SECOND EDITION

TECHNICAL COMMUNICATION

IN THE TWENTY-FIRST CENTURY

SIDNEY I. DOBRIN
UNIVERSITY OF FLORIDA

CHRISTOPHER J. KELLER
UNIVERSITY OF TEXAS–PAN AMERICAN

CHRISTIAN R. WEISSER
PENNSYLVANIA STATE UNIVERSITY, BERKS

Prentice Hall
Upper Saddle River, New Jersey
Columbus, Ohio

Library of Congress Cataloging in Publication Data
Dobrin, Sidney I.
Technical communication in the twenty-first century / Sidney I. Dobrin, Christopher J. Keller, Christian R. Weisser. -- 2nd ed.
p. cm.
Includes index.
ISBN-13: 978-0-13-503174-2
ISBN-10: 0-13-503174-5
1. Rhetoric. 2. Communication. I. Keller, Christopher J. II. Weisser, Christian R., 1970- III. Title.
P301.D63 2010
808′.066--dc22
2008048842

Editor in Chief: Vernon Anthony
Senior Acquisitions Editor: Gary Bauer
Editorial Assistant: Megan Heintz
Development Editor: David Ploskonka
Project Manager: Rex Davidson
Copy Editor: Mary Benis
Production Coordination: Shelley Creager, Aptara®, Inc.
Art Director: Diane Ernsberger
Cover Designer: Candace Rowley
Senior Operations Supervisor: Pat Tonneman
Director of Marketing: David Gesell
Marketing Manager: Leigh Ann Sims
Marketing Coordinator: Alicia Wozniak

Chapter opener photos: iStock

Photo credits are on pages 763–765.

This book was set in Garamond by Aptara®, Inc. It was printed and bound by Courier/Kendallville. The cover was printed by Phoenix Color Corp.

Pearson Education Ltd., London
Pearson Education Singapore Pte. Ltd.
Pearson Education Canada, Ltd.
Pearson Education—Japan
Pearson Education Australia Pty. Limited
Pearson Education North Asia Ltd., Hong Kong
Pearson Educatión de Mexico, S.A. de C.V.
Pearson Education Malaysia Pte. Ltd.

Prentice Hall
is an imprint of

www.pearsonhighered.com

10 9 8 7 6 5 4 3 2 1
ISBN-13: 978-0-13-503174-2
ISBN-10: 0-13-503174-5

This one's for Asher, Shaia, Will, Cole,
Carson, and Greta Jane—the *TCTC* kids.
Long may you run.

SID DOBRIN

CHRIS KELLER

CHRISTIAN WEISSER

BRIEF CONTENTS

TO *TCTC* STUDENTS AND THEIR TEACHERS

FROM THE AUTHORS

Welcome to the second edition of *Technical Communication in the Twenty-First Century (TCTC).* Our goal is to provide an accessible and useful set of materials to prepare students to be successful writers and readers of technical communication, regardless of their career path. *TCTC* covers the standards, conventions, and expectations of primary technical communication genres so that students develop the appropriate knowledge and skills to produce the proficient and ethical documents required by their professions.

TCTC continues to use innovative approaches to learning technical communication and more important, evaluating one's own role in and relationship to the production and interpretation of technical communication.

Responding to changes in the field of technical communication, to technological advances, and to comments from a wide variety of users of the first edition, we have made many changes in this edition. Most notably, the second edition includes the following:

- A new focus on **transnational** and **transcultural** workplace communication throughout the book. Chapter 5, "Workplace Writing in a Transnational World," is a new chapter devoted specifically to communicating with transnational and transcultural audiences. *TCTC* is the first technical communication textbook to provide this level of coverage of transnational and transcultural communication.
- The addition of 45 **new case studies** to the book and Companion Website that provide a variety of writing scenarios for students to develop and apply writing and rhetorical thinking skills.
- Expanded coverage of the relationships between **technology** and **communication**, including the most current workplace communication technologies such as wikis, instant messaging, and blogging.
- Many new and challenging **writing scenarios** added to the end of each chapter.
- A variety of new **interesting** and **thought-provoking examples** throughout the chapters that engage readers in critical thinking, writing, and discussion.

Another major enhancement for the second edition is the availability of **MyTechCommLab**, which can be packaged with new copies of the textbook. Students will find an interactive textbook integrated with a variety of multimedia technical communication resources, including interactive document examples, website and visual rhetoric tutorials, grammar and mechanics exercises, diagnostic and instructional tools, and online research tools.

Ultimately, the goal of *TCTC* is to prepare students to efficiently and effectively accomplish the communication tasks they will face in the workplace. We know that instructors will find the information in this book useful to students today in class and later on the job.

SIDNEY I. DOBRIN, University of Florida

CHRISTOPHER J. KELLER, University of Texas–Pan American

CHRISTIAN R. WEISSER, Pennsylvania State University, Berks

KEY THEMES AND FEATURES OF *TCTC*

REAL PEOPLE, REAL WRITING

MEET BETH KUZMINSKY • Nursing Curriculum Specialist

What kinds of writing do you do as a health-care professional?
I write every day in a variety of ways. I create or reply to approximately forty to sixty professional e-mails each day. I also write regularly for local and national newsletters, develop curriculum for nursing courses, and develop presentations to share at forums throughout my health-care system. When working with outside guests, thank-you notes and descriptions of speaker performances are also my responsibility.

How is ethics addressed in your profession?
As we work closely with patients, families, and health-care providers, ethics is at the core of all interactions. My medical center has an ethics and compliance program, and we depend upon a guidebook entitled "Workplace Ethics." The guidebook highlights our foundation of honesty and integrity in the workplace. It is the obligation of staff to ensure that we always act honorably and appropriately, since the decisions we make impact our patients every day.

What are the most important ethical issues you face in writing documents in your workplace?
Patient safety is most important. As we implement new patient-care strategies in the workplace, it can be challenging to transition from "the way we've always done it" to the best, most efficient, and safest way for patients and providers. It is important to reinforce the good work being done in health care by the frontline providers while also testing new ways to serve our patients.

Can you describe an example of a recent ethical issue in your workplace and the way you dealt with it through writing?
I led our medical center's new Condition Help program, which allows patients and families to call our rapid-response team during emergencies. Ethically, we believe this is right for patients and families. My role has been to promote the program, and I have written extensively about how Condition Help is beneficial for patients even though the program sometimes challenges the traditional health-care delivery system.

What is the relationship between ethical and legal approaches to issues in your profession?
In the health-care profession, they are closely related. Our medical center has an Ethics and Compliance Advisory Committee to provide guidance as needed. This committee consists of a number of people with expertise in ethics and legal issues. If I encounter a legal or ethical problem, I refer to this committee before making any decisions.

Does technology affect the way you address ethics?
Yes, technology has an impact on ethics in my workplace, particularly with online communication. Patient confidentiality must be maintained at all times. In references to a patient in e-mail, names are not included, or only initials are used. When case studies are used as an educational strategy and shared electronically, patient names are never included.

What general advice about workplace writing, do you have for the readers of this textbook?
Remember that your readers come first. If you put their best interests first and share this through writing, you are almost always doing the right thing. Consider that each person is a unique human being with a background that might be different from your own. Keep the message appropriate, meaningful, and applicable.

REAL PEOPLE, REAL WRITING

To emphasize the human side of technical writing, each chapter begins with a brief, original interview with a professional who speaks credibly about the particular issues covered in that chapter. These interviewees include individuals who do a good deal of writing in their respective workplaces but were not hired as technical communicators per se. Instead, the interviews convey that all workers, regardless of their positions, interpret and produce technical documentation.

RHETORICAL APPROACH

Technical communication is pragmatic writing intended to communicate crucial information to its audiences. To develop a clear sense of audience, writers in the workplace must think *rhetorically.* In this text, *rhetoric* means, essentially, the study and practice of using language to influence others' thoughts and actions, and *thinking rhetorically* in workplace writing means thinking about how one's words—written or spoken—will affect an audience, as well as how they will reflect one's credibility.

In addition, thinking rhetorically involves important choices in invention, arrangement, style, and genre, as well as ethos, emotion, and logic. *TCTC* emphasizes rhetorical thinking as critically and intrinsically necessary in *every* phase and context of technical communication.

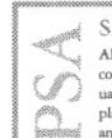

Situation and Perspective

All workplace and rhetorical problems occur within specific situations that provide the context in which they can be solved. Ethical choices are no different. The context of a situation determines what the appropriate professional and personal response can be. Workplace writers can ask themselves a number of questions to better understand the situation and thus to better enable themselves to address the ethical issues of the situation.

- What is my reason or purpose for writing this document? What is the exigency?
- Who is the audience for this document, and how will that audience be affected by the choices I make in writing the document?
- Who, other than the audience, does the information in this document affect?
- What authority do I have in conveying this information? Does that authority affect how the information will be understood?
- What are the ramifications of conveying the information in this way in my document?
- Where does this information fit in the larger context?
- Are there limits to what my writing can and cannot convey? How do I determine those limits? What or who imposes those limits?
- Will this document lead to any environmental effect?
- What responses are this document, and my choices, likely to evoke?

PROBLEM-SOLVING APPROACH (PSA)

Technical and professional documents solve problems when they inform, when they persuade, and when audiences can act on them. Tied directly to rhetorical thinking, the problem-solving approach (PSA) encourages students to solve document-related problems—who should receive the document, what type of document to create, what the document should contain, what tone the document should take, and whether to include visuals. The PSA icon in the margin indicates that the PSA is being used or reviewed.

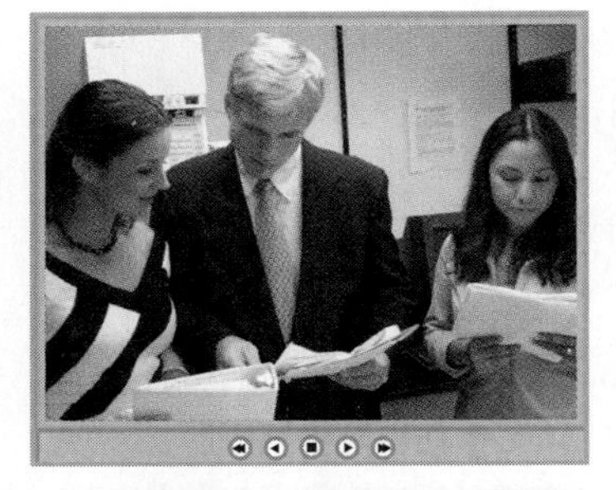

CASE STUDY APPROACH

Case studies offer an effective way to understand the workplace contexts in which technical communication takes place. *TCTC* incorporates both real-world and constructed case studies, which students analyze through discussion questions and writing prompts. Each chapter presents two case studies in the text, and the Companion Website provides at least four additional cases for each chapter.

TCTC Video Cases Provide Visual Context for Writing

Video cases, integrated into the text, allow students to watch how people in four different workplaces interact and how those interactions affect their writing. Each video case presents a rhetorical situation similar to those students will experience during their careers. The videos can be used in class for discussion or as assignments. They are available on the Companion Website and instructor DVD.

ETHICAL APPROACH

Ethics is a key component in understanding how technical communication works, and *TCTC* makes it a cornerstone of rhetorical thinking, from invention through revision to final production.

TCTC addresses the kinds of ethical issues that workers are likely to encounter in workplace environments, including patents, privacy laws, confidentiality, intellectual property, copyright, software licensing, nondisclosure agreements, billable time, and use of clients' equipment and facilities. Practical guidance and advice are given about how to analyze and resolve the kinds of ethical dilemmas encountered by today's professionals.

Professionalism, Community, and Personal Responsibility

TCTC addresses professionalism, community, and personal responsibility as foundational components of writing in the workplace. In particular, the book draws considerable attention to how workplace texts function as extensions of writers' professional selves and identities. It makes clear that writing often is used by colleagues and supervisors to assess a writer's professionalism, and, ultimately, his or her status in the workplace.

TCTC also integrates discussions of community to show student writers how technical documents can affect more people than just those in the immediate audience. An understanding of how writing affects numerous communities inside and outside a company or organization enables students to see the relationships between writing and personal responsibility. Workplace writers are seen as real people with real decisions to make—decisions that have real consequences for different communities of people.

COLLABORATIVE APPROACH

Whether in class or in the workplace, collaboration is a crucial part of producing technical and professional documents. Most professionals collaborate in one form or another: gathering information from experts, asking a colleague to proofread or verify information, consulting with intended audiences, or conducting usability tests. *TCTC* includes numerous discussions of collaboration, addressing the pragmatics of collaboration at various stages of writing, the role technologies play in collaboration, and the role of professionalism, ethics, and community in collaboration.

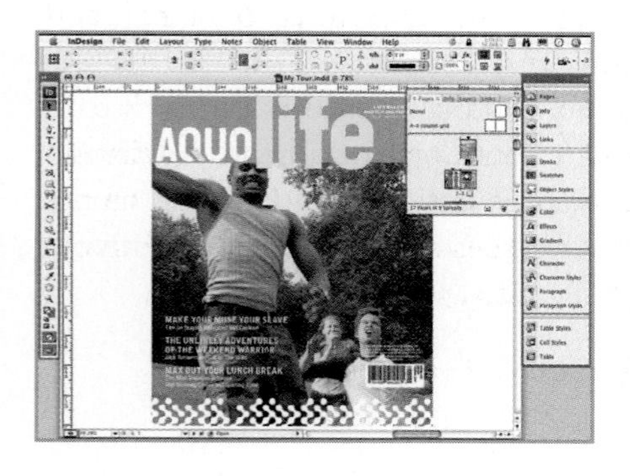

FOCUS ON TECHNOLOGY

TCTC asks students to evaluate how technology, particularly computer hardware and software, affects the way that workers write in various workplaces. Students should understand technology as an integral part of rhetorical thinking and writing that shapes how and what they write. Students and instructors are encouraged to examine critically the ways that computers influence how workers think about, read, and write documents.

Chapter 3: Technical Writing and Electronic Technologies

Although we devote Chapter 3 specifically to the role of electronic technologies in the workplace, students gain both critical awareness and practical knowledge of technology in every chapter.

FOCUS ON TRANSNATIONAL AND TRANSCULTURAL COMMUNICATION

Because technical communication in the twenty-first century often involves communication with transnational and transcultural audiences, *TCTC* offers detailed considerations of international and cross-cultural audiences at all stages of technical communication.

Chapter 5: Workplace Writing in a Transnational World

TCTC offers a full chapter about transnational and transcultural audiences. This groundbreaking chapter is a first for technical communication textbooks and establishes the need for attention to transnational and transcultural topics in almost all workplace writing. These topics are addressed throughout *TCTC*, making it the first technical communication textbook to attend to these issues as a central focus for all workplace writers, not just a secondary possible consideration. Provocative examples and challenging exercises reinforce the concepts discussed.

FOCUS ON VISUAL RHETORIC

Because of recent striking developments in visual technologies, technical communication relies more than ever on visual elements to convey meaning. *TCTC* emphasizes the important role of visual elements in technical communication. From discussions of how visuals affect the design of various documents to more sophisticated theoretical and practical analyses of the relationships among images, written text, and vocal communication, the text stresses that workplace writing often requires writers to integrate visuals effectively into their work.

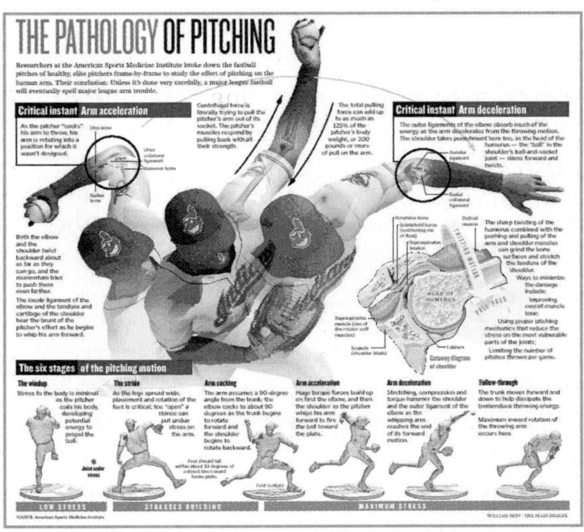

Chapter 8: Visual Rhetoric and Technical Communication

Chapter 9: Layout and Design

Students are shown that visual rhetorical choices in technical communication are inseparable from technological, professional, and ethical considerations; and concepts of visual rhetoric and the role visuals play in technical communication are introduced.

Layout and design are addressed as part of the visual rhetorical choices writers make, from page design to typography to the material qualities of a document. Issues of transnational and transcultural visual considerations are also introduced.

FOCUS ON USABILITY

One of the most significant recent developments in technical writing has been the attention paid to usability testing. In addition to promoting collaboration between writers and potential audiences, usability ensures that technical documents are readable, user friendly, and accurate. *TCTC* devotes a chapter to usability guidelines, tests, and protocols and then reinforces that foundation in subsequent chapters with information on the role of usability testing in developing readable, feasible, and usable documents and the role it plays in larger issues of community, professionalism, and personal responsibility. Chapter projects and writing scenarios at the end of chapters reinforce this material.

Chapter 11: Usability

This chapter describes how to develop usability testing, working from planning usability testing sessions to collecting and analyzing data.

IN-TEXT LEARNING AIDS

Numerous features are designed to guide students in their study of technical communication.

Marginal Information

Margin comments throughout the book provide students with four highlighted types of information:

- **Key terms** are defined.
- **Web references** direct students to the Companion Website when additional information can be found there.
- **Cross-reference information** directs students to additional information within the text.
- **Experts Say** quotations from famous and workplace writers give students insight into how writers think about particular workplace writing issues.

In Your Experience

These boxes stimulate students to think about chapter topics as they relate to their own lives and past experiences.

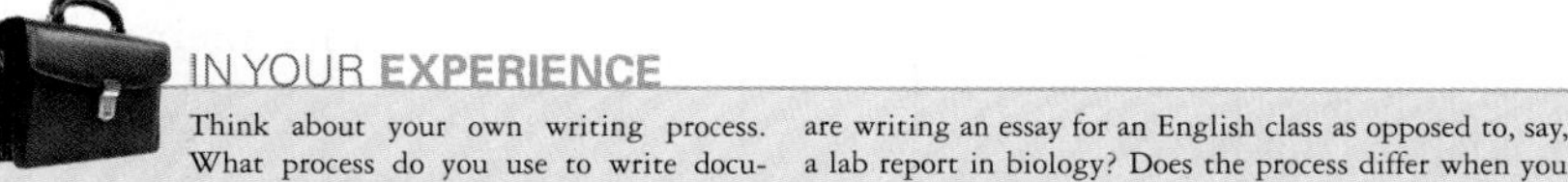
IN YOUR EXPERIENCE

Think about your own writing process. What process do you use to write documents? Does the process differ depending on the kind of writing you are doing? For example, does it differ if you are writing an essay for an English class as opposed to, say, a lab report in biology? Does the process differ when you are writing documents for school as opposed to work or personal communication? Explain.

Explore

These boxed exercises encourage students to look beyond the text and conduct further investigation to learn more about particular subjects.

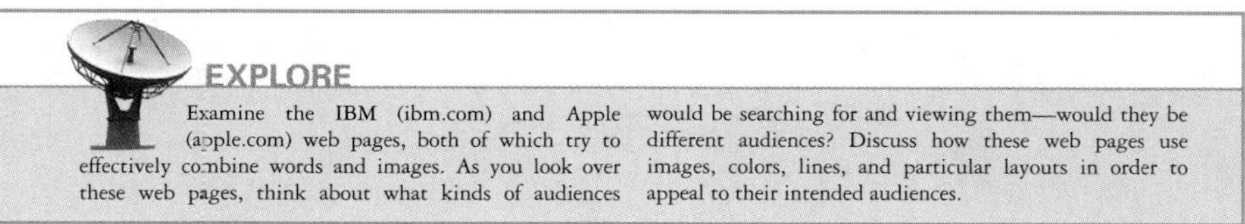
EXPLORE

Examine the IBM (ibm.com) and Apple (apple.com) web pages, both of which try to effectively combine words and images. As you look over these web pages, think about what kinds of audiences would be searching for and viewing them—would they be different audiences? Discuss how these web pages use images, colors, lines, and particular layouts in order to appeal to their intended audiences.

Analyze This

These boxed exercises engage students in rhetorical analysis of a range of technical documents in order to better understand the subject matter at hand.

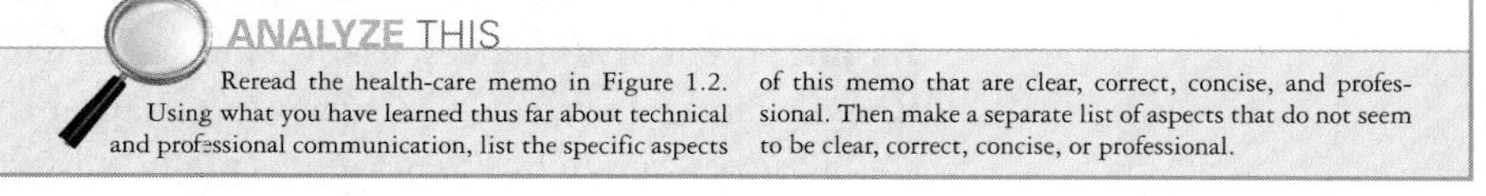
ANALYZE THIS

Reread the health-care memo in Figure 1.2. Using what you have learned thus far about technical and professional communication, list the specific aspects of this memo that are clear, correct, concise, and professional. Then make a separate list of aspects that do not seem to be clear, correct, concise, or professional.

BUILD ANALYTICAL WRITING SKILLS

CHAPTER SUMMARIES

Each chapter provides a concise review of the primary information presented in that chapter. These summaries are designed both to emphasize key points and to provide students with an overview of the chapter.

CONCEPT REVIEW

A series of review questions ask students to consider the key elements of each chapter. These questions reflect the Learning Outcomes of each chapter and are designed to ensure that students have met those outcomes.

WRITING SCENARIOS

Each chapter provides a number of scenarios that ask students to think about and produce longer, more substantial writing projects. In genre-based chapters, the Writing Scenarios ask students to write about the issues discussed in the chapter and to write the kinds of documents addressed in that chapter. Such assignments encourage students to consider issues that relate to audience, rhetorical choices, genre, style, form, and function, as well as professionalism, technology, community, and ethics.

CASE STUDIES

Two case studies appear at the end of each chapter. Each case study is detailed, complete with learning objectives, case information and prompts for class discussion, and writing assignments. In addition, each chapter offers at least two long cases and two short case studies on the Companion Website. Some of the more than 110 case studies put students in the role of workplace writers, some ask students to analyze and respond to workplace writing scenarios, and some ask them to conduct further research into particular workplace writing issues.

VIDEO CASE STUDIES

A video case study appears at the end of each chapter, along with discussion and writing prompts.

TOOLS FOR LEARNING

FOR THE STUDENT

COMPANION WEBSITE

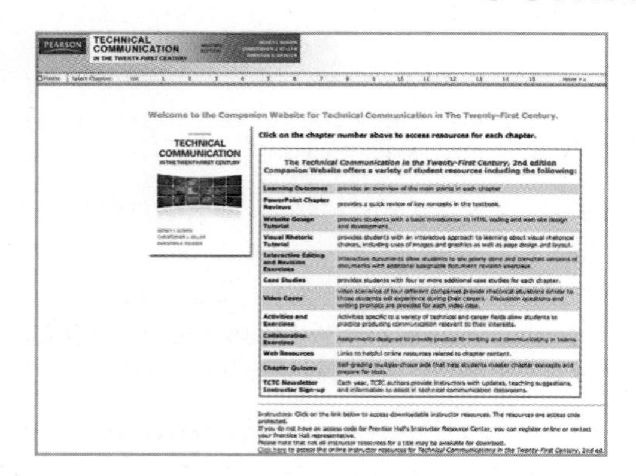

Go to www.prenhall.com/dobrin

The Companion Website offers a variety of student resources, including the following;

- **PowerPoint Chapter Review** provides a quick review of key concepts in the textbook.
- **Website Design Tutorial** provides a basic introduction to HTML coding and website design and development.
- **Visual Rhetoric Tutorial** provides an interactive approach to learning about visual rhetorical choices, including use of images and graphics as well as page design and layout.
- **Interactive Editing and Revision Exercises** allow students to see poorly written and corrected versions of documents along with additional document revision exercises.
- **Case Studies** provide four or more additional case studies per chapter to allow variety in assignments.
- **Activities and Exercises** specific to a variety of technical and career fields allow students to practice producing communication relevant to their interests.
- **Collaboration Exercises** provide practice for writing and communicating in teams.
- **Web Resources** provide links to helpful online resources related to chapter content.
- **Chapter Quizzes** are self-grading, multiple-choice aids that help students master chapter concepts and prepare for tests.
- ***TCTC* Newsletter Instructor Sign-up**—Each year, *TCTC* authors provide instructors with updates, teaching suggestions, and information to assist in technical communication classrooms.

YOUR ONE-STOP SOURCE FOR TECHNICAL COMMUNICATION RESOURCES

MyTechCommLab for *TCTC*, 2e

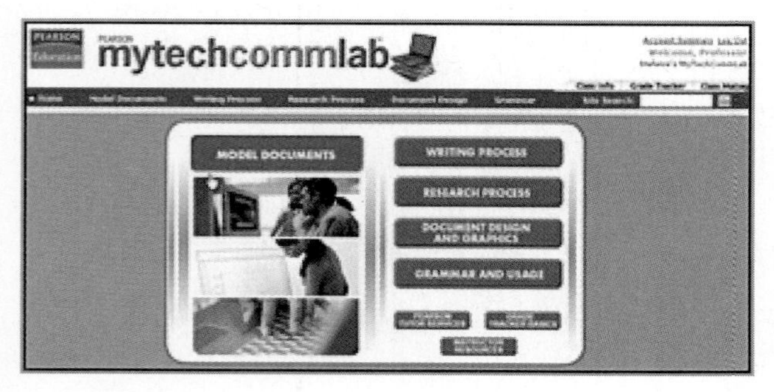

This dynamic, comprehensive resource can be packaged at no additional cost with the purchase of a new text. MyTechCommLab comes in two versions: a generic version that requires no instructor involvement and an e-book version that includes an instructor gradebook and classroom management tools. Both versions provide a wide array of multimedia tools, including all content from the Companion Website—all in one place and all designed specifically for technical communicators.

Model Documents in MyTechCommLab

- **More than 80 model documents** in ten categories (memos, reports, proposals, definitions and descriptions, letters, and more) offer samples in all the important technical communication categories.
- **More than 50 interactive documents** include rollover annotations highlighting purpose, audience, design, and other critical topics.
- **Quizzes** for each category of model document let students practice what they have learned.

Writing Process Resources in MyTechCommLab

- **Writing Process Tutorial** leads students through each stage of the writing process, from prewriting to final formatting.
- **A tutorial on Writing Formal Reports** offers step-by-step guidance for creating one of the most common document types in technical communication and working with sources.
- **Activities and Case Studies** provide more than 65 exercises, all rooted in technical communication and many document-based, including 3 new case studies on usability.
- **Weblinks** offer an annotated list of relevant websites for both technical communication in general and writing in particular.
- **An online reference library of e-books** includes PDF files for books on visual communication and workplace literacy.

Research Process Resources in MyTechCommLab

- **Pearson's exclusive Research Navigator** offers access to four credible databases, including EBSCO's ContentSelect database of more than 25,000 articles from a wide spectrum of academic journals
- ***The New York Times* archives**
- ***The Financial Times* Search-by-Subject archives**
- **Link Library**, which provides access to thousands of links for discipline-specific key terms.
- **A step-by-step tutorial on the research process** is also included in Research Navigator.
- **Citation style guides**, including MLA and APA.
- **Avoiding Plagiarism Tutorials** are self-paced and self-scoring tests, exercises, and tutorials on how to recognize and avoid plagiarism.
- **Document Design Resources**
- **Visual Rhetoric Tutorial** leads students step-by-step through the process of designing an effective document, including basic document architecture, choosing colors and typefaces, using images, and putting it all together.
- **Web Design Tutorial** provides basic instruction for novices on building a Web page and creating hyperlinks.

Grammar Resources in MyTechCommLab

- **Two diagnostic tests** allow instructors and students to identify individual weaknesses and create study plans.

- **More than 3,000 practice items** cover all major topics of grammar, style, and usage, include sentence and paragraph editing exercises, and provide basic, assignable review for students who need it.
- **Weblinks** provide an annotated list of appropriate websites.

MyTechCommLab is also available with book-specific resources and course management tools!

MyTechCommLab is available in Pearson's proprietary CourseCompass online course management system for use with specific introductory technical communication texts from Pearson. CourseCompass versions include a complete e-book as well as a wealth of book-specific resources, including quizzes, PowerPoints, activities, case studies, and much more. Contact your Pearson representative for more information.

To order *TCTC*, 2e, with the generic MyTechCommLab access code, order ISBN: 0136098916

To order TCTC, 2e, with the MyTechCommLab with e-book in CourseCompass access code, order ISBN: 0136098932

A standalone access code can be purchased online at www.prenhall.com.

To preview MyTechCommLab, go to www.mytechcommlab.com.

TOOLS FOR TEACHING

FOR THE INSTRUCTOR

To access supplementary materials online, instructors need to request an instructor access code. Go to www.pearsonhighered.com/irc, where you can register for an instructor access code. Within 48 hours of registering you will receive a confirming e-mail that includes an instructor access code. Once you have received your code, locate your text in the online catalog and click on the Instructor Resources button on the left side of the catalog product page. Select a supplement, and a log-in page will appear. Once you have logged in, you can access instructor material for all Prentice Hall textbooks.

INSTRUCTOR'S MANUAL WITH NEW INSTRUCTOR TOOLKIT

The detailed instructor's manual accommodates the needs of instructors with any level of experience. For first-time instructors or experienced hands, this tool contains a wealth of teaching materials, including teaching tips; sample syllabi for 10-, 12-, 15-, and 16-week classes; chapter outlines; and chapter notes.

PRENTICE HALL TESTGEN

This computerized test-generation system gives maximum flexibility in preparing tests. It can create custom tests and print-scrambled versions of a test at one time, as well as build tests randomly by chapter, level of difficulty, or question type. The software also allows online testing and record keeping and the ability to add problems to the database.

POWERPOINT LECTURE PRESENTATION PACKAGE

High-quality lecture presentation screens for each chapter are available online and on the Instructor Resources DVD.

ANNUAL *TCTC* NEWSLETTER

The *TCTC* newsletter contains additional teaching suggestions and new case material. Instructors can sign up online at the Companion Website to receive the newsletter via e-mail.

ADDITIONAL RESOURCES

DICTIONARY AND THESAURUS

A dictionary and thesaurus can be packaged with this textbook at very low cost. To order this package, instructors should contact your Pearson sales representative.

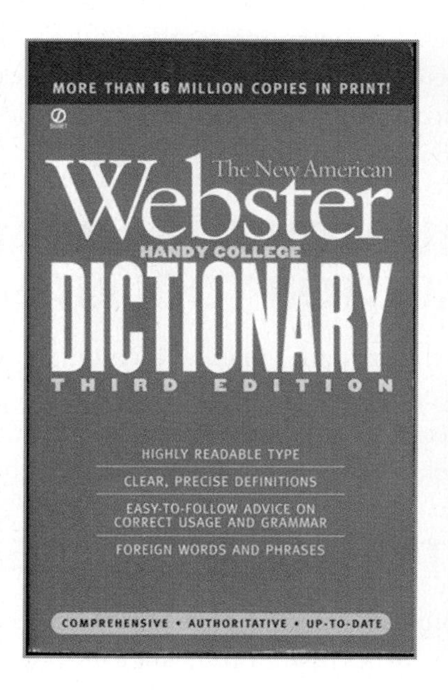

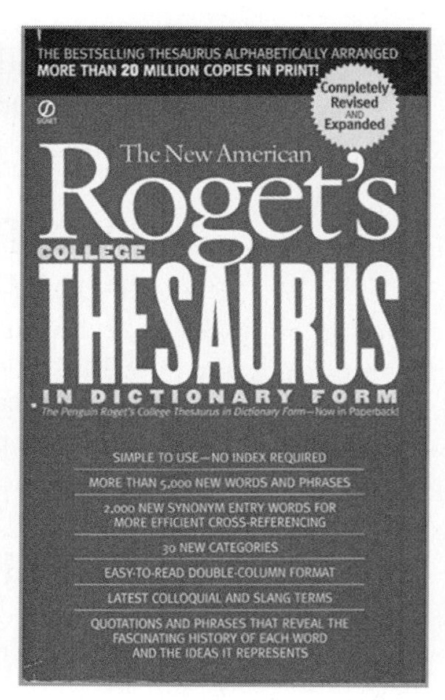

MYWRITINGLAB ACCESS CODE

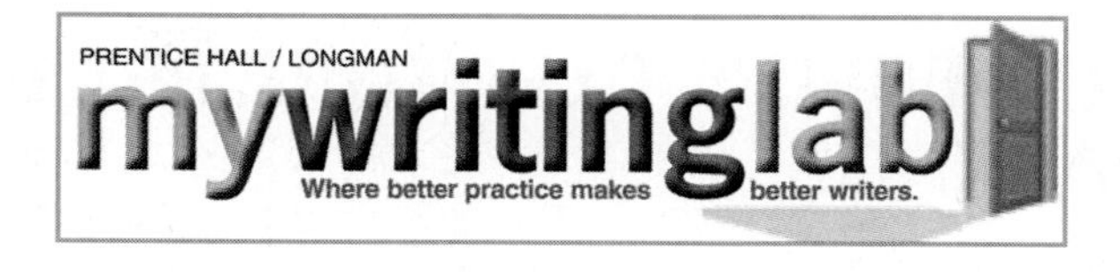

MyWritingLab is an online learning system that provides better writing practice through diagnostic assessment and progressive exercises to move students from literal comprehension to critical thinking and writing. With this practice model, students develop the skills needed to become better writers. An access code to MyWritingLab can be packaged with this text at very low cost.

TCTC, 2e, packaged with MyWritingLab (ISBN: 0205668976)

ACKNOWLEDGMENTS

MANUSCRIPT REVIEWERS

Comments and insights from reviewers were invaluable in helping us refine our ideas and presentation. We are grateful for their candor and useful suggestions.

Kaye Adkins, Missouri State College
Jennifer Adkinson, Idaho State University
Mark Amdahl, Montgomery County Community College
J. D. Applen, University of Central Florida
Cindi Baker, Belmont Technical College
Alice Benham, Bowling Green Technical College
Howard Benoist, Our Lady of the Lake University
Magdalena Berry, Southwest Missouri State University
Margery Brown, State University of New York, Farmingdale
Lillie Busby, Sam Houston State University
Rocky Colavito, Northwestern State University
Nancy Coppola, New Jersey Institute of Technology
Helen Correll, Metropolitan State University, St. Paul
Ken Cox, Florence Darlington Technical College
Natalie Daley, Linn Benton Community College
Bradley Dilger, Western Illinois University
Marc Donadieu, Oregon State University
Angela Eaton, Texas Tech University
Crystal Edmonds, Robeson Community College
Margaret Ellington, Georgia Southwestern State University
Douglas Evman, Michigan State University
Joyce Fisher, Henry Ford Community College
Sandy Friend, East Carolina University
Jose Gallegos, University of New Mexico, Belen
Daniel Gonzalez, University of New Orleans
Judy Grace, Arizona State University
Chris Grooms, Collin County Community College
Angela Haas, Michigan State University
Judith Hakola, University of Maine
D. Alexis Hart, Virginia Military Institute
Barbara Heifferon, Clemson University
Jurgen Heise, Texas Tech University
Julia Helo, North Carolina State University
Groves Herrick, Maine Maritime Academy
Erma Hines, Southern University at New Orleans
William Hooton, Texas A&M University
Michael Hricik, Westmoreland County Community College
Kathleen Hurley, Minnesota State University, Mankato
Leslie Janac, Blinn College
Trish Jenkins, University of Alaska, Anchorage
Marjorie Justice, Governors State University
Michael Kerley, New Jersey Institute of Technology
Marguerite Krupp, Northeastern University
David Kuipers, South Georgia Technical Institute
Lisa Lebduska, Wheaton College
Gail Lippincott, University of Central Florida
Mark Mabrito, Purdue University, Calumet
Barry Maid, Arizona State University, Polytechnic Campus
David Major, Austin Peay State University
Michelle Manning, University of Northern Carolina, Wilmington
Donna Marino, Macomb County Community College South
Jodie Marion, Mount Hood Community College
Phillip Marzluf, Kansas State University
Lisa McClure, Southern Illinois University, Carbondale
Candie McKee, University of Central Oklahoma
Brooke McLaughlin Mitchell, Wingate University
Rick Mott, New Mexico Tech
Marshall Myers, Eastern Kentucky University
Harriet Napierkowski, University of Colorado at Colorado Springs
Gail Nash, Oklahoma Christian University
Carol Nelson-Burns, University of Toledo
Gian Pagnucci, Indiana University of Pennsylvania
Shelley Palmer, Rowan Cabarrus Community College
Anne Papworth, Brigham Young University
Dee Pruitt, Florence Darlington Technical College
John Rothfork, Northern Arizona University
Sumita Roy, Louisiana State University
Marilyn Seguin, Kent State University
Carol Senf, Georgia Institute of Technology
Janet Yagoda Shagam, University of New Mexico
Herbert Shapiro, Empire State College
Sim Shattuck, Louisiana Tech University
Todd Solomon, University of Maryland
Karen Stewart, Norwich University
Jan Strever, Spokane Community College
Carol Sullivan, Delaware Technical & Community College
David Tillyer, City College of New York
Lowell VerHeul, Ohio University
Wanda Worley, Indiana University–Purdue University Indianapolis

CLASSROOM TESTERS AND STUDENT USERS

Over a two-year period, early versions of this text were used by faculty and teaching assistants at the University of Florida. We thank them for their patience and feedback.

Teachers who classroom-tested early versions of *TCTC* include Angela Schlein, Derrick Hoeben, Sean McCartin, Andrew Wislocki, Leah Carroll, Lyndsay Brown, William Scherban, Christine Bertrand, Adam Nikolaidis, Lisa Dusenberry, Janel Simons, David Ramsey, Georgia Gelmis, Brittany Luck, Melissa Garcia, Rachel Slivon, Anna Donovan, Antionette Baker, Marianne Kunkel, Wylie Lenz, Julie Christenson, Margaret Franklin, Linsey Maughan, Brandon Hartley, Saara Raappana, Lindsay Skorupa, Carrie Crumrine, James Addcox, Peter D'Ettore, Eric Bliman, Brittany Parkhurst, Joshua Miller, Christina VanHouten, Delores Amorelli, Megan Leroy, Charity Burns, Tarah Dunn, Kate Sweeney, Daniel O'Malley, Lee Pinkas, Laurel Czaikowski, Christe Duncan, Michelle Lee, Cosme Caballero, Curtis D'osta, Hayden Draper, Sharon Lintz, Maggie Powers, Kelly Dunn, Randy Romano, Chris Cowley, Erin Bohannon, Velina Manolova, Felice Lopez, John Hart, Chris Shannon, Emily McCann, Richard Paez, Rania Williams, Josh Coonrod, Elizabeth Femiano, and Sean Fenty.

Student users who made suggestions for revision include Dustin Reynolds, Ruben Orduz, Josh Caltagirone-Holzli, Jimmy Nguyen, Yissel Cabrera, and Harold Rodriguez.

CONTRIBUTORS

This project truly reflects the notion that it takes a village to create an outstanding text and teaching package. We gratefully acknowledge and thank those individuals whose assistance led to the successful completion of this text and the accompanying supplements:

Kaye Adkins, Missouri Western State College, helped to prepare the instructor's manual.

Jennifer Dareneau, Penn State University, helped to prepare the test item file.

Sean Morey, University of Florida, helped to prepare the instructor's manual and case studies on the Companion Website.

Scott Reed, University of Georgia, created the PowerPoint Lecture Presentation Package.

Melanie Rosen Brown, St. Johns River Community College, provided material for the Companion Website.

Clay Arnold created the Web tutorial and assisted with the creation of the Visual Rhetoric tutorial.

Dr. Tim Morey of the University of Florida gave the use of his lab in the video cases.

Case Study contributors were Brenda Maxey-Billings, Lloyd Willis, Laurie Taylor, Sean Morey, Lindsey Collins, Carol Steen, Linda Howell, Jeff Rice, and Richard Paez.

Real People, Real Writing interviewees were Aprille Ericcson, Darren Barefoot, Kim Zetter, Margie Musser, Heejong Haas, Beth Kuzminsky, Max Lally, Robert Harrell, Heidi Machul, Lori Booth, Robert Uy, Dave Ferron, Jeff Rodanski, Matt Keffer, Stephanie Hinson, Orlando Lamas, Clint Hocking, Jen deHann, Diane Stielstra, Stormy Brown, Devin Maxey-Billings, Etzer Darout, and Monty Hansen.

Video Case Study talent and crew: High Noon Productions and Computer Wiz. The crew: M. Adam Ball, Alison Muckenfuss, Terrance Wissel, Skyler Slade, Desha Marshall, Megan Forrest, Vincent Holt, Ayelet Gaito, Matt Grosswald, Jared Thomas, Diedra Kelley, Zach Randall, Michael Fong, Stephen Reale. The cast: Alison Muckenfuss, Robinson Moore, Albert Lin, Kristin Samuelson, Donna Wright, Valerie Denise Jones, Tequila Q. Brooks, Marcellus "Chello" Davis, Kenneth Matthews, Evelyn King, Milorad Djomlija, Scot Davis, Peaches Joy Brown, Stephen Reale, Mark Redd, Ruth Buehrig, Melissa Pamela Jegers, Keegan Carter, Megan Forrest, Vicki Hall, Matthew A. Eunice, Richard Dunlop, Krystof Kage, and Mercy Fiallo.

We offer special thanks to our families for their patience over the past several years. To our wives—Teresa, Cindy, and Traci—and our kids—Asher, Shaia, Will, Greta Jane, Cole, and Carson—we offer our love and gratitude for their endless support. We also thank Cleo, Evey, Emmanuel, and Marianna Martinez, and Chris's dog, Chloe.

Finally, we acknowledge and thank the editorial and production staff at Prentice Hall for their faith, patience, and diligence. We acknowledge and praise the work of our editors, Gary Bauer and David Ploskonka, our copy editor Mary Benis, and all of the members of the Prentice Hall production team.

ABOUT THE AUTHORS

SIDNEY I. DOBRIN is Associate Professor of English at the University of Florida. He served as Director of Writing Programs in the Department of English and Coordinator of the technical writing program in English for ten years. He has published a dozen books about writing and is completing several others, including *The Prentice Hall Technical Writer's Handbook.*

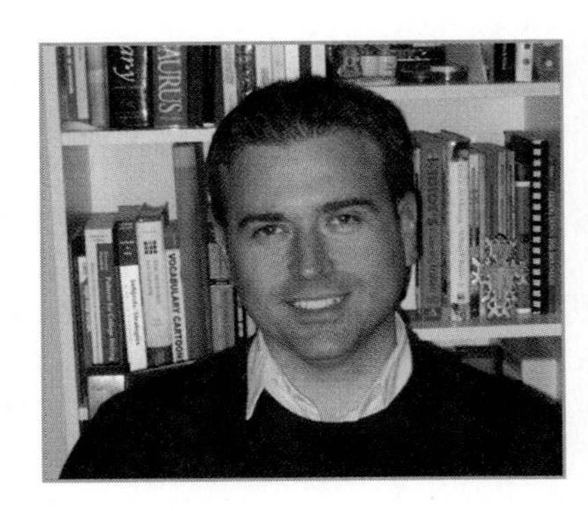

CHRISTOPHER J. KELLER is Associate Professor of English at the University of Texas–Pan American, where he teaches courses in technical communication, composition theory, and American literature.

CHRISTIAN R. WEISSER is Associate Professor of English at Penn State University, Berks, where he coordinates both the professional writing program and the writing across the curriculum program. He is the author and editor of numerous books and articles about writing and currently serves as the editor of *Composition Forum*, a peer-reviewed scholarly journal in rhetoric and composition.

CONTENTS

CHAPTER 1 TECHNICAL AND PROFESSIONAL COMMUNICATION IN THE WORKPLACE

CHAPTER 2 RHETORIC AND TECHNICAL COMMUNICATION

CHAPTER 3 TECHNICAL COMMUNICATION AND ELECTRONIC TECHNOLOGIES

CHAPTER 4 ETHICS AND THE WORKPLACE WRITER

CHAPTER 5 TECHNICAL COMMUNICATION IN A TRANSNATIONAL WORLD

CHAPTER 6 RESEARCHING AND EVALUATING SOURCE MATERIAL

CHAPTER 7 ORGANIZING AND DRAFTING DOCUMENTS

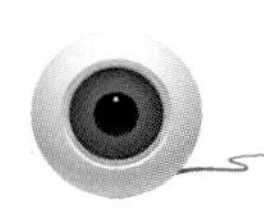

CHAPTER 8 VISUAL RHETORIC AND TECHNICAL COMMUNICATION

CHAPTER 9 LAYOUT AND DESIGN

CHAPTER 10 REVISING, REWRITING, AND EDITING

CHAPTER 11 USABILITY

CHAPTER 12 E-MAIL, E-MESSAGES, AND MEMOS

CHAPTER 13 LETTERS

CHAPTER 14 FINDING AND OBTAINING EMPLOYMENT

CHAPTER 15 TECHNICAL DEFINITIONS

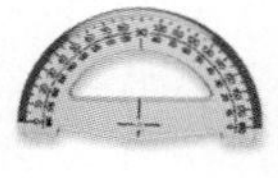

CHAPTER 16 TECHNICAL DESCRIPTIONS AND SPECIFICATIONS

CHAPTER 17 WEBSITES AND ONLINE ENVIRONMENTS

CHAPTER 18 TECHNICAL INSTRUCTIONS

CHAPTER 19 MANUALS

CHAPTER 20 PROPOSALS AND REQUESTS FOR PROPOSALS

CHAPTER 21 INFORMAL REPORTS

CHAPTER 22 FORMAL REPORTS

CHAPTER 23 PRESENTATIONS

TECHNICAL COMMUNICATION

IN THE TWENTY-FIRST CENTURY

1

Technical and Professional Communication in the Workplace

CHAPTER LEARNING OUTCOMES

After completing this chapter, you will be able to do the following:

- Define **technical and professional communication**
- Realize that all **workplace employees** produce technical and professional communication
- Recognize technical communication in your **everyday life**
- Understand the significance of communication in all **workplaces**
- Know the basic characteristics of technical and professional communication: **clear and correct**, **concise**, **professional**, **aware of audience**, **rhetorical**, **ethical**, **technology oriented**, **visual**, and **research oriented**
- Understand that writing is **an activity and a process**
- Know that writing and communication involve **problem solving: workplace problems and rhetorical problems**

DIGITAL RESOURCES

On the Companion Website www.prenhall.com/dobrin:

- Case 1: Writing in the Workplace: Apalachicola Zoological Sanctuary
- Case 2: Ethical Considerations of Writing in the Workplace: Absolm and Brown
- Case 3: Writing in the Workplace: Gemshine Marine Hardware
- Case 4: Shipping Problems: A Lubricant Mix-Up
- Video Case: Writing the Lab-Use Recommendation Report
- PowerPoint Chapter Review, Test-Prep Quiz, Exercises and Activities

REAL PEOPLE, REAL WRITING

MEET APRILLE ERICCSON • Aerospace Engineer for NASA Goddard Space Flight Center

What types of writing do you do as an aerospace engineer?
I write a variety of documents, such as short internal memos, longer technical letters that record the obligations of groups that work for us, detailed specifications documents, justification documents for purchases, and even proposals and reports. I also write technical papers for conferences and international meetings. For the most part, I write every day.

Can you describe an important document you recently created at work?
We've been developing the James Webb Space Telescope, which is a replacement for the Hubble Space Telescope, and we plan to launch it in 2011. Because it is such a large effort, it requires partnerships with space agencies worldwide, and we've been developing a written plan describing who will be involved, what tasks they will perform, what their deadlines are, and other factors.

Who are the typical readers of your workplace documents, and how do different audiences influence the style and content of your writing?
When I work with government employees, the information remains fairly technical, specific, and standard. However, if we were writing for business, we'd give more overview and explain points more simply. The audience definitely influences the style and content of the document.

What role does technology play in the writing you do?
Technology plays a huge role, from the way I create information, to the way I transfer information, to storage and backup methods. I've used floppy disks, CD-ROMs, and, more recently, memory sticks. PowerPoint plays a big role in my presentations.

What role does ethics play in the writing you do?
When factors such as cost or time become important in decision making, I often need to consider how much or how little to emphasize these aspects. Sometimes it is necessary to emphasize certain positive details (such as the benefits of a component or a plan) to persuade a group of readers, rather than emphasizing details that might be seen as negative. That involves an ethical choice in terms of the information to include in a document so that it is clear and balanced.

What advice about workplace writing do you have for the readers of this textbook?
Communication is one of the most important tools that you have to accomplish your tasks. Writing is an important skill you need to master, so think critically and carefully about what you want to communicate.

INTRODUCTION

If asked to imagine a typical work or office environment, you would probably picture people sitting behind desks, engaged in some act of communication. For instance, you might imagine a person talking on the phone, dictating a letter or message to a secretary or assistant, typing a report on a computer, e-mailing a memo to a colleague, reading a marketing report, or maybe even searching the Web to gather information. Such scenes are common in workplaces around the world; businesses and other organizations often succeed or fail on the basis of how well they are able to communicate information.

As an example of how workplace success or failure depends on communication, we can look at one of the most significant scientific endeavors in recent history: NASA's quest to reach the planet Mars. In May 2008 NASA's Phoenix spacecraft landed successfully on Mars to collect data and dig into the Martian soil in search of ice and organic matter. However, NASA's missions to Mars have not always had such positive outcomes. In 1999 NASA lost a $125 million spacecraft on a mission to Mars. The craft—the Mars Climate Orbiter—was meant to orbit Mars, collect data about the planet's climate (e.g., its water history and the potential for life on the planet), and send the data back to Earth for analysis. But the spacecraft came about fifteen miles closer to the planet than was originally planned. This caused the craft's propulsion system to overheat, which probably "stopped the engine from completing its burn, so Climate Orbiter likely plowed through the atmosphere, continued out beyond Mars and now could be orbiting the sun."[1]

The main collaborators in the mission to Mars—NASA; the Jet Propulsion Laboratory (JPL) in Pasadena, California; and Lockheed Martin—spent a great deal of time reviewing this catastrophe. After much debate and conversation, they found the major culprit that caused the loss of the Mars Climate Orbiter: English measurement units. One group working on the project made all of its technical calculations using English units (such as inches, feet, miles), whereas the other groups used metric units (like millimeters, meters, kilometers), and no one recognized this mismatch of data until it was too late. In other words, the real problem had to do with a lack of communication.

There's an old, sarcastic saying that many of us use when we refer to an easy task: "It doesn't take a rocket scientist to . . . ." In this case we see a large space project that probably involved hundreds of rocket scientists whose project failed, not because of their lack of expertise in building rockets, their lack of knowledge about physics or space travel, or their lack of dedication to their work. The project failed in large part because of an inability to communicate technical information across group boundaries. Maybe it doesn't take *just* rocket scientists to build successful rockets but skilled communicators as well.

Effective technical and professional communication is crucial in organizations like NASA, but this book is not written exclusively for those who have a future in engineering and rocket science. Rather, *Technical Communication in the Twenty-First Century* presents all students with practical information about communicating in workplace environments in fields such as accounting, marketing, engineering, computer science, agriculture, biology, architecture, medicine, and more. Learning to communicate effectively is vital to success in all fields. This book describes technical and professional communication in the workplace as follows:

Communication about complex, highly detailed problems, issues, or subjects in the professional world, which helps audiences visualize and understand information so that they can make informed and ethical decisions or take appropriate and safe actions.

---

[1] Robin Lloyd, "Mars Craft Probably Destroyed," *CNN.com*, September 23, 1999, http://www.cnn.com/TECH/space/9909/23/mars.orbiter.04/.

Technical and professional communication may seem unfamiliar to those with little or no work experience. However, even if you have never worked in an environment where technical and professional communication has been produced, you have still been influenced greatly by it. Most people in the world read, write, or communicate some form of professional and technical information every day, and many do so without recognizing it.

- Did you drive to school or work today and read street signs to help guide your way? If so, you were informed by a type of technical communication.
- Have you ever read the Nutrition Facts on the back of a drink or food product, such as a Coca-Cola can, a Rice Krispies cereal box, or the wrapper of a Snickers candy bar? If so, you have read a document that communicates technical information to consumers.
- Have you ever taken money out of your bank account from an automatic teller? If so, you followed technical instructions on the automatic teller's screen.
- Did your instructor for this or any other course provide you with a syllabus? If so, your instructor produced a technical document for you and your classmates.
- Have you ever written someone a note, letter, or e-mail that gave directions to your house or apartment or instructions on how to prepare a recipe? If so, in some respects you were writing like a technical communicator.

In all of these instances, someone provided information that an audience needed—selecting the right amount of content information, arranging it in a certain way, and presenting it coherently and effectively for the audience. These are the basic ingredients of technical and professional communication.

EXPLORE

Locate a food package and look closely at all the information that it conveys to consumers. What information is most prominently displayed? What information is least prominently displayed? Given what you know so far about technical and professional communication, what makes the information on the food package technical and professional? Be sure to review the previous definition before responding to these questions.

Because we are surrounded with so much technical and professional information, we often do not notice that we are reading or writing technical and professional documents unless they are inadequate. For example, when someone gives you directions to get somewhere and you get lost, you might notice that the instructions were incomplete or inaccurate. When you cannot find refund or contact information on a company's website, you notice that the layout and organization are ineffective. Or when you try to assemble a bookshelf but cannot figure out how to put Peg B into Slot C, you realize that the diagrams in the assembly manual are vague and confusing. Unfortunately, poor technical and professional communication is all too prevalent, and both good and bad technical communication does affect your life and the lives of others daily.

The yellow diagram in Figure 1.1 is an excerpt from an instruction manual about using a pellet gun properly and safely. A quick glance at these instructions alerts you to the fact that the document is poorly produced and highly unprofessional. A reader would hope that whoever produced this manual did not also produce and assemble the actual pellet gun.

Writers who produce technical and professional communication may also work on much more lengthy and sophisticated documents that take a great deal of time, energy, and expertise to produce. Consider for a moment the owner's manual that came with your computer, the many different kinds of reports and specifications needed to develop a spacecraft for NASA, or the lengthy proposals and contracts that circulate between two corporations that plan to merge. All of these are examples of technical and professional communication, too.

ANALYZE THIS

Review the short definition of technical and professional communication presented earlier in this chapter. In keeping with this definition, how does the pellet gun diagram function as technical and professional communication? Explain.

FIGURE 1.1 Pellet gun safety warnings

Many individuals are employed as technical and professional communicators. For instance, some technical and professional communicators work with computer scientists and programmers to help them produce technical manuals for software programs. Others work closely with mechanical engineers to help them explain the workings of, say, a complex airplane engine. However, it is important to note that *not all* people who write technical and professional documents are officially deemed "technical" or "professional" communicators. A great many are simply employees whose jobs require them to write memos, manuals, reports, and other documents that are forms of technical and professional communication. An accountant, for example, may write financial reports, or a web designer may provide technical information on a website. In this book we refer to those who produce technical and professional documents in any capacity as workplace writers. This includes anyone who has to write in his or her job—and that is just about everyone.

IN YOUR **EXPERIENCE**

Think about past jobs you have had. What kinds of communication did these jobs require of you? What were the most challenging aspects of this communication? Did you have to communicate technical information to customers, clients, or other audiences? Develop a list of the kinds of technical communication you had to produce on the job. What characteristics did any documents have in common? In what ways did the documents differ?

GENRES OF TECHNICAL AND PROFESSIONAL COMMUNICATION

This book covers different types of workplace writing with the understanding that these different documents are composed by workplace writers who often must meet different needs:

- Helping to negotiate and create the knowledge that users need
- Creating texts that interact with the various audiences involved
- Making complex information understandable to the audiences who need it
- Serving as a bridge between various groups of individuals, both inside and outside the workers' organizations
- Reducing internal costs of production, support, and development
- Adding credibility and competitiveness to the workers' companies

Therefore, it is not surprising that the kinds of documents produced in workplaces differ in many ways from the documents that students create in their college careers. Instead of essays, term papers, and final exams, workplace writers compose memos, letters, reports, proposals, feasibility studies, websites, and manuals, among others. Each of these different document types—or *genres*—is associated with certain formal characteristics or conventions, such as cover pages, particular writing styles, lengths, fonts, visuals, and professional appearance. Chapters 12 to 23 provide specific information about each of the following document types:

- E-mails and memos (chapter 12)
- Letters (chapter 13)
- Job search documents (chapter 14)
- Technical definitions (chapter 15)
- Technical descriptions (chapter 16)
- Websites (chapter 17)
- Technical instructions (chapter 18)

- Manuals (chapter 19)
- Proposals and requests for proposals (chapter 20)
- Informal reports (chapter 21)
- Formal reports (chapter 22)
- Presentations (chapter 23)

As you make your way through this book, you will learn a great deal about how to produce these different kinds of documents. You'll notice that there are many differences among them; however, you'll also see very quickly that there are some underlying characteristics common to all of them. For instance, Figure 1.2 is an example of one type of professional and technical document—a memo.

See chapter 12 for more information on writing memos.

Memos, in particular, are common workplace documents composed by workers at all levels and produced every day. Composing an effective memo requires knowing spe-

To: Francis Stafford
From: Richard Humber R.H.
Subject: Health-care service rules updates
Date: January 12, 2009

The health-care service rules updates for 2009 are expected to be effective on or about February 11, 2009. The rule changes for 2009 will do the following:

1. Adopt 2007 CPT and HCPCS for coding purposes.
2. Use 2006 RVU data and a conversion factor of $48.49 (1.5% increase).
3. Create a **payment methodology for freestanding surgical outpatient facilities.** Refer to R 418.10923B for billing instructions and R 418.101023 for reimbursement rules.
4. Language in R 418.10912 **requires use of generic medication** or documentation of medical necessity if brand is written for.
5. Increase from $5,000 to $20,000 the threshold when a carrier is required to provide professional review (R 418:101205).
6. Create a **broader fee schedule for laboratory services.** Lab procedures will be paid at 10% over Medicare; pathology services will continue to be paid by RBRVS. The lab and pathology codes will no longer be published in the rules but will be contained in the manual.
7. Update several rules to reflect the department name change from CIS to DLEG and Bureau of Workers, Compensation to Worker's Compensation Agency; also to note the change of HCFA to CMS 1500.

The health-care rules, manual, and fee schedules will be posted to the website when the effective date is known. The updates are effective seven days after the rule package is filed with the secretary of state, and we currently believe that the package will be filed on 2/2/2009.

FIGURE 1.2 An example of a memo

cific formatting traits, such as clear headings with the writer's name, the audience, the date, and a subject line. Memos also include the writer's initials next to his or her name, they begin with short introductory paragraphs, and they end *without* a closing line such as "Sincerely" or "Best wishes." Such formatting traits are specific to memos and are not part of all technical and professional genres. However, memos do share certain characteristics with other genres; memos are, among other things, aware of audiences, and they are clear, ethical, and problem-solving documents. The following section provides a more detailed look at these characteristics of all technical and professional documents.

CHARACTERISTICS OF TECHNICAL AND PROFESSIONAL COMMUNICATION

Technical and professional documents in the workplace—memos, e-mails, reports, and manuals, for example—serve their audiences by incorporating certain necessary and expected characteristics. This section outlines each of these characteristics and explains why workplace writers should strive to produce documents with these characteristics.

Rhetorical

Technical and professional communication is about solving problems and compelling readers to *act*. The action you want them to take may be something overt, such as hiring you for a job, buying your product, or showing up for a meeting on time; or it may be more subtle, such as considering a long-term plan or thinking of your organization in a more positive way. Because technical and professional communication is meant to persuade readers to act, it requires rhetorical thinking and problem solving. We define *rhetoric* as, essentially, the study and practice of using effective language—written, spoken, or visual—to influence others' thoughts and actions. Thus, thinking rhetorically in workplace writing means considering how documents solve problems, affect an audience, and ultimately reflect one's credibility as a worker.

See chapter 2 for more on rhetoric and rhetorical thinking.

For instance, if a corporate accountant is writing a fiscal report for her superiors, she is thinking rhetorically when she considers not only the information that will go into the report but also the purpose of the report, the problems it will solve, and the expectations of her audience for this report, including its layout, design, style, and wording. She also considers how her report will reflect on her integrity and reliability as a worker for that company. Rhetorical thinking, then, involves one's words, one's audience, and one's own role in the production of technical communication.

In many ways a workplace writer solves problems by integrating a variety of perspectives and ideas into clear and workable documents, presenting useful results that are comfortable to readers and users. One of the authors of this textbook, for instance, once assisted more than seventy scientists who had studied the flow of silt in the Atlantic Ocean for the Department of Energy. He had to take their data and compile a single, comprehensive report, integrating the information each scientist provided and using language that both scientists and nonscientists could easily understand. Thinking rhetorically involves making choices about how to best reach readers and users.

Audience Centered

Those who communicate technical and professional information—whether through written, spoken, or visual discourse—must be highly aware of their audiences and the needs of those audiences to *use* the information contained in technical documents. Your own success and effectiveness as a writer will be based on how well you understand and address the needs of your audiences. You have probably heard the realtor's adage about the three most important aspects of a property: location, location, location. In professional and technical communication there is a parallel adage: audience,

See chapter 2 for more on audience awareness.

audience, audience. For this reason, audience awareness appears throughout this book as a central aspect of effective technical communication.

When people write in professional workplaces, they usually write for one of two audiences: an internal audience of colleagues and coworkers inside their organization or an external audience of readers outside their organization. These audiences determine many factors in the document, including the genre, style, level of formality, level of detail, format, design, and length.

This book addresses some specific genres of writing that are considered exclusively internal—such as interoffice memos—and some that are considered exclusively external—such as résumés and cover letters. However, many of the documents discussed here can be *either* internal or external, and some can be *both* internal and external. The significant point is that the technical and professional documents you compose will generally be classified as internal or external, and that classification will play a big role in what the document will say and do. As you read more about the various document genres covered in this book, consider the degree to which they are internal or external and the way that difference affects the documents themselves. In some instances, internal and external audiences may be from other countries and regions of the world or from different cultural backgrounds, requiring workplace writers to develop strategies for writing to transnational audiences. Chapter 5 focuses on writing for transnational audiences, and throughout the book we'll consider how transnational audiences affect workplace writing.

See chapter 5 for more on transnational audiences.

Think about the different kinds of documents you have written during the last few years. In which were you most aware of your audience and the needs of the audience, and in which was the audience most unclear and hard to envision? How did these differences in audience affect how you wrote the documents? Explain in a few short paragraphs.

Technology Oriented

Evolving technologies heavily influence the ways individuals produce and disseminate communications in and from their workplaces, which are often multimedia environments. This means that workplace communication isn't always produced by an individual who sits down to create a document on a typewriter or a personal computer. Instead, technical communication often comes about through conversations via e-mail, text and voice messaging, or tele- and videoconferences. And it might derive from research done on the Internet or with the help of electronic databases. Furthermore, writers in the workplace must often produce documents that are not circulated as hard copies—such as paper brochures, manuals, or reports—but may be electronic in orientation, made for websites, e-mail distribution, or other media forms such as videos and multimedia presentations. Technology decisions depend on the particular project and audience needs.

See chapter 3 for more on technology in technical and professional communication and chapter 17 for more on web design.

Ethical

Although the term *ethics* may conjure up images of philosophers meditating on the nature of good and evil, ethics in technical communication is often a practical, tangible matter dealing with specific issues and circumstances. Ethics is important because communicators work and act in the *real* world with *real* consequences to their actions and decisions. Although *ethics* is a difficult word to define, many people understand it as simply making right or wrong decisions.

See chapter 4 for more on ethics in technical and professional communication.

For technical communicators ethics is a part of their daily work environment. As Mike Markel writes in *Ethics in Technical Communication*, "Presented with inaccurate and incomplete information, assessing unclear and inadequate options, torn between competing claims and goals, {writers} must make decisions."[2] Technical and professional communicators produce, shape, and convey information to audiences that use that information for varying purposes. Writers must seek to provide information that can be used not just efficiently and successfully, but also safely. Producing documents involves making choices, often within gray areas of ethical consideration. Appropriate professional behavior is not always clear-cut.

ANALYZE THIS

As a class or in small groups, discuss the ethical principles that you believe should guide your writing for this class. You might also address the ethical expectations you have for the documents your instructor will give you. Use this exercise as an opportunity to clarify ethical responsibilities. Once you have thought about ethical principles, write a definition of *ethics* for the technical writing classroom.

Research Oriented

In many cases clear and efficient documents are produced only after writers have carried out the necessary research, which may not be the kind of library research that students are accustomed to doing in college. Research in technical and professional fields is multifaceted and context specific. This means that different kinds of documents and different kinds of audience needs will dictate the types and depth of research needed. Writers may conduct many types of research, from Internet searches to interviews, to field and site visits, to databases, to conversations with groups and individuals, both inside and outside the company. Again, the type of research is dictated by the purpose of the document being produced.

See chapter 6 for more on research.

Professional

Because most technical and professional documents are audience and user centered, workplace writers must strive to make their texts abundantly clear and accessible to audiences. *Accessible*, in this case, means (a) that the documents describe every aspect and provide every detail about the subject that an audience needs and (b) that the documents follow the conventions of correct grammar, punctuation, usage, and style. Ignoring such conventions often confuses audiences and diminishes the communicator's credibility. In technical and professional communication, perhaps more than in any other kind of communication, *form* and *content* must be exact. In additional, the tone of documents should be professional and formal in most cases; workplace writers should not typically strive for humor or a casual and laid-back style. The primary goal of most workplace documents is not to entertain, but to inform. As with business attire, if you're not sure of your audience, it is generally better to be a bit more formal and professional in your technical writing than to be overly casual and informal.

See chapters 7 and 10 for more on composing clear and correct professional documents. You may notice that just about every chapter in this book is about producing "professional" documents.

Visual

All writing in the workplace employs visuals and graphics to some degree in order to provide clear, readable, and pragmatic information to the audience. Texts that consciously use visual elements have greater meaning and clarity for readers. Visual

[2] Michael Markel, *Ethics in Technical Communication: A Critique and Synthesis* (Westport, CT: Ablex, 2001), 6.

elements ranging from typographical choice (i.e., fonts) to choice of design and medium, to the inclusion of images and graphics, all affect how information is conveyed to readers. More and more, readers expect dynamic visuals to help them take meaning from a document, and workplace writers must be able to address visual components of documents with the same level of professionalism and accuracy as they use with the written content of a document.

See chapter 8 for more on incorporating visuals into documents.

Many professional writers are adept at using words to provide clear and cogent information; however, far fewer are trained and skilled in using visuals to provide that information. But in our current technology-centered society, linking visuals with writing is crucial for professional and technical communication, partly because many texts and images are competing for our attention and creating something visually dynamic and memorable has definite advantages. Learning how to use the visual dimensions of documents and considering how readers read visuals are part of *visual rhetoric*, which addresses the choices a writer makes in using visual elements.

EXPLORE

Examine the IBM (ibm.com) and Apple (apple.com) web pages, both of which try to effectively combine words and images. As you look over these web pages, think about what kinds of audiences would be searching for and viewing them—would they be different audiences? Discuss how these web pages use images, colors, lines, and particular layouts in order to appeal to their intended audiences.

Design Centered

In most work environments documents are expected to be neat, professional, and formatted to the standards of the particular organization or profession. Related to this is *document architecture*, which refers to the way words and visuals appear on pages and screens. Workplace writers must consider not just what they write but also how that writing will look to audiences. For example, writers must make choices about font types and sizes, spacing, colors, headings, placement of visuals, paper types, and bindings. These choices relate specifically to the most effective and efficient way to provide intended audiences with the information they need.

See chapter 9 for more on designing documents effectively.

Concise

Workplace writers must recognize that their audiences are likely busy and wish to receive as much information as possible in the smallest amount of time and space. Therefore, writers must eliminate superfluous words, phrases, and sentences. A good workplace writer is *efficient* and gets the most out of the fewest words.

Try to think of your sentences as if they are athletes in training. A good athlete will have lots of necessary muscle, but relatively little unnecessary fat. If your sentences are overweight, put them on a strict regimen of diet and exercise, sweating off those unnecessary words and phrases, and you'll have them back in shape in no time. These two examples illustrate the importance of concise writing:

See chapter 10 for more on writing concise sentences.

A wordy (unhealthy) statement:

In the event that an operator is occupied with the operation of the vehicle, it is recommended for reasons of safety that the operator discontinue use of any communication device in order to ensure that focus and attention are paid to the operation of the vehicle and not to the communication device.

A concise (healthy) statement:

Do not use your cell phone when driving.

ANALYZE THIS

Reread the health-care memo in Figure 1.2. Using what you have learned thus far about technical and professional communication, list the specific aspects of this memo that are clear, correct, concise, and professional. Then make a separate list of aspects that do not seem to be clear, correct, concise, or professional.

THE ACTIVITIES OF TECHNICAL AND PROFESSIONAL WRITING

Learning about the characteristics of technical and professional documents requires workers to also think carefully about the processes and activities of actually producing these documents. *Technical Communication in the Twenty-First Century* will teach you *how* to compose different documents and will provide strategies for doing so in workplaces.

Learning to produce technical and professional documents in the workplace requires students to see writing as a process, a series of activities that lead to the production of a document. Although the process may vary from document to document, it is important to consider the most appropriate and useful steps in creating a successful document. The following activities are addressed in great detail in chapters 6 through 11:

Planning	Designing	Editing
Researching	Integrating visuals	Testing
Organizing	Revising	
Drafting	Rewriting	

You are likely familiar with some of these writing activities but unfamiliar with others. This text will show you how crucial these activities are to the production of technical and professional documents in the workplace and how important these activities are in solving complex and sophisticated problems that arise in all of our careers.

IN YOUR EXPERIENCE

Think about your own writing process. What process do you use to write documents? Does the process differ depending on the kind of writing you are doing? For example, does it differ if you are writing an essay for an English class as opposed to, say, a lab report in biology? Does the process differ when you are writing documents for school as opposed to work or personal communication? Explain.

SOLVING PROBLEMS THROUGH TECHNICAL AND PROFESSIONAL COMMUNICATION

See chapter 2 for more on problem solving in workplace documents.

Technical and professional documents solve problems when they inform and persuade and when audiences can act on them. Typically, we think of a *problem* as an unfavorable or disagreeable situation. However, workplace problems are more accurately viewed as questions to be considered or addressed; they need not always have a negative dimension. For example, many workplace documents address the problem of how best to market a product and yield the highest possible sales or how to handle sales when demand is high.

Writers in workplaces must consider two different kinds of problems when they produce documents. First, workplace writers face problems in the real world that must be solved by taking action. These problems require that writers understand the situation facing their company and have a plan or strategy for dealing with it. Second, workplace writers face problems in creating the most effective documents to address the real-world problems. Related issues include who to write to, what type of document to create, what the document should contain, what tone the document should take, and whether to include visuals or graphics.

In other words, writers face two sets of related problems: the *workplace problems* out in the world that need to be solved and the *rhetorical problems* that arise in producing the documents to help solve the external problems. An external problem might be a company's need to determine what kind of water treatment facility it should build. A rhetorical problem might be a writer's need to determine who the actual audience of a feasibility report will be and what information that audience will need.

- Workplace problems require information and action to solve.
- Rhetorical problems require choosing the best approach to communicate information.

Professionals solve field-specific workplace problems all the time, but knowledge about the solution to a problem is different from the ability to communicate that solution to others. Thus, it is common for specialists in different fields and careers to know a great deal about problems and solutions but to have difficulty articulating those problems and solutions to other audiences. Unfortunately, there is no single, direct formula for communicating information properly; most forms of communication must be driven by the context of the problem, requiring workers to develop creative, effective, feasible documents relevant to the particular situation or problem at hand.

Technical Communication in the Twenty-First Century asks you to take up your writing and communication endeavors in the workplace with a sophisticated five-step guide. Our problem-solving approach (PSA) is designed to address the entire process of solving workplace problems through communication, from understanding the problem to releasing the document. The PSA is detailed here:

PSA

Plan

Define or describe the real problem or reason for writing
Establish goals and purposes for writing
Identify stakeholders and what they want or need
Consider the ethical choices involved with the problem
Consider document formats and delivery methods
Identify what information you have and what information you need
Choose technologies that will best assist you and your audience

Research

Determine the types of information necessary and how to get them
Conduct research and gather information
Organize the information you gathered
Evaluate the information and decide if further research is necessary

Draft

Confirm your goals and purposes, document format, and delivery methods

Organize and draft the document

Design and arrange the document

Create and integrate visuals that help communicate the information, if appropriate

Review

Test the usability of the document

Solicit feedback and response from peers and colleagues

Revise or rewrite the document based on feedback

Edit the document to ensure correctness

Distribute

Include related documents or attachments when applicable

Confirm appropriate means to transmit the document to the intended audience

Transmit the document to the intended audience

Follow up to ensure that your audience received the document

Assess the document's outcomes and decide if further correspondence is necessary

The PSA is not always linear; that is, entry points into the process will vary according to the task, and there is no set beginning and end to the process. Writers may conduct only a few of the tasks, they may repeat tasks, or they may approach the tasks in varying orders. In fact, writers may often find it necessary to go back to earlier stages if they have difficulties at any step along the way.

To understand how the PSA might help solve a workplace problem, imagine this scenario: A geophysical contractor is hired by a large suburban city on the East Coast of the United States to explore one of the city's problems—that landfills in the city are becoming overstressed and overburdened because of the city's increasing population. The geophysical contractor is employed to study the problem carefully and to propose to the city a course of action that will lead to a feasible solution. In particular, he must determine whether old landfills should be expanded or new ones should be created elsewhere. Thus, he has a difficult workplace problem to solve (i.e., what to do about the city's landfills) as well as a difficult rhetorical problem (i.e., how to explain his findings in a way that is clear, ethical, and helpful to the audience of city councillors so that they can make appropriate decisions). The contractor might begin with a *plan* to ensure that he understands the problem and is prepared to address it accurately. Next, he might survey the land and do *research* on the area to study the problem. He might then employ a diverse team to help him *draft* and design the report for the city councillors. Then he might *review* it, soliciting feedback from fellow geophysical contractors to determine whether his solution is feasible and effective and revising the report based on their feedback. Finally, he might *distribute* the document to the city councillors by e-mail or in person.

This chapter serves merely as an overview of technical and professional communication in the workplace, so the details of the PSA will develop as you read and study further.

Although the PSA may appear somewhat detailed, it is nevertheless a straightforward approach to problem solving that covers all the steps a workplace writer might need to take to solve a problem with workplace documents. Chapter 2 provides more information about the PSA and allows you opportunities to practice and master this approach. Other chapters look at the PSA in the context of various features, formats, and genres of workplace writing in an effort to make the problem-solving approach both useful and familiar. Whenever you see the PSA icon—which appears to the left of this paragraph—you should consider the ways in which problem solving is tied to the creation of successful workplace documents.

IN YOUR EXPERIENCE

Which aspects of the PSA do you already use when you write? With which are you familiar and with which are you not? Explain your answers in detail, listing specific examples of past essays and other writing projects.

SUMMARY

- Technical and professional communication is communication about complex, highly detailed problems, issues, or subjects in the professional world that helps audiences visualize and understand information so that they can make informed and ethical decisions or take appropriate and safe actions.
- Technical and professional communication includes various kinds of document genres: e-mails, memos, letters, résumés, definitions, descriptions, websites and text messages, instructions, manuals, proposals, informal reports, formal reports, and presentations.
- Technical and professional communication is characterized as rhetorical, audience centered, technology oriented, ethical, research oriented, professional, design centered, visual, and concise.
- Producing technical and professional communication is an activity that involves planning, researching, organizing, drafting, designing, integrating visuals, revising, rewriting, editing, and testing.
- Technical and professional communication involves solving two different but related types of problems: workplace problems and rhetorical problems.
- The problem-solving approach incorporates five steps that guide the process: plan, research, draft, review, and distribute.

CONCEPT REVIEW

1. What is technical and professional communication?
2. What is a workplace writer?
3. Which fields and careers involve written and spoken communication?
4. What does it mean to write clearly and correctly?
5. What does it mean to write concisely?
6. What does it mean to say that workplace writing should be professional?
7. Why should workplace writers concentrate on audience awareness?
8. What does *rhetorical* mean in terms of technical and professional communication?
9. What is ethics?
10. Why is ethics important in workplace communication?
11. What are visuals?

12. Why is research a necessary component of workplace communication?
13. What does it mean to say that writing is a process?
14. What is the problem-solving approach?
15. How does the problem-solving approach function as a guide for writers?
16. In the PSA what does it mean to plan?
17. In the PSA what does it mean to research?
18. In the PSA what does it mean to draft?
19. In the PSA what does it mean to review?
20. In the PSA what does it mean to distribute?

CASE STUDY 1

Mismeasuring: Poor Technical Communication and the Mars Climate Orbiter

Here is a news story about the mission to Mars:

Mars Climate Orbiter Team Finds Likely Cause of Loss

A failure to recognize and correct an error in a transfer of information between the Mars Climate Orbiter spacecraft team in Colorado and the mission navigation team in California led to the loss of the spacecraft last week, preliminary findings by NASA's Jet Propulsion Laboratory internal peer review indicate.

"People sometimes make errors," said Dr. Edward Weiler, NASA's Associate Administrator for Space Science. "The problem here was not the error, it was the failure of NASA's systems engineering, and the checks and balances in our processes to detect the error. That's why we lost the spacecraft."

The peer review preliminary findings indicate that one team used English units (e.g., inches, feet, and pounds) while the other used metric units for a key spacecraft operation. This information was critical to the maneuvers required to place the spacecraft in the proper Mars orbit. "Our inability to recognize and correct this simple error has had major implications," said Dr. Edward Stone, director of the Jet Propulsion Laboratory. "We have underway a thorough investigation to understand this issue."

Two separate review committees have already been formed to investigate the loss of Mars Climate Orbiter: an internal JPL peer group and a special review board of JPL and outside experts. An independent NASA failure review board will be formed shortly. "Our clear short-term goal is to maximize the likelihood of a successful landing of the Mars Polar Lander on December 3," said Weiler. "The lessons from these reviews will be applied across the board in the future."

Mars Climate Orbiter was one of a series of missions in a long-term program of Mars exploration managed by the Jet Propulsion Laboratory for NASA's Office of Space Science, Washington, DC. JPL's industrial partner is Lockheed Martin Astronautics, Denver, CO. JPL is a division of the California Institute of Technology, Pasadena, CA.[3]

Imagine that you're a NASA administrator who worked on the mission to Mars project and who will work on the next mission to Mars. Write a short, informal letter, note, or memo (depending on your teacher's instructions) to the various members

---

[3] Douglas Isbell, Mary Hardin, Joan Underwood, "Mars Climate Orbiter Team Finds Likely Cause of Loss," Mars Polar Lander, September 30, 1999, http://www.iki.rssi.ru/jp/mirror/mars/msp98/news/mco990930.html.

of your team, highlighting the importance of using just one unit of measurement. Using what you learned in this chapter, be sure to remind your team of the importance of the different characteristics of good technical writing as applied to the mission to Mars scenario.

CASE STUDY 2

You'll Shoot Your Eye Out: Rewriting Pellet Gun Safety Warning

You have recently been hired in the quality control division of the Marathon Toy Company, which manufactures the pellet gun and produces the pellet gun instructions presented in Figure 1.1 earlier in this chapter. Many complaints have arisen about the gun and the safety booklet that accompanied it. In the past, Marathon has outsourced the writing of such documents but has decided now to produce them in-house. You have been asked by your supervisor to look carefully at these pellet gun instructions and come up with a detailed plan to update them.

Considering all that you learned about technical communication in this chapter, compose a memo or essay that explains in *detail* exactly how the instructions should be revised. For example, how should the wording, visuals, and design be revised to improve the instructions? Be sure to look at all aspects of the instructions that you believe need changing. The document you compose will be sent to the company's art department, where the instructions will actually be rendered and published. To complete this task effectively, you'll need to consider what you have learned about effective, efficient, and ethical technical communication. In addition, the document you actually write should conform to what you have learned in this chapter. Your instructor will identify what document type you should use for this case study.

CASE STUDIES ON THE COMPANION WEBSITE

Writing in the Workplace: Apalachicola Zoological Sanctuary

This case looks at the rhetorical contexts for producing a brochure for visitors to a small zoological park, which is home to several dozen alligators as well as numerous turtles, rabbits, armadillos, snakes, and birds, all protected in a natural setting.

Ethical Considerations of Writing in the Workplace: Absolm and Brown Accounting

This case asks you to assume the role of an accountant with a client who reveals some information that puts you in a difficult position. You will need to consider how confidential information affects the way you write certain internal documents.

Writing in the Workplace: Gemshine Marine Hardware

This case involves Gemshine Marine Hardware, a company that produces metal fixtures and equipment for use in boats and guarantees its products against corrosion. However, a new system of electropolishing isn't protecting the equipment as it should, and the company must notify customers about the problem. You will address the rhetorical issues that Gemshine is facing.

Shipping Problems: A Lubricant Mix-Up

In this case you will analyze the particular problems facing a warehouse manager trying to rectify a technical miscommunication. You will need to consider how to address multiple employees while maintaining the principles of good technical communication.

VIDEO CASE STUDY

Cole Engineering–Water Quality Division

Writing the Lab-Use Recommendation Report

Synopsis

Chris Key has been assigned to the Simulations Lab team. Following a meeting in the first week, Chris is assigned to write two documents: a short report determining which research team should have first access to the Simulations Lab and an internal memo report summarizing his findings.

WRITING SCENARIOS

1. Consider the words *rhetoric* and *rhetorical*. Where else have you heard these terms? In what contexts? What connotations do they usually hold? How do your previous impressions of *rhetoric* or *rhetorical* differ from the ways these terms are presented in chapter 1? As a group, write a short summary about why rhetoric is important in workplace writing.
2. Locate a visual on your campus—for example, in the parking lot, in a building, or in your classroom—and discuss the purpose of this particular visual in its context. In what ways can or cannot this visual be considered technical and professional communication? Review this chapter's definition of *technical and professional communication*.
3. In a small group, compile a list of all the types of writing you have done in school, on the job, and in your personal lives. This should be a fairly lengthy list of the different documents you all have produced in these different environments. Go through the list and decide which can be considered technical and professional documents and which cannot. Explain your answers in detail.
4. On the Internet or in your local library, spend some time researching your future profession. In particular, pay attention to the kinds of problems that workers in this profession often face. What are they? What kind of writing and communication is expected in this profession? Who will you have to communicate with? In what capacity? What sorts of documents will you have to produce? How do you think these documents will help solve the problems faced by those in your future profession? To help answer these questions, you might interview someone who already works in your future profession and ask what kinds of problems he or she solves and how communication and writing help solve them.
5. Locate a document that you believe communicates technical information, such as a brochure, a newsletter, a manual, a set of instructions, or a report. Which aspects of this document do you think came from research? As far as you can tell, what kind of research do you think the writer conducted in order to put the document together? Explain briefly in a short, informal essay.
6. Go to your college or university's library website. Browse around the site for ten to fifteen minutes, trying to get a good sense of how the website works. Then, in a short writing assignment, explain what you discovered. What kind of research can you conduct at this library? What kinds of sources would you find? Which are most likely to contribute to research in technical and professional communication? What difficulties did you have navigating the website? What questions do you have about using your library (and its website) as a research tool?

7. On a daily basis we are surrounded by texts and images—on TV, radio, the Internet, billboards, magazines, newspapers, textbooks, and signs. Many social and cultural critics suggest that we are often manipulated by the texts and images we see—consciously and subconsciously. We might ask whether we live in a world of unethical messages that try hard to persuade us to do things. Write a short essay that discusses the kinds of everyday texts we encounter that could be considered unethical, and support your position with specific ideas and examples.
8. Write about a time in your life when you witnessed or were the victim of someone who communicated information unethically. In detail explain what happened and why you believe the act of communication could be considered unethical. If possible, write about a situation involving technical or professional communication.
9. What kinds of problems have you solved in the past by using effective communication? Think of one example, and explain in a short essay what the problem was and how you used a form of communication to solve it. Look carefully at the PSA. Were there any facets of the PSA that you used while you were working to solve this problem? Be detailed and specific in your response.
10. Review chapter 1. In what ways is this chapter a technical document, based on the characteristics of technical communication discussed here? In what ways is this chapter audience centered, and how does it cater to its users (i.e., you, the student)? Write a short, informal essay to a classmate or your instructor, explaining how chapter 1 itself is a technical document.
11. In what ways do technical and professional documents differ from the following kinds of texts: literary texts, such as novels or short stories; academic essays that you might write for, say, a history or sociology course; or news stories on the front page of a newspaper such as the *New York Times* or *USA Today*? Are there any similarities between technical and professional documents and these other kinds of texts?
12. What kinds of documents do you think are usually not written clearly and concisely (e.g., legal documents)? Locate any technical or professional document that you think is not written clearly and concisely, and explain in a short essay which parts of the document are not clear and concise.
13. Look closely at your favorite website—especially its homepage—and study it as a type of technical document. Given the text, the images, the colors, the fonts, and the kinds of links available, what kinds of audiences are being catered to? Does the website seem more accessible (or friendly) to certain kinds of users than it does to others? How does or doesn't the website fit the characteristics of technical and professional writing described in this chapter—clear and correct, concise, professional, aware of audience, ethical, rhetorical, visual?
14. Technology plays an important role in technical communication. As thoroughly as you can, construct a timeline showing how communication technologies have changed in the past twenty years. You might discuss, for example, how pens, typewriters, computers, word processors, e-mail programs, or PDAs have influenced technical communication.
15. In a college-level dictionary or etymological dictionary, look up the words *technology* and *technical*. Where do these words come from, and what were their original meanings? How do these definitions help you better understand the role and purpose of technical writing in the workplace? How do these words

relate to other words and phrases in our society, such as *technician*, *technique*, or even *technical foul* (as in a basketball game)? In your opinion, is technical communication about subject matters that are technical, or is there something about the writing itself that is technical?

16. What other blunders have happened because of communication errors in the workplace? Do some basic research on an Internet search engine to locate other examples of technical communication mistakes that have led to blunders like the mission to Mars. Try using keywords such as *technical communication mistakes*, *bad technical communication*, and so forth.

2

Rhetoric and Technical Communication

CHAPTER LEARNING OUTCOMES

After completing this chapter, you will be able to do the following:

- Know the differences between **workplace problems** and **rhetorical problems**, as well as how they relate to one another
- Understand what it means to **think rhetorically**
- Recognize the different components of the **problem-solving approach**
- Know how the PSA ties into **rhetorical thinking**
- Understand **exigency** and its relationship to problem solving in workplaces
- Know the different purposes of workplace writing, including **informing**, **defining**, **explaining**, **proposing**, and **convincing**
- Recognize that audiences have **expectations** and **attitudes**
- Recognize that audiences **use** documents in the workplace
- Recognize that **multiple audiences** often read documents
- Recognize how problem solving is influenced by **contexts** and **constraints**
- Understand the ways writers achieve **credibility**, particularly by producing **correct** documents, showing their **experience** and **expertise**, showing **goodwill**, and demonstrating **identification** with audiences

DIGITAL RESOURCES

On the Companion Website www.prenhall.com/dobrin:

- Case 1: Elegance Limousines: The Difficulties of Cross-Cultural Communication
- Case 2: The Baby Comfort 5.2: The Whistleblower's Dilemma
- Case 3: Carolina Construction: Meeting the Needs of Multiple Audiences
- Case 4: Is the Customer Always Right?
- Video Case: Planning the Shobu Automotive Proposal
- PowerPoint Chapter Review, Test-Prep Quiz, Exercises and Activities
- TCTC Visual Rhetoric and Design Tutorial

REAL PEOPLE, REAL WRITING

MEET DARREN BAREFOOT • "Head Geek" at Capulet Communications

What types of writing do you do in your workplace?
I do a broad variety of writing for my company—everything from technical white papers to website text to advertising copy. I contract out most of our pure software documentation but generally edit the content myself.

Can you describe an important document you recently created at work?
I recently completed a business blogging strategy document for a large software company here in Vancouver. It was a twenty-five-page blueprint for the company's blogging strategy over the next twelve months. This document was unusual in that I've got some expertise in this area, so I didn't have to do much research. For me, writing is usually 80 percent research and preparation and 20 percent actual writing. This project was the reverse of that.

Who are the typical readers of your workplace documents?
Because I mostly write technical marketing materials for high-tech companies, I generally divide my audience into two groups: the suits and the geeks. For the suits my writing will emphasize the business benefits and go easy on the technical detail. For the geeks I'll pitch the technical angle.

What are your primary purposes or goals in writing?
Most of my writing is meant to help promote or persuade. I do my best never to overstate a product's capabilities—I'm aware that most marketers lie at least some of the time. This is easier to do with software, for which claims are relatively easily proven or disproven.

How do you establish a bond or connection with your readers?
I always try to write with an appropriate tone and diction. For example, the suits speak a different language than the geeks. The first step in establishing trust with readers is using language that they're comfortable with. After that, it helps if you tell them the truth.

Is collaboration an important part of your writing?
I'll collaborate with our clients to do research and establish the basic goals and parameters (e.g., audience, length, tone, format) for the document. In addition, most of the documents I write are reviewed by a colleague here at Capulet. The more marketing focused a document is, the more I'll work with others (sometimes including the audience itself) to achieve the appropriate message.

What advice about workplace writing do you have for the readers of this book?
Whatever you're writing, readers can tell when you're faking it—that is, when you haven't done enough research or don't understand the subject matter. Listen to the voice in your head when it tells you to go back to the drawing board—it'll save you a lot of pain in the long run.

INTRODUCTION

We're probably all familiar with the phrase "Houston, we have a problem." These words, uttered by Apollo 13 astronaut Jim Lovell, are now used to connote any dire situation. Sometimes we use the phrase ironically, applying it to situations that seem to have no easy or immediate solution or that are beyond our control. In 1970 the Apollo 13 lunar mission suffered a series of problems: an explosion aboard the spacecraft, leaking oxygen tanks, power failures, and navigational problems, among others. Most people now credit those working in the mission control center in Houston, Texas, with devising the solutions to all these problems and getting the crew back safely. The mission control crew had to devise and discuss many different options.

Almost forty years later, we know that these problems were solved with ingenious solutions, solid teamwork, and good communication skills. Those in mission control, for example, had to explain in detail to the crew how to put together various improvisational and technical gadgets that would save their lives, including a "mailbox" device made of spare parts that served as a kind of vacuum to clear the spacecraft of poisonous gas. To this day, the Apollo 13 mission is viewed as one of the best examples of innovative problem solving.

Our lives require problem-solving activities each day, even if they are not as dramatic as those of Apollo 13. We may face relatively simple problems, such as what clothes to wear for a job interview, or more complex problems, such as how to complete a twenty-five-page research paper or how to help our company fight a hostile takeover from a corporate giant. It is important to recognize that each of these problems has many solutions or strategies for dealing with it. Many different clothing combinations are appropriate for job interviews, a twenty-five-page paper can be written in a number of satisfactory and impressive ways, and a corporate takeover can be successfully fought by various means. In other words, when this book talks about problem solving, there are usually many possible solutions to any given problem. The point is to find, explore, and create the best solutions among these possibilities. The purpose of this chapter is to give you the tools and techniques to engage different kinds of problems and solve them effectively.

A problem is any matter that poses an obstacle that needs to be addressed to promote harmony, improvement, and/or agreement for stakeholders. This chapter discusses two separate but often-interrelated kinds of problems: workplace problems and rhetorical problems. Workplace problems are simply those that present obstacles to a particular workplace environment, such as a company, institution, agency, or office. Not all workplace problems are negative, some arise because of positive factors, and some deal with how to maximize or emphasize the positive. In the following list of workplace problems, note that the third example can be attributed to a positive situation:

- A manager of a large grocery store finds that employees continually show up late or miss work altogether.
- A pharmaceutical company is sued for millions by a group of consumers who claim that a particular drug directly causes heart attacks.
- An Internet-based real estate agency sees its client list grow tenfold in a one-year time period and must figure out how to effectively handle this sharp increase.

Rhetorical problems relate to the difficulties or troubles involved in the production of a text or document. Because solving workplace problems often involves producing documents along the way, the two kinds of problems are frequently connected. As chapter 1 noted, rhetoric is the study of the ways that texts—written, spoken, and/or visual—work to create effects on and solve problems for intended audiences. Solving rhetorical problems, then, means finding ways to produce a text or document that affects problems and audiences in a desired manner. The following are examples of rhetorical problems that connect with the previous workplace problems:

- The grocery store manager must make the point to employees that lateness and missed days are not acceptable. She must decide whether to communicate this to employees verbally or through a written reprimand.
- The pharmaceutical company executives must speak and write to consumers, lawyers, medical experts, and company employees to ensure a successful outcome to the trial. One can easily imagine various meetings, notes, memos, and reports involved in such a scenario.
- The Internet-based real estate agency must research the causes of its growth and its customer needs and expectations and develop an updated website that can handle these various customer needs.

All of these rhetorical problems would pose challenges for those involved because there are multiple ways to produce appropriate documents, whether written or spoken. Just as options exist for solving workplace problems, options also exist for solving rhetorical problems; rarely is there only one way to handle either a workplace or a rhetorical problem.

WORKPLACE PROBLEMS

In *Strategies for Creative Problem Solving*, H. Scott Fogler and Steven E. LeBlanc discuss the importance of defining problems, and they provide the following example:

> A student and his professor are backpacking in Alaska when a grizzly bear starts to chase them from a distance. Both start running, but it's clear that eventually the bear will catch up with them. The student takes off his backpack, gets his running shoes out, and starts putting them on. His professor says, "You can't outrun the bear, even in running shoes!" The student replies, "I don't need to outrun the bear; I only need to outrun you!"

Fogler and LeBlanc provide this story to make a distinction between a real problem and a perceived problem. The *perceived* problem was initially to outrun the bear; however, after just a few seconds the student recognized that the *real* problem was not to get caught by the bear. In short, real problems are not always easy to recognize and define at first because perceived problems can sometimes get in the way. Thus, it's important that in the planning stages of the PSA, we spend time ensuring that we have adequately and accurately defined the real problem; otherwise we may be wasting time, energy, and money trying to develop solutions to problems we don't truly understand. Note that the planning stage of the PSA addresses this directly:

PSA

Plan

Define or describe the real problem or reason for writing
Establish goals and purposes for writing
Identify stakeholders and what they want or need
Consider the ethical choices involved with the problem
Consider document formats and delivery methods
Identify what information you have and what information you need
Choose technologies that will best assist you and your audience

Fogler and LeBlanc provide another real-life example of a workplace problem that is misunderstood because a perceived problem gets in the way. In the example, guests staying on the upper floors of a high-rise hotel continually complained that the elevators were too slow. The hotel manager then instructed his assistants to figure out how to speed up the elevators. The assistants called the elevator company and learned that nothing could be done, so the manager tried to solve the problem by building a new elevator shaft. However, before the new elevator was constructed, the real problem was discovered. The perceived problem was needing trying to speed up the elevators, whereas the real problem was minimizing the guests' complaints. To take the guests' minds off the wait time for the elevators, the manager installed mirrors on all floors in the elevator corridors. Guest complaints quickly stopped. The time guests spent admiring themselves in the mirrors was enough to keep them from thinking about the wait.

See chapter 6 for more information about researching and analyzing information as well as interviewing people who might have insight into the problem at hand.

To make sure that we are defining and describing the real problem, it is important to spend time studying the problem. This effort involves, when possible, and appropriate, studying the problem firsthand, researching and analyzing information and data, and communicating with others—internally and externally—who are familiar with the problem.

Defining and describing the problem may happen more than once while developing solutions to workplace and rhetorical problems. Workplace writers and problem solvers constantly encounter new issues that may require them to redefine problems. In other words, once a problem arises, it does not always remain static and unchanging. As workplace writers begin to understand problems, they may encounter new or related problems they hadn't originally envisioned.

IN YOUR **EXPERIENCE**

It is common for people to face workplace problems on a daily basis—some small and some large. Think of a workplace problem you have encountered in the past for which you perceived the problem in a way that may have covered up the real problem. If you have not had a job in the past or cannot think of a workplace problem, think of any problem in your life in which the real problem differed from the way you initially perceived the problem. In two or three paragraphs, write about this scenario.

RHETORICAL PROBLEMS

Rhetoric is a word used loosely in the media and elsewhere; many use it to suggest that a speaker or writer is up to some trick, skirting the truth, or outright lying. In short, *rhetoric* is often used as a pejorative term to label someone who uses words in an unethical or untruthful way, as the quotes about President George W. Bush and Senator John Kerry illustrate (see Experts Say).

EXPERTS SAY

"President Bush's recent claim that weapons of mass destruction have been found in Iraq highlights two disturbing trends in rhetoric from the White House."

BRYAN KEEFER
http://www.spinsanity.org/columns/20030612.html

EXPERTS SAY

"{John} Kerry's rhetoric on Iraq simply doesn't match reality."
SENATOR JOHN CORNYN
http://www.freerepublic.com/focus/f-news/1144096/posts

This view of rhetoric, despite being commonplace, is oversimplified. The term is actually more complex and has been defined in different ways by different people throughout history, most notably by ancient Greeks such as Aristotle, who defined *rhetoric* as the study of the available means of persuasion in any given situation. No singular definition is likely to be accepted any time soon; however, most scholars agree that rhetoric relates to how words persuade people, or how words work and affect audiences. In this sense rhetoric could refer to deception or falsehood, but it usually does not.

This chapter is not about definitions of *rhetoric* but instead presents rhetoric as a way of thinking about and approaching workplace problems that can be are solved with the help of documents. Rhetorical thinking and problem solving help you understand how writing works in contexts and how writing, writers, and audiences are all part of those contexts. This chapter will not provide tidy definitions of *rhetoric* but will help you to start thinking rhetorically. The problem-solving approach, which was presented in chapter 1, is the main tool used throughout this book to help generate rhetorical thinking. It is printed on the inside back cover of this book for easy reference.

An ancient rhetorician named Zeno once discussed the "open hand of rhetoric." He was referring to oral speeches rather than written texts, but one of his points was that rhetoric allows speakers or writers options. The open hand symbolizes reaching out to an audience to engage in an open discussion of the situation or subject. Each situation or subject can be viewed in different ways and, often, can be solved or addressed in different ways. Engaging in a dialogue is often the best way to approach a subject, since it invites options rather than a simple resolution, which may not, in the long run, be the most productive or useful.

Recall from chapter 1 that the problem-solving approach is not linear; that is, writers do not necessarily move through the steps in sequential order. Instead, different problems and writing situations necessitate that writers bounce around the PSA and address the steps most suited to their needs at the time. Good rhetorical thinking involves knowing how to move through the PSA appropriately for any given situation. Thus, the PSA allows you to be open with your words and ideas, to figure out how best to put them together in any scenario or situation. There is not just one way—each problem or situation is open to many possible solutions and responses.

EXPLORE

Use an Internet search engine to look for definitions of *rhetoric* or *rhetorical*. Try to find at least three or four different definitions, and read them carefully to explore their similarities and differences. In a few paragraphs summarize your findings, and discuss any additional ways you see the relevance of rhetoric to workplace writing.

Here are some different situations in which individuals encounter workplace problems that require the writing of documents as part of the solution. Of course, producing each document brings up its own set of rhetorical problems:

- The president of a large automobile manufacturing company wants an update on the status of a new braking system to be put in all new models. The president is not a technical expert in this area, so a mechanical engineer for the company writes a formal report to the president that details the new braking system in a way that is clear and understandable to him.
- The manager of a large department store sees an increase in sales that involved stolen credit cards. She must research current policies and laws on this issue, develop a new procedure for accepting and processing customer credit cards, and communicate the new procedures to all employees involved in customer sales.
- A web designer is asked to create a new site for a grassroots environmental organization, visited mainly by environmentally minded citizens in the southeastern United States. Before beginning, the designer must learn a great deal about the organization, the issues they are concerned with, and the intended audience for the website. Because it is a grassroots movement, there is little money to compensate the designer, who must figure out how to develop the site on a limited budget.
- A stockbroker recognizes a trend in the market and thinks she can use her insight to help her clients make a quick profit. She wants to communicate this information to old clients as well as use it to obtain new ones, so she must decide on the means to get this information across effectively and reliably.

- The head of human resources at a community college notices a large increase in the number of calls from employees who have questions about their benefits packages. Obviously there is a good deal of confusion about the paperwork involved. The head of human resources and his staff write an instruction booklet for faculty and staff that explains these benefits packages, such as the different health-care plans and instructions on filling out the applications and paperwork correctly. He must figure out the best way to deliver these instructions to the large group of employees who need them.

On the surface these different scenarios seem quite commonplace, hardly requiring a process such as rhetorical thinking or problem solving. The PSA, however, is designed for all such situations—those that call for long, complex documents as well as those that need only short, simple ones—because it guides writers through all aspects of composing documents: from discovering why communication and problem solving are necessary in the first place, to using the best and most effective means of transmitting the finished document to audiences. Writers must approach these situations with open minds that allow the processes outlined in the PSA to lead them to answers and solutions suitable to their audiences.

At its most basic, *thinking rhetorically* involves recognizing and understanding that problem solving and composing documents, regardless of the situation, usually have a few things in common:

- An *exigency* (i.e., a situation, event, or impetus) that causes a workplace problem that needs to be solved
- A *workplace writer* whose purpose is to produce documents that help solve problems
- An *audience* that can receive, understand, and use the document to help solve problems
- A particular kind of *document* that serves as the medium between the writer and the audience
- *Contextual factors* and *constraints* that influence writers, audiences, and documents during problem-solving activities

The rest of this chapter explains in detail that these elements are crucial and interrelated aspects of writing in the workplace. It also shows how the PSA helps writers sort through these components.

IN YOUR **EXPERIENCE**

Recall two different documents you have written in the last year or two. What were the exigencies, purposes, and audiences of these two documents? Which steps of the PSA do you specifically remember using? Why did you feel the need to use these particular steps? Explain in a short, informal essay.

Exigency and Purpose

Exigency is the impetus that requires some kind of writing or communication—a real-life problem to be solved, an issue to be addressed, or a situation that requires a response, explanation, or additional information. For example, customers may become suspicious that a product is not as safe as it could be, stockholders may be anxious that a company's profit margins are shrinking, or a corporation may need more sales representatives in South America. Each of these examples demonstrates that some form of communication is necessary to resolve the issue at hand.

If customers are suspicious about the safety of a product, the company might need to provide a written or spoken press release addressed to these consumers to alleviate their anxieties. If stockholders are fearful that profits are shrinking, the company might send them a report to explain the reasons for the losses and detail how profits will return in the future. If a corporation discovers that it needs more sales representatives in South America, it might send letters to its U.S. sales representatives, inquiring about their interest in transferring to South America. Many such exigencies arise in workplaces every day, new situations that need writers with strong rhetorical skills.

When the Plan stage of the PSA asks you to define or describe the real problem, part of the task involves understanding the exigency of the problem. Recognizing how and why a problem arises goes a long way in helping one define or describe the problem. This, in turn, leads to better understanding of the purposes for composing documents. Because each situation derives from a different exigency, writers in the workplace always have different purposes for writing.

Students in college who write essays for a technical communication course might do so to receive a good grade or to learn writing skills that will help them succeed after college. However, this chapter asks that you start thinking about purpose a bit differently, considering not just *your* purpose for writing but the purpose of the document itself. Looking at purpose in this way allows writers to think rhetorically—to consider what documents accomplish, whom they reach and affect, and how they reflect on the writers themselves.

Let's focus first on purpose in terms of what a document does or accomplishes. You know that a résumé, for instance, has a different purpose from that of a manual, a report, or a proposal. Each is designed to do something different and to reach a different kind of audience. Here are some different purposes that documents can serve:

- *To inform*—to present information or answer questions; for example, a memorandum written by a supervisor that informs his employees that new software is being installed on their computers
- *To define*—to elucidate or illustrate the characteristics of something, for example, an instruction manual written by software engineers that defines for users a new feature on a graphic-design program
- *To explain*—to clarify why and/or how something happened or will happen or how a process works; for example, a letter from an airline to potential passengers that explains the process by which they can enroll in a frequent-flyer program
- *To propose*—to recommend that a particular course of action be taken or a decision be made; for example, a report by civil engineers to company executives that proposes that more testing take place on a new alloy before it is used in a building design
- *To convince*—to persuade an audience to do something or to understand an issue in a particular way; for example, a memo from a manager at an accounting firm to her employees that convinces them of the necessity of coming in to work on Saturdays

Many documents have more than one purpose. As chapter 22 shows, a formal report might need to inform, explain, propose, and convince. Therefore, don't assume that a document serves only one purpose.

The Plan stage of the problem-solving approach allows writers to discover the purposes of their writing. The first two items in this stage deal specifically with problem solving and the purpose for writing. Yet problem solving and purpose cannot be fully understood without considering stakeholders, or audience, as well as ways to meet the purpose for writing in an ethical fashion (chapter 4 discusses ethics).

Audiences

If there is a guiding principle for workplace writing, it is that audiences determine how and what one writes; never underestimate how important audiences are to the creation of documents. To see that concept in real-life terms, begin this section with the Analyze This exercise.

ANALYZE THIS

Envision the following situation: You're waiting just outside your apartment one morning at 8:30. Your best friend is supposed to pick you up and give you a ride to campus so that you can make your 9:00 class. At 8:45 your friend still has not shown up, and you begin to worry that you will not make it to class in time. After repeated phone calls, the friend has still not answered. When you finally and begrudgingly decide that you'll have to walk to campus, it begins pouring rain. But before you start out in the thunderstorm, you decide to leave a note on your apartment door in case your friend finally shows up.

What might this note say to your best friend? Now, what if it wasn't your best friend who was supposed to pick you up but was instead your spouse or significant other? What would the note say then? What if it was your boss? a grandparent? a sibling?

In the Analyze This exercise, the content of the note is shaped by its purpose, your emotions, and the unfortunate situation (i.e., the exigency). However, the writing style, the choice of words and phrases, and the tone of the note are determined by the particular audience. Even though this is a simplistic example, it gives you a sense of how audiences affect writing, and it indicates how writers analyze their audiences in everyday life without really thinking much about it. Writing in the workplace is different in degree from this example, but it is not different in kind. Writers in the workplace must carefully analyze their audiences and understand how those audiences will interpret and use the documents being produced.

Audiences Vary Significantly

Audiences in the workplace include individuals in different departments and divisions and individuals who have different positions, backgrounds, and areas of expertise (i.e., internal audiences). Audiences in the workplace also include those who do not work for the company or organization (i.e., external audiences). Workplace writers classify audiences according to their knowledge about the subject, their expectations and attitudes, and their ability or power to use or act on the document. Such classification can become more complex when writing for transnational audiences.

See chapter 5 for additional information about transnational audiences.

Classifying audiences allows writers to determine which document genre to use, what information to include, which style of writing to adopt, what language or jargon to use, and which organizational strategies to follow. Writers often need to produce documents intended for multiple audiences and sometimes must produce multiple documents about the same subject matter, each intended for a different audience.

A civil engineer, for example, who writes a report about a new bridge design, produces a distinctly different report for other engineers than for his company's public relations department or for a company manager with no engineering background. The report for other civil engineers can include technical terms and concepts that the other individuals probably would not understand. Workplace writers know how to appropriately rework a document for different audiences. Look at the memorandum in Figure 2.1 and see what hints indicate the intended audience.

Memorandum

To: M. Miller
From: F. Scott
Subject: New SPAM Filter
Date: February 13, 2009

Because of recent problems employees have had with accumulated junk mail in their e-mail inboxes, the IT Department is starting a junk e-mail reduction program. Starting March 9, 2009, all junk e-mail will be blocked by the Thunder Clap E-mail Traffic Control System. Any e-mail identified as spam will be filtered by Thunder Clap. The following day employees will receive an e-mail from the system letting them know which e-mails have been quarantined. Links in the e-mail from Thunder Clap will allow employees to save or delete quarantined mail.

A short video explaining the Thunder Clap System can be found at the following URL: http://www.thunderclapsystems/email_traffic_control/training/video.html. The more employees use the system, the smarter it gets, recognizing which e-mails to filter and which to let through, based on users' instructions.

I believe Thunder Clap will reduce problems employees have with spam and allow them to better manage their e-mail accounts and their time. If you have any questions, please contact me by phone (555) 456-8750 or by e-mail, francine.scott@itechus.com.

FIGURE 2.1 A workplace memo

ANALYZE THIS

Reread the memo in Figure 2.1. Who is the audience for this memo, and what helps you determine that? What is the purpose of this document, and can you discover an exigency that likely gave rise to this memo? What additional information would you need to better understand the rhetorical situation? Respond to these questions with informal written responses, or discuss them in small groups.

Audiences Have Expectations and Attitudes

Audiences are not passive recipients of the information presented in workplace documents. In fact, audiences usually have predetermined expectations and attitudes about those documents and the information within them. Audiences typically expect documents to follow genre conventions, and they expect to receive information that allows them to do something. In addition, audiences usually hold attitudes about particular subjects—positive, negative, or neutral—and writers must effectively address those attitudes.

- *Positive attitude.* When writing a document for readers with a positive attitude about the subject, writers reaffirm that the readers' attitudes are correct.
- *Negative attitude.* When writing a document for readers with a negative attitude about the subject, writers must figure out what the readers object to and explain how their objections are unfounded or how benefits outweigh the objections. Writers should not approach the subject matter as if they are quarreling with the readers. Instead, writers try to bridge gaps between themselves and the readers by looking for common ground, areas where they all share attitudes or beliefs, and thereby increasing the likelihood that readers will be willing to change their attitudes about the subject.
- *Neutral attitude.* When writing a document for readers with a neutral attitude about the subject, writers persuade the readers to agree with them by showing the positive results of doing so and the negative consequences of not doing so.

Workplaces are filled with people who do not always get along; many ideas and beliefs are met with stiff resistance from coworkers and other audiences. Consequently, it is crucial for workplace writers to understand the ways their documents address audiences with varying expectations and attitudes and to see each writing endeavor as distinct. Learning how to formulate documents in diverse situations is a highly valuable skill for today's workers, no matter what their career field.

IN YOUR **EXPERIENCE**

Recall a time in your life when you tried to write or speak about a topic and your audience disputed your viewpoints. What was your position on the subject? What were your audience's objections to your position? What strategies did you use to get your audience to see things your way? Did these strategies work? Why do you think they did or didn't work? Respond to these questions in an informal writing, or discuss them in small groups.

Audiences Use Documents Differently

Not surprisingly, different audiences use documents in different ways, and how they use them should help determine how you write them. Readers might do any of the following with a document:

- Skim it quickly
- Read it partially
- Read it closely from beginning to end
- Revise it and return it to you
- Take a related course of action
- Make a related decision
- Use it in support of other documents

Because writers shape their documents according to how readers will use them, a strong correlation exists between a writer's purpose and readers' use of the document. Getting a feel for this relationship takes practice and experience in the workplace, although audience needs make certain writing strategies apparent. For instance, if a writer knows that readers will only skim the document, it is important to put pertinent information near the beginning or in easy-to-read sections with bulleted lists instead of in traditional paragraphs. Or, if a reader is using the document to make a decision, the writer must clearly explain what choices the reader has and clarify the consequences of each. The ways readers use a document also determine what genre or format the document follows. Later chapters in this book explain in detail what kinds of genres are used in particular problem-solving situations and for particular audience needs and uses.

Multiple Audiences Often Read Documents

In some cases workplace writers prepare documents that are read by multiple audiences; that is, more than one kind of audience will read the document. Memos, for instance, are often written for all workers in a company and use generalized, nonspecific language so that all readers can understand them. Writing for multiple audiences becomes more difficult when writers create longer, more sophisticated documents such as formal reports, which are discussed in chapter 22.

In these situations writers may include distinct sections, each of which is geared for a particular kind of audience. Such a report, for instance, might provide an overview or summary intended for a manager or supervisor who does not have time to read the report from cover to cover. On the other hand, the detailed body of the report might be written for experts or other interested readers who need to read through the entire document carefully. When writers are producing highly technical documents, they often attach appendixes to their reports to include specific kinds of technical data of interest only to technicians and other experts.

In short, writers cannot always assume that their work will be read by a single audience type or by a single professional community. When multiple audiences need to access documents, writers should enable them to effectively and efficiently find the information they need. Later chapters detail common multiple-audience documents, such as websites, manuals, proposals, informal reports, and formal reports.

EXPLORE

Go to two or three websites that you frequently visit. Look carefully at these sites, perhaps exploring links that you usually ignore. Explain the purpose of each website as well as the probable audiences for each. Are there multiple audiences for each, or do you think the audience is pretty homogeneous? Support your answers by giving details from the website itself. Respond to questions in a short writing, or discuss them in small groups.

Workplace Writers

Sometimes it's easy to forget that real people produce the documents we see. For instance, when we see billboards on the freeway, an instruction manual for a computer, or an application form for employment, we often forget that a writer or a team of writers actually sat down to produce those documents. Much of workplace writing, however, requires writers to put their names on their documents. Therefore, it is important that writers consider how their work makes them appear.

Good workplace writers always consider the ways their writing reflects their *ethos*—that is, their character, or credibility. Because a document always reflects the writer, workplace writers want this reflection to be one of correctness, experience, expertise, and goodwill.

Correctness

Writers achieve credibility when their documents are produced correctly:

- Using the correct genre for the particular writing situation. For instance, detailed financial information for a client might require a report instead of a letter or a memo.

- Using the correct conventions and standards of the genre. Formal reports, letters, and memos, for example, all have particular standards that must be followed because readers expect them.
- Editing carefully for spelling, grammar, and mechanical errors. Such mistakes are unacceptable in workplace environments.

See chapter 10 and appendix A for more on correctness.

Making sure that workplace documents are correct enhances a writer's reputation and trustworthiness and, by extension, enhances the reputation and trustworthiness of the company or organization that employs the writer.

Experience and Expertise

Writers want to assure their audiences that they have the experience and expertise to produce documents that solve problems. If a building contractor is putting together a proposal to complete a new wing on a local shopping mall, she can garner credibility with the audience by outlining past experiences with similar projects and explaining her qualifications for the job. Even those applying for jobs right out of college can write cover letters detailing how past experiences make them suitable for certain positions. In short, workplace writers build credibility when they show their audiences that they have the experience and expertise to deal with the situation at hand.

Goodwill

Writers must be sure not to focus too much on themselves; audiences are often turned off when documents concentrate too much on the writer or the writer's organization. On the other hand, writers achieve credibility when they show *goodwill*, that is, when they explain what they or their companies can do for the audience. The building contractor mentioned earlier must ultimately show goodwill; her proposal must focus on how the audience will benefit from her services. Writers who establish goodwill do not concentrate on themselves but on what they can do for the audience.

Identification

As writers establish their reliability, they should also show audiences that they share common ground, often that they identify with the readers' goals or perspectives. Think about what happens when you meet someone new. Most of the time the beginning of your conversation with this person is a search for similarity, for things you have in common. Where are you from? What is your major? What kind of music do you like? If you have a lot in common, you're more likely to get along and more likely to trust this person. Writing in the workplace is not much different. Writers want to gain credibility by showing their audiences that they share common beliefs, ideas, goals, and visions.

Trust

All of these elements—correctness, experience and expertise, goodwill, and identification—must be considered in relation to a writer's purpose. Writers always want to establish credibility, but different workplace situations call on them to use these strategies in different ways. In some cases writers need to focus on their experiences, whereas other situations necessitate more emphasis on goodwill. Being a good writer means being able to achieve credibility with diverse audiences in diverse writing situations. If audiences do not trust or believe writers, they will not take their writing or their companies or organizations seriously.

Writers must make sure their solutions are appropriate, their documents are formatted and arranged correctly, and those documents are revised and edited appropriately for their audiences. The drafting and reviewing components of the problem-solving approach deal in part with writers' credibility:

PSA

Draft

- Confirm your goals and purposes, document format, and delivery methods
- Organize and draft the document
- Design and arrange the document
- Create and integrate visual elements that help communicate the information, if appropriate

Review

- Test the usability of the document
- Solicit feedback and response from peers and colleagues
- Revise or rewrite the document based on feedback
- Edit the document to ensure correctness

ANALYZE THIS

Examine the e-mail provided in Figure 2.2, which was sent by a job seeker to the human resources department at a regional newspaper to inquire about an opening in the purchasing department. Given the discussions in this chapter about writers and their credibility with audiences, analyze the ways this writer might gain or lose credibility with this e-mail. Write your response in a short essay, or discuss this scenario in a small group.

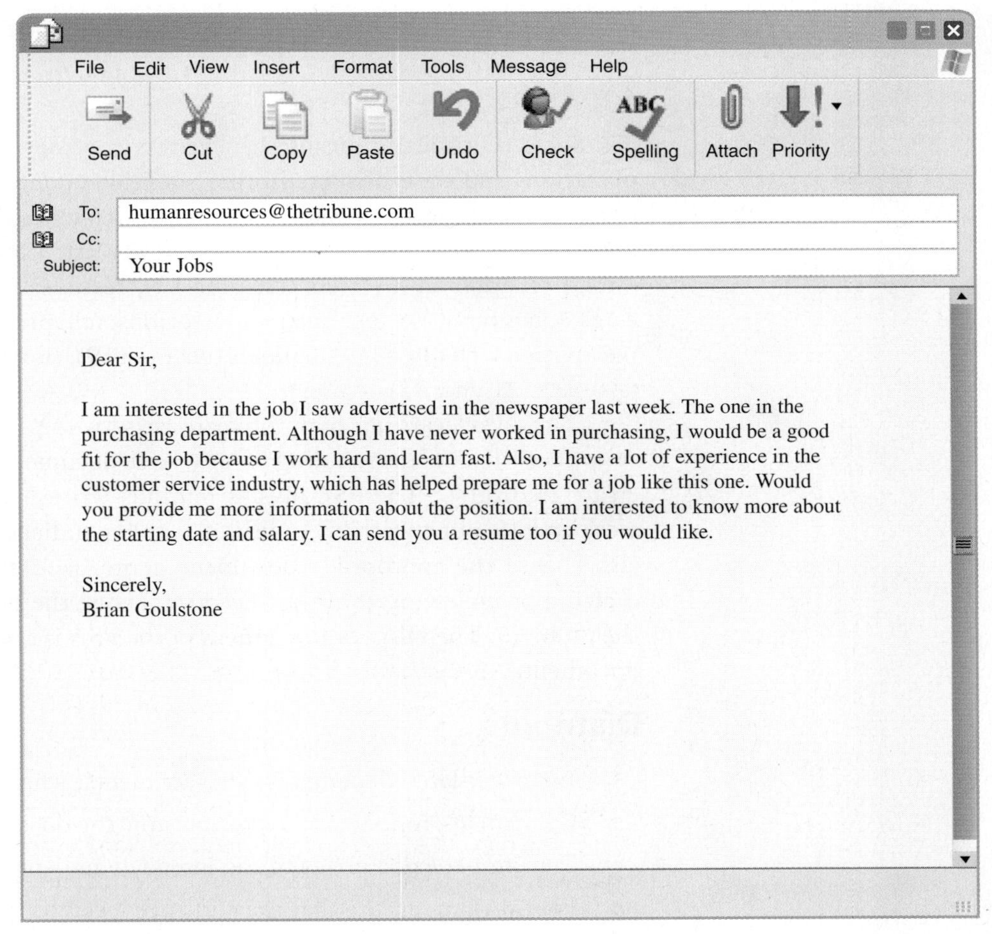

humanresources@thetribune.com

Your Jobs

Dear Sir,

I am interested in the job I saw advertised in the newspaper last week. The one in the purchasing department. Although I have never worked in purchasing, I would be a good fit for the job because I work hard and learn fast. Also, I have a lot of experience in the customer service industry, which has helped prepare me for a job like this one. Would you provide me more information about the position. I am interested to know more about the starting date and salary. I can send you a resume too if you would like.

Sincerely,
Brian Goulstone

FIGURE 2.2 Job seeker's e-mail

Contexts and Constraints

Writers, audiences, and documents do not exist in a vacuum, free from the various contexts in which problems circulate. Thus, it is important for workplace writers to fully understand problems in relation to all the factors that influence those problems and their solutions. For example, an international company with offices in New York and Buenos Aires may encounter similar problems with its computer network; however, because the locations are different, the contexts of the problems are also different. Perhaps the computer technicians equipped to handle the problem are better trained in one location; perhaps replacement parts for the network arrive in one location a week sooner than they do in the other; perhaps one location has a more pressing need for the problem to be solved.

Location, however, is only one contextual factor. Other factors, often termed *constraints*, put limitations on how a problem can be solved. Common examples of constraints are time and money. Many problems can easily be solved if time and money are not an issue, but in the real world most problems require quick solutions and monetary investment. For instance, a company that produces, distributes, and sells a toy microwave oven might learn that the toy is prone to overheating and starting fires. Thus, in addition to fixing the overheating problem, the company must recall all the toys that have already been sold. This seems obvious enough, but the company must act quickly to avoid liability. Time is an important constraint in this case because the company must get the recall notice out before any more fires occur.

When workplace writers use the Plan stage of the PSA, they must try to account for all the contextual factors and constraints that will have some bearing on their ability to solve the problem.

Documents

Thinking rhetorically means that writers consider their audiences' needs and expectations and their own credibility, and documents are the means through which those needs are addressed and that credibility takes shape. Much of the rest of this book teaches you how to create and use specific kinds of documents: memos and e-mail (chapter 12), letters (chapter 13), résumés (chapter 14), technical definitions (chapter 15), technical descriptions (chapter 16), websites (chapter 17), instructions (chapter 18), manuals (chapter 19), proposals (chapter 20), and reports (chapters 21 and 22).

However, learning how to create these workplace documents must always be coupled with an understanding of how these documents can serve audience needs and create credibility. The PSA always reminds writers to be aware of how documents work and are put together. As writers plan for audience needs and expectations, they also choose the appropriate document genre. And sometimes they decide during drafting or reviewing that they need to change the document genre to better solve the problem. The Distribute segment of the PSA is also important in thinking about documents.

Distribute

- Include related documents or attachments, when applicable
- Confirm appropriate means to transmit the document to the intended audience
- Transmit the document to the intended audience
- Follow up to ensure that the audience received the document
- Assess the document's outcomes and decide if further correspondence is necessary

Thus, documents are not finished when writers type the final word. Writers must also consider how to present and transmit documents to readers, and they must make sure that their documents have done what they are supposed to do. That is, they must make sure their documents have met audience needs and expectations so that they can be used appropriately. In many cases the end of one document leads to the beginning of others.

Technology can enhance writers' abilities to get their messages across to their audiences, no matter what the document's purpose. A report sent to presidential advisors outlining the perfect plan to end poverty will not be given serious consideration if it is handwritten in pencil or printed with an old dot-matrix printer. A job candidate on a telephone interview with a prospective employer is less likely to seem credible if he's talking on a gas station pay phone during rush hour. On the other hand, a sales presentation using PowerPoint and an overhead projector instead of handouts may appear more interesting and, therefore, more persuasive. There are myriad situations in which writers must make technological choices in producing and transmitting documents. Experienced workplace writers recognize these as rhetorical choices.

IN YOUR **EXPERIENCE**

What are the three most important documents you have produced since entering college? In what ways did you try to ensure that those documents created credibility? How did you transmit those documents to your audiences, and how did you decide whether each document met audience needs? In hindsight, are there other things you could have done to enhance your credibility with your audiences through these documents? For at least one of these three documents, write a short essay that responds to these above questions.

In short, workplace writers need to carefully consider the ways in which workplace problems and rhetorical problems go hand in hand. Learning to address and understand rhetorical problems helps writers deal with workplace problems, and vice versa. Understanding the situation and evaluating your choices is an important first step. We hope that your workplace problems are never as serious or life threatening as those encountered by the crew members of the Apollo 13 Lunar Mission described in the chapter introduction, but we do hope that your rhetorical choices are as effective and successful as theirs.

SUMMARY

- Workplace problems are difficulties or troubles that arise in a workplace environment and that must be solved with the help of documents.
- Rhetorical problems are those that relate to the composition of documents necessary to help solve workplace problems. Rhetorical problems and workplace problems are intertwined.
- Workplace writers produce documents by thinking rhetorically through the problem-solving approach.
- Exigency is the problem or impetus that gives rise to the need for a workplace document. The purpose of a document is related to its exigency.
- Informing, defining, explaining, proposing, and convincing are different purposes of documents, which may have more than one purpose.
- Rhetorical thinking involves consideration of audiences, writers themselves, and documents.
- Audiences vary significantly.
- Audiences have expectations and attitudes.

- Audiences use documents in different ways: they skim them quickly, read them partially, read them closely from beginning to end, revise and return them, use them to take some course of action, use them to make decisions, and use them in support of other documents.
- Multiple audiences often read documents.
- Writers must consider their credibility when producing documents. They do so by making sure their documents are correct, demonstrate their experience and expertise, show goodwill, and establish identification or common ground.
- Various contexts and constraints surround workplace problems and must be understood in order to create effective documents to help solve these problems.
- Documents are the means through which writers meet audience needs and expectations and establish credibility.

CONCEPT REVIEW

1. What is a workplace problem?
2. What is a rhetorical problem?
3. What is the difference between a perceived problem and a real problem?
4. What does it mean to think rhetorically?
5. What are the different components of the problem-solving approach?
6. How does the PSA tie into rhetorical thinking?
7. What is exigency?
8. What does it mean to inform in one's writing?
9. What does it mean to define in one's writing?
10. What does it mean to explain in one's writing?
11. What does it mean to propose in one's writing?
12. What does it mean to convince in one's writing?
13. What are audience expectations and attitudes?
14. In what ways do audiences use documents in the workplace?
15. How do writers strategize about documents read by multiple audiences?
16. What is credibility?
17. What does it mean to produce correct documents?
18. Why do writers want to show their experience and expertise?
19. Why do writers show goodwill?
20. What does it mean to demonstrate identification with audiences?

CASE STUDY 1

When Alligators Attack: Informing Tourists about Dangers in the Park

Recently, on two separate occasions, individuals who were walking by a small pond in a Florida state park were attacked by alligators. Fortunately, no one was seriously injured, but news coverage about the attacks has made town residents and tourists nervous about going to the park. Wildlife experts say that the timing of these attacks is merely coincidental, and they are not likely to continue. However, attendance at the park has declined dramatically since the attacks. The majority of park visitors are tourists from other parts of the country.

Park officials want to start an information campaign to warn park visitors about possible alligator attacks, but they do not want to create fear, so it is important to assure visitors that attacks are uncommon and can be prevented. They intend to hand out pamphlets to visitors as they enter the park and also install signs throughout the park that provide the necessary information.

As a public relations employee with the state parks, you are asked to help create the information in these documents—both the pamphlets and the signs. Use the problem-solving approach to begin to Plan for the situation; focus your attention on the first three steps—define the problem, establish your purpose, and determine what your audiences need. What is the real problem? Is there a peceived problem that might obscure the real one? What do you believe would be the audience's expectations and attitudes regarding such documents? What particular things must you do to ensure the park system's credibility when audiences read these documents? Look back through the rest of the problem-solving approach, and discuss other rhetorical concerns related to these two documents. Give all of your responses in a one- to two-page essay or memo.

PSA

Plan

- Define or describe the real problem or reason for writing
- Establish goals and purposes for writing
- Identify stakeholders and what they want or need
- Consider the ethical choices involved with the problem
- Consider document formats and delivery methods
- Identify what information you have and what information you need
- Choose technologies that will best assist you and your audience

CASE STUDY 2

Saving the Puma at Eastern Delaware University

You are a student at Eastern Delaware University and are approached by several members of your campus's environmental organization—Students for Environmental Activism (SEA)—who have recently learned that your university mascot (the puma) is actually an endanged species. The group wants to start a campaign to bring greater awareness about this issue to the campus community in order to help get the puma off this list. In particular, members of SEA want to begin by writing articles in the school newspaper about this issue, as well as a letter to the president of the university asking for financial support to help save the puma.

The head of SEA, Jenny Carlson, comes to you for help in planning these documents. The organization already knows the problem that exists and understands the purpose and goals for writing, but they are having difficulty composing these documents. After looking through the problem-solving approach, you tell Jenny that they now need to identify what information they have and what information they need in order to compose documents suited to their purpose and audience. A few days later Jenny returns with some data about the puma from the U.S. Fish and Wildlife Service, and she asks you to help the organization with the writing. You agree to read through this information and aid them in planning the documents.

After reading the U.S. Fish and Wildlife Service information about the puma (see Figure 2.3), you are to write a letter (or other document assigned by your instructor) to Jenny Carlson, explaining what additional information she and the organization will need in order to compose the *two* different documents they are working on. Focus on what information is already present in the U.S. Fish and Wildlife Service data that they can use and what other kinds of information they will need. You do not need to do the actual research for the organization; rather, you simply want to let them know what additional types of information are necessary to develop (a) an effective newspaper article that will be read by students and (b) a letter to be read by the university president. Think carefully about the two different audiences and the purposes for writing to each. The document you write to Jenny Carlson should detail the kinds of information needed and should explain why that information is needed. For example, in writing a newspaper article to students, it would probably be helpful to include information about how and why the puma became the university's mascot, whereas this information would probably be less important to the university president.

U.S. Fish and Wildlife Service

Eastern Puma (concolor couguar)

Taxonomy

Kingdom	*Animalia*
Class	*Mammalia*
Order	Carnivora
Family	*Felidae*

General Information

Eastern pumas are reddish brown-tan in color. They have white fur on the belly and under the chin. Black markings are apparent behind the ears, on the face, and on the tip of the tail. On average, they weigh between 65 to 130 pounds and grow to be 6 ft in length. Their western counterparts can grow up to 170 pound because of availability of larger prey. Many eastern pumas have an upward turn or kink at the end of the tail and a swirl or cowlick in the middle of the back. Puma cubs are pale ith sots and have rings around the tail. They lose their spots and rings at approximately six months of age.

States Where the Eastern Puma Is Known to Occur

Connecticut, Delaware, District of Columbia, Illionois, Indiana, Kentucky, Maine, Maryland, Massachusetts, Michigan, New Hampshire, New Jersey, New York, North Carolina, Ohio, Pennsylvania, Rhode Island, South Carolina, Tennessee, Vermont, Virginia, West Virginia

Critical Habitat

No critical habitat rules have been published for the eastern puma.

Conservation Plans

No Habitat Conservation Plans exist for the eastern puma.

No Safe Harbor Agreements exists for the eastern puma.

No Candidate Conservation Agreements exist for the eastern puma.

No Candidate Conservation Agreement with Assurances exist for the eastern puma.

No petition findings have been published for the eastern puma.

FIGURE 2.3 Information bulletin about the eastern puma

CASE STUDIES ON THE COMPANION WEBSITE

Elegance Limousines: The Difficulties of Cross-Cultural Communication

A cultural misunderstanding between Elegance Limousine Company and the Windsor Grand Hotel threatens their productive working relationship. You will consider how to write a document to ease the tension between the two companies.

The Baby Comfort 5.2: The Whistleblower's Dilemma

As an assistant marketing director, you become aware of some customer dissatisfaction and potential design flaws in a baby stroller your company produces. You will consider both rhetorical and ethical dilemmas.

Carolina Construction: Meeting the Needs of Multiple Audiences

The general contracting firm that you work for faces legal action for less-than-standard work that occurred before you joined the company. The president has assigned you to the problem; you'll have to consider public perception, fiscal feasibility, and legal issues.

Is the Customer Always Right?

Your company sells products to boost automobile fuel efficiency. A customer writes a complaint letter, demanding a refund, but your supervisor refuses to give it. You must figure out how to navigate rhetorically between the wishes of your boss and the demands of the customer.

VIDEO CASE STUDY

DeSoto Global

Planning the Shobu Automotive Proposal

Synopsis

Dora Harbin has been assigned to examine the possibility of securing contracts within the Japanese automotive industry. The case revolves around a proposal DeSoto Global is writing in response to a request for proposals (RFP) placed by Shobu Automotive, which seeks top-of-the-line electronic components for their luxury car line.

WRITING SCENARIOS

1. Develop a list of two or three problems that exist in the world today. These might be workplace or other local, national, or international problems. In an informal writing assignment or other kind of document required by your instructor, discuss who is involved in solving these problems. Also, discuss what forms of communication and what written documents they would use to help solve these problems. Explain in as much detail as possible.
2. Locate three technical documents written for multiple audiences; for example, memos, junk mail, reports, statistical graphs, automobile specifications, even road signs. In an essay or memo, describe each document's purpose(s) and the audiences it tried to reach. What tells you that these documents were intended for multiple audiences? Discuss the ways that these documents try to achieve credibility with the different audiences.
3. In a document format specified by your instructor, write about the contexts in which technology does or does not help make a message persuasive. For example, does a school essay or workplace report seem more persuasive if printed with a laser printer, as opposed to being handwritten or typed? Why or why not? Or in what circumstances might a text message on a handheld device seem more persuasive than a Post-It note? Discuss the persuasiveness of technology in other situations, explaining specifically what it does or does not accomplish in each context.
4. Consider a research essay you wrote for a college course. What kinds of strategies did you use to make yourself credible as a writer and a student? How did you use your research to enhance your credibility? Write a short analysis of your essay, focusing on the issue of credibility. For example, you

might expand the notion of correctness to include elements such as proper citation and documentation of sources.

5. The dean of student affairs at a large college has been asked by the college president to write a report that examines the possibility of constructing a new 500-space parking lot somewhere on campus to alleviate a campuswide parking problem. Thus, the purpose of the dean's report is to present the positive and negative factors associated with a new parking lot. What kind of information do you think the dean will have to find to satisfy the president's (i.e., his audience's) needs and expectations? Explain your responses in any kind of document allowed by your instructor.
6. Stakeholders and audiences are not always the same. In fact, quite often they are not. Read through the first case study at the end of this chapter, "When Alligators Attack." Note that the case study revolves around park officials producing a pamphlet for visitors and tourists, who serve as the intended audience for this pamphlet. Given this particular situation, who are all the stakeholders in the success of this pamphlet? Explain.
7. Consider again the research eassay identified in Writing Scenerio 4. Discuss the context of that essay and any constraints that surrounded your writing of it. Compose your response in a short, informal essay.
8. Word processor technologies allow writers to do things they could not have done in the past. Pull up your current word processing program, and explore all of its functions and features, even the ones you typically do not use. In a short, informal essay discuss whether any of these features could be used unethically or could be viewed as unethical altogether. For example, is there anything unethical about changing the font size from 12 to 12.5 to extend the length of a term paper to meet page requirements? Or is it unethical to use the thesaurus to incorporate words whose meanings you really don't know in order to sound more scholarly? What other features could be used unethically? Explain.
9. Documents are not simply passive mediums that carry information from writer to reader. Instead, documents are *used* by their readers and audiences. Think about two or three items or services that you purchased recently and for which you received a receipt. Even though we don't always think of them as such, receipts are technical documents. Find those receipts and study them carefully. Write a few short paragraphs that explain the purposes these documents serve, as well as how they can be used by those who receive them. Pay attention to the way these documents are designed, in addition to the content itself.
10. Choose a particular company's website that you would like to visit, preferably one that you are not already familiar with. Before actually going to that website, write down what your attitudes are toward that company and what your expectations are for the website. Then visit the actual website. In what ways did the website reinforce or change your attitudes? How did or didn't it meet your expectations? Do you believe that you are the intended audience for the website? Support your responses with details from the website's design.
11. What is your favorite website? What is it about this website that you trust; that is, what about the website creates credibility? List all of the characteristics (at least ten) that make the website credible to you. Then decide which of these characteristics derive from the website's words and which derive from the website's design (i.e., layout, images, sounds, colors, interactive features, and other applications). In what ways are these two areas connected? distinct?

12. E-mail technologies have helped people communicate but have brought them irritation as well—such as messages from solicitors all over the world willing to deposit millions of dollars in our bank accounts, if we'll just give them access. In a short essay, explain the ways that e-mail technologies have changed your reading and writing habits over the last five or six years. Do you now write more or fewer messages? Are those messages generally longer or shorter? Do you read more or fewer messages? Do you read more quickly or more slowly? Are you more trustful or distrustful of e-mail as a form of communication? Explain your responses.

13. Most documents we receive today are printed out or sent to us electronically; the days of handwriting with pens and pencils seem to be coming to a close. Many students even take class notes with PDAs or laptops. Yet in some instances it seems that handwritten messages are more persuasive. Describe three different rhetorical situations in which a handwritten message would be more persuasive than one printed or sent electronically. Explain why these handwritten messages would likely be more persuasive.

3

Technical Communication and Electronic Technologies

CHAPTER LEARNING OUTCOMES

After completing this chapter, you will be able to do the following:

- Understand the broad concept of **technology** and the way it has developed over time
- Recognize that **workplace writing** depends on computer technology
- Consider how the **problem-solving approach** incorporates technologies
- Consider how computer technology creates **new rhetorical situations**
- Recognize programs and applications used to create documents: **word processors**, **presentation software**, **graphics and imaging software**, **web-authoring software**, **desktop publishing software**, **help-authoring tools**, and **single-sourcing programs**.
- Recognize programs and applications used to communicate and collaborate: the **Internet** and **intranets**, **e-mail**, the **World Wide Web**, **electronic messaging**, **videoconferencing**, and **groupware**.
- Consider the many **ethical dimensions** of electronic communication
- Recognize some of the ways that computer technology will affect workplace communication in the **future**

DIGITAL RESOURCES

On the Companion Website www.prenhall.com/dobrin:

- Case 1: ADA Compliance: A Case of Corporate Responsibility
- Case 2: Human Relations and Technological Training: Francesca's New Warning System
- Case 3: Planning Ahead: Blossie Moore and Internal Documentation
- Case 4: Texting More than Just Engineering Data
- Video Case: Using Technology to Advertise the GlobeTalk Messaging System
- PowerPoint Chapter Review, Test-Prep Quiz, Exercises and Activities
- TCTC Visual Rhetoric and Design Tutorial
- TCTC Website Design Tutorial

REAL PEOPLE, REAL WRITING

MEET KIM ZETTER • Senior Reporter for WIRED News

What types of writing do you do on a regular basis at WIRED?

I write every day—sometimes nearly all day long—lots of e-mail and instant messaging in the office to communicate with my colleagues, interview subjects, and talk with experts. I write news stories on a regular basis, and that requires lots of communication with people in various technical fields. I also do some writing just for fun, and I recently published a restaurant guide to San Francisco.

What role does technology play in the writing that you do?

It plays a huge role, since all of my writing is done on a computer. I write it on my computer, and it is published online at WIRED News, so there really wouldn't be an outlet for my writing without technology. I use technology to do the interviews, to do resourcing, to gather information online—it's really invaluable.

Has technology changed the way that you write?

Oh, yes. The flow of thoughts is much easier on a computer. I do much more revision with computers, too, and I use cut-and-paste a lot to rearrange my writing. I can't imagine being a reporter using a typewriter, correction fluid, and things like that. Technology—especially the Web—gives me access to things I wouldn't normally have access to in terms of research, and that plays a big role in what I actually write.

Do you use technology to communicate with others on a regular basis?

I communicate with people outside my office through e-mail much more than I do by phone. We joke about this at WIRED, but we spend a lot of time instant-messaging each other rather than talking verbally—even though we're just a few desks away from each other. We communicate through instant message when we need quick answers to things, and that way we don't need to interrupt someone's train of thought. They can attend to our question or comment when they have the time.

Do you foresee any changes in the ways that writers use technology in the future?

I think we're using technology to a fairly high degree—it plays a huge role in our lives. I think it will become more efficient in the future, though. We have some problems with technology and written documents now in terms of different platforms, different software, and document sizes that are too large to be sent through e-mail and other electronic formats. Hopefully, technology will become more reliable and easier to use in the future.

What advice about workplace writing do you have for the readers of this textbook?

I believe that good writers *become* good—you don't start out that way. Writing is challenging, but that's what's rewarding about it. Understand that there are steps to everything, and don't try to truncate those steps. Sometimes your best work will come in revision, not necessarily in the first thing that comes out of you. Also, be willing to take advice from others during revision; lots of times, getting some feedback from someone else helps a writer see the subject from a new perspective.

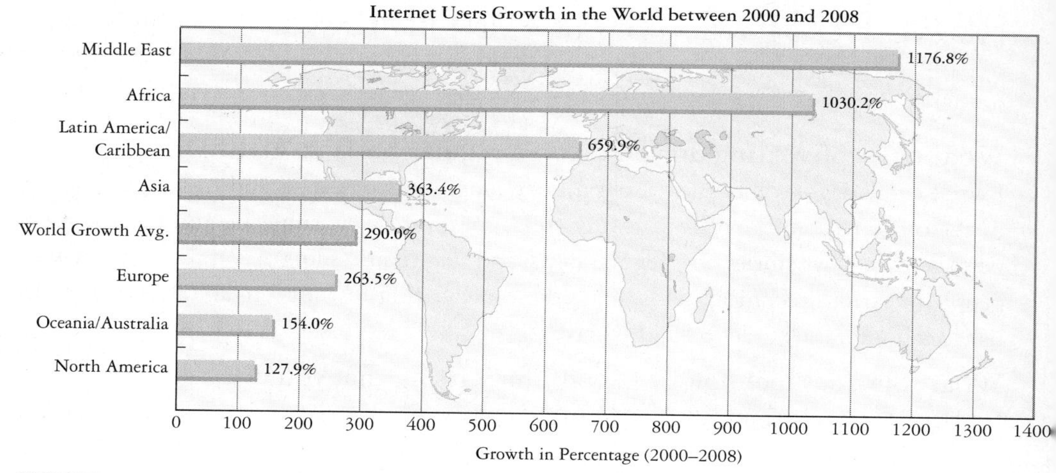

FIGURE 3.1 Worldwide growth in Internet usage

INTRODUCTION

The International Telecommunications Union (ITU) is an international organization within the United Nations system designed to coordinate global communications networks in both the private and the government sectors. In 2008 the ITU conducted a survey of worldwide Internet users. Their results were astounding. The survey revealed that from 2000 to 2008, the number of individuals using the Internet grew by more than 300 percent, to over a billion people worldwide (see Figure 3.1). The number of Internet users in the Middle East and Africa rose more than 1,000 percent in that eight-year period, increasing the spread of the Internet to continents in which usage previously had been very low. Internet users in North America more than doubled in that time, expanding an already-large number of users to more than 70 percent of the population.

What do these statistics tell us? They reveal that the Internet and other electronic technologies are spreading throughout the world at a rapid pace, creating new opportunities and new challenges for transnational communication. The statistics also reveal the extent to which our economies and our workplaces—global, national, and local—depend on technology. To put it plainly, no discussion of technical communication is complete without accounting for electronic technologies. The purpose of this chapter is to describe the current computer technologies used to create, revise, and disseminate workplace documents and to communicate and collaborate electronically. This chapter provides an overview of how computer technologies function in the workplace and addresses the ways these tools can be used to solve problems. However, because technology is central to workplace writing, our discussion of technology runs throughout the book. Regardless of your level of technological expertise, it is important to understand and recognize the significance of technologies in the workplace.

Chapter 5 provides more information about using electronic technologies to communicate with transnational audiences.

Nearly all workplace writers depend on various technologies to communicate through and collaborate on documents to solve workplace problems. As you will discover, a host of technologies and programs facilitate this process. However, don't let the many choices intimidate you; although all of the technologies addressed in this chapter are used by workplace writers, few writers use them all. But knowing that they are available is important, because computer technologies are an inevitable and necessary part of workplace communication.

PSA

Evolving technologies influence how individuals produce and disseminate information in and from their workplaces, affecting every phase of the problem-solving approach. In the planning phase, workplace writers consider which technologies are best suited to solving the problem at hand. And as Kim Zetter suggests in the opening interview, computer technology plays an important role in research, allowing writers to access and organize information in electronic formats. Most of the drafting done in workplace situations also relies on technology, and computers allow writers to revise their work more effectively and thoroughly. Finally, many workplace documents are distributed through some technological avenue, such as e-mail or the World Wide Web. In short, technology and workplace writing cannot be separated.

IN YOUR **EXPERIENCE**

Consider the tools you used in elementary school for writing and designing your school papers and other documents. Write a few short paragraphs describing what those tools were and what they were like to use. How have the tools you use for writing and designing documents changed in the last fifteen years? How have they changed in the last five or ten? Have there been any major changes in technological tools in the last year that have directly affected your writing? Explain.

The Impact of Technology

Although much of this chapter addresses the practical applications of computers in workplace writing, we hope that you examine the tools of technology as important subjects in and of themselves. Changes in computer technology affect not only the ways technical documents are produced and disseminated but also their content. Computers influence how workplace writers read, write, and think. For example, documents written for the Web are often arranged differently, include more images, and contain shorter paragraphs and more sections than those written for print-based distribution. Announcements, letters, and memos have generally become shorter and more casual as they've moved from print to e-mail. Even simple conversations have changed; wireless and instant messaging have given birth to new acronyms—such as IMO for "in my opinion," LOL for "laugh out loud," BTW for "by the way"—as well as *emoticons*—;-) or :-0—which are impossible to duplicate through traditional conversations.

See chapter 17 for more information concerning the differences between print-based and web-based documents.

Electronic technologies have led to important developments in workplace communication, and workplace writers must be proficient in using these technologies to create professionally designed documents. However, writers should also be able to critically examine the very technologies they use in that creation. Recognizing the ways that technology affects communication gives you a greater ability to make conscious choices about nearly every aspect of the production and dissemination of your writing. Thus, although we hope you will learn how to use technology to produce more sophisticated documents, we also hope you will inquire into how those technologies shape what you say and how you say it.

See chapters 12 and 13 for information about the differences between short electronic messages and short print-based messages.

Keeping Up with Progress

Like worldwide Internet usage, computer speed and power have also increased in recent years. Moore's law (named for Intel cofounder Gordon Moore) suggests that computer memory develops exponentially, doubling the speed and power of computers every few years. Figure 3.2 shows the increase in computer memory and processing power over time. Nearly all computer technology, including software, continues to develop at an increasing rate. This development allows computers to do more sophisticated things more quickly but can also make it difficult to keep up with changes and advances in the tools you'll use as a workplace writer.

Of course, individuals must learn to use the tools they'll need in a way that suits them best. Some like to learn by watching and talking with others. Some do-it-yourselfers like browsing the Web, using online documentation and preinstalled help

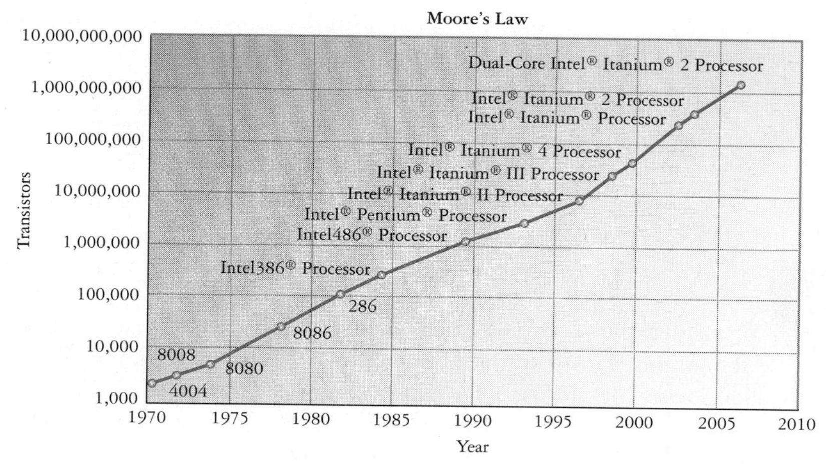

FIGURE 3.2 Moore's law

files, reading how-to books, or following the trial-and-error method. Others find intensive seminars and workshops to be the best approach to learning.

You'll have to decide which approach works best for you, but it is important for workplace writers to stay abreast of changes and developments in the hardware and software they use. Because of rapid changes in technology, this book won't give you step-by-step instructions in using a particular tool. Instead, it discusses the various types of applications and tools used by workplace writers, as well as how they can help you solve problems. Our purpose is to expose you to what can be done with technology rather than giving specific instructions on how to use it.

ANALYZE THIS

Look closely at Moore's law and the graph in Figure 3.2, which shows how the processing speed of computers has changed from 1970 to 2008. These developments in processing speed and memory have led to significant changes in our ability to use computers to communicate technical information. Discuss with classmates and your instructor how changes in computer technology have affected our general ability to communicate with one another. How do you think such technological changes have affected our ability to communicate technical information?

CREATING DOCUMENTS

Most of us have forgotten (or were born too late to know) what it's like to create professional documents without computers. As a result, we tend to overlook the revolutionary effect they've had on nearly every aspect of writing. Can you imagine being forced to use a typewriter for all of your writing today? Picture yourself hitting the carriage return after each line. Envision a world without Delete or Backspace keys, where each change or addition requires messy correction fluid or even a new sheet of paper, if not a completely new start. Or imagine having to dip a quill pen into an ink well every few words. Or having to chisel each letter into stone. Feeling frustrated yet?

Luckily, most of us have access to computers to create, revise, and edit our work, and with that access we are able to do many things. Word processing programs allow us to create and easily revise professional print-based documents. Presentation software enables workplace writers to create media-rich, portable, and professional slideshows. Web-authoring software makes it easy to create sophisticated web pages. Graphics and imaging programs can be used to create sleek images, graphics, and animations and also to manipulate, crop, and edit existing images. For sophisticated print-ready documents like magazines or newsletters, we can use desktop publishing programs. Help-authoring and single-sourcing programs make it easy to transfer content to various formats, including online tutorials and instructional manuals. In short, many tools are available to help writers create professional, sophisticated, and precise documents.

Word Processors

Word processors are tools that most writers use frequently—you have certainly used them yourself to create documents for your classes or in your workplace. Word processors allow writers to do many things:

- *Revise or change documents* at any stage in the writing process. Nearly all word processors allow writers to cut-and-paste chunks of information to a clipboard of some sort, where they can save the information for later use or insert it immediately somewhere else in the document.
- *Use templates and style guides* (i.e., patterns for composing a document in a preset format) to create consistent-looking letters and memos, reports, and even web pages. Templates guide writers through the steps and features found most commonly in a particular type of document (e.g., introduction, body, conclusion), whereas style guides apply a set of formatting characteristics to certain document features, such as titles, headings, or paragraphs.
- *Create tables or columns* to change the layout and appearance of text or graphics. Tables help organize and present information in cells; columns provide an arrangement similar to that found in a newspaper.
- *Use multiple windows* to compose various versions of a document, quickly shifting from one to another to combine, add, or subtract material.
- *Find and replace specific words or phrases* in a larger document, checking them or replacing them with more appropriate words or phrases.
- *Insert symbols, images, and charts,* which help create professional, visually appealing documents. Most word processing programs allow writers to add and manipulate images that have been created or downloaded, and many allow the input of data and information to create useful graphs, charts, and diagrams.
- *Edit for language, grammar, and style.* Editing programs let writers skim a document for practically every error associated with grammar and mechanics, easily correcting misspelled words, overused words, passive voice, clichés, and sexist pronouns. Such programs even check documents for readability—a concept discussed at length in chapter 10—analyzing the number of words per sentence, average paragraph length, and other features that might make a document confusing or difficult to read.
- *Share or post documents on the Internet.* Most new versions of word processing software allow writers to share drafts with others via web discussions and online meetings. Similarly, most of them easily convert print-based documents into web-ready HTML or XML formats.
- *Include active hypertext links.* Most programs now recognize web and e-mail addresses and format them so that they are active links when put on the Web.

Word processing software has developed from relatively simple, stand-alone applications to those that can perform a variety of tasks and operations. Most programs now allow you to convert from one application to another, such as from Word to Wordperfect, and most developers now produce versions of their software for both Windows and Macintosh operating systems. However, even though opening, saving, and translating documents in different formats has become easier, problems with translation still persist, so make sure you consult with others if you plan to share files. When working collaboratively, it is best to discuss and agree on the program and operating system beforehand, to save you from formatting problems later when you try to share your work. Most of this textbook was written collaboratively using files created with a word processor. Figure 3.3 depicts Microsoft Word, the most popular word processing program.

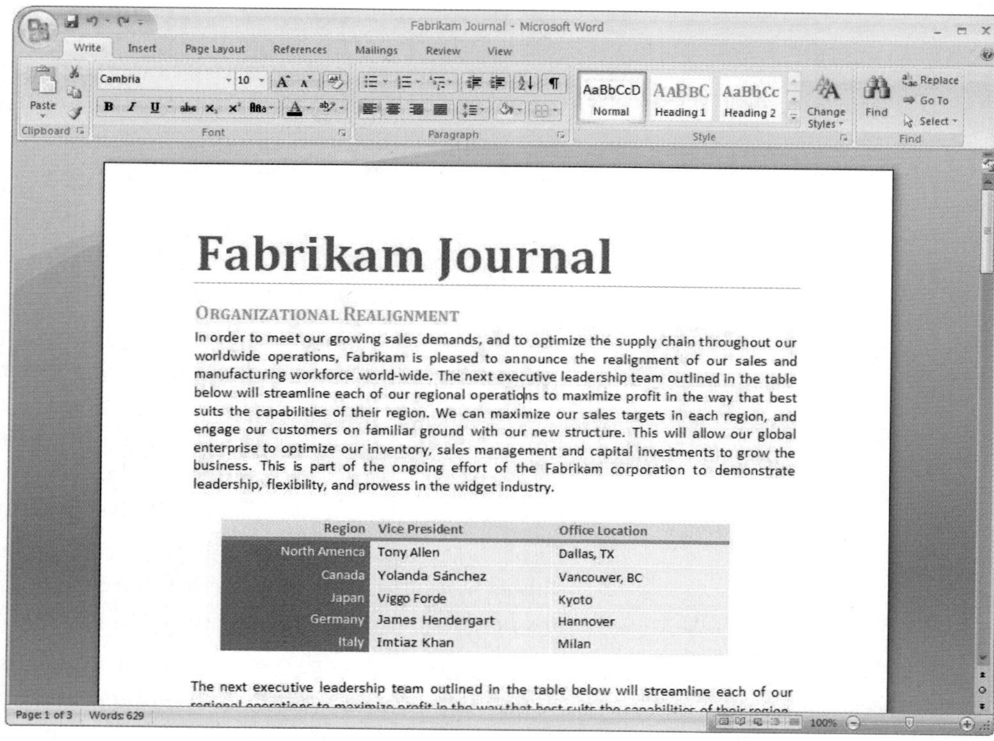

FIGURE 3.3 An example of Microsoft Word

EXPLORE

You may know about and use many of the features of word processors mentioned in this section; for most people the word processor is the most familiar and most often used software on their computer. However, you probably haven't used every feature that your word processor offers. Spend some time investigating the features of your word processor, and find one or two that you have never used. Familiarize yourself with their uses and functions, and then write a short document informing a classmate about the features, as well as how and why they might be used. As a class, share and discuss these documents.

Presentation Software

Chapter 23 provides more detail about presentation software.

As chapter 23 points out, computer technology has become indispensable in presentations; many audiences now expect to see presentations delivered through software such as PowerPoint or KeyNote. Such programs are easy to use and practically required in workplace communication. Studies show that audiences take a speaker more seriously and retain more information when it is delivered with multimedia equipment. Figure 3.4 is an example of PowerPoint, the leading presentation software program.

Most presentation software programs allow workplace writers to do the following:

- *Apply background and themes* to presentations to create audience interest and increase professionalism.
- *Add animations, timing, and transition styles* to presentations to alter the ways in which text, images, and slides appear in a presentation.
- *Insert audio, video, and graphic content* into slides to provide examples or detail in presentations.
- *Switch between different views* to arrange or move slides or to add notes to the presentation.

- *Share presentations* with users across different platforms and devices.
- *Upload presentations* to websites or intranets.

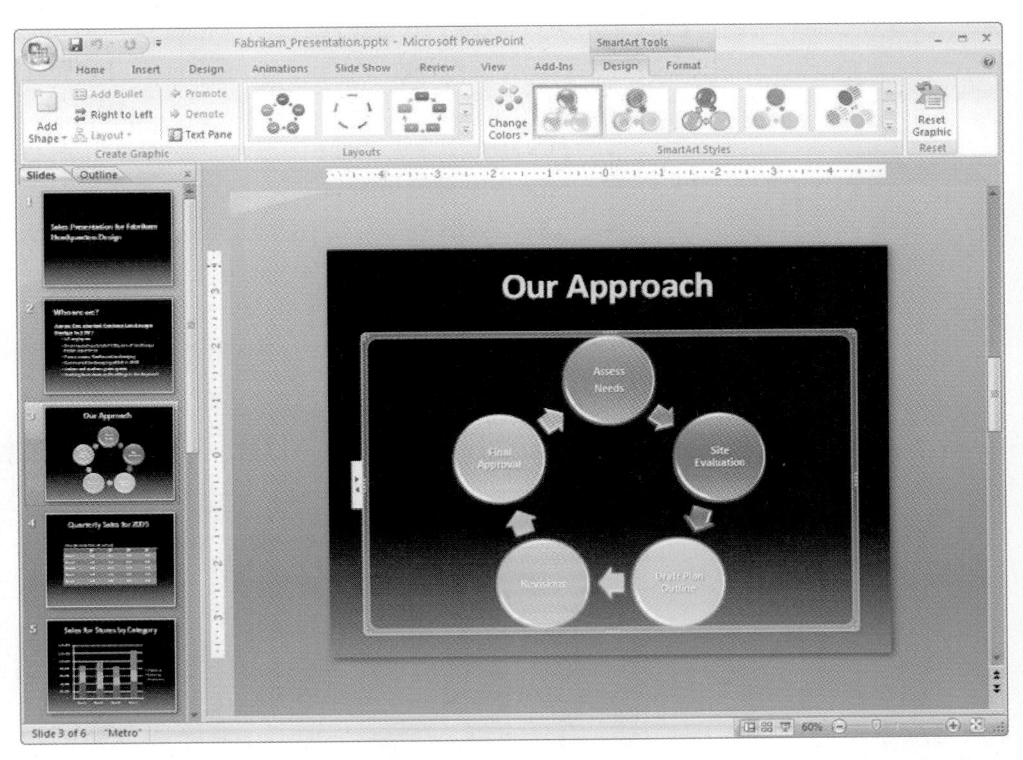

FIGURE 3.4 An example of Microsoft PowerPoint

Graphics and Imaging Software

Graphics and images are vital components of workplace writing. Many technical documents incorporate visuals to help market products; promote and represent organizations and companies; explain how processes work; identify instruction or assembly procedures; and provide further detail, examples, and explanation of objects, activities, and systems. Although advanced graphics and imaging may be outsourced to advertising firms or imaging technicians, the visual aspects of technical communication have increasingly become the responsibility of the workplace writer. Luckily, many of the programs designed for graphics and image development have become easier to use.

Chapter 8 provides more detail about creating and using visuals.

Graphics and imaging software allows you to manipulate, change, and edit photographs and create visuals. Quite often, visuals make the difference between a document that serves a basic purpose and one that is highly marketable, professional, and noteworthy. In fact, we believe visuals are so important to technical writing that we devote an entire chapter to them—see chapter 8, "Visual Rhetoric."

Because there are many graphics and imaging programs available and they often serve similar purposes, many technical writers develop a familiarity with just one or two of these programs and use them for a variety of functions. We recommend that you find one or two programs that match your skill level; they run the gamut from very basic and user-friendly to advanced and highly sophisticated. Figure 3.5 depicts one of the most popular graphics and imaging programs—Adobe Photoshop.

Regardless of which one you choose, most graphics and imaging programs allow you to do the following:

- *Edit and crop existing images,* such as digital photographs, to fit a particular size or space.

- *Erase or "paint away" part of a photograph,* leaving just the unpainted parts as a foreground image.
- *Manipulate or distort pixels in a digital photo* to touch up blemishes or smooth over areas that contain unnecessary or unwanted detail.
- *Create text objects and graphics* for letterhead and company logos.
- *Develop graphic images of objects,* such as exploded or cutaway diagrams.
- *Design images that show motion or progress,* such as graphs, tables, flowcharts, and other visual elements.
- *Change text appearance* by blurring, highlighting, or shadowing text.
- *Add sound or motion* to web-based and electronic documents.

FIGURE 3.5 An example of Adobe Photoshop

ANALYZE THIS

Visit one of your favorite websites, and consider for a moment how images are used in it. Do the images market or promote a product, service, or organization? Do they explain a concept or a process? Are they there to attract, engage, or entertain readers? Write a few paragraphs about the role or function of those images within the website. Then share your analysis—as well as the actual images—with your classmates.

Web-Authoring Software

Many technical documents produced in business and industry today are disseminated through the World Wide Web. In some instances the primary duty of a workplace writer is to design, create, and maintain web pages. Designing technical documents in a web-based format has many benefits for workplace writers, including the potential to embed graphics, links, sound, and video in their documents; the capacity to link to other documents and pages on the Web; and of course, the ability to connect with the more than one billion Internet users around the world.

Chapter 17 is devoted exclusively to web pages, addressing their creation and dissemination in detail.

However, composing documents for the Web is a bit more complicated than writing word-processed documents for print formats because web-based documents require hypertext markup language (HTML). HTML consists of special tags or commands, which specify how parts of a web page should look or behave when opened in a web browser (such as Internet Explorer or Mozilla Firefox). All HTML commands—as well as those for extensible markup language (XML), its more recent complement—are written inside angle brackets (i.e., less-than and greater-than signs), and many of the most basic commands use container tags, such as <title> and </title> to define where the title begins and ends. Many technical writers are familiar with these tags, although few know all of the tags and their uses. Luckily, nobody needs to learn HTML or XML coding to create sophisticated and useful web pages because web-authoring software allows writers to easily and quickly create web documents.

<TO CODE> OR </NOT TO CODE>

Even though many workplace writers use authoring programs to create web documents, there are still reasons for learning how to use HTML and XML tags. Knowing how they work will help you understand what they can and cannot do, allowing you to overcome limitations and problems you might have in getting program-authored pages to look the way you want them to look.

In fact, many professional web designers use authoring programs to do the time-consuming and tag-heavy tasks, such as tables and frames, but they edit and fine-tune their pages by hand, adding, deleting, and changing the code to make the pages uniquely their own.

If you anticipate doing a great deal of web authoring in your workplace, we encourage you to take this approach so that as you learn and experiment with authoring software to do the grunt work of your web-based writing, you'll be able to control how the finished product appears and behaves. You can learn how the code works and refer to it when necessary without the need to create all of your HTML and XML documents entirely by hand. Figure 3.6 is an example of Apple iWeb, a popular web-authoring software program.

Web-authoring programs allow you to do some important things:

- *Insert text, graphics, and multimedia objects* directly on a page. Most programs provide a WYSIWYG format—that is, what you see is what you get. In other words, the pages you create will look approximately the same in the program's window as they will when posted on the Web.
- *Switch between various modes* to see and edit web pages in a standard viewing mode, in HTML and XML format, and in various browser formats to test the hyperlinks.
- *Easily upload or post pages* to a web server without actually leaving the program.
- *Use drop-down menus* to create and manipulate text, graphics, tables, frames, and multimedia objects, including rollovers and Flash objects.
- *Create Cascading Style Sheets* (CSS), which serve as templates for multiple pages within a larger website. CSS allows writers to transfer stylistic choices (e.g., margins, fonts, positioning) from one page to another to maintain a consistent structure and appearance without manually re-creating these choices with each new page.
- *Insert internal navigation buttons and external links* for moving from page to page within a site as well as to other websites.
- *Automatically find and correct bad or unnecessary HTML and XML coding.* Bad hypertext coding may appear normal in the author's browser but may create problems for viewers using other platforms or programs. Web-authoring programs can eliminate unseen coding errors.
- *Create design notes* within files so that coauthors can remember what was done, what needs to be corrected or updated, and how they plan to change or renovate the web pages in the future.

FIGURE 3.6 An example of Apple iWeb

Desktop Publishing Software

Desktop publishing (DTP) software is used by some workplace writers interested in creating their own publications. Traditional DTP software was used primarily to create print-ready booklets, magazines, manuals, newsletters, and even calendars, cards, and tickets. Today's software combines many of the features of word processing, web-authoring, and graphics and imaging software in one package. DTP users can create page layouts with text, graphics, photos, and other visual elements. For small jobs a few copies of a publication might be printed on a local printer. For larger jobs a computer file can be sent to a vendor or publisher for high-volume printing. Although DTP software still provides extensive features necessary for print publishing, modern word processors have publishing capabilities beyond those of many older DTP applications, blurring the line between word processing and desktop publishing. Figure 3.7 is a screen shot of Adobe InDesign, a widely used desktop publishing program.

Help and E-Learning Authoring Tools

Some programs allow workplace writers to create online documents that instruct, train, or educate readers. These programs have a variety of applications and are used primarily by those working with computer software and with online education. For instance, some workplace writers are in charge of explaining how to use, navigate, or search through extensive online sites, files, or documents. These writers are often referred to as *documentation professionals*, and they may use help authoring tools, or HATs, to create and organize online help systems. If you've ever searched a website for troubleshooting advice about a software program, chances are good that a documentation professional created that website using a HAT.

One growing trend in both business and education is to deliver training and instruction through interactive, web-based environments. These environments allow learners to participate in a class or training seminar without being physically present. In some instances, the instructor and the learner communicate in real time; in other instances, the content is available to the learner at any time. Various e-learning authoring tools help workplace writers create these instructional environments and add content to them.

Figure 3.8 shows an instructional program created with *Course Lab*, an e-learning authoring tool. Help and e-learning authoring tools allow workplace writers to do the following:

- *Create and organize information* to help guide readers through a process or problem.
- *Develop searchable, easy-to-navigate menus* of information about a particular application, program, product, or website.

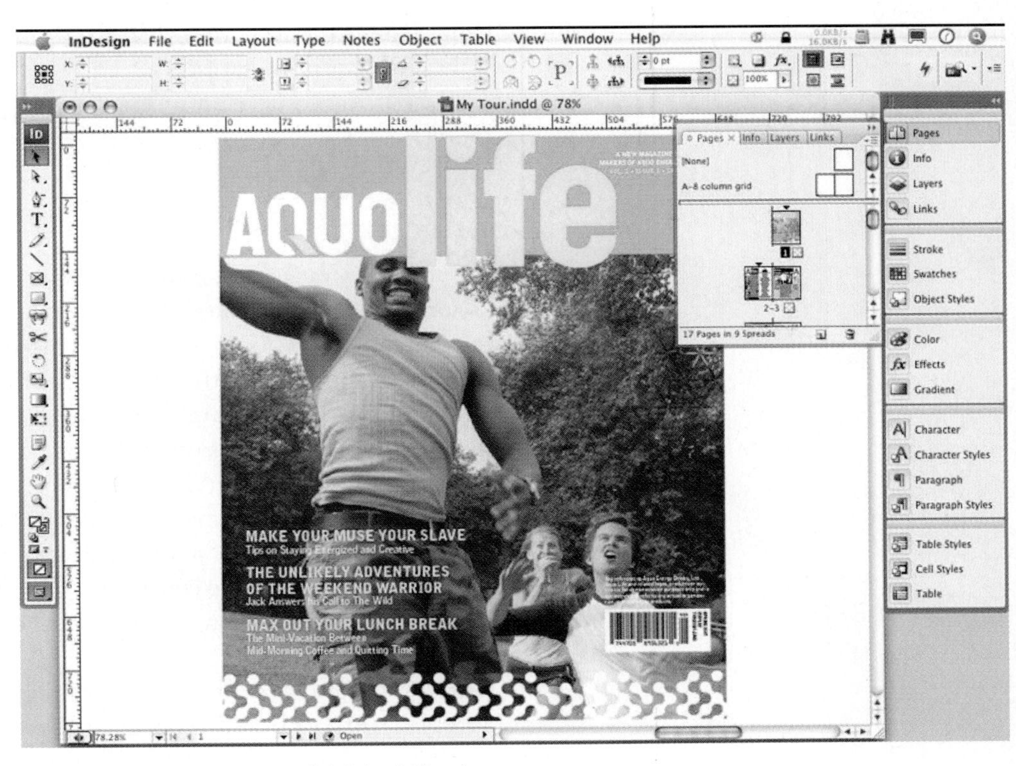

FIGURE 3.7 An example of Adobe InDesign

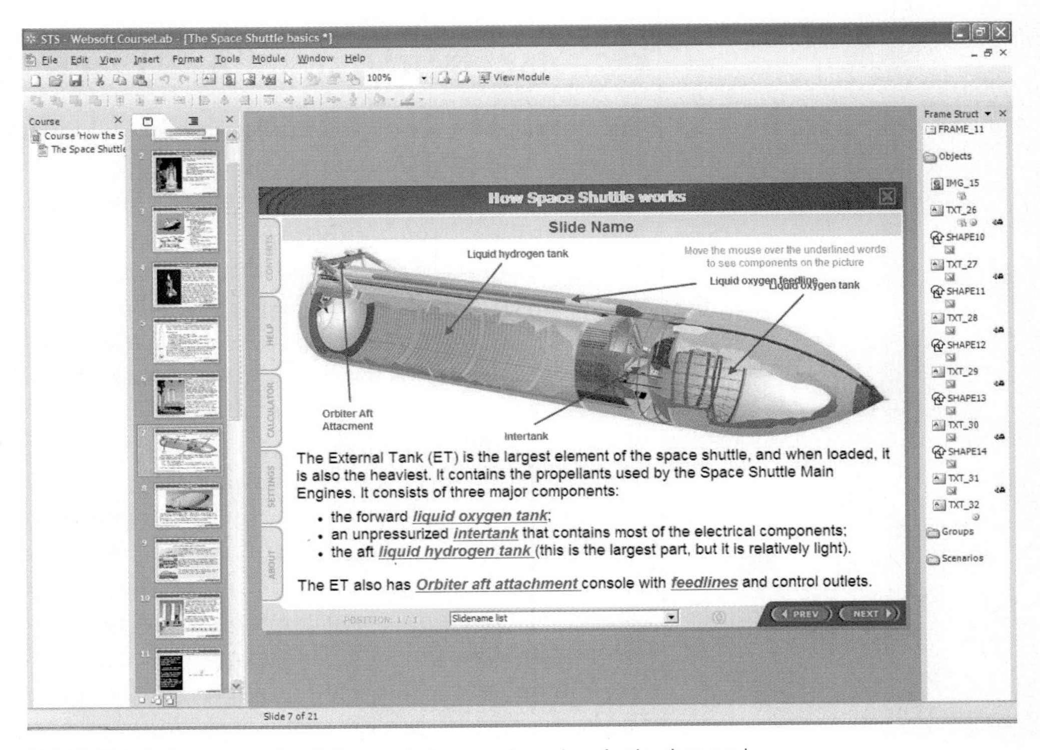

FIGURE 3.8 An example of CourseLab, an e-learning Authoring tool

- *Prioritize and sequence help information* so that users follow a logical progression through information.
- *Include various search functions* that allow users to search for help information by keyword, topic, or category.
- *Create glossaries, appendixes, and tables of contents* to organize and prioritize information.
- Create help systems that can be *viewed in a variety of web browsers* and that are transferable across various platforms.
- *Create and use a variety of navigational aids,* such as forward, back, and home buttons.

Single-Sourcing Programs

Companies that produce a variety of similar products often reuse or duplicate many of the documents, or parts of documents, that correspond to their products. For example, a software company might use a similarly structured users manual for its many different software programs. Rather than rewriting or re-creating the manual for each program, the workplace writer in charge might use a single-sourcing program, which would allow her to store just one version of the information to use in creating a range of documents in various platforms, styles, and types. In other words, single sourcing allows the creation of multiple documents in various formats from the same content. This capability not only saves time and money by avoiding duplicated efforts but also helps to eliminate errors and improve consistency. In addition, single sourcing allows for easier sharing of documents among colleagues and coworkers because it relies on one standard source.

See chapter 19 for more on single sourcing.

Single sourcing has much in common with help authoring; in fact, two of the best-selling HATs—AuthorIT and RoboHelp—are often used for their single-sourcing capabilities. Other programs designed specifically for single sourcing exist as well.

COMMUNICATING AND COLLABORATING

The image of the writer as an isolated, solitary individual is far removed from the realities of modern technical communication. Today, workplace writers spend much of their time interacting, collaborating, and communicating with others. In fact, as we note in chapter 1, one of the primary duties of workplace writers is to interact with others; workplace writers often work with and among technicians, designers, marketers, advertisers, executives, and those who will ultimately use, assemble, or operate a product. Workplace writers are often part of collaborative teams assigned to work on a project at various levels and in different stages of the production process.

At one time most of this collaboration occurred during face-to-face meetings, in phone calls, and through documents mailed back and forth. Needless to say, this sort of collaboration was rife with problems. Bringing individuals together for face-to-face meetings can be difficult, especially if members of the team are scattered across vast distances. Phones can help but still require collaborators to interact at the same time and have access to the same documents at that time. Postal mail can be time-consuming and limits collaboration since participants might not have the same version of the document at the same time.

Computer technology has significantly reduced many of these difficulties. Today, workplace writers can share documents through e-mail and the Web, communicating in milliseconds regardless of distance or location. Collaborators can interact in real time through instant messaging and can communicate from the field through wireless messaging. Workplace writers can even hold virtual meetings in cyberspace through webcams, virtual meeting software, and networked writing environments. Computer technology has broken down many barriers of time and space in collaborative technical communication.

The Internet

As you probably already know, the Internet consists of a huge network of interconnected computers. Strictly speaking, the Internet is a collective, global, government-subsidized network of both public and private networks linked together through the Transmission Control Protocol/Internet Protocol suite. Many of the tools workplace writers use to communicate with others and conduct research are connected through the Internet, including e-mail, Listservs, web pages, discussion lists, newsgroups, and electronic messaging.

Even though the Internet is a global network for sharing information, some companies use their own *intranets*, or private, local networks, to share and collaborate internally. These intranets generally require some sort of authentication to be accessed, such as a password, whereas others can be accessed only from a computer inside the company. Like the groupware programs mentioned later in this chapter, intranets can help users collaborate and communicate on a variety of workplace problems.

The World Wide Web

The World Wide Web is part of the network of networks that is the Internet; it is made up of Internet servers that support documents (i.e., web pages and sites) formatted in HTML and similar hypertext languages. In other words, the World Wide Web consists primarily of the web pages that are stored on servers (i.e., larger computers that archive and transmit web pages) that you can access through a browser like Netscape or Internet Explorer. The World Wide Web is not the Internet, although it does make up a large portion of the Internet that is visible to most of us. It is believed that as of 2008, popular search engines like Google can access nearly 50 billion indexed web pages, and most researchers believe this index excludes many web pages that are hidden, private, or not yet indexed.

The Web is currently a primary medium through which technical information is published, transmitted, archived, and accessed; and it will likely play an important role for the rest of your working lifetime. For this reason our consideration of the Web includes detailed discussions of ethics and the Web in chapter 4; research and source material in chapter 6; visual rhetoric, graphics and visuals on the Web in chapter 8; layout and design of websites in chapter 9; and web-based presentations in chapter 23, as well as a complete chapter devoted to creating and posting Web pages in chapter 17. Workplace writers use the Web for many purposes:

Access. Technical writers often use the Web to access and retrieve documents that other writers have posted to public or private servers; in fact, most technical documents contain information from other sources. As we will discuss in chapter 6, the Web can be a powerful tool for finding useful information to support a position, clarify context and background material, and show trends and patterns.

Storage. The Web can be a great place to store technical documents. Each web page is given a specific address known as a uniform resource locator (URL)—for example, http://www.prenhall.com/dobrin—which makes it easy to find that page at a later date. Web pages can also be stored in files to keep them organized and easily accessible.

Multimedia use. Unlike print-based documents, web pages allow workplace writers to embed multimedia components—such as interactive images, video, sounds, and hyperlinks to other websites—within a document. As access, bandwidth, and software developments continue to expand, it will become easier for writers to include more advanced visual, textual, and aural media in their documents.

Transmission. Posting documents to the Web can make it easier for others to access information. For this reason many workplace writers post longer, detailed documents on the Web and then send the URL and a short note to readers in an e-mail message. Because large files can take a long time to download and many e-mail programs have limits on the size of files they can accept, posting documents to the Web offers a useful alternative and allows readers to access documents at their leisure.

Collaboration. Many technical writers choose to collaborate on documents through the Web. Although collaborative writing through the Web may require a password and other information to make changes, the process allows readers at different locations to add to, delete, and revise the content of a web page so that readers at other locations can see the changes instantly and make their own as well. For instance, when we shared the draft versions of this chapter among our offices in Florida, Texas, and Hawaii, we relied on posting the large draft files on the Web for all of us to access.

E-mail

E-mail is accessed and delivered through the Internet and is probably the most widely used tool for collaboration and communication between workplace writers today. In fact, according to the International Data Corporation, Internet users in the year 2007 sent approximately thirty-five billion e-mails per day—a staggering twenty-one *trillion* e-mails per year. Because e-mail is so important, chapter 12 addresses it in depth.

See chapter 12 for more on e-mail.

Workplace writers send text messages back and forth among various recipients and also use e-mail to send technical documents to others as attachments, that is, documents created in another program and attached to the e-mail so that the recipient can open and view them. An attachment can be a text document created by a word processor; a picture created by a graphics program; a video or audio file created with a camera, microphone, and some editing software; or a computing application that can be run on a Mac or PC. Many workplace writers also subscribe to specific Listservs or newsgroups that automatically distribute e-mail messages to members who have subscribed to the group.

IN YOUR EXPERIENCE

In a few paragraphs write about all the different kinds of e-mails you compose on a monthly basis. What sorts of people do you write to, what kinds of information do you convey, and in what sort of writerly style do you convey this information? What do these different audiences, kinds of information, and varieties of writing styles that we all use tell us about the Internet and e-mail as a medium through which we communicate?

E-mail has many advantages over traditional print-based forms of communication:

Speed. E-mail messages can be delivered much more quickly than even the fastest bike messenger or courier could accomplish. The distance between sender and receiver can affect the speed of transmission—but not by much. Even messages going halfway around the world can be delivered in seconds. The connection speed of the computers being used can slow down the e-mail, especially if one or more are using a slow dial-up connection. And the size of the e-mail can affect the speed of transmission, especially if the document contains a large attachment. However, even a large file sent to a remote location by way of a slow connection will usually arrive in a short time.

Price. Although computer hardware can be expensive, the cost of sending e-mails—especially if you send lots of e-mails—is cheap. Current pricing allows most users to send a message to one recipient or a thousand for the same price. Although there has been some talk of charging senders per e-mail, this change doesn't appear likely to be become a reality anytime soon. E-mail saves money not only in terms of shipping and mailing fees but also on photo copying and printing costs. For this reason some offices and businesses are striving to become paperless, eliminating as much paper usage as possible. Even though original hard copies are often necessary for legal reasons, e-mail has greatly reduced the amount of printing and copying required for many technical documents.

Convenience. Quite simply, e-mail is easier to use than postal mail. E-mail allows writers to compose a message and send it without fussing with paper, envelopes, postage, and other inconveniences. In addition, most e-mail programs automatically download messages into an inbox, where they can be read at the recipient's convenience. And it is easy to forward messages to other recipients, although we should all avoid sending chain e-mails to colleagues and clients.

Further, e-mail is *asynchronous*, which means "not at the same time," or not in real time, referring to the fact that the sender and receiver need not both be present during the transmission of information. With e-mail, people can send and receive messages at different times and read them at their convenience. This aspect of e-mail also allows individuals to carefully analyze the messages they receive and craft an appropriate and thoughtful response.

Organization. Most messages are automatically assigned a time and date, allowing the recipient to keep track of when each message was received. And most new, unread e-mails appear in bold or italics to differentiate them from those that have already been read. Many e-mail programs also allow easy storage of messages in files that mimic the filing cabinets found in most offices—but without the use of physical space and the frequent clutter. In addition, address books allow users to keep track of the names and addresses of those they e-mail, while also creating groups of recipients, such as a design team or an editorial staff, who can all be e-mailed with just a click.

Ethics and Electronic Communication

The quotation from famed broadcaster Edward R. Murrow suggests that although computer technology makes it easier for us to contact one another, we are still left with the same problem of communicating effectively, clearly, and appropriately. Like all other communication, communicating through electronic technologies requires an understanding of the appropriate and ethical behavior in each situation. However, because the tools of electronic communication are both new and changing, it can be difficult to determine how to communicate effectively in each situation. The fact that writer and reader are rarely both present when they communicate through technology can lead to misunderstanding and frustration.

EXPERTS SAY

"The newest computer can merely compound, at speed, the oldest problem in the relations between human beings, and in the end the communicator will be confronted with the old problem of what to say and how to say it."

EDWARD R. MURROW

ANALYZE THIS

Look at any one of the e-mails you have written in the last day or two. Write a brief analysis of the rhetorical situation of this e-mail; that is, provide a brief analysis of your audience, yourself as a writer in this situation, and the information you needed to get across. Consider why you chose this medium as an effective way to communicate in this scenario, how you were able to achieve credibility, and how you persuaded your audience of something. Think back to our discussion in chapter 2 about rhetorical situations.

netiquette informal guidelines for behavior on the Internet

To help alleviate some of the confusion, Internet users have adopted a loose collection of guidelines known as **netiquette**. Netiquette is network etiquette, the dos and don'ts of online communication; it encompasses both common courtesy and the informal "rules of the road" of cyberspace. Since e-mail is still the primary method of electronic communication, most of these guidelines are focused upon that medium; however, these same guidelines can be applied to nearly all forms of electronic communication.

Workplace writers should follow these guidelines to ensure that they communicate professionally and effectively:

- *Choose an appropriate e-mail address.* Whenever possible, use a work-related e-mail address, preferably something like yourlastname@workplace.com. Your e-mail address gives your readers important clues about you before they read your message, and it will be difficult for readers to take you seriously if you have an e-mail address like studmuffin#1@yahoo.com.
- *Use an appropriate subject line.* Try to make the subject line as specific as possible by referring to the exact content of your message. Many of the individuals you'll communicate with will receive dozens—perhaps hundreds—of e-mails each day, and they might ignore or delete your e-mail if you don't tell them precisely what it contains. Keep those subject lines short, though—four to eight words should do it in most cases.
- *Respect others' bandwidth.* Bandwidth refers to the amount of information-carrying capacity of the wires and channels that connect everyone in cyberspace. There's a limit to the amount of data that any piece of wiring can carry at any given moment—even for a DSL, or broadband, connection. Some servers, in fact, block the reception and transmission of large files and attachments to prevent the influx of worms and viruses. For these reasons as well as general efficiency, keep your messages short and to the point. Don't include the complete text of an e-mail that you're replying to unless it seems necessary. Instead, paraphrase its contents or briefly quote from it.
- *Lurk before you leap.* If you are new to a discussion, especially if it is of a technical nature or includes many different individuals, listen to the conversation before you jump in. If an archive of previous messages exists, read through it before posting a response or asking a question; someone may have already addressed it in a previous message. *Lurking* may seem like a negative word, but on the Internet it's a good thing.
- *Polish your writing.* Even though e-mail is less formal than print-based writing, you should still edit and proofread messages before sending them. Remember that your words represent you. If your e-mails appear sloppy and careless, your readers may think the same of you.
- *Watch those caps.* Capitalization seems to be a particular problem on the Internet. Some writers use all capital letters in e-mails, which gives readers the

impression they are shouting. Other writers refuse to capitalize anything, which makes it more difficult to determine where one sentence ends and another begins. Use the same rules of capitalization you'd use in writing a formal, print-based technical document.

- *Use attachments appropriately.* Attachments are great to preserve the formatting of a document, but they do require recipients to go through several steps to open them. If the text is fairly short and doesn't require formatting, put it in the body of the message rather than including it as an attachment. If you must send an attachment, be sure to let your readers know what it contains and which software program they'll need to open it. Also, remember that some users have limited server space for e-mail messages, so contact them in advance if you'll be sending a particularly large attachment.
- *Back it up and save it.* Most e-mail programs allow you to save material in folders, which can be helpful in keeping track of past correspondence. However, computer hard drives (and often the servers that store e-mail messages) are prone to viruses and other problems that can wipe out these files. Remember to save copies of important files, either on a disk or some other electronic storage device or in a printed format.
- *Remember that when it's out there, it's out there.* This last guideline is less a matter of netiquette and more a matter of self-preservation. Because nothing is ever private on the Internet, think before you send anything that could be illegal, inappropriate, or offensive. That crack about your coworker might be funny for the moment—until he becomes head of your division and has access to your e-mails.

ANALYZE THIS

Working in a small group, analyze the two versions of the message shown in Figure 3.9. What specific differences do you see between the two? How do they follow or fail to follow the basic guidelines of effective communication and netiquette described in this chapter? Record your analysis in a short, informal document, and share your impressions with your classmates.

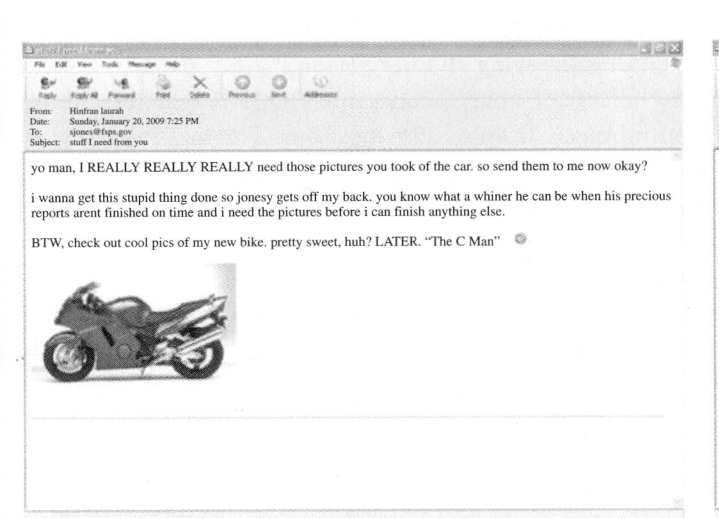

From: Hinfran laurah
Date: Sunday, January 20, 2009 7:25 PM
To: sjones@fsps.gov
Subject: stuff I need from you

yo man, I REALLY REALLY REALLY need those pictures you took of the car. so send them to me now okay?

i wanna get this stupid thing done so jonesy gets off my back. you know what a whiner he can be when his precious reports arent finished on time and i need the pictures before i can finish anything else.

BTW, check out cool pics of my new bike. pretty sweet, huh? LATER. "The C Man"

(a) An unprofessional e-mail

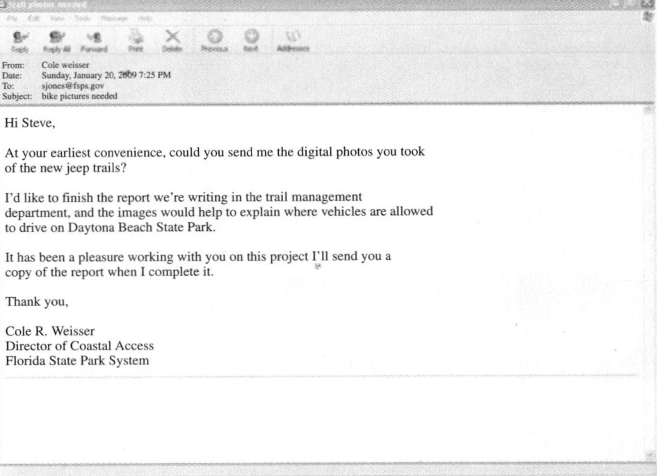

From: Cole weisser
Date: Sunday, January 20, 2009 7:25 PM
To: sjones@fsps.gov
Subject: bike pictures needed

Hi Steve,

At your earliest convenience, could you send me the digital photos you took of the new jeep trails?

I'd like to finish the report we're writing in the trail management department, and the images would help to explain where vehicles are allowed to drive on Daytona Beach State Park.

It has been a pleasure working with you on this project I'll send you a copy of the report when I complete it.

Thank you,

Cole R. Weisser
Director of Coastal Access
Florida State Park System

(b) A professional e-mail

FIGURE 3.9 Two versions of one e-mail message

Electronic Messaging

Over the past few years, electronic messaging has become widely used among workplace writers through computer-based instant messaging and chat programs, as well as through wireless devices such as "smart" cell phones and personal digital assistants (PDAs). Most people who use the Internet on a regular basis are familiar with software such as AOL Instant Messenger, Yahoo! Messenger (see Figure 3.10), ICQ, or one of the other Internet messaging programs. Many of these programs use simple pop-up software that appears on the computer, allowing the user to communicate through short text-based messages with someone else. Typically, the program displays a split screen, with a scrolling conversation on the top and the user's own typed text below. Such software is extremely easy to use; in most cases the writer simply types some text and selects the Send key, enabling all of the people in the conversation to see the message. Users can set up lists of contacts and, with one click, ask everyone in a group to join the conversation.

Many workplace writers use Internet messaging for short discussions, follow-ups, and immediate questions. Messages are sent instantly and must be read synchronously—in real time. However, like most Internet applications, the software works with the same relative speed, regardless of where participants are located. In fact, as Kim Zetter suggests in her opening interview, this type of software is so quick and easy to use that many people use it to communicate with coworkers just down the hall or even in the next office or cubicle. Although most writers rely primarily on textual messages, most programs allow users to send files and visuals as well. Some can also connect directly to the user's e-mail and the Web, sending the user a message when new mail has arrived or providing new reports and alerts. As bandwidth among typical users continues to grow, we can expect such features to increase and improve.

Chat rooms serve a similar function. The primary difference between chat rooms and other messaging software is in their use; chat rooms are electronic spaces, often devoted to particular topics, where people meet for conversation, discussion, or online socializing. People who have similar interests often come together to interact in chat rooms even though they may not know each other in real life. For example, technical writers might meet in chat rooms such as http://www.attw.org to discuss technical writing programs, http://www.freelancewriting.com to talk about issues writers face, or http://jobsearchtech.about.com to find a job as a technical writer. Other messaging software is typically used by individuals who already know each other and have an established relationship; in many cases messaging requires knowing another person's screen name.

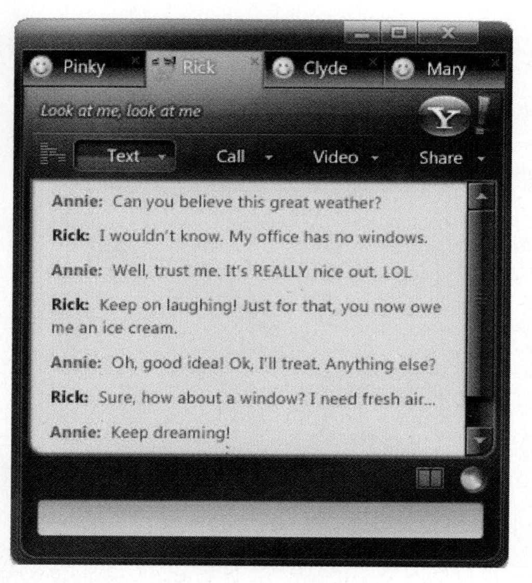

FIGURE 3.10 An example of instant messaging

The primary benefit of smart phones and PDAs is that they allow individuals to communicate from remote locations. The messages sent to and from these wireless devices are typically short and text based (without visuals) because typing, or "texting," from them is more cumbersome than from a computer. Nonetheless, wireless electronic messaging has become more widely used by workplace writers, and the benefits of communicating in practically any location cannot be overlooked. These forms of electronic messaging are addressed in further detail in chapter 12, which includes an analysis of the rhetorical situations created by these technological developments.

Videoconferencing

Videoconferencing is another recent technological development with applications for workplace writers. Like messaging software, videoconferencing

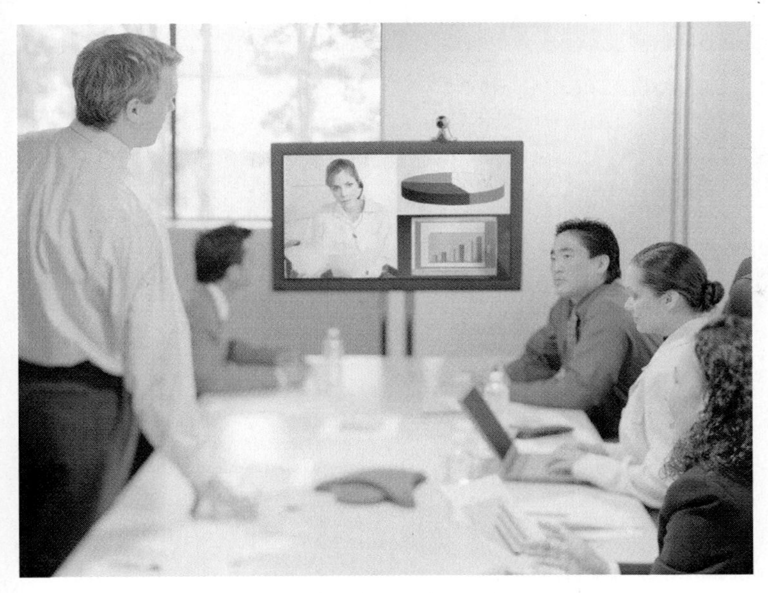

FIGURE 3.11 An example of videoconferencing

allows individuals at different locations to interact in real time. However, whereas electronic messaging is generally text based, videoconferencing is image based. Users generally see a video image of those they are conferencing with and have the option of seeing themselves as they appear to the other person(s). Videoconferencing software can be quite sophisticated and expensive but is also available to everyday users in the form of webcams.

Many companies with offices, partners, and customers in various geographic locations use videoconferencing to hold detailed discussions, organizational gatherings, and sales meetings. On a smaller level, webcams can be useful to interact with one or two others on a collaborative project. Workplace writers might discuss a product with a large, diverse group through an advanced videoconferencing program, or they might work from remote locations with their colleagues through one or two linked webcams. Videoconferencing software can allow for multiple windows showing individuals at multiple sites, as well as documents, graphics, and other multimedia applications (see Figure 3.11). Workplace writers can expect this method of collaboration and interaction to develop in complexity, usage, and quality as bandwidth and software improve in the future.

See chapter 23 for more on videoconferencing.

Groupware

Collaboration software, or groupware, often brings together many of the features of videoconferencing, document sharing, and the World Wide Web in one complete package. Some companies use programs like WebEx, Lotus, or SharePoint to work collaboratively. In fact, you may be using a form of groupware in your technical writing class—perhaps WebCT or Blackboard. Many of these groupware programs can perform the following tasks:

- Help people in one or more locations to *plan, draft, revise, and manage a document.*
- *Develop sophisticated presentations, spreadsheets, and documents* within the groupware program.
- Allow individuals to securely and privately *share documents and files* with group members without the risk that those files will be seen or downloaded by others.
- Support *instant messaging, videoconferencing, or blogging* with others through the groupware.
- Send *e-mail and text messages* within the program.
- Use *whiteboards* to "draw" on a screen that everyone can see, regardless of location.
- Upload or post *web pages* or other documents within the program.
- *Manage a list of contacts*, group members, or collaborative teams.

Groupware programs differ in their complexity, features, and uses. Some are designed for specific purposes, such as WebCT, which is primarily used in education. Most offer features that use both synchronous and asynchronous features of the Internet, allowing users to communicate and share information at the same time or at their convenience. Figure 3.12 shows an example of a groupware program.

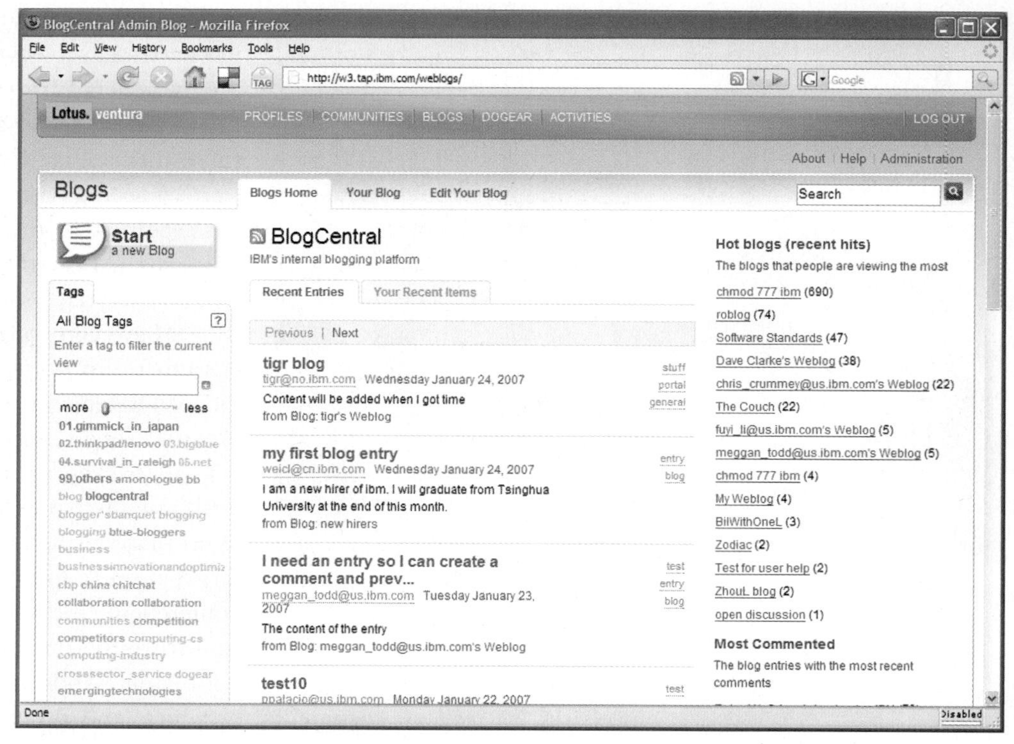

FIGURE 3.12 An example of Lotus Connections, a groupware program

Reprint Courtesy of International Business Machines Corporation, copyright 2008 © International Business Machines Corporation.

EXPLORE

You are probably familiar with some of the technologies used for communicating and collaborating, but you probably are not familiar with all of them. Do some research into one of these technologies with which you are relatively unfamiliar. Then write a short description of the technology, focusing on how it might be useful for you in your intended profession.

THE FUTURE OF WORKPLACE WRITING AND COMPUTER TECHNOLOGY

If we could accurately predict developments in computer technology, we wouldn't be textbook authors—we'd be retired millionaires. However, a few things do appear to be likely in the future:

- *Wireless technology,* known as Wi-Fi (wireless fidelity), will continue to improve and proliferate. Wi-Fi allows a person to connect to the Internet from remote locations without the need to be connected to a hard-wired office or computer terminal. Wi-Fi-enabled computers send and receive data indoors and out, anywhere within the range of a base station. In fact, your campus or workplace may have Wi-Fi zones, where you can access the Internet through your laptop or handheld device wirelessly.

  Wireless technology is beginning to have widespread ramifications for workplace writers, enabling them to stay connected via laptop computers and PDAs in the field, at home, in airports—just about anywhere. All of the collaborative tools of the Internet—e-mail, web pages, instant messaging, and so on—are becoming available anytime and almost anywhere. This allows workplace writers to work more closely with designers, engineers, and production and marketing teams without the need to return to an office, cubicle, or other wired location.

- *Bandwidth and connection speed* will continue to improve, allowing workplace writers to develop more sophisticated, interactive multimedia documents. The current amount of bandwidth available to the typical reader of a workplace document places limits on the amount of sound and video that can be included in a technical document. We envision a time in the near future when online technical documents can contain more detailed video and audio components, such as step-by-step video instructions complete with sophisticated video and audio, rather than the simple printed documents with standard images that are typical today.
- *Integration and portability* of components and software are constantly increasing. Phones are already becoming smarter and better equipped, PDAs are being used for web searching and multimedia viewing, and televisions are coming equipped with web browsers. Bluetooth and other connectivity programs allow for easy synchronization between different devices without any physical connections through wired ports. In other words, the tools that workplace writers use will become a greater part of mainstream society and will become more readily available than ever before. Workplace writers are increasingly able to carry in their pockets various devices with the capacity and capability of today's computers, as well as other tools for technical communication.
- *Prices* for computer technology will continue to drop, and more and more products will include electronic documentation, instructions, and other types of technical writing that have traditionally been print based. Many products are already sold with preinstalled programs or are equipped with DVDs, and documentation of this type will continue to drop in price and increase in availability. In the future many technical documents will come in electronic formats that are easily viewable and much more detailed. In fact, wafer-thin, foldable electronic paper will soon be available, which could contain video and audio files embedded within it. Many of the benefits of the World Wide Web will become available in formats similar to that of traditional printed text.
- *Transnationalism* will continue to influence computer technology, as computers and software respond to the demands of workplace writers for tools that will better allow them to work with colleagues and customers around the globe. This will require software programs that work seamlessly with programs at other locations, as well as developments in translation software and document portability. Chapter 5 addresses in greater detail the relationship between computer technology and transnational communication.

See chapter 5 for more on technology and transnational communication.

With these changes in technology, workplace writers will have more freedom to create interactive, highly complex documents, but they will also continue to have the responsibility to write clearly and precisely. Thus, workplace writers today need to learn not only how to write efficiently and clearly, but also how to use computer technology to communicate with others through more advanced electronic documents.

IN YOUR **EXPERIENCE**

What technological changes do you anticipate experiencing as a workplace writer? What changes are happening right now? Write a few paragraphs about current and future changes in the technologies used by workplace writers. You can focus on how our predictions are already occurring, you can focus on developments we did not mention, or you can speculate further into the future about technologies that will be used. Discuss or share your writing as a class.

SUMMARY

- Technology and technical communication are inextricably linked.
- Word processors allow you to revise or change documents at any stage; use templates and style guides; create tables or columns; use multiple windows; find and replace specific words or phrases; insert symbols, images, or charts into documents; edit for language, grammar, and style; share or post documents on the Internet; and include active hypertext links.
- Presentation software allows you to create slideshows with sound, video, and graphics; apply design themes and templates; use different views to insert notes or arrange slides; and share presentations with others.
- Graphics and imaging software allows you to edit and crop existing images; erase or "paint away" parts of images; manipulate or distort images; create text objects and graphics for letterheads and company logos; develop graphic images of objects and processes; design graphs, tables, flowcharts, and other visual elements; change text appearance by blurring, highlighting, or shadowing; and add sound or motion to web-based and electronic documents.
- Web-authoring software allows you to enter text, graphics, and multimedia objects directly on a web page; switch between various modes to see and edit web pages; easily upload, or post, pages to a web server; use drop-down menus to create and manipulate text, graphics, tables, frames, and multimedia objects; create Cascading Style Sheets; insert internal navigation buttons and external links; find and correct bad or unnecessary coding; and create design notes within files.
- Desktop publishing software can be used to create print-ready publications, such as magazines, newsletters, and flyers.
- Help and e-learning authoring tools can be used to create online documents that instruct, train, or educate readers.
- Single-sourcing programs allow individuals to create multiple files or documents using a single source document.
- Technology makes collaboration more efficient.
- The Internet increases communication efficiency. E-mail, in particular, is more efficient than postal mail in terms of speed, price, convenience, and organization.
- When communicating online, workplace writers should choose an appropriate e-mail address, use an appropriate subject line, respect others' bandwidth, lurk before they leap, polish their writing, pay attention to capitalization, use attachments appropriately, back up and save files, and remember that once it's out there, it's out there.
- The World Wide Web is part of the network of networks that is the Internet; it is made up of Internet servers that support documents formatted in HTML and similar hypertext languages.
- Technical communicators use the Web for access, storage, multimedia use, transmission, and collaboration.
- Instant messaging and chat rooms provide space for people to talk in real time.
- Wireless messaging allows individuals to communicate while in the field.
- Videoconferencing can bring workplace writers together using sound and video from remote locations.
- Collaboration software, or groupware, often combines many of the features of workplace writing and communication technology.
- The future of workplace writing and computer technology will probably include the following changes: proliferation of wireless technologies, improvement in bandwidth and connection speed, greater integration of technology into everyday life, declining prices for computer technology, and an increasing attention to the needs of transnational communication.

CONCEPT REVIEW

1. How can we define *technology*?
2. How is computer technology a part of the various phases of the problem-solving approach?
3. How does computer technology affect the content of documents?
4. How does computer technology create new rhetorical situations?
5. What is Moore's law, and why is it significant for workplace writers?
6. What features of word processors might be useful for workplace writers?
7. What features of presentation software might be useful for workplace writers?
8. What features of web-authoring software might be useful for workplace writers?
9. What features of graphics and imaging software might be useful for workplace writers?
10. What features of desktop publishing software might be useful for workplace writers?
11. What features of help and e-learning authoring tools might be useful for workplace writers?
12. What features of single-sourcing programs might be useful for workplace writers?
13. How can workplace writers use the Internet to communicate and collaborate?
14. How can workplace writers use e-mail to communicate and collaborate?
15. What are the basic guidelines of appropriate and ethical electronic communication?
16. How can workplace writers use the World Wide Web to communicate and collaborate?
17. How can workplace writers use electronic messaging to communicate and collaborate?
18. How can workplace writers use videoconferencing to communicate and collaborate?
19. How can workplace writers use groupware to communicate and collaborate?
20. What can workplace writers expect to see in the future regarding computer technology?

CASE STUDY 1

Addressing Student Technology Needs on Campus

Imagine that you are a member of a student advisory board on your campus. Your college president recently attended one of your advisory board meetings, and the subject of student technology needs became the focus of the discussion. Nearly everyone had some technological need to mention; some student members discussed specific hardware needs, others discussed changes in e-mail and website policies, and still others talked about the types of software they'd like to have available on campus computers. Rather than trying to write down all of these specific needs, the president asked the advisory board to create a one- to two-page document outlining the most pressing technological changes necessary to facilitate learning for the general student body.

Because your advisory board members know that you are taking a technical writing course and have been studying workplace technologies, they feel you are best suited to research this important topic and write the short document.

For this exercise do some research into the existing technologies available to the general student body on your campus. You may wish to begin by looking at your university's website and speaking with someone in the computing or information technology division or office. Through interviews, polls, and formal and informal discussions with your fellow students, try to determine the general technology needs of students on your campus.

After you have gathered the necessary information, write a one- to two-page document to your college president, outlining the most pressing or important technological needs.

CASE STUDY 2

Dillard Auto Parts Goes Paperless

You have recently taken a job as a purchasing manager for Dillard Parts, Inc., a national auto parts distributor. At a purchasing meeting with other managers in your division, your supervisor, Dave Bender, notes the growing trend among auto manufacturers toward more environmentally friendly cars and services, including hybrid autos, recycled tires and interiors, and other "green" products. Dave notes that many of the current best-selling auto parts in your warehouse are those that are more environmentally responsible; he goes on to suggest that he'd like to follow this trend by examining Dillard's own use of resources.

Since you've just begun working for the company, Dave asks if you'd like to do some research into ways in which Dillard's internal communication could be more environmentally responsible. Dave suggests that one benefit of using computer technology to collaborate and communicate in the workplace is that it can reduce the amount of paper that is used. In fact, some businesses strive for a paperless workplace, where all or nearly all printed documents are eliminated and documents are stored in electronic files, backed up on CDs, DVDs, flash drives, and other storage devices. However, he is not sure if using these technologies is actually better for the environment, and he wants you to find out.

Do some research into the materials and energy that go into producing printed documents as opposed to saving them on storage devices, and write a short report for Dave, identifying the better choice in terms of environmental impact. Be sure to carefully examine all of the potential effects and ramifications of your recommendation, including possible expenses, production times, and marketing values, as well as the impact of pollution, recycling values, and use of raw materials.

You might begin by looking at websites such as http://www.coopamerica.org, http://www.ecopaperaction.org, http://www.thegreenpc.com, the website on CDs and DVDs at http://store.yahoo.com/worldwise/reccdsanddv.html, and the BBC news article entitled "Chips Cost Environment Dear" at http://news.bbc.co.uk/2/hi/technology/2444675.stm.

CASE STUDIES ON THE COMPANION WEBSITE

ADA Compliance: A Case of Corporate Responsibility

A blind coworker points out that your building consulting company is not in compliance with the Americans with Disabilities Act. As an IT director, you'll address issues of access and the implication of noncompliance.

Human Relations and Technological Training: Francesca's New Warning System

Your supervisor has asked you to develop a proposal detailing possible new ways to conduct employee training on data security. You must figure out what technologies best suit the needs of this project.

Planning Ahead: Blossie Moore and Internal Documentation

As the new project manager in the corporate office of Blossie Moore's Kitchens, you learn that no documentation exists explaining internal company processes. You must find a way to document those processes in order to develop and manage projects effectively within the corporation.

Texting More than Just Engineering Data

This case looks at how new writing technologies, specifically text-messaging devices, might be used inappropriately in the workplace. As a technical writer at an engineering firm, you must develop documents that provide official policy on how messaging devices should and should not be used.

VIDEO CASE STUDY

GlobeShare Wireless

Using Technology to Advertise the GlobeTalk Messaging System

Synopsis

Three GlobeShare employees are developing an initial marketing and advertising campaign for GlobeTalk, the company's new wireless messaging system. Because they have no budget for advertising at this stage, the GlobeShare employees assigned to develop the campaign must be creative in the documents and approaches they use.

WRITING SCENARIOS

1. As a collaborative exercise, use an instant messaging program to discuss and analyze this chapter with one or two of your classmates. Then write a short document to your instructor that examines the strengths and weaknesses of this form of studying in comparison with face-to-face discussions.
2. This chapter mentions five primary purposes for which workplace writers use the World Wide Web: access, storage, multimedia use, transmission, and collaboration. In a group, discuss the benefits of each of these purposes. Which seems most significant for the typical workplace writer? Which seems least important? Rank the five purposes, and provide a few sentences explaining why each purpose is ranked as it is.
3. This chapter discusses many different technologies used by workplace writers; some may be familiar, whereas others may be unfamiliar. Working in a small group, discuss the technologies you use most as a student. Then write down your results, naming the technologies you feel are most useful for students and describing why they are useful.
4. Some writing scholars have noted the ways in which technology reflects the cultural values and worldview that surround it. For example, as Cindy and Richard Selfe note in "The Politics of the Interface," the Macintosh desktop replicates a reality familiar to white, upper-class, corporate America with such objects as manila folders, files, documents, telephones, fax machines, clocks and watches, and desk calendars. It does not, they argue,

   > represent the world in terms of a kitchen counter top, a mechanic's workbench, or a fast-food restaurant—each of which would constitute the virtual world in different terms according to the values and orientations of, respectively, women in the home, skilled laborers, or the rapidly increasing numbers of employees in the fast-food industry.[1]

---

[1] Cynthia L. Selfe and Richard J. Selfe, Jr., "The Politics of the Interface: Power and Its Exercise in Electronic Contact Zones," *College Composition and Communication* 45, no. 4 (1994): 480–504.

Examine a computer application or platform with which you are familiar (e.g., AOL software or Windows XP), and analyze the cultural values, mind-set, and worldview it exemplifies. Does it reflect a world familiar to one race, class, or gender to the exclusion of other perspectives? Why do you think the creators of this application or platform designed it in such a way? If you were to redesign this application or platform to reflect a different cultural perspective, what would it look like?

5. Choose a piece of software with which you are familiar, and visit the website that tells about that program. How does the website discuss the program? Does it mention features with which you are not familiar? Does it overemphasize the strengths of the program or fail to mention any weaknesses? Write a short document comparing the software and its representation on the website.
6. This chapter addresses many of the positive aspects of computer technology for workplace communication. In a short document, address some of the pitfalls and negative aspects of computer technology.
7. This chapter discusses netiquette in the context of workplace communication. Study the guidelines listed in the chapter, and write a short document analyzing whether these same rules apply to your class or university. Are there any guidelines listed that do not apply to this different context? Are there others that do apply to your class or university that are not listed?
8. Consider the ways in which your own economic, social, or educational background has influenced your familiarity with computer technology. Did you have great access to and experience with computers for much of your life? Or did you have minimal access to and experience with computers? How did this experience place you at an advantage or disadvantage in your course work? In what ways might this advantage or disadvantage influence or affect you as a workplace writer? Write a short document addressing these questions.
9. Imagine what it would be like to communicate technical information using more primitive forms of technology, such as by writing on a cave wall or papyrus or even using a simple printing press. How have computers and other advanced writing technologies changed the writing process? Do you think these changes primarily affect the form of writing, or have they changed the content as well? In other words, do you believe that technology can change the substance of a technical document as well as its appearance, layout, and design? Write a short document addressing these questions.
10. This chapter (as well as others in this book) provides screen shots of various e-mail programs. Compare the layout and design of one of those e-mail messages with that of the e-mail program you use. Which seems easier to use and navigate? Which has more features? Which seems more suited for workplace communication? Write a short document addressing these questions.
11. Skim through this book and try to determine which of the technologies addressed in this chapter might have been important in the book's overall design and appearance. Write down your responses, and discuss them as a class.
12. Find a free or trial version of a software program with which you are not familiar but which might be useful for you in your intended career. Write a short document analyzing it and assessing whether it is worth purchasing and learning to use.
13. Write a short letter to incoming freshmen in your discipline or field, describing the types of computer technology with which they should be familiar in order to be successful in their courses in that discipline.

14. Visit a large local store that has a comprehensive selection of computer software, and examine the shelves containing business and workplace software. Which of the programs listed in this chapter appear on those shelves? Which do not appear? Do you see other programs that might be useful in the workplace? Write a short document describing what you saw and examined.

15. Examine the history of writing technologies. For your research you may wish to explore one of these books:

    - Steven Fischer, *A History of Writing* (London: Reaktion Books, 2001)
    - J. David Bolter, *Writing Space*, 2nd ed. (Mahwah, NJ: Erlbaum, 2000)
    - Andrew Robinson, *The Story of Writing* (New York: Thames and Hudson, 2003)
    - Anne-Marie Christin, *A History of Writing: From Hieroglyph to Multimedia* (Paris: Flammarion, 2002)

    After you have completed your research, participate in a class discussion or write a short analysis. Which writing technologies had the largest impact on society and culture? How have writing technologies developed or built on each other? What were some of the material, social, or cultural factors that aided this development?

Ethics and the Workplace Writer

CHAPTER LEARNING OUTCOMES

After completing this chapter, you will be able to do the following:

- Consider the relationship between **rhetorical choices** and **ethical choices**
- Understand that **ethics** is often a difficult subject matter
- Make better ethical **decisions**
- Understand ethics in **communication** with others
- Understand that **conflict and disagreement** are part of making ethical decisions
- Understand different **definitions** of ethics
- Understand how and why we **study** ethics
- Consider the relationship between **legal and ethical positions**
- Identify the role of ethics in **workplace writing**
- Recognize legal considerations in making ethical choices, including the role of **liability laws, environmental laws, copyright laws, patent laws, trademark and service mark laws,** and **contract laws**
- Account for the role of **honesty** in workplace writing
- Account for the role of **confidentiality** in workplace writing
- Understand the role of **context and situation** in making ethical choices
- Recognize that **editing and revising** are closely tied to making ethical choices
- Identify and analyze **codes of ethics**
- Address issues of ethics and technology, or **cyberethics,** in writing, including e-mail, websites, and visuals
- Consider the role of **environmental ethics** in workplace writing
- Apply **strategies** to write more ethically, such as avoiding deceptive or evasive language, avoiding obscuring the issue, properly using jargon, properly suppressing or emphasizing information, ethically employing visual rhetoric, and identifying inaccuracies and bad information

DIGITAL RESOURCES

On the Companion Website www.prenhall.com/dobrin:

- Case 1: Instant Messages: Personal Space in the Professional Environment
- Case 2: Privileged Information
- Case 3: Product Recall: Complaint
- Case 4: Ethics and the Workplace Writer
- Video Case: Agritechno's S-194 Bt Corn: Informing Investors and Growers about a Potential Problem
- PowerPoint Chapter Review, Test-Prep Quiz, Exercises and Activities

REAL PEOPLE, REAL WRITING

MEET BETH KUZMINSKY • Nursing Curriculum Specialist

What kinds of writing do you do as a health-care professional?

I write every day in a variety of ways. I create or reply to approximately forty to sixty professional e-mails each day. I also write regularly for local and national newsletters, develop curriculum for nursing courses, and develop presentations to share at forums throughout my health-care system. When working with outside guests, thank-you notes and descriptions of speaker performances are also my responsibility.

How is ethics addressed in your profession?

As we work closely with patients, families, and health-care providers, ethics is at the core of all interactions. My medical center has an ethics and compliance program, and we depend upon a guidebook entitled "Workplace Ethics." The guidebook highlights our foundation of honesty and integrity in the workplace. It is the obligation of staff to ensure that we always act honorably and appropriately, since the decisions we make impact our patients every day.

What are the most important ethical issues you face in writing documents in your workplace?

Patient safety is most important. As we implement new patient-care strategies in the workplace, it can be challenging to transition from "the way we've always done it" to the best, most efficient, and safest way for patients and providers. It is important to reinforce the good work being done in health care by the frontline providers while also testing new ways to serve our patients.

Can you describe an example of a recent ethical issue in your workplace and the way you dealt with it through writing?

I led our medical center's new Condition Help program, which allows patients and families to call our rapid-response team during emergencies. Ethically, we believe this is right for patients and families. My role has been to promote the program, and I have written extensively about how Condition Help is beneficial for patients even though the program sometimes challenges the traditional health-care delivery system.

What is the relationship between ethical and legal approaches to issues in your profession?

In the health-care profession, they are closely related. Our medical center has an Ethics and Compliance Advisory Committee to provide guidance as needed. This committee consists of a number of people with expertise in ethics and legal issues. If I encounter a legal or ethical problem, I refer to this committee before making any decisions.

Does technology affect the way you address ethics?

Yes, technology has an impact on ethics in my workplace, particularly with online communication. Patient confidentiality must be maintained at all times. In references to a patient in e-mail, names are not included, or only initials are used. When case studies are used as an educational strategy and shared electronically, patient names are never included.

What general advice about workplace writing, do you have for the readers of this textbook?

Remember that your readers come first. If you put their best interests first and share this through writing, you are almost always doing the right thing. Consider that each person is a unique human being with a background that might be different from your own. Keep the message appropriate, meaningful, and applicable.

INTRODUCTION

On August 29, 2005, Hurricane Katrina, one of the five most deadly hurricanes ever to hit the United States, made landfall in southeast Louisiana and southwest Mississippi. More than 1,800 people died as a result of the storm. Five hours after Katrina struck the coast, Michael D. Brown, Undersecretary of the Federal Emergency Management Agency (FEMA), sent a memorandum to Michael Chertoff, Secretary of Homeland Security, requesting that one thousand Department of Homeland Security personnel be deployed to the Gulf Coast regions devastated by Katrina, which Brown identified in the memo as a "near catastrophic event." In an attachment included with the memo, Brown asked that Homeland Security personnel assigned to the affected areas "convey a positive image of disaster operations" (see Figure 4.1). Many critics argued that the memo was unethical in encouraging Homeland Security personnel to convey less-than-truthful information about the situation along the Gulf Coast. A few suggested that the proposed deception was justified, since a positive image of the disaster might have reduced panic and other unproductive reactions. Regardless of whether the memo was ethical or unethical, it brings up many questions about ethics in workplace communication. This chapter is specifically about the choices writers make in conveying accurate, honest information, which we call "ethical choices."

Role of Assigned Personnel:
Establish and maintain positive working relationships with disaster affected communities and the citizens of those communities.
Collect and disseminate information and make referrals for appropriate assistance.
Identification of potential issues within the community and reporting to appropriate personnel.
Convey a positive image of disaster operations to government officials, community organizations and the general public.
Perform outreach with community leaders on available Federal disaster assistance.

FIGURE 4.1 Excerpt from a FEMA Memo

EXPERTS SAY

"Practically all professional and academic organizations involved with technical communication have developed guidelines or codes of ethical conduct."

PAUL DOMBROWSKI, *Ethics in Technical Communication*

In chapter 1 you learned that workplace writing involves the dissemination of information to diverse readers under various conditions, and you learned that there are many kinds of writing for many different workplace situations. In chapter 2 you read that workplace writing requires careful rhetorical choices in order to solve problems. Chapter 3 showed you the role of technology in solving workplace writing problems. We hope you have noticed several key factors: (a) all workplace writing requires writers and readers; (b) writing in the workplace requires writers to make careful, informed choices to coherently and concisely inform readers; and (c) those choices are affected by professionalism, technology, and ethical concerns.

Workplace writing, like other forms of communication, relies on the standards or conventions of particular contexts. What is considered appropriate for one audience or situation can be inappropriate for another. For instance, the choices a nurse makes in writing a report may differ greatly from those made by an electrical engineer in writing a proposal. And just as the nursing and engineering communities may have different standards for writing reports, so, too, do different professional communities develop different standards of ethical practices. As a worker and writer, it is your responsibility to attend to ethical concerns. Not only is ethics a standard consideration for every profession, but you are also personally responsible for engaging in ethical practices and behaviors. We cannot stress enough the importance of ethics in the daily activities of workplace writers.

The Charles Schwab investment firm recently developed an advertising campaign with the recurring slogan "I need information I can trust." Most people in the work world operate with a similar philosophy: they believe they should give and receive trustworthy information. Because we live and work in a culture of trust, each day as we leave our homes and commute to school or work, we trust that others on the commute are going to stop at red lights. We trust the woman at the coffee shop to return with our change. We trust that our teachers are going to show up in the classroom. We trust that the food we buy won't be contaminated. All day we trust that others will do us no harm and take no (or little) advantage of us and that we will be safe as a result of that trust. But what happens when that trust is violated? What happens, for instance, when we or someone we know is harmed because that trust failed? What happens when a company we have worked for over many years goes out of business because it has misrepresented financial information? What happens when a politician promises to lead us in a particular way but then uses that leadership role for personal gain? What happens when we read a document that disseminates information we later find to be inaccurate, untrue, or manipulative? Each of these questions involves ethical concerns.

Each year countless cases involving ethics and workplace writing are heard in the nation's court system. Many make headlines, but others do not. However, whether an ethical choice is likely to end up in the court system should not influence our decisions and behaviors. Ethical decisions require asking the difficult but direct question: Is this right? Sometimes unethical decisions result in injury or death or in financial loss or damage, but in other cases the results are less drastic. A product's documentation may make claims the product doesn't fulfill—"Guaranteed to cure hair loss!" Or the documentation may claim that certain products are included when actually they are not—"All parts included!" It is not just the *job* of workplace writers but their ethical and professional *responsibility* to be aware of any ethical concerns related to the production of any workplace document.

PSA

The Plan phase of the problem-solving approach includes this: "Consider the ethical choices involved with the problem." Although this is the only part of the PSA that mentions ethics directly, nearly every aspect of the PSA requires workplace writers to frame their problem-solving activities with ethical questions. For instance, planning also requires writers to determine what information is needed; these are ethical decisions about what to report and what not to report. And the Research phase requires gathering and organizing that information, tasks that require ethical choices about what to gather, where to look, and how to organize it. Drafting itself is an act of ethical problem solving: how to organize the document, how to persuade, which visuals to choose. Moreover, in reviewing, getting feedback from readers and colleagues or editing the words of the document requires ethical choices. And, of course, distributing information relies on ethical decisions about how and when to release the document. The entire PSA is grounded in ethical choices.

IN YOUR **EXPERIENCE**

Soon you'll read about ethics and various definitions of *ethics*, but before you do, it is a good idea to get a handle on your own preconceptions about what *ethics* means. Take a minute now to write a short definition of *ethics*. What does the word mean to you? Where does ethics come from? How do we learn ethics? How do we act ethically? Why would a technical communication textbook devote an entire chapter to ethics? Why does ethics in technical communication matter?

Paul Dombrowski's words in the Experts Say box at the beginning of this introduction identify how important ethics is in workplace writing. Beyond acknowledging the importance of ethics, however, it is crucial to understand what ethics is, where ethical principles come from, and how workplace writers make ethical choices at all stages of their writing. This chapter begins with a general treatment of ethics and moves to more specific ethical issues in workplace writing, such as the role of patents, privacy laws, confidentiality, intellectual property, copyright, billable time, and the use of clients' equipment and facilities. Throughout our discussion, this text advocates a number of core values that are central to writing in the workplace—honesty, privacy, legality, social responsibility, teamwork, avoiding conflicts of interest, cultural sensitivity, and professional growth. Here are some introductory points about ethics:

EXPERTS SAY

"You shouldn't just do something because you can; there ought to be some information that disseminates wisdom about why and what happens if you do."

DIANE STIELSTRA
User Assistance Lead, Microsoft Corporation

- Discussing ethics is difficult because it requires asking difficult questions about beliefs and roles in larger communities.
- This book cannot force you to be ethical. It can teach you only what ethics is, why you should think and write ethically, and how ethics affects your life and your workplace. Ultimately, the decision to act ethically, professionally, and responsibly is yours.
- The only way to fully understand ethics is through communication with others. Even though we each make our own ethical choices, those choices affect and are affected by others.
- Because ethics depends on individual decisions within communities, discussions of ethics and individual ethical choices reveal conflicts between individuals and their communities. Because disagreement and conflict are part of ethical decisions, ethics can be uncomfortable.

WHAT IS ETHICS?

ethics a code of conduct that helps individuals determine what is right and what is wrong

Simply put, **ethics** is about right and wrong. It is the principles one chooses to live by within communities, yet ethics is by no means universal, consistent, constant, or even always easily identifiable. Sometimes making an ethical decision is difficult; as we all know, what is right in one situation is not always right in another. Nonetheless, we all make choices about what we consider "right" countless times each day. Making ethical choices is much like making rhetorical choices: we depend on our available tools (i.e., the ethics we have learned), and we situate those tools within contexts. In fact, making ethical choices and making rhetorical choices are intertwined. Sometimes we can't explain exactly why we made a particular choice; it just seemed to be the most appropriate at the time.

For as long as humans have questioned what is right and wrong, they have studied ethics. Ethics has, in fact, become a field of study in its own right, affecting many other disciplines, such as writing, communication, business, medicine, engineering, criminology, law, computer science, allied health, and science. The study of ethics involves the study of moral philosophy. It investigates the ways that concepts of right and wrong are defined, presented, defended, justified, systematized, taught, learned, controlled, and so on. Those who study ethics define three primary types or categories of ethics:

metaethics the study of where ethical ideas come from and how they develop

normative ethics the study of ethics concerned with classifying what is considered right and wrong

applied ethics the study of particular ethical issues, problems, and circumstances

- The first type of ethics addresses the origin of ethical ideas and asks where they come from, how they were formed, and what they mean in the contexts in which they are applied. Are ethical concepts universal, or are they socially defined? Are they linked to emotions? Are they related to religious beliefs? This type is called **metaethics**, which is the study of where ethical ideas come from.
- The second type of ethics addresses how we arrive at ethical standards, establishing categories for what is considered right or wrong. How should we behave? How should we respond to those who don't behave within an ethical code? How might we regulate ethical codes? This type is referred to as **normative ethics**, the branch of ethical studies that is concerned with classifications of right and wrong.
- The third type of ethics focuses on particular case studies to better analyze the contexts in which ethics is debated. These studies are generally called **applied ethics**

because they examine real-world decisions and the ramifications of an ethical issue. For instance, applied ethics might study the ethical dilemmas that arise in debates about controversial subjects in order to understand why some see different rights and wrongs. Applied ethics also studies how particular events or issues change the ways in which we define right and wrong.

Because applied ethics relies on metaethics and normative ethics, it is often difficult to identify where one consideration ends and another begins. For instance, to consider the ethical dilemma regarding the effects of hazardous waste dumping on populations of children, we must also ask why we protect children, how our definitions of "pollution" have changed over time, and how we balance the needs of business with those of the environment.

In many ways, then, these three approaches ask not just why a particular action might be right or wrong, but why that action is considered right or wrong to begin with. For those attentive to how workplace writing is produced, this is a crucial distinction because recognizing ethics as more than simple choices requires an analysis of contexts and classifications in order to make informed ethical/rhetorical decisions. Throughout this text you will be asked to consider the ethical implications of decisions you will make as a workplace writer. In many instances you will consider specific cases and their ethical implications. You will also examine how others establish ethical guidelines and how you can develop your own ethical guidelines. In fact, one goal of this book is to help you develop a set of professional and personal standards to use as a workplace writer. None of the conventions of workplace communication that you will learn in this book will be worth much unless you understand your ethical, professional, and personal obligations as a worker, writer, problem solver, and human being.

Ethics and the Workplace Writer

In workplace writing it is important that you not only make good choices about what is right or wrong but also understand why those choices are worth consideration. The book *Ethics in Technical Communication: A Critique and Synthesis* details three primary reasons that writers in the workplace should study ethics:

1. Studying ethics can help us think more clearly and more sensitively.
2. Studying ethics enables us to articulate our views to others.
3. Studying ethics enables us to advance in our ethical thinking.[1]

Studying ethics can do more than simply help you decide what is right or wrong; it can help you to become a more effective, successful workplace writer and communicator. By examining ethics, both in theory and in context, we better understand why ethical decisions are necessary and can then better explain those decisions to others. In the workplace you may find yourself in situations that require not only ethical and rhetorical decisions but also explanations of those decisions to company superiors, clients or customers, or even courts of law. As we learn more about ethics, we become more sophisticated in our thinking about ethics. That doesn't necessarily mean that individuals become more ethical by studying ethics, but it does mean that their ethical decisions might be better informed, and in turn, they may see ethical implications they never considered before.

For many years the relationships between ethics and workplace writing were not addressed; only recently have we begun to realize just how crucial ethics is for workplace writers. Paul Dombrowski explains:

> This awareness is due in part to the recognition of many important ethical lapses in recent years involving communications about technology. The major technological

---

[1]Michael Markel, *Ethics in Technical Communication: A Critique and Synthesis* (Westport, CT: Ablex, 2001), 30.

> disasters of recent years seem to be linked in various ways to problems in communication in this country—from the dangers of charred O-rings in the *Challenger* to the danger from leaking silicone breast implants. In other countries we have learned of inadequate safety documentation at Bhopal and botched technical procedures at Chernobyl.[2]

In short, we have identified that many disasters are the result of poor documentation and inadequate attention to ethical considerations. Thus, it stands to reason that more accurate and ethical workplace writing should reduce these disasters, both large and small.

EXPLORE

The case of the Ford Pinto is one of the most famous cases in ethics and workplace writing. The Pinto, a small car produced by Ford Motor Corporation from 1971 to 1980, was found to burst into flames in low-speed rear-end collisions because of a design flaw in the placement of the fuel tank. In 1978 Ford lost more than $125 million in lawsuits over injuries and deaths caused by the car, despite the company's consistent argument that there was no design flaw in the vehicle. However, in October 1979 a memo was discovered indicating that Ford had been aware of the design flaw but had not addressed it in any of its documentation about the car. Ultimately, Ford recalled the vehicles and stopped manufacturing the car, explaining their action as a result of the Pinto's reputation, not an admission of the design flaw. Both the infamous memo acknowledging the design flaw and the documentation that never acknowledged the flaw were prominent issues throughout the case.

Take some time to conduct an online search into the Ford Pinto case to see what role technical documents played in the case and what those documents show about how ethical decisions contributed to the case. Also, take some time to explore the Web for examples of other newsworthy events that can be attributed to ethical decisions made in producing technical documents.

ETHICAL GUIDELINES FOR WORKPLACE WRITERS

To help you think carefully about ethical issues you may face when writing, this chapter offers three basic guidelines to consider.

Guideline #1: Ethics and Laws Are Not the Same

People often equate ethical with legal. Rather than asking what is right, they may ask what is legal. However, ethics and legality are not the same thing: what is ethical may not be legal, and what is legal is not always ethical. Think, for instance, about laws that have allowed practices such as slavery, the dumping of hazardous waste, the exclusion of certain people from jobs, the production and distribution of unsafe products like tobacco, the presentation of inaccurate advertising, and more. We may believe something is right or wrong, yet the laws that govern that subject may differ from our beliefs. Certainly, workplace writers want to be aware of the laws that govern their industries, but they also need to consider the ethical implications of those laws. A good rule to live by is "Just because I can doesn't mean I should."

Laws regulate much of what we do in the work world. There are laws that regulate what writers can write, what they can't write, and what they must write in documentation. Think, for instance, about an insurance policy—a specific kind of technical document. That document must detail exactly what the insurer is legally responsible for; and the language must be specific, reflecting local and federal (perhaps even international) laws regarding insurance policies. When legal cases develop over what insurance policies say and mean, ethical representation may not be the same as

[2]Paul M. Dombrowski, *Ethics in Technical Communication* (Boston: Allyn & Bacon, 1999), 1.

legal representation. It is up to the writers who design those documents to ethically express the legal conditions contained in them.

Writers must always be exact in the legal facets of their documents. For instance, writing contracts is a specialized form of technical communication, one that has serious legal and ethical implications. Contract laws dictate how companies and clients must respond to one another, based on the agreements they signed to establish a working relationship. Good contracts can prevent disputes, but breaches of contract can result in lawsuits, loss of contracts, and loss of client or customer faith. Contracts can come in a variety of forms, ranging from formal contracts to warranty information (i.e., a company's contract with its consumers). However, writers who produce contracts of any type should be attentive to the legal and ethical implications of the wording.

Workplace writers often see the law as an easy way out of ethical dilemmas. Think, for example, about natural supplements that purport to enhance performance or promote weight loss. Because many of these products are not approved by the Food and Drug Administration, there are no legal restrictions on what they can and cannot claim on labels or in documentation. Consequently, some manufacturers of supplements push the ethical limits in their documentation and then hide behind legalities that allow them to make such claims. Again, just because something is legal doesn't mean it's ethical.

You should also be aware that laws vary from place to place; what may be legal in one place may not be legal in another. For writers this is critical because their documents may be read and used in places other than those where they were written. Therefore, writers must be aware of the laws in all places where a document might be read or used.

ANALYZE THIS

Locate an over-the-counter medical or cosmetic product: a nutrition supplement, cold remedy, pregnancy test, makeup, shampoo, or deodorant. Then read the label and look for the claims about what the product can do. What language is used? What disclaimers are provided? How does the technical documentation provided demonstrate legal compliance and ethical attention to product claims? You may also want to visit the website of a company that sells enhancement or nutrition supplements and examine the ways in which that website presents information.

Many laws that affect workplace writing protect those who rely on that writing. In particular, many laws address safety and liability issues, ensuring that those responsible for a document are held liable if that document contributes to injury, financial loss, or other harm. The opposite holds true, too. When workplace writers do act ethically and legally but the information they provide is misused and harm results, the writers are protected by those same laws. Workplace writers need to consider a number of specific kinds of laws when producing documents.

Liability Laws

Liability laws protect individuals from defective products, accidents, and inappropriate actions by an individual or company. *Liability* refers to the condition of being responsible; therefore, liability laws dictate who can be held responsible for particular actions. For example, numerous laws dictate the extent to which a manufacturer is responsible when a consumer misuses a product and is injured, as well as when a consumer is injured because of a product defect.

Environmental Laws

All workers should be concerned about whether their work harms local or global environments. The choices of some workplace writers may indeed harm the environment—for example, a report that hides information about the environmental impact of building a new gas station could result in groundwater pollution or habitat destruction. Thus, professionals should always consider the impact of the information they disseminate. Environmental laws address issues such as energy use, waste disposal, land use, natural resource use, and water rights. Later in this chapter we will address environmental ethics in greater detail.

Copyright Laws

Copyright and trademark laws are among the most important laws for workplace writers. Simply put, copyright laws give owners of written works the right to display, publish, reproduce, license, and distribute those works. In addition, any work contracted to be written is considered work for hire; if a workplace writer produces a report, manual, or memo as part of regular work duties, the copyright to that document is owned by the worker's employer.

Continuous technological developments have led to reinterpretations of copyright laws, which now understand writing as more than just traditional written documents. Currently, copyright laws protect audio recordings, films, software, graphic arts, and even architectural designs.

Patent Law

Much like copyright laws, patent laws protect the rights of inventors and ensure them "the right to exclude others from making, using, offering for sale, or selling" the invented product or idea in the United States. Simply put, patent laws prevent individuals or organizations from profiting from another's invention. Such laws do *not* grant the owner the right to produce, sell, or offer the product or idea for sale, but they do exclude others from doing so. Patents are generally granted for twenty-year periods.

Trademark and Service Mark Laws

Trademarks are names, symbols, devices, or words that accompany goods and indicate that those goods are distinctly different from other goods. For instance, both Coca-Cola® and Pepsi-Cola® register their names and logos as trademarks, indicating that these seemingly similar products are distinctly different. A service mark is essentially the same as a trademark, but it identifies a service. Companies such as Visa or Century 21 use service marks, since what they provide is primarily a service. The word *trademark* is used to refer to both trademarks and service marks.

Trademark laws are designed to prevent other individuals and organizations from using the same or similar representations to identify a product or service. Names, symbols, or words that are similar to those used by other companies or products are not allowed by trademark law. Trademark laws do not, however, prevent someone from making a similar product, only from representing it in a similar way. For instance, a company can produce and sell adhesive bandages but cannot market them under the name Band Ayd or Bang-Aid.

EXPLORE

- For more information regarding environmental laws, visit the EPA's website at http://www.epa.gov/, or read about the National Environmental Policy Act at http://www.epa.gov/compliance/nepa/index.html.
- You can learn more about copyright laws or access the entire U.S. Copyright Code by visiting the U.S. Copyright Office web pages at http://www.copyright.gov/. Be sure to pay particular attention to Section 106 of the U.S. Copyright Code, which specifically addresses what kinds of documents are protected by copyright law.
- You can learn more about patent and trademark laws by visiting the U.S. Patent and Trademark Office at http:// www.uspto.gov/.

Contract Laws

Contracts are legally binding written agreements between parties. They are written when one party makes a promise to another; they record the details of the promise. For a contract to be legally binding, the promise detailed in the contract must be offered in return for some benefit or detriment that is a reasonable exchange for the promise. Nearly every business and organization relies on written contracts to get work done. Contracts can range from the agreement a worker has with an employer to an agreement between companies regarding their business interactions. Contract laws provide solutions to disputes when those agreements have been breached.

Writing contracts requires particular knowledge of contract laws, as well as an understanding of both previously signed contracts and negotiations for future contracts. For instance, an engineer might submit a proposal detailing the use of a leased storage facility as a test site for a particular project. However, if that engineer is not aware of the details of the contract signed with the leasing agency, the proposal may violate the agreed-upon conditions of that lease.

Guideline #2: Be Honest

This one should be easy: be honest. It's that simple. Not only is the dissemination of false information illegal, but it is downright unethical. And this includes presenting information in a way that obscures the truth or vaguely represents it or asks others to do the same (think about the FEMA memo discussed at the beginning of this chapter). Workplace writers have an ethical responsibility to present information in a clear, concise, and coherent manner. Remember that rhetoric is a powerful tool that allows you to decide which information to present and how to do so. A skilled communicator can tell the truth while still obscuring important details, and such practices may appear to make good business sense in the short term. However, such rhetorical decisions are deeply dishonest and may have longer-term ramifications. Consider, for instance, how the tobacco industry's documentation minimized the health risks associated with smoking, depicting smoking as cool and fun—and then consider the $350 billion lawsuit they had to settle in 1997.

EXPERTS SAY

"There is no such thing as an employee who doesn't steal. . . . The thing that is most often stolen from the workplace is time."

DR. LEONARD E. DOBRIN
Criminologist

Another aspect of honesty requires that workplace writers always identify the sources that inform their documents. If they rely on, reference, or borrow someone else's work, they must obtain permission to use that work and then acknowledge that permission.

The quotation from criminologist Leonard E. Dobrin in the Experts Say box refers to yet another aspect of honesty. A worker is employed to do a particular job at a particular time and is paid for that time. For instance, if you are paid seventeen dollars an hour to enter data from hard-copy files into a computer file, your seventeen dollars is contracted for a full hour's work. If you take a moment to make a doctor's appointment or to check the price of a new car, you have spent part of that hour not doing the work for which you are being paid. Hence, you have stolen time if you accept the full seventeen dollars for that hour's work. Of course, most employers understand that some personal business must be conducted during working hours, and most have no problem with that, as long as the employee is productive. When such theft of time becomes excessive, however, it becomes an issue. Part of being an honest worker is using time ethically.

The same holds true for theft of materials, services, and supplies. Often workers solve personal problems using company equipment—for example, using a company copy machine to make flyers about a garage sale or a lost kitten, taking a company pen or software home, or using company tools to fix personal things. Like the theft of time, most employers overlook the theft of small items. Many, in fact, might not even consider the use of an office printer or the removal of a few pens to be theft. Again, this type of theft is often an issue only when the theft becomes excessive.

Why is it sometimes seen as OK to steal something small but wrong to steal something large? One of the primary considerations has to do with how that theft affects others—what we might refer to as *the public good.* Taking a pen or two will probably not damage your company or your colleagues to any great degree, but stealing an office computer certainly could. Because human beings rely on one another for things to function, when anyone denies honesty as fundamental to human interaction, then the strength of that interaction begins to weaken. As workplace writers we must all be aware of our contributions to the public good, which is also our own good. An honest, functioning public benefits us all. In the long run, honesty comes down to thinking about the welfare of others as equal to our own.

IN YOUR **EXPERIENCE**

Have you ever taken anything from a workplace? Many workers take things like office supplies or other small items. Some workers use company phones to make personal calls or use company computers or copy machines to produce personal documents. Other workers may take advantage of services provided by a company; for example, a person who works for an automotive repair station may fix his own car on company time.

Chances are good that every one of us who works has at one time or another taken something from our job. But how much of this taking is considered OK? That is, at what point does taking something become a violation of ethical responsibility, and at what point is it just an accepted practice? Consider your own boundaries. Where do you draw the line? Discuss these issues with your classmates.

Guideline #3: Respect Confidentiality

Confidentiality, sometimes referred to as privacy, is both a legal and an ethical issue. Some people even consider it a right. Workplace writers often access information that should be discussed or communicated only in prescribed ways. If information is sensitive, any inappropriate transmission of it could cost a company contracts, money, jobs, or reputation. Computer technology contributes to confidentiality issues, since personal or confidential information is often stored or transmitted electronically, making it easier to access or send than ever before.

Good workplace writers are clear about what information can be, or should be, made public, and they obtain documented permission to release information when necessary. Workplace writers have an obligation to protect the privacy of employers, employees, clients, partners, or anyone else with whom they maintain professional relationships, even if those relationships have been terminated. Only when workplace writers have received releases from those involved should they share sensitive information. Many companies ask employees to sign confidentiality agreements, which are usually legally binding contracts.

ETHICS IN CONTEXT

Making ethical choices and solving ethical problems can be difficult because there are no formulas, no definitive rules. Ethical choices can be made only in the context of a specific situation, and even within that situation the parameters of the ethical options may be unclear. You may respond in a particular way to a situation one day and in another way to a similar situation the next day. You may have to make difficult or even uncomfortable choices. Certainly, Codes of Ethics (discussed in the next section) and guidelines can offer us frameworks within which to work, but ultimately, an ethical decision is an individual, professional choice made in a particular context.

PSA

Situation and Perspective

All workplace and rhetorical problems occur within specific situations that provide the context in which they can be solved. Ethical choices are no different. The context of a situation determines what the appropriate professional and personal response can be. Workplace writers can ask themselves a number of questions to better understand the situation and thus to better enable themselves to address the ethical issues of the situation.

- What is my reason or purpose for writing this document? What is the exigency?
- Who is the audience for this document, and how will that audience be affected by the choices I make in writing the document?
- Who, other than the audience, does the information in this document affect?
- What authority do I have in conveying this information? Does that authority affect how the information will be understood?
- What are the ramifications of conveying the information in this way in my document?
- Where does this information fit in the larger context?
- Are there limits to what my writing can and cannot convey? How do I determine those limits? What or who imposes those limits?
- Will this document lead to any environmental effect?
- What responses are this document, and my choices, likely to evoke?

Keep in mind that the rhetorical choices you make represent your ethical positions. Those rhetorical choices are always tied to the context of the situation and your own interests and beliefs. For instance, consider the concept of taking a life and the rhetorical choice among words like *murder*, *kill*, or *euthanize*. Each of these words conveys a different ethical stance dependent upon the situation. *Murder*, for instance, implies that a crime has been committed, whereas *kill* doesn't necessarily carry that same baggage, and *euthanize* can even have a positive connotation. Consider how an animal rights activist might describe what a hunter does to a deer as compared with how that hunter might describe the act. Because the two operate from different ethical positions, their rhetorical choices are apt to be different.

Circulation

At the beginning of this chapter, we told you about a memo that FEMA Undersecretary Michael D. Brown sent to Michael Chertoff, Secretary of Homeland Security, in the aftermath of Hurricane Katrina. In the past, interdepartmental memos like this would probably not have reached the public eye. But contemporary technologies provide the opportunity for documents to be shared worldwide and ensure that information is made public quickly. This means that when someone recognizes a document as potentially unethical—like the FEMA memo—there is a good chance that the document will circulate to a wide audience rapidly. Because of this capability workplace writers need to consider the ramifications of widespread distribution of their documents, both in terms of their own accountability and in terms of what others can do with that documentation. Competing companies and even individuals can access and use technical documentation in ways such documentation was never intended to be used. For instance, Apple introduced the iPhone in the summer of 2007 as exclusively tied to AT&T, but when the phone and its documentation were released, it didn't take long for hackers to make the iPhone compatible with other networks. The hackers then used the Internet to distribute their adaptations, undermining Apple and AT&T's contract agreements. Such a situation raises questions about the ethics of hacking, the corporate ethics of preventing cross-network use, and the ethics of using publicly available documentation for ulterior purposes.

Revision and Editing

Throughout *TCTC* we stress the importance of revision and editing—the task of crafting a document into a professional, readable, accessible communication. Even though the ultimate goal of revision and editing is to make a document "better," writers must be alert to the ethical considerations that accompany editing and revision activities. Because editing and revising can alter a document in form and content, writers need to be aware that even the slightest changes can alter or obscure the information being conveyed. Workplace writers must be acutely aware of how their revisions and edits change a document, particularly when offering revision suggestions to someone else. The choices made in editing and revising are themselves ethical choices.

See chapter 10 for more on revision.

Revising and editing the work of others requires acting in their best interests—an ethical concern. The most important principle in revising or editing someone else's work is to be clear about your goal before you begin. Make sure you both understand what and how much you will do and how you will do it. You may wish to agree on a particular method of editing or revision beforehand. If the original author wants to see the changes you suggest, you may wish to use a track-changes feature in your word processor or use a colored font to designate the new or changed information. On the other hand, if the author expects you to make the changes yourself, tracking them might create unnecessary work for both of you.

CODES OF ETHICS

Many groups, organizations, and companies maintain their own codes of ethics, which address issues such as harassment, grievances, employee relations, and basic business practices. Codes of ethics help employees with ethical decisions by providing company policies and other forms of information to steer workers through thorny ethical issues. Figure 4.2 shows the code of ethics for the Institute of Electrical and Electronics Engineers.

ANALYZE THIS

Like the Institute of Electrical and Electronics Engineers, many other organizations have codes of ethics. Locate two other codes from professional organizations. Then compare the two codes, and make a list of what kinds of things each covers and what differences there are between the two. You might try looking for codes like the Software Engineering Code of Ethics, the American Psychological Association Code of Ethics, the Association for Computing Machinery Code of Ethics, or the American Library Association Code of Ethics.

ETHICS AND TECHNOLOGY

Because technical communication relies on and is enmeshed with technology issues, it is crucial to consider the relationships between ethics and technology, or *cyberethics*. The ethical questions related to technology are mostly about use: How do we use technology ethically to convey information? And how do we act ethically in cyberspace?

In *Code and Other Laws of Cyberspace*, Larry Lessig explains four principal factors that regulate our behaviors in cyberspace:

1. Laws—the rules that governing bodies impose on how we use cyberspace
2. Social norms—the sense of normalcy that a community imposes on how we use cyberspace

3. The market—which creates standards for what can be bought and sold on the Internet
4. Architecture—the way the Internet is constructed to let us get to some places and not others.[3]

Workplace writers must be aware of how these factors influence their writing and work.

We, the members of the IEEE, in recognition of the importance of our technologies in affecting the quality of life throughout the world, and in accepting a personal obligation to our profession, its members and the communities we serve, do hereby commit ourselves to the highest ethical and professional conduct and agree:

1. to accept responsibility in making decisions consistent with the safety, health and welfare of the public, and to disclose promptly factors that might endanger the public or the environment;
2. to avoid real or perceived conflicts of interest whenever possible, and to disclose them to affected parties when they do exist;
3. to be honest and realistic in stating claims or estimates based on available data;
4. to reject bribery in all its forms;
5. to improve the understanding of technology, its appropriate application, and potential consequences;
6. to maintain and improve our technical competence and to undertake technological tasks for others only if qualified by training or experience, or after full disclosure of pertinent limitations;
7. to seek, accept, and offer honest criticism of technical work, to acknowledge and correct errors, and to credit properly the contributions of others;
8. to treat fairly all persons regardless of such factors as race, religion, gender, disability, age, or national origin;
9. to avoid injuring others, their property, reputation, or employment by false or malicious action;
10. to assist colleagues and co-workers in their professional development and to support them in following this code of ethics.

Approved by the IEEE Boards of Directors
February 2006

FIGURE 4.2 The Institute of Electrical and Electronics Engineers Code of Ethics

E-mail

The first thing to remember about e-mail is that it is never private. Any message you send or receive can be accessed by unintended readers or even later through archives. Consequently, you want to think carefully about sending e-mails that may be slanderous, may cause harm to another, or may convey false information. In addition, you should confirm information you receive before forwarding it to others. Frequently,

[3]Larry Lessig, *Code and Other Laws of Cyberspace* (New York: Basic Books, 2000).

e-mails that make claims about company practices, political actions, or newsworthy events are sent to large numbers of people. But as often as not, these widely circulated e-mails are false. A writer in the workplace should recognize that forwarding false or inaccurate information is as unethical as originating it.

See chapter 12 for more on e-mail.

One of the most significant problems with the wealth of information available on the Internet is that just about anything can be posted—true or false. Often you may receive unconfirmed information through e-mail, claiming something that seems outrageous and unbelievable. For example, you may have seen the e-mail claim that Bill Gates and Microsoft were giving away free money or that Starbucks had closed its stores in Israel for political reasons. These stories circulated quickly and widely, even though neither was true. Several websites—see www.snopes.com, for instance—can help you track and confirm the veracity of such stories and sometimes their origins.

Websites

"It must be true; I read it in the newspaper." That used to be a commonly accepted belief, reflecting our trust in published information. However, as the World Wide Web grows and as more private users publish their works on the Web, relying on published information becomes more complicated. It is crucial for workplace writers to pay close attention to the information they gather from web pages, as well as the information they publish on web pages. Web pages can be made to look official but may carry no legitimate information. They can also allow writers to make false information appear real. Ethically minded workplace writers must be conscious of how information becomes public on the World Wide Web.

See chapter 17 for more on web pages.

(a)

(b)

(c)

FIGURE 4.3 Image (a) is a dramatic manipulation created by blending image (b) with image (c).

Visuals

In the past, workplace writers relied on graphic artists to produce any needed visuals, and the technologies for including visuals in documents were limited. However, current technologies allow just about anyone to create visuals, and writing and publishing software allows writers to produce visually dynamic documents. Although these new technologies provide workplace writers with many opportunities for including visuals in their documents, such technologies also present the opportunity for workplace writers to manipulate images to their own ends. Anyone can fake an image now. For example, the first image in Figure 4.3 was reported to have been taken off the coast of South Africa and was nominated as Photo of the Year. However, the picture was proven to be a case of digital manipulation of the second and third images, as reported by *National Geographic*. The image of the helicopter was actually taken in San Francisco with the Golden Gate Bridge in the background.

But image forgeries aren't the only ethical concern that workplace writers must consider in their use of visuals, nor are they even the primary concern. All workplace documents include some visual component, whether that is images and graphics or simply the design and layout of the document. Chapters 8 and 9 address visual rhetoric and document design in detail, but for now it is important to understand that choosing how and when to use visual elements in a technical document is an ethical decision. Consider this example: In March 2005 one of the headline news stories in the United States was a court decision to remove the feeding tube from Terri Schiavo. In covering the story, the Cable News Network (CNN) published a chart depicting the results of a CNN/*USA Today*/Gallup poll that had asked, "Based on what you have heard or read about the case, do you agree with the court's decision to have the feeding tube removed?" The poll showed that 62 percent of Democrats polled agreed, 54 percent of Republicans polled agreed, and 54 percent of Independents polled agreed. With these numbers in mind, consider the way the first bar graph in Figure 4.4 reports the information from the poll.

In evaluating this graph, compare the visual representation of the poll results and the actual difference in percentages. The poll found a total difference of 8 percent

Results by party

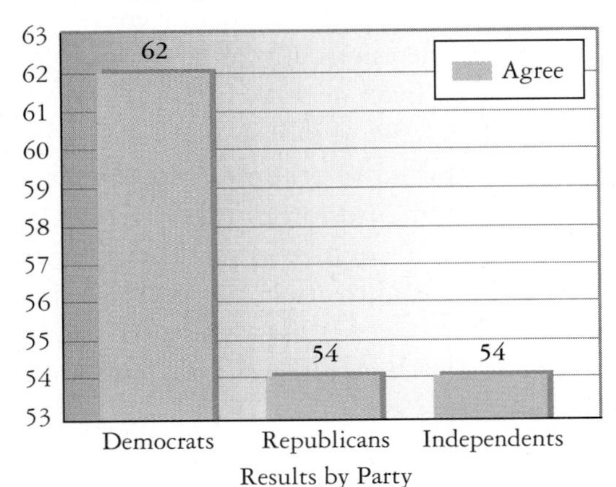

Question 2: Based on what you have heard or read about the case, do you agree with the courtś decision to have the feeding tube removed?

SAMPLING ERROR: +/– 7% pts

SAMPLE: Interviews conducted by telephone March 18-20, 2005 with 909 adults in the United States.

(a)

Results by party

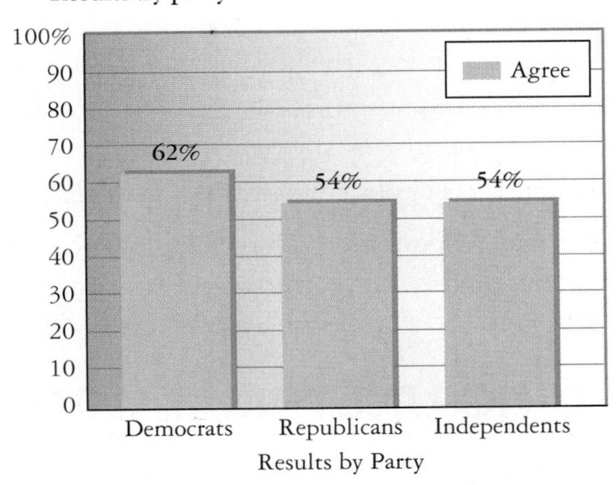

Question 2: Based on what you have heard or read about the case, do you agree with the court's decision to have the feeding tube removed?

SAMPLING ERROR: +/– 7% pts

SAMPLE: Interviews conducted by telephone March 18-20, 2005 with 909 adults in the United States.

(b)

FIGURE 4.4 (a) Graph of CNN/*USA Today*/Gallop poll from March 18–20, 2005, as originally published by CNN and (b) the updated version published after CNN received a number of complaints about the misleading graph.

between the number of Democrats in agreement and the number of Republicans or Independents in agreement. And the poll had a sampling error of +/−7 percent. The second bar graph in Figure 4.4 shows a revised, more visually accurate representation of the poll numbers. This example demonstrates how visual representations can lead readers to see information in particular ways. Ethical writers consider how the use of visuals in a document can skew the perception of information.

ENVIRONMENTAL ETHICS

You might find it odd to see a segment about environmental ethics in a technical communication textbook, but ethical/rhetorical choices can have important environmental ramifications. Simply put, what we do in the workplace affects the world around us. Therefore, good workplace writers develop a sense of environmental ethics in conjunction with their personal and professional ethics. *Environmental ethics* suggests ethical choices about how we behave in relation to our environments. You might infer that this means natural environments, and in many ways it does. But environments include all the spaces and places with which humans interact—community environments, business environments, academic environments, electronic environments, and so on. Our ethical decisions affect the natural world as well as the communities in which we work and live.

Earlier we cited Paul Dombrowski, who identifies Chernobyl as an example of a disaster attributed to technical documentation. The failure to produce reliable technical documentation in this instance led to an ecological disaster. Thus, one of the things to recognize as workplace writers is that the documents we produce have the potential to affect not only the lives of other people but also the environments in which we all live. And there are countless other examples in which technical documents have affected local and global environments, both negatively and positively.

It is the responsibility of writers and communicators to be aware of how their documents might impact Earth's environments. The success of an engineering project or a business deal should not be separated from its impact on the natural environment, for example. Environmental ethics asks important questions about these relationships:

- How do our actions affect the world in which we live?
- How do our actions in regard to our environments affect the quality and the length of the lives of the people who inhabit those environments?
- What are the compromises we make to protect human interests over environmental interests?
- Do other parts of our environments have rights, and if so, what are they?

Asking these sorts of questions is crucial for workplace writers in order to consider the implications of information conveyed in their documents. Many documents have ramifications beyond their immediate context.

AVOIDING UNETHICAL WRITING

Ethical concerns in workplace writing often revolve around the ways information is presented: Does the writing obscure, hide, or evade the truth? Does the writing emphasize or de-emphasize key information? Does the writing misrepresent data or other findings? Does the writing intentionally oversimplify or overcomplicate the information? The words that are chosen, the organizational patterns, the tone, even the design of documents are all bound up in ethical issues.

Don't Use Deceptive or Evasive Language

Language allows writers to clarify as well as conceal information; language can hide and obscure meaning and, therefore, truth. An ethical workplace writer does not evade the truth through deceptive language that attempts to mislead the reader. For instance, sentences that use ambiguous subjects can hide information about responsibility for a given act. In most cases obscure, deceptive language is used to cover up an error or an oversight. Consider this example:

See chapter 10 for more on evasive language and ambiguous subjects.

> Action had to be taken because certain personnel did not meet department expectations.

This sentence is unclear about what was done, which individuals were involved, and who was responsible for the action. Instead, good writing should provide specific details:

> The director had to fire twelve civil engineers from the highway division who had approved the construction of an unsafe bridge.

At times it may be appropriate to obscure who is to blame, particularly when it is a customer or client who has made a mistake. For example, you wouldn't write to a customer stating, "You ordered the wrong part, but we will ship you the correct one anyway." Placing blame in this way could create negative feelings. Instead, you might write, "We noticed that the wrong part was ordered, so we are sending the correct one."

This next example is also unclear and potentially misleading:

> Numerous patients contracted a staph infection after being treated by the seventh-floor nurses.

The word *numerous* does not provide accurate information about the number of cases reported. It suggests a large number, perhaps even an indeterminable number, even though it is unlikely that the number of infected patients is that great. A writer filing a report containing this sentence might be considered deceptive, encouraging readers to believe the number of infected patients is greater than was actually recorded. Or the opposite might be true with this example:

> A few patients contracted a staph infection after being treated by the seventh-floor nurses.

To be accurate and ethical, writers should provide nondeceptive data:

> Seventeen patients contracted a staph infection after being treated by the seventh-floor nurses.

Similarly, evasive language can intentionally, or unintentionally, alter readers' interpretations of documents. Evasive language skirts the truth, avoiding accurate information without outright lying or misrepresentation. One of the most common examples of evasive language is the use of the passive voice, which again can avoid responsibility. Consider this example:

See chapter 10 for more on the use of the passive voice.

> The accident was caused by an oversight.

In this sentence no subject accepts responsibility for the accident. The subject of the sentence, "accident," actually receives the passive action of the verb, "was caused." To ethically and clearly report this information, a true subject should be identified, and the accident should be reported as the direct object of an action verb. The following is a nonevasive, clearer statement of what occurred:

> An oversight caused the accident.

Even clearer and more informative is the following:

> An engineering oversight caused the accident.

Don't Obscure or Misrepresent the Issue

Just as we can use language to present information clearly, we can also use it to obscure or misrepresent information, to cloud a reader's ability to accurately interpret information. To avoid either intentionally or unintentionally obscuring or misrepresenting issues or information, writers should pay careful attention to both their word choice (i.e., diction), especially avoiding abstract wording and unnecessary use of jargon, and the various ways they present information.

Abstract Language

Abstract, as opposed to concrete, language refers to concepts that cannot be confirmed by sensory perception; that is, they are not words that are understood through seeing, hearing, smelling, tasting, or touching. Words like *truth, freedom,* and *happiness* are easily identifiable as abstract. The workplace writer, however, must be attentive to a broader range of abstractions. Consider this example:

> The ideal employee will have experience, confidence, intelligence, and the ability to collaborate on important projects. As a coworker the ideal employee should be able to participate in the coordination, implementation, and execution of tasks.

This description of an ideal employee may sound appealing, but it leaves many questions: What kinds of experience? How will confidence be shown? What constitutes intelligence? How much should the employee participate? By using so many abstract words, the writer provides an opportunity for many different and potentially inaccurate interpretations.

Jargon

Jargon can also keep some readers from understanding a document. Jargon is simply specialized language that conveys specific meaning to audiences familiar with the terms. Some refer to jargon as the slang of a professional community, and sometimes it can be the most effective and efficient means of communicating because it carries specific meaning without long explanations. However, because jargon is understood only by specialized audiences, it can be confusing to readers unfamiliar with it.

A good workplace writer is careful about when and how to use jargon. By intentionally using familiar jargon, a writer can often save time in conveying information. For instance, many readers can understand jargon terms like *downsizing* or *downtime* because they have become familiar to mainstream audiences. But what about terms like *quant* or *hotboxing*?

See chapter 10 for more on jargon.

The decision to use jargon depends on the audience. A good general approach is to avoid using jargon in an external document. In an internal document jargon may be acceptable if the audience uses that same jargon. But workplace writers must always avoid intentionally using jargon to confuse or mislead readers. Doing so is an unethical practice.

Emphasis or Suppression of Information

The famous author George Orwell explained that all writing is political, meaning that all writing attempts to convince an audience of something. Certainly, all workplace writing is used to convince readers of something; even the objective presentation of data is meant to help readers solve a problem. And the way writers present information affects the way readers respond to it. What a writer emphasizes or suppresses can determine a reader's reaction.

Emphasizing or suppressing information can be used to a writer's advantage, but workplace writers must recognize the ethics surrounding such behavior. Does a document overemphasize a point in order to distract a reader from other information? Does a document neglect information a reader should have in order to make an informed decision? A writer who intentionally manipulates information to affect a reader's decision making is acting unethically.

Visual Rhetoric

See chapters 8 and 9 for more on how visual rhetoric affects information in technical documents.

Visual rhetoric also allows writers to emphasize or de-emphasize information. Therefore, workplace writers must consider not only how the wording of a document stresses particular information but also how the visual appearance of a document emphasizes or de-emphasizes the conveyed information. In their decision making, workplace writers are determining whether the visuals in the document are used ethically.

Plagiarism

See chapter 6 for more on plagiarism.

Plagiarism is using someone else's words or ideas without identifying the source. Simply put, presenting someone else's work as your own is unethical. Conscious plagiarism amounts to stealing, and even inadvertent failure to credit a source can be construed as plagiarism. Whether intentional or unintentional, plagiarism misleads readers.

Use of Inaccurate Information

Workplace writers sometimes find themselves gathering information that others have produced and, on occasion, discover or suspect that part of that information is inaccurate or somehow flawed. An ethical writer does not simply transmit questionable information and expect the original source to be blamed if the problem is detected. Workplace writers are responsible for the accuracy of the information they communicate; inaccurate information can have greater ramifications in the future if not corrected. To be safe, workplace writers should get into a habit of double-checking everything.

THE ETHICAL WRITER'S CHECKLIST

- Is the information presented in my document honest?
- Is the information presented in my document legal?
- How does the information in my document impact local or global environments?
- Have I used technology to obscure or alter meaning?
- Does my use of visuals accurately represent the truth?
- Does my language obscure the truth in any way?
- Have I used vague, ambiguous, misleading, abstract language in my document?
- Is my document jargon filled?
- Have I confirmed all information I've presented?
- Have I credited all information I have borrowed and used in the document?
- Have I identified and made known any inaccuracies in the information?
- Have I violated anyone's confidentiality?
- Have I accurately represented my company or organization?

SUMMARY

- Ethics is about right and wrong.
- Understanding ethics is crucial for the workplace writer.
- Discussing ethics can be difficult because we must consider difficult questions.
- No one can make you ethical, but learning more about ethics can help you make better-informed decisions.

- The only way to truly understand ethics is through communication with others.
- Ethics can be studied from three perspectives: metaethics, normative ethics, and applied ethics.
- Studying ethics helps us think more clearly about issues, provides us with the tools to explain our ethical decisions, and advances our own ethical thinking.
- Ethics and the law are linked, but what is legal may not necessarily be ethical, and vice versa.
- Workplace writers should be attentive to the law, honesty, and confidentiality.
- Ethical choices, like rhetorical choices, are driven by the contexts in which they occur.
- Workplace writers should be alert to the ramifications of rapid and widespread circulation of documents.
- Workplace writers must consider how editing and revising a document might alter the meaning of a document.
- Some organizations develop codes of ethics, which assist them in making decisions based on local policy.
- Ethics must be considered when addressing issues of technology, particularly e-mail, web pages, and visuals.
- Environmental ethics should concern all workplace writers.
- Workplace writers should avoid using deceptive or evasive language or obscuring the issue through abstract wording, inappropriate jargon, or manipulating information or visual rhetoric.
- Plagiarism is unethical, whether intentional or unintentional.
- Workplace writers should confirm all information and report any inaccuracies.

CONCEPT REVIEW

1. In what ways are ethical and rhetorical choices related?
2. Why can ethics be difficult to discuss?
3. Why can't this text make you ethical?
4. How can studying ethics help you make better ethical decisions?
5. Why is communication the best way to come to understand ethics?
6. Why might conflict and disagreement be part of making ethical decisions?
7. What is ethics?
8. Why do we study ethics?
9. How is ethics classified for study?
10. What is the relationship between ethical and legal thinking?
11. What role does ethics play in workplace writing?
12. What legal considerations do workplace writers have to take into consideration when making ethical decisions?
13. In what ways do liability laws relate to ethics?
14. In what ways do environmental laws relate to ethics?
15. In what ways do copyright laws relate to ethics?
16. In what ways do patent laws relate to ethics?
17. In what ways do trademark and service mark laws relate to ethics?
18. In what ways do contract laws relate to ethics?
19. What role does honesty play in workplace writing?
20. What role does confidentiality play in workplace writing?
21. How does the context of a writing situation affect the ethical choices a writer must make in that situation?
22. What is the relationship between editing and revising and ethical choices?
23. What is a code of ethics?

24. What is the relationship between technology and ethics in technical communication?
25. What should you consider in relation to e-mail and ethics?
26. What should you consider in relation to websites and ethics?
27. What should you consider in relation to visuals and ethics?
28. What role does environmental ethics play in technical writing?
29. What are some strategies for writing ethically?

CASE STUDY 1

Kansas City Hyatt Regency: A Case of Ethical Considerations

The Hyatt Regency in Kansas City opened its doors to the public in July 1980. The forty-story building had taken four years to design and build and was a major addition to Kansas City's hotel scene. The building housed entertainment facilities, restaurants, bars, an atrium, and, of course, guest rooms.

One of the hotel's featured constructions was three walkways suspended from the atrium ceiling. Connected to the ceiling by 32-millimeter tension rods, the walkways were suspended in a stack: the third- and fourth-floor walkways were suspended from the ceiling, and the second-floor walkway hung from the beams of the fourth-floor walkway, directly beneath that walkway. The walkways spanned the distance from the hotel tower to the function halls, a distance of just over 121 feet.

Despite the planning, the atrium ceiling collapsed during construction. The collapse was attributed to improperly installed steel-to-steel concrete connectors and improper movement in an expansion joint. Following this initial collapse, hotel owner Crown Center Redevelopment Corporation, a subsidiary of Hallmark Cards, hired an engineering firm other than the firm that had designed the building to assess the cause of the collapse. In consultation with the construction firm, the second engineering firm found no structural flaws in the design or the roof construction. Consequently, construction was resumed, and the hotel opened to the public.

On July 17, 1981, a local radio station held a dance competition at the Kansas City Hyatt. An estimated 1,500 to 2,000 people attended the event, filling the atrium floor and the walkways above. At 7:05 that night the fourth-floor and second-floor walkways collapsed. Witnesses reported hearing a loud crack that resonated throughout the atrium as the two walkways gave way. The collapse killed 114 people, injured more than 200 others, and is considered one of the worst structural-failure disasters in U.S. history. This event was reported by numerous major news sources, including television, radio, magazine, and newspaper. To supplement what you now know about the case, conduct a simple web search to locate specific articles about the disaster.

The National Bureau of Standards investigated the accident and determined that the collapse was caused when the rod hanger pulled through the mounting box on the ceiling. Investigations showed that the new construction did not meet building codes, was improperly constructed, and used inadequate building materials. The original design for the box-beam loading system, which would attach the walkways to the ceiling, was to use a single rod (see Figure 4.5). Preliminary documentation identified both the single-rod construction and the use of rods designed for a specific strength of 413 Mpa. However, notes for this construction were not included in the final documentation and drawings, and the construction crew used rods designed for a strength of 248 Mpa.

During construction the contractor determined that the original design was impractical in the actual building, so he modified the design (see Figures 4.6 and 4.7). The engineer then approved the modified design without reviewing the altered sketches and documentation and without testing the new design. The modification, which was ultimately identified as the cause of the collapse, shifted the burden of the weight of the two walkways to a single nut under the fourth-floor beam.

What Happened at the Hyatt?

Peter McGrath with Donna Foote in Kansas City

Flags flew at half mast throughout Kansas City last week, and funeral processions would through the streets. Outside the Hyatt Regency Hotel, where 111 people died in the collapse of two aerial walkways two weeks ago, ìNo Trespassing" signs barred the curious and the morbid. Inside, a fine dust covered the floor, and a few balloons clung wanly to the ceiling. But the wreckage was gone, trucked to a nearby warehouse, and the sole remaining "sky bridge" had been dismantled. With investigators arriving daily and lawyers, lining up to file suits, the city was beginning to come to grips with a tragedy that may not have ended yet: 81 victims still lay in hospitals, 8 of them on the critical list.

The biggest question–how a year-old structure could fail so spectacularly–remained unanswered. Investigators said it might be a year before they knew the precise cause. One reason for the delay was that the hotel's owner, Crown Center Redevelopment Corp., a subsidiary of Hallmark Cards, Inc., began restricting access to the debris: its own investigators got first crack, while others representing the Hyatt Corp., the architects and various victims' attorneys waited in line. After the first few days. "the door pretty much slammed shut," complained Michael J. Davies, editor of the *Kansas City Star* and [Kansas City] The Times, which are conducting independent investigations. Crown Center promised eventual cooperation, but, said Bernard Ross, president of Failure Analysis Associates, a firm retained by the hotel's architects, "We're stymied ... Whether they altruistically give up parts and pieces for examinations in our labs without court orders remains to be seen."

In the absence of facts, theories abound. An early favorite is that one or both of the walkways buckled from "harmonic" vibrations set up by people swaying or dancing, each wavelike motion reinforcing the one before until the stress became too violent for the structures to endure. But witnesses disagree on whether there was dancing on either walkway; moreover, says Roger McCarthy of Failure Analysis, there are telltale signs in the skeleton of a structure well before it suffers vibration failure. "I haven't seen the hallmarks of that here," he says. Another theory is that the walkways were overwhelmed by sheer weight; Hyatt president Patrick Foley, howevere, says that they were designed to hold "wall-to-wall people"–more than were on them at the time.

Double Stress: Wayne G. Lischka, a structural engineer hired by the Kansas City Star, has discovered a design change that "would result in double stress" on the fourth-floor walkway's steel box beams. Originally, six steel rods were to be run from the ceiling through the upper bridge to the lower one on the second floor. But the actual construction used twelve rods, six suspending the top bridge from the ceiling, the other six hanging the lower bridge from the upper. The result: the fourth-floor bridge was subjected to increased loads from two directions. While declining to say that this could have caused the collapse, Lischka does call it "significant." Other investigators say that two of the washers intended to spread out the stress where the rods met the beams appeared to be missing, and there is some suspicion that they had never been installed.

There also questions about the credibility of the investigations. While Mayor Richard Berkley has asked the National Bureau of Standards for help on the case, most other investigators represent people with interests at stake. And there is skepticism about Crown Center's conduct, especially after records of the Occupational Safety and Health Administration revealed that during construction a large section of the lobby roof had caved in. This was discovered only because OSHA inspectors were at the site to look into an unrelated accident. Then there is the removal of the remaining walkway, for both safety and legal reasons. Noting that it would have been useful to test the structure in place, Berkley says that the move, coming as it did in the middle of the night, "doesn't give a very good impression." A city with a reputation for community cooperation suddenly seems at odds with itself even as it struggles to live with its sorrow.

FIGURE 4.5 A *Newsweek* article about the Hyatt disaster.

A number of legal cases evolved from the Kansas City Hyatt Regency walkway collapse. During one trial, testimony from the architect, the technician, and the fabricator identified that they had each discussed the structural integrity of the new box-beam connection. According to their testimonies, the engineer assured each of them that the connection was structurally sound and that he had confirmed the changes and checked the new connection. The trial revealed that he had, in fact, *not* examined the new connection and had never calculated the safety and load capacity of the new connection.

Given the information presented in this case, what ethical considerations do you think affected the overall situation? Was there an ethical discrepancy in how the walkways were designed? in how those designs were documented and reported to those constructing the walkways? In what way do you imagine that documentation and

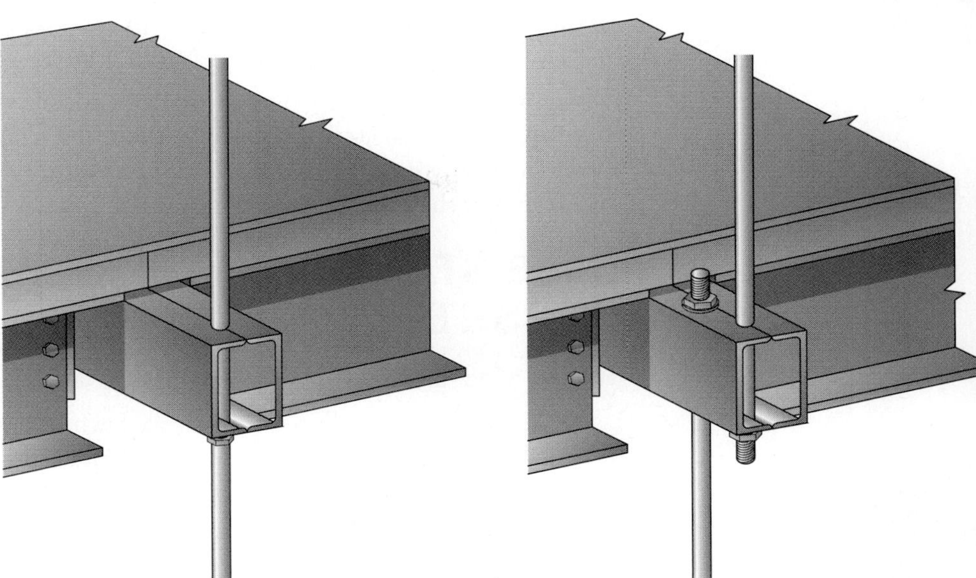

FIGURE 4.6 Original design for box-beam loading system

FIGURE 4.7 Modified design for box-beam loading system

technical writing played a part in this situation? Use the information that is provided here, and write an assessment about what this case can teach us about ethics and technical writing. If you want to know more about the case, there are numerous websites that can provide more details. However, what is included here should provide sufficient information for thinking about ethics and technical writing.

CASE STUDY 2

Ethics Documentation in the Tobacco Industry

For more than thirty years, the tobacco industry has been accused of unethical advertising and marketing practices geared toward attracting a younger market of smokers without acknowledging the health risks of smoking. In the late 1990s much of this criticism came to the fore as tobacco industry documents were made public when a number of states sued the tobacco industry for deliberately concealing information about how dangerous cigarettes are. Company documents revealed an active and direct marketing approach to attract a younger population of smokers. A 1984 memo from R. J. Reynolds Tobacco Company—the fifth largest tobacco company in the world—stated: "If younger adults turn away from smoking, the industry must decline, just as a population which does not give birth will eventually dwindle."

Ten years earlier, R. J. Reynolds official J. W. Hind wrote in an internal company memo, "To ensure increased and longer-term growth for Camel filter, the brand must increase its share penetration among the 14–24 age group, which have a new set of more liberal values and which represent tomorrow's cigarette business." That same year Executive Vice-President of Marketing C. A. Tucker echoed the information in the Hind memo: "They represent tomorrow's cigarette business. As this 14–24 age group matures, they will account for a key share of the total cigarette volume for at least the next 25 years." In this same document, Tucker went on to write, "This suggests a slow market share erosion for us in the years to come unless the situation is corrected. . . . Our strategy becomes clear for our established brands: 1. Direct advertising appeal to the younger smokers." Of course, this approach was not limited to R. J. Reynolds but was widely adopted by most of the tobacco industry.

Many of these documents are housed in the Legacy Tobacco Documents Library; visit the LTDL at http://legacy.library.ucsf.edu/. Take the time to explore some of the documents in the LTDL pertaining to marketing to younger smokers. Then, consider the context in which these documents were written. In what ways do these documents pose ethical dilemmas? Choose four specific documents, and analyze them for ethical concerns. Be sure to consider how the rhetoric of the documents reveals or conceals the key ethical issues. Then, write an assessment of the ethical situation as you see it manifested in the documents you examined.

CASE STUDIES ON THE COMPANION WEBSITE

Instant Messages: Personal Space in the Professional Environment

Lemonsee Corporation allowed workers to download an instant messenger service onto its network in order to speed up interdepartmental communications. Several supervisors have had to admonish employees about personal use of IM during business hours.

Privileged Information

Mayfield Software's small-business management software is sold in office supply and large chain electronics stores. Royal Allen, Mayfield's software development editor, learns that their latest tax software contains a programming error that makes it incompatible with two major operating systems. Mayfield will not issue a recall but will address individual complaints. Royal's brother Dana is a buyer for Computer Town, a large chain of computer and electronics stores.

Product Recall: Complaint

Many professions have established codes of ethics, but how do these apply to real-world rhetorical situations? This case deals with ethical responsibility and the dissemination of information when legal action may follow.

Ethics and the Workplace Writer

You are a new employee at an environmental engineering consulting firm. Your boss has made efforts to make the company more "green," but you notice many areas for improvement. How can you convince your boss that move must be done without hurting his pride?

VIDEO CASE STUDY

Agritechno

Agritechno's S-194 Bt Corn: Informing Investors and Growers about a Potential Problem

Synopsis

In this case Agritechno geneticist Lev Andropov is working in his lab with another geneticist, Teresa Cox, and his two lab assistants, Poncho and Lefty. As the four are conducting their work, Donna Holbrook from Marketing, Stephan Girard from Accounting, and Jaylen Castillo from Product Development come to the lab with some discouraging news: they've been getting early reports from some growers in the South that some of the caterpillar-resistant corn that was planted this year is failing in areas with higher-than-normal rain falls. The group must decide how to report that information to growers and to investors.

WRITING SCENARIOS

1. Write a list of the twenty most important things workplace writers need to think about in terms of ethics and the documents they produce. Then, as a class, compare everyone's lists, and create a single list of the most commonly agreed-upon ethical issues for workplace writers.

2. Nearly every college and university has some type of ethics code that indicates what ethical guidelines students are expected to follow. In some cases this is called a code of conduct; in others, an honor code or an academic conduct code. Take some time to examine, analyze, and discuss your school's code of ethics. What does it cover? What kinds of details does it include? Write a summary of that code, explaining how it gives students parameters for making ethical choices.
3. Workplace writers are likely to face different kinds of ethical choices in different fields and different contexts. Consider a discipline in which you are interested. Using the resources available to you on your campus—Internet, library, career services, faculty—spend some time learning about the kinds of ethical decisions that workers in your chosen field may face. (Your campus may even have specific courses that address ethics in your field; reading lists and assignments from the courses may help here.) Keep in mind that a particular discipline may have a range of professions within it. For instance, even though pharmacists and physicians both work in the broad area of medicine, the ethical choices they have are quite different. Once you have gathered your information, write a memo to your instructor explaining what you have discovered about the ethical choices in your chosen field.
4. This chapter introduced you to the Ford Pinto case and the Kansas City Hyatt Regency case. Locate another prominent case that evolved from poor ethical choices in documentation. After reading about it, write a summary of the events and an explanation of the role technical writing and ethics played in those events.
5. Using a local, state, or federal government web page, search for "ethics" and explore how that government body addresses the topic. Then, in a letter to your instructor, explain what you found, and identify any shortcomings in the way ethics was handled.
6. Take a moment to consider your role as a professional, a student, and a citizen in your classroom and your college. What are the guiding principles that drive your ethical decision making? What is of ethical importance to you in your class in technical communication? Develop and write a code of ethics for students in your class. What kinds of ethical issues come into play—legal issues, technological issues, environmental issues? Address at least ten points of ethical decision making.
7. What role does personal responsibility play in workplace collaboration? How do your own individual ethical decisions affect a company document? Write an explanation that details the role individuals have in making collaborative ethical decisions in workplace writing.
8. Imagine that you work as a civil engineer for a company that has just been contracted to develop a new industrial park on the edge of a growing city. The park will provide warehouse, office, and some retail space. The city anticipates that the space will be filled rapidly by local businesses wanting to expand. Two major department store chains have also expressed interest in securing space in the development to use as regional distribution centers. The city anticipates that the industrial park will create several thousand new jobs and produce revenue for the city.

   In the preliminary development plans, your company supported an environmental impact study to ensure that the land designated for the project was environmentally viable. The initial report from the environmental engineers, which was submitted only to your company and not to the city planners, found that building on the designated land itself posed no direct environmental risk but the proposed road system that would be needed to provide access to the warehouses would require that a wooded area between the main highway and the industrial park be cleared. These woods, the study showed, provided a buffer zone between the high-traffic highway and a local wilderness area. The study recommended that the road system be designed to come in from another direction in order to preserve the woods.

Your company estimates that the proposed change would cost approximately $15 million more and would delay the project by at least six months. Your boss has asked you to write a progress report for the city planners, explaining that everything is going as planned, he has told you specifically not to mention the environmental impact study. If the project can be launched as it stands, the city will have to stick to the original plans, no matter what the environmental impact study shows.

What would you do in this situation? Write a letter to your instructor explaining how you would handle this situation.

9. Design choices are rhetorical choices; they determine how a reader will access information. For instance, if you write a letter that conceals critical information until the very end, your design has manipulated the information flow. Thus, design choices are also ethical choices. Make a list of ten things that you think document designers need to consider about ethics when designing their documents.
10. Locate a company or business web page that interests you. Look carefully at how the page is designed. In what ways does that design reflect ethical choices the designer/writer had to make? Where are the images? How easy is the page to navigate? What kinds of information are emphasized? de-emphasized? Write an assessment of the ethical choices you believe were made in designing the page.
11. Many product labels are required by law to report certain information that doesn't really promote the product or shine a positive light on it. For instance, cigarette and alcohol packages are required to carry health warning labels. Likewise, food labels are required to display certain nutritional information. Select a product with a label that presents some less-than-favorable information. Analyze the design of the label. Where is the bad news presented? What does the label emphasize? Write a letter to your instructor explaining why you think the label was designed as it was and why some information is given prominence while other information is subdued.
12. As a student, one of the most obvious ethical issues that you are likely to face is the decision to plagiarize or not to plagiarize. Many professors around the country argue that plagiarism is on the rise because technology makes it easy to find work that students can pass off as their own. A quick web search of nearly any topic can locate written papers, websites, and databases, all readily available. In addition, the number of online clearinghouses for college papers is growing. At the same time, however, efficient search engines and plagiarism databases (e.g., the famous www.turnitin.com) make it easy for teachers to find out whether student papers have been plagiarized. In a letter to your instructor, address how technology makes plagiarism an easy act, how technology makes identifying plagiarism even easier, and what role ethical choices play in the technology/plagiarism situation.
13. Take some time to consider the fundamental ethical issues related to technology that workplace writers must take into account. Write a code of ethics for workplace writers who rely on technologies like e-mail, the Web, and text messaging. Be sure to include what to do and what not to do when using these technologies to convey professional, technical information.

14. In 2005 a student of one of the authors of this text turned in a paper that the teacher suspected was plagiarized. When the teacher conducted a simple web search, he found the paper in a paper clearinghouse, from which it had apparently been purchased and then turned in as the student's own work. However, when confronted and accused of plagiarism, the student produced a receipt, showing that she was the one who had sold the paper to the clearinghouse. She said that she had written the paper, submitted it for class credit, and then sold it to the clearinghouse (for thirty-five dollars, by the way).

    At this point the teacher had to acknowledge that the student had not plagiarized her paper but had made it available for other students around the world to plagiarize. Regardless of whether students should have the ability to sell their writing for someone else to plagiarize, the teacher realized he could not prevent them from doing so.

    However, still unconvinced that the student had been the original author, the teacher continued to search for the paper elsewhere and discovered a database of student papers that had been archived online by a professor at another university. The database contained hundreds of student papers the professor had used in conducting research about student papers, and within that database the teacher found the original paper his student had submitted. She had plagiarized the paper from the database and then sold the plagiarized paper to the online clearinghouse in order to be able to produce a receipt indicating that the paper was originally hers.

    Consider this situation from both technological and ethical standpoints. What role did technological access play in this situation? What were the student's ethical choices? What were the teacher's ethical dilemmas? Write two letters, one addressed to the student involved, responding to her actions and explaining what you would have done if you had been the teacher. The second should be addressed to the teacher, explaining your perspective on the ethical dilemma of students being able to sell their own writing to online clearinghouses.

15. As a private construction contractor, you often submit proposals and bids in response to requests for proposals from your local city planner, and over the years you have developed an amiable yet professional relationship with the personnel in that office. After you have submitted a proposal to build a new public parking garage, one of the clerks in the office forwards you a copy of an e-mail the city planner has sent out indicating that he wants the garage contract to go to another construction company, which he has been favoring recently. The e-mail is dated three days before the deadline for submitting proposals.

    Consider how you might respond to this situation. What might be the ramifications of forwarding the city planner's e-mail to other local contractors, government officials, and other citizens of the city? Specifically, who would you want to send the e-mail to? How might widespread circulation affect reaction to an internal government e-mail? What might you write to accompany the forwarded e-mail in order to provide readers with context and suggestions for understanding the original e-mail and its ramifications? Write an e-mail that would accompany the forwarded e-mail explaining the context of the original e-mail, the ramifications of the original e-mail, and the reasons you are forwarding the e-mail.

5

Technical Communication in a Transnational World

CHAPTER LEARNING OUTCOMES

After completing this chapter, you will be able to do the following:

- Recognize and solve problems associated with writing for **transnational audiences**
- Account for **difference between transnational and transcultural** audiences
- Address the **roles of language** in transnational communication
- Understand the **roles of technology** in transnational communication
- Examine the roles of differences in **approaches to education** among transnational audiences
- Understand how differences in **politics and law** affect transnational communications
- Address **economic differences** in transnational communication
- Consider how **religious differences** affect transnational communication
- Avoid **stereotypes** in transnational communication
- Understand the **roles of translation** in transnational communication
- Understand reasons for and processes of **localization**
- Understand the goals of **internationalization**
- Understand the ramifications of **globalization**
- Develop strategies for **verbal communication** with transnational audiences
- Address **ethical issues** associated with transnational communication
- Establish **guidelines** for writing to transnational and transcultural audiences

DIGITAL RESOURCES

On the Companion Website www.prenhall.com/dobrin:

- Case 1: Translating Video Games to Foreign Markets
- Case 2: Big Owl's Goes Global: Internationalizing a Restaurant Chain
- Case 3: Learning about Chinese Business Culture
- Case 4: The Hawksbill Turtle Society: Securing International Protection for an Endangered Species
- Video Case: Collaboration and Writing a Request for Proposals
- PowerPoint Chapter Review, Test-Prep Quiz, Exercises and Activities

REAL PEOPLE, REAL WRITING

MEET HEEJONG HAAS • Transnational Business Consultant

How do you define transnational communication?

Transnational communication takes place between people and countries all around the world; it seeks to move beyond cultural and language barriers. Within the U.S. business community, a writer has one language with a common cultural and business background to communicate within. In transnational communication, that writer will be exposed to many different cultural aspects and different languages, which can create challenges.

What types of writing do you do as a transnational business consultant?

Like most workplace writers, I write a lot of e-mails every day. I also write many reports for my clients, most of which strive to develop successful relationships between Korean businesses and their partners around the world.

What is the most important thing writers should consider when writing for transnational audiences?

Using humor can be problematic when communicating with transnational audiences. What makes American audiences laugh may confuse or offend people from other countries. I think, as a general rule, writers should not use humor or jokes because the risks of miscommunication are high, whereas the chances of getting a positive response are low.

How can workplace writers learn to account for cultural differences when writing for transnational audiences?

The best thing they can do is to learn to be clearer writers. A good way to do this is to edit their writing to take out cultural references (like lines from TV shows, sports references, etc.), idiomatic expressions, "corporate-speak" (most nonnative speakers of English are not up-to-date with buzzwords), and ambiguous sentences. Most workplace writers need to consider that English is a second language for transnational audiences, so clarity is important.

What role does ethics play in transnational communication?

Ethics is an extremely important matter when it comes to transnational communication. The language barrier you often face means the trust level must be very high. Without trust in transnational communication, things can easily go wrong.

How do you address issues of document design or the use of visuals when writing for transnational audiences?

Using visuals in documents is a good thing, but writers must be aware of different cultural factors. There are certain colors and elements in every culture that have religious, cultural, or historical meaning. Therefore, when designing a document or preparing visuals, writers should always research the target audience's culture to make sure that the design and visuals are not offensive.

What advice about workplace writing do you have for readers of this textbook?

My general rule in communication is to be simple, to the point, and easy to understand. Most importantly, respect others and keep dignity in your communication. I often read e-mails, proposals, or reports that contain lots of buzzwords or big words with lengthy sentences. What these writers do not understand is that this kind of writing bores and frustrates readers. Clear and simple is always best.

INTRODUCTION

More than thirty years ago, Xerox established its Multinational Customer and Service Education organization (MC&SE). This division of Xerox was created to translate written documentation for worldwide use. The MS&CE writers and translators developed a system called Multinational Customized English to establish a set of standards for how Xerox documents would be translated. These standards were used to assist both machine and manual translations; the objective was consistency and accuracy in document translation. The MC&SE group also developed an Employee Development System, which worked to train Xerox employees in two key skills for global communication: "writing for translatability" and "global design."

The creation of MC&SE (and similar divisions in other companies nationwide about this time) signified an important threefold shift in technical and professional communication. First, it acknowledged that commerce had become a global endeavor; no longer were businesses limited by national borders. Second, it recognized the role of technological advancement in opening lines of communication between international organizations, as well as the ways those technologies expedite international communication. Technology allows trade to happen farther and faster. Third, MC&SE marked a shift *away* from relying on International English (more on this later) as the standard language of international communication and a shift *toward* working in and with local languages, cultures, and customs.

You have probably heard phrases like "global economy," "international business," or "transnational commerce." Business and commerce now commonly function in a global market economy. Companies not only exchange goods and services with companies in other countries, but many larger firms now have offices in other countries, employing linguistically and culturally diverse populations within one company. These different branches and offices must communicate with each other, and that transnational and transcultural communication affects internal and external documents.

EXPERTS SAY

"International technical communication can be used by any audience that is culturally, linguistically, and technologically variant from the audience in the source country."

NANCY L. HOFT, Author of *International Technical Communication*

This chapter uses the words *transnational* and *transcultural* to encompass concepts like global, international, multicultural, and cross-cultural, although later we address the specific terms *internationalization* and *globalization* as facets of transnational communication. In other words, we say that transnational and transcultural communication encompasses global, international, multinational, and cross-cultural audiences, although there are important distinctions among these terms. For instance, *multinational* suggests definitive borders between countries, whereas *transnational* envisions the global community without national borders. *Transnational* and *transcultural* suggest a sense of movement, particularly the flow of ideas, goods, and services and the flow of communication and information. *Transnational* and *transcultural* suggest connectivity despite differences.

Writing documents for transnational and transcultural audiences presents a number of problems for workplace writers, and this chapter is designed to help you consider those rhetorical and workplace problems and to provide approaches for efficiently solving these problems. For example, consider the situation that recently arose when Digibox, a growing digital imaging company based in New York City, acquired Digishare, a digital imaging company based in Bogotá, Columbia. As Digibox accountants began to take stock of and create documents about Digishare's holdings, the accountants often called the Bogotá employees to ask for verification and explanation of information conveyed in Digishare's financial records. In one conversation the accountants asked Digishare representatives to prepare tables to organize financial information. The Digishare personnel were confused by the request and never provided the data needed. After investigation the accountants realized that their use of the word *table* caused the problem because of varying linguistic and cultural understandings of that word. Look carefully at the three visuals in Figure 5.1, which represent ways

North American Workplace

	Jan	Feb	Mar	Total
East	7	7	5	19
West	6	4	7	17
South	8	7	9	24
Total	21	18	21	60

European/North American

South American (mesa)

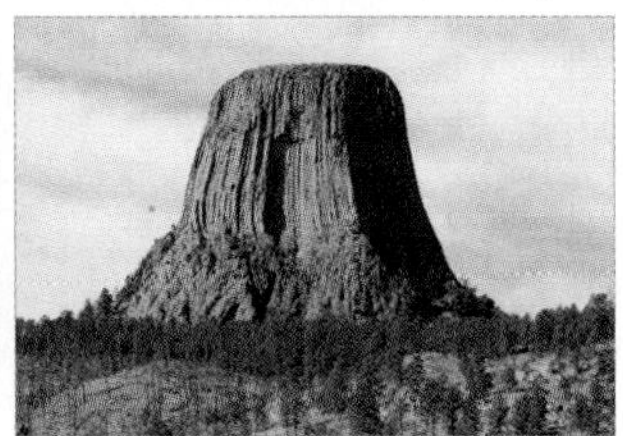

FIGURE 5.1 Various interpretations of *table*

different audiences have interpreted the term *table*. The first shows how workplace writers in North America might understand the term; the second shows how many European and North American general audiences are likely to understand the term; and the third shows how many South American audiences understand the term, as happened in the Digibox/Digishare situation.

This example shows that writing for transnational audiences can challenge even our basic assumptions about how diverse groups communicate and understand information. What may seem a common understanding—like the definition of the word *table*—may in fact be interpreted and understood in different ways among varied cultures and nationalities. This chapter provides strategies for better understanding the role of communication and writing in transnational and transcultural workplace situations. All writing requires careful consideration of audiences, but writing for transnational and transcultural audiences presents particular challenges for workplace writers.

EXPLORE

The U.S. Commercial Service (USCS) is a government agency that promotes global business. Its web pages offer advice about conducting business in many other countries. Read through the U.S. Commercial Service web pages (http://www.buyusa.gov), and locate discussion of suggestions for doing business with a specific country—you pick the country. Then write a short summary of the advice the USCS offers for communicating with individuals and businesses in the country you have selected. (You can find information quickly by typing into your search engine "US Commercial Service" followed by the name of the country in which you're interested.)

LEARNING ABOUT DIFFERENCES

Chapter 2 explained that understanding audiences is a crucial step in producing effective workplace documents. This chapter explains that understanding transnational and transcultural audiences is difficult. Nonetheless, learning how to analyze transnational and transcultural audiences is crucial to reaching global audiences. There are seven key points of difference that writers should account for when writing to transnational and transcultural audiences: language, technology, education, politics, economics, society, and religion. The approaches suggested here are adapted from Nancy L. Hoft's *International Technical Communication*; our simplified version is designed to help you begin thinking about transnational and transcultural audiences within the broader scope of workplace writing.

You might think of these seven areas of focus as components of culture, which can be defined in many ways. Some say that *culture* is how a group of people think about things together; it might be a combination of the group's approaches to religion, politics, education, ethics, economics, and art. Others say that culture is really the same as a group's method of communication, or at least that you can't tell the difference between culture and communication. Still others think of culture as a system of deeply held, shared principles among a group of people. Regardless of the definition chosen, workplace writers must recognize that culture greatly influences nearly every aspect of communication. But learning about those influences can be tricky because not all are visible or recognizable. Some aspects of culture may be evident, but others are unspoken or even inherent—simply the natural way of doing things.

One of the most important points in learning about cultures is understanding that you can never get the full picture. Each time you engage a culture, you learn more, and all of those moments compound into a better and better long-term view of how the culture functions. Consequently, it is important to develop a system of taking notes about a culture and updating those notes with each encounter. Because workplace writers' needs and expectations change and because transnational and transcultural audiences' needs and expectations also change, it is important to keep up with the shifts and the ways that cultures react to and account for those changes. The seven points of difference can help you organize your growing understanding of transnational and transcultural communication.

Language

Understanding the language requirements of a country or cultural group is important because it provides direction in how to write documents and offers insight into how audiences will read those documents. But language also affects how *they* write documents and *we* read them. Transnational and transcultural interaction entails more than just "writing for them"; it entails learning about how others write technical documentation, as well. Here are five key aspects of language to consider:

1. *The target language:* Most transnational communications experts and language experts agree that language is the key to how individuals express themselves and come to know and understand the world around them. For workplace writers, then, it is important to understand as much about a target language as possible, because understanding the language provides an understanding of the target audience. As an added benefit, it also helps writers understand more about their own language. Collaborating with a translator is one of the best ways to learn more about a language. Other resources are introductory language textbooks (like a first-year Spanish, French, or Japanese textbook) and audio lessons that teach languages (like Russian or Chinese on CD or MP3).
2. *Official national languages:* Many countries identify their official languages. Resources like IBM's *National Language Support Reference* identify both official and nonofficial languages. In many countries, numerous languages and dialects are commonly used, officially and unofficially. For instance, Singapore identifies four official languages: Chinese, Malay, Tamil, and English. It is important to be aware of a country's official language for legal reasons, as many laws require documentation to be presented in the official language. Also, knowing a country's official language can reduce the number of languages into which you translate a document.
3. *International English:* For many years English speakers could assume—or even require—that the business they conducted internationally would be conducted in English, which was considered the language of international business. International English is an attempt to standardize English as a global language for commerce. However, a true International English can never be attained; worldwide users will adopt any language to their own needs. Thus, International English appears in many dialects even though it works toward a single English to be used in all transnational communication. Realistically, we can no longer

expect International English to be the language that drives world commerce. For instance, less than 5 percent of the Japanese population speaks English. Although English is certainly used with great frequency throughout the global business world, it has become more common—and perhaps more appropriate—for transnational communication to occur in the local languages of the organizations involved in the exchange of information.

ANALYZE THIS

Do a web search for sites that address Global English or International English. Consider how they talk about these kinds of English? Write an analysis of some of the primary discussions about Global or International English.

4. *Text directionality:* Although most of us are used to writing left to right and thinking of a document as beginning in the top left corner and proceeding across a page and then down, some languages do not follow this directional path. Text directionality is not universal; Arabic and Hebrew, for instance, are written right to left, so that documents like manuals and books are designed to open and be read right to left. Some forms of Japanese used in marketing are read in columns from right to left instead of in horizontal rows. Thus, workplace writers accounting for transnational audiences must be sure to consider text directionality in designing documents.
5. *Writing style:* Like other facets of writing, writing styles vary from country to country, culture to culture. American English, for instance, tends to be more repetitive than French is. It is often a good idea to work with a translator to translate documentation into more appropriate target-country writing styles while maintaining accuracy in information dissemination.[1]

EXPLORE

Within the borders of a single country, workplace writers may speak a variety of languages. For instance, the United States does not have an official national language, even though 82 percent of the population speaks English. Approximately 337 languages are spoken in the United States, and knowing where various languages are spoken can be quite useful. The Modern Language Association's language map helps track the languages used in the United States. Take some time to explore the language map to see how diverse language usage is in the United States. You can find the map at http://www.mla.org/map_single.

Technology

The first issue to consider regarding technology is whether the target country has access to the same technology you are using. For instance, software developers need to know whether their applications can be used on existing technology platforms in the target country or whether they need to adjust the software to meet local needs—or whether the industry should make the technology available that supports their software. To understand the technological culture of a target country, you should first identify commonly used technologies and then assess the reliability of that technology.

[1]Nancy L. Hoft, *International Technical Communication: How to Export Information about High Technology* (New York: John Wiley and Sons, 1995).

Machine Translation

One way technology helps to overcome differences is through the use of machine translation software, which automatically translates documents into different languages. However, machine translation programs are limited in their capabilities. Because they use formulaic translation schemes, these programs rely on restrictive grammars, such as the standardized grammars called controlled English. These forms of English grammar are designed to aid in automatic translation programs but have not yet proven to benefit transnational readers. Controlled English is a rule-bound English that does not allow writers many options in how they convey information; restrictive grammars limit what can be said and how it can be said. One primary objective of using machine translation is to reduce the amount of editing needed for the translated language, but this goal has not been reached. Machine translation software is expensive and thus beyond the budgets of most companies, so few companies use it.[2]

Education

Education systems and approaches vary greatly from country to country, and understanding those systems and approaches greatly influences how documents are written. Three facets of education should be considered in learning about a target country:

1. *Literacy:* Literacy is calculated differently from country to country, and many countries inflate their announced literacy rates. Nonetheless, having a sense of a country's literacy rates can influence how you approach documentation. A number of online resources present international literacy rates, including the United Nations Educational, Scientific, and Cultural Organization (UNESCO) and Brigham Young University's *Culturgrams*.
2. *Common body of knowledge:* Understanding what most people in a country know is important. If your research into literacy rates reveals that most people in the target country have, say, seven years of education, it would be valuable to learn what people with seven years of education in that country are expected to know. Good sources for researching this information are professors of education and professional education journals. You should also check to see what information is available through the target country's official educational body on the national level. For example, in the United States this would be the U.S. Department of Education; in Greece it would be the Ministry of Education; in Japan, the Ministry of Education, Culture, Sports, Science, and Technology; and in France, the National Ministry of Education.
3. *Learning style:* Identifying a target country's learning styles can be difficult, particularly since you may find many approaches and learning styles within one group. But identifying how people are taught to learn can provide a good deal of insight into how you should write technical documents, especially documents that include instructional material like tutorials, manuals, or technical instructions. You should explore the methods of education, the role of teachers and technology, the role of apprenticeships and mentoring, and the role of credentials and degrees in different learning styles.[3]

---

[2]Nancy L. Hoft, *International Technical Communication: How to Export Information about High Technology* (New York: John Wiley and Sons, 1995).
[3]Ibid.

Politics and Law

All target audiences are influenced by the laws and governments of their countries. When writing for transnational audiences, workplace writers should examine the differences and similarities between the legal and governmental influences on the writer and those on the target audience. Workplace wirters must consider and account for the following:

- *Trade issues*, such as import/export practices
- *Legal issues*, such as laws governing copyrights, trademarks, intellectual property, liability, and fraud
- *Political traditions and symbols*, such as forms of government, voting rights, politically significant dates (e.g., most American companies are closed on July 4), flags (e.g., the colors and symbolism of a country's flag, as well as cultural conduct in respecting it), national documents, and national symbols

To research these kinds of differences, it is a good idea to check not only with international legal and political experts, but also with any professional organizations that have developed lists of standards.[4]

Economics

Accounting for economic differences and similarities is important in understanding transnational cultures, and also central to effective transnational business practices. Along with the basics of currency value and exchange rates, it is helpful to understand how transnational audiences assign monetary value to products and services. For instance, in February 2008 a businessman from the United Arab Emirates purchased a vanity license plate for $14 million. In many countries that would be considered excessive; in others it may be taken as a sign of status, conveying important information about the owner of that vehicle. Such differences underscore the ways economic values vary from one culture to the next. One way to document such differences is to create a comparative chart that examines the costs of common items (e.g., grocery items, real estate, technology) and the relative value they have in different cultures, taking into consideration who has access to those items and what owning them says about the owner's role in the culture.[5]

Society

In most countries and cultures there are noticeable differences in various aspects of life. For instance, how one functions in the workplace may be very different from how one functions in family settings. Thus, to investigate social factors, it is important to consider different components separately, while understanding that they all contribute to a cultural atmosphere. Here are three of the factors you should be attentive to:

- *Age*—A culture's attitude toward age can impact some business practices and can shed light on other cultural attitudes, such as those toward education.
- *Business etiquette*—Business etiquette varies not only from culture to culture, but also from industry to industry. Understanding how various transnational audiences approach business etiquette can enhance your ability to interact

---

[4]Nancy L. Hoft, *International Technical Communication: How to Export Information about High Technology* (New York: John Wiley and Sons, 1995).
[5]Ibid.

with them, as well as with members of your own organization from different cultural backgrounds.

- *Family and social interaction*—Many factors contribute to family and social interaction within a culture. Considering the roles of marriage, funerals, and birth ceremonies, for instance, can give insight into how a particular culture envisions family life.[6]

Religion

In many countries religion plays a significant role in all aspects of cultural life, including government and law. To better understand a transnational and/or transcultural audience, it is important to examine the target countries' indigenous religions, most of which can be studied through books and online research. It is a good idea to understand which religions are prevalent and what the basic tenets are of those religions. You should know the significance of a religion's symbols and colors; reproducing some religious images is considered offensive. In addition, many religions influence dietary habits. For instance, some religions forbid eating pork, and in others eating lamb or fish is customary on particular days. It is important to respect another culture's religious practices, although you should not feel compelled to sacrifice your own beliefs in the process.[7]

AVOIDING STEREOTYPES

A difficult problem to overcome in transnational and transcultural communication is stereotyping. Traditionally, when textbooks and other instructional resources provide lists of cultural issues to consider in working with specific transnational audiences, that information might be presented like this: "When writing for a Japanese audience, you should always. . ." or "Spanish readers tend to. . . ." This kind of cultural-guidebook approach promotes typecasting and stereotyping by reducing differences to simple formulas. *Effective* awareness of cultural, linguistic, and national differences cannot result from such formulaic approaches; important differences exist within seemingly monolithic cultural units.

Ethical and responsible workplace writers must be diligent in developing accurate understandings of their transnational audiences. A number of methods exist for becoming culturally aware, but all require dedicated research. None of the suggestions here can provide a nonnative with a perfectly accurate vision of another culture, but actively pursuing these approaches can prepare workplace writers to make more effective rhetorical choices and solve transnational communication problems in productive ways.

See chapter 6 for more on research strategies and methods.

Avoid Assumptions

A common mistake in learning about other cultures, languages, and nations is to make assumptions about them. For instance, people often assume that an entire cultural group can be addressed in the same way, that everyone within a given culture will respond similarly. But such a culture-specific stereotype does not allow for intracultural differences. Likewise, one cannot make assumptions about individuals within

---

[6]Nancy L. Hoft, *International Technical Communication: How to Export Information about High Technology* (New York: John Wiley and Sons, 1995).
[7]Ibid.

a cultural group. For example, one should not assume that a woman who wears the traditional clothes of her culture is or is not fluent in English, that she is or is not a professional, that she does or does not work outside the home, that she is or is not technologically literate, or even that she has certain beliefs about or ways of conducting business. In short, assumptions don't lead to an accurate understanding of other cultures, languages, or nations.

Ask Questions

The best way to learn about differences is to ask questions. Whenever possible, workplace writers should find answers to their questions before they meet a new transnational or transcultural client, customer, or colleague. However, if they cannot anticipate or ask their questions beforehand, they should develop relationships with individuals in the target audience whom they can comfortably approach with questions.

It is important to ask culturally sensitive questions, being careful that the phrasing or context of a question is not offensive in itself. Good questions can create connections between you and your audience, rather than highlight differences or manifest biases. In addition, you should remember that you are asking for one person's perspective and should not consider his or her response to be a universal or final answer on the subject, nor should you treat the person as a representative of the entire country or culture.

Collaborate with the Translator

Most professional translators are not only proficient in a language but are also well-versed in the culture. For instance, it is likely that a translator of French is familiar with the differences between French and French Canadian cultures, as well as with the linguistic differences between the two populations. It is important to see the role of the translator as a collaborator on transnational documents, someone who is able to explain linguistic and cultural nuances that can assist you in crafting effective documents for specific groups. Most translators gladly provide such guidance. And the more your writing reflects audience awareness, the easier it is for good translation to occur.

ENHANCING TRANSLATION

The difficulty with translation is that most workplace writers aren't translators, so they must hire translators. Thus, workplace writers must anticipate that their documents will be translated by someone else, and translators must work with documents that others have prepared. This can create a difficult division of labor. In addition, few translators are able to translate in more than one language direction, for instance English to Japanese. And since documents often require translation into multiple languages—perhaps a company distributes software to a number of European countries—companies often hire translation agencies that employ translators in many languages. Few workplaces are able to staff dedicated translators.

EXPLORE

Many translation companies maintain websites that explain their approaches to translation and the kinds of work they do. Locate three translation company web pages, and compare and contrast their approaches. Then write a letter to your instructor that explains the similarities and differences among the companies' approaches to translation.

The responsibility of most workplace writers is not to translate documents but to prepare documents that can accommodate translation—that is, documents that are attuned to the countries, cultures, and languages of the audiences and that adhere to the guidelines explained here. Such documents make the job of the translator more efficient and assure that the documents will be understood as intended.

The first step in producing such documents is to become aware of your own language. Learning how English, for instance, differs from other languages provides a sense of how your language use might be interpreted by nonnative speakers and how it might be translated. The following four elements are central to developing a sense of your own language use.[8]

Terminology

Pay attention to how you use specialized terminology and jargon in your documents because specialized terms usually don't translate directly. One method of accounting for terminology is to maintain an annotated glossary of specialized terms, jargon, and new words used in your documents. Such a glossary can include definitions, contexts, and even the frequency of use of the specialized terms. Glossaries like this serve two purposes within a document: First, they provide a method of keeping track of the specialized words used—including how and how frequently you use them. By understanding the contexts in which you use specialized language, you will learn more about how your writing functions. Second, glossaries of specialized terms can be given to translators as a resource to explain how you use specialized terminology. Thus, a glossary can help ensure that the translator understands precisely what you mean and can prepare a more accurate translation.

For more about jargon use, see chapter 10; for more about writing definitions, see chapter 15.

Clarity

Documents that will be translated should be written clearly and concisely, as all workplace writing should be, but this characteristic is especially important with transnational and transcultural audiences. Translators that must work with poorly written, unclear language have difficulty producing accurate and effective translations. Keep in mind that the translator's job is to translate, not to rewrite.

Specifically, you should avoid linguistic features such as idioms, acronyms, ambiguous antecedents, use of synonyms, modifying phrases (i.e., adjective or adverb phrases), gerunds, shifts in person, and dropping *that* at the begining of clauses. All of these can lead to ambiguities that are difficult for translators to work with.[9]

See chapter 10 for more on clear, concise writing.

Cultural and Rhetorical Differences

Workplace writers cannot know everything about all cultures and languages. However, you should remember that idiomatic phrases, references to popular culture, and even humor can be problematic for translators. You should actually avoid humor in *all* professional workplace documents since you can never know how an audience will respond. But it is especially important to avoid humor with transnational and transcultural audiences since humor does not usually translate well. What you may think is funny—even a simple joke—might be construed as inappropriate or insulting by a transnational reader. Similarly, idiomatic expressions might not make sense to a nonnative speaker and might not be easily translatable into other languages.

---

[8]Bruce Maylath, "Writing Globally: Teaching the Technical Writing Student to Prepare Documents for Translation," *Journal of Business and Technical Communication* 11 (1997), 339–352.
[9]Ibid.

Design

According to translation research, most languages require 30 percent more space than English requires to convey the same message.[10] Furthermore, some languages like Arabic, Hebrew, and certain forms of Japanese don't follow the left-to-right directional flow of English. Because translators focus on language translation and not necessarily on design, workplace writers need to attend to design in order to accommodate shifts in space requirements.

Design architecture becomes important in organizing information clearly and guiding transnational and transcultural readers through a document. Documents that are deductively organized and that provide clear headings, topic sentences, and summaries are easier to translate.[11]

See chapter 9 for more on document design and design architecture.

ACCOMMODATING TRANSNATIONAL AUDIENCES

Localization

localization adapting or creating a document for a specific target country or market.

Localization refers to adapting a product and/or translating a document for a specific local audience. The key to effective localization is to begin with a high-quality original document. According to Hoft, there are two degrees of localization:

1. *General localization*, which addresses superficial cultural differences like language, currency, and date and time formats. General localization often requires little more than adjustments made in translation.
2. *Radical localization*, which focuses on more substantive cultural differences that affect how readers and users think, feel, and act, including learning approaches and culture-specific examples. Radical localization can drastically change the language, design, and approach of a document.[12]

EXPERTS SAY

"If the technical information is poorly written in the original language, even with the best translation and design in the world, the information will still be poor."

CHARLENE NAGY, President of NCS Enterprises, LLC., a full-service global communications company in Pittsburgh, PA

Localization offers four key benefits:

1. *Localized products improve sales.* Localization is a form of marketing; it makes products and documents more applicable to a transnational or transcultural audience and thereby makes selling the product or distributing the documents more effective.[13]
2. *Localized products overcome cultural differences.* If a target country's cultural differences are understood, products can be adapted to local needs in ways that make them more attuned to cultural needs. Hoft offers an interesting example of a wood products industry. In the 1950s, Japan, which has limited wood resources, approached U.S. and Canadian wood manufacture's about supplying Japan with wood. One of the restrictions was that Japan would import only wood that had been cut to meet traditional Japanese building customs, including specific sizes and shapes (the 2x4 is not a universal building size). The U.S. wood industry was shortsighted, did not foresee

---

[10]Bruce Maylath, "Writing Globally: Teaching the Technical Writing Student to Prepare Documents for Translation," *Journal of Business and Technical Communication* 11 (1997), 339–352.
[11]Ibid.
[12]Nancy L. Hoft, *International Technical Communication: How to Export Information about High Technology* (New York: John Wiley and Sons, 1995).
[13]Ibid.

Japan's becoming an economic force, and rejected its proposals. Canada, on the other hand, localized wood production for Japan and now dominates the wood suppliers industry.[14]

3. *Localization helps overcome inherent resistance.* As an example of this benefit, Hoft explains that when American-based McDonald's decided to expand its worldwide business into India, it met with an inherent cultural resistance to its primary product: the all-beef hamburger the Big Mac. India, which has the world's second largest population and, thus, a massive consumer market, has a significantly large percentage of vegetarians. Cows hold a culturally different position in some Indian cultures than they do in the United States. So, to overcome this inherent resistance, McDonald's developed the meatless Big Mac.[15] McDonald's also offers localized variants of the Big Mac, which originally used lamb instead of beef but now use chicken. These versions are called Maharaja Mac. Likewise, because Jewish dietary laws restrict the mixing of meat and dairy products, McDonald's in Israel offers a Kosher version of the Big Mac that does not include cheese. McDonald's has also made efforts to localize the Big Mac to cultural attitudes toward nutrition. The Big Mac offered in Australia, for instance, contains 20 percent fewer calories and 24 percent less sodium than the Big Mac sold in Mexico.
4. *Localization is a good approach to being the first to reach world markets and entering global niche markets.* By ensuring that company products and documentation are localized from early stages of development, companies can more rapidly reach markets not yet tapped by their competitors.[16]

Despite these benefits, localization can be expensive, time-consuming, and legally complicated. The Boeing Company, for example, has stopped translating and distributing maintenance manuals with the airplanes it exports because of the liability Boeing would assume if the translation of the manuals was incorrect or misleading. Many countries now require standardized product documentation written in the official language of the target country.[17]

Internationalization

Localization often requires adapting documentation to specific target audiences, which may mean identifying the writer's embedded cultural markings and shifting them to localized approaches. This can be a difficult and complex process.

For instance, think about the simple act of alphabetizing a list of names. If a U.S. writer was documenting the list, the order would follow the U.S. alphabet; but if the writer was alphabetizing the same names in Spanish, the order would be different because the Spanish alphabet includes the letter "eñe" (*ñ*) between *n* and *o*. Similarly, translating the list into languages such as Russian, Greek, Hebrew, Japanese, Mandarin, Arabic, or Tamil would require translating or transliterating the names to fit those alphabetic needs and then reconfiguring the order based on those languages.[18] In addition, not all languages rely on alphabets. Cherokee, for example, uses a syllabary, a system that uses a set of written characters to represent syllables.

---

[14]Nancy L. Hoft, *International Technical Communication: How to Export Information about High Technology* (New York: John Wiley and Sons, 1995).
[15]Ibid.
[16]Ibid.
[17]Ibid.
[18]Ibid.

Because of countless cultural factors like these, it is often difficult to localize documents that are written for native English speakers and are already embedded with cultural markings. **Internationalization** is a process of developing software that can be easily adapted to various transnational audiences without changing the core of the software design. Internationalization (which is often referred to as *i18n*—the first letter in the word, the number of letters between the first and last letters, and the last letter in the word) can also be adapted to technical communication in general. We can apply the goals and approaches of internationalization to the process of writing, rewriting, designing, and redesigning documents so that they can be more easily localized to any transnational audience.[19]

internationalization the process of writing documents so they can be easily localized for transnational audiences

Most workplace writing is comprised of core information that is constant among various documents, despite their differences in design or audience. *International variables* are those parts of documents that can be localized, and internationalization is the process of identifying the international variables that come into play in specific documents and determining which parts must be localized.[20]

Internationalization is necessary for companies that work transnationally; because it anticipates localization, it helps these companies save time and money in their localization processes. Internationalization also compounds its efficiency over time by continually adding to a company's database of localization strategies and standardizing many of those processes.

Globalization

As Hoft points out, internationalization is part of a two-step process: internationalization of a transnational document and then localization of the document for each specific context. Both internationalization and localization increase the amount of time and resources that must be dedicated to each cultural variation of a document beyond the planning, researching, drafting, reviewing, and distributing of the original document. When a company invests in localizing documents for many target audiences, the associated cost will likely outweigh the return on the investment. As a result, the company may have a difficult time justifying the cost of internationalization and localization.[21]

The ultimate goal for writers working with transnational target audiences is to develop a single document that can be read and understood by any audience. The question, of course, is Can there be such a document? Business professionals talk about global products, which can be used by anyone anywhere, but those conversations typically identify the near impossibility of developing such items. A paper clip might be a global product, but only if all cultures agree on its function and only if all have a use for it.[22] The Onge tribe is a seminomadic tribe living in the Andaman Islands in the Bay of Bengal. Is there a need for a paper clip in Onge tribal life?

Global is a relative term, and we must understand that a truly global document is probably not a possibility. For practical purposes a global document is one that can be read and understood by as many transnational and transcultural audiences as possible. One way that many workplace writers try to produce global documents is by increasing the use of visual representations to account for linguistic differences, but visuals are also culturally influenced, as chapter 8 explains, so writers must make their visual rhetorical choices very carefully.

---

[19]Nancy L. Hoft, *International Technical Communication: How to Export Information about High Technology* (New York: John Wiley and Sons, 1995).
[20]Ibid.
[21]Ibid.
[22]Ibid.

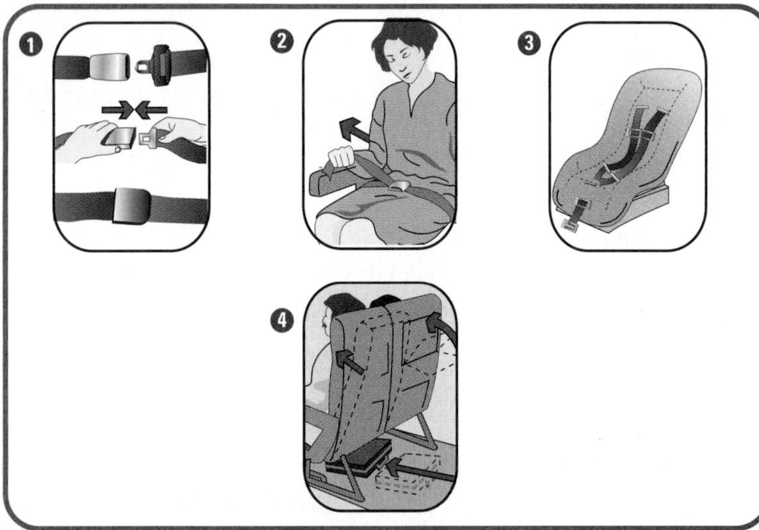

FIGURE 5.2 A safety information card using visuals to convey global information

One example of a global visual document is an airline safety card (see Figure 5.2), which is generally provided in multiple languages and with visuals designed to convey information even to those who cannot read the available languages. Despite these efforts, however, most experts agree that airline safety cards are not intuitive for many readers and attempt to convey too much information in one document. Notice, for instance, in Figure 5.2 the use of numbers to direct readers—numbers that are not used in all languages. Notice, too, the need for a preliminary understanding of what a seatbelt is used for, what a cigarette is, and what the red circle with a slash through it indicates. Finally, notice the visual depiction of the plane's passengers.

Despite the obstacles that hamper total globalization, workplace writers with transnational audiences should still strive for global documents.

Verbal Communication

Translation, localization, internationalization, and globalization can all help to adapt workplace writing for transnational audiences, but workplace writers often find themselves needing to communicate verbally, too, with their transnational and transcultural audiences, perhaps even with nonnative speakers of English within their own multinational company. In these instances it is important to remember that native speakers often speak rapidly, making it difficult for nonnative speakers to understand. In addition, cultural differences can come into play, and native English speakers may inadvertently offend nonnative speakers through their body language, tone, or humor.

In an article for the Society for Technical Communication, Bill Gruener offers nineteen guidelines for communicating with nonnative English speakers, guidelines he devised after polling multilingual colleagues:

- Speak slowly and enunciate clearly.
- Have face-to-face conversations whenever possible because body language adds meaning (a point stressed by every respondent).
- Avoid noisy locations where multiple conversations are taking place.
- Keep conversations one-on-one in order to avoid potential distractions.
- Remember that phone calls are difficult because of the lack of face-to-face contact.
- Pause often and ask if you've been understood.

- Speak slowly when giving a phone number, address, or some other number.
- Use simple sentences, especially when writing (e.g., "Click OK" instead of "Select the OK button and press").
- Use widely accepted international business terms, such as *invoice*.
- Use common English words found in small, simple dictionaries (e.g., Please read the specification" instead of "peruse the specification").
- Choose a single common verb that explains the action of a sentence. (e.g., "Analyze the profit margins in the ACME report" rather, than "Give me your take on the margins").
- Avoid two-word verb phrases ending with *off*, *up*, or *on* (e.g., "She quit" instead of "She gave up").
- Be careful with verbs that have varied meanings, such as *get*, *look*, *take*, or *put*.
- Avoid verbs used colloquially (e.g.,"Deliver the work order to Bob in manufacturing" instead of "Hoof it over to Bob").
- Avoid everyday jargon and local expressions.
- Avoid idioms, idiomatic phrases, and metaphors (e.g., "He was robbed" instead of "He was taken to the cleaners").
- Avoid words adopted from other languages into English (e.g., use "appetizers" instead of "hors d'oeuvres").
- Avoid humor based on knowledge of the language.
- Assume that your listener does not understand everything you've said.[23]

TRANSNATIONAL ETHICS

See chapter 4 for more on ethics.

International business ethics and the ethics of economic systems have become important to ethicists and to the international business community. However, because culture determines ethics, workplace writers must know that ethical understandings don't always cross borders; in other words, what you may see as an ethically appropriate choice may not appear so to other audiences.

A number of ethical issues exist that should be considered in writing for transnational audiences. The following issues arise from a broad perspective:

- Are there values that transcend national or cultural borders that might be thought of as universal ethics?
- How do different countries and cultures tend to address ethical issues, particularly in relation to business and workplace communication?
- Does the target country or culture traditionally hold nonnative business people in a different regard than it does native business people?
- Do various target countries or cultures have different ethical traditions? How might workplace writers account for such discrepancies?
- Do religious perspectives inform a target country's or culture's ethical approaches?
- How do certain business practices—fair trade agreements, for instance—interact with a target country's ethical understanding of business practice?

---

[23]Bill Gruener, "Communicating with Nonnative Speakers," http://www.stcsig.org/itc/articles/0309-nonnative_speak.pdf.

- Does the target country have certain industry-specific ethical expectations—for example, in pharmaceutical industries, food industries, or biotechnology industries?
- How do localization, internationalization, and globalization of documents as well as work toward Global English affect cultural imperialism?
- Do target countries or cultures have varying expectations of and ethical approaches to labor issues, such as child labor?
- Does working with transnational target audiences open the door to taking advantage of cultural differences? (Think about how companies outsource production, like clothing manufacturing, or service support, such as call centers.)
- Are there ethical concerns with engaging in commerce with countries that are at political odds with the source country?

Other issues arise from a specific perspective:

- Does my language in any way obscure information in the document?
- Have I accurately translated the document?
- Has the document accounted for the cultural differences of my target audiences?
- Has my document obscured or accurately represented liability issues?
- Have I used visuals to highlight or distort information?

GUIDELINES FOR WRITING FOR TRANSNATIONAL AUDIENCES

Even though we cannot offer approaches for specific cultural, linguistic, or national audiences (because to do so would encourage stereotyping), we can offer a series of general guidelines for writing for transnational and transcultural audiences that encompass the strategies discussed in this chapter.

Write Clearly

See chapter 10 for more on how to write clearly.

Translators can more accurately and easily translate documents that are written clearly and concisely. Likewise, nonnative readers of English are more likely to understand clearly written documents.

Use Correct Punctuation

See chapter 10 for more on punctuation.

Ensure that your documents are punctuated correctly. Misuse or lack of punctuation can confuse nonnative readers and, more important, can lead to misinterpretation of information. Also be sure that translators are working with the most accurate version of a document; punctuation errors can result in translations that differ from the original intent.

Include Definite Articles

Leaving definite articles out of sentences can confuse nonnative readers. For instance, although it may sound appropriate to say, "Products will be shipped Tuesday," it is more complete to write, "The products will be shipped Tuesday."

Avoid Using Pronouns

Native English writers tend to use pronouns that refer to nouns preceding the pronouns. Often, this backward reference can confuse nonnative speakers. In addition, because other languages use and position pronouns differently than English does, exact translation can be linguistically difficult. Instead, repeat the specific noun to clarify. This practice is also helpful in writing for mechanized translation programs.

ANALYZE THIS

The pronoun *you* can be particularly tricky for translation. Explore the ways in which the pronoun *you* is used in German, French, and Spanish. Then write an explanation about why it might be difficult to translate the U.S. use of *you* into these three languages.

Use Terminology Consistently

Consistency helps nonnative speakers contextualize information. Use of a single term throughout a document also helps translators convey meaning. Even though English may offer multiple synonyms for a concept, other languages may not accommodate such word shifts. Consistency in terminology also ensures that transnational audiences understand that you still mean the same thing; shifts in words can suggest nuanced shifts in meaning.

Avoid Idiomatic Language

Remember that idioms do not translate well, so you should avoid idiomatic expressions. The same is true for acronyms, abbreviations, jargon, and slang. If you must use specialized language, provide definitions of the terms early in the document. Glossaries can also be of use.

Avoid Comparatives

Comparing things makes sense only if the two things are culturally understood as similar. Promotional materials often use comparatives to show how products or services compare with other products or services. In the United States, for instance, it is common for advertising or promotional materials to indicate that Product A is better than Product B or that Product A has 25 percent more meaty goodness than Product B. Such comparatives rely on familiarity with Product B, which you cannot be sure transnational audiences have. And in some countries, such as Germany, it is illegal for advertising and promotional materials to use these kinds of comparatives.

Localilze Your Writing

Recognize Alphabetic Differences

Not all languages use the same alphabets; in fact, not all languages use alphabets. Consequently, alphabetized lists may not translate exactly, and using letters as organizational markers—as in an outline—may not provide readers with directional information they understand.

Use Local Numbers

Remember to localize numeric information like dates, time, currency, measurements, telephone numbers, and addresses. Also, confirm that the target audience uses traditional Arabic numerals (i.e., 1, 2, 3…).

Be Alert to Time Differences

Time zones shift between most countries, and you don't want to miss deadlines or meetings because of time differences. This may also be a factor if you provide time-sensitive information or time-restricted support—say, for instance, a help desk that is not open 24 hours a day. In addition, you should recognize that many cultures do not use forty-hour, five-day work weeks.

Avoid References to Holidays

Remember that not all cultures or countries celebrate the same holidays, nor do they celebrate holidays on the same dates or in the same ways. Furthermore, certain celebrations in one culture or country may be seen as offensive to another. Be sure to avoid using visuals that reference holidays for the same reasons.

Avoid Cultural References

Most transnational and transcultural audiences will not understand references to culturally specific items. Many things that you know well—like eating or cooking utensils, sports equipment, or holidays—are probably not common reference points for most of the people in the world. Similarly, references to celebrities, movies, or music should be avoided because they do not always cross borders. In addition, you should not make cultural references to a culture that you are unfamiliar with; doing so risks displaying your ignorance about the culture and perhaps insulting or offending your audience.

Avoid Humor

For the most part, humor doesn't translate universally and risks insulting or offending target audiences. You should also not try to use another culture's humor unless you are well versed in the target language, culture, and humor. The result could be even more offensive than a poorly translated joke.

Account for Visual and Auditory Perceptions

Consider Visual Interpretations

Remember that visual rhetorical choices are culturally informed; images can convey different meanings to different people. For instance, in the United States an image of an owl (adopted from Native American lore) might symbolize wisdom. However, in parts of Central and South America, an owl might represent the blackness of night, implying death, black magic, or witchcraft.

Workplace writers should also avoid using religious symbols. From 1996 to 2004, the web page askjeeves.com (now simply ask.com) used a cartoon version of a butler as its logo. But when the company began to move into Asian markets, it was noted that the butler image would not translate with any real meaning. One suggestion was to replace the Jeeves butler with an image of a monk—another figure one might ask for answers. However, research showed that such a depiction would be considered offensive to several cultures.

For more about visual rhetoric, using visuals, and transnational use of visuals, see chapter 8.

In addition, you should be sure to understand cultural interpretation of color variations.

Avoid Images of People and Hand Gestures

Because body language and gestures are culturally defined, it is best to avoid using images of people or gestures; those images may not be interpreted as you intended. For example, you might use the icon 👌 as a bullet or to indicate a correct step. In the United States that hand sign communicates correctness or that everything is OK. However, in France that same gesture or icon would likely be interpreted to mean zero or that something is worthless.

Reevaluate Design Elements and Principles

Be sure that your document designs—for both print and electronic documents—are easy to access and that navigation is clearly explained. Because not all languages are

See chapter 9 for more on design elements.

read left to right, what you consider to be a logical document flow may not seem logical to a target audience. Also, keep in mind that most languages require about 30 percent more space than English does.

In addition, don't forget that computer access varies worldwide. Individuals may have limited time to work with documents online because of shared work spaces and varying phone rates and Internet availability. Since it is common for transnational recipients to print hard copies of online documents, you would be wise to produce small files that require minimum download time. Remember that large graphics and image files slow downloads.

Account for Differences in Sound Interpretation

Like visuals, sounds can mean different things to different people. Because multimedia documents often include sound, workplace writers should account for cultural differences in sound interpretation, just as they would account for visual interpretation. For example, in the United States we are comfortable with a computer beeping to indicate an error; however, in Japan the same sound is considered embarrassing because it calls attention to a mistake.

SUMMARY

- Transnational and transcultural communication occurs internally and externally for many companies.
- Transnational and transcultural communication must account for how nonnative speakers will read documents written in English and how documents will be translated from English into other languages.
- Learning how to analyze transnational and transcultural audiences is crucial in order to reach global audiences.
- It is important to learn as much about a target language as possible.
- It is also important to identify official national languages.
- Workplace writers can no longer anticipate that International English will be the language of world commerce.
- Some languages do not follow left-to-right directional flow.
- Workplace writers must account for various levels of technology in target countries.
- Machine translation software automatically translates documents. It relies on controlled English, but the software is unreliable.
- Educational systems and approaches vary greatly from country to country.
- Workplace writers must account for differences in laws and politics in target countries.
- Economic factors and perceptions affect workplace writers who write transnational and transcultural documents.
- Family and business life must be kept separate when considering transnational and transcultural audiences, but they both contribute to cultural atmospheres.
- In many countries religion plays a significant role in all aspects of cultural life.
- When writing for transnational and transcultural audiences, writers should avoid stereotyping.
- It is the responsibility of most workplace writers not to translate documents, but to prepare them for translation.
- Localization is the process of adapting or preparing a document for a specific local audience.

- Internationalization is the process of writing, rewriting, designing, and revising documents so that they can be more easily localized.
- Globalization is the process of preparing documents to be readily understood by as many transnational audiences as possible.
- Workplace writers often find themselves having to communicate verbally with transnational and transcultural audiences and nonnative speakers within their own company.
- Because culture determines ethics, workplace writers must realize that ethical understandings don't always cross borders.
- It is important to develop guidelines for writing for transnational and transcultural audiences.

CONCEPT REVIEW

1. What are some problems associated with writing for transnational and transcultural audiences?
2. What factors contribute to national and cultural differences?
3. What linguistic differences contribute to transnational and transcultural communication?
4. What role does technology play in differences among transnational and transcultural audiences? How does technology contribute to approaches to transnational and transcultural audiences?
5. How do educational differences affect how you would write a document for a transnational or transcultural audience?
6. What are some issues of politics and law that workplace writers must account for in writing transnational and transcultural documents?
7. What economic factors affect how workplace writers should think about transnational and transcultural audiences?
8. How does religion affect transnational and transcultural communication?
9. What are some methods of avoiding stereotypes?
10. What issues pertaining to translation should workplace writers account for?
11. What is localization, and how are documents localized?
12. What are the goals of internationalization, and how are documents internationalized?
13. What are the goals of globalization, and how are documents globalized?
14. What are some strategies for communicating verbally with transnational and transcultural audiences?
15. What ethical issues are associated with transnational and transcultural communication?
16. What are some guidelines for writing for transnational and transcultural audiences?

CASE STUDY 1

Nola-Cola Goes Global

The Coca-Cola Company has a long history of working with transnational audiences. In the 1920s, Coca-Cola executive Robert W. Woodruf initiated an effort to market Coca-Cola internationally. By the 1930s, Coke had established bottling and distribution centers in France, Guatemala, Honduras, Mexico, Belgium, Italy, Peru, Spain, Australia, and South Africa. By the time the United States entered into World War II, Coke was being bottled in forty-four countries. During the war Coke established more than sixty bottling plants worldwide to supply American troops with Coke. According to the Coca-Cola Company, this effort began after General Eisenhower requested bottling equipment for his North Africa base of operations. After the war the Coca-Cola Company converted many of its military bottling plants into civilian bottling centers, positioning the company for rapid growth in the international market. Sales in the United States also increased as soldiers returned with a taste for Coke and created a larger demand for it.

Like any other company, the Coca-Cola Company faced a number of challenges in bringing its product to transnational audiences. It has long been rumored that when Coke was introduced to China, the transliteration of the Coca-Cola name was expressed in characters that sounded like "ke-kou-ke-la," which means "bite the wax tadpole" or "female horse stuffed with wax," depending on the dialectal pronunciation. However, these bits of industry lore have also been shown to be nothing more than urban legend.

Consider that you have recently been hired by Nola-Cola, a New Orleans-based soft drink manufacturer seeking to expand into international markets. Nola-Cola has been in operation for more than fifty years, but its products are not recognized outside the southern Louisiana region. The newly acquired executive officer believes the time is right to extend sales nationally and internationally of Nola-Cola's three best-selling products: the premiere product, Nola-Cola; Tangeraide, a citrus soda; and Joggle, a double-caffeine, double-sugar energy soft drink that gained some attention when riders in the Tour de Louisiana cycling event began drinking it in place of other caffeinated colas.

The Nola-Cola CEO is familiar with the Coke-in-China legend and is also well aware of Coke's success. The CEO has asked you to learn what you can about the Coca-Cola Company's move into international markets, including the controversies that surrounded some of its business dealings. Your task is then to use what you learn about Coke to devise an internal manual or guidebook that can be used in localizing the Nola-Cola products for three large international markets: China, India, and Brazil. You will need to conduct some research to learn what you can about each of these soft drink markets (hint: the American Beverage Association web page can be of use here), about the Coca-Cola Company's experiences in these markets, and about localization. You should then create a series of guidelines to direct Nola-Cola through the localization processes for these three markets. Be sure to consider everything from product name to advertising campaigns to cultural positioning of soft drinks.

CASE STUDY 2

Emergency Evacuation on Peligo

The island nation of Peligo relies on tourism as its primary economic industry. The island's vast beaches and calm, clear water on its south side and the heavy, rolling surf on its north side attract tourists from the United States, Europe, Asia, and India. Saltwater enthusiasts of every kind find that Peligo offers just about any activity they could desire: snorkeling, scuba diving, flats fishing, offshore fishing, kayaking, surfing, sailing, spear fishing, and more. Peligo's warm climate draws tourists year-round, and its beaches and top-rated resorts and restaurants provide world-class facilities. Despite Peligo's paradise-like qualities, however, its location in the Caribbean leaves it susceptible to hurricanes and tropical weather systems.

Peligo's geography, like that of many other small islands, is dominated by its shoreline. The entire country is only thirty miles long and six miles across at its widest point. With the exception of a two-mile stretch of coastline along the northwest shore that is a mix of rock and mangrove, most of Peligo's shoreline consists of broad beaches. And lining those beaches are rows of resorts, restaurants, and shops. Before tourism brought investors to build the resorts, Peligo's coastline was home to the island's lowest economic classes. Peligo's wealthier residents, many of them of European descent, built homes in the central part of the island, in the mountains away from the coastline, where storm surges and floods were less likely to wash their homes away. As is the case in many island nations, Peligo's wealthiest residents took the safest locations, forcing its poorer citizens to occupy the more dangerous regions along the coast. But Peligo's tourism industry then forced the residents away from the beaches as hotels, resorts, and shops were built to appeal to tourists from around the world. Finally, Peligo's poor were forced to move to an intermediate zone between the tourist beaches and the mountains.

In 2004 Peligo was devastated by three hurricanes: Charlie, Francis, and Ivan. The island suffered extreme property damage and a large number of fatalities. Many of those that died were native Peligans whose homes could not withstand the storms, who had no other place to go, and who could not afford to leave the island before the storms arrived. A number of tourists were also stranded and died when Peligo Air could not shuttle people off the island fast enough. Because Peligo serves tourists from all over the world, one of the difficulties with emergency evacuation was an inability to communicate evacuation procedures to all tourists.

In the wake of these tragedies, the Peligo Tourism Ministry has insisted that new evacuation plans be developed and that evacuation procedures be presented in ways that will be easily understood by tourists, no matter their place of origin. In addition, the Tourism Ministry has created new policies to ensure that Peligo's citizens are provided with accessible evacuation plans and that the Peligo tourism industry provides the resources for evacuating resident Peligans when necessary.

The Peligo Tourism Ministry has contracted with the U.S.-based company Safety International (SI) to develop the evacuation plans and information cards and brochures that explain evacuation procedures. SI currently provides safety information cards for a number of airlines worldwide and has also developed evacuation plans and resources for companies, militaries, and governments around the world.

As one of SI's safety strategy engineers, you have been assigned to work on the Peligo plan. For this assignment you are charged with developing a plan to account for all of Peligo's needs and detailing it in a written report. You will need to create a list of the kinds of documents that should be developed to publicize the evacuation procedures. Then, write a description of the kinds of transnational and transcultural concerns SI will have to account for in writing those procedures.

CASE STUDIES ON THE COMPANION WEBSITE

Translating Video Games to Foreign Markets

This case focuses on Lateral Designs Unlimited, a company that specializes in re-releasing successful U.S. video games in foreign markets. Your team of technical writers prepares game documents for localization to target markets outside the United States.

Big Owl's Goes Global: Internationalizing a Restaurant Chain

As a technical writer for a U.S.-based chicken wing franchise, you have been asked to research the feasibility of opening a chain restaurant in several Middle Eastern countries. You will need to determine how to best transfer your product to a different market in a culturally sensitive manner.

Learning about Chinese Business Culture

As an employee of THC Industries, you find yourself having to make an extended business trip to Beijing in order to negotiate a number of new contracts. You decide to do a bit of a preliminary research to familiarize yourself with some of the basic cultural issues that you will need to consider when working with Chinese business people.

The Hawksbill Turtle Society: Securing International Protection for an Endangered Species

In this case you will consider the problems faced by the Hawksbill Turtle Society as the members attempt to convince legislators in other countries to adopt better protection measures for this endangered species. The economic and social role of this turtle in certain cultures creates complex rhetorical problems.

VIDEO CASE STUDY

DeSoto Global

Collaboration and Writing a Request for Proposals

Synopsis

DeSoto Global's automobile navigation systems are popular with North American markets because they all feature DeSoto Global's patented Tracker1 system, which allows DeSoto Global to locate stolen units through a GPS tracker chip embedded in each unit. Dora Harbin reports that Japan has launched an aggressive campaign to combat increased car theft. Of greatest interest to DeSoto Global is the campaign's desire to use more efficient vehicle-tracking devices—like DeSoto's Tracker1 system. The snag for DeSoto is that, in order for the Tracker1 system to work, it needs to be tied into a local security company with connections to local law enforcement agencies.

WRITING SCENARIOS

1. Listed here are thirty-eight idioms commonly used in American business communication. First, select fifteen of these terms, and learn their origins/etymologies. How did these come to mean what they do? Then create a glossary that explains what each of the expressions means in contemporary communication so that a translator who encounters any of them will be able to provide a more direct translation.

across the board
at a loss
bail a company out
ball park figure
bang for the buck
banker's hours
bean counter
big wheel/cheese/gun/wig
black and white
bottom fell out
bottom line
budget crunch
buy out
captain of industry
carry the day
cold call
company town
cut and dry
cut back
cut your losses
deliver the goods
face value
finger in the pie
give the green light
heads will roll
in the long run
in the red
jack up
kickback
number cruncher
on the chopping block
piece of the action
red ink
sell like hot cakes
strike while the iron is hot
take a nosedive
throw money at a problem
write off

2. In a 2008 letter to potential clients, the Sphere Group Distributors included these two sentences: "Although updated figures have not been confirmed, ballpark figures suggest a 34 percent increase over last quarter" and "It would seem that with the QuadOne notification process, we've hit a homerun." Of the 2,500 copies of the letter that were distributed, more than 750 were sent to potential European clients. Why might these sentences be problematic for a European audience? How might these two sentences be rewritten to better reach a European audience?
3. In the United States, written contracts between companies, individuals, and organizations are considered binding, legal documents and are regarded as critical to most professional relationships. Generally, no work or trade takes place until contracts have been agreed upon and signed, and they are then highly regarded as defining the relationship. Writing contracts is itself a massive industry in the United States, as is contract law. However, contracts do not have the same kind of legal or social status in many other countries. For instance, in China, contracts are much less important than developing a relationship between organizations, and many business arrangements never require contracts.

   Consider this situation: A small U.S. gaming software company is attempting to establish its newest line of social-space games in Asian markets. Because it does not have the resources or the local knowledge to effectively place its product in local retail locations, the company must work with local distribution companies. After a number of telephone conversations with a Chinese software distribution company, the U.S. company agrees that the distribution company meets its needs and verbally acknowledges that it would like to work with the distribution company. The U.S. company then suggests that the two companies should reach an agreement and develop a contract. The Chinese representatives, however, seem less concerned about a contract; their words suggest that they accept the conversation as contractual. In fact, their final statements indicate that they will begin including the gaming software in their distribution catalogs and that they are looking forward to receiving the first 10,000 units for distribution.

   Given that contracts are important to U.S. business and culture and are less so to Chinese counterparts, how would you explain to the distribution company the gaming company's need for a contract? Write a document that explains the need for the contract and then attempts to explain this need to the distribution company executives.
4. Select a food item that is made and distributed in the United States. Pick something that is culturally situated (and perhaps marketed) as "American": hot dogs, frozen dinners, pies, cake mixes, individually wrapped slices of cheese food, potato chips, candy bars, and so on. Take some time to analyze the package in which the food item is distributed. What text is included? What visuals are used? How is it designed? After you have analyzed the package, imagine that your company handles the item and has asked you to look into marketing and distributing it in three countries: Ukraine, Moldova, and the Kyrgyz Republic.

   The item you are working with has not been previously distributed in these countries, and there has been little attempt to create a market for it. Because of the political situation in these countries, only recently have they been opened to U.S. trade. Your company's venture into these market areas will prompt consideration of them as potential long-term markets, but the uncertainty has caused your company to offer only limited resources for the project. You will not be able to develop three new packages and three new market strategies; your supervisors want a single, global approach to distribution.

Develop a document that outlines what your company needs to know about each of these three countries in order to market the item you have selected. Identify factors like language, religion, political history, and so on that might affect how you market the item. Consider specifically how the kind of food you have chosen fits within the countries' culinary cultures and what advantages or disadvantages this might cause. Be sure to identify factors that will lend themselves to a single market approach, such as common customs, languages, and visual approaches.

5. Recently, your Chicago-based company has been negotiating with a South Korean company to enter into a long-term agreement that could open up new Pacific-rim markets for your company. You have been central in these conversations. Last Friday morning when you arrived at your office, you found an e-mail in your inbox from your counterpart with the South Korean company, asking you to call him Monday morning at 10:00 to plan the details of a phone conference the two companies have scheduled for Tuesday. The relationship you have developed with your counterpart has been positive and, you believe, partly responsible for the success of the negotiations. You respond to the e-mail, agreeing to call.

   On Monday when you call, there is no answer. You call back fifteen minutes later, still no answer. Assuming that your counterpart has been delayed by other business, you decide to wait for an e-mail. On Tuesday morning when you arrive at the office, there is a terse e-mail waiting for you from your South Korean counterpart. He is insulted by your not calling, and the conference call has been postponed.

   Why did he assume that you did not call him? What mistake did you make? Write a letter of apology to the South Korean representative, explaining what happened (the clues are in the assignment), apologizing for the situation, and attempting to reschedule both the independent and the conference calls.

6. Using English as culturally neutral as possible, write a step-by-step set of instructions for a global audience on how to use a paper clip.

7. Pick a country other than the United States. Conduct a bit of research, and then create a chart that outline's what you have learned about that country's approaches to politics and law, economics, society (including attitudes toward age, business etiquette, and family and social interaction), religion, education, technology, and language. Once you have finished the chart, write a list of guidelines based on what you've learned that would assist someone who is writing to that target country.

8. Many industries have developed networks of professional conversation about transnational communication within their industries. Many have also begun to develop industry standards for transnational communication. Select an industry in which you might be interested in developing a career. Then research how that industry approaches transnational and transcultural communication. Are there established standards? If so, what are they, and who maintains them? Who in the industry is talking about transnational and transcultural communication? Are there companies and/or organizations that are leading these conversation? Write a document explaining how the

industry you have selected talks about transnational and transcultural communication. Don't limit your research to simple web searches; do talk with professionals in the industry.

9. Writing documents for audiences in third world countries (also known as developing countries or the global south) taps into specific kinds of economic, technological, and educational issues that differ from those of first and second world audiences. Think about what those issues might be, and develop a set of guidelines for writing technical documentation for third world target audiences.

10. When U.S. businesspeople work in countries outside the United States, they often find local business customs so different from what they are accustomed to that they have difficulty adjusting to the new working conditions. For example, many U.S. workers—particularly managers, professionals, and upper-level executives—are accustomed to having their own offices. But in some countries where space is at a premium and social practices emphasize community over individuals (e.g., Japan), those workers often find themselves working at desks in large rooms with other employees in the same room. U.S. workers often find such conditions distracting. As another example, business meetings in some countries often begin with informal conversations that U.S. workers might consider small talk. Actual discussion about business matters is delayed until each person present has had the opportunity, for instance, to diligently inquire about the health of the families of others in attendance. And in some countries it is considered impolite to begin addressing business matters until everyone present has had a chance to take refreshment, like a cup of tea.

    Conduct some research to learn about the business culture of three other countries that differ from the United States in practices and expectations. Then write a document that explains these practices and offers strategies for how U.S. workers might best acclimate to such practices should they find themselves working in those countries.

11. This chapter explains the problems of stereotyping that might arise from guidebook approaches to communicating with particular target audiences. Despite these concerns there are endless resources available whose strategies for doing business or communicating with various target audiences rely on methods that can be construed as stereotyping. Locate a resource that offers advice on how to communicate with others in a specific target country other than your own. Analyze the advice, and determine whether any of the suggested strategies contribute to stereotypes. Then write an analysis of the resource, explaining what you found.

12. This assignment comes from Bruce Maylath:

    > Many translation companies report that, while the largest segment of their work is for documentation heading in and out of the United States, the fastest growing segment of their business is for documentation staying within the United States. Part of the growth is a result of the North American Free Trade Agreement. Go to any store in the United States, and you'll see many package labels in English, French, and Spanish for the joint U.S., Canadian, and Mexican market. The labels are just as useful for French speakers in Maine, northern New York, and Louisiana and for Spanish speakers in Florida, Texas, and California as they are for French Canadians and Mexicans. More striking, perhaps, is the increase in translation into Navajo, Chinese, Somali, Hmong, and other languages for communities within the United States. In what ways will the emphases on niche marketing, multiculturalism, and reaching

populations in their native languages affect the way American firms conduct business both at home and abroad? In what respects is the situation today different from 1900, when quite a few documents were translated into German, Swedish, Yiddish, Italian, and other languages for immigrant groups settling in the United States?[24]

Write a detailed letter to your instructor explaining your answers to these questions.

---

[24]Bruce Maylath, "Translating User Manuals: A Surgical Equipment Company's 'Quick Cut,'" in *Global Contexts: Case Studies in International Technical Communication*, ed. Deborah S. Bosley (Boston: Allyn and Bacon, 2001), 79.

Researching and Evaluating Source Materials

CHAPTER LEARNING OUTCOMES

After completing this chapter, you will be able to do the following:

- Recognize the importance of **research** in writing and problem solving
- Understand how research involves **asking and answering questions**
- Distinguish between **primary and secondary sources**
- Find and **evaluate** sources from different locations: **World Wide Web**, **blogs and wikis**, **databases**, and **intranets and archives**
- Find and select **visuals** to incorporate into your research
- Know how to conduct primary research through **observation**, **inspections**, **experiments**, **interviews**, **surveys**, and **focus groups**
- Take effective notes from sources: **research narratives**, **double-entry journals**, and **problem-solution logs**
- Understand the ways that **note taking** is a form of **drafting**
- Know how to **document** sources correctly
- Avoid unethical research practices and **plagiarism** by correctly **quoting**, **paraphrasing**, and **summarizing**

DIGITAL RESOURCES

On the Companion Website www.prenhall.com/dobrin:

- Case 1: Northside Urgent Care Goes Shopping: Searching for the Right Office Machine
- Case 2: IPD Body Armor: Planning a Major Research Project
- Case 3: Street Sports Market Research: Conducting Primary Research through Focus Groups
- Case 4: Going Global: Transnational Management Accounting
- Video Case: Planning the Shobu Automotive Proposal
- PowerPoint Chapter Review, Test-Prep Quiz, Exercises and Activities

REAL PEOPLE, REAL WRITING

MEET MAX LALLY • Police Officer with the Village of Tequesta, Florida

What types of writing do you do as a police officer?

About 90 percent of my job involves writing, and the types of writing are varied. I write incident reports, crash investigation reports, summaries of interviews with witnesses to crimes and accidents, as well as lots of e-mail. Typically, my e-mail messages are with other officers and supervisors; for example, if residents ask me for more patrols around their neighborhood while they are away, I'll e-mail the request to other officers so they have the information.

What role does research play in your writing?

There are many types of research I do as a police officer. The Internet is a primary source for investigating changes in laws and procedures and for finding out information about other departments and officers in our area, especially if I need to investigate an issue or problem that covers a wide area. We also use our CAD (computer-aided dispatch) system to look up individuals to determine whether they've made previous reports or have a valid license or a police record. I research laws and procedures through printed sources as well, such as state statute books and case law books, although many of these are also available online.

What are the sources for your research?

Aside from printed and Internet sources, I also rely on individuals to gather information. I conduct interviews with witnesses both in person and over the phone to help establish a chronology of what happened. Accurately recording the details and the time frame is crucial in an incident, so I try to conduct and write up these interviews as quickly as possible. In addition, I consider other police officers as good sources for research. New officers undergo four months of field training with a seasoned officer, and this provides lots of background information about how we proceed.

Do you ever need to cite the sources of your research?

Most of the incident reports I write involve some sources, whether they are interviews with witnesses or specific laws and cases that apply to the situation. Citing these sources is especially important if the incident goes to court, because lawyers and judges need to know where the information came from. Sources are typically referenced at the bottom of our reports, and I try to include as much information about the source as possible in case someone needs to follow up on it.

What advice about workplace writing do you have for readers of this textbook?

Take as many writing and communication courses as you can, because writing will be a part of your job—no matter what you do. Many police officers attend additional report-writing seminars, because clear writing is so important to the job. Keep in mind that what you write represents you to many different people; part of being a professional is communicating accurately and effectively.

INTRODUCTION

research a means to find information as well as to solve complex problems

Research plays an important role in solving problems in the workplace and in the world in general. Nearly every major decision made in business, industry, and government involves detailed research to discover the best possible approach or solution. Research is also often conducted after a solution has been implemented to confirm that it truly addresses the problem or issue.

You may not immediately recognize the ways in which research influences the world around you, but it does. For example, the Transportation Security Administration (TSA) regularly conducts research to determine whether airports are safe and secure (see Figure 6.1). In a recent TSA field experiment, a covert testing agent was equipped with a fake bomb placed inside a back brace. The agent was able to get through airport security, indicating that some airport screeners may be hesitant to thoroughly search passengers with medical problems. Other tests involve inserting digital images of guns, knives, and explosives into x-rays of luggage to see if screeners will detect them. The TSA also conducts regular inspections of x-ray machines to determine whether they are operating properly. All of these activities are types of research, intended to look for weaknesses in airport screening that might require new procedures, equipment, or training. This research is integrated into a variety of reports, memos, and other documents that influence airport security measures. It is also made public through newspapers, magazines, and other public media, particularly if it will affect the general public. Research like this is a necessary component of complex problem-solving activities—many of which have no easy answers or quick solutions.

FIGURE 6.1 TSA research to improve airport security

Although you may never conduct covert research into airport security, most careers require research to solve workplace problems. Some research activities entail simply the retrieval of information, whereas other research involves addressing difficult problems, conducting your own primary research, and writing sophisticated documents over long periods of time. This chapter demonstrates that research in the workplace shares similarities with research in school and in our daily lives, but the chapter also shows you how to employ particular methods and strategies to solve the kinds of complex problems that you might encounter in your career.

PSA

In order to solve problems, workplace writers must know how to distinguish among different kinds of research, know which sources are relevant for the problem at hand, and know where to find those sources, how to collect and assess them, and how to incorporate them into workplace documents—all of these activities are covered in this chapter. The research that workplace writers conduct depends on the particular variables of the situation—that is, the problem at hand, the purpose for researching and writing, the stakeholders, and the type of document that must be produced. The problem-solving approach is particularly important in producing documents that require research. In fact, as you should recall, one of the phases of the PSA is Research.

Remember that the problem-solving approach is not a linear process; each rhetorical situation calls for a unique approach that uses some, but usually not all, of the PSA strategies. And each rhetorical situation requires writers to determine the order of steps that best suits their needs and the needs of their audience.

Let's return for a moment to the TSA example. The TSA's goal is to determine how best to ensure the safety of airline passengers. In this situation the researchers already know the problem (threats to airline security), the purpose of researching and writing (to ensure public safety and domestic security), and the stakeholders (airline passengers, airline employees, the U.S. government, and others). However, the researchers may not know the most appropriate and most ethical solutions to the problem; the research needs to balance the goal of airline safety with passenger civil rights. In this case the entry point would probably be under Plan, to "identify what information you have and what information you need," and then move to Research, to "determine the types of information necessary and how to get them." Each rhetorical situation you encounter in the workplace will require you to determine the appropriate route through the PSA.

IN YOUR **EXPERIENCE**

Write a short summary of the last research-based document that you wrote—for school, work, or personal use. Explain the steps you took in researching and writing this document, particularly which aspects of the problem-solving approach you used. Include your purpose, your audience, and the way you went about researching.

Answering Questions for Specific Audiences

Experienced researchers and writers know that most research and writing begins with *questions*, not answers—the questions that arise from problems and situations in the workplace and the new questions that arise from the research itself. These problems often come from particular groups with specific interests or goals, and the ways in which the problems are addressed will vary with each audience. Researchers usually do not do research until they have encountered questions that need to be answered, problems that need to be solved. For example, a biologist who notices a rise in infectious diseases in areas where habitats have been destroyed might be interested in posing a research question about this problem: How does habitat destruction influence infectious diseases, particularly deadly viral diseases such as hemorrhagic fevers? The biologist who poses this research question does so because she sees habitat destruction and infectious diseases as real problems that need to be solved through research and subsequent actions. Thus, she articulates a problem to be understood and solved—that is, her purpose. And because many others in the world would be interested in hearing more about this problem and solution, the biologist has a large potential audience.

The nature of that large audience will help the biologist decide how to present her findings. The audience may very well be split into two main groups—for example, scientists who specialize in this area of research and politicians who hold the power and means to address the problem. If the biologist writes for the scientists, she might write a technically sophisticated report or a scientific article; for politicians, she might write a report and a proposal that suggest some particular course of action. The two documents shown in Figures 6.2 and 6.3 demonstrate how different audiences shape the ways in which research is discussed within a document. The first, an excerpt from an article about a particular strain of hemorrhagic fever, was found on the Center for Disease Control (CDC) website. This article contains specific scientific research geared toward an audience of experts. The second, which is from the World Health Organization (WHO), explains to general readers the dangers of a different strain of hemorrhagic fever. This sort of document might convince politicians to enact laws or policies to deal with the problem.

In each example the audience shapes the type of research as well as how it is presented and discussed. Pay particular attention to the different audiences being

Crimean-Congo hemorrhagic fever (CCHF) is an acute illness affecting multiple organ systems and characterized by extensive ecchymosis, visceral bleeding, and hepatic dysfunction; and it has a case-fatality of 8% to 80%. CCHF virus (CCHFV) (genus *Nairovirus*, family *Bunyaviridae*) is transmitted to humans by bites of infected ticks (several species of genus *Hyalomma*). CCHFV has also been transmitted to patients or viremic livestock through contact with blood or tissue. Epidemics of CCHFV have previously been reported from Eastern Europe, Africa, and central Asia. Many cases have been reported from the countries around Turkey, including Albania, Iran, Iraq, Russia, and the former Yugoslavia. Although serologic evidence indicated the existence of CCHFV in Turkey several decades ago, no clinical cases have been documented. We describe 19 patients from the eastern Black Sea region with hemorrhagic fever compatible with CCHF, who were admitted to Karadeniz Technical University Hospital during the spring and summer of 2002 and 2003.

Several patients in May through July 2002 and 2003 were referred from surrounding county hospitals to our hematology unit with varying degrees of fever and hematologic manifestations. All of the patients had similar clinical and laboratory findings, including fever, petechiae, headache, abdominal pain, nausea, vomiting, liver enzyme elevations, and cytopenia. Bone marrow aspirations and routine serologic tests excluded hematologic malignancies and known viral or bacterial infections. Serum samples from several patients admitted in 2003 were stored at −80°C for further diagnostic testing for a possible hemorrhagic fever agent.

The most commonly encountered signs and symptoms were malaise, fever, abdominal pain, myalgia, nausea, vomiting, petechiae, and bleeding from gingiva, nose, vagina, or gastrointestinal system. Complete blood counts showed thrombocytopenia in all patients (median 15×10^3/UL, range 1–87 $\times 10^3$/UL), leukopenia in 15/19 (median 1,700/UL, range 700–5,200/UL), and anemia in 5 of 19 patients (median 13.8 g/dL, range 6.1–17.3 g/dL). Serum aspartate aminotransferase (AST) (median 693 U/L, range 178–5,220 U/L), alanine aminotransferase (ALT) (median 248 U/L, range 66–1,438 U/L), and lactate dehydrogenase (LDH) (median 1,601 U/L, range 650–20,804 U/L) levels were elevated in all patients. Coagulation tests showed prolonged prothrombin time (PT) (median 13.4 s, range 12.1–18.5 s) and activated partial thromboplastin time (aPTT) (median 34.9 s, range 30.2–59.1) in 7 of 19 patients. Fibrinogen was decreased and D-dimer was elevated in one patient with suspected CCHF, which indicated disseminated intravascular coagulation. Fibrinogen and D-dimer levels were normal in other patients. Creatine phosphokinase (CPK) levels were elevated in 14 of 19 patients (median 568 U/L, range 81–2,500 U/L). Blood urea nitrogen and creatinine (median 0.8 mg/dL, range 0.5–6.2 mg/dL) were found to be elevated in 2 of 19 patients. Hematologic malignancies were excluded after bone marrow aspiration smear and trephine biopsy in 14 patients.

FIGURE 6.2 A research article written for a specialized audience

addressed, as well as the ways each document presents its research findings. Each of these documents is research based, but the two are significantly different from each other. Even if you do not understand their content, you can recognize that they have different audiences, purposes, and problems to solve.

primary sources those that present information in its original form

secondary sources those that repackage primary sources by interpreting, evaluating, summarizing, describing, or commenting on them

Understanding Primary and Secondary Sources

Figure 6.2, which relies on clinical research conducted by the writers themselves, is an example of a text that uses primary research, whereas Figure 6.3 was composed by writers that used secondary sources. Basically, **primary sources** are those that present information in its original form; **secondary sources** repackage other sources by interpreting, evaluating, summarizing, describing, or commenting on their information. In the world of science, the results of a laboratory test would be considered a

As of 11 April, 231 cases of Marburg haemorrhagic fever have been reported in Angola. Of these cases, 210 have died. Uige Province, with 202 cases and 190 deaths, remains the most severely affected area.

The present outbreak of Marburg haemorrhagic fever is unprecedented in its size and urban nature, and its dimensions are still unfolding. Although surveillance to detect cases has improved, it remains patchy. In Uige, where daily mobile teams are active, surveillance continues to be largely concentrated on the investigation of deaths and collection of bodies. The security of teams remains a concern. More vehicles are needed, and WHO is making the necessary arrangements on an urgent basis. To bring the outbreak under control, the detection and isolation of patients needs to be much earlier, but this will not happen until the public understands the disease and the high risks associated with treating patients in homes. Infection control needs to improve in heath care settings, and WHO is continuing to supply effective personal protective equipment, for both national and international staff, adapted to conditions in African countries.

A welcome development is the decision by the International Federation of Red Cross and Red Crescent Societies to strengthen its presence in Uige. Volunteers from these societies are part of a group of workers mobilized to conduct a door-to-door public information and education campaign in collaboration with community and church leaders and traditional healers. Today, workers received specialized training from experts in social mobilization and medical anthropology drawn from the Global Outbreak Alert and Response Network. These workers have been rapidly deployed to deliver public talks at markets and schools.

The International Federation of Red Cross and Red Crescent Societies has extensive experience in responding to emergencies in Africa and has been instrumental in bringing large outbreaks of Ebola under control. Because of this experience, workers from the Federation are usually viewed by communities as welcome help. WHO anticipates that this added and welcome support to response activities will help create greater acceptance of control measures and reduce high-risk behaviours.

All currently available data indicate that casual contact plays no role in the spread of Marburg haemorrhagic fever. Transmission requires extremely close contact involving exposure to blood or other bodily fluids from a patient who will most likely be showing visible signs of illness. The disease can also be transmitted following exposure to items, including bedding and clothing, recently contaminated by a patient.

In addition, transmission can occur in hospitals lacking adequate equipment and supplies for infection control and training in their proper use. The hospital system in Angola has suffered from almost three decades of civil unrest, and several cases of Marburg haemorrhagic fever have occurred in health care staff exposed during the treatment of patients in Uige.

To date, WHO is not aware of any cases of Marburg that have occurred in foreign nationals other than those involved in the care of cases in Uige. WHO does not recommend restrictions on travel to any destination within Angola but does advise some precautions. Travellers to Angola should be aware of the outbreak of Marburg haemorrhagic fever and of the need to avoid close contact with ill persons. Persons with existing medical conditions who might require hospitalization should consider deferring nonessential travel to Angola, particularly to Uige Province.

FIGURE 6.3 A research article written for a more general audience

primary source; a subsequent report about the implications of the laboratory test would be a secondary source. There are examples of primary and secondary sources in all disciplines and fields of study.

Writers and researchers in the workplace typically consult both primary and secondary sources when producing problem-solving documents. Some documents require primary sources, some secondary sources, and some both—depending on the problem, purpose, audience, and solution. Often workplace writers produce the actual research they'll use. In most cases the following documents can be considered primary:

- Scientific journal articles that report research results
- Lab reports
- Technical reports
- Statistical data, such as a census report
- Interviews, surveys, and fieldwork reports
- Diaries and journals
- Letters and other correspondence
- Speeches
- Photographs and works of art

In most cases the following documents can be considered secondary:

- Encyclopedias
- Handbook and data compilations
- Histories
- Summaries
- Review articles
- Instructional texts
- Treatises
- Biographies

It is not always possible to tell whether a source is primary or secondary based solely on the genre or kind of document. For example, one website could present primary information (e.g., statistical data gathered from a survey), whereas another website could serve as a secondary source, interpreting and analyzing the statistical data. Thus, it is important to consider what a source does and how it works. Primary sources are original materials on which other documents are based. They may present original thinking, report discoveries, or share new ideas; they are often the first formal appearance of results in print or electronic documents. Primary sources present information in its original form, not evaluated, summarized, or interpreted. Secondary sources describe, interpret, analyze, and evaluate the evidence and ideas in primary sources. Secondary sources are usually removed, in time, from the original event or information. They often rely upon primary sources but often combine them with other sources or apply them to new situations.

The more proficient writers become, the more they recognize when to incorporate primary sources in their texts and when to use secondary sources. In fact, many workplace documents are primary sources themselves, particularly when those documents address new problems in the workplace. You may conduct original research yourself, incorporate primary sources into your own research, or rely on secondary sources to explain or address workplace problems. The nature of the problem, the purpose of the writing, the audience needs, and the genre of the document all help determine what kinds of research are necessary to compose an effective document.

EXPLORE

Visit three of your favorite websites, and look at each carefully. As you explore these websites, pay attention to which information on these sites can be considered a primary source and which information can be considered a secondary source. Construct a list in which you discuss these sources for each of the three websites.

FINDING AND EVALUATING SOURCE MATERIAL

The real trick for writers and researchers is not finding sources but evaluating them for trustworthiness and accuracy.

The world is full of sources of information—friends, relatives, coworkers, magazines, books, journals, websites, TV shows, movies, posters, road signs, maps, billboards, Post-it notes stuck to refrigerators. Typically, we encounter so many sources of information in our daily lives that we fail to recognize them as sources at all. In fact, the difficulty is rarely in *finding* sources of information, but often in finding useful, credible, and reliable sources. This is the real trick for writers—not just finding sources of information but finding trustworthy sources of information and, of course, knowing the difference!

Evaluating the sources you have found is a significant part of the research and writing process. Unfortunately, learning to evaluate sources for reliability and trustworthiness is not easy. Many texts do tip us off that we would be wise not to believe them; the *Weekly World News* magazine in the grocery store checkout lane is a good example (see Figure 6.4). That scientists have learned to create diamonds from dog poop is not likely to be viewed as reliable information by most of us. However, many sources that look professional, sound trustworthy, and seem reliable are not. Thus, the workplace writer must become a discerning judge of source materials, and one of the first steps is knowing where to look; there is a connection between where you find things and the likelihood of their being reliable.

Scientist Creates Diamonds From Dog Poop!
By VICKIE YORK

It's doggone amazing! A scientist has invented a method for deriving synthetic diamonds–from dog doo!

Dr. Florence Gurnley, founder of Caninegems, Inc., says her machine can convert a few pounds of pooch poop into a diamond of perfect color and clarity.

"We can create a 2-carat diamond suitable for an engagement ring that might cost $8,000 in a jewelry store–for as little as $50." Dr. Gurnley brags.

Gemologists who've examined the man-made stones confirm that even the most experienced jeweler would be unable to distinguish them from high-quality natural diamonds.

"These diamonds are absolutely dazzling," marvels one top gemologist. "But I made sure that I washed my hands after I handled them."

Dr. Gurnley refuses to divulge details of the technology that transforms piles of dung into jewels fit for a queen's tiara. She would only say the process is similar to a cutting-edge technique for creating industrial diamonds known as chemical vapor.

FIGURE 6.4 A questionable source of information

A Workplace Example

Let's explore information sources through a real-world example: The director of patient support services at the Health Services Clinic (HSC) at a large university must find ways to improve student access to the clinic. In particular, these are the problems that need to be resolved:

- Students have prolonged waiting times on hold or get a busy signal when they call HSC. All medical or illness-related questions are referred to one nurse. Thus, many messages are taken, and often the patient is not available by the time the nurse is able to call back.
- The long waits for the nurse mean that untrained receptionists often deal with student medical issues.
- More students come in for care without appointments because they can't get through on the phone.
- The waiting areas become overcrowded.

The director of patient support services recognizes that these problems can be resolved by redesigning the process by which students make appointments and employees meet the volume of student needs. The director suggests the following changes:

1. Remodel the call-center area where staff members answer phones.
2. Redefine the roles and expectations of staff members who answer calls.
3. Reconfigure the automatic call distribution system (ACDS) to include deeper queues and a modified cascade so that callers can select a general information prompt and be routed to a staff member who can answer specific questions.
4. Increase staff by three to meet the volume of calls.
5. Reassign staff to meet the projected volume of calls coming in at peak call times.

Although these suggested changes will be highly beneficial, the director and her staff are not exactly sure how to go about making them. The director will need to research the five suggested changes in order to figure out how each can best be put into operation. And before any real changes are made, she will need to write a report and a proposal for the university's vice president for student affairs.

Given the variety of changes that the director hopes to make, there is no one place to find all the necessary information. The director and her staff will need to look at websites, databases, and internal information (e.g., budgets) and even conduct some of their own primary research.

The World Wide Web

For most of us today, the World Wide Web is central to our daily lives. We check the weather and sports scores, read the news, order goods and services, send and receive e-mails, and do much more. In general, the Web is an important tool for conducting research—that is, for finding various kinds of information. However, it is not always the best research tool available.

Search engines such as Google, Yahoo!, or Ask.com can provide starting places for research, but they are not always completely accurate and often provide information based on corporate sponsorship rather than validity or applicability. In addition, savvy web designers now employ strategies to increase the likelihood that their website will appear early in search engine results. And because of the sheer number of websites, it is sometimes necessary to sift through dozens of sites before finding the right information. This is not to say that the Web should not be used as a research tool in the workplace; it does provide certain kinds of information that can be highly useful. But those who conduct research on the Web should be aware of its advantages and disadvantages.

EXPERTS SAY

"Search engines, with their half-baked algorithms, are closer to slot machines than library catalogues. You throw your query to the wind, and who knows what will come back to you?"

DAVID ROTHENBERG
"How the Web Destroys the Quality of Students' Research Papers"

Advantages of the World Wide Web

1. **Ease of use:** Search engines are relatively easy to use. Keyword searches, for instance, bring a wealth of information in mere seconds. In addition, most professional websites constantly undergo usability testing to make them more and more functional for users.
2. **Access:** Web searches bring users a good deal of information that they would otherwise not have access to. Those who use the Web as a research tool can find documents and information from companies and organizations around the globe, as well as information, ideas, and opinions from people whose voices would previously have gone unheard.
3. **Portability:** Researched information from the World Wide Web can be easily saved, stored, and transported from one place to another. It's easy to download words, images, and web pages, and it's usually convenient and simple to cut and paste information for later use. Saving information from a useful website is now as easy as bookmarking the site on your web browser.

4. **General information:** The World Wide Web is a good research tool for those who are looking for general or topical information about a particular subject matter. Many websites provide basic or introductory information and often will identify where more specific information can be found. Someone looking for information about the history of technical communication in the United States would be sure to find websites that provide an overview, but detailed information might be found only in books and specialized journal articles.
5. **Products and services:** The World Wide Web is a tremendous resource for those who are doing research in order to purchase a product or service. In fact, the dot-com (.com) part of web addresses is actually an abbreviation for the word *commercial*. Thus, we can't overlook the fact that most websites exist in order to sell us things. However, because a good many websites do mean to serve clients, we can expect to discover important information, such as the lowest prices on goods and services and specific information about the companies that sell those goods and services. Informative websites can effectively lead us from the first stages of purchasing to the last—from browsing for information to having a product or service delivered.

See chapter 17 for more on web pages and usability testing of web pages.

Disadvantages of the World Wide Web

1. **Lack of trustworthiness and reliability:** The World Wide Web grants us access to all sorts of information from one corner of the globe to the other, but this is also one of its downfalls. Because so much information is attainable from the Web and because almost anyone with a minor amount of computer know-how can design a website and post information on it, we can't always trust what we see on the Web. Many sites do not list the names of designers and authors of information, so that viewers have no sense of who is providing the information. Most print-based documents are carefully reviewed by experts before they are published—a process known as **peer review**. However, unless a website is affiliated with a group of experts (such as a scientific organization or a scholarly review board), it may contain unverified or unconfirmed information. There is much good, reliable information on the Web, but it is often wedged between other sites that are not trustworthy or credible. Deciding what is and is not reliable is not always an easy task.

peer review the process of verifying information by a group of experts before the information is published

2. **Irrelevant information:** Search engines can't read our minds. They might bring hundreds or thousands of websites to our attention after we type in a few keywords, but these sites are not always relevant. Search engines work well if users know precisely what information they are looking for but don't work so well if users are a bit uncertain. For instance, a search engine will likely lead you where you want to go if you're looking for the state of Vermont's Department of Education or a corporate website. However, searching for information written by geological experts about the latest studies of soil erosion in Eastern Europe will likely yield few worthwhile results.

For more on how search engines work, see chapter 17.

3. **A fee for the best:** The best, most sophisticated websites are those that sell things, not those that give things away. This should help you understand the kind of information available through the Web. If you consider that scholars, writers, and other experts who produce documents usually expect to get paid for their work, it makes little sense that they would want it placed on the Web for individuals to read and download for free. Notice that textbooks are not available online for free. Therefore, although you can expect to find some helpful information on the Web, often the most helpful and valuable information can be found only in books, magazines, journals, DVDs, and websites that users subscribe to.

Thus, our director of patient support services at the Health Services Clinic may find *some* help on the World Wide Web to make her suggested changes.

But it is doubtful that she will find there all the information she needs unless she's willing to pay for access to certain websites.

Blogs and Wikis

Researchers can also use web-based blogs and wikis for quick, up-to-date information. As we suggest in chapter 3, both forms can be problematic because the information contained in them is not always verified or confirmed by experts. However, a benefit of both blogs and wikis is that they are generally current, often containing information that is updated daily, hourly, or even to the minute.

blog a website where entries are made in journal style and are recorded in reverse chronological order

Blogs, short for weblogs, are websites where entries are recorded in reverse order, with the newest information appearing at the top of the page. Most blogs include daily or weekly entries about issues of importance to the author. Some take on the style and content of personal diaries and may not be useful for serious research. Others, however, are written by experts for public use and can contain analyses of current events and technologies, descriptions of business and industry trends, and discussions of political, legal, or social developments—which could be useful sources of research information. Blogs are generally easy to find: databases of blogs can guide readers to the blogs that suit them. For example, those interested in scientific advancements might visit http://www.scienceblog.com; those who want to research business trends might use http://www.iblogbusiness.com; and those looking for political and legal analysis might go to http://www.watchblog.com. One large directory of blogs can be found at http://www.technorati.com, which claims to archive 112 million blogs.

wiki a website that allows visitors to add, remove, and sometimes edit the available content

Wikis are web-based pages of information that visitors can edit in real time. They depend on numerous participants working under the premise that the contributors will add comments and information and build the wikis into bodies of knowledge. In this sense they are the opposites of blogs, which are generally written by just one person. Some wikis, such as the popular Wikipedia, contain information that has not been validated by experts and can be edited or modified by just about anyone (see Figure 6.5). These wikis can be useful sources of summary or background information, much like encyclopedias, but they should not be relied on as trustworthy

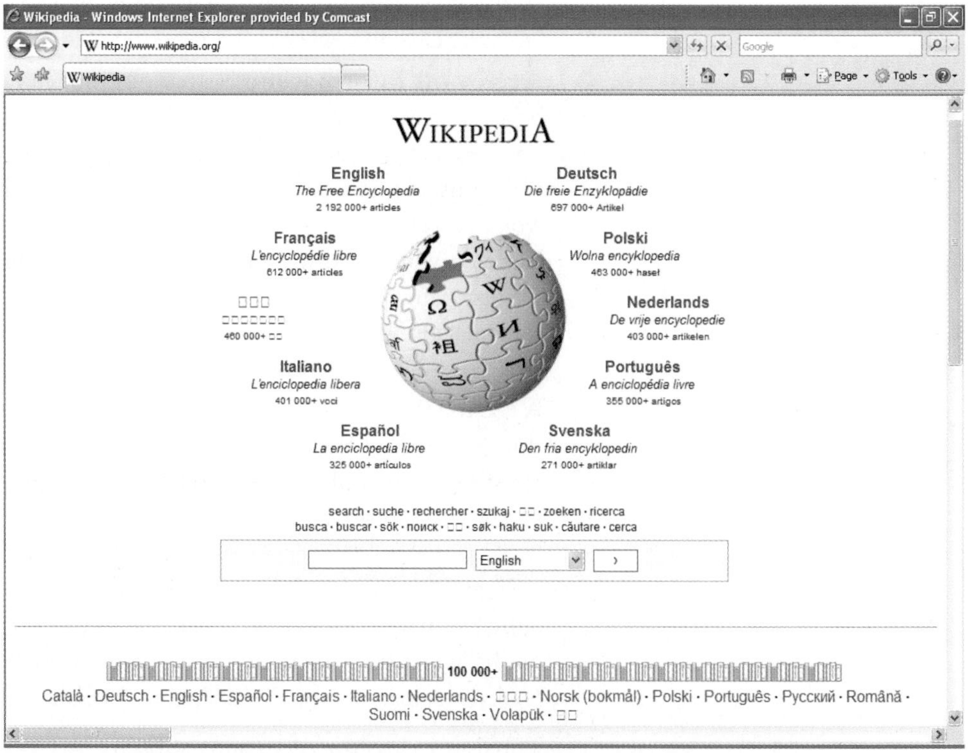

FIGURE 6.5 An example from Wikipedia, a web-based, collaborative encyclopedia

sources, nor should you expect to find great depth or specificity in the information they provide. Other wikis are moderated—that is, contributors are invited or supervised. Some of these are more sophisticated and contain information written or verified by experts; for example, http://www.openwetware.org provides a site for sharing current research in biological sciences and engineering.

The director of patient services in our example may review some student blogs to find out what patients are saying about their experiences in the Health Services Clinic; many universities provide students with space and software to develop their own blogs. She may also visit different wikis to learn about the history of health clinics or to get an extended definition of an unfamiliar term dealing with call distribution systems. However, she probably would not base any important decisions on information gathered through either blogs or wikis, since their accuracy is often in question.

ANALYZE THIS

Review the five proposed changes to the Health Services Clinic (page 136). Given the advantages and limitations of finding information on the Web, which of the five changes might best be researched on the Web? Which of the changes would be least likely to be successfully researched on the Web? Could any useful information about these changes be found on a blog or a wiki? Write a short analysis of how these changes could or could not be researched using information on the Web.

Databases

Databases are indexes or catalogs that contain information about published works, such as books, journals, newspapers, and government documents. Most databases are now available in electronic formats, are searchable, and lead users to specific and reliable sources of information. Many databases, in fact, do not just lead users to information about sources but also lead them to full-text sources online. Some databases, like those for the *New York Times* or the *Wall Street Journal*, contain archives of all or many of the articles published in those sources; other databases are geared toward broad subject areas, like engineering or medicine, and contain articles from a variety of sources (see Figure 6.6). Because many databases charge a subscription fee, most

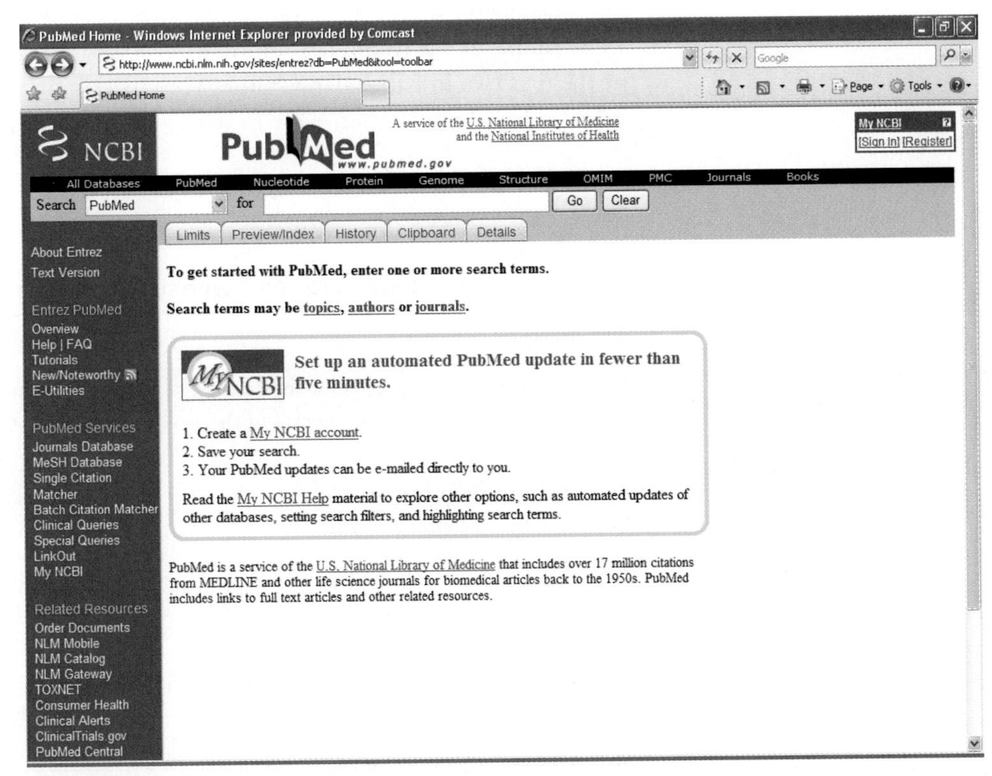

FIGURE 6.6 An example from PubMed, an important database of biomedical research

users access them in city, county, state, federal, or university libraries. You've probably accessed a database in your university's library. Many corporations, institutions, and other organizations subscribe to the databases most relevant to their needs.

Different databases cater to users looking for certain kinds of information. Listed here are examples of different disciplines and the databases that correspond to them:

Agriculture
Agricola
CAB Abstracts

Chemistry
SciFinder Scholar

Electrical Engineering
IEEE Xplore

Biology
Biological Abstracts
Knovol

Medicine
PubMed

Psychology
PsycInfo

Although the names of these databases may sound strange, writers and researchers quickly get used to them and use them efficiently and effectively. As a college student, you may already be familiar with certain databases commonly used for research in your major—JSTOR, EBSCO, or Lexis-Nexis, for example.

Even though those who use databases for research usually do so through a computer interface, it is important to note that online databases are distinct from websites. Table 6.1 outlines some of the important differences.

Most libraries provide an alphabetical list of databases and the kinds of subjects covered in each. However, not all libraries subscribe to all databases; because many databases are quite expensive, libraries have to pick and choose which databases best serve the needs of their users. The library at a small liberal arts college, for instance, might subscribe to many databases devoted to the humanities but not to databases devoted to agriculture or business if students at the college do not major in the latter two fields.

Databases would be helpful to the director of patient support services in our earlier example as she researches information about the five suggested changes. Let's look at the first two in particular. Change 1, to redesign the call-center area, would require information about the best ways to structure that physical space. The director has some

TABLE 6.1
A comparison of databases and websites used for research

Databases	Websites
Contain information about published works written by professionals and experts in their fields	Can be written by anyone, regardless of expertise
Provide all the information necessary for users to cite the sources they find	Do not always provide the information necessary to cite sources
Help users narrow their topics and find specific kinds of information in specialized sources	Are not organized to support specialized research activities
Always include the publication dates of sources and are typically updated frequently	Do not always provide the dates they were created or updated and are not always updated

ideas but is not herself an architect or designer; she needs help from experts in room design and layout for professional offices. Change 2, redefining the roles of those who answer calls, requires information about the tasks of particular employees. The director has a great deal of experience in how to run the HSC, but she needs to learn more about how best to delegate authority to individual staff members and manage the labor pool. Furthermore, she needs to research the most up-to-date kinds of technological systems that allow staff members to discuss medical issues with patients by phone.

The director finds that her university's library subscribes to many health and medicine databases. An initial search with keywords *nursing*, *technology*, and *telephone* leads her to hundreds of possible sources of information, including the following:

"Callers' Ability to Understand Advice Received from a Telephone Health-Line Service: Comparison of Self-Reported and Registered Data," *Health Services Research* 38, no. 2 (2003): 697–710.

"Harnessing Technology to Improve Patient Care," *Nursing Times* 100, no. 10 (2004): 26–27.

"Human Resources Issues in University Health Services," *Journal of American College Health* 50, no. 1 (2001): 43–47.

"Do You Give Phone Advice to Patients? Learn the Risks," *ED Nursing* 6, no. 10 (2003): 120–21.

"Nurses Help Members Navigate System: Callers Get Information on Conditions," *Case Management Advisor* 14, no. 6 (2003): 67–68.

"The Experience of Decision Making among Telephone Advice/Triage Nurses," University of San Diego Press, 1998.

Although these are only a few of the hundreds of possible sources on this subject, a quick glance at the titles of these articles and the sources of their publication—*Health Services Research*, *Nursing Times*, *ED Nursing*—shows that such information is not floating around freely on the Web. These sources can be researched through databases, some of which are available in full text, whereas others are not.

Advantages of Databases

1. **Trustworthiness:** Databases provide users with specific information about sources that are usually written by experts and are published in respected journals, books, and newspapers. Even though users still have to search diligently for the information they need, they usually do not have to worry about the credibility of the information they find.
2. **Frequent updates:** Most databases are updated at least once a year to make sure users are searching through some of the most current information. Not everything published will immediately appear in the proper database, but most databases include information within a year or two of initial publication.
3. **Full texts online:** The majority of databases now can be searched through a computer interface, usually in a library, and a good many databases connect users to the sources they seek through full-text versions. Thus, rather than gathering information and having to track down sources on library shelves, users are directly linked to an online version of the source, in either HTML or PDF formats. This tends to be a remarkably convenient way to do research.
4. **Amount of information:** The sheer volume of sources and information available through databases can be staggering. Databases open up sources of information that users would never be able to find elsewhere—a universe of information brought to the user in an orderly and indexed fashion.

5. **More relevant results:** Using a database is not all that different from using a web search engine: users type in keywords, titles, and/or authors' names in order to retrieve information. However, because databases do not sift through commercial websites, they provide more relevant and credible sources and fewer sales and advertisement sources. Someone looking in databases for information about computer chips will find specialized information about how they began, how they are designed and marketed and sold, how they have changed the world, and how they will be designed in the future. Someone who looks on the Web for information about computer chips might find corporate websites, books for sale, or a link to the web project about computer chips that was built by Mrs. Johnson's fourth-grade students in Cedar Rapids, Iowa.

Disadvantages of Databases

1. **Expensive access:** Databases are often cost prohibitive for individuals and for many companies. Even though most libraries—from small city and county libraries to larger city, state, and university libraries—now make databases accessible to almost anyone, traveling to a library is certainly less convenient than using a resource in the comfort of your office or home.
2. **Full-text inconsistency:** Many databases provide users with links to full texts of articles, chapters, and reviews, but other databases provide only a citation for a source. Researchers who become accustomed to full texts may neglect to search out other sources, which might contain new and desirable information. Thus, the convenience of databases that provide full texts can inhibit researchers from considering all sources of information, encouraging them to use only those sources that are readily available.
3. **Longer learning curve:** Those who use the Web for research quickly find a comfort zone with search engines and enjoy a one-stop-shopping approach to research. However, because there are so many databases, most of which have their own field of specialty, users often find that it takes more time to learn how to navigate through databases in an efficient manner. Users have to learn which databases are more likely to index the kinds of information they are seeking, which function better with keyword or subject searches, and which provide the most comprehensive amounts of information. In addition, most researchers want to learn which databases are more likely to contain full-text documents.

EXPLORE

Go to your school's library homepage, and find the links that take you to the databases. Note who is allowed access to the databases and who is not. How can you, as a student, access them? Once you enter the databases, explore them and take notes about what you discover. Is it easy to navigate the databases? Why or why not? Which databases did you browse? Which databases were you most comfortable with, and why?

Intranets and Archives

intranet a private computer network used within a company, university, or organization

Intra means "internal" or "within," so an **intranet** is actually an internal or private Internet. It is used strictly within the confines of a company, university, or organization, for example, by individuals employed by or affiliated with that entity. An intranet is an online environment—a series of web pages usually accessible through the World Wide Web—that is closed off to the general public, available for use only by those with permission to access it. For instance, a company, organization, or institution might wish to allow certain users the convenience of accessing vital information through the Web but might not want everyone to have access to this information. Thus, intranets can often be accessed only by those with clearance to do so, usually through a username and password. For researchers a company intranet can be highly valuable because they might find records, documents, and transcripts of information vital to their current research projects. In some cases an intranet contains current documents and information as well as archived documents.

Let's review three of the changes our director of patient services wants to make at the Health Services Clinic:

- Redefine the roles and expectations of staff members who answer calls.
- Increase staff by three to meet the volume of calls.
- Reassign staff to meet the projected volume of calls coming in at peak call times.

To research these three potential changes, it would be helpful to look at records and documents from the past: What were the previous roles and expectations of staff members? How many staff members were there? How many calls were they taking during an average workday? Which staff members most efficiently and effectively dealt with the calls to the Health Services Clinic? The director would probably have access to such information in the center's employee archives, which might or might not be available through an intranet interface. In particular, the director would want to research staff members' work experiences, their specific duties, their particular areas of expertise, their work hours, the number of calls they handled, and the way the calls were handled (i.e., recommendations made). Looking through such records would help the director determine when and how the clinic was run effectively and when and how it was running ineffectively. Such information would help her make decisions about the future direction of the Health Services Clinic.

More and more, companies, institutions, and organizations keep such records online, accessible to those with the proper clearance. Whereas twenty years ago the director would have had to sift through piles of papers, forms, and folders, such information would likely now be available in a more ordered online format, one that is easier to search. Those who research and write in workplaces often need to consult these sorts of archived information as they create memos, reports, proposals, and other documents.

Advantages of Intranets

1. **Private information:** Users with access to intranets are often privy to information that is unavailable elsewhere, valuable information that they might not otherwise have, even though users are often able to use this information only for internal documents.
2. **Ease of use:** Companies, organizations, and institutions that put files, records, and other documents on a secure intranet allow users an easier approach to research. Researchers can access most of these items through a computer interface, allowing research to be accomplished more quickly and efficiently. Such information is easy to download, add to documents, and transport elsewhere.

Disadvantages of Intranets

1. **Restricted information:** If private information is an advantage in some situations, it is a disadvantage in others. Many intranets do not make it easy

for users—even company employees—to access information or to use it once it is accessed. In other words, there may be many hoops to jump through in order to gain access to information on an intranet.

2. **A difficult search:** An intranet might be accessed through a computer interface and might look like a website, but that does not mean that it functions like a website or that it can be searched like the World Wide Web. An intranet may have a menu or an index providing information about available information, but new users might not know which information is found in which places. And because an intranet may not have a search engine, users may be forced to take a trial-and-error approach to finding the information they need.
3. **Incomplete records:** Transferring information to either the Internet or an intranet is a time-consuming process. Imagine a large company that wishes to upload all of its crucial files from the last twenty years onto an intranet. Scanning all of these documents into the intranet could take an enormous amount of time. And then the intranet should be constantly maintained and updated, something that may not always happen. As a result, users must be aware that not every document they need will be available.

IN YOUR EXPERIENCE

Recall the last time you conducted research at school or work and found a source that you did not trust. Write a short document explaining what this source was, how you evaluated it, and why you decided that it was not trustworthy.

Visuals

Along with textual research, you may also need to find visuals to add to your documents. Like text references, visuals and their sources can add credibility and professionalism to your own ideas. Because there are many ways to find visuals, finding an appropriate graphic or image should be relatively easy.

Before conducting your research, you should first determine what you need. There are millions of visuals available in a variety of sources, and you could spend all day searching for one image if you don't know what you want beforehand. You should decide whether you need a photograph, an illustration, a diagram, or some other type of visual. You should also consider what you want to accomplish with the visual, since different types of visuals convey different ideas. Do you want an artistic visual? a technical visual? Will you need a high-resolution visual? Do you want color or black and white? Questions like these will help you to narrow your search to find the best visual in the shortest amount of time. In addition, you'll want to consider the appropriate search terms for the visual you need; you should use specific terms, categories, types, and styles whenever possible. For example, conducting a search for "airport" will give you lots of different sources, but researching "airport security photograph" will provide more specific results (see Figure 6.7).

After you've determined what you need, you'll begin your search for a visual. Some visuals are available for free, although the most professional visuals with the best quality often require a fee. You'll need to consider the importance of the document as well as the medium. You probably wouldn't want to pay for a visual for a short,

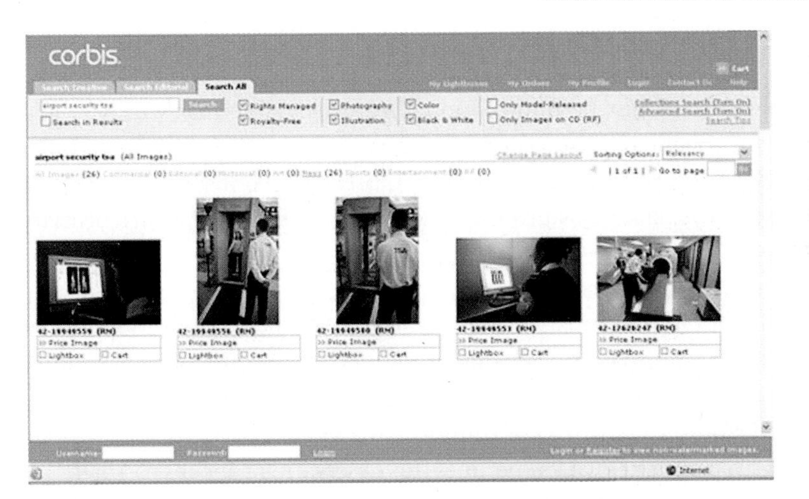

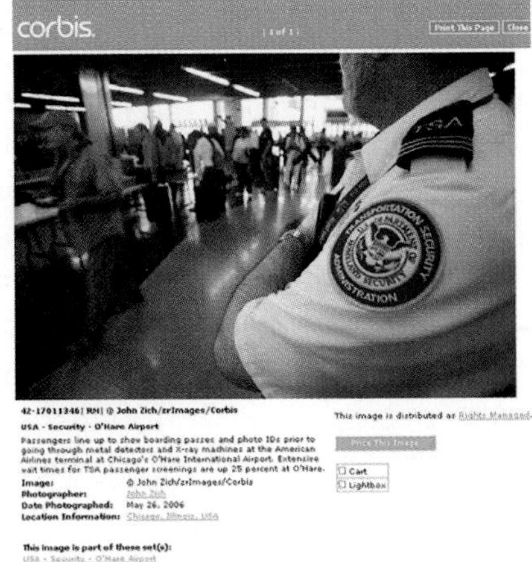

FIGURE 6.7 Commercial databases provide high-resolution visuals for a fee

internal memo to your peers. On the other hand, if you were preparing a PowerPoint presentation for a large group of potential investors, you might use a high-resolution, commercial visual for clarity and professionalism. The following electronic sources are worth considering:

1. **Licensed databases:** Most universities and some businesses subscribe to databases of visuals. Your university library probably subscribes to a number of visual databases you could use for research. Many of these are collections of particular types of images or graphics, allowing you to search according to a specific discipline or field. For example, an art historian might search for an early watercolor at Yale University's Manuscripts and Archives Digital Image Database; an architect might browse sketches of contemporary skyscrapers at the Architectural Archives of the University of Pennsylvania; or a medical researcher might look for a diagram of the inner ear at the Health on the Net Database.
2. **Free web resources:** Many good visuals can be found for free online. The number of these visuals increases every day, but you should consider the quality and resolution of the visual before you incorporate it into a document. Google's Image Search contains more than a billion images; other sources like AltaVista Images and Freefoto.com are also worth examining. Keep in mind that some of these images may be copyright protected, where as others may not be. Most of these resources allow you to search by file, size, color, and other parameters.
3. **Commercial databases:** If you want high-resolution, professionally rendered visuals, you often have to pay for them. Corbis is a large database of visuals, most of which have been created by professional photographers and graphic designers. Getty Images contains creative and editorial images as well as film footage and music. Both of these services charge a fee for materials and allow you to purchase visuals for any number of purposes. Many of the visuals in commercial databases are watermarked until you've paid for their use.

After you've found an appropriate image through one of these electronic sources, you can download it or cut-and-paste it into your document. If you want higher resolution, it is best to download the visual and insert it into your document, since cut-and-pasted images can often lose resolution. Be sure to cite the source of your visual, just as you would when referencing a written work.

CONDUCTING PRIMARY RESEARCH

Sometimes there is just no information already available that will help a writer solve the problem at hand or create the necessary documents. Thus, writers sometimes need to conduct primary research themselves. Observations, inspections, experiments, interviews, surveys, and focus groups are common methods of doing primary research.

Observations

One important type of primary research is the simple act of observation. Workplace writers often observe a situation, location, or process in order to understand and evaluate it. Observations can be the most exciting kind of research, since you are dealing with real people and places, can experience things firsthand, and can come up with new and useful information on your own. Sometimes this observation happens naturally and may even be the cause of the research that follows. For example, our director of patient support may simply *observe* long lines in the Health Services Clinic, which motivates her to conduct further research about the causes of and solutions for the lines. Observations can lead to other forms of primary research; our director may decide to interview a group of patients after observing them waiting in line at the clinic. At other times observations may take place after a problem is identified, when a researcher visits a location to consider the problem firsthand. Either way, observations involve making careful assessments of who is involved and what, when, where, and why something is happening.

Inspections

Inspections are similar to observations in that you visit locations firsthand and use your senses to gather information. However, inspections are generally more active; conducting an inspection often requires the researcher to operate devices, test equipment, or manipulate the environment to collect data. Many inspections are carried out by experts—inspectors—who have specialized training or expertise in a particular area. Inspectors use their experience and professional judgment to determine whether a location, object, group, or document is safe, correct, or effective. Some inspections are scheduled at regular intervals whereas others are done randomly. Those done regularly might assess whether equipment is operating correctly or needs maintenance; random inspections might assess whether employees are doing their jobs effectively or appropriately. Inspections can sometimes be quite complicated and can result in further types of primary research or in sophisticated documents such as inspection reports.

For more information on inspection reports, see chapter 21.

As noted in our introductory example, the Transportation Security Administration conducts regular inspections of baggage screening equipment in airports. TSA officers, who have experience and training with the equipment, examine and test it carefully to make sure it is operating as it should and then write inspection reports identifying the effectiveness or ineffectiveness of the equipment in detecting dangerous objects in passenger luggage (see Figure 6.8).

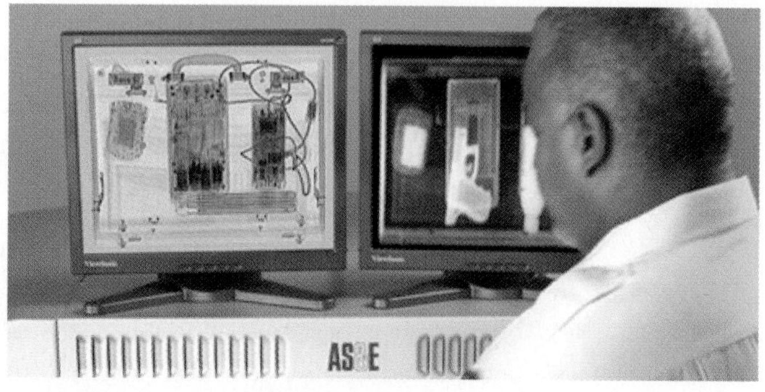

FIGURE 6.8 Inspection of baggage screening equipment at an airport

Because observations and inspections are similar, they are often carried out in similar fashion. Here are some guidelines to follow in both observations and inspections:

- *Conduct some background research before you observe or inspect.* Understanding the situation will allow you to focus your research on what is most important.
- *Observe or inspect what is typical or normal* if you want to know how something normally operates or functions. In these cases conduct your research on something that represents the norm, whether it is a typical day, piece of equipment, or employee. Our patient support director would probably want to observe health clinic waiting lines on a typical day in the semester, rather than on a weekend or during the summer break.
- *Observe or inspect what is atypical or unusual* to test the limits or shortcomings of something. In these cases it may be best to look at extremes. For example, TSA officers might inspect the oldest baggage screening equipment or observe screeners with the least training to determine weaknesses in airport security.
- *Consider working with others.* If your observation or inspection addresses something big or complex, ask others to help you. It might be useful to assign your roles according to experience or expertise.
- *Confirm availability.* Check on hours and days of operation if you are visiting a field location, to be sure it is open when you plan to visit. If you hope to observe or interview individuals, make sure they will be there then—not on vacation or out sick.
- *Take careful notes.* Observation and inspection rely on keeping track of what you see or encounter, since you may need to write about or discuss the details at a later time. Note taking is covered in more detail later in this chapter.
- *Have all necessary equipment with you.* Observation and inspection often require the use of specialized tools to conduct research; make sure you have everything with you to avoid an unnecessary second visit. Also, consider taking backup or related equipment—extra batteries, film, test kits, or other supplies that may save you a repeat visit.

Experiments

Experiments allow us to test possible solutions before implementing more permanent changes. We typically think of experiments as activities that take place in a regulated space with specific steps and equipment, and that is the case with many scientific or medical experiments. However, some experiments take place in the field, where researchers try new things to see what will happen. Strapping a fake bomb to a TSA agent is a type of field experiment done to test airport security. Some experiments require a great deal of training or expertise in order for someone to conduct them accurately and safely; if you are involved in an experiment that could put you or someone else in danger, you must be sure to follow all protocols and safety procedures carefully. Other experiments can be simple and easy to conduct, such as adding a second receptionist to a health clinic to see if that reduces wait time for patients. The complexity of the experiment depends on the type of information to be gathered.

Most experiments follow a basic procedure:

1. *Determine what you want to know.* Start with a basic question or series of questions you want to answer, and then figure out how to proceed. Consider which steps or activities will answer your questions.
2. *Conduct your experiment.* Make sure that you follow all of the steps you've outlined, and never assume an experiment will have certain outcomes. The purpose of an experiment is to confirm *or* to contradict.
3. *Analyze the information.* Spend time studying the information you obtain through your experiment, looking for outcomes, trends, and new questions to consider.

4. *Discuss or report the information.* In most cases the information you gather through experiments will affect others. Present your information and findings in an appropriate manner, whether that is a report, memo, website, or some other document.

Interviews

Workplace writers also conduct research through interviews with people who have a certain expertise or knowledge relevant to the problem at hand. Interviewing people for information may seem like something that happens only on news or celebrity shows, but it is actually a common way for workplace writers to get detailed information from the people who are involved with or affected by the subject being researched. Interviews can be important in various aspects of employment, too: you might conduct informal interviews to find out more about jobs in your field, and a hiring committee or supervisor will certainly interview you when you apply for such a job. A good way to prepare for an interview is to consider who, what, when, where, and how.

- *Who?* Researchers who conduct interviews must be certain that they have chosen an interviewee with relevant information about the subject matter. Some initial research about the interviewee should verify the person's authority and credentials. In many cases writers interview individuals whom they have read about in books or journals; in other cases they may interview individuals known from work, home, or the local community. The type of research project and writing genre, the purpose, and the audience all help determine which people should be interviewed.
- *What?* After identifying a good interviewee, the researcher needs to decide what questions to ask. Questions should be short and direct and should refer specifically to the interviewee's areas of expertise and the problem being addressed in the particular research project.
- *When, where, and how?* These cues refer to the logistics of setting up an interview. The time and location depend in large part on the schedule of the interviewee. Someone who works down the hall is likely going to be easier to schedule than a scientist who lives in Belgium. Be aware, however, that interviews do not have to be conducted in person; they can be done on the phone, through e-mail, and, increasingly, through interactive videoconferencing.

Surveys

Like interviews, surveys begin by defining what you want to know and deciding which people have that information. Rather than talking with one person face to face, however, surveys allow you to ask numerous individuals a series of questions about a particular subject. In addition, because surveys usually ask closed-ended questions—typically multiple-choice questions—those who conduct surveys typically see themselves as collecting data, that is, quantitative information, rather than the qualitative information derived from interviews. **Quantitative information** can be tabulated or measured, whereas **qualitative information** reveals subjective reasons, motivations, or experiences.

quantitative research research that relies on statistical or numerical data

qualitative research research that relies on the impressions or opinions of a group or individual

There are many ways to design surveys, a task often performed by experts with experience in statistics and survey design. However, the typical workplace researcher and writer can conduct surveys on an informal level. Here are some guidelines to follow:

- Keep the survey short, about six to ten questions.
- Ask one question at a time.

- Avoid biased or loaded questions, such as "Do you believe it is wrong to execute criminals in harsh, dehumanizing ways such as lethal injection?"
- Ask clear, fair questions. Avoid any questions that could be easily misinterpreted or considered offensive.
- Ensure that your participants understand how to respond. If you use numerical ratings, make sure participants know what is best and worst.
- Use round numbers whenever possible. It will be easier to quantify your data if you survey one hundred people rather than eighty-seven.

Surveys, like interviews, can be conducted in many different ways. We typically think of surveys as printed forms filled out by users, but surveys can be conducted by phone, in person, and, increasingly, online. Large-scale surveys that try to reach a broad cross-section of society are typically handled by companies that specialize in such surveys—surveys about presidential candidates, new products, or political issues, for example. A small-scale survey can be conducted by an individual or a small team of workers.

Our director of patient services at the Health Services Clinic would benefit from the direct feedback of her employees, as well as that of the students who visit the clinic. Remember that the director believes that (a) the call-center area needs to be remodeled, (b) employees' roles need to be redefined, (c) the automatic call distribution system (ACDS) needs to be reconfigured, (d) the number of staff members needs to be increased, and (e) staff members need to be reassigned. To discover how to implement these changes, the director needs to get information directly from those at the forefront—the students who use the HSC and the staff who interact with them. Interviews and surveys would both be appropriate.

Interviewing student clients and staff members could be accomplished simply by sitting down with them, asking them directly about their experiences, and soliciting suggestions for implementing the changes. Interviews could bring forth a wealth of helpful information. However, interviews are time-consuming and somewhat difficult to schedule. Therefore, the director might prefer to develop two short surveys: one for students and one for employees. A survey could be completed at the students' or employees' convenience, wouldn't require the director to be present, and could also provide her with a wealth of data. In addition, because most surveys are done anonymously, the director is likely to get more honest results from a survey than from a face-to-face interview; there would be no perceived repercussions for saying something negative. Figure 6.9 is an example of a helpful employee survey.

The director might wish to conduct further primary research—interviewing employees and students, developing additional surveys, or speaking with and interviewing other experts in the field. But whatever method(s) she chooses, her

ANALYZE THIS

You have read a great deal about the director of patient support services and the problems she wishes to research for her report and proposal. Given the details of her situation, what might a survey of student clients of the HSC look like? What kinds of questions should the survey contain? What would be the benefits of such a survey? What would be the drawbacks? What other people might the director wish to interview as she researches her problems and proposed changes? Write a short analysis of these issues, and then create a one-page student survey she might use to gather information.

Employee Survey

This survey contains six questions about your position at the Health Services Clinic. The results of the survey will be incorporated into a strategic report that outlines future changes to be made to the center for the benefit of both student users and employees. Please answer as honestly and accurately as possible. Thanks for your time and energy.

General Information:

Position title: ____________________

Job responsibilities: ____________________

Number of years employed: ____________________

Questions: (Circle the letter that best applies.)

1. How well defined are the responsibilities and requirements of your job?
   a. Extremely well defined
   b. Well defined
   c. Somewhat well defined
   d. Not well defined
   e. Not at all defined
2. Overall, how would you rate your satisfaction with your job at this time?
   a. Extremely satisfied
   b. Very satisfied
   c. Somewhat satisfied
   d. Not very satisfied
   e. Not at all satisfied
3. How satisfied are you with your job supervisor?
   a. Extremely satisfied
   b. Very satisfied
   c. Somewhat satisfied
   d. Not very satisfied
   e. Not at all satisfied
4. How satisfied are you with your physical work environment?
   a. Extremely satisfied
   b. Very satisfied
   c. Somewhat satisfied
   d. Not very satisfied
   e. Not at all satisfied
5. How satisfied are you with the quality of your work equipment?
   a. Extremely satisfied
   b. Very satisfied
   c. Somewhat satisfied
   d. Not very satisfied
   e. Not at all satisfied
6. What aspects of your position hinder your job performance and ability to serve our student clients?

FIGURE 6.9 An example of an employee survey

situation—like that of many workers—requires primary research. So often the information we need comes from those closest to us and cannot be found in print or electronic media forms.

Focus Groups

Simply put, focus groups are group interviews, usually conducted with six to fifteen participants and overseen by a moderator. Focus groups are designed to generate conversation within the group about particular subjects in order to gain a general sense of participant response. Focus group research is used extensively in market research, particularly research for new products, but can be applicable to other workplace scenarios as well. Focus groups can provide information about customer satisfaction, consumer problems, potential problem solutions, product failures and successes, packaging issues, or any number of other concerns. Focus groups provide qualitative information and can be a useful tool for evaluating services, testing new ideas, and getting consumer feedback. However, for focus groups to provide useful information, the moderators of the groups need to do some careful planning, facilitating, and following up.

Preparing for a Focus Group

Here are some guidelines for preparing for a focus group session:

1. ***Know exactly what you want to accomplish during the focus group session.*** Identify the primary objective of the session; that is, have a goal in mind, and design the session around gathering the information you most want to have.
2. ***Develop a series of questions for the group.*** These questions should both guide the conversation and elicit the information you want. Because most focus groups last only an hour to an hour and a half, you should limit your questions to about six; you can ask about that many questions during a one-hour session without rushing participants in their answers. In a longer session, participants may lose interest, become tired or bored, or feel they are repeating themselves. All questions should specifically work toward your purpose, and no questions should be used that could be interpreted as aggressive, misleading, entrapping, or otherwise unethical.
3. ***Select participants carefully.*** As we mentioned, focus groups are usually comprised of six to fifteen participants plus a moderator. Generally speaking, the participants should share something in common in relation to the subject of the focus group. For instance, if the director of patient support services decides to hold a focus group about student perceptions, she might select students who have filed complaints or who have failed to receive prompt help on more than one occasion. It is also a good idea to select participants who want to participate. Those who seem resistant or hesitant may not provide feedback as freely.
4. ***Have a detailed plan for the session.***
   - *Schedule:* Be sure the session is scheduled to take place long before you'll need to report the data you gather. Consider the time of day: some participants may have difficulty focusing early in the morning, and others may be tired immediately following lunch or in the late afternoon. Participants tend to be most alert during mid-morning sessions.
   - *Setting:* Secure a location in which participants will be comfortable and free from distractions. Configure the room in such a way that all participants can see each other; use a conference table or place chairs in a circle. Eye-to-eye contact encourages participants to join the conversation. Also consider having refreshments available.
   - *Ground rules:* Because you have only limited time to gather the information you need, consider setting some ground rules to keep everyone on task. You might establish limits for how long participants can respond to a given question, in order to ensure that all participants get to voice their responses. You might also develop a system of etiquette to shape how participants respond to another participant's comments. In addition, you might define a set order in which participants respond, again ensuring that all participants actually participate.

- *Agenda:* Be sure to have a set agenda for the session; without a set agenda you are less likely to stay focused on the objective.

5. ***Send a formal follow-up letter of invitation to all participants.*** Thank them for their willingness to participate; provide a proposed agenda; identify the date, time, and location of the session; and explain that you will provide them with a copy of your findings after completion of the project.
6. ***Plan how you will record the session.*** You cannot rely on remembering all that is said during the session. You may decide to take notes, involve a cofacilitator to act as a recording secretary, or record the session on audio- or videotape. If you choose to tape, you will need to secure written permission from each participant. And if you are using recording equipment, be sure that it works, that it is already set up in the room before the session, and that you have sufficient tape for the session.

Facilitating a Focus Group

In order to collect useful information, the moderator must carefully facilitate the session. Here are some suggestions for doing so:

1. ***Introduce yourself and any cofacilitator who might be present.*** Be sure to give not only your name(s), but also the position(s) you hold in the organization conducting the focus group. In that way you establish authority within the group and let participants understand who wants the information they will be providing.
2. ***Explain thoroughly the purpose of the session.*** Your participants should know why they are there, what you are looking for, and what will be done with the information they provide. Think of this segment as you would any good piece of writing: make sure your audience understands the purpose.
3. ***Explain how the session will be recorded and how the data will be processed.*** Participants will be interested in the process, and it is important that they understand how their voices will be recorded.
4. ***Stick to the agenda.*** Departure from the agenda can keep you from getting the information you need and may convey a sense of disorganization to your participants. Remember that the participants will leave the focus group with a particular perception of you and the organization you represent; maintaining an agenda adds to your professionalism.
5. ***Stick to the questions you have planned.*** Be sure to give each participant time to think about the question before responding. In some instances you may want to provide participants with paper and pen or pencil so they can write notes as others speak and thus remember what they want to say.
6. ***Ensure that each participant has an equal opportunity to respond.*** Encourage everyone to speak, and specifically call on those participants who are either shying away from the conversation or being silenced by more dominant participants. Steer the conversation toward those who are participating less. One good strategy for balancing participation is to keep a time limit on responses.
7. ***Review responses to be sure that what you heard was what participants intended.*** After all participants have responded to a given question, you do not have to repeat specific responses but might provide a summary of the group's general and combined responses. Identify particular

discrepancies or highlighted points that may have developed during the conversation.

8. ***Thank participants for their time and energy.*** Remind them that you will send them a copy of the results or findings of the focus group, and identify when they should expect that document.

Following Up after a Focus Group

Immediately after participants have left the focus group setting, confirm that your recording system actually recorded the session. Take a few minutes to write any notes to yourself that you'll need later to remember what transpired during the session. You may want to clarify notes you took hastily or explain details that won't be clear from the recorded transcription—for example, consultation between participants or subtle factors such as anger or nervousness. You may also want to note any unexpected events that may have occurred or any important information that was brought up.

NOTE TAKING

Research requires careful note taking to produce problem-solving documents.

When workplace writers conduct research, they may be doing many quite distinct activities, all of which are conveniently included in the phrase "conducting research." Despite their differences, however, these activities all generally require researchers to take notes about what they find, in order to produce the documents necessary to solve their particular problems. Taking notes is an important step in making sure that workplace writers document their sources accurately and correctly. In addition, those who take notes in certain ways can actually see themselves as beginning to create drafts of their documents.

IN YOUR EXPERIENCE

How do you take notes as a student? What are your methods and procedures? Do you find it easy to review your notes after you have taken them? What strengths and flaws do you recognize in your current note-taking ability? Do you use different note-taking strategies for different courses (e.g., English, math, science, history)? If so, why do you think you take notes differently for these different courses? Write a short document addressing these questions.

Strategies for Note Taking

There is no one right way for researchers in the workplace to take notes. Most people develop their own means of note taking through years of experience, a means that best suits their own writing habits. You may already have developed an organized and effective system of taking notes in research. Nonetheless, there are a few basic concepts that all workplace writers should consider:

- *Save everything.* Once you have found sources that look important, be sure to save them in one way or another. This might mean photocopying, printing them out, or bookmarking them on an Internet browser. It is also beneficial to develop some kind of folder or filing system to organize your sources.

- *Read sources carefully.* When you find sources that look beneficial to your project, be sure to read them carefully. Because it's difficult to take notes from sources that you do not fully understand, taking the time to read carefully will ensure that you get the most out of your sources.
- *Forget the highlighter.* Many have grown accustomed to reading and highlighting valuable information. A highlighter marks relevant passages but does not allow you to actually write notes. Reading with a pen in hand allows you to write notes while you read, rather than having to return later and reread highlighted passages.

Researchers need to develop forms of note taking that allow them to produce documents as effectively and efficiently as possible. Your project may require books, journal articles, newspaper articles, surveys, and interview transcripts—that is a lot of information. Therefore, it is important to take notes in a way that helps you codify all of your source information. We discuss here a few possible strategies, all of which require, not surprisingly, some form of *notebook.*

The Research Narrative

This form of note taking is like writing a diary entry about your research findings. It is an informal way to jot down summaries, opinions, and criticisms of sources. You might also outline how you plan to use your sources—all in a narrative format. In some ways a research narrative is similar to writing a letter to yourself about sources of information. Here is an excerpt from a research narrative written by our director of patient support services, referring to the journal article "Human Resources Issues in University Health Services," which she found in a nursing database:

> Philip Meilmanís article "Human Resources Issues in University Health Services" discusses ways to provide first-rate services to students, arguing that college health services need the best possible staff. Managers and supervisors play a critical role in guiding the work of their employees to enhance performance. Reference checks for new employees and regular performance-appraisal dialogues for ongoing employees are important tools in this process. Meilman discusses these issues and suggests formats for reference checks and performance appraisals. He even provides an excellent chart for managers and supervisors to use in their performance appraisals of staff members.
>
> As I put together my report and proposal for suggested changes at the Health Services Clinic, this article will come in handy. In particular, I will use Meilman's section on "Toxic Behavior" (pp. 44–45) to consider the working relationship of current staff members at the HSC. I really like this: "Of course it is important to select staff carefully whenever openings occur. Individual competence is a natural prerequisite for satisfactory job performance; in some ways, however, it is not the most important factor. By the time a prospective employee has made it through the screening process, review of credentials, and intial reference checks, competence is rarely an issue with candidates who are serious contenders for a position. It is important, therefore, to think about personality traits and work habits. In-depth reference checks may be helpful here."

A research narrative is a means of writing notes in a casual, informal, and fluid way—almost like freewriting. The main benefit of a research narrative is that writers can concentrate on their thoughts and ideas rather than on the form and structure of their notes, providing a sense of freedom. The drawback to this approach, however, is that it is somewhat difficult for researchers to later return to their notes and easily mine them for relevant information. That is, the research narrative does not lend itself to rigid order and organization.

Double-Entry Journal

The double-entry journal uses a more formal structure than the research narrative. Most researchers divide pages in their notebooks with a central, vertical line that runs from the top of the page to the bottom, dividing the notebook page in half. On the left side of the page, researchers note passages from their sources that they believe will be important to their writing projects. These are quotes, summaries, and paraphrases—the author's words and ideas. The right-hand side of the page is reserved for the comments, criticisms, and ideas of the researcher, responding in whatever way is deemed necessary.

There is a wide range of questions that researchers can address during their note taking. Of course, a researcher should address only those questions that are relevant to the given purpose and audience. If the director of patient support services had used a double-entry journal with the article "Human Resources Issues in University Health Services," it might have looked like this:

Notes from "Human Resources Issues"	Responses to "Human Resources Issues"
P 43—Employee relations and human resources issues—including morale, staff attitudes, and work behavior—must surely affect the quality of care provided in health centers.	This is central to the problem we have at HSC, that human resources issues, particularly our staff members' training, are affecting how we deal with our patients in the clinic and on the phone. I wonder if Meilman has had any specific experiences dealing with automated call distribution systems.
P 43—Problematic are staff members who have idiosyncratic ways of looking at patients making ethical decisions, interpreting interpersonal interactions, and focusing not on their rights as employees but on their responsibilities.	This is key! As we reevaluate ways to define the roles and responsibilities of staff members, we'll have to provide training that better standardizes how staff members approach patient needs. This will need to be a separate section in the report.
P 44—When employees are preoccupied with intrastaff problems or adverse interactions within their departments, they are unable to give their full attention to the students sitting in front of them. They may miss a keyword or sentence or miss some clues that would help them diagnose and treat. They may also miss the opportunity to develop the kind of rapport that makes a student feel truly attended to.	I wonder what kinds of intrastaff problems in particular Meilman is referring to here. There are many different kinds of problems that fall in this category. Meilman's point here is a good one, but it is also dealt with a bit too simplistically. I'd like to see some more specific examples to see if they are similar to what we are experiencing at HSC.

The double-entry journal is a more time-consuming approach to note taking than the research narrative. However, the double-entry journal allows researchers and writers an easier time of sifting through their notes because the notes are more organized; responses are tied directly to summaries of sources in a clear-cut manner. Nonetheless, some individuals feel constrained by the format of a double-entry journal and prefer a less rigid approach to note taking.

Problem-Solution Log

This form of note taking is more straightforward than the research narrative and the double-entry journal. It simply asks that researchers list the problem that they wish to solve in a clear statement or that they phrase their problem in the form of a question. If they are working to solve more than one problem, they should present all problem statements or questions in a numbered list. Then as the researchers read and evaluate their sources, they jot down beneath the appropriate problem statement or question the information that helps them solve that problem. Depending on how much research is needed, a problem-solution log could go on for many pages; each problem could reference many different sources of information. This approach helps researchers keep an eye on the problem(s) that make their research necessary in the first place. If the director of patient support services had chosen this method of note taking, a portion of her problem-solution log might have looked like this:

1. How do we remodel the call-center area where staff members answer phones?
   After interviewing staff members who work in the call-center area and Tim Smith, the architect who specializes in health-care facilities, I think the best thing to do is to design an area with more space for each staff member, as well as small offices in the call-center area for each staff member. This allows more privacy and less background noise for those taking calls.
2. How should we redefine the roles and expectations of staff members who answer calls?
   Philip Meilman's essay "Human Resources Issues in University Health Services" suggests that staff members should be assigned very specific and well-defined roles. On page 45 he suggests that staff members' roles should be reappraised every six months, giving me the ability to keep up with any changes taking place.
3. Should we increase the staff by three to meet the volume of calls?
   While browsing through the HSC records for the last two years, I noticed a large increase in the number of student users (an average increase of 19 percent each year), particularly those who came in for minor health issues that could have been addressed on the phone. This seems to necessitate more staff members to answer calls. My survey of employees also suggests addressing the needs of the students.

Researchers can accomplish any of these note-taking strategies with pen and paper or with a word processor or portable electronic device. As you become more experienced with research note taking, you may develop a note-taking method that combines several approaches, or you may devise a new method altogether that fits your own style.

PSA

Note Taking as Drafting

Experienced researchers do not see themselves as simply taking notes; they recognize that they are also beginning to draft the documents they wish to write, part of the Drafting portion of the PSA. In the three different note-taking strategies, workplace writers are also responding to, critiquing, and figuring out how to use their sources in the documents they will create. For instance, in the double-entry journal example, the second entry summarizes some information but also notes that this information should be treated in a separate section of the report. Thus, as the director was taking notes, she was also thinking about how her report would be structured. Remember what the PSA makes clear: writing does not always work in a straightforward, linear fashion.

See chapter 7 for additional information about drafting and organizing documents.

IN YOUR **EXPERIENCE**

In a paragraph or two, explain how you currently take notes in your college classes and how you use these notes for studying and writing. Explain what purposes these notes serve. Then compare your note-taking experiences with the three kinds of note-taking methods described here. What similarities and differences do you see between your current note-taking style and these new approaches?

ETHICAL CONCERNS

Documentation shows that you have given credit to your sources.

Documentation

As a college student, you have probably had some experiences with research writing and, along with that, some experiences with documenting sources. You may have been shown a particular format and told to document your sources in that way. However, proper documentation is not just a matter of following a format carefully; it is an important ethical concern. Documenting sources properly shows that you have used them in an ethical manner—you have given credit to others for their ideas, and you have made clear to your readers how you have used sources in your document.

Different styles or systems of documentation and formatting have been developed for particular disciplines or areas of study by their respective governing associations. These different styles are outlined in various manuals, which explain how to cite a source in the body of a text, create a works-cited page or bibliography, use endnotes and footnotes, write an abstract, and use particular forms of punctuation, spelling, and sentence structure. The primary style manuals are listed here with their specific disciplines; you should become familiar with the style manual related to your chosen field of study.

Biology—*CSE Style Manual: A Guide for Authors, Editors, and Publishers in the Biological Sciences*

Chemistry—*The ACS Style Guide: A Manual for Authors and Editors*

Engineering—*Information for IEEE Transactions and Journal Authors*

English—*MLA Style Manual and Guide to Scholarly Publishing*

Geology—*Geowriting: A Guide to Writing, Editing, and Printing in Earth Science*

Government—*The Complete Guide to Citing Government Information Resources: A Manual for Writers and Librarians*

Information sciences and computer science—*Scientific and Technical Reports: Organization, Preparation, and Production*

Law and legal studies—*The Bluebook: A Uniform System of Citation*

Management—*The AMA Style Guide for Business Writing*

Mathematics—*A Manual for Authors of Mathematical Papers*

Medicine—*AMA Manual of Style*

Physics—*Style Manual for Guidelines in the Preparation of Papers*

Psychology and other social sciences—*Manual of the American Psychological Association (APA)*

Various arts and sciences—*The Chicago Manual of Style*

For information regarding the use of CSE, MLA, and APA styles, see appendix B.

Other disciplines in the humanities, sciences, and social sciences have their own style manuals, which you can consult as needed. Your school library website should provide further information about them.

As you write more in upper-level courses and in the workplace, you will become accustomed to the particular documentation style and guidelines that best suit your writing and audience needs. Using the proper style for your particular field is important because those who do not use the proper guidelines are likely to lose credibility with their audiences, who may view the writers as ignorant of accepted conventions and perhaps as unethical.

Plagiarism

plagiarism the act of passing off as one's own the ideas or writings of someone else

There is one main ethical strand that runs through all style manuals: *avoid plagiarism at all costs*. **Plagiarism** is defined simply as the act of passing off as one's own the ideas or writings of someone else. Writing in college and in the workplace is expected to be the result of the writer's own ideas and/or research. Writers are guilty of plagiarism when they submit documents as their own that borrow ideas, structure, organization, or wording—that is, anything—from other sources without properly documenting what has been borrowed.

EXPLORE

Visit http://www.plagiarism.org, a website that explains the growing problem of Internet plagiarism, as well as the new technologies used on campuses and elsewhere to help track plagiarism. As you explore this website, be sure to visit the link for "statistics," which provides research information about the percentages of students who cheat and plagiarize. As you browse the statistics, consider what kind of research was done to come up with these numbers. Write a short analysis of the website.

PSA

Rewriting or resubmitting documents or parts of documents composed by someone else is plagiarism. When the words of another writer are used, the borrowing writer must put quotation marks around the passage and indicate the origin of the source. If the words are paraphrased, the borrowing writer must still indicate their origin. Avoiding plagiarism is part of the Draft segment of the problem-solving approach, particularly as you think about the arrangement and organization of words, sentences, and paragraphs. Making slight changes to a passage but leaving its organization, structure, and content intact is plagiarism as well. Allowing another person to write or revise your work is also seen as a kind of plagiarism, particularly in university settings. In the workplace those employed to help with writing should be acknowledged. Nearly all style manuals discuss plagiarism, although the ways in which it is addressed reflect the mindset of each particular discipline. Here are statements on plagiarism from three of the most widely used style guides:

Council of Science Editors (CSE) statement on piracy and plagiarism:

> Piracy is defined as the appropriation of ideas, data, or methods from others without adequate permission or acknowledgment. Again, deceit plays a central role in this form of misconduct. The intent is the untruthful portrayal of the ideas or methods as one's own.
>
> Plagiarism is a form of piracy that involves the use of text or other items (figures, images, tables) without permission or acknowledgment of the source of these materials. Plagiarism generally involves the use of materials from others but can apply to researchers duplicating their own previous reports without acknowledging that they are doing so (sometimes called self-plagiarism or duplicate publication).[1]

From the style manual of the Modern Language Association (MLA):

> In short, to plagiarize is to give the impression that you have written or thought something that you have in fact borrowed from someone else, and to do so is considered a violation of the professional responsibility to acknowledge "academic debts" ("Statement on Professional Ethics," *Policy Documents and Reports*, 1984 ed., Washington: AAUP, 1984, 134).

---

[1]Available online at http://www.councilscienceeditors.org/editorial_policies/whitepaper/3-1_misconduct.cfm.

The most blatant form of plagiarism is reproducing someone else's sentences, more or less verbatim, and presenting them as your own. Other forms include repeating another's particularly apt phrase without appropriate acknowledgment, paraphrasing someone else's argument as your own, introducing another's line of thinking as your own development of an idea, and failing to cite the source for a borrowed thesis or approach. The penalties for plagiarism can be severe, ranging from loss of respect to loss of degrees, tenure, or even employment. At all stages of research and writing, you must guard against the possibility of inadvertent plagiarism.[2]

From the style manual of the American Psychological Association (APA):

Quotation marks should be used to indicate the exact words of another. Summarizing a passage or rearranging the order of a sentence and changing some of the words is paraphrasing. Each time a source is paraphrased, a credit for the source needs to be included in the text.

The key element of this principle is that an author does not present the work of another as if it were his or her own work. This can extend to ideas as well as written words. If an author models a study after one done by someone else, the originating author should be given credit. If the rationale for a study was suggested in the discussion section of someone else's article, that person should be given credit. Given the free exchange of ideas, which is very important to the health of psychology, an author may not know where an idea for a study originated. If the author does know, however, the author should acknowledge the source; this includes personal communications.[3]

ANALYZE THIS

Read the three statements on plagiarism from the CSE, MLA, and APA. How are they similar or different? How do their definitions imply different ideas or ways of thinking about intellectual property? Which seems most useful to you, and why? In a one-page document, compare and contrast these three statements.

Ethical Use of Sources

Knowing how to quote, paraphrase, and summarize research will enable you to include information in your documents in ethical, professional ways.

Quoting

Quoting refers to using someone else's exact words in your document. A writer may choose to quote someone else in order to accurately and fairly represent the information or to ensure that the original information is presented with the same emphasis or expression that the originator intended. Here are a few guidelines to keep in mind when quoting other people's work:

1. Use only brief quotes in technical documents; try to avoid quotes of more than three sentences. If you need to represent more than three sentences of information, try to summarize and/or paraphrase the material.
2. Don't quote just to show you've looked at other sources. Be sure the research you quote fills a need in your writing.

---

[2] *MLA Handbook for Writers of Research Papers*, 6th ed. (New York: MLA, 2003).

[3] *Publication Manual of the APA*, 5th ed. (Washington, DC: APA, 2001).

3. The primary content of your document should be your own writing, your own argument, your own point. Quoted research should support your position; your document should not be a collection of quotes.
4. Be sure you understand appropriate punctuation for direct quotes: quotation marks to identify the beginning and the ending of quoted material, ellipsis points to show when something has been left out of a quotation, brackets to explain or clarify something within a quotation, and appropriate documentation.
5. Develop a varied list of ways to introduce quoted material so that you don't overuse the same introductory clause, such as "The report states." To add to the readability of your document, try some of these verbs: *says*, *notes*, *explains*, *argues*, *contends*, *identifies*, *explains*, *articulates*, *declares*, *announces*, *comments*, *remarks*, *observes*, *clarifies*, *details*, *describes*, *claims*, *maintains*, *recognizes*.

Paraphrasing

Paraphrasing refers to representing someone else's work in your own words. A good paraphrase captures the author's intent and information but presents it in your own words. A paraphrase is not necessarily shorter than the original, although the style and wording are usually different. Here are some basic guidelines for paraphrasing:

1. Be sure you understand the original thoroughly, so that you accurately convey the information of the original.
2. Include all of the ideas, concepts, or data from the original work. Don't compress the information by leaving out critical details. Deleting or altering the information can misrepresent the original research. There is a dual concern here with ethics and accuracy.
3. Give credit to the original author. Identify the source from which you have adapted the information you are presenting.

Summarizing

Summarizing is much like paraphrasing: you represent someone else's work in your own words. However, in summarizing, you abbreviate the information without compromising the original meaning. Many forms of workplace writing require summaries—for example, abstracts, executive summaries, and letters of transmittal. In addition, writing conclusions in just about any genre requires you to summarize your own work. Here are a few guidelines for summarizing:

See chapters 20, 21, and 22 for more on abstracts, executive summaries, and letters of transmittal.

1. Do not omit any of the central or relevant information in the original.
2. Be sure the summary identifies all key or central ideas of the original.
3. Identify what you are summarizing, giving full credit to the original writer(s).

Different occasions will call for different strategies for including outside information in your documents. Quoting conveys a sense of authority and exactness; paraphrasing allows you to capture an author's meaning while using your own wording and style; summarizing is useful for condensing a long passage down to its essence. Here is an example of a statement from the Transportation Security Adminstration that has been quoted, paraphrased, and summarized:

> ***Quotation:*** In their *Civil Rights Policy Statement*, the TSA notes that "TSA employees, applicants for employment, and the public we serve are to be treated in a fair, lawful, and nondiscriminatory manner, without regard to race, color, national origin, religion, age, sex, disability, sexual orientation, status as a parent, or protected genetic information.

Paraphrase: The TSA's *Civil Rights Policy Statement* states that employees, potential employees, and members of the public are to be treated fairly, lawfully, and without discrimination for any reason.

Summary: The TSA's *Civil Rights Policy Statement* contains the agency's policies regarding nondiscrimination, sexual harassment, equal opportunity employment, and other important civil rights issues.

ANALYZE THIS

Ford Motor Company's website explains that the company is in the process of developing an "alternative vehicle" that is powered by a hydrogen combustion system. The website explains:

> When petroleum-based gasoline is used in a traditional internal combustion engine, hydrocarbons and carbon dioxide emissions are produced, which contribute to environmental pollution. However, the hydrogen combustion process produces neither hydrocarbon nor carbon dioxide emissions because there are no carbon atoms in hydrogen fuel. With further research to reduce tailpipe output of potentially smog-forming emissions to below-ambient conditions in many cities, the air leaving a hydrogen internal combustion engine's tailpipe could actually be cleaner than the air coming into the engine. In addition, a hydrogen engine can attain overall efficiency of 38 percent, approximately 25 percent better than a gasoline engine. Designing a gasoline engine to burn hydrogen fuel has typically resulted in significantly lower power output—until now. Ford researchers have shown that with supercharging, the hydrogen internal combustion engine can deliver the same power as its gasoline counterpart and still provide near-zero-emissions performance and high fuel economy. The centrifugal-type supercharger provides nearly 15 pounds per square inch (psi) of boost on demand.*

Rewrite this passage from the Ford website by paraphrasing. Try to put as much of it as possible into your own words and organization, although you may find it necessary to quote small portions.

*Ford Motor Company, "Building a Hydrogen Transportation System," http://www.ford.com/en/innovation/technology/hydrogenTransport/default.htm.

Ethics in Primary Research

Because primary research requires writers to create new kinds of information, the ethical issues involved are a bit different. Instead of plagiarism, the ethical concern is with treating human subjects properly. When we create primary research—everything from lab experiments to interviews to surveys—we often use other people to help us gather information and data, and they must be treated fairly. Those in the sciences who actually experiment on humans are bound by a strict set of ethical codes; see, for example, the "Regulations and Ethical Guidelines" for the protection of human subjects of research, written for the National Institutes of Health and found at http://ohsr.od.nih.gov/guidelines/belmont.html.

Even though the average workplace writer does not usually expose anyone to physical harm, all types of research require ethical behavior.

- *Confidentiality:* When researchers conduct interviews or surveys, they should either keep private the identities of those interviewed or surveyed or get written permission from them to use their words or disclose their identities.
- *Unbiased questions:* In interviews or surveys researchers should ask questions that are unbiased, that is, that do not purposely lead their subjects into particular responses.
- *Reporting results:* Researchers have an ethical duty to report the results and data gathered from interviews or surveys fully and fairly. They should not exclude information harmful to their agenda, nor should they fabricate or falsify any information provided by interviewees or respondents.

Surveys, in particular, are difficult to create, even when researchers are trying their best to be fair and ethical. Consequently, many workplace writers have professional survey makers develop their surveys to meet the particular needs of a research project. Many companies, organizations, and institutions also have their own ethics offices, which can answer many questions that workplace writers have regarding ethical issues in research.

SUMMARY

- Research is conducted to provide information that helps solve problems.
- Workplace writers should be attentive to the exigency of their situation, the purpose or reason for conducting the research, the audience to whom the research will be presented, and the genre in which the research will be presented.
- Good research answers questions for specific audiences.
- Primary sources present original information; secondary sources present interpretations, evaluations, summaries, descriptions, or comments related to primary research.
- Being able to locate sources is important, but being able to evaluate them is even more important.
- The World Wide Web has advantages: ease of use, access, transportability, general information, and products and services. It also has disadvantages: lack of reliability, irrelevant information, and a fee for the best information.
- Blogs and wikis can be useful for current information or summaries, but their reliability is questionable.
- Databases have advantages: trustworthiness, frequent updates, full texts online, lots of information, and more relevant results. They also have disadvantages: expense, full-text inconsistency, and a longer learning curve.
- Intranets and archives have advantages: access to private information and ease of use. They also have disadvantages: restricted information, search difficulty, and incomplete records.
- Primary, or qualitative, research can be conducted through observation, inspections, experiments, interviews, surveys, and focus groups.
- Useful note-taking strategies include the research narrative, the double-entry journal, the problem-solution log, and seeing note taking as drafting.
- Documenting research information is an ethical component that is crucial in avoiding plagiarism.

CONCEPT REVIEW

1. How does research for personal use differ from research in workplace writing?
2. Why is research an important part of problem solving in the workplace?
3. What are some examples of situations in the workplace when research is necessary?
4. In what ways does research help writers both ask and answer questions?
5. What are the differences between primary and secondary sources?
6. What are some examples of primary sources?
7. What are some examples of secondary sources?
8. What does it mean to evaluate a source?
9. What are the advantages and disadvantages of using the World Wide Web?
10. What are the advantages and disadvantages of using databases?
11. What are the advantages and disadvantages of using an intranet?
12. When might a workplace writer need to conduct primary research?

13. What are some of the ways to conduct interviews?
14. What are some of the ways to develop and use surveys?
15. What are some of the ways to conduct focus groups?
16. What is the relationship between note taking and research?
17. What is a research narrative?
18. What is a double-entry journal?
19. What is a problem-solution log?
20. How does note taking function as a type of drafting?
21. What does it mean to document sources correctly?
22. What is the relationship between documenting sources and ethics?
23. What is plagiarism?
24. What do you need to know about quoting to avoid plagiarism?
25. What do you need to know about paraphrasing to avoid plagiarism?
26. What do you need to know about summarizing to avoid plagiarism?

CASE STUDY 1

What to Buy: Researching Personal Electronic Devices

Over the past few years the popularity of personal entertainment devices has increased dramatically: consider the iPod, the Microsoft Zune, or the Sony Digital Walkman for listening to music; Microsoft's Xbox 360, Nintendo's Wii, and Sony's PS3 for playing video games; and a variety of different portable players for watching DVDs. There has also been an increase in the popularity of cellular phones, personal digital assistants (PDAs, such as the Blackberry or Palm), digital cameras, and video. You may own one or more of these kinds of devices, and if you purchased the device yourself, you likely did a little research before purchasing it. There is a lot of information available: friends, relatives, and salespeople make recommendations, and corporate advertisements make claims. Discerning what information is useful and what is not takes some careful attention and analysis.

Select any one of these types of devices, and identify at least three different brand versions that are available for purchase. Then locate three sources of information about each that might influence your decision to purchase it. Reflect on the research you find, and evaluate its usefulness in your decision making. In a double-entry journal write analyses of each of the documents for each of the brands. Remember that the left side of the journal is reserved for important information you have found in the documents (i.e., quotes, summaries, paraphrases), whereas the right-hand side is reserved for your comments, criticisms, and ideas. Finally, write about which of the devices you would buy, and explain how the journal entries helped you come to that conclusion.

CASE STUDY 2

We Brake for Research: Public Statement about a New Auto Safety Feature

The Insurance Institute for Highway Safety is an independent organization dedicated to reducing deaths and injuries from crashes on U.S. highways. The IIHS conducts a wide range of research to solve these problems, including the testing of new automobile safety features. The IIHS released this public statement about one of its research studies:

> About half of the 28,000 fatal passenger-vehicle crashes that occur each year involve a single vehicle. Equipping cars and SUVs with electronic stability control (ESC) can reduce the risk of involvement in these crashes by more than 50 percent. The effect on all single-vehicle crashes (fatal and nonfatal) is somewhat less (about 40 percent), and

> the effect on multiple-vehicle crashes is much less. These are the main findings of a new Insurance Institute for Highway Safety study comparing crash rates for cars and SUVs with and without ESC. ESC is an extension of antilock brake technology, which has speed sensors and independent braking for each wheel. For ESC, additional sensors continuously monitor how well a vehicle is responding to a driver's steering input. These sensors detect when a vehicle is about to stray from the driver's intended line of travel (that is, lose control), which usually occurs in high-speed maneuvers or on slippery roads. Then ESC brakes individual wheels automatically to keep the vehicle under control.*

Write a brief analysis of the problem that this statement addresses, the purpose of the statement, and the various stakeholders who might need this information.

CASE STUDIES ON THE COMPANION WEBSITE

Northside Urgent Care Goes Shopping: Searching for the Right Office Machine

Northside Urgent Care was outfitted with second-hand and leased equipment to reduce start-up costs. Some of that aging equipment is now failing, and the office manager, who must investigate the possibility of buying some new equipment, needs to determine the two best options for replacement.

IPD Body Armor: Planning a Major Research Project

The chief of police of the Indianapolis Police Department has begun to wonder if National Body Armor still provides the best product for his officers. This case examines how to conduct research to determine which body armor the department should purchase.

Street Sports Market Research: Conducting Primary Research through Focus Groups

This case focuses on Joel Fogelman, the CEO of Pinnacle Sports, Inc., one of the nation's largest sports apparel and equipment companies and Street Sports, a subsidiary of Pinnacle Sports. The final piece in the Street Sports debut product line is its signature basketball, the StreetBall 1.0, which is in need of testing.

Going Global: Transnational Management Accounting

This case asks you to assist an international management accounting firm as it looks to expand into South American markets. By identifying valid sources and documenting information, your research will help First Global Management decide in which countries to open branch offices.

VIDEO CASE STUDY

DeSoto Global

Planning the Shobu Automotive Proposal

Synopsis

Senior sales analyst Dora Harbin has been assigned to examine the possibility of securing contracts within the Japanese automotive industry. The case revolves around

---

*Insurance Institute for Highway Safety, "Electronic Stability Control Found Effective," October 28, 2004, http://www.iihs.org/news/2004/iihs_news_102804.pdf.

a proposal DeSoto Global is writing in response to a request for proposals (RFP) placed by Shobu Automotive, which seeks top-of-the-line electronic components for their luxury car line. Dora asks Bill Kemble and Tosh Takashi to write the proposal. Bill, a mid-level operations manager at DeSoto Global, has a great deal of experience putting together external proposals like the one Dora needs, and Tosh has previous sales and marketing experience in Hawaii, Japan, and Korea, as well as personal experience with Japanese culture.

WRITING SCENARIOS

1. In a small group, decide on one social or ecological issue that is relevant to your campus, city, or state. Individually gather two to three sources that deal specifically with the issue, and write a brief summary of each. Then reconvene as a group, and discuss everyone's summaries. Finally, discuss what additional questions are raised by the research you have talked about.
2. Explain your future career goals in two or three detailed paragraphs. Then search both the World Wide Web and databases of your choice to find further information about this career: how one gets into this career, what the benefits and the drawbacks are, what responsibilities and duties come with this career, and whether this career requires any kinds of research.
3. Imagine that you have been hired by a mid-sized newspaper to serve as the head of purchasing. The job requires you to make decisions about purchasing supplies for company employees—everything from coffee and paper clips to computers and sophisticated equipment. Your first task is to look into purchasing six new computers for the reporters who write for the sports section of the paper. You must do some research to make sure you get the best computers for the reporters at the lowest cost. In a few paragraphs describe what your research strategy would be. Explain in detail the kinds of primary and secondary sources you would need to find or develop.
4. Find a total of five sources (no more than three from the Web) about the new Freedom Tower, which will replace the fallen twin towers in New York City. For each source describe the kinds of research the author(s) conducted to compose the document. Provide specific examples to support your answers.
5. In a series of informal notes, discuss the different ethical dilemmas the director of patient support services is likely to face as she researches, writes, and tries to implement her five changes at the HSC.
6. Write an ethical (i.e., plagiarism-free) summary of the following passage from Dell's 2005 fiscal report to investors (available at http://www.dell.com/downloads/global/corporate/annual/2005_dell_annual.pdf):

> Diversity is essential to maximizing our customers' experience and creating a winning culture that fosters excellence and continuous improvements throughout the company. Today, more than half of our U.S. employment is represented by women and minorities, and more than one-third are managers. We provide online diversity and multicultural training, networking groups, mentoring, career coaching and work/life effectiveness programs to ensure we're recruiting and retaining the best and brightest talent in the industry.
>
> Dell remains committed to supplier diversity. Since 2001, Dell has steadily increased its spending with Minority and Women Business Entrepreneurs (MWBE) by more than 66 percent. Last year, we estimate $1.5 billion in spending with MWBEs and ranked No. 7 on the Div50 listing, the

highest ranking among technology companies. We continue to partner extensively with diverse organizations such as Congressional Black Caucus Foundation and the Congressional Hispanic Caucus Institute; historically black colleges and universities; National Urban League; United Negro College Fund; Catalyst; National Society of Black and Hispanic MBAs; National Council of La Raza; Out and Equal; and the Asian, Native American, Hispanic and African-American Chambers of Commerce. For more information on Dell's diversity initiatives, visit: http://www.dell.com/diversity.

Globally, Dell employees are dedicating thousands of hours and donating millions of dollars every year to organizations such as International Red Cross, Habitat for Humanity, United Way, Second Harvest and Earth Share. Through the company's One Dell: One Community program, employees combine team building and volunteerism to benefit local communities. Last year, the program culminated with Global Community Involvement Month when nearly 18,000 employees from 28 countries volunteered time in their communities. During the year-end holidays, U.S. Dell employees donated 545,000 pounds of food, which provided more than 716,000 meals for hungry families. Employee food-bank support helped earn the company the Group Volunteer of the Year Award from America's Second Harvest, the nation's largest hunger relief organization.

The Dell Foundation, the company's corporate giving division, offers several grant opportunities which provide funds for health and human services, education, literacy and technology access programs for youth around the world. More information about Dell's community activities can be found at www.dell.com/dellfoundation. (p.24)

7. Explain in short informal notes how the ethics of citing research in your school or workplace writing relates to each of the five different parts of the problem-solving approach.
8. Decide which one of the documentation and style manuals discussed in this chapter is most relevant to your discipline and future career. Locate a copy of this manual—either online or in your school library. As you review the manual, take notes in a double-entry journal on what you discover.
9. Compose a research narrative based on Dell's 2005 Fiscal Report to Investors, which is available at http://www.dell.com/downloads/global/corporate/annual/2005_dell_annual.pdf. Read through this report carefully, paying close attention to both words and images. In your research narrative discuss which aspects of this report could be viewed as researched data and information. Is this research primary or secondary? In what ways is this report itself a source of information that was researched by its intended audience?
10. Design a survey for students in your technical communication course. Your purpose is to solicit feedback about the layout and the structure of your classroom. What is advantageous and what is not? Distribute your survey to classmates, and then review their responses to determine whether your questions were well posed. Ask your classmates whether there was any part of the survey that should be revised for clarity.
11. After completing Writing Scenario 3, briefly compare the sources you found on the World Wide Web and those you found in the databases. Were the sources different or similar? Which ones were more helpful? Given what you were researching, why do you suppose one resource was more helpful than the other?
12. Describe the ways that current technologies make research easier for workplace writers, as well as the ways that these technologies make research more

difficult. You may base your answers on real-life experiences and observations if you wish.

13. How do you define *technology*? What kinds of hits would you expect if you typed the word *technology* into a search engine like Google? Now go to this search engine, and type in the word *technology*. What kinds of hits did you get? Were they different from what you expected? How does this search engine technology sort out the word *technology*?

7

Organizing and Drafting Documents

CHAPTER LEARNING OUTCOMES

After completing this chapter, you will be able to do the following:

- Understand how **drafting and organizing** work within the problem-solving approach
- Understand the role of **front matter**, **body**, **and end matter** in larger drafting approaches
- Use a number of **drafting strategies** in writing documents
- Develop strategies for writing the **body** of a document
- Employ a variety of **organizational strategies** in putting information into a document
- Better understand how **visuals** can clarify information in documents
- Develop strategies for writing the **conclusion** to a document
- Develop strategies for writing the **introduction** to a document
- Consider the positive and negative aspects of using **templates and wizards** to draft documents

DIGITAL RESOURCES

On the Companion Website www.prenhall.com/dobrin:

- Case 1: Streamlining Communications in the South Carolina Department of Transportation
- Case 2: Time for an Upgrade: Mark Simmons Confronts the Problem of Obsolete Computers at Quality Care Nursing Home
- Case 3: Beacon Biological: Retracing and Remedying Six Months of Business Disaster
- Case 4: Designing the Fighting Platypi Little League Baseball Team Web Page
- Case 5: Software Woes: Training Employees in How to Use HXaune 3.0
- Video Case: Collaboration and Drafting the Oxygen-Bleach Report
- PowerPoint Chapter Review, Test-Prep Quiz, Exercises and Activities

REAL PEOPLE, REAL WRITING

MEET ROBERT HARRELL • Technician with the U.S. Department of Agriculture Biological Science Laboratory

What kinds of writing do you do as a laboratory technician?

Much of my writing consists of protocols (i.e., action plans), procedures, updates, and data reporting concerning biological experiments. As a technician with the U.S. Department of Agriculture Biological Science Laboratory, I conduct the experiments and run assays (i.e., analyses) that produce results that I communicate to the primary investigator and to other technicians. Generally, I want things done the same way each time, so I have to maintain written protocols and procedures.

What are some of the guiding principles that determine how you organize data in your documents?

Because of the kinds of documents I write, I generally rely on a chronological organizational strategy or an order of priority. When I'm writing to people that are working for me, I may want them to watch out for particular events, so I write down the important things and how I want them to respond to those things in an order of importance.

What are some general principles you use in drafting your workplace documents?

My primary investigator often goes on business trips and may want to know what's going on in the lab when he's away. It may even be a month between the times that we talk, and I may have five or six things going on in the lab. I keep track of information in data tables, which provide updated information of what has happened since we last spoke. I often use older drafts as a starting point for newer drafts and will then include data from the tables in newer drafts. So part of my drafting is keeping a running chronological report of what's going on in the lab.

What are some of the issues of professionalism that you must take into consideration as a laboratory technician?

You always want to be honest and truthful about everything that you do because you may have to go back and conduct an experiment again. You need the information you report to be as accurate and detailed as possible. So a professional makes certain to write down all of the information from the first time the assay was conducted so that it's conducted consistently the next time or the twentieth time. This keeps the data from being skewed.

What advice about workplace writing do you have for the readers of this textbook?

Know that you are going to be writing. Write as clearly as you possibly can because things can be misconstrued easily. Make sure to write as much as you can when gathering information because you think you will remember all of the information, but you won't. In addition, you will work with other people, and if they go back and look at what you've written, unless you've been very clear in your writing, they may not understand what to do, or they may miss a crucial bit of information.

INTRODUCTION

We're probably all familiar with the scene in the movie or TV show in which the hero has to defuse the bomb and the sidekick is reading the instructions as the hero's sweaty hands work the wire clippers. "Cut the red wire leading to the detonator," the sidekick reads aloud to the hero. Click; the hero cuts the red wire. "But only after you've cut the blue wire," the sidekick quickly adds. The hero turns to the sidekick with a you've-got-to-be-kidding-me look. "Run," says the hero.

The order in which information is presented can be crucial. Although most of us will never have to defuse bombs (thankfully), nor will we likely have sidekicks (unfortunately), we will present information that can be misconstrued or misunderstood if it is not organized effectively. Organizing information well is a matter of presenting that information in ways that will make sense to the audience, will be accessible to readers, and will solve the problems the document works to solve.

PSA

Writing in the workplace often begins when writers gather information during the Research phase of the problem-solving approach. This chapter discusses how to organize that information and craft it into usable, efficient workplace documents. Both the Research and the Draft phases of the PSA encourage writers to strategize about ways to organize information; this chapter helps you do that and then write drafts of documents. Organizing drafts is just one of the numerous aspects of making documents, from developing ideas to the final stages of composition, which are discussed in this part of the text.

IN YOUR EXPERIENCE

Make a list of the kinds of documents you most frequently write, separating them into two categories: those for which you write various drafts, or versions, and those for which you write just one draft from start to finish. These don't necessarily have to be technical or workplace documents. Then spend some time thinking about or discussing these two lists. Are there any similarities among the kinds of documents for which you write drafts? Are there similarities among the documents you write from start to finish? Who tend to be the audiences of these different kinds of documents?

In other writing courses you may have learned about the writing process, but there really is no such thing as *the* writing process. There is no one correct way to move from gathering information to writing early drafts to producing final documents. This is the reason that the problem-solving approach is not a linear process. Each writer develops his or her own methods for moving through writing processes in different contexts and situations. For instance, a workplace writer would probably not use the same writing process to develop and write a formal proposal for corporate expansion into international markets as she would to write an e-mail to coworkers about a new company policy. The proposal would require more research, visuals, collaboration, and rough drafts, whereas the e-mail might require little research, no visuals, no collaboration, and a single draft.

In this sense writing in the workplace is a bit like hitting a baseball. All boys and girls are taught the fundamentals of hitting when they first learn the game: keep the feet spread apart, the bat off the shoulder, the chin up, and so on. Yet if we watch a college or professional game, we see that each hitter has developed a unique batting stance: some keep their feet close together, and others spread them far apart; some rest the bat on their shoulder, and some raise it high in the air; some hold their heads up, others down. The point is that these players have developed their own styles after years of practice and experience. They all started with the same fundamentals but altered them over time to suit their own purposes. Writing is similar. Most writers learn the same fundamental processes of writing, but with time and experience they develop their own methods and processes for composing documents effectively and efficiently.

Other factors also influence both baseball players and workplace writers: the contexts and situations in which their activities take place. Depending on how the game is going, how many runners are already on base, and how the pitcher is throwing, batters will adjust what they do. In addition, the first inning of the first game of the season is quite different from the ninth inning of the last game of the championship series. Writers, too, are affected by the contexts and situations that surround their documents. For example, a press release announcing the recall of a potentially dangerous product might be written more quickly than the manual that originally accompanied that same product. Unique situations dictate the processes of writing.

This chapter cannot offer a specific step-by-step process to guide you through organizing and drafting all documents in all situations. However, it provides an overview of how to organize materials and formulate drafts, and it encourages you to develop appropriate methods for producing diverse kinds of documents. It suggests practical strategies and helps you discover your own strengths as a writer.

IN YOUR **EXPERIENCE**

Consider the last time you wrote a document that you considered important, such as a job application letter, a résumé, or a term paper. How would you describe the process by which you put together your first draft? And once you had that draft, how much attention did you pay to revising, editing, or just cleaning up that document? Now, consider the last time you wrote a document that you didn't consider important, such as an e-mail to a friend, directions to your home, or even a school writing assignment. Did your writing process differ from how you wrote the more important document? Did the differences in those processes have any impact on the effectiveness of the documents? Compare the two documents, their effectiveness, and the reasons that one needed a draft and the other did not, and write a comparison of your approaches with each.

PREDRAFTING STRATEGIES

Workplace writers do not usually sit down and write first drafts of documents without having already begun the writing process in other ways. In fact, all five phases of the problem-solving approach contribute to how writers draft documents. Consider the following strategies as you prepare to draft a document:

Confirm Your Purpose

PSA

As the Plan portion of the problem-solving approach shows, you must always be certain about the purpose of your document. If you begin to write without knowing exactly what the goals of the document are, you may find that your finished product does not accomplish what it needs to or that you have completed a document that was not necessary. Understanding the purpose allows you to control the document, write it efficiently, and organize it clearly and effectively. To be sure you understand your document's purpose, ask these questions:

- What should the document accomplish?
- What should readers do when they have finished reading the document?
- What information should the document convey?

Analyze Your Audience

Along with purpose, the audience of the document determines how and what you write. Again, the Plan phase of the problem-solving approach includes understanding who the stakeholders are. In thinking about your potential audience, consider these aspects:

- Level of expertise
- Level of education
- Cultural differences
- Attitudes
- Expectations
- Context in which the document will be read

Gather Your Information

PSA

Chapter 6 and both the Plan and Research parts of the problem-solving approach show that gathering and researching information is a crucial part of the predrafting process. As you collect information, recognize that what you write down will become a part of the first draft of the document. Be sure to record as much information as you can in a clear, careful manner way so that when the time comes to write the first draft, you will be able to recall the details of the information from your notes and will be able to incorporate your notes directly into your draft.

Develop Ideas about the Information

Once you have gathered the data, you will need to generate ideas about what you wish to say about that material. There are a number of strategies that workplace writers use to generate ideas, but writers also develop their own approaches.

Collaboration and Discussion

By the time you sit down to produce the first draft of a professional or technical document, chances are good that you have already developed some familiarity with the information to be included. However, you may not know all there is to know about the subject, and you may not have thought about all the facets of that information. Taking the time to talk with others about the information allows you to articulate what you already know; sometimes writers forget particular details that resurface as they discuss the subject. Ideas that writers suggest to others in conversation often spark new ideas. In addition, talking with others about the document encourages colleagues to ask questions, and answering such questions helps you better understand how the document should be put together.

Listing

This strategy asks you to write out on paper as many ideas about the subject as you can. Don't worry about order of importance as you create a list of ideas; you can organize your ideas at a later point. Simply writing ideas down can lead you to new ideas that might become important in the document. Lists also allow writers to identify ideas that are not useful.

Freewriting

Freewriting strategies involve writing about the subject at hand without any planned agenda, simply jotting down whatever comes to mind. The writer writes for a given amount of time, usually fifteen to thirty minutes, without stopping to revise, edit, or adjust what has been written. The point is to get as many thoughts and ideas out as possible. The writer can pose questions or state ideas—what is written does not matter. At the end of the session, the writer reviews the freewriting document to identify useful elements that are pertinent to the project.

IN YOUR **EXPERIENCE**

You have probably already spent some time thinking about what kind of career you would like to have and perhaps what kinds of writing it might require. Take fifteen minutes now and freewrite about the kinds of writing you anticipate—the writing it will take to secure the job, the writing the job will regularly require, and the special writing tasks that might occasionally arise. Remember to write consistently for fifteen minutes, anything that comes to mind. Once you have finished the freewriting, go back through your writing, and highlight or underline any ideas, phrases, or sentences that you think might be useful in describing the type of writing you might do in the career you have chosen.

Clustering

Clustering is a method of taking inventory of what a writer already knows about a subject and beginning to organize that information. It also provides an opportunity for writers to identify what information is lacking and still needs to be acquired before the draft is completed. You may find it useful to ask a series of standard questions—beginning with *what*, *where*, *when*, *who*, *why*, and *how*—to identify the gaps in your information.

Workplace writers often use this strategy because it offers a visual component. Clustering uses diagrams—most frequently simple circle drawings—to show the relationships among pieces of information. Clustering works like this: First, you write the primary subject of your document in the middle of a piece of paper and draw a circle around what you have written (see Figure 7.1). Then you add secondary circles around the primary circle to identify what you see as the main issues surrounding the subject. Connect those circles to the primary circle by drawing lines between them (see Figure 7.2). Next, you add more circles around the secondary circles, identifying facts, ideas, examples, or other pertinent information that supports or explains the circle to which it is attached (see Figure 7.3). You can repeat this process many times, adding as many circle details as necessary.

Organize Your Information

organization the coherent arrangement of information in ways that make sense to and are useful to readers

Organization is the way a writer arranges information within a document in order to present that information coherently. Good organization can help readers better understand information and better access it. To create useful workplace documents, writers must know how to control their information and integrate it into a well-organized body. Information control requires attention to several key factors, two of which we reviewed at the beginning of this chapter:

- *Purpose*—A writer organizes a document differently to persuade a group of investors that a new product is safe rather than to persuade that same audience that the product will be financially viable.
- *Audience*—A writer organizes pertinent information differently for an audience of experts than for an audience of novices, for example. For the experts, basic information might appear in an appendix, but for the novices, early in the document.

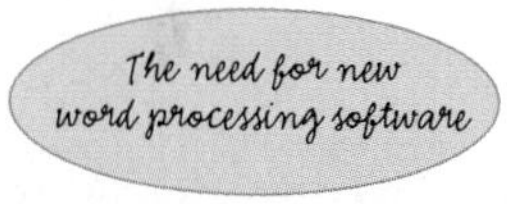

The need for new word processing software

FIGURE 7.1 Step 1 of clustering

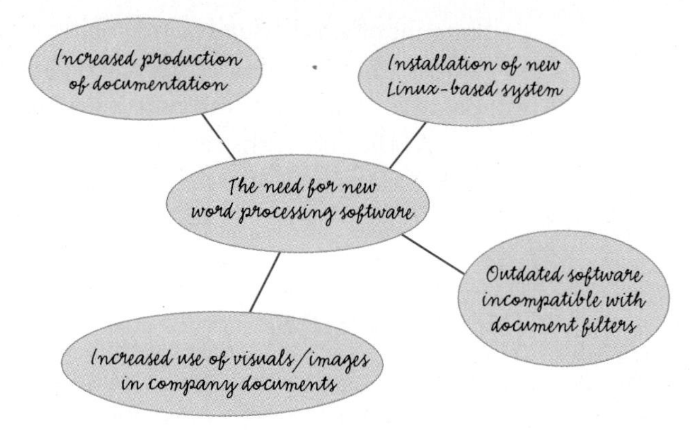

FIGURE 7.2 Step 2 of clustering

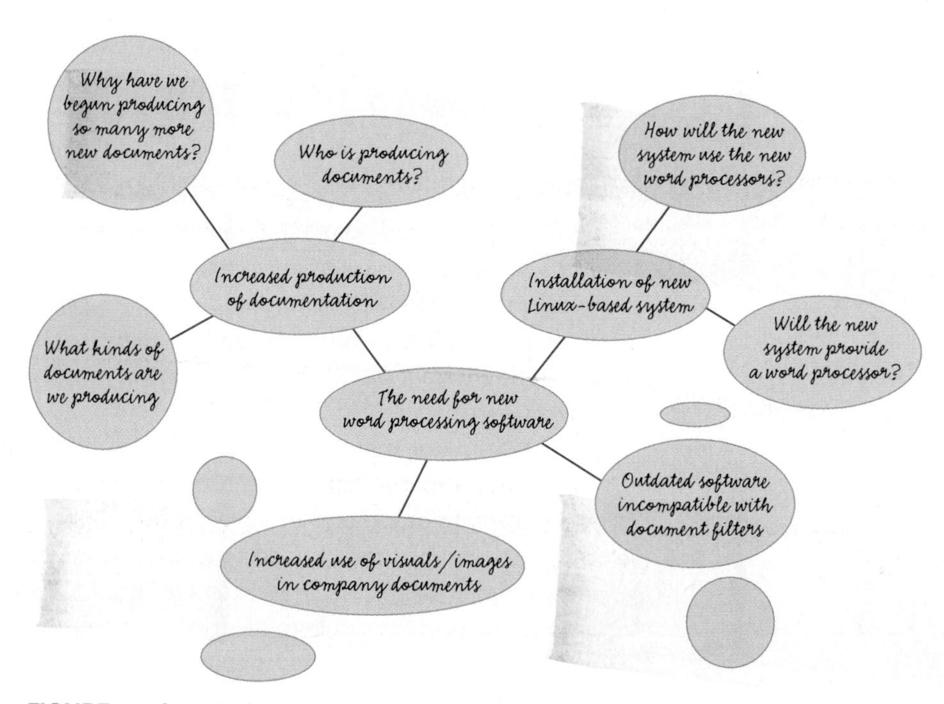

FIGURE 7.3 Step 3 of clustering

- *Logic*—A writer chooses to arrange information along a logical path in order to lead the audience to the proper conclusion rather than to a wrong or misleading one.
- *Ethics*—A writer organizes a document to emphasize or de-emphasize particular information in order to guide the audience to a particular conclusion—a process with ethical significance. For example, a writer would need to consider whether it is ethical to list all of the benefits of a particular service in the body of a proposal and then list all of the drawbacks in an appendix.

organizational strategy the method by which a writer arranges the materials that comprise a given document

In addition to controlling how their information is presented, writers need to decide on an **organizational strategy** before they begin the first draft. This chapter addresses nine possible organizational strategies with which writers can experiment to determine what will work best: sequential, chronological, order of importance, general/specific, division, classification, cause and effect, comparison/contrast, and spatial. Sometimes writers develop hybrid approaches, blending several strategies to create the most effective documents. For instance, a comparison/contrast approach might be combined with a division or classification strategy. And as they draft their documents, writers may shift from one strategy to another. Writers should not feel compelled to stick to just one strategy but should be flexible.

SOME HINTS FOR THINKING ABOUT ORGANIZATION

In addition to organizing every document according to the four factors listed here, you should always

1. make your organizational approach explicit
2. be consistent in your organization within a given document
3. consider how readers will need to access the information
4. organize information in ways that present it ethically
5. identify the differences between primary and secondary points of information
6. be sure task-based information is presented in the order of action

Sequential

Sequential organization places information in the order that it progresses or should progress. The sequential approach asks readers to move through information from beginning to end in a linear fashion. For instance, in a set of technical instructions (see chapter 18), the body of the document consists of the steps a reader is to follow. These steps are organized sequentially, one after the other in a particular order. This organizational strategy is usually simple for readers to follow; it moves them through a process or a procedure. In some cases it's a step-by-step approach (see Figure 7.4).

SOME HINTS FOR USING A SEQUENTIAL STRATEGY

To move readers through a sequence, you should consider using the following:

1. Numbered lists to guide readers through the sequence
2. Transitional words to identify movement through the sequence, such as *first, next, then, after, finally*
3. Sequence guide words—for example, *step, part, phase,* or *segment*—accompanied by an identifier such as a number and sometimes by descriptions of the individual parts (e.g., Phase One: Inverting the Lens)
4. Images to clarify the sequence, often corresponding to written descriptions

How to Change the Screen Saver in Windows Operating System

First, click the Start menu button in the lower left-hand corner of the screen.

Second, click Control Panel.

Next, double-click the Display icon to open the Display Properties window.

Then, click the tab labeled Screen Saver to open the screen saver window, which allows you to make choices about your screen saver.

- Click the screen saver list to locate screen savers installed on your computer; then click the screen saver you wish to use.
- Click the Settings button to adjust the screen saver commands, such as speed, color, and size.
- Click the Preview button to look at the screen saver before enabling it. If you do not like what you see, go back to the Settings button, and adjust the commands. You can also go back to the screen saver list and select a different screen saver.
- Next, determine how long you want your computer to wait before running the screen saver.
- Finally, click OK when you are satisfied with your choices.

FIGURE 7.4 An example of sequential organization

Chronological

The chronological organizational strategy also moves readers through a sequential process, but this sequence is related to time. Like the sequential, the chronological strategy moves from a beginning to an end. Just as technical instructions and other forms of directions (see chapter 18) use a chronological strategy, so, too, do some reports (see chapter 21), such as lab reports, which guide readers through an experiment; incident reports, which detail the sequence of events that led to an accident or a customer complaint; and travel reports, which detail an employee's travel expenses and completed work while on a trip. Figure 7.5 uses a chronological strategy to organize historical achievements.

SOME HINTS FOR USING A CHRONOLOGICAL STRATEGY

Because a chronological strategy is a form of sequential strategy, you should consider the hints offered there as well as those found here. To move readers chronologically through a sequence, you should consider using the following:

1. A time line to identify the events in order.
   8:00 AM: The on-site crew notified the field manager of the accident.
   8:15 AM: The accident victim was transported to the hospital.
   8:35 AM: The field manager arrived at the accident site.
   Monday, July 16, 2007, RFP posted
   Wednesday, July 25, 2007, Bid from Falcon Group submitted
   Friday, July 27, 2007, Falcon Group bid returned for failure to supply all required documents
2. A flowchart to see the sequence of events before you describe them in writing

EXPLORE

Using your web browser, conduct a simple search for the terms *time line* and *chronological order*. What kinds of documents did your search reveal? Look at several of those sites and determine the similarities and differences in how information is reported chronologically. List five of the sites you found, and describe how they use chronological order as an organizational strategy.

1936 Konrad Zuse develops the Z1 computer, the first freely programmable computer.

1942 John Atanasoff and Clifford Berry start the first computing company: ABC Computer.

1944 Howard Aiken and Grace Hopper develop the Harvard Mark I computer.

1946 John Presper Eckert and John W. Mauchly, using more than 20,000 vacuum tubes, develop the ENIAC 1 (the electronic numerical integrator and computer), the first electronic digital computer and the prototype of the modern computer. It was designed to compute ballistic firing tables during World War II.

1947–48 John Bardeen, Walter Brattain, and William Shockley develop the transistor. Although not a computer, the transistor would influence further development of computer technologies.

1951 John Presper Eckert and John W. Mauchly develop the UNIVAC computer (universal automatic computer), the first commercial computer. It was used to predict the winner of the 1952 presidential election.

1953 International Business Machines (IBM) opens.

1954 IBM and John Backus develop FORTRAN computer programming language, the first successful programming language.

1958 Jack Kilby and Robert Noyce develop the integrated circuit, the first computer chip.

FIGURE 7.5 A chronology of the early development of modern computers

Order of Importance

The order-of-importance strategy is commonly used because it permits writers to present information in either an increasing or decreasing order, allowing them to emphasize or de-emphasize particular information. In an increasing order of importance, writers begin with the least important bits of information, followed by increasingly important items until the last piece of information, which is the most important. This strategy is used when writers want to hold the most important piece of information until the end of the document, which emphasizes the most important point by allowing the less important pieces to build up to the conclusion. Placing the most important information last keeps it fresh in the audience members' minds when they finish reading.

On the other hand, workplace writers may find a decreasing order of importance useful in documents such as letters, memos, and reports. With this strategy writers provide the most important information up front and then offer progressively less important information. This strategy is useful with readers who do not have time to carefully read documents from beginning to end.

Consider the memo from the programming manager to his staff, regarding a critical meeting (see Figure 7.6). The body of the memo offers six pieces of information in decreasing order of importance. This strategy ensures that a reader is immediately made aware of a document's primary purpose and information.

SOME HINTS FOR USING AN ORDER-OF-IMPORTANCE STRATEGY

Both increasing and decreasing orders of importance can be effective, depending on the context of the document. Consider using these approaches:

1. An increasing order of importance in PowerPoint and other presentations, particularly when you do not leave your audience with a printed copy of the presentation or supporting documentation to which they can refer later
2. A decreasing order of importance if you are presenting numerous points and want to ensure that your reader is attentive to your most important information
3. An alert in the introduction of the document to the order-of-importance strategy you have chosen:
   "In the following report, information regarding the most critical aspects of the changes in our employee training program will be addressed before the numerous less-critical issues that have evolved this year."
4. Numbered lists to guide your readers through the specific points, whether in increasing or decreasing order
5. An explanation to your reader of why you have decided that a particular point is more important than another

General/Specific

The general/specific strategy can also be used in two ways: progressing from general to specific information or from specific to more general information. Both ways provide a balance of abstract ideas and concrete details within the document.

In some instances a writer may need to present general information before providing specific details; the general information provides background, scope, and context for the more specific information. This approach allows a writer to make a general statement about a particular topic and then support that statement through specific examples. Some workplace genres, such as formal reports, include abstracts or executive summaries early in the report to give readers a general overview of the information housed in the report before moving into the details of the report. Scientific reports, too, often rely on this strategy because many scientific methodologies, such as taxonomy, move from the general to the specific: kingdom, phylum, class, order, family, genus, and species, for example. This kind of organizational strategy derives from deductive reasoning, the process of moving from a general principle to specific examples.

MEMORANDUM:

TO: Programming Staff

FROM: Lev Azar, Programming Manager

DATE: 9 November 2004

RE: Critical Meeting

There will be a mandatory meeting on 11 November 2004 at 9:00 am in the Conference Room. This meeting will address the problems that have developed regarding the TCP/IP protocols in regard to the recent shift from Windows to Linux servers. The meeting should last about an hour, so please plan accordingly. Coffee and donuts will be served.

The most important piece of information is that there will be a meeting on November 11.

The second most important piece of information is when the meeting will be held.

The third most important piece of information is the location of the meeting.

The next most important piece of information is the subject of the meeting. It is more important that people attend the meeting than that they know what the meeting will be about.

The next piece of information lets readers know how long the meeting will be, but since the meeting is mandatory, the length is less significant. This information is mostly courteous.

This piece of information is really of no importance, although it may be of interest to readers.

FIGURE 7.6 A memo using decreasing order of importance

In other instances writers may know that readers need specific information first and then general information after they have read the specifics. This sort of organizational strategy is common in many forms of journalistic writing, but it is also useful in workplace settings. For example, an engineer who investigates the structural flaws found in a suspension bridge might choose to address the most specific flaw—say, the steel-plate girder-deck stiffening system—before addressing the general flaws common to suspension bridges. This organizational strategy derives from inductive reasoning, the process of moving from specific examples to more general principles or theories.

Deciding when to use either of these strategies depends on audience needs: do they need a general background before they can understand specific information, or is the specific information what they need first?

SOME HINTS FOR USING A GENERAL/SPECIFIC STRATEGY

Moving readers from general to specific or from specific to general information can have a serious effect on how readers interpret your documents. When choosing these organizational strategies, consider using these elements:

1. A flowchart to help you better organize your information before drafting it into your document

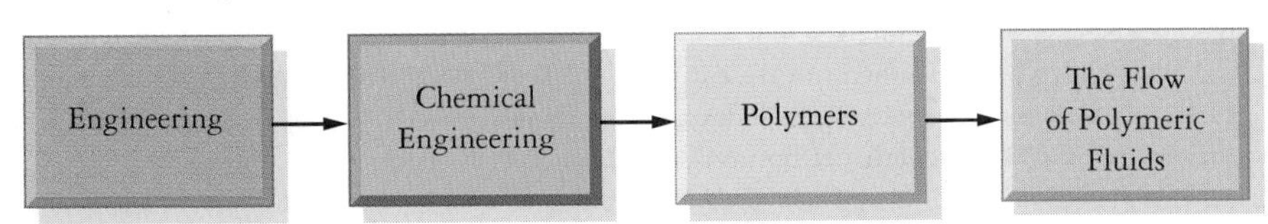

2. Language that specifically identifies your pattern: "We examined heating unit operations in general, heating unit problems similar to our own, and nine common solutions employed in situations similar to our own. Our findings show that retrofitting the heating unit is the most feasible solution."

IN YOUR **EXPERIENCE**

The general/specific strategies are quite common in a variety of documents. Addresses on letters, for instance, begin with the specific, the name of the addressee, and move to the general, the state and sometimes the country of the addressee:

Maurice Richards
Laboratory Assistant
Endocom Corporation
756 Coffins Court
Portsmouth, NH 00210

Spatial theorist George Perec played with this idea of moving from the specific to the general in addresses and expanded on it. He might have written the previous address like this:

Maurice Richards
Laboratory Assistant
Wooden Desk, Last Office on Left, Fourth Floor
Endocom Corporation, Langley Building
756 Coffins Court
Portsmouth, Rockingham County, New Hampshire, New England
United States, North America, World, Universe

Using Perec's approach to addresses, how would you write the address for where you are right now? Write that address, moving from as specific a detail as you can identify to the largest, most-general location.

Division

The division strategy is based on the idea that some things can best be understood by treating them as a series of smaller parts. This organizational strategy allows a writer to divide—and even subdivide—a whole idea, object, or phenomenon into its various components. This strategy is particularly relevant in documents that include parts lists or describe the parts of a process, procedure, policy, or event. Division is specifically useful when addressing physical items. Figure 7.7 uses a division strategy to describe a search engine.

SOME HINTS FOR USING A DIVISION STRATEGY

Dividing information into parts of a whole requires writers to understand the relationships among the parts and the way they make up that whole. When using a division strategy, consider using these elements:

1. Images to depict the ways in which the parts come together to make up the whole
2. Lists to identify the parts of the whole
3. A classification system within the divisions to group similar parts together

> Web search engines are systems that retrieve information from the World Wide Web by mining information from databases, web pages, images, and other file types. Although search engines operate on complex algorithms, there are three key parts of search engines you should understand: spiders, indexes, and search legs. Spiders are programs that methodically browse through the World Wide Web to keep the search engine supplied with up-to-date information. They are primarily used to make copies of the web pages they visit so that those pages can be mined and cataloged in the search engine's index. Indexes collect, sort, and store information found on the World Wide Web in local caches. By storing information in an index, search engines can optimize their search time by searching the index for relevant, up-to-date information rather than having to literally search whole documents or the entire World Wide Web. The search leg can be thought of as the link between the browser, the index, and the results of a search. This manual explains how these three parts work together to optimize search engine efficiency and accuracy.

FIGURE 7.7 An example of division strategy

Classification

The classification strategy groups items, ideas, phenomena, or events together according to their similarities. This strategy often uses categories to classify various items; scientific taxonomy is a prime example. Classification strategies are particularly useful in feasibility studies that present multiple solutions, which can be presented by category. Figure 7.8 employs visuals to clarify the differences among the plane classifications and provide readers with cues.

SOME HINTS FOR USING A CLASSIFICATION STRATEGY

A classification strategy is useful only when the data being presented contain a number of items that should be classified. With this strategy consider using the following:

1. A well-designed and well-thought-out order of categories that is logical and useful to your readers
2. Parallel structures for your classifications
3. All elements of the classification strategy, without leaving anything out
4. Keywords to identify not only the various classifications but also the rationales behind them

Cause and Effect

The cause-and-effect strategy is frequently used to explain the relationships between events or the reasons that something has happened or will happen.

- A study might be commissioned to determine the causes of an accident.
- A report might detail what environmental impact is apt to result from particular construction activities.
- A memo might explain how a discussion at a meeting led to a policy change.
- A set of instructions might explain what will happen if a product is misused.

The tricky part about this strategy is that in many cases it is difficult to prove that certain causes do or did indeed lead to certain effects. Therefore, the cause-and-effect strategy requires writers to understand that they are probably reporting plausible, not definitive, causes. Figure 7.9 shows an internal report that uses the cause-and-effect strategy.

SOME HINTS FOR USING A CAUSE-AND-EFFECT STRATEGY

One of the most difficult parts of using a cause-and-effect strategy is ensuring that readers can identify the relationship between the cause and the effect. Be sure that readers do not have to make great leaps of faith to accept the causal relationship. Consider using the following:

1. Only evidence that is directly related to your conclusion; no evidence that is tangentially linked
2. Evidence that represents more general data or a representative overview of the data about the causes, and not necessarily *all* of the causes
3. Only plausible data that your audience can accept
4. A chart or diagram to organize the information to better show the relationships

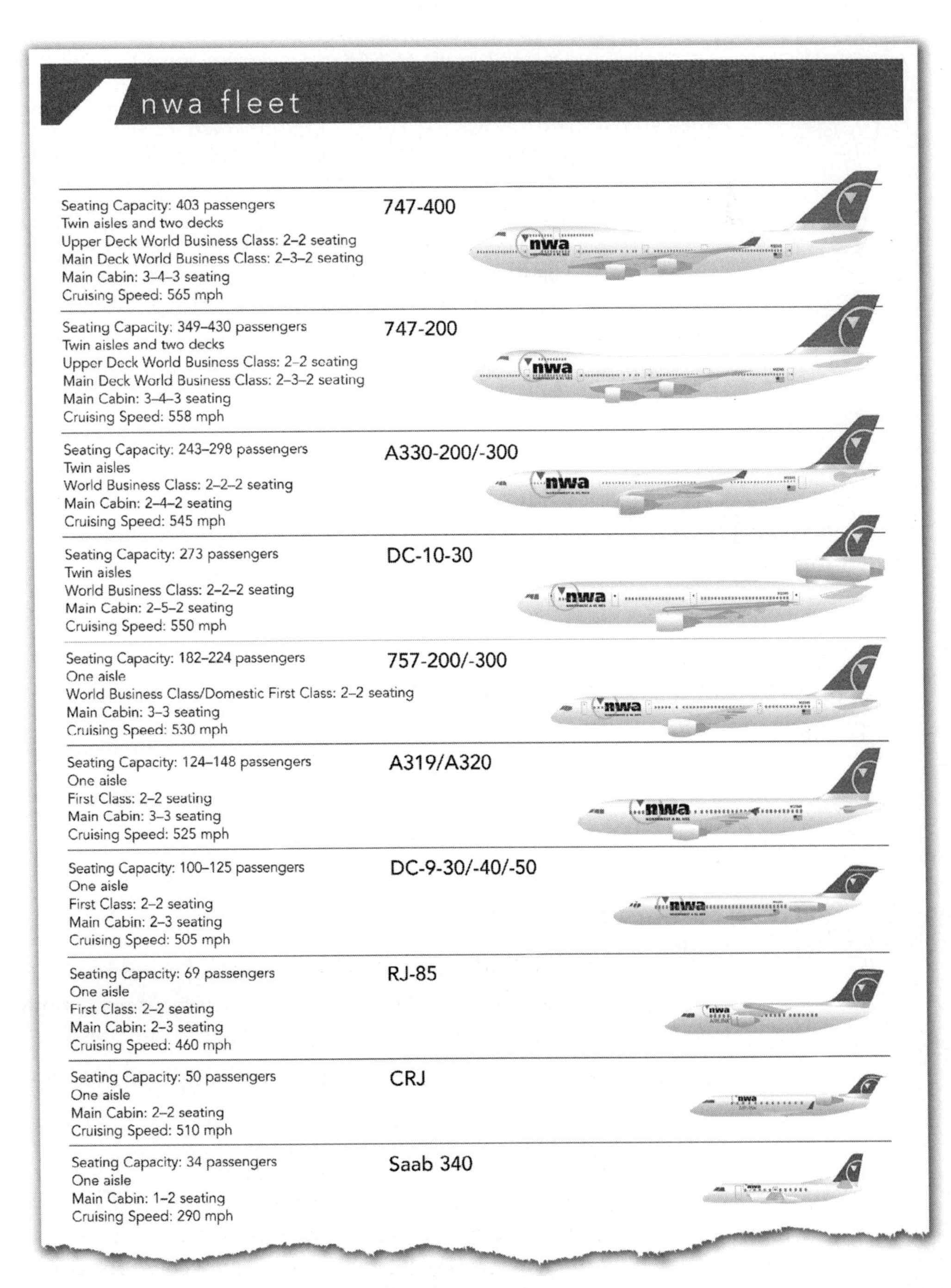

FIGURE 7.8 An example of a classification strategy

November 14, 2006

On July 14, 2006, SAB's Account Division transferred all payroll accounts to Citizens Flexible Systems (CFS), which required a number of upgrades in order to accurately handle SAB's diverse accounts. Soon after the upgrades were installed, at an additional cost to SAB of nearly $200,000, the company's Payroll Division began receiving reports of errors in employee paychecks and initiated an investigation. SAB accountants discovered that an error in the CFS system had caused nearly sixty SAB employees to be underpaid for four pay periods.

On the advice of CFS, SAB owners decided to invest another $100,000 to purchase patches for the payroll system to correct the problems. CFS has assured SAB that the system will be upgraded in time to accurately disburse payroll checks for the next pay period.

On November 10, 2006, SAB owners issued a statement indicating that reimbursement for the underpayment to employees would be delayed for two months to allow the company to recover from the unexpected $300,000 outlay for the new system. In response, SAB union leaders met to discuss these actions and have called for a general walkout tomorrow at 3:00 PM to protest the company's disregard for its employees and its apparent contract violations. If SAB does not respond to union demands within forty-eight hours, both sides anticipate a call for a full strike, which would further strain relationships between SAB owners and SAB employees.

FIGURE 7.9 An example of a cause-and-effect strategy

Comparison/Contrast

The comparison/contrast organizational strategy is commonly used in professional and technical writing because it lets readers consider how items relate to one another; they can compare the similarities and/or differences among possible solutions to a problem. For instance, a document organized with a comparison/contrast approach might do the following:

- Compare the proposals of two companies to determine which to accept
- Compare and contrast designs submitted by engineers for a new foundational molding
- Compare the credentials of several colleagues to determine who gets a promotion

A comparison/contrast approach requires that writers understand the scope of the problem and the scope of the solutions; they must recognize all the criteria to be considered. The criteria involved can be understood as the conditions or contexts in which the items are assessed.

The comparison/contrast strategy can be presented in two primary ways. The first is a comparison of the wholes, in which all of the criteria of the first item are provided and then all of the criteria of the second. This approach is particularly useful if the writer wants the reader to see the whole picture before considering the options. However, the order in which items are presented can influence how a reader assesses the information and makes a choice. For instance, an option that is presented first might be viewed as being superior. On the other hand, the item presented last may be the one the reader remembers most. An ethical writer should be attentive to such decisions.

The second approach to the comparison/contrast strategy is to present the parts of the whole using a point-by-point consideration, providing one aspect of all the options, then the next aspect, then the next, until all of the parts have been covered. Figure 7.10 uses a point-by-point comparison of the qualifications of five colleagues who are possible candidates for promotion. With this strategy writers must also decide the most efficient method of ordering both the criteria and the options to be compared. For example, should the colleagues be consistently presented in the same order, or should they reflect their ranking within each criterion?

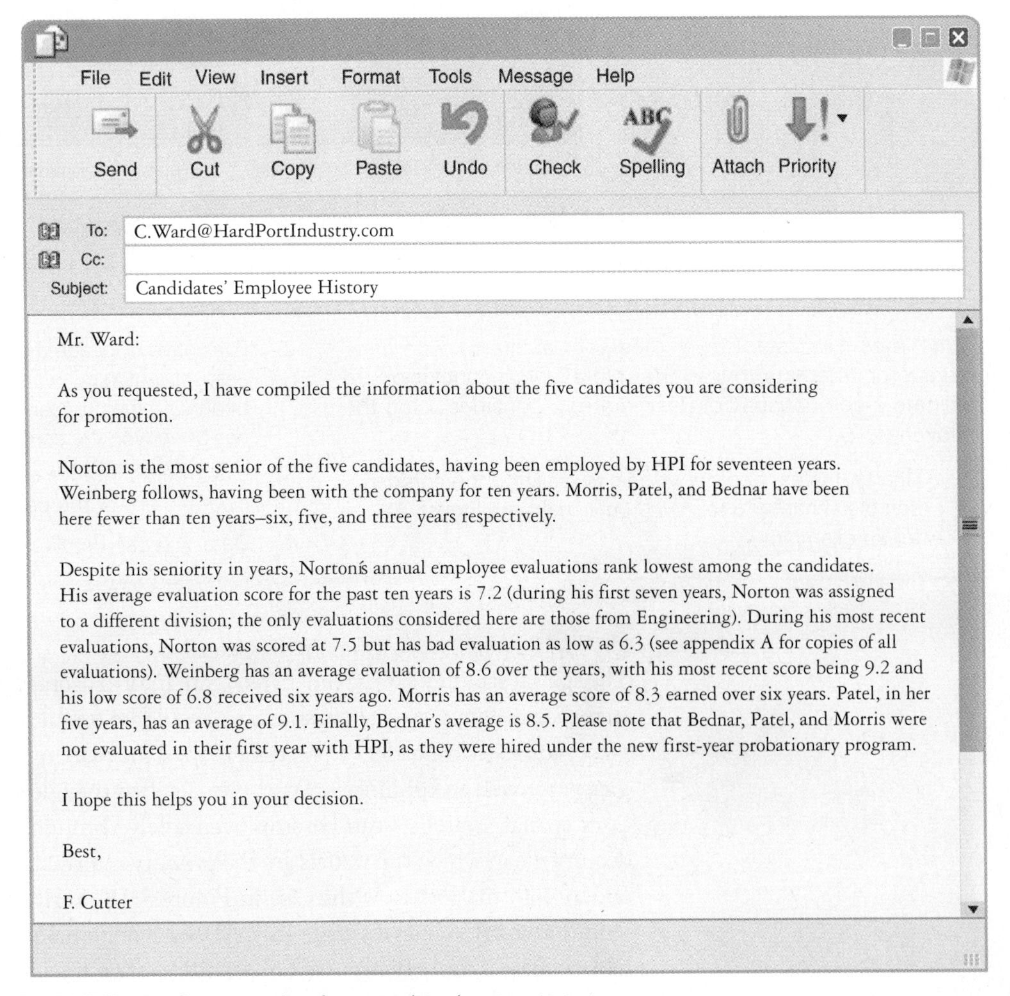
File Edit View Insert Format Tools Message Help

Send Cut Copy Paste Undo Check Spelling Attach Priority

To: C.Ward@HardPortIndustry.com
Cc:
Subject: Candidates' Employee History

Mr. Ward:

As you requested, I have compiled the information about the five candidates you are considering for promotion.

Norton is the most senior of the five candidates, having been employed by HPI for seventeen years. Weinberg follows, having been with the company for ten years. Morris, Patel, and Bednar have been here fewer than ten years–six, five, and three years respectively.

Despite his seniority in years, Nortonś annual employee evaluations rank lowest among the candidates. His average evaluation score for the past ten years is 7.2 (during his first seven years, Norton was assigned to a different division; the only evaluations considered here are those from Engineering). During his most recent evaluations, Norton was scored at 7.5 but has bad evaluation as low as 6.3 (see appendix A for copies of all evaluations). Weinberg has an average evaluation of 8.6 over the years, with his most recent score being 9.2 and his low score of 6.8 received six years ago. Morris has an average score of 8.3 earned over six years. Patel, in her five years, has an average of 9.1. Finally, Bednar's average is 8.5. Please note that Bednar, Patel, and Morris were not evaluated in their first year with HPI, as they were hired under the new first-year probationary program.

I hope this helps you in your decision.

Best,

F. Cutter

FIGURE 7.10 An example of comparison/contrast strategy

A document using the comparison/contrast approach can contain numerous items. For instance, in the example of the colleagues up for promotion, there might have been ten or even a hundred candidates. A report summarizing the qualifications of each of the candidates would need to consider the best way to present them all. A chart or a table providing a visual representation of the comparisons/contrasts might be useful, like the one in Figure 7.11.

Ethics plays an important role with the comparison/contrast strategy. A professional and ethical writer needs to present all of the options in the same light so as not to skew the reader's decisions. If a document is designed to offer a choice, then it is the writer's ethical responsibility to present the information as clearly and honestly as possible without overemphasizing any option or criterion.

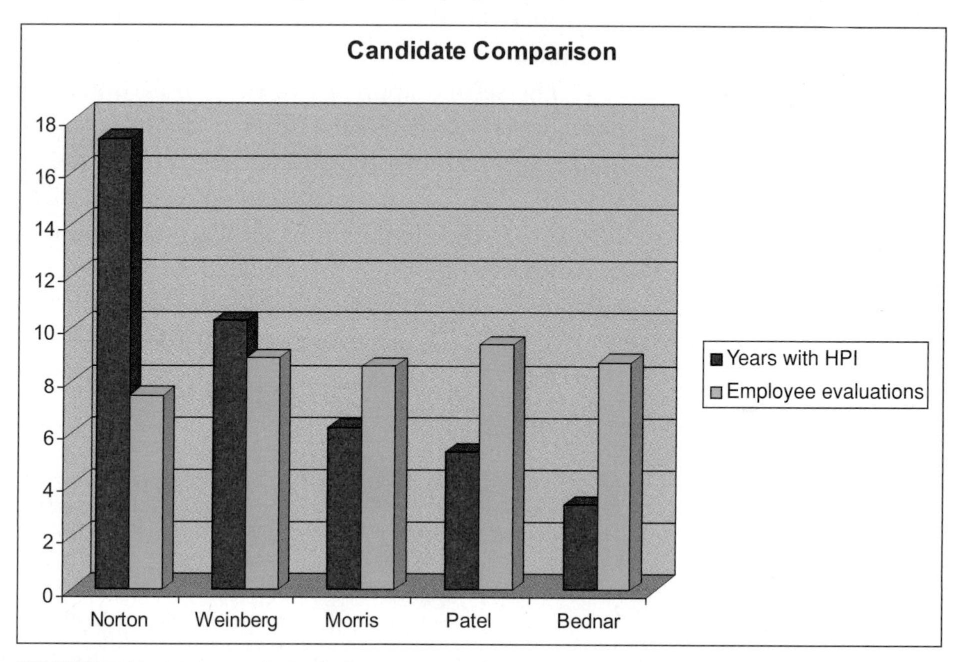

FIGURE 7.11 A bar graph depicting a comparison/contrast strategy

SOME HINTS FOR USING A COMPARISON/CONTRAST STRATEGY

Sometimes it is useful to provide your audience with clear and appropriate data to lead them to a logical conclusion. To facilitate a comparison/contrast strategy, consider using the following:

1. Charts, tables, or other visuals to facilitate the comparison (see chapter 8 for more on visuals and Figure 7.11 for an example)
2. Evaluative language that leads your audience in their assessment (e.g., instead of saying that Weinberg has been with the company for ten years, saying that Weinberg has been with the company significantly longer than . . .)
3. Only representative evidence or evidence that gives a clear picture of the pertinent data
4. Data that can be plausibly compared

Spatial

The spatial strategy helps readers navigate information pertaining to physical space or objects. For instance, spatial organization would be useful for providing information about a place, whether the reader can access a map of the place or not. Spatial organizational strategies work well in conjunction with visuals, but they don't always require them. In other cases spatial strategies can be employed solely through visuals—like schematics—or in documents in which the visuals are the primary information and the written text is the secondary information, as is the case in Figure 7.12. Written descriptions and visual representations can cover a vast area—say, Alaska—or a smaller area, such as the layout of a bank lobby where a robbery occurred or an office space that is available for rent. In the case of the bank robbery, the police report might be organized spatially to explain where the robbers entered the bank, how they moved through it, and where they ended up. Such

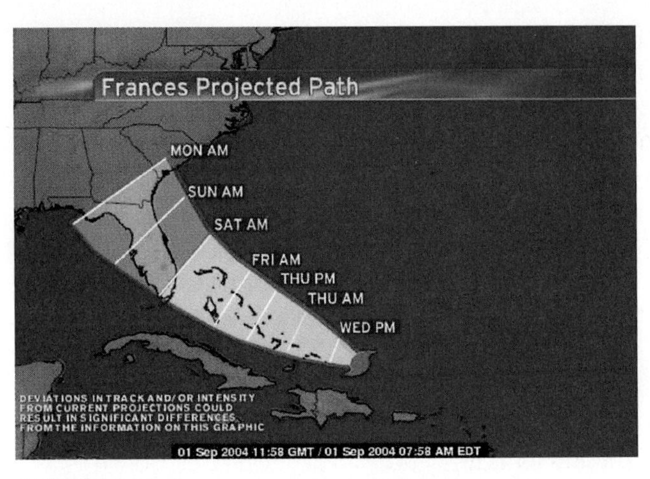

FIGURE 7.12 An example of a spatial strategy with a weather graphic

descriptions guide readers through the space and describe the role of the space in the event. A spatial strategy can also be useful to guide readers through a process using a particular object or piece of equipment.

But spatial organization is more than merely describing spaces. Spatial approaches to organization can be employed to guide readers through information in relation to the spaces affected by that information. For example, Figure 7.13 shows the introduction of a county commissioners' report about residential fire rates. As the introduction indicates, the report is organized by geographical regions and by spaces in common home and apartment construction where fires often start. The report is designed to convey information about fire rates by way of the places where fires occur.

A Report to the Daniels County Commission

February 28, 2007

Introduction

The U.S. Fire Administration, a division of the Federal Emergency Management Agency, reports that in 2006 there were 412,500 residential fires in the United States. Those fires caused 2,620 deaths, 12,925 injuries, and nearly seven billion dollars in damages. Locally, 27 residential fires were reported in Daniels County during 2006; of these, 13 were ruled to be acts of arson. Although 19 people were injured in those fires, residential fires resulted in no fatalities in 2006. (From 2001 through 2005 Daniels County averaged three fire-related deaths per year.) Estimated financial loss from 2006 residential fires in Daniels County is approximately 2.5 million dollars.

The Daniels County Fire Department (DCFD) reports that in 2006 55% of all residential fires occurred in the northeastern quadrant of the county, 19% in the southeastern quadrant, 16% in the southwestern quadrant, and 11% in the northwestern quadrant. Given that Jefferson Township, which houses the largest population in the county, is located in the northeastern quadrant, it is expected that the most residential fires would occur there. Likewise, as the northwestern quadrant is primarily zoned agricultural with small pockets of light industry, there are significantly fewer residences in that area. The DCFD further reports that within the northeastern quadrant and Jefferson Township, rates of fire occurrence can be identified within specific regions of the area, most often identified with neighborhoods developed during particular construction phases and shifts in county fire and construction codes.

The following report, commissioned by the Daniels County Commissioners, examines residential fire risks in each of the four county quadrants, beginning in the northeast quadrant and moving clockwise around the county. The analysis of the northeast quadrant considers five regions within the quadrant: Oak Grove region, the oldest residential area of Daniels County, with homes ranging in construction dates from circa 1830 until 1935; the Deer Run area, which was developed immediately following World War II; the University Park area, which is comprised primarily of apartment complexes built in the late 1970s; the Larchmont district, consisting primarily of single-family homes built in the late 1980s and early 1990s; and the Fox Pond area, an area growing from the recent development expansion. Within each of these regions, common construction approaches are examined, along with rates of fire initiation, specifically in kitchens, common rooms, bedrooms, and out buildings such as garages and barns. By considering specific physical spaces and rates of fire damage to those spaces, this report hopes to make recommendations about future county building codes in order to decrease risks of residential fires.

FIGURE 7.13 Introduction to a report that follows a spatial organizational strategy

SOME HINTS FOR USING A SPATIAL STRATEGY

When using spatial strategies, consider the following guidelines:

1. Show how the areas of the covered space are tied to the information in the document. For instance, the robbery report would likely not describe places not affected by the robbery itself.
2. Describe the necessary details of the space to the extent that those details are pertinent to the information you are conveying. Do not include unnecessary information. For instance, the robbery report would likely not include the color of the carpeting in the bank lobby.
3. Organize information in ways that explicitly indicate connections between the various parts of a space.

A spatial strategy often relies on images to clarify spatial information for readers. When incorporating visuals with a spatial strategy, consider using the following:

1. Subdivisions within larger, more-detailed images to allow readers easier access to information
2. Images that do not cram large quantities of spatial information into small visual spaces, such as listing all of the parts of a complicated device in a single image
3. Corresponding numbered or lettered lists to direct readers between images and written text

Outline Your Important Ideas

Regardless of the organizational strategy you choose, you may find it useful to develop an outline, particularly if you are working with a long, complex, and detailed document. Outlines ensure that you include all necessary information and encourage you to think about the following:

- How the information flows logically
- How you divide and label parts of the document
- How you develop transitions between parts of the document

When writing an outline, you should approach it not as a rigid mechanism that defines what you *must* do but as a general guide that you can adjust as the document demands. There is no universal outline format; many times an informal, rough outline may serve a writer's needs, whereas at other times a detailed and formal outline is necessary. A formal outline serves two purposes: an organizing guide for your document and a draft of your table of contents. Figure 7.14 gives a standard format.

Title

Purpose of document (think of this as your thesis statement)

I. First primary idea or argument

  A. First secondary idea

    1. First example, clarification, explanation, illustration of first secondary idea

      a. Detail supporting example

      b. Detail supporting example

    2. Second example, clarification, explanation, illustration of first secondary idea

  B. Second secondary idea

    1. First example, clarification, explanation, illustration of second secondary idea

      a. Detail supporting example

      b. Detail supporting example

    2. Second example, clarification, explanation, illustration of second secondary idea

II. Second primary idea or argument

And so on ...

FIGURE 7.14 A standard outline format

WRITING THE DRAFT

Parts of a Document

Nearly every workplace document can be divided into three primary parts: the front matter, the body, and the end matter. The *front matter* refers to all information provided before the core substance of a document; it gives introductory information that readers need to access and understand the document. Front matter often includes these materials:

- Title
- Cover image
- Date
- Byline
- Table of contents
- List of figures
- Executive summary/abstract
- Materials, parts, or tools lists
- Inside and return addresses
- Definitions
- Alerts and warnings
- Introduction

The *body* includes all of the core information that needs to be conveyed to the reader. The format and content of this segment are determined by the purposes of the document and the needs of the reader. The body of a document can be as simple as a one-line memo announcing a meeting or as complex as proposals and manuals spanning hundreds, or even thousands, of pages. The body of a document can include these details:

- Procedures
- Data (both summaries and actual data)
- Steps

The *end matter* encompasses all conclusions, all suggestions, any final synthesis of data, and any additional information that the reader might need. End matter helps the reader reach specific conclusions and closure, based on the information contained within the body of the document. The end matter might incorporate these elements:

- Conclusions
- Additional information
- Troubleshooting suggestions
- Additional warnings or alerts
- Recommendations
- Indexes
- Appendixes
- Glossaries
- Contact information
- Follow-up information

Of course, not all documents need all of these parts; some documents combine parts, and some eliminate them altogether. Chapters 12 to 23 in this book provide specific details about the different genres of workplace writing, and each chapter explains the kind of front matter, body, and end matter needed for that particular genre. Our purpose here is to discuss the ways writers approach drafting the front matter, body, and end matter of documents.

A Nonlinear Process

Workplace documents are rarely successful when writers use a simple, linear writing process. Instead, workplace writers usually draft larger, more detailed documents by piecing together bits of information and then organizing and revising that information into useful, readable documents. And workplace writers often collaborate with other researchers and writers, all of whom may offer input during the drafting of the document. Even with less-detailed documents—such as letters, memos, or e-mails—most writers engage in *some* writing activity prior to the formal writing of the document. The writer may have notes that direct the focus of a letter or may refer to another document that served as the catalyst for writing the new document.

PSA

A draft refers to the first version of a document that includes all of its parts. Drafting documents—part of the Draft phase of the PSA—is closely related to revising and editing documents. Although most writers revise and edit to some degree as they produce the first draft, this book addresses revision independently, simply as a matter of convenience. However, just as research can affect and blend with drafting, so, too, can drafting affect revision. Often, as writers move from research phases through drafting phases, they revise their approaches, their plans, their notes, or even their overall objectives. Thus, research, drafting, and revising are all closely connected.

When writing your draft, be sure to include the following:

- All the parts the document requires—front matter, body, and end matter
- A strong statement of purpose to ensure that your audience understands the function of the document
- Information that logically and concisely supports the purpose of the document
- Strong conclusions and recommendations

Drafting the Body

When you draft a document, the body should be the first part you write, even though it may not be the first part your audience reads or even the part that your audience reads all of. Introductions and conclusions, front matter and end matter, are dependent on what is presented in the body, which is the primary part of the document. The body is often the biggest part of the document simply because it contains the most information.

It is where you present the information that will solve the problem(s); it is how you will convince your readers. By doing your work in the body section, you will gain command of the document and control over the information.

Even though we often think of revising as something we do after writing the draft, drafting the body of a document is closely tied to revising the document. Drafting the body helps writers see clearly how the information plays out, exposing the details of the information they have and do not have and making explicit the work they still have to do to make their information useful and accessible to their readers.

Every kind of workplace document requires a different approach to writing the body of the document. For instance, the content, style, and format—not to mention quantity of information—vary greatly among, say, a memo, a letter of inquiry, a proposal, and a formal report. Yet despite such differences there are some general guidelines that can help you think through writing the body of any document.

Coverage and Length

Every document has a certain number of things that must be presented in order for the document to be effective; the document body must include all of the parts that inform the whole, all of the necessary details within the coverage area. One of the best ways to ensure that you include all necessary information in the body is to develop detailed notes and an outline during your research and predrafting phases. Notes and outlines can help you see the structure of the body and give you a sense of direction in writing it. Outlines especially can help guide you through the information you need to present. As we have said before, research, note taking, and predrafting strategies are all inseparable from the actual drafting.

As you focus on what needs to be communicated to your audience, don't think about length. Thinking about how long the body of your document will be puts an emphasis on quantity over quality. Don't limit yourself to explaining difficult information in limited space, and don't force yourself to overwrite information just to fill space. Let the clear, concise presentation of information guide the length of the body; don't let assumptions about length shape the information.

Organization and Access

Because some of the information you present in the body may be more important than other information or because you are writing to emphasize or de-emphasize particular information, how you choose to organize the body will have great effect on how your audience understands the information you are presenting. Drafting gives you the opportunity to try various organizational strategies to determine which best suit your needs.

In addition to considering how the organizational strategies you choose will make information available to your readers and influence how they understand that information, you need to make choices about how you will make that information accessible to them. For instance, will your body contain subheadings that direct readers to specific kinds of information, as the body of a formal report might? Or is your body going to be brief and self-evident in its presentation? Regardless of the length or complexity of your document, remember that *all* readers are helped by clear topic sentences that effectively announce the focus of the section or paragraph.

Drafting the Conclusion

After organizing and drafting the body of a document, it might seem logical to next draft the introduction, the information readers first encounter. However, introductions provide information that is later presented in a document's conclusion; therefore, drafts of conclusions are often written before drafts of introductions.

Drafting a conclusion can be a difficult task because it is influenced by the information presented in the body of the document, giving the writer many options to consider. A good conclusion requires careful decision making and can serve a variety of different purposes, including the following:

- A summary of information
- Analytic predictions based on the information in the body
- Recommendations about how the reader should respond or act
- A judgment about the information

A well-written conclusion both ends the document and draws something definitive from the information contained in it. Many conclusions end not by offering significant closure to the information in the document, but by calling for further action or further research. Figure 7.15 presents a conclusion from a document called "A Technical Commentary on Greenpeace's Nanotechnology Report," which was prepared by the Center for Responsible Nanotechnology in September 2003. As you can see from the conclusion, the commentary does not reach any specific conclusion but instead urges further discussion about nanotechnology policy. You may want to examine the entire report at http://www.crnano.org/Greenpeace.htm.

VI. Conclusion

The Greenpeace report correctly notes that molecular nanotechnology appears to be possible, and could have significant negative impacts. However, their analysis is based on an early understanding of MNT, and does not take into account the limited MNT that has been proposed more recently and developed in more detail. LMNT would be much simpler and cheaper to develop, and powerful enough to be extremely attractive to a variety of interests. If there is not already a targeted LMNT development program somewhere in the world, there probably will be soon.

Although some of the consequences of traditional MNT, such as self-replicating nanobots, become less significant with LMNT, other potential consequences remain areas of considerable concern. The sudden discovery of an LMNT project nearing completion would not allow time for formulating and implementing good policy. It should be emphasized that the final stages of LMNT development are likely to be the easiest and most rapidly accomplished. Hurried or panicked policy would likely be both oppressive and inadequate to prevent the negative consequences, including geopolitical instability, economic disruption, and a variety of unfortunate products and capabilities being widely accessible.

However, cautionary discussions should not ignore the fact that MNT, including LMNT, could be a strong positive assets. If administered well, the existence of cheap, clean, local, easy-to-use manufacturing capability (even limited to diamondoid products) could go a long way toward reducing poverty and underdevelopment, as well as alleviating current environmental impacts. Whether suitable adminisration can be developed depends largely on how soon the policy process begins.

FIGURE 7.15 A conclusion that urges further discussion

Drafting the Introduction

Documents usually contain detailed introductions unless audiences are already familiar with the subject. An introduction often functions as the most important section of a document because readers usually pay attention to what they read first. An introduction alerts readers to what follows.

The form of an introduction depends on the type of document being written—that is, the specific genre—and the needs of the readers of that document. Later chapters of this text focus on the unique conventions and requirements of each type of document and explain how to write an introduction for each of the genres. However, despite their differences in form, most introductions contain the same kinds of information.

We must note here that web pages do not rely on traditional introductory formats; instead, they use more visually constructed introductions. And instead of explicit sections, readers may gain information from images, bulleted lists, and linked phrases. Web pages can and do include the standard elements of an introduction, but web design provides writers with different opportunities and rhetorical choices. Most introductory information on web pages appears on a homepage, or on the first page readers access. The sections that follow pertain to written documents.

Purpose/Objective

The introduction identifies specifically why the document was written; a purpose is usually determined before any writing occurs. The purpose might evolve from a specific task, such as a supervisor's asking you to find out which software is most applicable to your company's needs and write a report detailing your findings. The purpose or objective of a document is usually one of the following:

- To convince an audience (i.e., to persuade)
- To provide direction (i.e., to instruct)
- To offer information (i.e., to report)
- To initiate action (i.e., to motivate)

Introductions often begin with specific statements of purpose; and in some cases, such as reports, the purpose of a document is identified by a specific heading:

> **Purpose**
>
> The purpose of this manual is to provide standards and guidelines for representing LitCom at international trade shows.

> **Objective**
>
> This report details the analytic techniques used to determine the viscosity and resistance of six compounds.

In other instances the objective of a document may not be identified by a heading:

> I am writing to let you know that we have reviewed the data you sent and have serious questions about their validity.

Scope

The scope of a document spells out specifically what is and is not covered or what the document does and does not do. This is crucial for readers because it allows them to determine the relevance of the document to their own needs. Sometimes writers begin documents with a particular scope in mind but adjust it as they gather data, synthesize the data, and reach conclusions. Therefore, the scope should not be written until the body has been completed.

Like other elements of the introduction, the scope may or may not be identified by a heading, depending on the type of document:

> **Scope**
>
> This document provides an overview of hypertext markup language (HTML) and a general reference for coding HTML documents. This document is not to be considered a tutorial for creating HTML documents but is an introduction and reference for beginners working with HTML.

> In this letter I attempt to give you a short summary of what has occurred to date but will not burden you with the mundane details of the problem.

Because the scope can include information relevant to the statement of purpose, sometimes writers choose to combine these parts into one segment.

Statement of the Problem

Workplace writing involves various kinds of problem solving: How should we market a product? What design will best suit our needs? What information is available to us? How should we proceed? Why did this event occur? Many writers provide readers with specific information about the problem at hand as part of the introduction. This approach brings readers immediately to an understanding of the document's focus. The problem may or may not be explicitly identified in a heading: longer, more formal documents such as reports and manuals are likely to use heading indicators, whereas letters and memos are not.

Statement of the Problem

Lack of attention to software needs and upgrades during the past two years has caused FireCorp to need massive upgrades in software compatibility. These needs are of three types: immediate needs (i.e., within the fiscal year), short-term needs (i.e., anticipating development in the next three to five years), and long-term needs (i.e., projected needs of five or more years). FireCorp has traditionally outsourced its software development; however, increased demands for rapidly available updates and local customizations require that FireCorp now examine the possibility of answering its growing software needs in-house.

Dear Mr. McHenry:

I am writing to thank you for your recent evaluation of FireCorp's software needs. As you know, FireCorp has traditionally not addressed its software problems in a timely or proactive manner.

The statement of the problem brings readers into the larger context of the document so that they can better understand its content.

Relevant Information/Background

Introductions often contain other details that help readers understand information in context. Writers sometimes include relevant background information to indicate how a current problem arose. That information allows readers to gain a better perspective on the information and understand why certain parts of it are relevant. Readers can also become more interested in the presented information as they identify their own investment in the situation. Like other parts of the introduction, the background may or may not be presented with a heading that explicitly alerts audiences to it.

Key Terms

Some introductions define key terms to help clarify information. Writers want to make sure that key terms are not misinterpreted or misunderstood, leading to a possible misunderstanding of the entire document. Unlike glossaries or other end-matter dictionaries, the key term segment of an introduction addresses *only* the primary terms that might cause confusion. Not all documents require key term sections, but using one might eliminate interpretation problems. Figure 7.16 gives an example of a key term section used in the introduction of a document.

Overview of Organization

Introductions announce information that lies ahead, and they show readers how that information will be presented. Introductions—particularly in long, formal documents such as manuals, reports, and feasibility studies—often show the format, structure, and organization of the document. This approach allows readers to anticipate how they will receive information and how they can best understand and use it.

Organization

This document is organized in five parts, excluding this introduction. Four of the parts contain detailed data from four resonance tests conducted over the course of the experiment. These parts are organized chronologically according to the dates when the resonance procedures were initiated. The final part provides an analysis of the data from the four resonance tests.

Summary

Audiences sometimes become distracted or may simply not have the time to read an entire document. Therefore, many writers include a summary of the document in the

Introduction

This introductory note seeks to provide a basic—but not an exhaustive—overview of the key terms employed in the United Nations Treaty collection. . . .

Agreements: The term "agreement" can have a generic and a specific meaning. It also has acquired a special meaning in the law of regional economic integration.

(a) *Agreement as a generic term:* The 1969 Vienna Convention on the Law of Treaties employs the term "international agreement" in its broadest sense. . . . The term "international agreement" in its generic sense consequently embraces the widest range of international instruments.

(b) *Agreement as a particular term:* "Agreements" are usually less formal and deal with a narrower range of subject-matter than "treaties.". . . Nowadays by far the majority of international instruments are designated as agreements.

(c) *Agreements in regional integration schemes:* Regional integration schemes are based on general framework treaties with constitutional character . . . but the subregional instruments entered into under its framework are called agreements.

Conventions: The term "convention" again can have both a generic and a specific meaning.

(a) *Convention as a generic term:* Art. 38(1)(a) of the Statute of the International Court of Justice refers to "international conventions, whether general or particular" as a source of law. . . . The generic term "convention" thus is synonymous with the generic term "treaty."

(b) *Convention as a specific term:* Whereas in the last century the term "convention" was regularly employed for bilateral agreements, it now is generally used for formal multilateral treaties with a broad number of parties. . . .

Charters: The term "charter" is used for particularly formal and solemn instruments, such as the constituent treaty of an international organization. The term itself has an emotive content that goes back to the Magna Carta of 1215. Well-known recent examples are the Charter of the United Nations of 1945 and the Charter of the Organization of American States of 1952. . . .

FIGURE 7.16 An example of key terms in an introduction

introduction to provide an overview of the key data and conclusions found in the document. Summaries ensure that readers anticipate the information that follows and are guided toward a particular conclusion. In addition, summaries enable readers to get the key information whether they read the entire document or not.

More often than not, summaries are placed at the end of the introduction; however, some proposal and report formats (see chapters 20, 21, and 22) require the front matter to include a separate executive summary, which details the key information, conclusions, and recommendations of the document. Executive summaries are designed for executives who have little time to read documents but must make decisions based on them nonetheless.

In standard summaries it is better not to use obvious phrases such as "To summarize this report . . ." or "In summary we recommend. . . ." Instead, allow your language to guide your reader through the summary: "In this report the primary investigators show how . . ." or "Based on the acquired data, it is recommended that. . . ."

ELECTRONIC TEMPLATES AND WIZARDS

Chapter 3 examines the role electronic technologies play in producing workplace writing and offers advice about how to best use available technological resources. It is important to consider template software and wizards in relation to drafting. A template is a preestablished format for a document. Many word processing programs include templates for documents such as memos, résumés, and letters. Users need only to fill in the blanks with key information, and the software formats the information into a standardized document. Wizards are interfaces that lead users through a series of dialog boxes in order to accomplish a particular task. Wizards can ask writers/users to provide basic information that can then be used to automatically produce formatted documents such as web pages. More advanced wizards can ask a writer questions about content arrangement, design elements, and stylistic approaches and then adjust documents according to the user's responses. Wizards can also be used as problem-solving tools that ask users to provide yes/no answers to basic questions until the Wizard narrows the user's problem to a few possible solutions. Many kinds of help and electronic troubleshooting documents use wizards in this way.

See chapter 18 for more on writing help pages and troubleshooting sections.

Templates are easily available through word processing software, like Microsoft Word, but many companies and organizations also develop their own custom-designed templates. Custom templates offer workplace writers a number of advantages, including a framework for documents that are used frequently by a company—for example, weekly progress reports. Such frameworks can reduce the amount of time a writer spends on a document by showing the writer exactly what to include. Custom templates also help companies establish a consistent look to their documents, a "brand look." Companies may even provide document templates when they request documents from outside contractors. For instance, a company seeking proposals from multiple bidders might provide a proposal template to ensure that it receives the same kinds of information from each bidder and to expedite its review process by standardizing where certain kinds of information are in each proposal.

See chapter 20 for more on proposals and requests for proposals.

Although templates and wizards save time and are useful tools for beginners unfamiliar with document design, standard formats, or standard content inclusions, they also have some flaws. First, because some template software is distributed to large populations (such as everyone who has Microsoft Word on his or her computer), many users create documents based on those templates. Consequently, when readers encounter a template document that they have seen numerous times before, they may become bored. More important, in recognizing the document as a template document, readers may make judgments about the writer's initiative, professionalism, knowledge, and ability. In addition, because templates are designed to answer the needs of many generic users, rarely are they able to answer the specific needs of a specific rhetorical situation. Templates also limit design possibilities; writers lose much control over their documents by relying on a template. By writing the documents themselves, they can be more certain that information is presented as they wish.

Despite their benefits, we generally recommend avoiding template approaches to writing workplace documents. However, many companies and organizations develop their own templates, boilerplates, and formatting standards for particular types of documents to ensure consistency and inclusion of particular information. In such cases you should work within the parameters established by your company; those templates serve a particular institutional need.

SUMMARY

- Predrafting strategies include confirming your purpose, analyzing your audience, gathering your information, developing ideas about your information, and organizing your information.
- Organization is controlled by the document's purpose, its audience, logic, and ethics.
- Organizational strategies include sequential, chronological, order of importance, general/specific, division, classification, cause and effect, comparison/contrast, and spatial.
- Generally speaking, technical documents can be divided into three primary parts: front matter, body, and end matter.

- Writing the end matter and the front matter often requires first writing the body.
- The body is the core of the document and may contain these elements: procedures, data, and steps.
- The conclusion creates closure and may summarize the given information, provide analytic predictions, make recommendations, or issue a judgment.
- The introduction alerts the reader to what is contained in the document and may include these elements: purpose/objective, scope, problem identification, relevant information/background, key terms, organization, and summary.
- Web page introductions use similar parts but may do so with a more visual approach.
- Templates and wizards may be useful tools but have several disadvantages.

CONCEPT REVIEW

1. Where does drafting fit in the problem-solving approach?
2. What are general strategies that should precede drafting documents?
3. Why might a writer need to use a specific kind of organizational strategy to draft a document?
4. How does the sequential strategy present information?
5. How does the chronological strategy present information?
6. How does the order-of-importance strategy present information?
7. How does the general/specific strategy present information?
8. How does the division strategy present information?
9. How does the classification strategy present information?
10. How does the cause-and-effect strategy present information?
11. How does the comparison/contrast strategy present information?
12. How does the spatial strategy present information?
13. In what ways might visuals help clarify information in a document, regardless of the organizational strategy employed?
14. What elements of a document are contained in the front matter?
15. What elements of a document are contained in the body?
16. What elements of a document are contained in the end matter?
17. What are some strategies for drafting the body of your document?
18. Why would a writer need to write the body of a document before writing the conclusion or introduction?
19. What strategies apply to writing conclusions?
20. What strategies apply to writing introductions?
21. What are positive and negative aspects of using templates and wizards in drafting documents?

CASE STUDY 1

Organizing Information about Search Engine Options

As you probably know, each of the popular search engines on the World Wide Web—such as Google or Yahoo!—works a bit differently, offering users a variety of ways to search the Web for specific information. Imagine that you have been asked to prepare a document that will help your company's executive board determine which search engine to recommend as your company's web tool. They have asked that you present information about at least three commonly used search engines and address the following capabilities: Boolean tags, case sensitivity, search fields, default search, proximity searching, and truncation.

First, identify the three search engines you wish to study—don't feel obligated to use the three we've named; there are many others that offer good search tools. Then gather the data, organize the data, and determine what organizational strategy would best suit your document. Create a draft that begins to organize your information in the way you think would best present your findings. Then write a letter to your instructor explaining why you have chosen to present the information in the way you have.

CASE STUDY 2

Organizing an Informal Field Report

Figure 7.17 shows a short, informal report prepared by Duane Warren, an environmental field technician with Pacific Innovations Northwest Electric (PINE). As part of his regular duties, Warren is required to submit informal reports about the field sites he visits. His reports advise field supervisors about how they should install high-voltage wiring in buildings. Warren's job is to find the best way, with the least environmental impact, to install wiring from high-voltage towers and junction boxes. Warren's reports are internal documents, but often his conclusions guide the field supervisors who write proposals, bids, and reports for external audiences. What would Warren need to have known before writing the introduction or the conclusion to this report? Could he have made recommendations or offered conclusions before he conducted his study? Why or why not? Could the introduction have been written before the body? Why or why not? Write a short evaluation of how Warren reaches his conclusions and what he uses as a basis for his introduction.

To: Rachel Orr, Field Supervisor

From: Duane Warren, Environmental Field Technician

Subject: Filbert Fiber Optic, SW K St., Grants Pass, Oregon

Date: July 9, 2008

Cc: Karl Branyon, Field Supervisor
Danielle Bohlke, Field Supervisor

On July 7, 2008, I visited the job site located on SW K St. in Grants Pass, Oregon, for which PINE is considering issuing a bid to rewire Filbert Fiber Optics, which has recently acquired the building located at 3605 SW K St.

This report examines the potential environmental impact of bringing new conduit and wiring from the three primary PINE junction boxes within a two-mile radius of the job site:

Junction Box #34511-13 at the intersection of SE 11th Ave. and SE J St.

Junction Box #34511-32 at the intersection of SE 11th Ave. and SE Riverside Ave.

Junction Box #34521-34 at the intersection of SW K St. and SW Pine St.

After examining the routes needed to bring wire in from each of these three sites either underground or via poles, I recommend that PINE offer in its bid to bring wire in from either junction box #34511-13 or #34511-32.

Even though junction box #34521-34 is closest to the work site and offers the most direct route to the work site, bringing wiring in from junction box #34521-34 would require spanning poles and wire across Skunk Creek, a site that is currently being reviewed by local authorities in an ongoing environmental impact study (see Grants Pass, Oregon Tier 2 Study #6089-675).

Although both junction boxes #34511-13 and #34511-32 require the use of more materials, these sites should be considered closely in a bidding for this contract, as they are the only two sites from which PINE could feasibly bring wiring into the job site without delays imposed by the environmental impact study.

I recommend that the PINE bid offer both junction boxes #34511-13 and #34511-32 as possible sources for the rewiring connections since both would provide the job site with the needed power, while giving PINE some room to maneuver in figuring overall cost when the job is undertaken.

Neither of these sites would involve any environmental impact.

FIGURE 7.17 An informal field report

CASE STUDIES ON THE COMPANION WEBSITE

Streamlining Communications in the South Carolina Department of Transportation

The Director of Highway Engineering for the South Carolina Department of Transportation officially supervises every highway construction project in the state. Currently, regional directors must submit weekly reports on the work being done in their districts. The director realizes that the engineering department could be more efficient.

Time for an Upgrade: Mark Simmons Confronts the Problem of Obsolete Computers at Quality Care Nursing Home

Mark Simmons, a customer relations specialist for a software company, works extensively with clients once they have purchased his company's software. He helps them install the software, teaches them how to use it, and provides customer support if problems develop.

Beacon Biological: Retracing and Remedying Six Months of Business Disaster

The founder of Beacon Biological no longer manages the day-to-day business, and the company's profit margin has dropped 25 percent over the past two quarters. Beacon's employees fear cutbacks and layoffs, and they need to be reassured that these are not part of Beacon's plans for the future.

Designing the Fighting Platypi Little League Baseball Team Web Page

In this case you will analyze the concerns of a little league coach who is considering designing a website for his team of 9- to 12-year-old boys and girls. Your task includes proposing a website that satisfies the desires of the team—which wants a networking website where the members can communicate and develop the team's history—and the needs of the parents, who are concerned about the children's online safety.

Software Woes: Training Employees in How to Use HXaune 3.0

In this case you will consider the problems an IT training coordinator confronts in developing a company's new software training materials. The rehetorical problems include analyzing what technical information the employees need and how best to organize that information in a professional document.

VIDEO CASE STUDY

Cole Engineering–Water Quality Division

Collaboration and Drafting the Oxygen-Bleach Report

Synopsis

After years of work, Harry Lonsdale and Betty Childs have developed a new oxygen bleach. Now the two have to share their findings with others at Cole, including Cole's legal representatives, a public relations representative, a marketing director, two chemical engineers, and a safety engineer. Each of the members of the product development team will also be required to produce documentation about their aspect of product development. This case focuses on a conversation between Harry Lonsdale and Betty Childs about the documents they will have to produce to initiate the procedure.

WRITING SCENARIOS

1. Locate the nutrition information on a food package. Read and analyze the package, and discuss which parts of the packaging (e.g., health warnings or ingredient lists) might be considered the front matter, the body, and the end matter. Then write an analysis of the relationship of the information found in each of the parts.
2. Using an order-of-importance strategy requires that the writer determine which information is most important, least important, and somewhere in between. Such determinations are not always easy. Consider the information listed here, and organize it according to an order-of-importance strategy. Then discuss with your class what difficulties impacted your decisions. This information relates to a work project that is temporarily suspended while a fatal accident is investigated.
   - The site visit was intended as a preemptive inspection of a gas line intersection.
   - All project engineers are required to attend the meeting.
   - Moses Hightower was making a site visit.
   - The information is to be transmitted on May 20, 2006.
   - The accident occurred on the project site.
   - There will be a meeting on May 24, 2006, to discuss the investigation.
   - The site visit had been unannounced.
   - The information is to be directed to the project engineers.
   - The project will now be delayed during an investigation.
   - The information concerns an accident that occurred on May 18, 2006.
   - The project coordinator had anticipated a real inspection on May 25, 2006.
   - Moses Hightower was killed when the gas line intersection ruptured.
   - Work on the project is temporarily suspended.
   - Moses was a good person and an asset to the company and the project.
   - The information is from the project coordinator, Lynn Hopkins.
   - On May 18, 2006, Moses Hightower was killed in an accident.
3. At one time or another, you have probably written a document that required collaboration. Write a short explanation of how you went about organizing and drafting those documents. For example, you might want to address how the time spent researching, organizing, and drafting the information was divided and coordinated. Then create a table with two columns. In the first column write down all of the positive aspects of collaborative writing. In the second column write down the negatives or potential drawbacks of collaborative writing. Discuss your results, weighing the pros and cons of organizing and drafting collaborative documents.
4. Find a web page for a company or professional organization that does work in a field that you might pursue. Analyze the web page, and identify how it presents its introduction, purpose, and organizational strategy. Write a document that explains how the web page conveys key points of information.
5. It is likely that your career will be in a subfield of a larger discipline. There may also be within your area of expertise even smaller divisions of specialization. Write a short document employing a classification strategy to describe a field in which you might be interested.
6. Scientific taxonomy serves as a prime example of a classification strategy. According to its taxonomy, the genus and the species of a sailfish are *Istiophorus platypterus.* Using a web search engine, locate the family, order, class, and phylum of the sailfish. Then continue your research to find out what characteristics apply to each of these levels of classification. Finally, write a document that

explains the sailfish classification, identifying all levels of taxonomic classification and the characteristics of each.

7. In what ways is deciding on an organizational strategy also an ethical decision? How can solving organizational problems also solve ethical problems? In an essay or a letter to your instructor, explain the relationship between ethics and organizational strategies.
8. Consider the ethical implications of writing an introduction or a conclusion before writing the body of a document. Write a short essay or letter to your instructor that explains the ethical ramifications of drafting the body of a document before its conclusion or introduction.
9. One of the important aspects of using a comparison/contrast strategy is to be sure that you are comparing like items. In what ways can you imagine a writer using a comparison/contrast strategy unethically? Write a short explanation that details how a comparison/contrast strategy might be used unethically.
10. The steps for burning CDs are usually straightforward; most instructions use a sequential strategy. Imagine that a friend is trying to use your computer to make a copy of a CD but can't find the instructions. Write a sequential list of steps for your friend to follow, beginning with turning the computer on and ending with turning it off.
11. Locate two of your favorite music CDs and pull out the liner notes. Analyze these different documents: What does each contain? Are lyrics included? production information? images? copyright information? writing credits? music credits? advertisements? acknowledgments? Once you have analyzed the two documents, use a comparison/contrast strategy to compile two different lists that could be used to draft a document. In the first list use a comparison of wholes, and in the second list use a comparison of parts.
12. This text can be thought of as an extended series of technical instructions about how to successfully write workplace documents (see chapter 18 for more on technical instructions). Select any chapter in the book, and consider the method used to organize the information in that chapter. What strategies have we used? Are they effective? What other methods of organization might we have employed? Explain your responses in detail.
13. Identify the templates provided with the word processor you use. Are there templates for letters, reports, memos, or manuals? Select one template and analyze it in relation to what this chapter has said about organizational strategies and drafting. In what ways does the template conform to the chapter information, and in what ways is it different? In an informal essay, compare and contrast the features of the template with the chapter suggestions.
14. Many instruction manuals for electronic devices are organized using a division strategy, which allows the documentation to address various components of the device—for example, the processor, the monitor, and various input devices of a computer. Consider the computer you use to write most of your class assignments. List the key components you would address if you were explaining the various parts of your computer. Then list what you would explain for each of those primary divisions.
15. Consider the advantages that technological applications like word processors, e-mail, and the World Wide Web offer writers in the drafting phase of their writing. How might each of these technologies benefit workplace writers? How might each work to the disadvantage of workplace writers? Write a short document that details the advantages and disadvantages of technological applications in drafting.

8

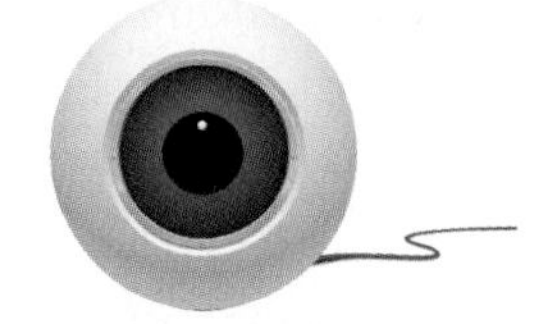

Visual Rhetoric and Technical Communication

CHAPTER LEARNING OUTCOMES

After completing this chapter, you will be able to do the following:

- Know what **visuals** are
- Understand the role of visuals in **documents**
- Discern the difference between **images and graphics** and the uses of each
- Make informed **visual rhetorical choices**
- Identify and employ the **various forms** of images and graphics
- **Find, capture, create, and format** visuals
- Use **color** effectively
- Understand **when and where** to use visuals in your documents
- **Revise** visuals while revising an entire document
- Recognize the **ethical implications** of using visuals

DIGITAL RESOURCES

On the Companion Website www.prenhall.com/dobrin:

- Case 1: Visuals with Teeth: The Ethics of Graphically Representing Shark Attacks
- Case 2: Visualizing Performance: Designing Graphs for Cyclists
- Case 3: Identity Theft: The Ethics of Logo and Image Manipulation
- Case 4: Making Bikes Green
- Video Case: Revising the Cole Engineering Website
- PowerPoint Chapter Review, Test-Prep Quiz, Exercises and Activities
- TCTC Visual Rhetoric and Design Tutorial

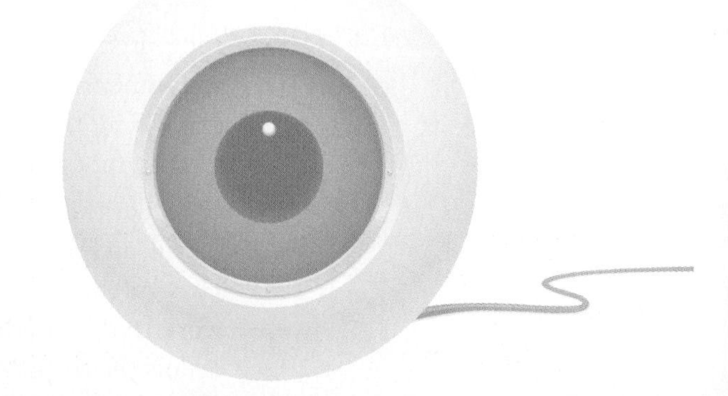

REAL PEOPLE, REAL WRITING

MEET HEIDI MACHUL • Architect with Richard Taylor Architects

What types of writing do you do as an architect?

When I sit down with a client, the first thing I do is to write a program outlining the size of the building, the budget, and the different spaces, as well as descriptions of how those spaces will be used. I also write many transmittals explaining what drawings are being sent and at what stage they are. Throughout the entire process of a project, we document our meetings with meeting notes. Even when visiting a construction site, we document the condition of the job site and the status of the project. Having the ability to track the history of a project from start to finish is very important, and documentation makes it possible.

Who are the typical readers of the documents you create?

The majority of the writing that I do is for clients. Because I work as a representative of the client, I must have a complete understanding of what the client is trying to achieve in order to convey that to the engineer, product suppliers, and contractor. Taking notes and documenting everything in writing allows the client to confirm that we have understood what is being said. The better the communication among all of the parties involved, the more successful the project will be.

What types of visuals do you incorporate in your writing?

We use a variety of visuals in an architect's office to help our clients envision what their projects are going to look and feel like once they are built. We start each project with a site analysis so that the client can see how the building will be oriented on the site, to take advantage of its positive features and minimize the negative. We then create sketches to illustrate how the spaces can be laid out and what they might look like. This process continues until the sketches match what the client has been envisioning. Construction drawings include floor plans, elevations, sections, and details; however, a structure cannot be built with drawings alone. Structural notes, construction notes, and specifications for the products to be used must also be included.

Another graphic tool that we use to help a client envision the project is 3-D modeling. It is difficult for the majority of our clients to look at a two-dimensional floor plan or elevation and be able to understand what the structure is going to look like once it is built. With a client that is not able to visualize the design, we create a 3-D model to visually show how the space is going to feel. Sometimes we build a cardboard model so that the client can get a better idea of how the building will actually look.

What is the relationship between text and visuals in the documents you create?

The relationship between text and drawings is very important; they have to work together in order to be successful. Notes in our drawings identify and explain items that cannot be explained by graphics alone. All of the drawings, sketches, and details that we create are supplemented by the name of the image and the scale to which it has been drawn. Sometimes words carry more meaning than graphics; however, without the graphics it would be very difficult to understand how a space is laid out.

What advice about workplace writing do you have for readers of this textbook?

Writing must be clear and accurate. Documenting information is very important in any line of work and must be taken seriously to avoid confusion and errors. Learn to be a great listener; you will always learn more by listening than by speaking. My favorite thing to hear from a client is "Wow, you really listened to me." That lets me know I'm doing a good job because clients are getting what they have been dreaming about.

INTRODUCTION

On September 16, 2007, Senator Barack Obama, along with Governor Bill Richardson, Senator Hillary Clinton, and other Democratic presidential hopefuls, attended a fund-raising event sponsored by Iowa Senator Tom Harkin. The candidates stood on the stage while a vocalist sang the national anthem.

FIGURE 8.1 Senator Barack Obama, Governor Bill Richardson, Senator Hillary Clinton, and Ruth Harkin standing during the national anthem

A photograph of the candidates at the event appeared in the October 1, 2007, edition of *Time*. That magazine article stated that the photo had been taken while a vocalist sang the national anthem, not during the Pledge of Allegiance. However, the photo quickly became the focus of a controversy in which Senator Obama was accused of refusing to place his hand over his heart during the pledge (see Figure 8.1). The photo appeared on a number of websites, in a series of chain e-mails sent to millions of people, and on major news programs. Many of these messages or broadcasts used the picture to question Obama's patriotism, some suggesting that the image offered "proof" that he was unsuitable to be president.

Rumors about the photo and Obama's patriotism continued for many months. Senator Obama discussed the photo during political rallies and on television interviews, stating that "the photo is very frustrating and has been misidentified. It's simply not true. I've been pledging allegiance since I was a kid." Obama later noted the impact that such photos can have, suggesting that "there's only so much you can do about this kind of misinformation."[1] Amidst all of the debate about the image, one thing was clear: this image conveyed meaning and evoked a response. Many people saw the same image, but not everyone took away the same meaning.

Ours is a visual culture. Visuals like the photo of Senator Obama surround us every day, providing information, initiating conversation, and demanding our interpretation and understanding. Even visuals that are not tied to news stories or political campaigns play a crucial role in how we access and transmit information every day. Think, for instance, about the icons on your cell phone, the hyperlinks or program icons on your computer, or the images on the screen of your MP3 player—all provide us with information.

Because visuals convey information, we can say that they are rhetorical. How they are used to convey information is a matter of rhetorical choice—that is, *visual rhetorical* choices. Was the photo of Senator Obama used to advance a particular political agenda? Was the photo intentionally misidentified for an effect? How did moving the photo from the context of the *Time* magazine article change the ways in which the image was interpreted and received? How did the different mediums of magazine, Web, e-mail, and television affect the image's reception? Were there ethical choices in publishing and distributing the image? Does the information in the photo reach us in ways that a written description cannot?

For more on typography, see chapter 9, "Layout and Design."

The term *visual* refers to a pictorial representation used to convey meaning and information to an audience. Of course, written text itself is a visual representation, but this chapter considers visuals to be depictions other than text, including elements that affect the presentation of text, such as using a **bold,** *italic*, or color type. This chapter addresses two kinds of visuals: graphics and images. Technically speaking, these terms, along with the word *visual*, can be used interchangeably and often are. However,

[1]http://www.cbn.com/CBNnews/265357.aspx.

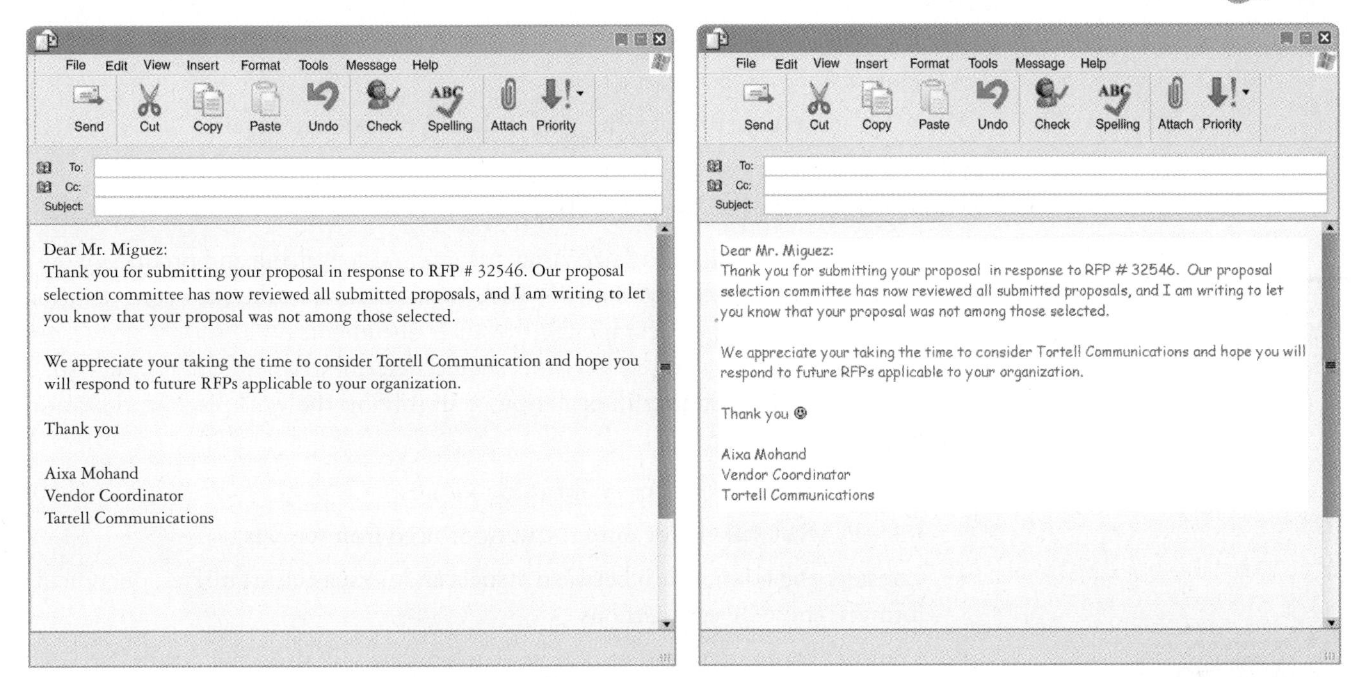

File Edit View Insert Format Tools Message Help
Send Cut Copy Paste Undo Check Spelling Attach Priority
To:
Cc:
Subject:

Dear Mr. Miguez:
Thank you for submitting your proposal in response to RFP # 32546. Our proposal selection committee has now reviewed all submitted proposals, and I am writing to let you know that your proposal was not among those selected.

We appreciate your taking the time to consider Tortell Communication and hope you will respond to future RFPs applicable to your organization.

Thank you

Aixa Mohand
Vendor Coordinator
Tartell Communications

File Edit View Insert Format Tools Message Help
Send Cut Copy Paste Undo Check Spelling Attach Priority
To:
Cc:
Subject:

Dear Mr. Miguez:
Thank you for submitting your proposal in response to RFP # 32546. Our proposal selection committee has now reviewed all submitted proposals, and I am writing to let you know that your proposal was not among those selected.

We appreciate your taking the time to consider Tortell Communications and hope you will respond to future RFPs applicable to your organization.

Thank you ☺

Aixa Mohand
Vendor Coordinator
Tortell Communications

FIGURE 8.2 Two visually different versions of the same e-mail

for the sake of clarity and ease, we treat graphics and images separately here. The next chapter also addresses visuals but focuses on elements of document layout and design. These four components—graphics, images, layout, and design—make up the visual elements of workplace documents.

Graphics refers to visuals that appear to be rendered or drawn, such as tables, maps, graphs, and diagrams. The term derives from the graphic arts, a field in which graphic designers create art for use in the production of documents. *Images* refers to photographic or realistic-looking visuals, such as photographs, screen captures, or even moving images like video clips or animations. The term relates to media technologies that incorporate visuals taken from real conditions, such as photographs of a piece of machinery. Graphics and images each have advantages and disadvantages, which this chapter discusses in detail. But for now it is important to recognize that all workplace elements require a consideration of visual elements. Even the most basic documents—like the short e-mails depicted in Figure 8.2—rely on visual elements, and those elements affect how readers interpret the documents. Consider how the visual difference between the e-mails in Figure 8.2 alters how you read the e-mails. Workplace writers must recognize that visuals are central to all of their documents. Deciding when to use visuals, which ones to use, and where to place them is an important part of rhetorical thinking.

VISUAL RHETORIC

visual rhetoric the ways visual elements communicate meaning to readers

Visual rhetoric refers to the ways visuals communicate meaning to readers. Workplace writers must choose how to design and produce visual elements to communicate effectively and clearly. Workplace writers must ask the following kinds of questions:

- Should I use a visual?
- What kind of visual should I use?
- What should the visual look like?
- Where should the visual be placed in the document?

- How should visual and written elements relate to one another?
- How should the document look?
- What form (e.g., print or electronic) should the document take?

Using the Problem-Solving Approach with Visuals

PSA

Workplace writers should recognize visuals as integral to writing and producing documents. Throughout each phase of the PSA, writers should take into consideration how they will address visual components of their documents.

In the Plan phase, workplace writers should determine whether and how visuals would be useful in explaining information or in solving the workplace or rhetorical problems at hand:

- Establish the goal or purpose for using visuals
- Identify what stakeholders might want or need from visuals
- Consider the relationship between visuals and persuasion strategies, document formats, and delivery methods
- Consider appropriate and ethical ways to employ visuals in the document
- Identify what information should be presented visually
- Choose technologies that will best assist the production of visuals

During the Research phase, writers should do the following:

- Determine the types of information and sources necessary to produce visuals
- Conduct research and gather information needed to produce visuals

During the Draft phase, writers should do the following:

- Reconsider appropriate and ethical ways of presenting information visually
- Confirm the persuasion strategy, document format, and delivery mechanism from a visual standpoint
- Organize and draft the visuals needed
- Design the document and arrange its components in an appropriate order
- Integrate visual elements that help communicate the information

During the Revise phase, writers should do the following:

- Test the usability of the visuals
- Solicit feedback and response from peers and colleagues
- Revise the visuals, based on feedback

During the Distribute phase, writers should

- Confirm that files with visuals are not too large to distribute

The visual problem-solving approach helps workplace writers make careful rhetorical decisions.

Communicating with Readers

Appropriate visuals can help audiences in many ways; a single visual can accomplish a number of functions all at once. However, writers must consider which functions they are attempting to accomplish and then select the best visual to solve the given problem.

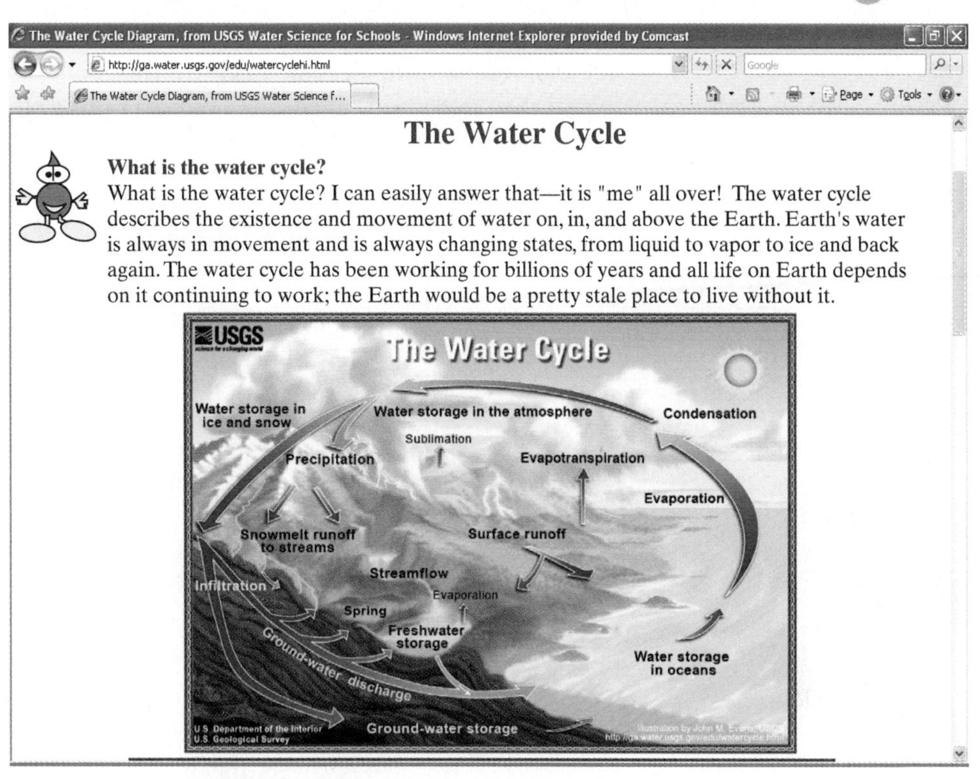

FIGURE 8.3 A visual used to clarify

Increasing Comprehension

In many instances visuals can be more efficient in conveying information than words. Visuals can elaborate details to help readers understand and can improve readers' retention and recollection of information.

Clarification Visuals often clarify difficult concepts, abstract information, and detailed processes and procedures by providing an easy-to-see reference. Examine the example found in Figure 8.3, which depicts the water cycle. Certainly, the writer of this document could have described the water cycle in various levels of technical detail, but the visual helps to readily clarify the process for readers. In some documents visuals provide examples for readers, which allow them to see what is described and to build a mental frame of reference.

FIGURE 8.4 Apple CEO using visuals to illustrate as he introduces the iPod video

Illustration Visuals can also illustrate key information for readers. Illustrations are generally linked directly to written texts to shed light on the written words. People who make presentations (see chapter 23) often use visuals to illustrate points because audiences are more likely to remember information they see. Visuals often depict relationships between pieces of information. For instance, Figure 8.4 shows Apple Chief Executive Officer Steve Jobs making a 2005 presentation in California, introducing three versions of the new iPod video. The visual he uses is designed to illustrate the differences in memory capability between versions of the video iPod. Likewise, our inclusion of this visual is designed to illustrate how visuals can illustrate.

Organization Visuals such as charts, tables, white space, and bullets help organize information into clear, accessible forms. Sometimes it is just easier to present a visual than to describe a lengthy process or piece of information with words. As the old adage says, "A picture is worth a thousand words": people learn and understand information best by seeing it.

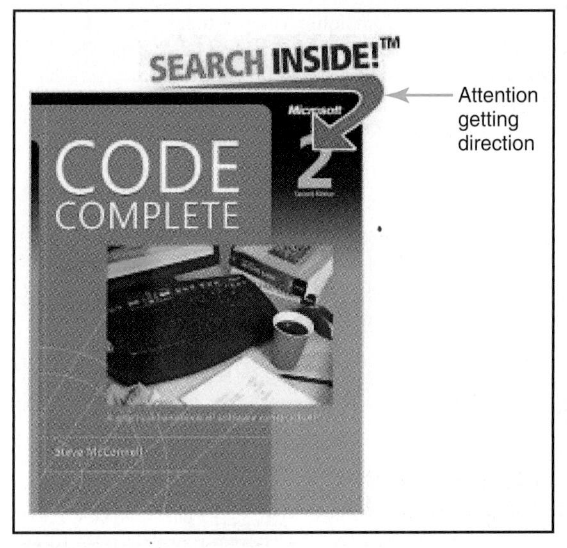

FIGURE 8.5 Visual used to draw attention

Gaining Attention

Visuals get our attention: think about how traffic signs get our attention, or flashy magazine covers, or a product package. Even the design and layout of a document can gain readers' attention. Visuals can be used in a number of ways to gain attention.

Emphasis Visuals can effectively emphasize key information—a usage that carries ethical implications. Notice, for example, the ways that a shift in the dimensions of text (i.e., typography) adds emphasis: *italics*, **bold**, underline, color, size. Or consider a combination of these ***effects.***

Graphics or images can also emphasize information by drawing a reader's attention to it. Notice, for instance, how the online bookseller Amazon.com uses visuals to let you "look inside" a book before purchasing it; the graphic is superimposed over the book cover, drawing attention to it (see Figure 8.5).

Highlighting In addition, visuals can highlight specific information. For instance, Figure 8.6 shows retail spending per capita in different provinces of China. The use of color highlights certain provinces in order to both identify their location and highlight specific information about spending trends.

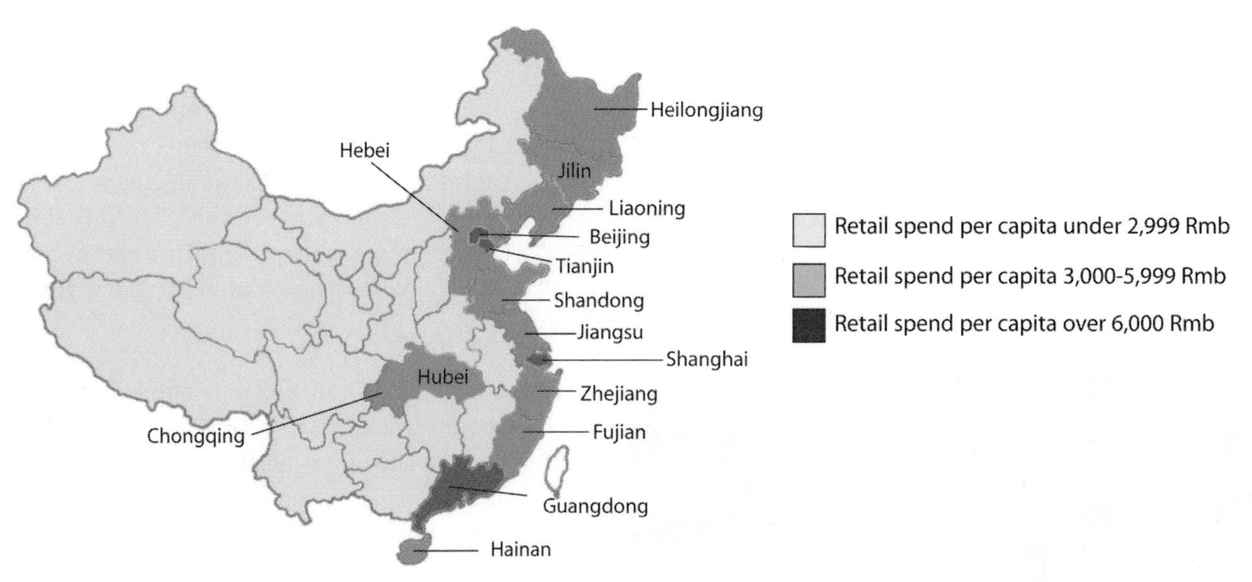

FIGURE 8.6 Diagram that highlights information

Establishing Authority

When disseminating information, writers often find that certain visuals, such as graphs and charts, can evoke a sense of authority and professionalism. Because such visuals are sometimes associated with the kind of quantifiable, statistical information that experts or scholars might generate, these visuals are often assumed to represent expert information. Some national newspapers like *USA Today* use graphs and charts to convey information and project authority through visuals that look professional, accurate, and reliable. Visuals can help establish credibility in the minds of readers.

IN YOUR EXPERIENCE

Visuals gain our attention every day. As we drive, our eyes are drawn to traffic signals and signs. Brightly colored store signs vie to draw us in as customers. Restaurant menus and store advertisements use visuals to draw our attention to specials and sales. Flyers and brochures use visuals to get us to read the contents. List ten rhetorical moments in which a visual has gained your attention.

Communicating with a Broader Audience

See chapter 5 for more on transnational and multinational writing.

As we explain in chapter 5, many workplaces now employ, serve, or communicate with workers and customers around the world, but global communication can lead to difficulty in disseminating information among groups with different languages and cultural backgrounds. Visuals can assist in communicating with a broader, more diverse audience. Consider the number of products individuals in the United States purchase that are produced in countries where English is not the primary language. Those manufacturers must produce, among other documents, instruction manuals for English-speaking customers; but translating documents can be time-consuming and costly, and many smaller organizations can afford neither the time nor the money. In such cases visuals can help bridge the gap between writers and readers. However, visuals are not necessarily universal; that is, not all visuals have the same meaning for readers in every culture. Consequently, writers must make culturally aware choices when using visuals. Later in this chapter we address guidelines for using visuals when writing for transnational and multinational audiences.

IN YOUR EXPERIENCE

Have you ever worked with a set of instructions that were written in a language other than your native language? If so, did the instructions include a visual element? Did the visuals help you better understand what the instructions were trying to accomplish? Have you ever worked with a set of instructions that were poorly translated from another language into the language you use? If so, what was the effect of the poor translation on you as a reader and a user of that document? Were visuals included? Did they help?

TYPES OF GRAPHICS

Icons

The term *icon* derives from the Greek word *eikon*, which literally means "image." Icons are visual metaphors because they denote more than they literally present; they are generally simple for readers to interpret and are designed to rapidly convey information. Many companies and organizations rely on icons as their trademarks or logos, which rapidly evoke connections to their organizations or companies. In visual rhetoric, then, logo icons serve as visuals that prompt readers to think about the products or concepts the logos represent. The Nike logo or icon, for example, prompts readers to think about the products Nike produces. Religious symbols like a cross or a star of David are icons that prompt readers to think about or make connections with the religions they represent. We are all used to seeing logos; in fact, most of the clothes we wear display logo icons to further promote those companies' products. As writers, we regularly use icons in the letters and numbers we choose; we even use emoticons to rapidly express emotions in word processors, instant message programs, e-mail, text messaging, and web production ☺.

In workplace writing, icons convey particular meanings. They may be used in bulleted lists, as alerts (see chapter 18 for more on alerts), as headings and dividers for highlighting elements, as buttons on web pages. Their primary purpose is to draw readers' attention to particular information.

Guidelines for Making Icons

Since an icon is a visual metaphor, you must first understand what the icon needs to represent before you are able to create it. Often, it helps to simply identify the words that the icon should evoke; you might develop a list of all of the terms. For instance, if you were creating an icon to be used in a print document to direct readers to a website for further information, your list might look like this:

link	further
information	connection
web page	computer

You can learn a lot by looking up terms in a dictionary. Then, when you have your key terms and concepts, the best way to make an icon is to sketch ideas on paper. After that, you can begin to use a graphics program, like Photoshop or Fireworks, to generate your icon.

ANALYZE THIS

The use of icons is so prevalent in Western culture that we rely on them for everyday information. Think, for instance, about the icons of the stop sign or the walk light at street corners. Think about corporate icons, like the McDonalds M or the Nike swoosh. For the next week keep an icon journal, making a list of every icon you see and describing what each looks like. If you can, include a picture of the icon—you can sketch it or photograph it or capture an image from the Web. Explain what each icon represents and why you think the people who made it made the choices they did.

Graphs

A graph is a diagram that represents the relationship between two or more kinds of quantifiable information. Generally, graphs depict information along two axes, with each axis representing one component of information. For instance, along one axis a graph might represent the year in which data were recorded, and along the other axis, the number of products sold. Graphs are excellent visuals for showing the relationships among many connected pieces of data.

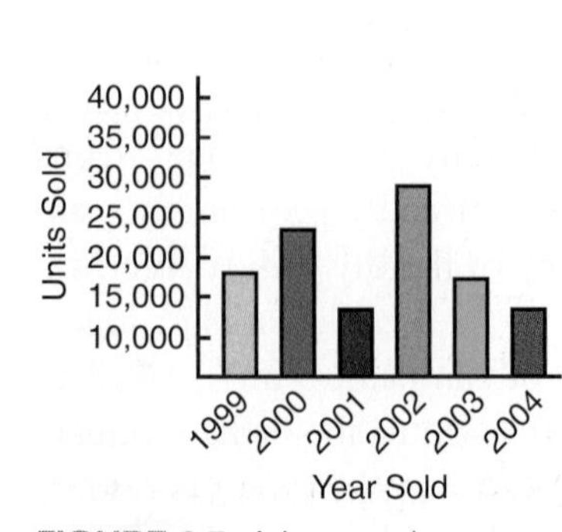

FIGURE 8.7 A bar graph

Bar Graphs

Bar graphs, also known as histograms, are the most commonly used graphs. Most often they emphasize quantity and are used with numeric information, such as sales figures or statistics. Bar graphs generally offer two types of information, each depicted on one side of the graph. For instance, in Figure 8.7 the vertical axis (i.e., the left side) of the graph represents the number of units sold, and the horizontal axis (i.e., the

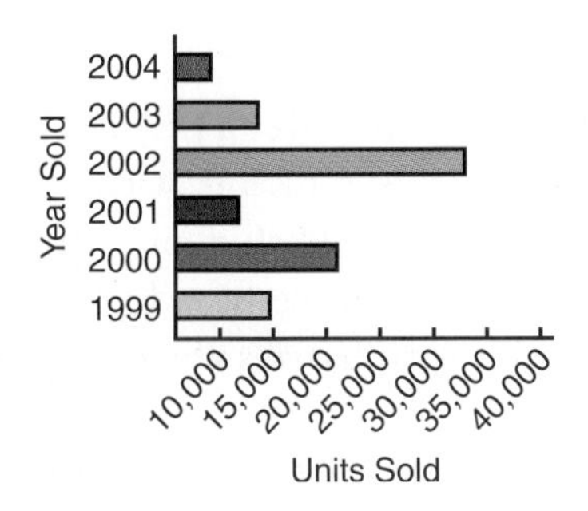

FIGURE 8.8 A bar graph with horizontal presentation

bottom) represents the years in which the units were sold. The basic bar graph can be modified in three ways.

Horizontal Presentation This modification is a simple rotation of the graph to present the information horizontally rather than vertically (see Figure 8.8). All bar graphs can be presented either vertically or horizontally; deciding which alignment to use is a matter of visual rhetorical choice.

Deviation from a Norm This modification provides for information that deviates from a set norm within the information. For instance, Figure 8.9 shows the same information presented in Figures 8.7 and 8.8 but includes sales figures from 2008 and identifies which years broke with the norm of selling at least 10,000 units per year. Deviation graphs often use a different color to depict the deviating data.

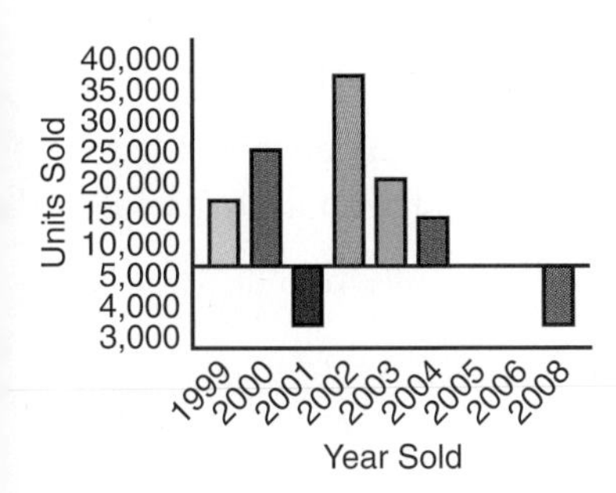

FIGURE 8.9 A bar graph showing deviation from a norm

Grouping This modification provides comparisons of information in groups. For instance, Figure 8.10 depicts not only the number of units sold and the years in which they were sold but also compares total sales with sales from the West Coast sales force and the Midwest sales force. Notice in this example how color is used to visually distinguish the data.

Guidelines for Making Bar Graphs Because bar graphs give readers visual keys to help them understand information, you must not misrepresent or confuse the information in bar graphs. Recall Figure 4.4, the bar graph that depicted opinions regarding the Terri Schiavo case; it depicted a significant difference of opinion despite the relative agreement that the poll numbers indicated. When designing and producing bar graphs, you should follow these guidelines:

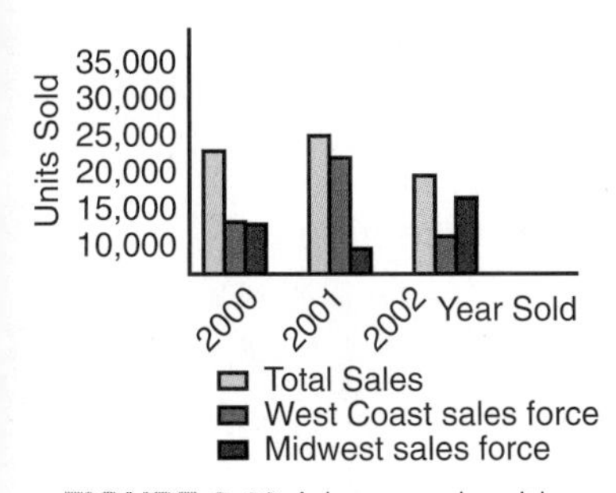

FIGURE 8.10 A bar graph with grouping information

1. *Depict data fairly and accurately.* Do not distort or manipulate numbers or information to influence readers.
2. *Begin graphs from a numerical point of zero or a comparable starting point whenever possible.* Beginning from a logical starting point familiarizes readers with the data. Examine the difference in representation between the two bar graphs in Figure 8.11. In the first graph Group 1 appears to have participated only minimally, whereas Groups 2 and 3 participated significantly more. The second chart, however, shows that there is really only a minor difference among the levels of participation of the three groups.
3. *Make sure your graphs are clear.* Use grids, marks, and other visual cues to help readers locate information easily. Notice the use of line marks in the second graph of Figure 8.11 to help identify the numbers of participants.
4. *Clearly label all items of information in the graph.* Identify the information represented along each axis, and if you are using multiple colors, provide a legend to identify what each color represents. Also, be sure to provide a nearby caption, heading, or title for each graph so that readers know what the graph represents and can connect it to any references. Placement depends on the design of your document, but it is common and logical to place the title just below the graph or beside it.
5. *Arrange data in a logical, readable manner within the graph.* Although chronological sequences seem evident, other data patterns may not be as evident. Consider how you want your readers to interpret the data, and use a visually logical arrangement.
6. *Use spreadsheet programs.* Programs like Microsoft Office Excel enable you to automatically convert spreadsheet data into graphs.

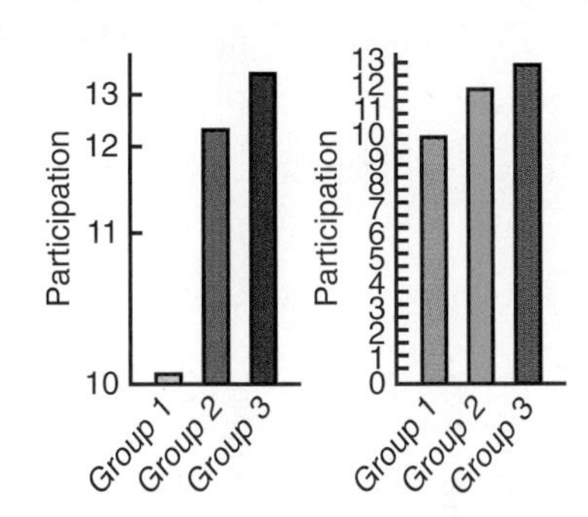

FIGURE 8.11 Different representations of the same data

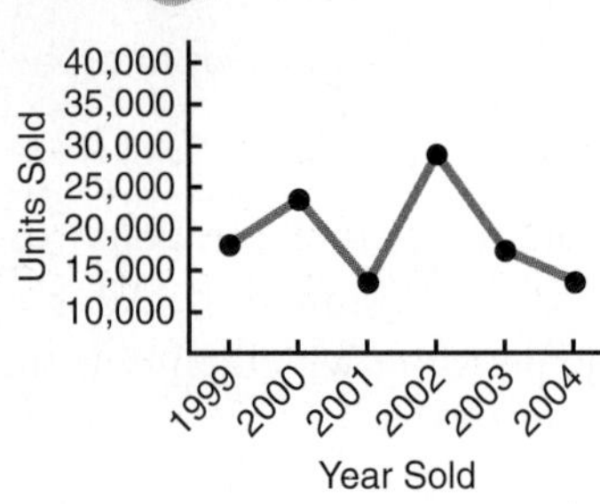

FIGURE 8.12 A line graph with a single entry

Line Graphs

Line graphs, also known as frequency polygons, often depict shifts over a period of time. For instance, Figure 8.7 depicts sales of a given unit over a period of time, highlighting the number of units sold. A line graph could be used to depict the same data, emphasizing the shifts in sales (see Figure 8.12). Notice how Figure 8.12 emphasizes the change over time, whereas Figure 8.7 emphasizes the quantity of units sold.

Line graphs can also represent multiple pieces of similar data to compare changes over time in more than one situation. Figure 8.13 represents the same sales data found in Figure 8.10; notice the shift in emphasis from quantity to change over time.

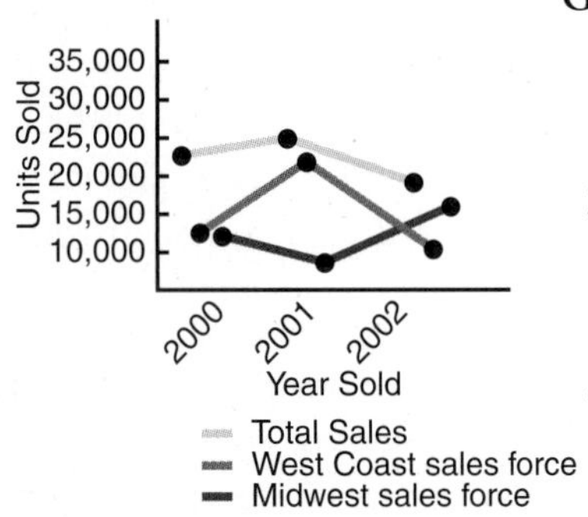

FIGURE 8.13 A line graph with multiple entries

Guidelines for Making Line Graphs

1. *Present data in a logical, easy-to-read manner.* Keep the information as simple as possible, and deliver it in a way that most readers would expect to see it.
2. *Use a caption or title for the line graph.* Readers can more easily understand what the graph represents when it is captioned or titled. This is particularly useful for busy readers.
3. With multiple lines in a single graph:
   a. *Use different colored lines for the different data.*
   b. *Use contrasting colors.*
   c. *Make separate graphs for each line if the lines cross often.*
   d. *Make the graph large enough so that all the details can be read and interpreted.*
4. *Provide horizontal and/or vertical grid marks.* These marks help readers more precisely locate information on detailed line graphs.
5. *Label each point of change to clarify values for readers.* This is particularly important when the plotted line changes frequently (see Figure 8.14).

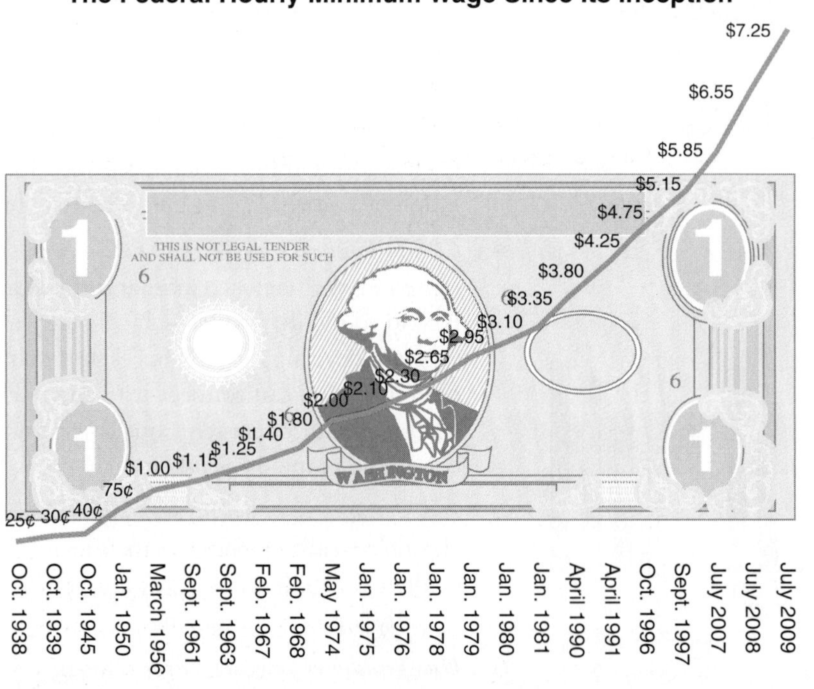

FIGURE 8.14 A line graph with points of change labeled

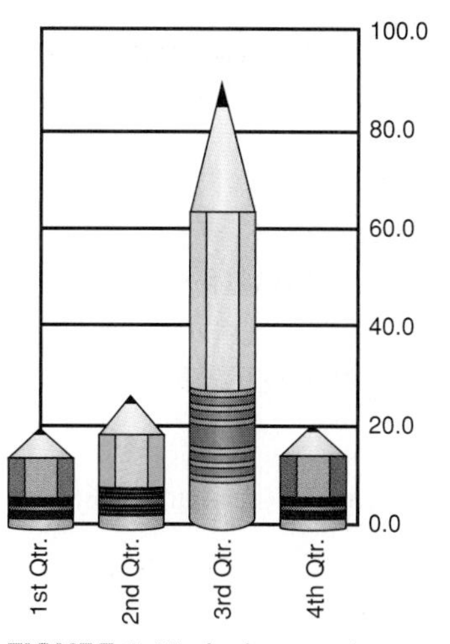

FIGURE 8.15 A pictograph

Pictographs

Pictographs use pictures or icons to represent numeric information (see Figure 8.15). Pictographs are usually easy to read and function much like bar graphs. They are most often used in documents that address general readers rather than specific audiences.

Guidelines for Making Pictographs

1. *Make your graphic clear, simple, and direct.* Clean graphics are always best.
2. *Make multiple graphics distinct from one another.* If they are similar, consider using complementary colors or styles.
3. *Be fair in your representations.* Do not distract from, distort, or misrepresent the information.
4. *Use clip art for making pictographs.* Clip art is simple and uncomplicated, which is good for pictographs. In fact, pictographs are one of the few professional applications of clip art.
5. *Use graphing programs to convert information and graphs into pictograms.* Microsoft Office Excel is a commonly used program that can easily convert information into pictographs.

Charts

Charts are closely related to graphs in that they represent information visually. The major difference is that graphs often present time as a numerical value. Time can be a factor in some charts, such as flowcharts, but it is not generally quantified. In workplace documents, charts make it easier for readers to understand data and the relationships among data.

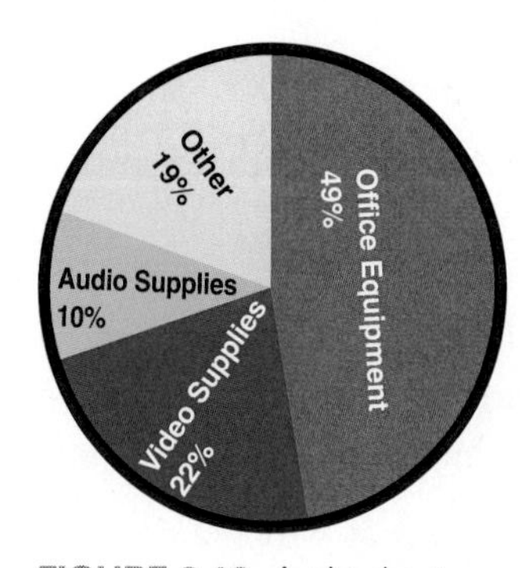

FIGURE 8.16 A pie chart

Pie Charts

Pie charts are circular charts that represent divisions within a whole. They most often represent percentages by dividing a whole circular area into representative sections. Within a pie chart each section's arc, angle, and volume are proportional to the quantity that the sector represents. For example, Figure 8.16 shows a pie chart depicting the percentages of expenses spent on office equipment, video supplies, audio supplies, and other expenditures.

Guidelines for Making Pie Charts

1. *Limit a pie chart to no more than seven sections.* Too many sections can be confusing or difficult to read.
2. *Use contrasting colors, particularly for those segments next to each other.* Use color to highlight or emphasize a given section. For example, the pie chart in Figure 8.16 shows a large section in a bright red color, which draws readers to that section.
3. *Label each segment clearly.* Choose one of three approaches: labels directly in or on each segment (e.g., see Figure 8.16), a corresponding legend, or outside labels that are visually linked by clear association or call-out lines. Call outs are textual highlights to clarify or draw attention to information in a visual. They are often used if a visual shows an exploded view to depict parts.
4. *Make the percentages in the chart equivalent to those in your data.* Always confirm that your percentages total 100 percent.
5. *Provide a caption for the chart.* Captions should explain the function of the pie chart and the information it depicts.

6. *Don't add too many effects to a pie chart.* Doing so may make the chart too cluttered or may distort the information. Avoid 3D effects that can distort information.

Flowcharts

Flowcharts depict a process or a procedure. Originally developed to help computer programmers think through program processes, flowcharts have become standard tools not only for illustrating but also for problem solving. Flowcharts use a set of symbols to identify various points within a process:

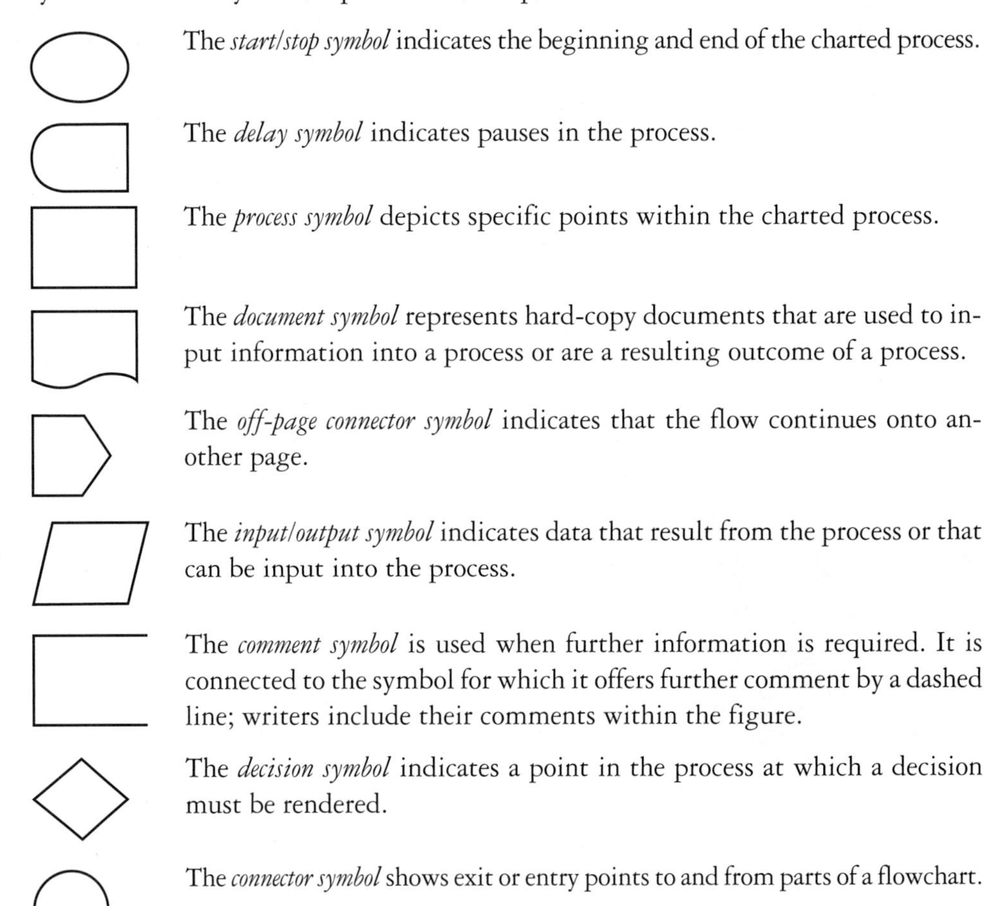

The *start/stop symbol* indicates the beginning and end of the charted process.

The *delay symbol* indicates pauses in the process.

The *process symbol* depicts specific points within the charted process.

The *document symbol* represents hard-copy documents that are used to input information into a process or are a resulting outcome of a process.

The *off-page connector symbol* indicates that the flow continues onto another page.

The *input/output symbol* indicates data that result from the process or that can be input into the process.

The *comment symbol* is used when further information is required. It is connected to the symbol for which it offers further comment by a dashed line; writers include their comments within the figure.

The *decision symbol* indicates a point in the process at which a decision must be rendered.

The *connector symbol* shows exit or entry points to and from parts of a flowchart.

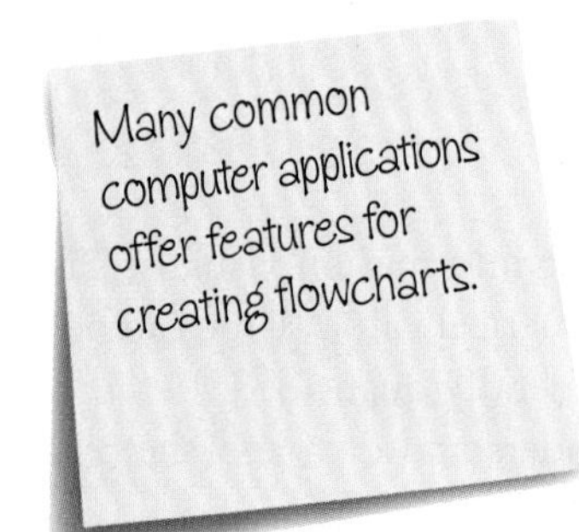

There are four primary types of flowcharts:

- *Top-down flowcharts:* These start at the top and present the primary steps of the process drawn vertically. Details of each primary step are then listed in order under each step. Generally, top-down flowcharts show a one-way flow and do not offer key decision points that might result in returning to previous points in the process. Figure 8.17 depicts a simple top-down flowchart.
- *Work-flow diagrams:* These illustrate the flow of work and are generally less detailed than standard flowcharts. They offer overviews of business processes, identifying information such as who a company's suppliers are, how products are distributed, who provides feedback to the company, and so on (see Figure 8.18).
- *Deployment flowcharts:* These show not only how a process flows, but also who is responsible for each part of that process. These charts also identify when collaboration is necessary or when one participant depends on the work of another before beginning a part of the process. Generally, deployment charts depict the names of the participants horizontally across the top of the chart. Their tasks and responsibilities are then listed vertically below their names,

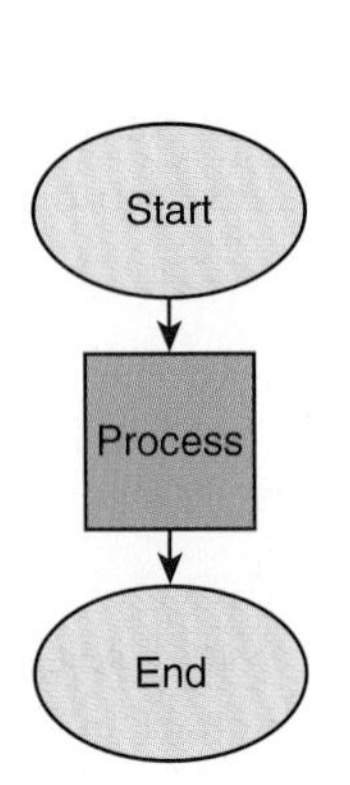

FIGURE 8.17 A simple top-down flowchart

and flowchart symbols show the relationships among the various participants' tasks. Figure 8.19 shows a simple deployment flowchart.

- *Detailed flowcharts:* These depict all the steps and sequences in a process, including the relationships among various internal stages. Figure 8.20 shows a detailed flowchart.

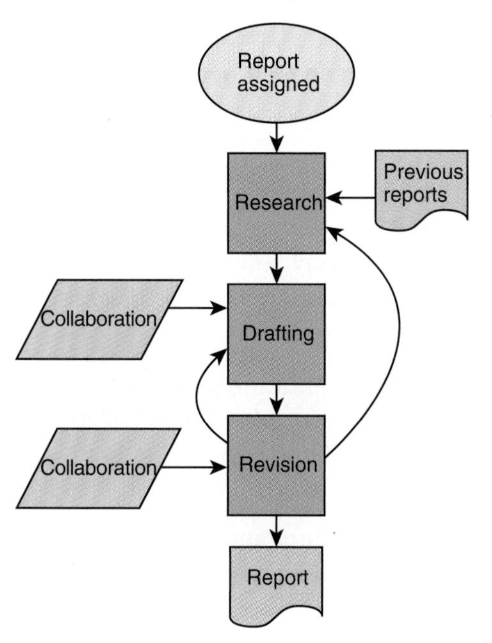

FIGURE 8.18 A work-flow diagram

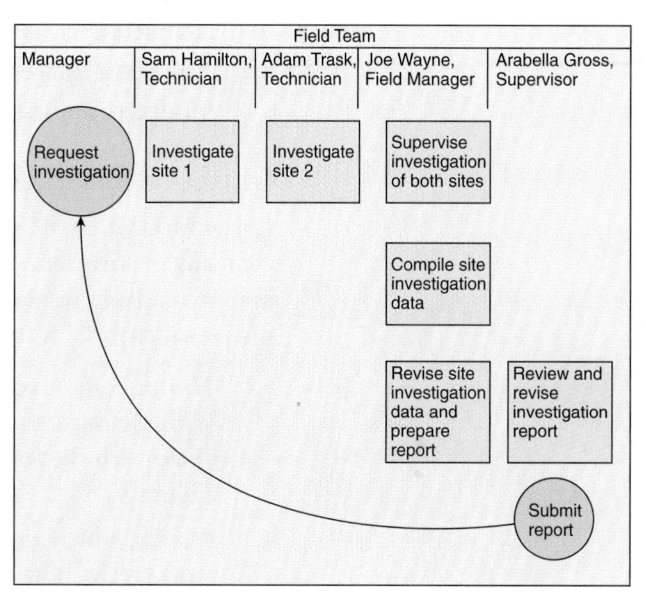

FIGURE 8.19 A deployment flowchart depicting an investigation procedure

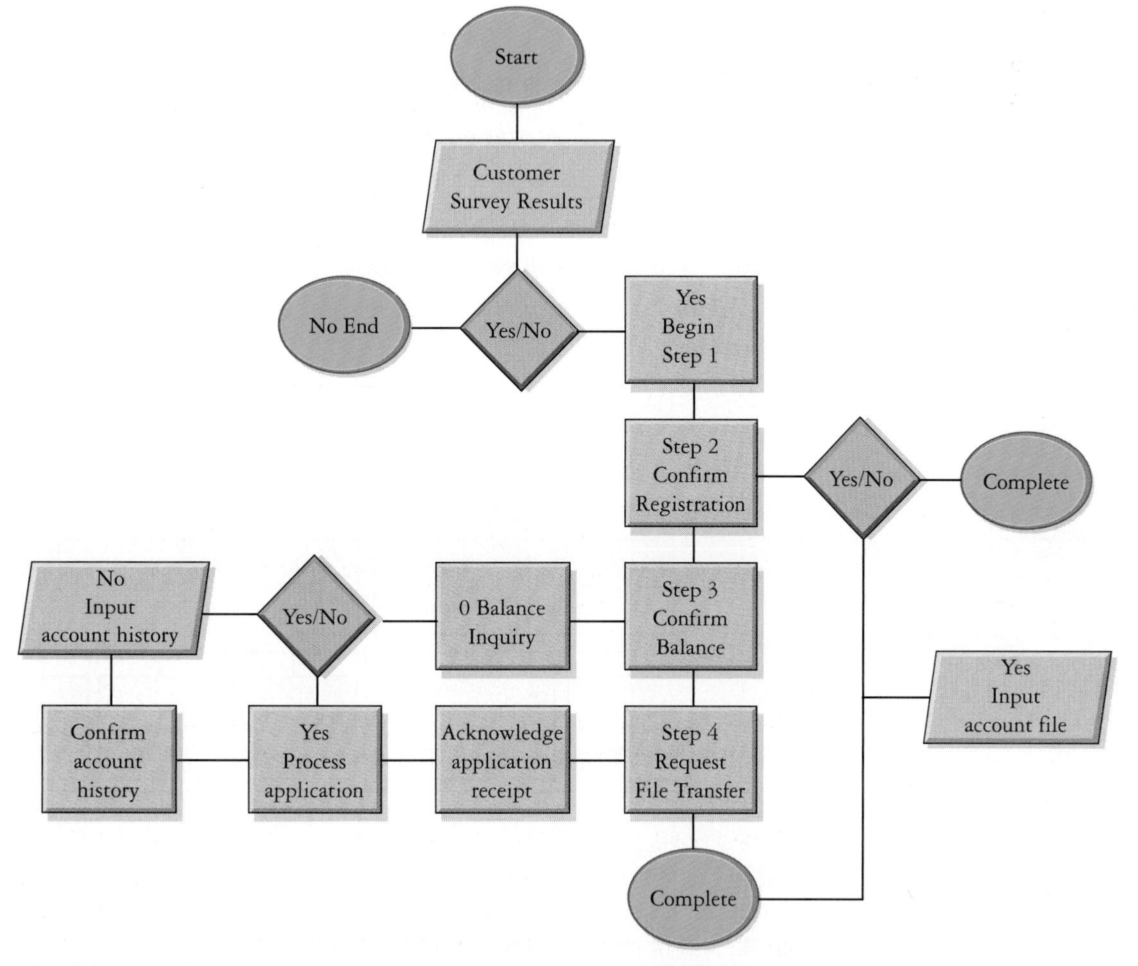

FIGURE 8.20 A detailed flowchart

Guidelines for Making Flowcharts

1. *Use standard flowchart symbols* when you can.
2. *Include all stages of the process,* including decisions, input, and documents that may affect or result from the process.
3. *Make sure the flowchart specifies a beginning and an end,* if there is one. Show a complete cycle when applicable.
4. *Avoid overlapping lines when possible.* If lines intersect, try to make them cross at 90-degree angles.

Organizational Charts

Organizational charts show the administrative, managerial, and staff structures of a company, institution, or organization. They display relationships among levels and identify paths of responsibility and authority. There are three primary types of organizational charts:

1. *Hierarchical*, which depicts a top-down organizational structure in which those with the most power—like a CEO—are identified at the top of the chart and those with the least power—like an intern—are at the bottom of the chart (see Figure 8.21).
2. *Matrix*, which emphasizes working relationships among various levels of workers by showing a web of connections among various divisions, groups, and individuals. Matrix organizational charts are usually beneficial to organizations that emphasize collaboration and interaction.
3. *Flat*, which depicts organizations in which almost everyone carries the same level of authority.

Guidelines for Making Organizational Charts

1. *Know and show the power structure of your organization.*
2. *Include all divisions and components of your organization.*
3. *Use color to distinguish among various levels of authority.*
4. *Include brief descriptions of the responsibilities of each division, if necessary.*
5. *Use solid lines to indicate direct relationships and broken lines to indicate functional relationships.*

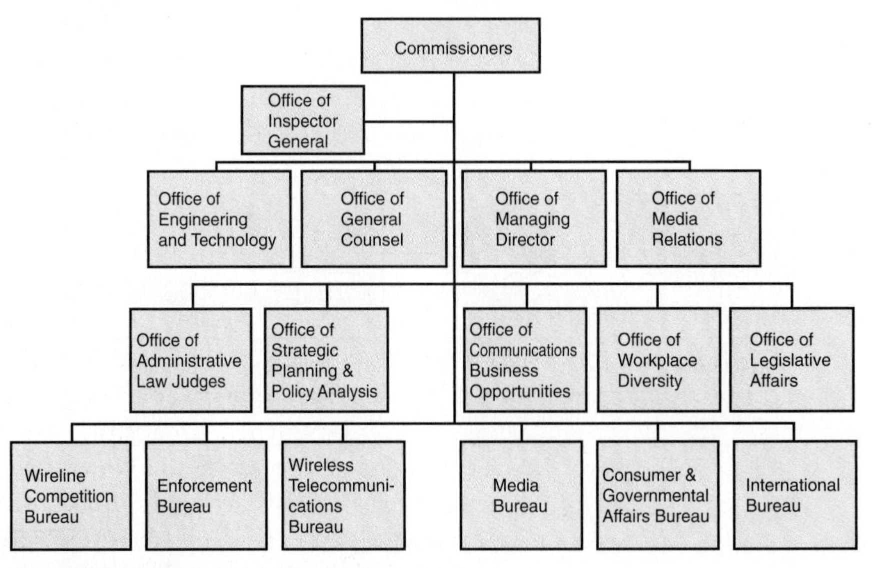

FIGURE 8.21 A hierarchical organizational chart

TABLE 8.1
A basic table

2006 Monthly Expenditures (in whole dollars)

	June	July	August	September	October
Office expenses	$1,600	$1,500	$2,600	$800	$4,500
Video production	2,700	2,900	1,300	12,300	3,900
Audio production	900	1,000	0	4,200	3,900
Advertisement	0	500	0	3,000	2,700
Travel	0	12,000	5,500	0	9,500
Total	$5,200	$17,900	$9,400	$20,300	$24,500

Tables

Tables are simply lists of data presented in a system of rows and columns. Rows are horizontal lines of information; columns are vertical lines. A cell is an individual box within the table. Tables provide readers with an easy way to access information that might otherwise be difficult to find and understand in relationship. For instance, if you were writing a document that reported monthly expenditures, you might use a table so that readers could easily access the data in a number of ways: by date, type of expenditure, or amount (see Table 8.1).

Imagine trying to write out all of the data in Table 8.1 in a traditional paragraph form. It would be difficult not only for writers but also for readers. Most spreadsheet software and word processing applications now include tools for easy table construction. Microsoft Office Word 2007, for instance, includes in its Insert tool bar a simple drop-down menu for creating tables. And spreadsheet software like Corel Quattro Pro X4 and Microsoft Office Excel 2007 allows users to input data directly into tables.

Guidelines for Making Tables

1. *Clearly identify your table with a caption.*
2. *Include a label for each column and row.*
3. *Use color for the text or the background to highlight information.*
4. *Include a source note if you gathered data from another source.*
5. *Identify units represented in the table.* If your numbers refer to a single type of measurement—say, dollars—then indicate that in the caption. If the values are different, use column and row labels to indicate the values.
6. *Present your data in an easy-to-understand, logical manner.* Remember that organization affects how readers interpret information.
7. *Be sure that your math is correct.*

Line Drawings

Line drawings are simple drawings used to represent objects. Line drawings are useful in documents like instruction manuals because they show specific parts or steps within a larger process or object. Figure 8.22 shows how line drawings can be used to isolate, emphasize, and simplify part of a visual. The example shows detailed color line drawings, highlighted bubbles, and then simplified line drawings for emphasis. Adobe Photoshop and Macromedia Fireworks make creating line drawings simple. Professional applications like computer-assisted design (CAD) allow professionals to integrate drawings into their documents. Schematics, a particular kind of line drawing, are technical diagrams that display the parts of a system or a process. They are most often used to display circuitry, as in Figure 8.23.

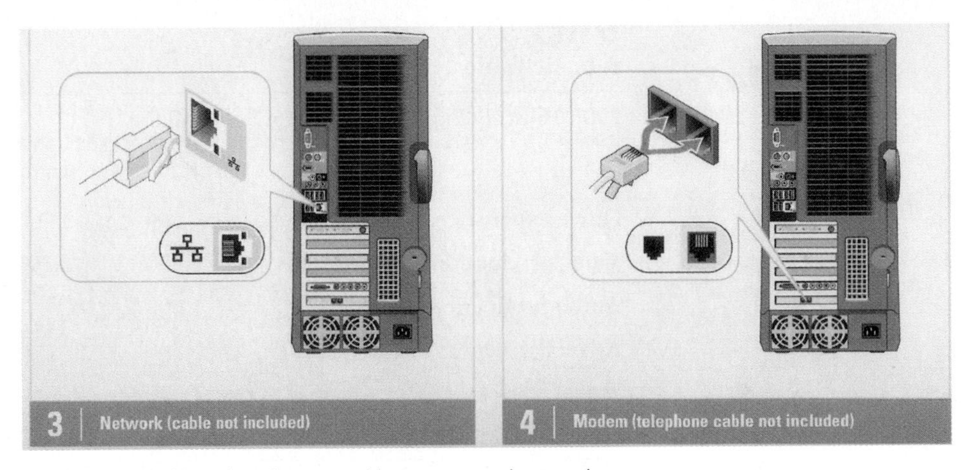

FIGURE 8.22 Line drawings used in computer instructions

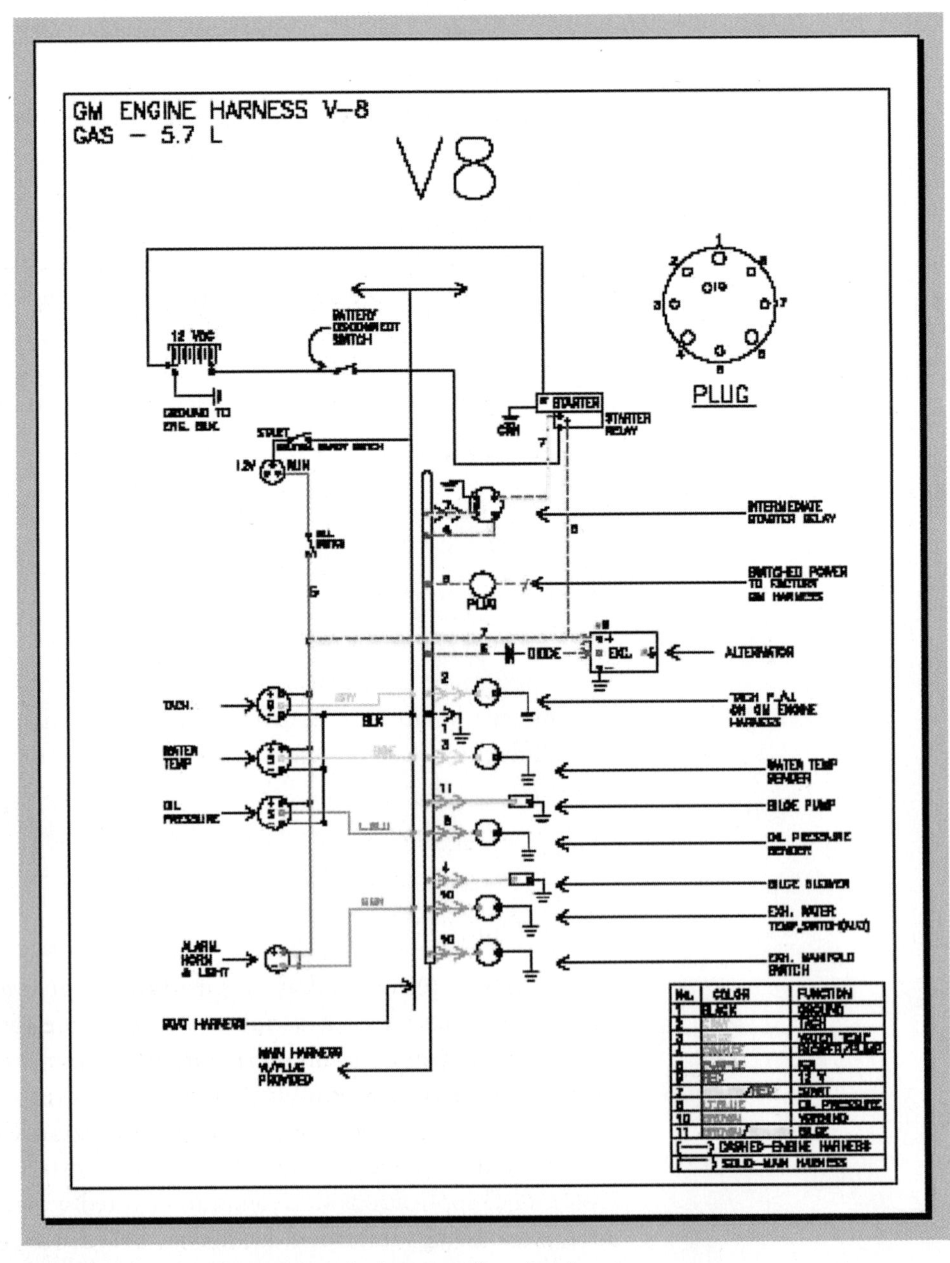

FIGURE 8.23 A wiring schematic that uses line drawings

TYPES OF IMAGES

Like graphics, images convey information and help persuade readers, but they have more photographic, realistic qualities. Workplace writers use many kinds of images in their writing.

Photographs

Contemporary technologies make including photographs in documents relatively simple. Photographs can be effective in manuals and instructions, showing the exact objects used in a process. They can also be useful in reports, such as accident reports or field reports, because they depict a scene as the viewer sees it. Photo manipulation applications, like Adobe's Photoshop, allow users to customize photographs, changing them in many ways—cropping, adjusting color, adding call outs or captions, adjusting contrast or brightness, zooming in on and highlighting particular parts of a photograph.

Guidelines for Using Photographs in Documents

1. *Make sure the size and resolution of your photograph fit the document you are writing.*
2. *Be sure your photograph shows just what you need to show.*
3. *Remember that altering photographs to misrepresent information is unethical.* Even basic cropping can change the meaning of an image.

Screen Shots

Screen shots are useful when describing a computer process or showing features of a particular kind of software or application. Many online computer manuals and help files use screen shots to guide users through on-screen processes.

Guidelines for Making Screen Shots

To make a screen shot on a PC, follow these steps:

1. Be sure that what you want to appear in the screen shot actually appears on your screen.
2. Press the Ctrl, Alt, and Print Screen keys at the same time.
3. Open the image editing or word processing software in which you wish to work with the screen shot.
4. Click the Paste icon in your tool bar. The captured screen shot should appear on your page.

Video

Many electronic documents, such as web pages and CD or DVD manuals, now employ film clips or streaming video as part of their visual production. Video clips provide a good method of showing readers/viewers exact processes. For instance, an online training manual might include video clips of particular work activities that employees need to know how to perform. Many software companies use streaming video to teach software owners how to use their applications, either in manuals or help pages. Similarly, many websites now include streaming video components to quickly convey information. Web writers know that viewers need video viewing plugins like Flash, Quicktime, Windows Media, or Real Player to view streaming video on the Web.

Animation

Animation generally refers to a series of computer-generated images that progress one frame at a time, giving the illusion of movement. Animation can be used to depict things that can't be shown or don't yet exist, such as chemical reactions at the molecular level or animated "walk-throughs" of potential buildings or structures. It is also widely used in marketing, since animation adds an eye-catching and professional touch to web pages and e-manuals. Workplace writers can now efficiently create animation through many software applications, such as Macromedia's Flash. Like video, however, animation requires that readers/viewers have the plug-ins to view the animation.

FINDING, CAPTURING, CREATING, AND FORMATTING VISUALS

In the past, visuals were added to workplace documents by graphic designers or graphic artists. Although these professionals are probably more adept than most workplace writers at making visuals or creating visually appealing documents, contemporary imaging, graphics, publishing, and word processing applications provide everyone with the tools to incorporate professional-caliber visuals into documents. In addition, a number of easily accessible resources are available to locate visuals.

Databases

Numerous searchable databases house millions of images that you can use in your workplace documents. Some databases make images available for free, whereas others charge fees; but if you need a particular image, it is likely that one of the many databases will have what you need. Professional, fee-based databases include the following: Alamy, Bridgeman Art Library, Corbis, Getty Images, Image Works, Index Stock Imagery, Landov, Magnum Photos, Masterfile, Peter Arnold, PhotoEdit, Photo Researchers, Picture Quest, Super Stock, Visuals Unlimited, and AP Images. Remember that images in these databases are copyrighted, and you must gain permission and pay fees, when applicable, to use them.

Web Searches

The World Wide Web continues to be one of the best resources for locating visuals. Keep in mind, though, that many visuals on the Web are also copyright protected. If you use a picture found on the Web, it is your responsibility to seek permission to use it, and you must acknowledge its source. Copyright laws do allow usage in some educational situations (see the U.S. Copyright Code Fair Use Doctrine); therefore, you might use some images in class work without needing to obtain permission but not in professional work.

There are a number of useful image search engines that help locate visuals. The common sources include Alta Vista Image Search, Ditto.com, and Google Image Search. Symbols.com is useful for locating symbols; and Gifworks, Animation Factory, and Web GFX are good for finding graphics.

If you are using a PC, capturing an image from a web page is simple: just place your cursor over the image you wish to copy, right-click the mouse, and select Save Image As (see Figure 8.24). Then be sure to note the full URL, the date you captured the visual (many web pages are changed or updated periodically), and the author/owner of the page from which you took the visual.

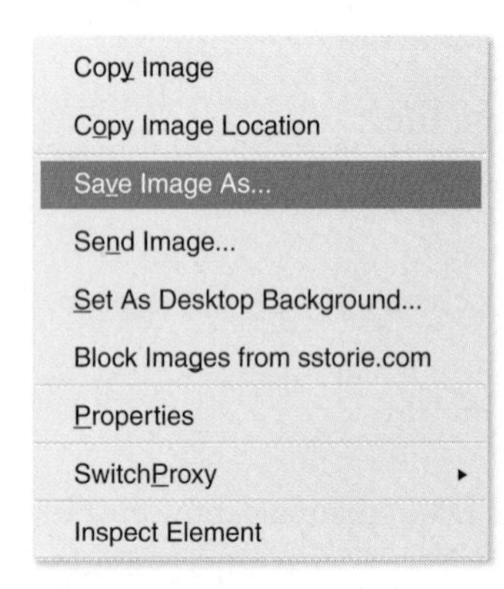

FIGURE 8.24 Drop-down menu for capturing an image from a web page

Clip Art

Clip art refers to pictures that you clip or copy from previously printed material. The term refers to physically cutting a picture from a paper document to use in another context, but in the digital age it also refers to precrafted graphics that are made available for public use. Nearly every word processing and publishing application contains libraries of clip art, and many businesses and industries maintain context-specific clip art libraries. Even though a number of commercial clip art galleries do sell their images, electronic clip art was developed to provide nonartists with simple graphics that are not bound by copyright and licensing restrictions. Most clip art is considered to be in the public domain, but you should confirm that the clip art you are using does not require permission. Although electronic clip art is easy to find and use, it usually conveys an amateur look.

Your Own Visuals

The best way to obtain context-specific visuals may be to make your own. Taking digital photographs is simple, and the hardware—a digital camera—is inexpensive and easy to use. In addition, most computers now contain simple paint and draw programs that allow users to make simple graphics, and many word processors and desktop publishing applications contain basic draw tools as standard features. Specific graphics programs like Photoshop or Fireworks offer advanced capabilities. Furthermore, external hardware, such as an electronic drawing tablet, allows users to draw on a flat screen surface and have images digitized and transferred to electronic files. Some computers can be equipped with convertible tablets that allow users to draw directly on the screen with a special pen and have pen strokes converted into electronic graphics. Despite this relative ease, workplace writers must understand a number of key points when making their own visuals.

Print and Electronic Visuals

Visuals that appear on computer screens are physically different from visuals that appear in print documents; electronic documents emit light, whereas printed visuals reflect light. Because of this difference you should take the following aspects into account:

- *Color type:* Print colors are subtractive; that is, they depict color through the absorption of light. They also reflect less white light; they are made from cyan, magenta, yellow, and key (black) and are known as CMYK colors. Electronic documents display color by emitting light. They use a red, green, blue (RGB) color scheme that is additive; that is, colors are formed by adding various amounts of the three primary colors to form the desired color. Nearly all image and graphic software allows you to select CMYK or RGB colors, depending on the format of your document.
- *File size:* Although higher-resolution visuals convert into higher-quality print documents, larger-file-size visuals may make an electronic document too large for easy download. Web pages, for instance, that contain multiple high-resolution visuals may be too large for computers with low-bandwidth connections to download efficiently.
- *Pixel density:* One of the things that affects file size is pixel density—the number of pixels per inch in a given visual. Higher pixel density can translate into cleaner print versions of visuals, but greater pixel density requires more electronic memory, causing an overall increase in document file size. Also, because most computer screens cannot project pixel density greater than 100 pixels per inch, high pixel density cannot improve what viewers actually see; the quality of the image is limited by monitor ability.

Software

Graphics software and image-editing software now make rendering visuals convenient for anyone with access to the software. Applications that create or edit visuals are generally known as either paint or draw programs and produce two kinds of visuals:

bitmap and vector. The kind of visual you need to produce is determined by the purpose and the design of the document.

Bitmap Bitmap, also called raster graphics, refers to a method of storing an image bit by bit, or pixel by pixel. Bitmap images are mapped by pixel onto a series of rows and columns, and each individual pixel is assigned its own color. If you zoom in on a bitmap graphic or image, you can see the individual pixels. The quality of bitmap visuals is determined by the number of pixels—called *resolution*—and the amount of color information stored in each pixel. The greater the resolution and the greater the amount of color, the sharper the visual. Bitmap, or raster, visuals are usually stored in GIF, JPEG, TIFF, BMP, PICT, PCX, and DIB files.

Graphics software and image-editing software that allow you to create bitmap files are referred to as paint programs. These programs are easily available and offer tools to do many things:

- Edit selected parts of a visual, working at the pixel level
- Add color, texture, and gradient color to a visual
- Change or match colors using a color pallet
- Determine whether the visual is rendered in RGB, CMYK, or another color scheme
- Control size, depth, and contour of paint lines using brush tools
- Insert text into a visual, and alter its typography
- Edit photographs for color, imperfections, and size
- Create multiple layers in a single visual
- Apply various effects (e.g., blurring) and styles (e.g., drop shadows) to visuals
- Select the format in which the visual will be saved
- Change the file format of a visual

Vector Vector graphics is a method of storing visuals based on a concept of paths, which refer to basic geometrical shapes like lines, curves, circles, polygons, and points. Vector graphics is also called object-oriented graphics because it works with objects rather than individual pixels. Vector visuals provide crisper, cleaner-looking visuals because they are not pixilated. They are stored in smaller files than bitmap files and require less bandwidth for transferring files. And because vector visuals are based on a mathematical formula, they do not distort when you zoom in to look more closely at them; they retain their proportion and scale. Vector visuals, therefore, do not depend on resolution; they remain the same no matter how you edit them.

The crisp, clean look of vector graphics makes them ideal for use in print documents. They are also useful in designing icons, line drawings, schematics, and other visuals that do not need to achieve the realism of photography. Many animators use vector graphics. However, because most output devices—printers and monitors, for instance—rely on pixilated output, they necessarily convert vector graphics to bitmap graphics in order to render the visual.

Software that allows you to create vector graphics is called a draw program and is considered more sophisticated and professional. CAD systems used by engineers and architects, for instance, produce vector graphics. Draw programs offer tools to do many things:

- Create composite drawings
- Edit objects within a visual through tools that skew, rotate, resize, stretch, mirror, and transform

- Combine basic objects into more complex objects
- Change the order of objects
- Create smooth surfaces, curves, and lines

Spreadsheet In addition to paint and draw applications, many spreadsheet applications provide simple steps for transferring data into graphs, charts, and tables. When using these features to produce visuals, remember to consider all visual rhetorical choices. The program may render the graph, chart, or table, but you need to define the type of visual, the use of color, and the organization of the data.

Word Processing Most word processing software now lets you create diagrams, auto shapes, and even basic drawings. In addition, these same programs allow you to insert visuals from clip art libraries or from stored files directly into your documents.

Thumbnails

Because many visuals require a large amount of memory, workplace writers often use thumbnail visuals as placeholders during the drafting process—as links to larger visual files or as overviews of the visual information. During drafting, many workplace writers include small thumbnails or quick sketches to show where visuals will be positioned and to identify what the image will accomplish. The actual images are added later. Thumbnails are also used as placeholders to direct readers to larger, more detailed versions of an image. Web pages, for instance, often use thumbnails to decrease the size of a web page, which makes loading it easier for readers. Files that contain large numbers of visuals often employ an index page of thumbnails linked to the larger files. Thumbnail index pages make it easy for readers to search through visual files to locate the ones they need without having to open all of the larger files. Most word processing applications, file managers, image editors, and graphics software allow you to search image files by selecting a thumbnail view, as is the case in Figure 8.25.

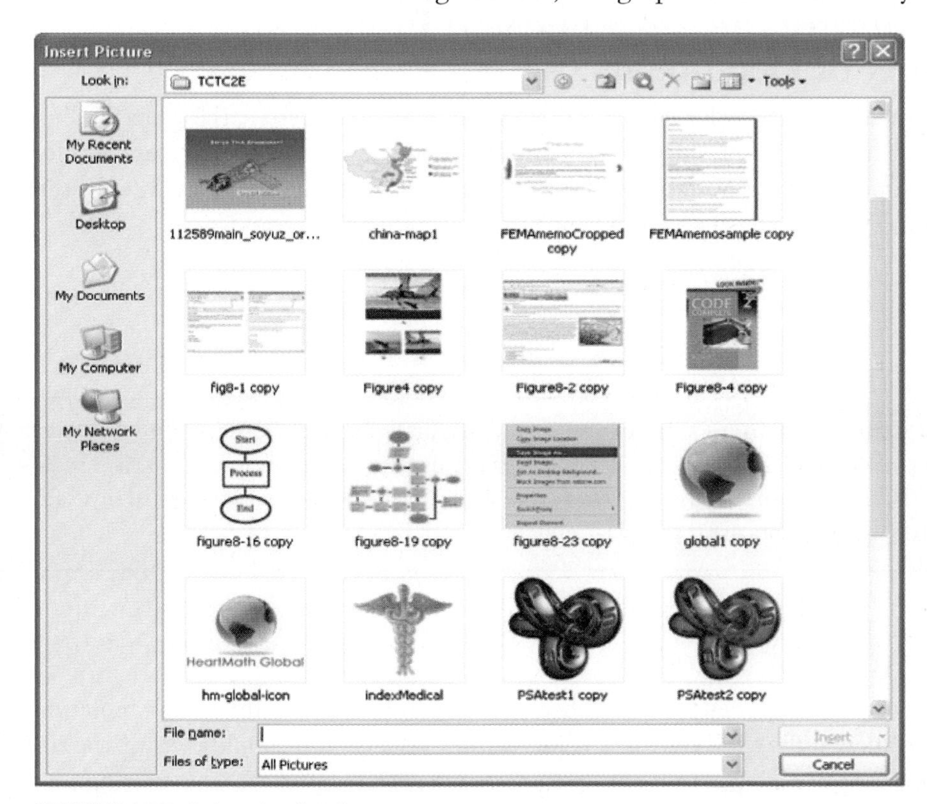

FIGURE 8.25 A thumbnail gallery

Many image editors and graphics software include automatic thumbnail features, which reduce the size of a visual but also reduce its quality. Generally speaking, thumbnails are sized at 80 to 200 pixels along their long side, although many organizations set their own standards for thumbnail size. The short-side measure equals the ratio of short to long sides in the original. The biggest mistake workplace writers make in creating thumbnails is failing to reduce the number of pixels and compacting them instead, thus keeping the actual size of the visual file the same. For instance, resizing a visual in a word processor does not reduce the size of the file; it resizes only the display. When making thumbnails, be sure to reduce the resolution and overall file size.

USING COLOR EFFECTIVELY

Color plays a crucial role in providing readers with information. Because producing color documents has become cost-effective for most companies and organizations, workplace writers now focus more on color in designing the visuals that accompany their documents. Thus, choosing colors is an important visual rhetorical choice.

To Identify Parts of a Document or Kinds of Text

Color can be used to identify parts of a document—for example, colored boxes, background page colors, and colored icons, all of which help guide readers through documents. Using a colored page background, for instance, might signal to readers the start or end of a chapter or a section.

In addition, color can be used to indicate particular kinds of text and thereby assist readers in understanding a document's organization. For instance, your document might include quotes from clients, which you could render in a different color to set them apart and to alert readers to the quotes. Furthermore, web pages and other electronic documents use color to indicate hyperlinks.

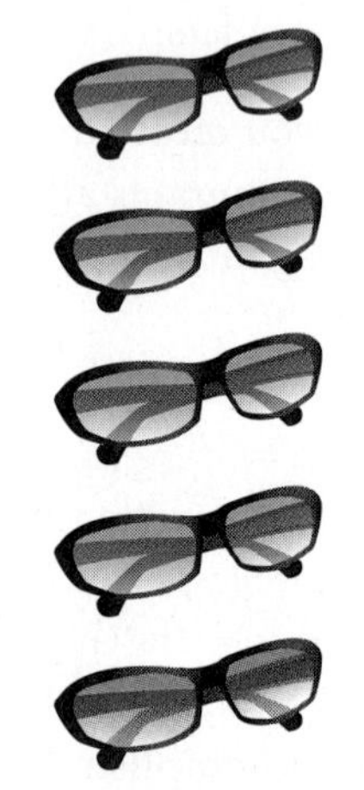

FIGURE 8.26 Drawing attention with warm colors

To Highlight or Draw Attention

Warm colors—red, orange, yellow—draw a reader's attention more than cool colors, such as blue and green. But not all readers respond to colors in the same ways; writers must be attentive to cultural differences in response to color. In Figure 8.26 notice how the change in color directs attention to the sunglasses at the bottom.

See chapter 9 for more on typography.

You can also use color to identify when your text shifts to a new heading or subheading. Consider, for instance, how the words "To Highlight or Draw Attention" in the title of this subsection compare with the larger section heading "Using Color Effectively."

To Improve the Aesthetic and Professional Quality of Documents

When used correctly, color simply makes documents look better; the proper use of color enhances their professional quality. Most readers equate the effective use of color in a document with its overall usefulness; documents with improperly used colors can appear unprofessional, sloppy, and confusing.

IN YOUR **EXPERIENCE**

By now you have become familiar with some of the physical properties of this textbook. What are some of the ways it uses color? What kinds of colors does it use? What colors are repeated? What colors are used as backgrounds, and what do those backgrounds signal? What colors are used in boxes, and what do they indicate? Are there any places where colors are used just for aesthetic enhancement?

How to Add Color

Most word processors, desktop publishers, graphics software, and image manipulation software allow users easy control of colors within visuals. You should consult the tutorial for the software you use, but here are some general approaches for coloring text and shapes in a word processor.

Coloring Text

Most word processors include tools for coloring text. In Microsoft Office Word 2007, for instance, you simply highlight the piece of writing you want to color and then click the font color icon in the Home tab toolbar. This will open a drop-down menu from which you can select the color you wish to apply to the text (see Figure 8.27). Coloring text can help demarcate divisions within documents or highlight information, but you should be sure not to overuse colors. Too many colors in documents can cause confusion.

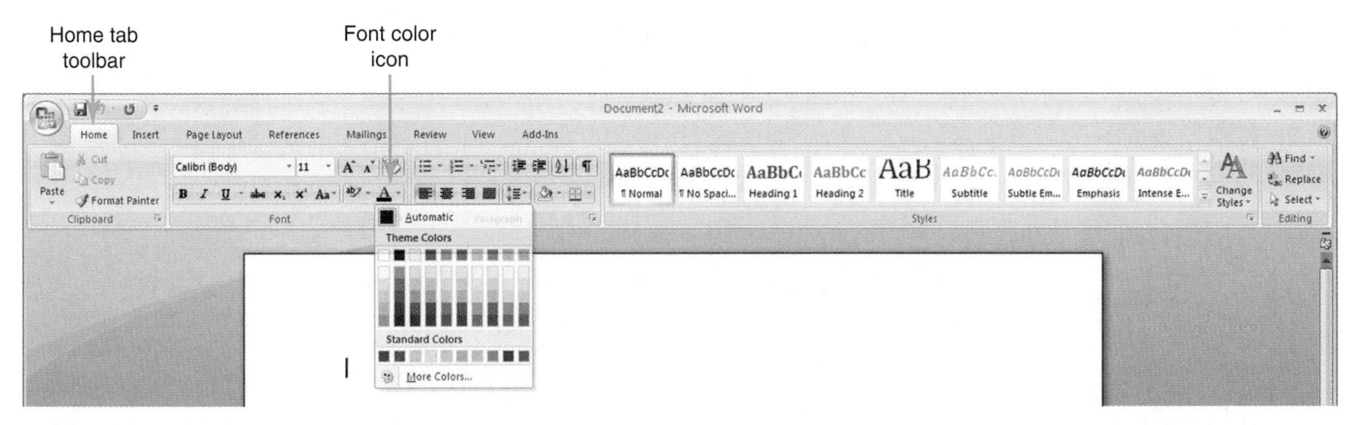

FIGURE 8.27 Drop-down menu for coloring text in Microsoft Office Word 2007

Coloring Shapes

Word processors also let users create simple shapes and drawings and then add color to those shapes. To color shapes in Microsoft Office Word 2007, simply click the Insert tab in your toolbar and then the Shapes icon. Once you draw your shape, the Drawing Tools Format panel, which includes the color selection panel, will automatically appear (see Figure 8.28).

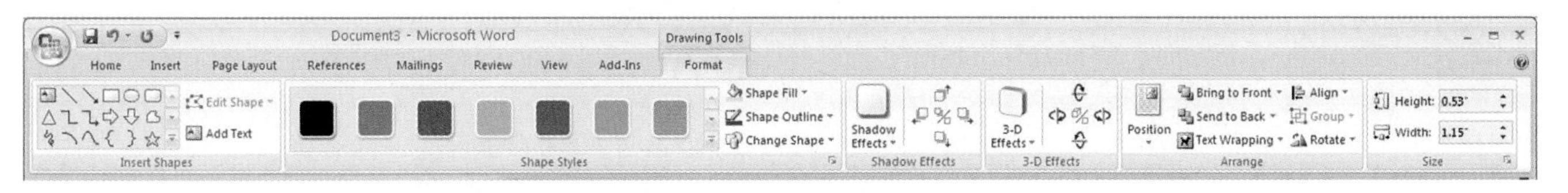

FIGURE 8.28 Shape Format toolbar in Microsoft Office Word 2007, including shape color selection

Coloring Text Boxes

Text boxes are visual elements that offset selected text by surrounding it in a box frame. The size and color of the box can be altered, as well as the line that forms the box frame, the color of the outline, the shape of the corners, and the color of the background. To create a text box in Microsoft Office Word 2007, follow these steps:

1. Select the Insert tab in the toolbar.
2. Click Text Box.
3. Select the format of the text box into which you want to place your text.
4. Either type or cut-and-paste your text into the text box.

Once you select your text box, the Text Box Tools Format menu will automatically appear, including color selection tools (see Figure 8.29). In this toolbar you can select tools to change the border color and size, as well as the text box fill color, or you can apply other visual effects like shadowing.

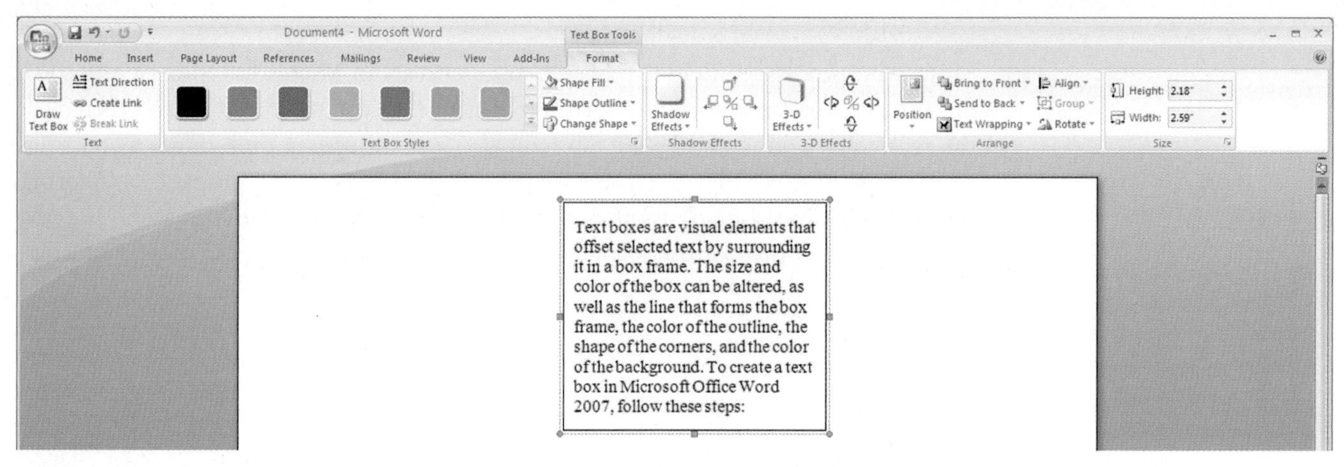

FIGURE 8.29 Text Box Format toolbar in Microsoft Office Word 2007, including text box color selection

Coloring Page Background

Word processors like Microsoft Office Word 2007 also allow you to set background or border colors or even to include watermark images for entire pages and thus visually demarcate different parts of a document. To set a background color, border shape and color, or watermark in Word 2007, simply select the Page Layout tab in the toolbar, and click the appropriate icon (see Figure 8.30).

1. Page Layout tab toolbar
2. Watermark tool
3. Page Color tool
4. Page Borders tool

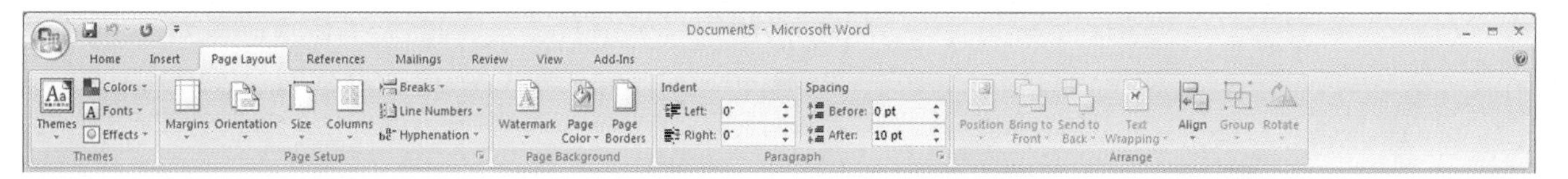

FIGURE 8.30 Microsoft Office Word 2007 Page Layout toolbar, including watermark, page color, and page border tools

Adding Color Appropriately

Color Scheme Choosing a color scheme—RGB or CMYK—depends on the kind of document you are producing. Visuals for electronic documents should use RGB color schemes; visuals for print documents should use CMYK color schemes. Fortunately, most graphics software and image manipulation software allow you to shift between these color schemes.

Contrast *Contrast* refers to a visual separation of parts of a text by differences in color; colors must have sufficient contrast so that readers can easily separate the different parts. Dark and light colors should be used in contrast, much like the standard of black type on a white paper background. You would not want a pale yellow text on the same white background. You also want to be careful with contrasting colors that are jarring to the eyes. In most contexts it is a good idea to use a light background and add darker elements like text on top of that background. Dark backgrounds with light visuals can be difficult for readers to focus on.

Consistency Colors used to signify headers, titles, boxes, pages, text, and shapes must be consistent. Using different colors for the same facet of a document confuses readers.

Limited Use Just because you *can* add color doesn't mean you *should*. Documents can become cluttered and confusing if they contain too many colors, and they can be tiresome for readers. Documents look most professional when they are clean and sharp; using too many colors looks messy and amateurish.

GUIDELINES FOR USING VISUALS

There are no rules about when to use visuals or what kinds of visuals to use; these are questions that require workplace writers to address rhetorical issues. Nonetheless, we can offer some guidelines to help.

Relevance

Visuals must be relevant to the document. Visuals *must* be used if their presence is necessary to the comprehension of the information. Visuals *should* be used if the information is clarified, supported by, or exemplified by the visuals. But visuals should *not* be used if the information can be easily understood without them, particularly when brevity or a reduced size best serves readers, as with a web page.

Size

Some research shows that readers respond more positively to large visuals than to small visuals. Consequently, when possible, present your visuals in larger formats, but do not compromise good visual-to-text relationships, document design (see chapter 9), or visual resolution. Also, when designing electronic documents, like web pages, remember that larger files may slow the download speed of the page, and a large visual may be too big for a viewer's screen. Find a balance between large visual size and the rest of the document.

Simplicity

Keep your visuals simple. Do not overwhelm readers with too much visual detail in a single image or with too many images. Whenever possible, eliminate any details that do not add to the readers' understanding of the subject or directly apply to the situation. Similarly, use visuals only when they serve a purpose. Too many visuals can distract readers from the real purpose of a document.

Transnational Awareness

If documents need to reach culturally diverse or multinational audiences, the meaning conveyed through visuals will be subject to various cultural contexts. Cultures, communities, and nations interpret visuals in different ways, and it is your responsibility to understand how your visuals will be perceived by your readers. If you are unsure how your visuals might be interpreted by a particular audience, conduct research to find out.

See chapter 5 for more on transnational communication.

Use Familiar Examples

Large numbers of visuals, particularly icons, have been proven successful in international documents. Locate those visuals for your field, and use them rather than attempting to create your own.

Direct Readers

Not all languages read left to right, and not all individuals read. The visual in Figure 8.31 was placed in South African mines to explain to miners unable to read that they should remove fallen rocks from the tracks. However, because the miners were not used to reading, they were not familiar with a left-to-right logic. As a result, many miners read the visuals from right to left and understood that they should place rocks on the track. Whenever possible, sequence visuals from top to bottom, and consider using arrows to direct readers through a sequence.

FIGURE 8.31 A wordless visual

Be Moderate

Because different cultures respond to formality and informality differently, it is better to strive for moderation so that you do not offend those at either end of the spectrum.

Provide Multiple Examples

Sometimes multiple visuals can clarify information for multiple audiences better than just one visual. Look at the approach taken in Figure 8.32.

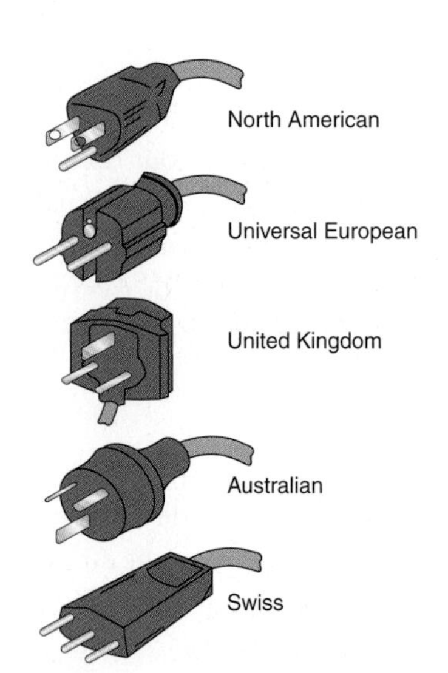

FIGURE 8.32 Multiple visuals used to relate to multiple audiences

Relationship to Written Text

Much of the effectiveness of visuals is determined by how they relate to the text they support. Consider the relationships between text and visuals, and design your documents accordingly.

Formal Relationship

A formal relationship between text and visual is established when the text directly refers readers to that visual. You may have noticed many references in this chapter to specific figures. If a pertinent visual cannot fit in the space adjacent to the text, the writer must establish a formal relationship between the two and direct readers to the visual. However, even when the visual can fit in the adjacent space, the writer may want to establish a formal relationship to highlight the importance of the visual. All formal visuals should include a figure, table, or illustration number and caption; and the numbering system should be consistent.

Informal Relationship

An informal relationship is created when the text does not direct a reader to a visual, but the visual, nonetheless, provides pertinent information for that text. Informal relationships should be self-evident. Such relationships can be useful if visuals support textual information but are not completely necessary to understanding it. However, informal relationships are most effective if the visuals are placed in close proximity to the relevant text; otherwise, readers may not understand the connection of the visuals to the text. Informal visuals may or may not include a figure, table, or illustration number, depending on the rhetorical context.

Identification

All numbered visuals should be listed in a table of figures or a table of illustrations in the front matter of a document, particularly in longer documents such as reports, manuals, and instructions. Such tables should provide enough detail about the visuals so that readers can easily identify them. Ambiguous information, like a simple list of figure numbers, does not help readers.

Fluff

Some visuals have no direct or relevant relationship with the information provided in the text. In most situations you should avoid fluff images because they do not help your document accomplish its task. However, in some contexts fluff images can improve the aesthetic appeal of documents and may add to the overall design. Generally speaking, you should not include fluff visuals because they take up space, increase the size of the document, and might actually confuse readers. However, if you do include fluff, you should not label it or give it a caption. If it requires a label or a caption, it may not be fluff after all.

Revision

PSA

One of the most common mistakes workplace writers make in adding visual elements to their documents is not revising those elements as they revise the rest of the document. Remember that revising is a crucial part of the problem-solving approach, and revising visuals is as crucial as revising any other part of your document. Here are some questions to guide you in revising your visuals:

For more on usability, see chapter 11.

- Does the visual depict what you intend it to depict? Consider either having a peer look through your visuals or asking a usability test group to consider them.
- Is the visual sized proportionally to your document? Consider the size of the visual in relation to the document and the role of the visual in the document.
- Is the visual of the caliber that your document requires? Sometimes workplace writers lower the overall quality of their documents by including lower-caliber visuals.

ETHICS AND VISUALS

See chapter 4 for more on image manipulation and misleading visual representation.

Workplace writers should be attentive to numerous ethical considerations when using visual components in their documents: manipulation, permission/rights, citation, accuracy, representation, obfuscation, concealment, clarity, and access. Each of these ethical concerns is addressed in detail here.

Manipulation

As you might recall from chapter 4, manipulating images digitally can result in exciting images that present false or inaccurate representations (see Figure 4.3). Such image manipulation can be used to misinform, alter opinion, and convey incorrect or inaccurate information. When manipulated images are used in the workplace without the readers' knowledge, they can have serious repercussions. There are a number of examples of such images leading to job termination and lawsuits against their creators.

In July 2007 the prestigious journal *Science* retracted an article it published a year earlier as a result of a manipulated visual. The article, written by researchers at the University of Missouri–Columbia, indicated that cells in a mouse embryo differentiated sooner than previously known, presenting significant implications for stem cell research. However, shortly after the article was published, questions arose about the authenticity of the mouse embryo photos included in the article. Three of the researchers who published the article signed a retraction, noting that the article "was founded at least in part on falsified or fabricated images." It was later discovered that a postdoctoral researcher in the group had altered the photos without the rest of the group's knowledge;

that researcher resigned from the university after the scandal broke. When questioned about the validity of the research, the lead author of the article stated that "the paper was written on the basis of those images. Everything has to be repeated."

These kinds of manipulations can have serious consequences not only for the originators, but also for others who use them as the basis of various types of decisions. When manipulations go undiscovered and are accepted as accurate, they can have a negative impact on those who trust them.

Permission

You may have noticed that many of the visuals we use in this text are taken from other sources. In order to use these visuals, we have had to seek permission from the copyright holders. Simply citing the sources of these visuals was usually not enough; we had to actively seek permission to use them. Copyright holders have the right to place restrictions on how their visuals can be used in order to ensure that they are not misused and do not discredit the maker or owner. In addition, copyright holders can charge a fee for the use of their visuals, particularly if the user stands to profit from that use. You may remember that earlier we discussed databases as sources for finding visuals and noted that most databases have set fees for use of their visuals. Using a visual without permission is not only unethical, but can also result in legal action.

See chapter 4 for more on copyright laws.

There are a number of situations in which you might be exempt from seeking permission to use a visual; many of these fall under fair use doctrines in copyright law. Fair use allows for use of others work without permission according to a four-factor balancing test. According to the United States copyright laws,

> The fair use of a copyrighted work, including such use by reproduction in copies or phonorecords or by any other means specified by that section, for purposes such as criticism, comment, news reporting, teaching (including multiple copies for classroom use), scholarship, or research, is not an infringement of copyright. In determining whether the use made of a work in any particular case is a fair use the factors to be considered shall include—
>
> 1. the purpose and character of the use, including whether such use is of a commercial nature or is for nonprofit educational purposes;
> 2. the nature of the copyrighted work;
> 3. the amount and substantiality of the portion used in relation to the copyrighted work as a whole; and
> 4. the effect of the use upon the potential market for or value of the copyrighted work.
>
> The fact that a work is unpublished shall not itself bar a finding of fair use if such finding is made upon consideration of all the above factors.[2]

Keep in mind, though, that just because something is legal does not mean it is also ethical. Workplace writers must be sure to consider the ethical implications of fair use and permissions, not just the legality.

For more on legal versus ethical concerns, see chapter 4.

Citation

Tied closely to issues of permission, citation is important in identifying the sources from which you have taken visuals that you have not made yourself. If you capture an image from a web page, for instance, and use it on your own web page, you

---

[2]US CODE: Title 17,107. Limitations on exclusive rights: Fair use.

should identify the source with a citation, even if you manipulate the image to suit your needs. Often workplace writers use visuals others have produced in order to show accurate and meaningful information, such as, say, a chart a local government has produced to depict industrial development. In such cases not only do workplace writers need to seek permission to use the visuals, but they must also acknowledge the image by citing its source. Failure to cite a visual source is the equivalent of plagiarism. Appendix B provides a detailed guide for citing various kinds of visuals.

See appendix B for more on source citation of visuals.

Accuracy

The visuals you present must accurately represent the given information. It is easy to manipulate visuals to intentionally (and unethically) misrepresent, but it is also easy to misrepresent unintentionally. Be sure to carefully evaluate your visuals during the Review phase of your to be sure that they accurately represent the information. One of the primary ways that workplace writers unintentionally present inaccurate information is by not confirming relationships between captions and visuals. Imagine, for instance, the difficulties that might occur if a lab report visual labeled "chemical formula following neutralization" depicted the formula prior to neutralization. The report would then, inaccurately, represent the neutralization process as ineffective. Be sure your captions and visuals correlate.

Representation

Workplace writers must consider that visuals represent not only information but also the company or organization for which the visuals are produced. Imagine that you were designing a catalog for a digital imaging supply company. How differently would an audience perceive the quality of the product you were selling if you used the visual in Figure 8.33(a) versus 8.33(b)? Does the quality of the image represent the quality of the product? Does it also suggest something of the quality of the company selling the product? Workplace writers must consider how they are representing their company through their use of visuals.

(a)

(b)

FIGURE 8.33 Varying quality of representation linked to choice of visuals

Obfuscation

In chapter 4 we noted that ethical writers avoid obscuring information by carefully considering the language that they use. So, too, can visuals be used to either clarify or obfuscate information. To avoid either intentionally or unintentionally obscuring issues or information, writers should pay careful attention to how they represent information in their visuals. You might recall how the first graph in Figure 4.4, depicting the CNN poll about the Terri Schiavo case, misrepresented information. For another example, consider the way in which the two bar graphs in Figure 8.34 represent the same information but present it in different ways. The first obscures the information by presenting it in a difficult-to-read three-dimensional graph that uses color and positioning of the bars to provide a distorted emphasis on the West division. Graph (a) also confuses readers by not beginning at a zero point on the vertical axis. Graph (b) accurately indicates the relationships among the three bars, does not distort the emphasis on any one bar, and provides clear information about both the graphing units and the units sold by each division.

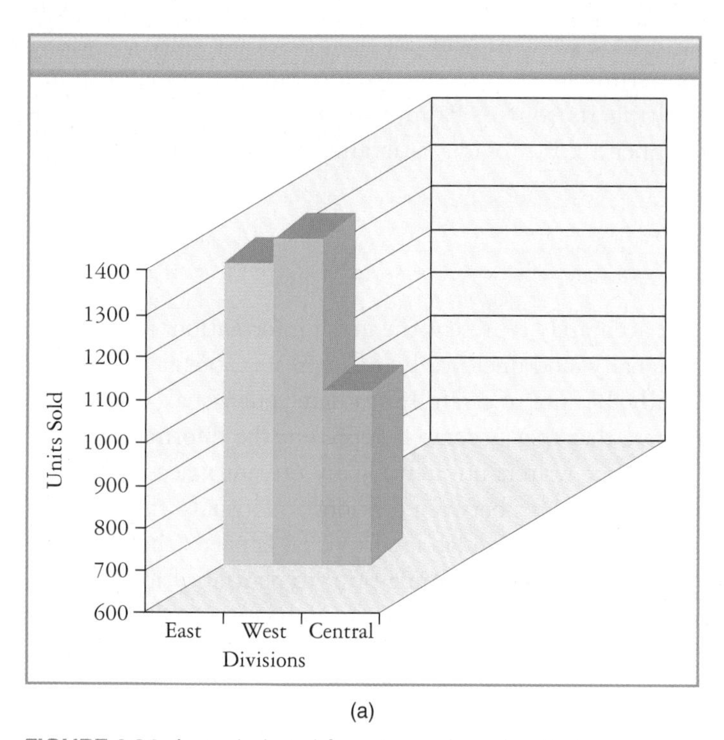

(a)

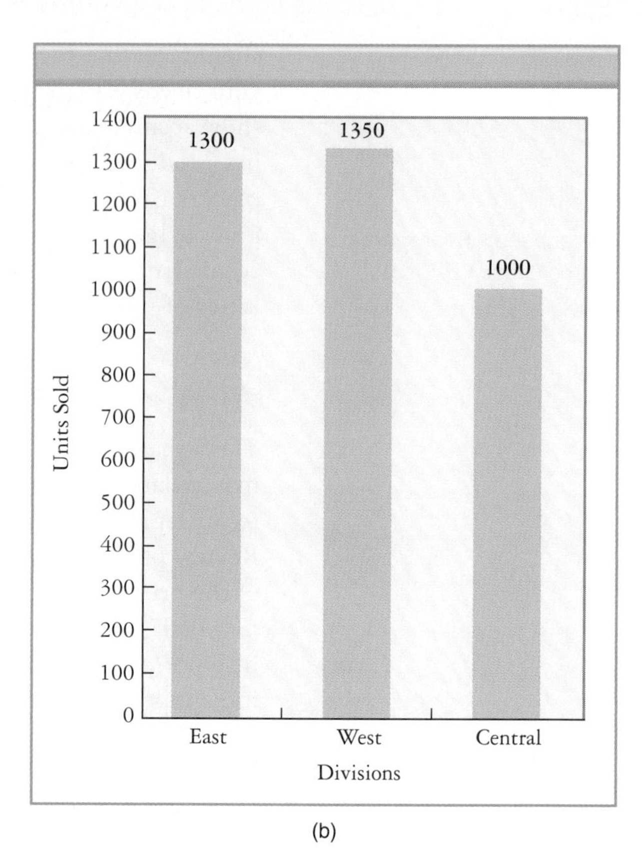

(b)

FIGURE 8.34 A graph that obfuscates and another that clearly presents the same information

Concealment

Visuals can also conceal information. One obvious example is that of cropping, mentioned earlier in this chapter. Sometimes cropping images can remove key information and thereby conceal the context of a visual or other important parts of the visual information. Consider Figure 8.35, which was purported to depict a polar bear resting near Arctic Village, Alaska. The original picture, shown in Figure 8.36, clearly indicates that the picture was taken in a zoo. In this case the image cropping and the caption used conceal contextual information in order to influence how readers perceive the visual.

FIGURE 8.35 Photograph that claimed to show polar bear resting near Arctic Village, Alaska

FIGURE 8.36 Original uncropped picture revealing more contextual information

Clarity

Just as workplace writing must be clear in order for readers to understand it, so, too, must visuals be clear in order for readers to get the appropriate information from the visual. Three primary factors affect visual clarity: size, readability, and resolution. Intentionally including visuals that are too small to be read can be an unethical way of hiding information from readers. Likewise, presenting images that are either difficult or impossible to read because of their placement in a document or because they are too cluttered can keep readers from getting a clear picture of the information. In addition, it is important to consider the resolution of visuals to be sure that they can be seen. Although print documents support higher-resolution visuals, electronic documents like web pages require lower-resolution visuals. Workplace writers must be sure that reduced resolution on visuals doesn't also render them unclear.

Access

In order for visuals to be of use to readers, those readers must be able to access the visuals. You should be alert to the formats and sizes of the visuals you use. Keep in mind that some servers may have difficulty downloading large visual files and some software applications may have trouble reading less-used image or graphic formats. Your visual file sizes and formats should be compatible with the kinds of applications your audience will most likely use to read the document and visuals. Presenting oversized or unusually formatted files might be interpreted as an attempt to keep information from readers.

SUMMARY

- Visuals are a crucial problem-solving tool that technical writers use to increase comprehension, gain attention, establish authority, and communicate with a broader audience.
- The term *visuals* refers to any pictorial representation other than written text that is used to convey meaning and information to an audience.
- *Visual rhetoric* refers to the ways in which visuals communicate meaning to readers.
- *Graphics* refers to any visual that appears to be drawn.
- Icons are visual metaphors that denote other things. Icons are designed to rapidly convey information to readers.
- Graphs are diagrams that represent the relationship between two or more kinds of quantifiable information. They generally depict information along two axes. Bar graphs, line graphs, and pictographs are the most common types of graphs.
- Charts make it easier for readers to understand data and the relationships among the data. Pie charts, flowcharts, and organizational charts are three common types.
- Tables are lists of data presented in a system of rows and columns.
- Line drawings are simple drawings used to represent objects.
- Images have more realistic qualities—for example, photographs, screen shots, videos, and animations.
- Searchable databases and the World Wide Web are good resources for locating visuals. *Clip art* refers to pictures that can be clipped or copied from previously printed material.
- The best way to obtain context-specific visuals may be to make your own.
- Visuals that appear on computer screens are physically different from visuals that appear in print documents.
- *Bitmap* refers to a method of storing an image bit by bit, or pixel by pixel. Vector graphics is a method of storing visuals based on a concept of paths.
- Thumbnails are smaller versions of larger visuals.

- Color can be used to identify parts of a document or kinds of text, to highlight or draw attention, and to improve the aesthetic and professional quality of documents.
- Much of the effectiveness of visuals is determined by how they relate to the text they support.
- One of the most common mistakes workplace writers make in adding visual elements to their documents is not revising those elements as they revise the rest of the document.
- Workplace writers must consider all aspects of ethical representation in visuals.

CONCEPT REVIEW

1. What are visuals?
2. What are graphics, and what are some examples of graphics?
3. What are images, and what are some examples of images?
4. What is visual rhetoric, and what kinds of rhetorical choices do workplace writers have to make?
5. What are icons, and what are some guidelines for making them?
6. What are graphs used for?
7. What are some variations of the standard bar graph?
8. What are some of the uses of line graphs?
9. What are pictographs?
10. What function do pie charts serve, and what are some guidelines for making them?
11. What are some kinds of flowcharts?
12. What are organizational charts?
13. What are tables generally used for?
14. What are line drawings?
15. In what circumstance might you use a schematic?
16. What are some guidelines for using photographs in documents?
17. How do you capture a screen shot?
18. What are some sources for premade visuals?
19. What is the difference between a bitmap and a vector visual?
20. What is a thumbnail used for?
21. What can color be used for?
22. How should you go about revising the visuals in your document?

CASE STUDY 1

The McDonald's Nutrition Icons

Background

In June 2007 Enterprise Language Solutions (ENLASO), a company that specializes in translation solutions, announced that it had developed for McDonald's Corporation a series of nutrition information icons designed to be understood worldwide. McDonald's objective in working with ENLASO in McDonald's Nutrition Information Initiative was to make nutrition information on its product packaging accessible to patrons all over the world. The primary problem for the initiative was to develop icons that could be understood with or without written explanation in 109 countries in the world. McDonald's contracted with ENLASO and its linguistic iconographers to develop icons that could be understood without being offensive to any of the cultures the icons would reach.

McDonald's wanted to provide more accessible nutrition information to its customers around the world and, through extensive research, identified that iconic representation was the most efficient way to do so. To develop a new "language" of nutrition that could reach a cross-cultural, multilingual audience, McDonald's and

ENLASO focused on five key nutrition areas: calories, fat, carbohydrates, protein, and salt. They then focused on four key questions that drove the project:

1. What visuals can communicate the desired nutrients?
2. Do the visuals work in 109 countries without evoking negative or socially/politically inappropriate connotations?
3. Will the visuals print or display well in all media, including packaging?
4. Does anyone else already own rights to the images that might prevent them from being used in this context?

For more on usability testing, see chapter 11.

After reviewing comments and feedback from over 13,000 cultural imagery experts around the world—a type of usability testing—McDonald's released its icons to the world, making them "freely available for unrestricted use within the restaurant and food industry."

Early in their work, ENLASO and McDonald's found that there were no already-established icons that could be used or revised; they would have to start from scratch. Much of their work involved creating icons and then seeking feedback from regional icon/language experts. Throughout this process the researchers found many icons that were unsuitable. For instance, the team identified four different icons for fiber that were considered to be high-risk icons based on responses they received about them (see Figure 8.37).

High risk: Associated with "scary alien" in 47 countries	High risk: Perceived as "slippery" or "curvy" road and "bird sanctuary"	High risk: Associated with marijuana because of three leaves	High risk: Associated with Christmas tree or a burning candle	Low risk: High acceptance from cultural analysis

FIGURE 8.37 Some possible fiber icons identified as high or low risk

ENLASO icon designers found that often their concepts of what seemed logical visually did not seem so to reviewers from other cultures. For instance, one designer proposed the icon for iron found in Figure 8.38. As the figure shows, however, a number of reviewers did not read the icon as the designer intended.

Croatia: "Looks like something heavy, therefore not good."
Argentina, China, Czech Republic, South Africa: "Looks like a lock, or a security device, or something that symbolizes security measures."
Australia, Brazil, Poland: "Looks like a handbag, a purse or a bell."

FIGURE 8.38 Different interpretations of a proposed icon for iron

ENLASO designers also ran into issues revolving around their use of color and the ways in which color is interpreted differently worldwide. For instance, one designer noted how frequently icons that were designed as black on yellow backgrounds were associated with traffic signals, something the designers had anticipated. Figure 8.39 shows one such icon and the responses it received.

Scotland: "Resembles orange and black signs used with Glasgow subway system."
Canada: "...yellow and black evoking road signs colors."
Ireland: "Colors resemble road signs."
Denmark: "...issue here could be that orange/black color combination is normally used on 'danger symbols.'"

FIGURE 8.39 Varied responses to particular icon colors

Ultimately, McDonald's and ENLASO settled on five icons. Figure 8.40 shows both early versions of the icons and the final versions.

	Calories	Protein	Fat	Carbs	Salt
Early icon					
Final icon					

FIGURE 8.40 Early and final versions of the McDonald's/ENLASO Nutrition Information Initiative icons

Assignment

McDonald's and ENLASO have released five key icons as part of their Nutrition Information Initiative, but those icons are by no means comprehensive in providing all of the nutrition information that a consumer might want. For instance, the carb icon does not offer any information about the differences between dietary fiber and sugars. Nor does the fat icon distinguish between saturated fat and trans fat. Likewise, the five icons do not offer any information about cholesterol.

Select one of these kinds of nutrition information that the McDonald's/ENLASO icons do not cover—dietary fiber, sugar, saturated fat, trans fat, or cholestrol—and create an icon for that information. You should have at least ten reviewers comment on your icon drafts before settling on a final icon. After you have designed and tested your icon, write an explanation about how you settled on that icon. Be sure to include an evaluation of the reviewers' comments and a description of how they influenced your revisions.

CASE STUDY 2

The *Charlotte Sun*–Altered Photograph

Background

On May 4, 2007, the Associated Press issued a mandatory photo elimination bulletin for a picture that had been digitally manipulated before being published in the *Charlotte Sun* newspaper. The executive editor of the paper, Jim Gouvellis, identified that the picture had been mistakenly published after the manipulation; there had been no attempt to publish a doctored photograph.

The photograph in question depicted a dog leaping over a pool of water during the Splash Dogs jumping challenge in the Oh boy! Oberto Redfish Cup in Punta Gorda, Florida. The photographer who submitted the picture to the *Sun*, Sarah Coward, noticed the alteration and contacted the Associated Press to request that the picture be deleted. According to Gouvellis, the photographer had made two versions of the picture. In an altered version a cord or leash had been digitally removed from behind the dog's leg. The altered version was intended to be sold to a private customer. The unaltered picture was to appear in the *Charlotte Sun*, but accidentally the altered version did.

The National Press Photographer Association (NPPA) ethics and standards chairperson questioned the mix-up. "If you are going to publish a news photograph one way, why would you sell it in a different way? It's still the newspaper's product." He went on to say "The newspaper still has to vouch for its integrity. Avoid the possibility of using the wrong photograph: don't create the wrong photograph to begin with." The Associated Press issued a mandatory photo elimination bulletin that read, "Please eliminate FL VEN101 transmitted May 4, 2007. The photo was digitally altered at the source." The AP offered no other explanation, leading many editors around the country to believe the photo had been intentionally altered. When questioned about the picture, photographer Sarah Coward—who had been a member of the NPPA since 2003—offered only that "my editors will be speaking on my behalf." Executive editor Gouvellis expressed his certainty that the photographer had not intentionally tried to deceive the newspaper or the public and that the situation was an honest mistake.

Assignment

Take some time to consider the ethical questions involved in this case. Read the NPPA Code of Ethics (codes of ethics are addressed in chapter 4). How do you respond to this situation? Was this an honest mistake as the *Charlotte Sun* executive editor claims? If it is, do honest mistakes excuse situations like this? Should Sarah Coward have altered the image in the first place? Write an assessment of the situation that considers the ethical implications of the case.

CASE STUDIES ON THE COMPANION WEBSITE

Visuals with Teeth: The Ethics of Graphically Representing Shark Attacks

Cathleen Bester is a web editor for the International Shark Attack File. She prepares visuals to graphically illustrate the relationships between shark attacks and other variables. Cathleen must revise some graphics in a way that ethically represents the data.

Visualizing Performance: Designing Graphs for Cyclists

You are a new computer engineer at VeloTrainers, a company that makes stationary bicycles. The bicycles connect to a computer that records speed, cadence, heart rate, time, and more. Many cyclists have complained that they would like to see the data represented graphically to make it easier for them to analyze and chart their progression.

Identity Theft: The Ethics of Logo and Image Manipulation

You are an assistant marketing director for a local car dealership, Bulldog Motors, named after the University of Georgia mascot. In the past most of the customers have been University of Georgia supporters and alumni. Your job is to help design a marketing campaign to attract this demographic to the dealership.

Making Bikes Green

Your boss wants to make his bicycle company look more "green." You must produce marketing materials that will please your boss but will also ethically represent the company to its current and potential customers.

VIDEO CASE STUDY

Cole Engineering–Water Quality Division

Revising the Cole Engineering Website

Synopsis

The Cole Engineering website is over two years old. Much of it is outdated and does not present the cutting-edge feel that the company wishes to portray to clients and customers. In addition, Cole Engineering has expanded. Because the Water Quality Division has grown more than any other division, this portion of the website is most in need of change. As a web designer, Dennis Blount can give the Water Quality Division a number of suggestions about the best way to present information on a website; however, he does not have the technical expertise in hydrology and other water-quality issues that would allow him to revise the website by himself. He needs their input and expertise to make sure that everything presented is technically accurate.

WRITING SCENARIOS

1. Identify a familiar way in which the public is given information visually. Analyze how that visual works and what visual rhetorical choices were made in order for it to function successfully or unsuccessfully. Then write a visual rhetorical analysis of that approach.
2. Conduct some research to find out how your school's administration is structured. This may involve web searches and talking with administrators. Then create an organizational chart that shows the relationships among the various levels.
3. Consider how we have used the various elements discussed in this chapter to produce the chapter's visual component. Make a copy of the chapter (or scan it to create an electronic version), and then use different colored pens to show how you would redesign the visual elements of the chapter. Be sure to address all aspects: text color, text box color, background color, graphics, photographs, line drawings, tables, charts, and so on. Write a letter to your instructor explaining the changes you would make and the visual rhetorical choices behind those revisions.
4. Locate four different documents that use visuals: a web page, a set of instructions, a manual, and a technical description. Analyze the use of visuals in each. Based on what you see in the documents, create a table that compares the features of the documents: (a) the purpose of visuals in each document, (b) the type of visuals in each, (c) the relationship of visuals to text, (d) distinguishing use of color, and (e) any ethical questions that the use of visuals might suggest.
5. Using a World Wide Web image search engine, locate a photograph, a line drawing, a graph, a piece of clip art, and a chart. Capture each of the visuals you locate, and save them to a file you can work with. Then use the word processor you use most often to place them in a single document. Resize them so that they all fit on a single page and are all readable. Add a caption to each, and identify the source information for each.
6. The National Cancer Institute maintains a web page called "NCI Visuals Online," which provides a searchable database of related images. These visuals are provided for public use as long as the user offers a citation and credit for them. Use the search engine feature to select a term for which you might find several kinds of visuals within the NCI database (e.g., something like "lung"). Locate two photographs, two line drawings, and two visuals that combine text and visuals. Download the visuals to a file from which you can work; be sure to retain citation information. Then create a document that includes all six visuals, with captions and descriptions of each.
7. Locate a digital picture—one you have taken or one you have found. Store the image in a file you can access. Then use image editing software to alter the image so that the manipulated version offers a different message from that of the original. Basic paint programs like Corel Paint Shop Pro or Microsoft Paint will let you manipulate images if those files are saved as bitmap files, such as BMP, JPG, or GIF. Use cropping, skewing, and/or color to manipulate the image. Then create a document that shows both the original visual and the manipulated version and explains the changes you made. Be sure to provide a source citation if you did not take the original photograph yourself.
8. Find an example of a document or a website from a group or organization that you believe supports an unethical position. Examine the use of visuals in that

document or website. In a memo or other document to your instructor, explain how the group or organization uses visuals to assert its perspective. Are the visuals distorted or cropped in a way that is misleading or unethical? Does the group use text or captions to explain or interpret the visuals in a way that is misleading or unethical? Or are the visuals accurate and ethical, despite the overall position of the group or organization?

9. Examine the websites of the two major U.S. political parties—www.democrats.org and www.gop.com. Write a memo or other document to your instructor in which you analyze the visual rhetoric used on these two websites. What is the first thing you see when you visit each page? How does each use colors to suggest a perspective or to persuade its audience? What types of visuals do they use, and are those visuals effective? Are the visuals accurate and ethical, or do they present a topic in a biased or unethical manner? If both websites address a common issue, do they use similar visuals, or are their visuals dramatically different?

10. In 2004 the United States imported $9.67 billion worth of cocoa, coffee, rubber, bananas, olives (including olive oil), spices, tea, and tropical oils. This accounted for 18.3 percent of all U.S. agricultural imports for 2004. In 2002 the United States imported $6.84 billion worth of these agricultural products; in 2003, $8.40 billion worth. Using these numbers, create a pictogram that depicts the dollar amounts of these agricultural products imported over these three years.

11. In 2006, as part of an argument for raising the county sales tax, the county commission of Shaia County, Florida, released a document explaining how taxpayers' money had been spent during the 2004–2005 fiscal year. According to that document, 56 percent of the money was devoted to school improvement, 13 percent to the county's general fund, 8 percent to various city and town governments within the county, 5 percent to nonincorporated districts within the county, 4 percent to development agencies within the county, 11 percent to parks and roads, and 3 percent to the county's emergency reserve. Create a pie chart that depicts this information.

12. Find an example of a document or a website from a major company or organization in your intended profession. How does the document or the website use visuals to present an image of the company or organization? What colors are prominent, and how do those colors assert a particular message or perspective about the company? Does this company or organization use visuals effectively? How could they be improved? Write a memo or other document to your instructor addressing these questions.

13. Using whatever paint program is available to you, create a line drawing of any piece of electronic equipment that you use regularly: an MP3 player, a DVD player, a laptop computer, a desktop computer, a cell phone, a digital camera, and so on. Your line drawing should focus on the external features of the equipment; do not include internal components. Be sure your drawing identifies all of the key physical features of the device, and render the drawing using color.

14. Using whatever paint program is available to you, create an icon/logo to represent a piece of electronic equipment that you use. Design the icon in such a way that it could be used either on a web page as a link to information about the device or in a print document as an alert to text about the device.

15. Visit the websites of two different graphic design or imaging software programs that you have heard of but have never used. Compare the features that each program offers. In a memo to your instructor, describe the two programs and their features, and recommend one of the two programs based on your research.

9

Layout and Design

CHAPTER LEARNING OUTCOMES

After completing this chapter, you will be able to do the following:

- Consider the role of **layout** and **design** in your documents
- Better integrate layout and design elements in your **problem-solving approaches**
- Understand how **document architecture** helps convey information to your readers:
    - Achieve visual **balance**
    - Employ visual **connection** to group elements together
    - Use **duplication** to assist readers in navigating your document
    - Make use of **variation** to emphasize contrast and difference
    - Create visual **flow** to move readers from one part of your document to another
- Make choices about the **typography** you use in your documents
- Design and place **titles and headings** in your documents
- Place effective **captions** with visuals
- Use **headers and footers** effectively
- Use **lists** to convey information to readers
- Control **line length**, **line spacing, and justification** to make documents readable
- Make effective use of **white space** to emphasize information or to make visual connections within your document
- Take into account the **physical properties** of your documents

DIGITAL RESOURCES

On the Companion Website www.prenhall.com/dobrin:

- Case 1: Publicizing Orton Construction's Upcoming Public Service Campaign
- Case 2: Crafting a Highly Detailed Document: Bruce Smithers and the Winged Elm
- Case 3: Managing Miscalculated Bills and a Public Relations Crisis at CPW
- Case 4: Putting Fish Online
- Video Case: Information for New Investors: Defining the Terms
- PowerPoint Chapter Review, Test-Prep Quiz, Exercises and Activities
- TCTC Visual Rhetoric and Design Tutorial

REAL PEOPLE, REAL WRITING

MEET LORI BOOTH • National Grants Manager for the Surfrider Foundation

What types of writing do you do at the Surfrider Foundation?

Documents related to obtaining and retaining grant funding—such as letters of intent, full proposals, interim and final reports—comprise the vast majority of the writing I do for the Surfrider Foundation. I also spend some time developing internal reports describing the status of grant funding for the organization.

How do you use layout and design in your documents?

Granting foundations often spell out their content and format requirements for various documents. Of course, those requirements dictate the layout and design of the document, sometimes right down to section headings, margin and paper sizes, and even preferred font style and size. When I have some flexibility in my writing, I try to use headings, lists, and white space to break up the information and to emphasize what is most important. This is essential in longer documents, where readers need breaks to enable them to digest or revisit various sections.

Could you describe the role images play in the documents you create at Surfrider?

I use images sparingly, but when I do use them, they play an important role. I use them in reports to highlight specific progress we've made; I also use them in proposals to emphasize the need for our work. I will also incorporate maps when appropriate. For longer documents I try not to have more than two consecutive pages of just text, so images and outlined text blocks can be very useful for providing breaks for the reader.

Do you use any particular font styles in your documents? Were these font styles chosen consciously for their effect?

I prefer Arial and Times New Roman. Even when a granting foundation has required a font style for a document, it is usually one of these two. Arial is somewhat less formal, and although one could argue that most of what I write is decidedly formal, I like this effect for most of my writing; it seems to fit my preferred tone. I will, however, look at a foundation's website or any past written correspondence from them to see what their preference is and then decide accordingly. I will also use Times New Roman for the document when I'm running up against space limitations.

What advice about workplace writing do you have for the readers of this textbook?

My main piece of advice is to PROOFREAD! Do not rely on spell and grammar checks because they are not foolproof. Always proofread the entire document before it goes out to your audience. It's even better if you can proof it with fresh eyes. Incorporate proofing time into your work plan so that it gets done. I can't remember the last time I proofed a document and didn't find something to correct or change. This one simple thing will improve the quality of your documents and reflect positively on you and your organization.

INTRODUCTION

layout the process and result of arranging the elements of a document—including medium, typography, spacing, graphics, and other visual elements—that give the document its shape and structure

design the process and result of composing the visual form of a document

One thing that distinguishes workplace writing from other writing is a focus on the **layout** and **design** of documents. Even though all good writing pays attention to the organization of information, workplace writing must carefully organize the placement and appearance of words and images.

As chapter 8 explains about visual rhetoric, workplace writing requires writers to pay attention to the visual elements of their documents, including the physical appearance of those documents. The design features discussed in this chapter allow you to communicate information in a variety of ways so that you can choose the methods most useful for your particular purposes and audiences. These layout and design elements apply to workplace documents in general and to both print and web-based documents.

Layout and design are visual rhetorical choices that are important in workplace writing for two reasons. First, readers of technical and professional documents are typically busy and often skim documents to find the information they need. Whereas the readers of a novel or a personal letter may read at their leisure, those examining a set of instructions, a research proposal, or a corporate website read quickly to find specific facts and details. Well-designed workplace documents allow readers to skim, jump from place to place, and refer back to information with minimal effort and time. Second, workplace writing often addresses complex, highly detailed subjects. As a result, workplace writers must make conscious choices about how to deliver the information in ways that readers will understand and follow. Workplace documents may not flow unless properly organized; layout and design choices become increasingly important as the complexity and quantity of information increase.

IN YOUR EXPERIENCE

As a class, spend a few minutes discussing pieces of writing you have read that communicated a complex, highly detailed subject. What specific layout and design features did the document use to present the information? Was the information clearly organized? Who was the intended audience for the document, and was it appropriately designed and structured to meet audience needs? After you've discussed these specific examples and their design features, discuss what you believe are general principles of effective layout and design.

Many of the layout and design features that we discuss in this chapter can be used to accentuate or suppress material, which can help you to create more persuasive messages. But as we emphasize in chapters 4 and 8, the ability to present information in sophisticated, rhetorically savvy ways comes with an ethical responsibility to present that information in an unbiased, fair manner. How you choose to employ visual elements in your documents affects the ethical presentation of information. You should not use design elements to hide facts your readers will need or to unfairly emphasize or distort details in any way. Visual rhetorical choices are also ethical choices.

See chapters 4 and 8 for more on ethics and the use of visuals.

Layout and design play an important role in many documents in the workplace and in public spheres. Organizations like Rock the Vote, an advocacy group focusing on political engagement among young people, make careful layout and design choices to increase the impact of their documents. Figures 9.1 and 9.2 offer two versions of a Rock the Vote flyer, providing an example of the importance of effective layout and design. Notice how Figure 9.1 provides relatively few navigational or organizational cues for the reader; the information is dumped into the document with little arrangement. Figure 9.2, on the other hand, guides readers from one point to another, carefully organizing and grouping information to make it easy for readers to skim, follow, and understand.

EXPERTS SAY

"Good document design enables people to use the text in ways that serve their interests and needs . . . the reader's needs should drive design activity."

KAREN SHRIVER
Dynamics in Document Design

Rock the Vote. Voting is easy. Voting is no harder than getting money from an ATM. Really.

Your vote counts. Florida was decided by 537 votes in 2000. New Mexico by 366. Every vote makes a difference.

Register to vote. You can fill out your form online at rockthevote.com. Just print it, sign it, lick it, and mail it.

Vote Early. Vote by mail. Your state may have early "in person" voting. You may also be able to mail in an ìabsentee" ballot. Find out and do it. Check rockthevote.com.

Vote from school. If you are a student, you may register and vote from your school residence—that is your right.

Where and when to vote. You will get your voting place location and hours in the mail. If you lose it, check rockthevote.com

Bring ID with you. States have new requirements for ID. Bring yours.

When voting, know your rights. If you mess up your ballot while voting, you can request a new one and do it again. If you are registered but for some reason not on the list, or if you have an ID problem when you get there, request a "provisional" ballot. There will be lawyers and poll watchers to protect you. Problem? Call 888 our vote.

Rock the vote. Join a street team and mobilize your friends at street.rockthevote.com.

Get campaign news on your cell phone. Signup by texting "rockme" to RTVMO (78866).

Rock the vote with your cash (or credit card) at donate.rockthevote.com.

FIGURE 9.1 A document without organizational cues

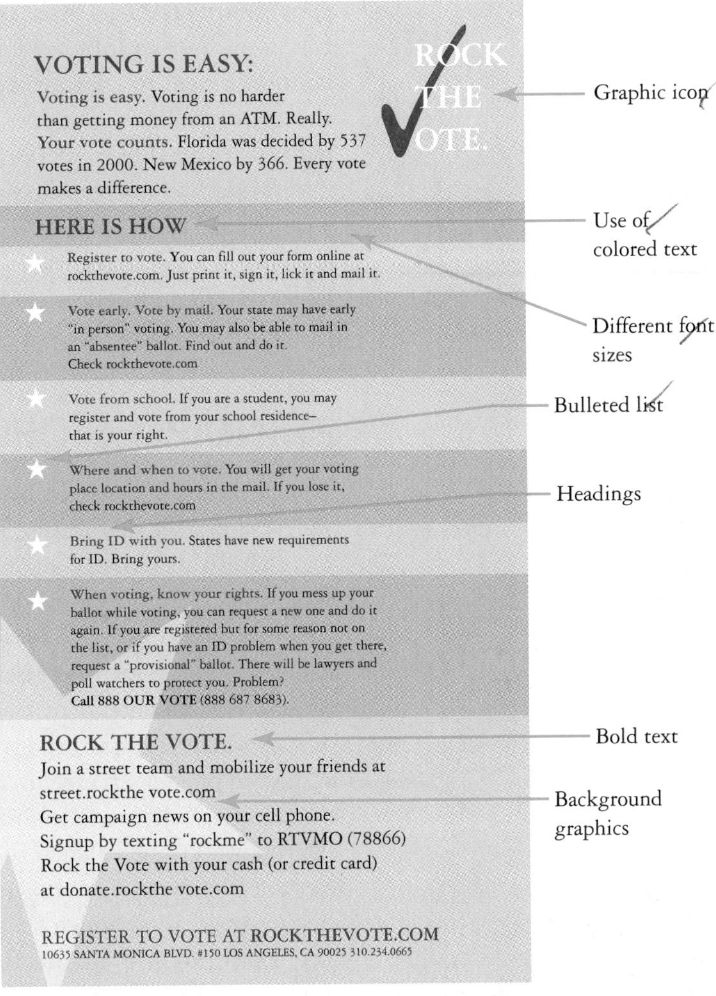

FIGURE 9.2 A document with organizational cues

ANALYZE THIS

Although the information contained in Figures 9.1 and 9.2 is essentially the same, the way that information is organized and transmitted is significantly different. Use the following exercise to investigate those differences: Draw a vertical line down the middle of a blank piece of paper. In the left-hand column write down all of the specific differences you find in Figure 9.2 as compared to Figure 9.1. In the right-hand column analyze the purpose or effect of each difference. In a small group or with the class, discuss the specific differences and your analyses of them.

LAYOUT, DESIGN, AND THE PROBLEM-SOLVING APPROACH

As you become more familiar with the problem-solving approach, you will realize that much of problem solving in workplace writing addresses two key aspects of document design: the content and the visual appearance of the document. As you already know, the Plan phase of the PSA encourages you to choose your persuasion strategy and format and to choose technologies that will best assist you. The Draft phase suggests that you design the document and arrange sentences, paragraphs, and sections in an

PSA

appropriate order and that you integrate any visuals or graphics that will help to communicate the information. And the Distribute phase proposes that you publish your document in appropriate form. All of these activities ask you to consider the visual components of your documents. In many ways the content of your document cannot be separated from the design of the document, and the design of your document depends on the content, audience, and purpose.

As you learn more about individual workplace genres, you will apply the general design information found in this chapter and in chapter 8 to those documents. Good design and good visual rhetorical choices can avoid or even solve many workplace writing problems.

PRINCIPLES OF DOCUMENT ARCHITECTURE

We use the term *document architecture* to explain how text and images can be positioned and designed for readability. In this sense document architecture refers to making visual rhetorical choices about the placement and layout of information on a page. Just as a physical architect makes conscious choices about where to position doorways, windows, and closets in a home to make it livable, a workplace writer must make conscious choices about where and how to position headings, visuals, lists, and other features to make a document readable.

Few rules exist in either type of architecture, but there are general principles that have been proven effective in both. For example, there is no one right way to position a window in a home, but windows are generally most effective when centered on an external wall rather than at the top of an internal wall. Similarly, there is no one right way to place a visual in a document, but visuals are usually more effective when placed close to the text to which they correspond rather than several pages later.

functional providing support for a particular purpose or use

Both physical architecture and document architecture are influenced by functional and aesthetic choices. In workplace writing, documents are primarily designed to be **functional**, and the various design elements assist documents in achieving this goal. For example, the headings or subheadings found in Figure 9.2 help readers more easily access specific information in the document. The headings are visually distinguished from the main body of text and provide readers with visual signals to locate parts of the document. In larger, more elaborate documents, functional elements like titles, headers, tabbed pages, page numbers, and tables of contents help readers locate information quickly.

aesthetic relating to beauty or visual appeal

Aesthetic elements are the parts of a document that make it look pleasing to the eye. Aesthetic choices are not as crucial as those that lead to function, but you should still think about the overall aesthetic quality of your documents. Aesthetic qualities often demonstrate a kind of professional refinement and help make documents more interesting to readers. Beyond the need for functionality, readers react positively to aesthetically pleasing documents. Figure 9.2 uses a colored background, text, and an icon to stimulate reader interest. However, you should remember that aesthetic appeal and functionality will not supersede a document's content. A good-looking document cannot mask poorly written or inaccurate information.

Balance

balance the alignment of elements on a given page or page spread to achieve visual equilibrium

Balance in a document refers to the way elements are arranged and aligned to create a unified whole. Effective documents should have an overall sense of symmetry between text and graphics. In many documents you'll want to place important elements in the *optical center* of your document, to emphasize them and provide balance with less important elements. However, the optical center of a document is not the physical center;

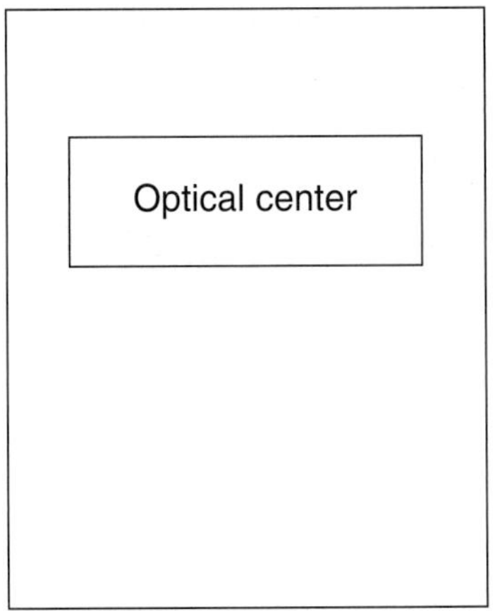

FIGURE 9.3 The optical center of a page

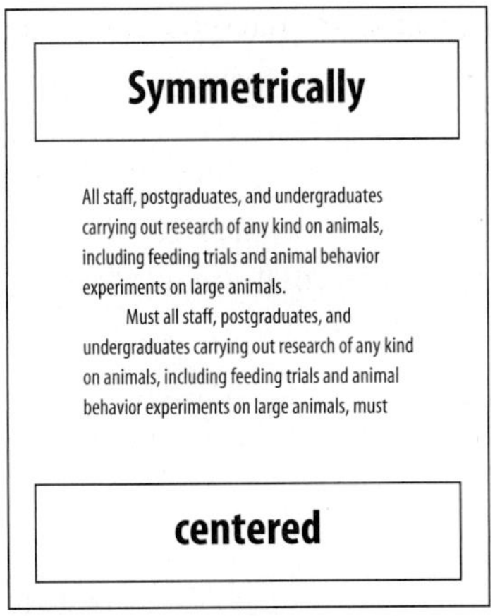

FIGURE 9.4 A symmetrically balanced document

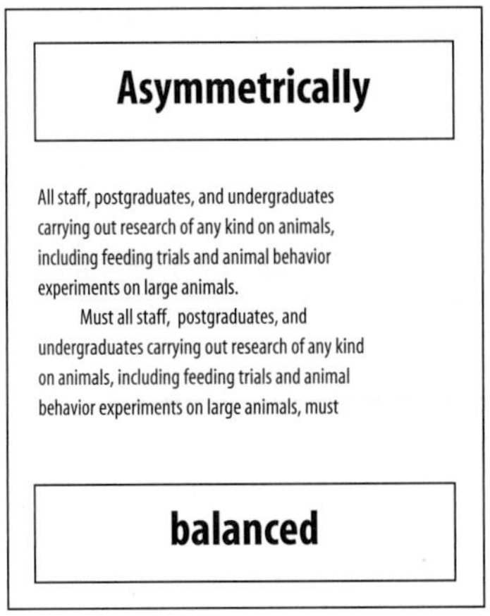

FIGURE 9.5 An asymmetrically balanced document

it is approximately one-third of the distance from the top of the page (see Figure 9.3). Because readers are naturally drawn to the optical center of a page, you'll want to carefully consider what they will see there.

However, you don't have to place everything important in the optical center. In fact, you'll want to consider whether to balance elements symmetrically (i.e., centered) or asymmetrically (i.e., off-center). Either choice can have a profound effect on how your document is read. *Symmetrical* balance has elements of equal weight on both sides of a document. This method is best when you want to convey a sense of stability, permanence, or formality (see Figure 9.4). Symmetrical balance is pleasing to the eye, but it can become boring and static when overused. *Asymmetrical* balance uses "heavier," more visually striking elements, which are offset by other less striking elements (see Figure 9.5). Asymmetrical balance is particularly useful to emphasize part of a document, like an image. However, it is important to include a counterbalancing element (e.g., white space or blocks of text) to maintain a sense of order and stability.

Just as the windows, doors, and rooms of a home are carefully balanced to be both functional and aesthetically pleasing, a document must be carefully balanced by aligning and positioning text, images, and white space to convey a sense of order. Consider how balance and symmetry are used in Figure 9.6, the web page for visitors to San Francisco. The title is placed near the top of the page, close to the optical center. The two large images are symmetrically balanced, as are the eight smaller images near the bottom, providing equal "weight" to both the right and left halves. The text in the middle of the page helps to break up the images, providing more balance. Even the text links at the bottom use a symmetrical alignment.

Connection

connection Creating a visual relationship by grouping elements together

Connection is a simple concept: if elements in your document are closely related to one another, you should emphasize this relationship by grouping them together. Readers easily interpret a relationship among closely grouped elements, and they assume that elements are unrelated if they are placed far apart. The principle of connection is particularly important when creating a document with a complex architecture; readers become confused when captions, headings, or descriptions are far away from the text to which they correspond. In Figure 9.7 the description and the price of each monitor are clearly connected to the image directly above. However, in

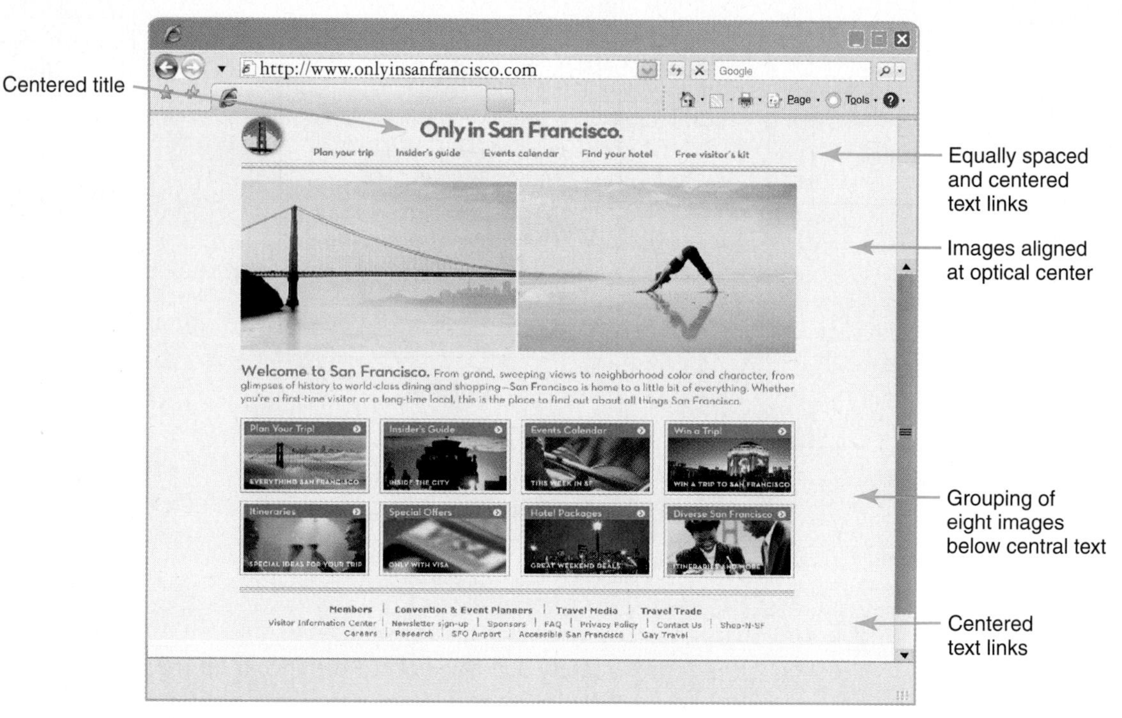

FIGURE 9.6 Effective use of balance and symmetry

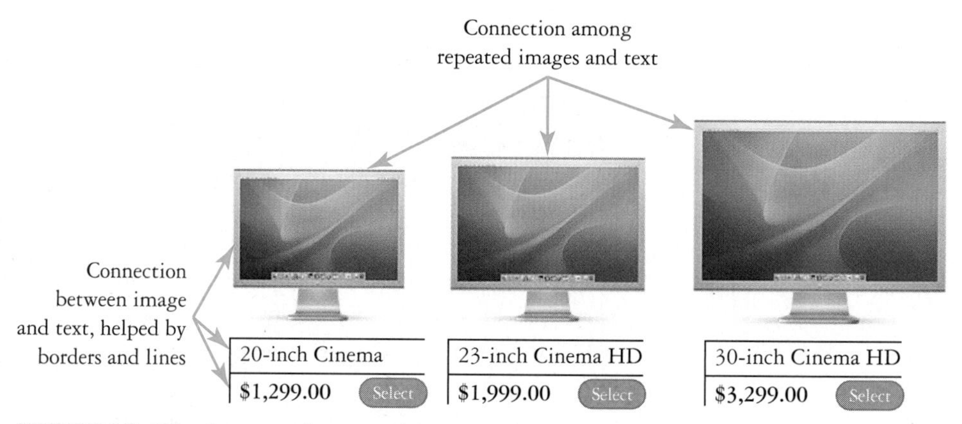

FIGURE 9.7 Effective use of connection

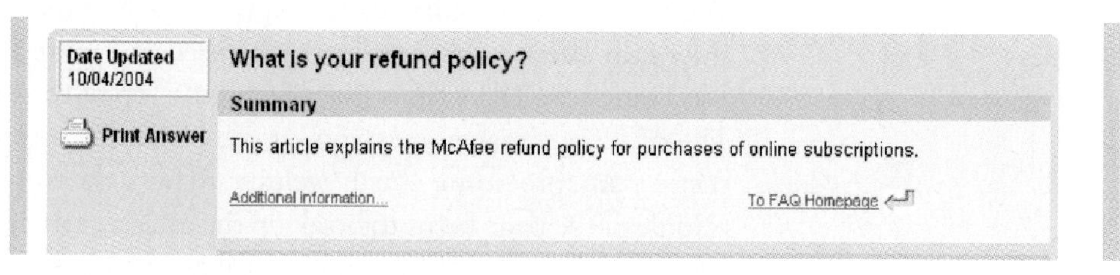

FIGURE 9.8 Ineffective use of connection

Figure 9.8 there is an unclear relationship between the article mentioned and its location on the page. Must customers print the article? Or is it under "Additional information" or on the FAQ homepage? In reality, none of the three links takes the customer to an article about refunds.

Duplication

duplication repeating visual elements consistently throughout a document

Duplication suggests that elements of the same type, importance, or magnitude should use the same basic format, layout, and design. Repeating design elements from page to page lets readers quickly determine the type of information they are

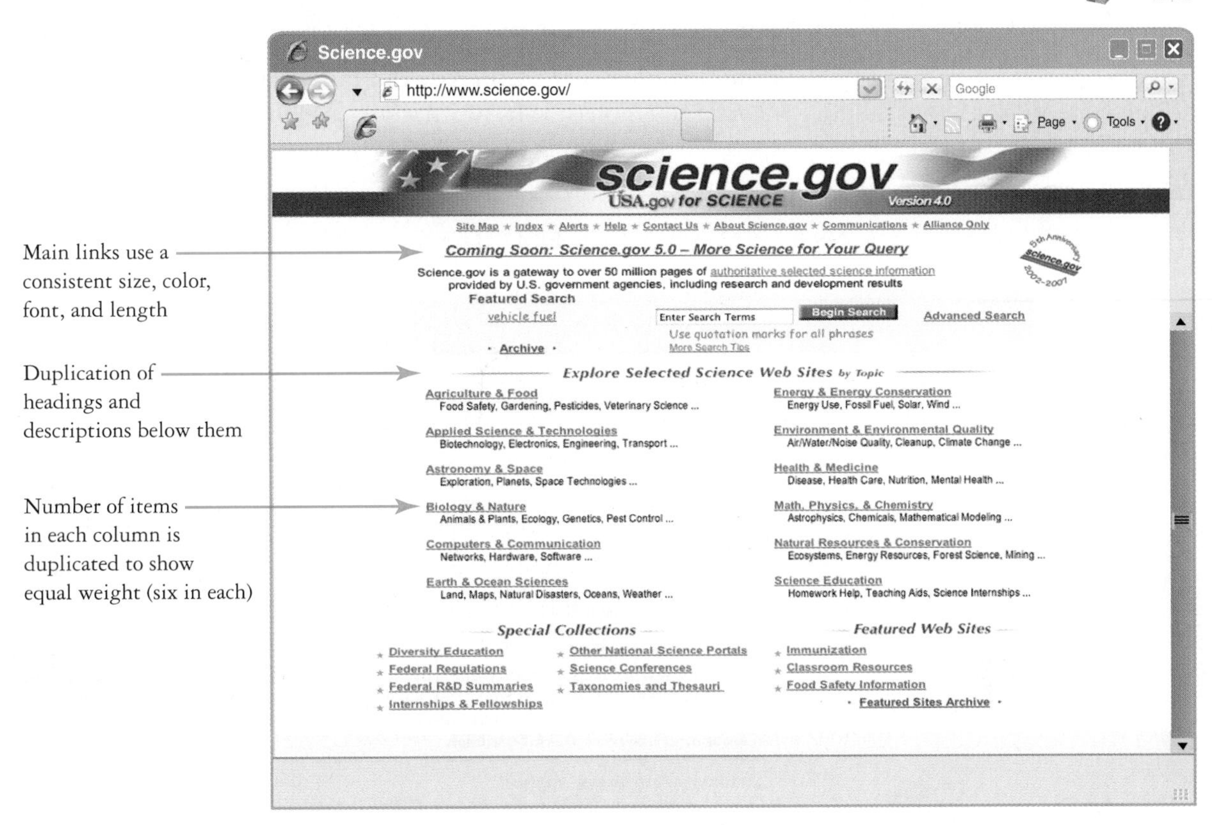

FIGURE 9.9 Effective use of duplication

reading, allowing them to learn the architecture of your document. Because readers instinctively ascribe greater importance to elements that are larger, darker, or set off with additional white space, you should coordinate elements of equal importance for consistency and clarity. For example, all section headings of the same weight or importance should duplicate the same font, font size, capitalization, spacing above and below, and so forth. Just as building architects often duplicate the size, shape, and style of doors or windows in a home to create a sense of harmony and coordination, you should apply this principle to your document architecture as well. Consider how the Science.gov website found in Figure 9.9 uses duplicated links, font sizes, colors, columns, and text descriptions to signal what kind of information is being presented.

Variation

variation changing or altering visual elements to distinguish differences

Variation allows you to show contrasts or differences among elements. Variation has a variety of purposes: it can show a hierarchy, focus a reader's attention, or signal a change or break. Building architects use variation and contrast to draw attention to various design features. For example, a home might use darker trim around a doorway or have a larger center window to draw attention. A document might use darker text or a larger font to do the same thing. Variation can be achieved with a variety of strategies:

- *Vary sizes* to show an organizational hierarchy, making more important elements larger and less important elements smaller. This can be done by varying the size of images, varying the font size of text, or using capitalization (see Figure 9.10).
- *Vary weights* to change appearance and signal importance. This can be done by adding italics, boldface, or different fonts to text.
- *Vary color* to emphasize some items. Typically, this is achieved by using a color opposite or nearly opposite the standard color of the document, such as using red text rather than black.

- *Vary position* to set one or more elements apart from other similar elements. If most elements are centered, you might align one to the left or right margin to emphasize it. Similarly, positioning an element with more white space surrounding it than other similar elements have can provide contrast.

Variation in font sizes to show hierarchy

Variation in use of boldface to distinguish details

Variation in font color to emphasize headings

Unprecedented Generosity

Donor Dollars At Work

(All figures as of July 31, 2007)

Emergency Assistance

More than 1.4 million families—more than four million people—received emergency assistance from the Red Cross. This helped hurricane survivors purchase urgently needed items such as food, clothing, diapers and other essentials. **Cost: $1.520 billion**

Food and Shelter

When hurricanes threatened the Gulf Coast, Red Cross disaster staff and volunteers prepared hundreds of evacuation shelters. The organization pre-positioned supplies including kitchens, prepackaged meals and emergency response vehicles and provided millions of people with food and shelter. **Cost: $229 million**

Physical and Mental Health Services

The Red Cross provided both physical and mental health services to hurricane survivors. Trained mental health professionals were available at Red Cross shelters and service centers to help survivors cope with stress, loss and trauma. Red Cross health care professionals delivered emergency first aid and attended to other health-related needs such as assistance with obtaining prescription medications to replace those lost in the storm. **Cost: $4 million**

Additional Red Cross Support

These funds enable the Red Cross to provide response and recovery resources to disaster survivors including coordinated damage and community needs assessment; deployment of trained workers and supplies; technology support for logistics, communications and information; and support of the disaster welfare inquiry system that helps families reconnect with one another. **Cost: $80 million**

Hurricane Recovery Program (HRP)

With offices in chapters along the Gulf Coast and in cities with large evacuee populations, HRP seeks to address the needs of survivors. From emotional and physical well-being to proactive case management and beyond, HRP's mission is to provide survivors with the tools they need to chart their path to recovery. **Cost: $50 million**

Fundraising Costs/Management and General Expenses

The Red Cross has managed an unprecedented number of contributions from generous donors who are helping meet the needs of people in this record-setting relief operation. These costs are associated with raising the funds that enable the Red Cross to respond to these and other disasters and to fulfill its mission. These costs include expenses such as finance and accounting, legal and auditing fees and public information outreach, all essential services in support of the Red Cross disaster relief effort. The fundraising costs/management and general expenses will be less than 6% of the total budget. **Cost: $80 million**

Funds raised: approximately $2.1 billion

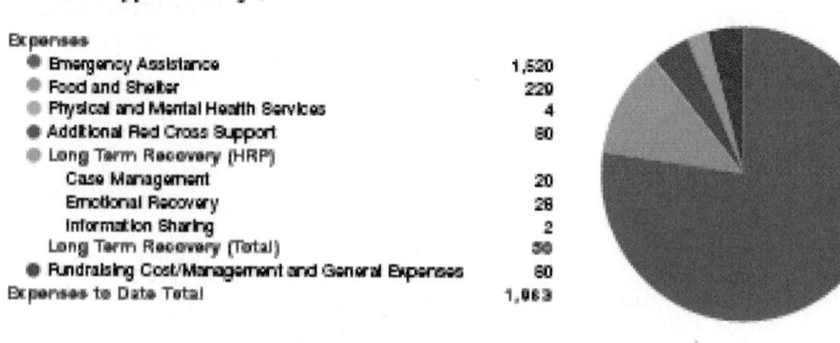

Expenses	
Emergency Assistance	1,520
Food and Shelter	229
Physical and Mental Health Services	4
Additional Red Cross Support	80
Long Term Recovery (HRP)	
Case Management	20
Emotional Recovery	28
Information Sharing	2
Long Term Recovery (Total)	50
Fundraising Cost/Management and General Expenses	80
Expenses to Date Total	1,963

10

FIGURE 9.10 Effective use of variation

IN YOUR **EXPERIENCE**

Look closely at the package of something you have recently consumed—perhaps a cereal box, a soda can, a candy bar wrapper, or a milk carton. Identify how the package uses visual strategies to convey information to its readers. Then, over the next week, be alert to how documents like food packages use balance, connection, duplication, and variation. Keep a list of the documents that catch your eye, and identify how they employ those strategies.

Flow

flow the use of visual elements to move a reader's attention fluidly from one element to the next

The **flow** of a document has to do with the reader's overall experience. Flow refers to the movement of the reader's eye from one part of a document to another. In an effectively designed document, the reader will get a clear and logical sense of movement from one section, idea, or image to the next. Effective flow is dependent on how the other four principles of document design are implemented. To develop flow in your documents, carefully place text and images in a way that leads the reader from one point to another, making ample use of white space and avoiding the overuse of borders and horizontal lines. Just as architects design homes to create a sense of flow by providing easy access and movement from one room to another, you should employ this same principle in your document architecture (see Figure 9.11).

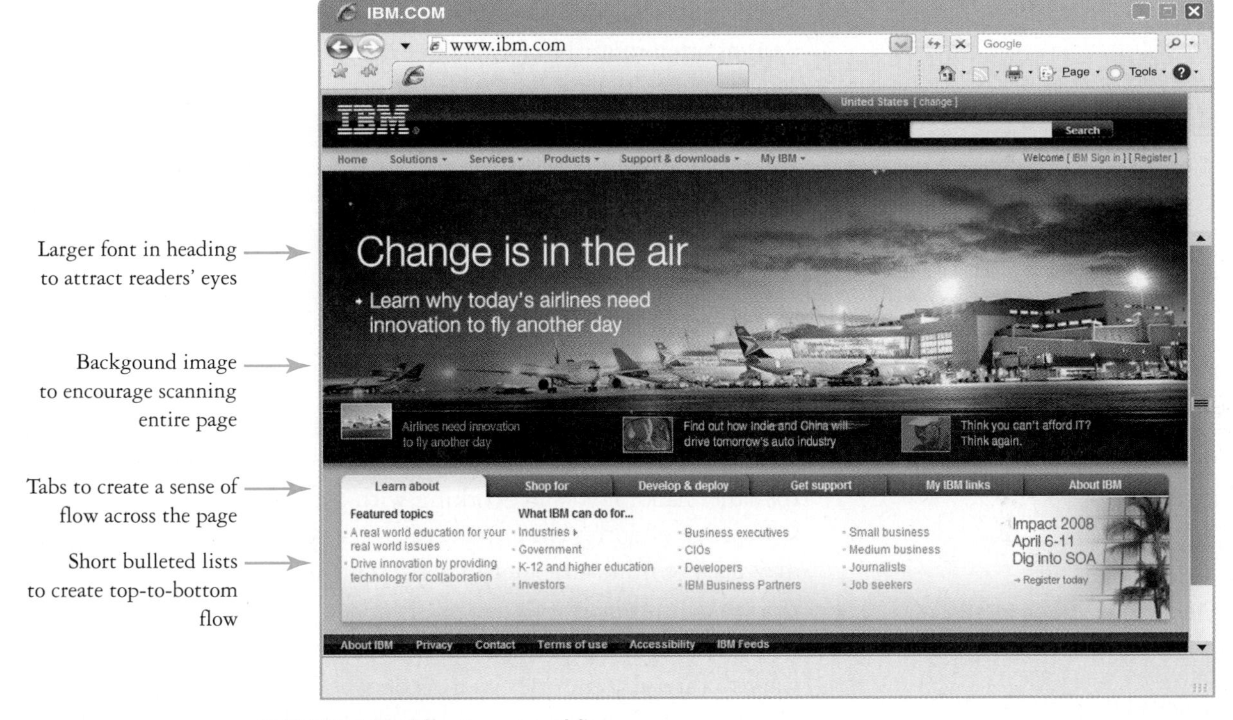

FIGURE 9.11 Effective use of flow

IN YOUR EXPERIENCE

Analyze the covers of two or three of your favorite magazines, and evaluate their use of balance, connection, duplication, variation, and flow. Then use your choice of word processing, web design, or graphics and imaging software to redesign one of those magazine covers to change the subject or the emphasis. Be creative—put yourself on the cover if you'd like—and feel free to visit the magazine's website to borrow images, logos, and typeface. For more information on working with visuals, see chapter 8.

ELEMENTS OF DOCUMENT ARCHITECTURE

The elements of a document consist of all the physical or visual components readers see—the bricks and mortar of the document. How you use those elements to create an effective document is shaped by several factors:

- *Space* is an important consideration in document architecture; you should carefully consider the relative size and space of the document you intend to create before you create it. Generally, more space in a document equates to a greater number and variety of elements. For example, a four-page report would probably have fewer sections and headings than a twenty-page report. Another space consideration might be the size and memory requirements of documents you plan to e-mail or post to a website.

See chapter 3 for more on netiquette and document size.

- *Time* can influence a document's architecture, particularly if you have little of it in which to create your document. A sophisticated design often requires additional time to produce, especially if you are creating your own visuals or web-based elements. In addition, documents with colorful visuals, unique paper sizes, or professional-quality binding may need to be outsourced for printing, which can take extra time. Consequently, if time is a factor, you may need to work with the basic tools available on your computer, as well as with a standard paper size and the capabilities of your office printer or copier.
- *Money* can affect your document's architecture as well, because sophisticated designs often result in more money being spent on creation and publication. Many companies have limited budgets and cannot afford professional designers, expensive publishing, or new equipment with special design features. Typically, employers will authorize more money to be spent on an external document, particularly a sales-related document, than on an internal document.
- *Equipment* can be a factor in your document's architecture, particularly if you have, or are familiar with, only basic word processing. If you have access to advanced graphics, web-design, or desktop publishing software, you may be able to create more sophisticated documents. Your printing equipment may also be a factor; a graphically complex document is pointless without the capacity to print in color.
- *Collaborators* may be able to help you create documents with a more elaborate architecture, particularly if they have expertise in graphic or web design. On the other hand, if you are working with collaborators who have only a basic knowledge of document architecture or who prefer a conservative design, you may need to minimize your use of certain architectural features in your writing.
- *Readers* will influence the architecture of your documents; in fact, they usually will be your primary concern in determining document architecture. You'll want to carefully consider your readers before beginning to create your document, and then you should think about how each choice in layout and design may affect their reception and understanding of your information.

See chapter 2 for a fuller discussion of the relationship between readers and documents.

IN YOUR **EXPERIENCE**

Think about the last big document you wrote for a class. How did the factors of space, time, money, equipment, collaborators, and readers come together to shape the document? Create a one-page table with a row devoted to each of these six factors. Write a few sentences in each row, describing that factor's influence on the final product.

Typography

typography the arrangement, style, appearance, and printing of text on a page

Typography refers to the arrangement, style, appearance, and printing of text on a page, and each aspect can affect readers in distinct ways. When most writers think about typography, if they think about it at all, they think almost exclusively about font. Font is indeed an important element of typography; it determines the style and size of the letters, numbers, and symbols used in a document. And in many cases the font is the most obvious typographical element in a document. However, typography consists of much more; it encompasses font style, font size, emphasis, case, spacing, and justification of text.

Font Style

Fonts differ greatly, and choosing the appropriate font is an important component of writing. That choice involves two main considerations: mood and legibility. The font style you choose can convey the overall mood and tone of a document through its appearance and shape. For example, a font style such as Times New Roman conveys a more formal, professional mood, whereas a font style such as Comic Sans conveys a more casual, creative mood. As with other writing decisions, you should carefully assess your audience, your subject, and the purpose of your document to select a font style that conveys the proper mood. If you are in doubt about the most appropriate font to use, you might want to examine previous documents written for similar audiences, purposes, or genres to determine whether there is a precedent or standard.

Although the mood you create with a particular font style is important, your primary concern should be its legibility. In most workplace writing you should choose a font style that is relatively simple and unadorned; avoid font styles that might distract readers with contrasts in line thickness or other stylistic features. Although you may occasionally use different font styles to create a contrast, you should not use more than two fonts in one document because multiple font styles can create a sense of disorder and disharmony.

In many cases you'll want to choose a font style that readers recognize, such as the following:

Arial	Century	Courier New	Garamond
Palatino	Tahoma	Times New Roman	Verdana

Another consideration is the presence or absence of serifs. *Serifs* are those small, decorative marks on the edges of type characters; they create a sense of flow from one character to the next (see Figure 9.12).

Studies show that serifed text is easier to read, particularly in longer blocks of text. However, some writers believe that the more complex shapes of serifed text look messy in very small and very large font sizes and that sans serif text has a cleaner, more modern look. It's common to use a serifed font to increase readability in body text and a sans serif font for a clean appearance in headings, labels, and notes. Most word processing programs offer an extensive selection of both serifed and sans serif fonts.

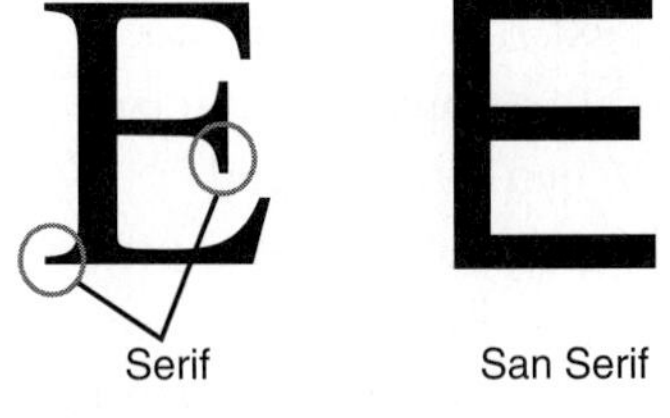

FIGURE 9.12 Serif and sans serif fonts

Font Size

Although it is easy to use your word processor's default font size, usually 12 point, you should carefully choose the size of your font. Ideal font sizes in paper documents usually range from 10 to 14 points. You want a font size that is large enough to be readable but not so large that it wastes space or appears awkward. One reason to carefully consider font size is that the same size specification differs from font to font. Look at the difference between the two common fonts shown in Table 9.1.

TABLE 9.1
Sample fonts and sizes

Size	Verdana	Times New Roman
10 pt.	Sample text	Sample text
12 pt.	Sample text	Sample text
14 pt.	Sample text	Sample text
16 pt.	Sample text	Sample text
18 pt.	Sample text	Sample text
20 pt.	Sample text	Sample text

Another factor to consider in choosing font size is whether your document will be read in electronic form, such as a web document, or in printed form. As screen resolutions increase, a given font size could appear smaller when read electronically, especially if readers have their monitors set on high resolution. Err on the large side if your text will be read in electronic form.

Varying font sizes within a document allows you to emphasize some sections of text and provide contrast. Larger font sizes, such as 14 point, are most commonly used for headings, although they can also be used to highlight certain blocks of text, such as quotations, cautions, or warnings. We recommend increasing font sizes in 2-point increments to produce noticeable changes. A single-point increment can be disconcerting to some readers and unnoticeable to others.

Devices for Emphasis

Sometimes you'll wish to place special emphasis on particular words, phrases, and even sentences, and a number of typographical features can help you do that.

- **Boldface** text contrasts in width and shading with the body text and is useful for headings, captions, or logos. Boldface is typically used for items that can be set entirely in bold, although sometimes it can be used as a stylistic device to draw attention to a word or phrase that you wish to emphasize. You should usually avoid using boldface text in the middle of a passage because the bolded text attracts attention and reduces the continuity of the passage itself.
- *Italicized* text attracts the eye because it is shaped differently than body text. Italics are often used to highlight the first use of a word or phrase that readers may not understand, such as a technical or foreign word. Because italics blend in more smoothly with surrounding text than boldface does, italics are more useful for indicating emphasis within a passage. Italicized text is also used for the titles of complete works. When writers used typewriters to create documents, they were unable to italicize titles and used underlining instead. Today, however, most writers have access to word processors and can italicize, making underlined titles obsolete. Underlining in conjunction with colored text is often used in URLs and e-mail addresses but rarely in titles anymore.
- Colored text can be used occasionally to emphasize words and phrases; it is especially valuable to alert readers to cautions or warnings. Colored text can also work well as a subtle means of distinguishing section heads. However, you should avoid putting colored text within a passage of text because readers might assume that the colored text is a hypertext link, even if they are reading a printed version of your document. If you do use colored text, choose dark shades of color that contrast with the page background, and avoid using colors that are similar to the default link colors of blue and violet. You should use colored text sparingly in workplace documents because it can quickly become overwhelming.
- CAPITALIZED text should be used rarely, only in short headings or for extreme importance or urgency. Capitalization is difficult to read in longer passages because text set in all capital letters forms rectangles with no ascenders or descenders (i.e., rise and fall of text height). We read more efficiently when we can use pattern recognition to guide us from word to word and sentence to sentence, but capitalized text forces us to read letter by letter, slowing down the reading.

See chapter 18 for more about cautions and warnings.

EXPLORE

Like many of the rules of writing, typographical choices change from one era to the next. Examine some books, magazines, and documents printed in earlier eras, focusing specifically on items one hundred years old or older. Such documents might be found in your library's special collections section, in another library's online collection of rare or antique manuscripts, or even through a commercial website such as www.antiquebooks.net.

As you examine these documents, pay particular attention to the font styles, devices for emphasis, and layout and spacing of text. What differences and similarities do you see in these older texts, as compared with newer documents? Which typographical features do you believe are influenced by the technology used to produce the documents? Which features are the product of cultural or social styles and norms of the era in which they were printed?

If possible, print a copy of the older texts you find, and discuss them in class. Be sure to jot down the dates on which the documents were published.

Titles and Headings

Titles and headings play an important role in page design because they forecast and introduce the subjects that follow. Effective titles and headings become increasingly important as the size and complexity of your document grow; a good title or heading lets readers know what to expect without having to read the entire document or section. Effective titles and headings are also important when you have multiple or diverse readers; they let readers know whether the information in a document or section applies to them, and they allow readers to easily find, reread, or refer to specific sections.

For example, imagine that a design team for a toy manufacturer has written a detailed proposal for a new playset. This proposal is comprehensive, containing sections on the playset's function, size, materials, intended market, production costs, advertising, and a host of other topics. Effective headings allow different readers to find the information that specifically applies to them. The production team, for instance, needs to read only the materials section of the proposal to determine whether the company has the capacity to produce the playset. And an effective title for the document lets members of the company who specialize only in board games know that they do not need to read the playset proposal at all.

Titles

Titles are the first—and most important—descriptive labels that readers see, and they should be displayed clearly and prominently. Titles appear on either a separate cover page or on the first page of text; they should be placed in the top one-third of the page for emphasis and should be centered rather than aligned to the right or left. To add further emphasis, you might consider presenting the title in boldface text in a font size at least 4 points larger than that of the body text. Thus, if you use a 12-point font size for body text, you should use a 16- or 18-point for your title, to differentiate it from body text and from headings, which will probably use a font size somewhere in between, such as 14 point. If you anticipate a long document with several levels of headings, you should consider using an 18- or even a 20-point font size for your title.

Headings

Headings are the descriptive labels that divide information into comprehensible sections within a larger document. When created properly, headings allow readers to understand the structure, order, and hierarchy of a document by quickly skimming through it. Headings are typically written as short phrases, although they may occasionally be complete sentences or questions. Headings should be parallel in structure and should follow a consistent word pattern, such as beginning with a gerund (i.e., an

-ing word) or being phrased as a question. As mentioned earlier, headings often use sans serif font styles to give a clean, orderly appearance.

Workplace documents often contain several levels of headings and subheadings, and you can designate the hierarchy of information in your document through the font size, font style, and placement of your headings. An effective and obvious way to distinguish one level from another is to vary the font size by at least 2 points. Stylistic choices, such as boldface or capitalization, can also be used to distinguish heading levels. In addition, the placement of headings on a page can help to clarify hierarchy. Major section headings and chapter titles in longer documents often begin on a new page; the additional white space and the prominence of the heading at the top of the page draw readers' attention. First-level headings in most professional documents should be aligned along the left margin; second-level headings may or may not be indented. Third-level headings may or may not be further indented, or they may appear on the same line as the text that follows them, called *run-in headings* (see Figure 9.13).

Document Title

This title uses 20-point Verdana, a sans serif font.

FIRST-LEVEL HEADING

This first-level heading uses 18-point Verdana font. All capital letters (in boldface) are used for this first-level heading to distinguish it.

Second-Level Heading

This second-level heading uses 14-point Verdana font, boldface with the first letter of each major word capitalized. It is noticeably smaller and less emphatic than the first heading level.

Third-Level Heading

This third-level heading also uses 14 point Verdana font, but it has been indented and italicized to differentiate it. The text following an indented heading should be indented accordingly.

Fourth-Level Heading These headings may be further indented to differentiate them. They often use the same font size as the text (although in boldface) and can be placed on the same line as the text.

FIGURE 9.13 Levels of headings

queuing using visual cues such as heading size and alignment to indicate the importance of information

Using various levels of headings—with varying font sizes and text alignments—contributes to **queuing**. Simply put, queuing is formatting a document to provide readers with visual cues that indicate the levels of importance of the information. Most often, documents are designed so that information under larger-sized headings is considered more important than information found under smaller-sized headings. Also, more important information is aligned to the left, with less important information indented.

Guidelines for Titles and Headings

Consider the following guidelines when creating titles and headings for your workplace documents:

- *Make titles and headings descriptive and specific.* They should contain proper nouns rather than generic nouns in most instances, as well as verbs that suggest action. For example, instead of using "Motorized Impeller," which contains little information about the content of the section, you might add descriptive words such as "Assessing Failure Modes in R 87 Motorized Impellers."
- *Keep titles and headings succinct.* Titles may occasionally extend to a second line of text, particularly when there is a subtitle, such as "California Bridge Safety: Assessing Structural Damage on Interstate Bridges from Los Angeles to San Francisco." Headings, on the other hand, should never be longer than a single line of text because they help readers skim through a document.
- *Omit all end punctuation except question marks in titles and headings.* Rhetorical questions (such as "What Is Our Budget?") can serve as effective titles and headings, but do not use other forms of end punctuation. When using rhetorical questions as headings, try to use them for all headings at the same level to maintain parallelism.
- *Capitalize the first letter of all major words in a heading, but not articles, prepositions, and conjunctions unless they are the first word in the heading.* Or you can capitalize the first letter of the first word in a title or heading and leave all other words uncapitalized. You should consistently follow whichever of these two conventions you choose.
- *Avoid using abbreviations or acronyms in titles and headings.* Even if you are sure your readers will know the abbreviation, it is considered informal to abbreviate in titles and headings.
- *Avoid underlining titles and headings.* This often obscures the descenders in lowercase letters, such as *g* and *y*.
- *Avoid stacking headings.* Place at least one line of text between consecutive headings if possible. This eliminates confusion in determining where the information belongs.

Captions

Captions are descriptive labels that identify or describe information in visuals. Captions follow the same principles of brevity and specificity as headings. Captions are usually placed directly below the visual to which they refer, although they may also be placed to either side of it. Captions are usually two font sizes smaller than the body text surrounding them and often use boldface text for emphasis.

Figures consist of diagrams or pictures, while **tables** consist of data arranged in rows and columns.

Material such as an image, chart, or other visual should be labeled "Figure" and given a numeral, sometimes followed by a period or a colon prior to the descriptive text—for example, "Figure 7. Exploded diagram of audiovox phone charger." Only tables should be labeled specifically as "Table" and should follow the same format—for example, "Table 2. Population density in Oklahoma." Captions in longer documents with chapters or discrete sections use two numerals, one to identify the chapter or section and the other to show the specific order of the visual within that section—for example, "Figure 9.3. Map of Everglades hiking trails." Most word processors allow you to easily create and insert captions with visuals; however, you should always consult the style guide used in your field for specific rules.

Headers and Footers

Headers and footers enable readers to easily find specific sections in a document. As their names suggest, a *header* runs along the top margin of each page in a document, whereas a *footer* runs along the bottom margin of each page. Pages occasionally include both. Headers and footers often use a smaller font size or lighter shading than the body text. They should be limited to one line so that they do not distract readers from the body text.

The primary purpose of headers and footers is to provide referencing information for readers—information such as the title of the document; the page number in documents longer than two pages; and sometimes the author's name, organization, or date. Headers and footers can also contain design features, such as a company logo or letterhead, a horizontal rule, or a graphic. Although header and footer information may seem relatively insignificant, it does involve rhetorical choices about what readers need to know. Figure 9.14 shows one configuration of headers and footers in a document.

Inserting headers and footers is fairly easy, and most word processors allow you to create textual headers of your choice. In fact, word processors include templates, themes, and styles that can format nearly every aspect of page design in your document, including headers and footers, headings, lists, and hyperlinks. You should use the header and footer functions in your word processor rather than inserting headers and footers manually; manually inserted headers and footers do not remain stationary on the page and can shift or move when new text or visuals are added or removed from the page or the page is reformatted in any way.

Some word processors include page numbering in the header and footer function; others have a separate option for page numbering. In many instances you'll want to suppress, or hide, the page numbering on the first page of your document. In addition, front matter, such as a table of contents, should use lowercase roman numerals to distinguish these sections from the body of the document. Your word processor will allow you to easily insert these numerals.

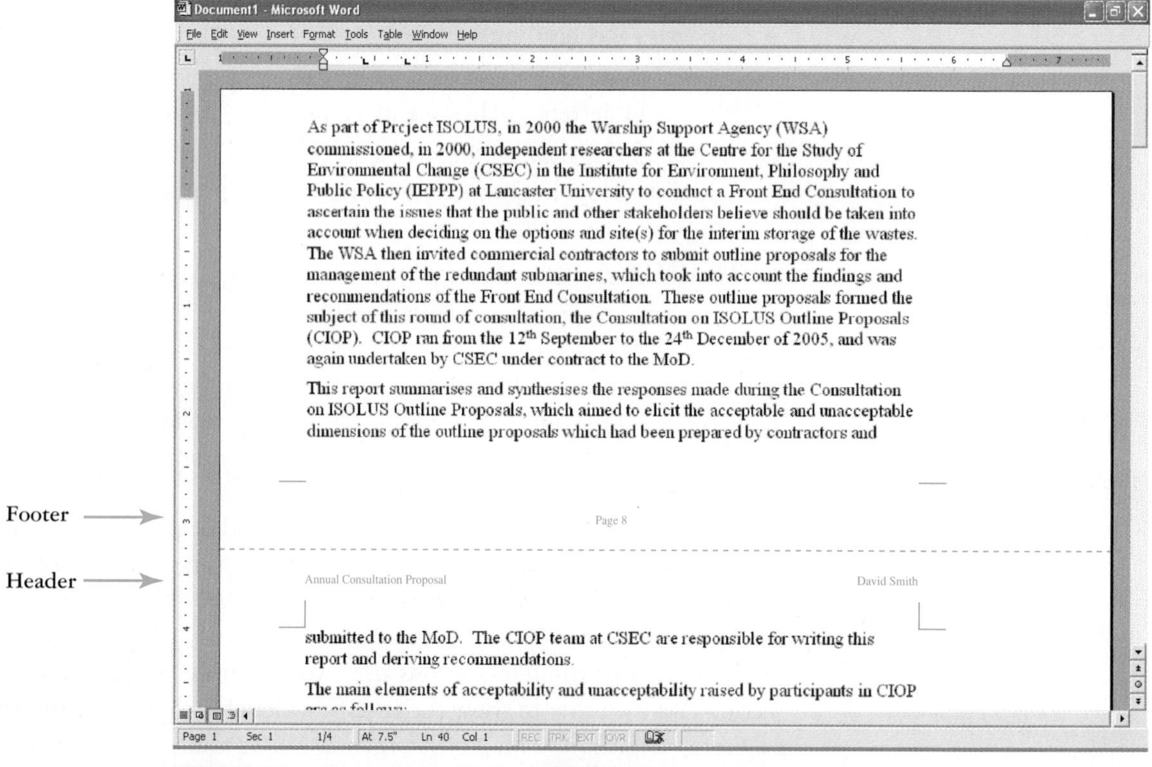

FIGURE 9.14 An example of a footer and a header

Lists

Lists emphasize words, phrases, or groups of items. They can also help summarize or review various types of information:

- Steps in a sequence
- Items, materials, or supplies
- Points of interest or importance
- Criteria for evaluation
- Conclusions or recommendations

Chapter 7 discusses organizational strategies that can help you develop appropriate lists for your documents.

Before creating a list, you should determine whether the information in it must follow a specific sequence, such as steps in a process, or whether sequence is unimportant, as in a list of materials. *Ordered lists* are used if a specific order is important; you can use numbers or letters to show a sequence, chronology, or order of importance among items. *Unordered lists* are used if the items are of equal value or require no specific order; you can use bullets, dashes, or check boxes to identify the items in the list. In formal documents you might consider using alphabetical order for items in unordered lists to convey a more orderly impression.

Lists often work best as individual words or phrases and do not generally have end punctuation. However, if you must use complete sentences, punctuate them accordingly. In addition, you should maintain consistency in length and style through all items in a list. And because most readers can remember only five or six items at a time, you should break up longer lists into smaller lists, or sublists, when possible. Sublists are indented below an item in the main list.

The short document found in Figure 9.15 shows ordered and unordered lists, as well as sublists. The list on the left side of the document outlines the collection route for a countywide recycling program. Because these locations will be visited in a sequential order, the list is numbered. The list on the right side contains the materials that can be recycled. Because the materials probably would not be organized in a particular order, the items are bulleted in an unordered list. However, because the recyclable materials should be grouped into categories, they are divided into sublists, allowing the materials to be collected and recycled more efficiently.

Lancaster County Recycling Program

The LCRP will collect recyclable materials every Monday (weather permitting). Because our collection times vary, we are unable to list a specific pick-up time. However, we will follow the same collection route each week. See the lists below for our collection route and for instructions on how to separate recyclable materials.

Recycling Program Collection Route:

1. Columbia
2. Mountville
3. Landisville
4. East Petersburg
5. Elizabethtown
6. Manheim
7. New Holland

We anticipate approximately 1–2 hours at each location. Please sort your materials according to the list on the right.

Recyclable Materials:

- Blue Bin Items
  - Plastic bottles
  - Aluminum cans
  - Steel and tin cans
  - Glass bottles and jars
- White Bin Items
  - Magazines
  - Newspapers
  - Office paper and mail
- Red Bin Items
  - Batteries
  - Motor oil (in sealed containers)
  - Hazardous materials containers

FIGURE 9.15 An ordered and an unordered list

Line Format

You'll also want to make choices about how lines of regular text appear in your document.

Line Length

The ideal line length is influenced by the design of the typeface, type size, line spacing, paragraph length, and even paper or screen size. Readability tests suggest that lines in a standard word-processed document (8.5 × 11 size) should have fifty-five to sixty characters (i.e., nine to ten words) for optimal readability. Readers find long lines of text difficult to read, whereas overly short line lengths break up the text and interrupt the flow of the document.

Unfortunately, because the default line length in most word processors is closer to eighty characters, a line length of fifty-five or sixty characters results in excessively large margins and wasted space. You should consider your subject, purpose, and audience when choosing line length and adjust lines accordingly. If you wish to create a document with a more balanced layout and fewer pages, you can stick with the default line length in your word processor.

We recommend experimenting with different line lengths in documents that will be read on a computer screen. You might ask several readers for feedback. If your document will be read through a web browser, you should consider increasing the font size and allowing the text to fill the screen, rather than using a fixed-width layout or a frame. This approach allows readers to structure their own view and adjust line lengths to fit their needs.

Line Spacing

Line spacing refers to the amount of white space separating either lines of text or a line of text and a visual. Line spacing is sometimes called *leading* (pronounced "ledding") and is measured in points. During the days of manual typesetting, printers inserted extra strips of lead between long lines of text to make them easier to read—thus, the term *leading*.

Most word processing and page layout applications let you adjust the leading in your documents. Manipulating the line spacing is a useful way to avoid **widows** (i.e., the last line of a paragraph appearing alone at the top of a page) and **orphans** (i.e., the first line of a paragraph appearing alone at the bottom of a page), both of which diminish readability. You can also manipulate the amount of space between individual characters, a process known as *kerning*, to improve readability or to avoid single-word widows or orphans. You should experiment with your word processor to find the line or character spacing that best suits your needs.

widow the last line of a paragraph appearing alone at the top of a page

orphan the first line of a paragraph appearing alone at the bottom of a page

> Too much line spacing can cause readers to skip a line when searching for the next one. It also conveys a sense of emptiness and fragmentation and increases the number of pages in a document.

> Too little line spacing can be uninviting to readers, giving a document an appearance of darkness and a sense of crowding. Too little line spacing can also cause readers to skip a line because lines are close together.

Line Justification

Line justification refers to the way words are aligned along the margins of the page. Most workplace documents use *left justification*, in which each line of unindented text is aligned along the left margin and unevenly aligned along the right margin, often called a *ragged right* margin. Studies show that left justification is easiest to read because the spacing between words remains the same. Consequently, most word processors use left justification by default.

Some documents use *full justification*, in which both the left and right margins are evenly aligned. Full justification is common in documents for which a more orderly, professional appearance is highly important, such as a sales brochure, a corporate report, or a publication like this textbook. Full justification is often used in conjunction with columns because the spaces between columns, sometimes called *alleys* or *gutters*, are more distinct and clearly defined when fully justified. Although full justification may convey a professional appearance, it is more difficult to read; the spacing between words is uneven, causing the eye to jump from one word to the next, often anticipating a sentence break where one does not exist. Some word processors also hyphenate words in irregular ways when full justification is used, again causing confusion for readers. Generally speaking, you should use left justification for documents that are not professionally printed, published, or distributed.

Examine the differences between Figure 9.16, a two-column document using left justification, and Figure 9.17, the same document using full justification. The left-justified document conveys a less orderly appearance, but the text is easier to read because the spacing between words is consistent. The fully justified document has a more professional appearance, but readability is sacrificed. As with all other layout and design choices, workplace writers must consider their own and their readers' goals when deciding on justification. Can you imagine a rhetorical situation for this exam-

Repowering Boats and Yachts

There are many reasons why repowering an existing boat may be necessary or desirable. To keep running costs down, a change from gas to diesel is often worthwhile. To avoid the vulnerability to theft of an outboard, an inboard engine may be the answer. Or it may be that the current engine is just worn out and needs replacement with a more modern unit. Sometimes greater performance is required, and again, this may result in a repowering operation being considered.

Our company can assist with all the technical aspects of a repowering project. We begin by inspecting the boat and assessing the feasibility, providing a report for the owner indicating the costs and potential benefits. Then a suitable power unit can be selected and a propeller chosen. Drawings will be produced for the new engine installation and sterngear arrangement. But before the engine is ordered the space is carefully checked out at the boat using full-size templates.

Typically, we then arrange for the work to be carried out at a suitable boatyard. When the engine arrives, its installation will be supervised by regular visits to the yard.

The engine will be installed on specially designed engine mounts with a drip tray.

FIGURE 9.16 An example of left justification

Repowering Boats and Yachts

There are many reasons why repowering an existing boat may be necessary or desirable. To keep running costs down, a change from gas to diesel is often worthwhile. To avoid the vulnerability to theft of an outboard, an inboard engine may be the answer. Or it may be that the current engine is just worn out and needs replacement with a more modern unit. Sometimes greater performance is required, and again, this may result in a repowering operation being considered.

Our company can assist with all the technical aspects of a repowering project. We begin by inspecting the boat and assessing the feasibility, providing a report for the owner indicating the costs and potential benefits. Then a suitable power unit can be selected and a propeller chosen. Drawings will be produced for the new engine installation and sterngear arrangement. But before the engine is ordered the space is carefully checked out at the boat using full-size templates.

Typically, we then arrange for the work to be carried out at a suitable boatyard. When the engine arrives, its installation will be supervised by regular visits to the yard.

The engine will be installed on specially designed engine mounts with a drip tray.

FIGURE 9.17 An example of full justification

ple in which left justification would be the better choice? Can you imagine a situation in which full justification would be the better choice?

White Space

White space may appear to be the absence of design, but the "empty" areas surrounding text and visuals are highly important. Effective use of white space can accomplish several things:

- Keep related items together
- Isolate and emphasize important elements
- Provide breathing room between blocks of information

White space in a document can be between lines of text, between paragraphs, between text and headings, between text and visuals, and along the margins of the page. Figure 9.18 shows the effectiveness of white space in keeping related items together.

chunking visually separating chunks of information with white space

Readers tend to understand and retain information better when it is presented in small chunks rather than large. **Chunking** refers to visually setting apart individual chunks of information by using white space. Most frequently, this means double-spacing, or adding an extra line of white space, between chunks of information.

White space is particularly important in addressing the five elements of document architecture: balance, connection, duplication, variation, and flow. As you design your document, consider how white space might increase the effectiveness of each element. You can use more white space around information that is highly important or complex; extra white space implies importance and will draw readers' attention. White space is equally important in showing relationships or connections among various items; additional white space before or after two or more related items establishes a connection between them.

10

There are many reasons why repowering an existing boat may be necessary or desirable. To keep running costs down, a change from gas to diesel is often worthwhile.

Boats

To avoid the vulnerability to theft of an outboard, an inboard engine may be the answer. Or it may be that the current engine is just worn out and needs replacement with a more modern unit.

Yachts

Sometimes greater performance is required, and again, this may result in a repowering operation being considered. Our company can assist with all the technical aspects of a repowering project. We begin by the

Unclear relationship between headings and text

10

ing an existing boat may be necessary or desirable. To keep running costs down, a change from gas to diesel is often worthwhile.

Boats

To avoid the vulnerability to theft of an outboard, an inboard engine may be the answer. Or it may be that the current engine is just worn out and needs replacement with a more modern unit.

Yachts

Sometimes greater performance is required, and again, this may result in a repowering operation being considered. Our company can assist

Clear relationship between headings and text

FIGURE 9.18 An example of white space and connection

ANALYZE THIS

This exercise invites you to analyze how different choices in the spacing and justification of a document influence its readability and clarity. Figures 9.19, 9.20, and 9.21 show three versions of the first page of a document outlining procedures for an ethics committee on animal experimentation. Figure 9.19 shows the document in its original form; Figure 9.20 shows the document with increased line length, increased line spacing, and increased white space; and Figure 9.21 shows the document with full justification, decreased line length, and decreased white space.

Analyze and discuss these three versions with a classmate. What is gained or lost with each of the changes? Which version would be best if document length was most important? Which version would be best if readability was most important? Are all three versions equally readable?

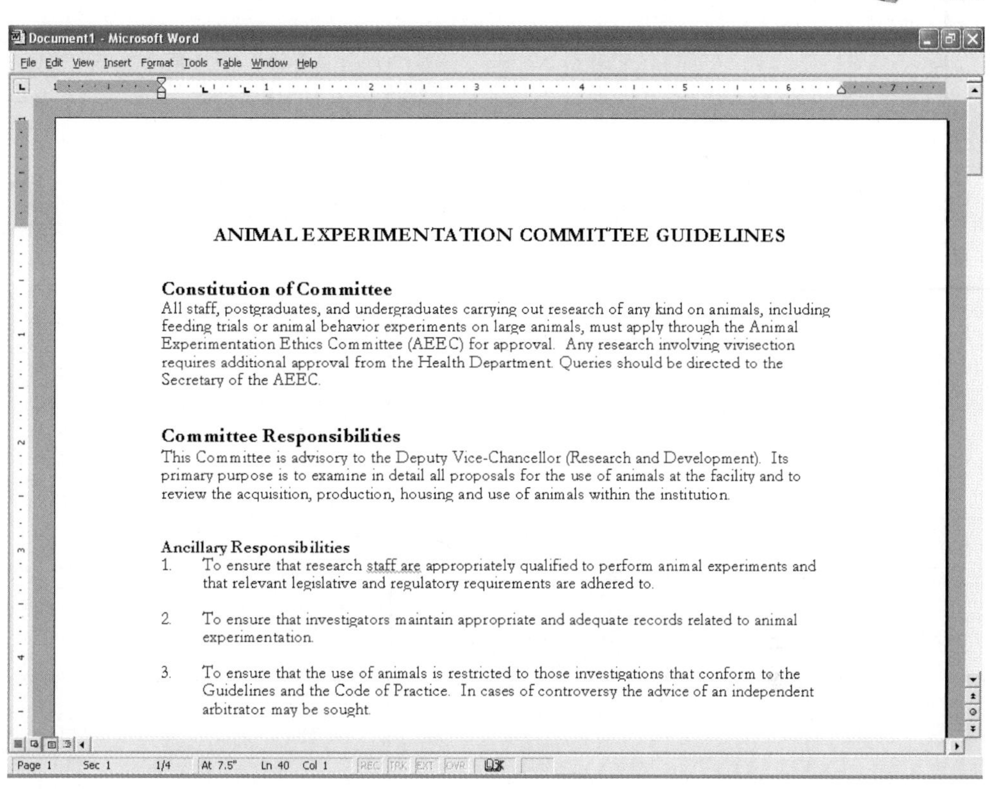

FIGURE 9.19 A document with standard formatting

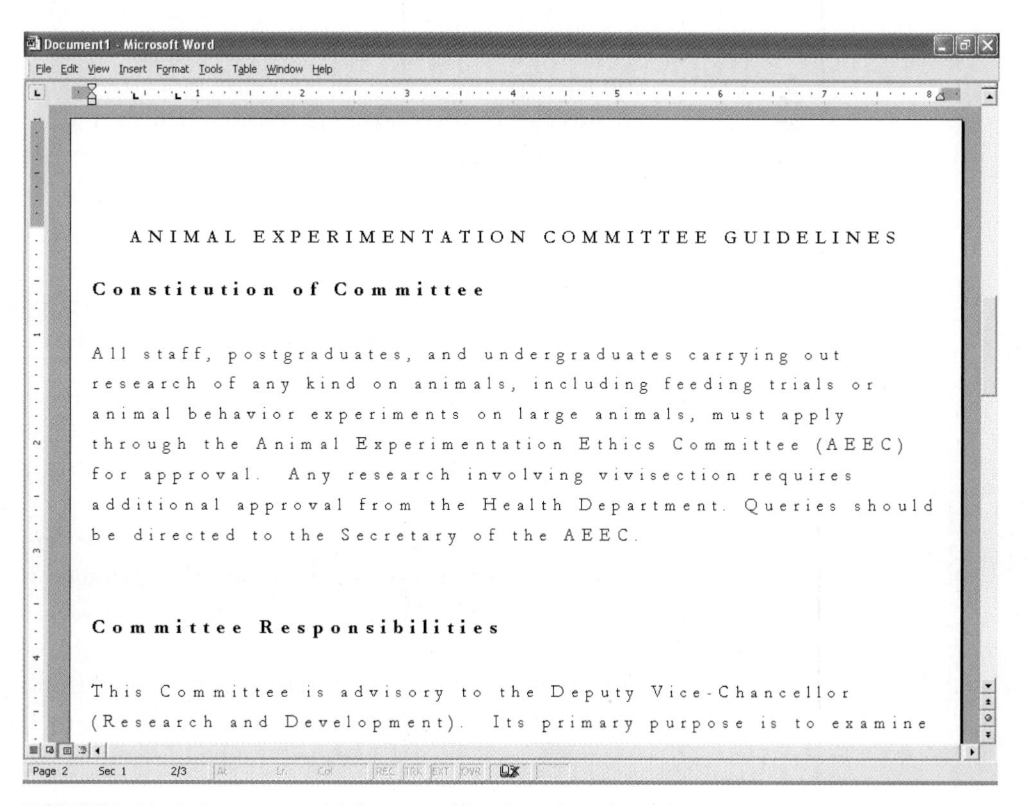

FIGURE 9.20 A document with increased line length and spacing

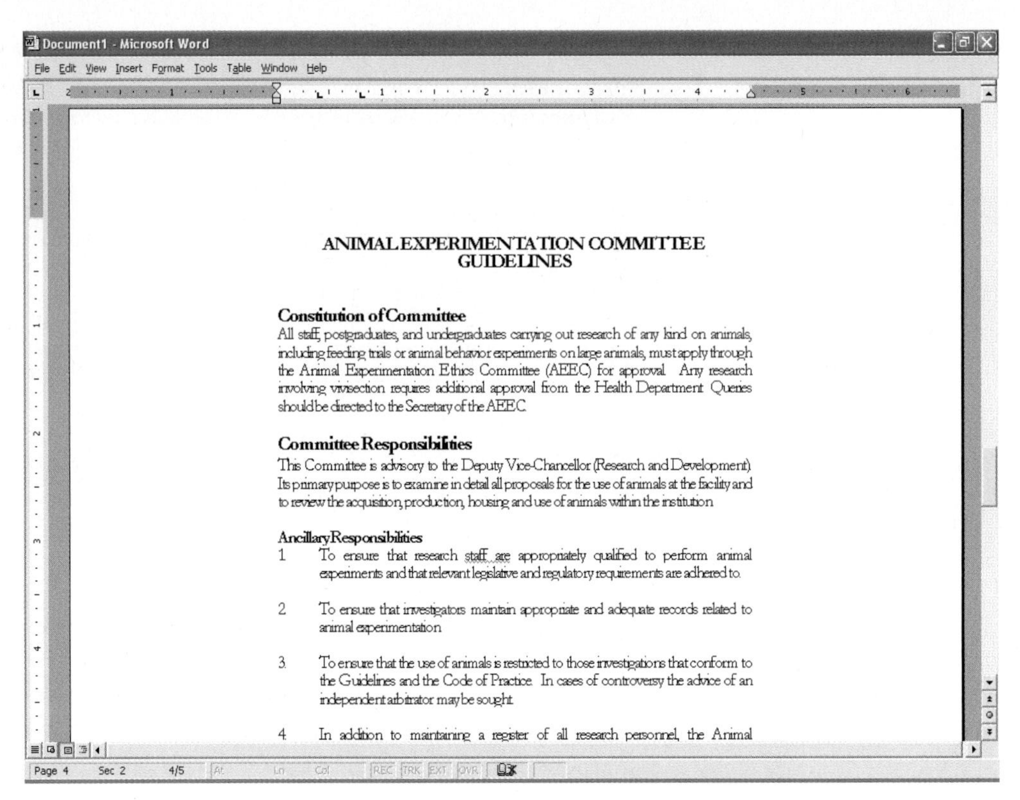

ANIMAL EXPERIMENTATION COMMITTEE
GUIDELINES

Constitution of Committee
All staff, postgraduates, and undergraduates carrying out research of any kind on animals, including feeding trials or animal behavior experiments on large animals, must apply through the Animal Experimentation Ethics Committee (AEEC) for approval. Any research involving vivisection requires additional approval from the Health Department. Queries should be directed to the Secretary of the AEEC.

Committee Responsibilities
This Committee is advisory to the Deputy Vice-Chancellor (Research and Development). Its primary purpose is to examine in detail all proposals for the use of animals at the facility and to review the acquisition, production, housing and use of animals within the institution.

Ancillary Responsibilities

1 To ensure that research staff are appropriately qualified to perform animal experiments and that relevant legislative and regulatory requirements are adhered to.

2 To ensure that investigators maintain appropriate and adequate records related to animal experimentation.

3. To ensure that the use of animals is restricted to those investigations that conform to the Guidelines and the Code of Practice. In cases of controversy the advice of an independent arbitrator may be sought.

4. In addition to maintaining a register of all research personnel, the Animal

FIGURE 9.21 A document with full justification and decreased line length and spacing

Physical Properties

When overseeing the construction of a home or a building, architects and construction supervisors carefully choose materials. As a document architect, you'll be faced with similar decisions about the physical properties of your documents. And like the architect or construction supervisor, you should guide your decisions by what your customers—that is, your readers—want and expect. With print-based documents, your decisions about physical properties will fall into two categories: paper and binding.

Paper

Paper is an afterthought for many writers, but in the professional and technical world, paper choices can be significant. You will want to consider several factors.

Paper Size Paper size is often influenced by the type of document you intend to produce. Reports, letters, and other traditional documents usually employ the standard paper size—8.5 × 11 inches. Manuals and reference guides are often smaller and more narrow because they are designed to fit in cramped workspaces, where they can be read while the reader is following the instructions or performing a task. Sometimes technical instructions or technical schematics are printed on oversized paper that opens to poster size or larger. Promotional materials and flyers can come in a variety of sizes and can employ creative shapes and forms, such as the shape of the product being described. When choosing paper size and shape, you should think about what the purpose of the document is, where readers will access the document, how readers will view or use the document, and what readers will find most familiar, useful, or appealing.

Page Count Page count can play a role in your document architecture, particularly if you are given a specific page range. Page count is usually determined by two factors: economic and psychological. Because paper and printing can be expensive, your budget may determine the number of pages, and thus the complexity of the information, in your document. And psychologically, most readers want their documents to be as short as possible; a busy reader generally prefers a ten-page report over a fifteen-page report, even if both reports contain the same information.

EXPLORE

When choosing paper for your documents, consider using recycled paper, which has improved in quality in recent years and is usually less expensive than other papers. Not only is recycled paper a good ethical choice, but it also conveys a positive, responsible image of you and your organization to potential employers or clients. To find out more about recycled paper, visit one of the following websites:

- http://www.recycledproducts.org/
- http://www.isri.org/
- http://www.nrc-recycle.org/
- http://www.epa.gov/epawaste/conserve/rrr/recycle.htm

Paper Quality Paper quality is also guided by economic and psychological factors; higher-quality paper is more expensive but conveys a greater sense of importance about the document. Choices include *newsprint*, an inexpensive, low-quality paper used for newsletters and other bulk publications; *stock*, a medium-quality paper used in photocopiers and laser printers; *bond*, a heavier paper used for letters and memos; *book paper*, a higher-quality paper that allows for better print resolution; and *text paper*, a high-grade paper used for formal announcements, brochures, and other important documents. Generally, higher grades of paper weigh more and are measured in grams per square meter, or gsm for short.

Along with these various grades of paper, you can also choose coated or uncoated paper, often categorized as glossy or matte. Coated paper gives documents a professional shine and increases durability and weight, but it is generally more expensive and has a greater negative impact on the environment because chemicals are part of the coating process. You should work with printing professionals to select the best paper to match your budget and the goals of your document. And remember that paper choice is both rhetorical (i.e., how it will be received) and ethical (i.e., how it will impact resources).

ANALYZE THIS

Examine the paper in one of the documents currently in your possession. What size and shape of paper are used? What type and quality of paper? Is the paper coated or uncoated? Was it recycled?

After you've carefully examined the paper, think about the process by which it was chosen. Were the publishers of the document more concerned with economics, professional appearance, or environmental ethics? What overall effect do the paper size, shape, and quality have on you as a reader? Would the document be as effective on a different size, shape, or quality of paper? If you had designed the document yourself, would you have chosen a different size, shape, or type of paper?

Share your documents as a class, discussing the possible motives and implications of each paper choice.

Binding

The binding of documents may be something you have not previously considered, but with professional and technical documents, the way that pages are bound can be a significant part of the design process. There are five common types of binding.

Plastic Grip Binding One simple method of securing pages is to use a binding bar and a clear vinyl cover. Plastic grip binding is relatively inexpensive and allows easy insertion and removal of pages. However, because it is less secure and holds fewer pages than other binding methods, it is not generally used for lengthy or permanent documents. Another disadvantage is that it is difficult to print anything on the spine if your document needs to be filed on a shelf.

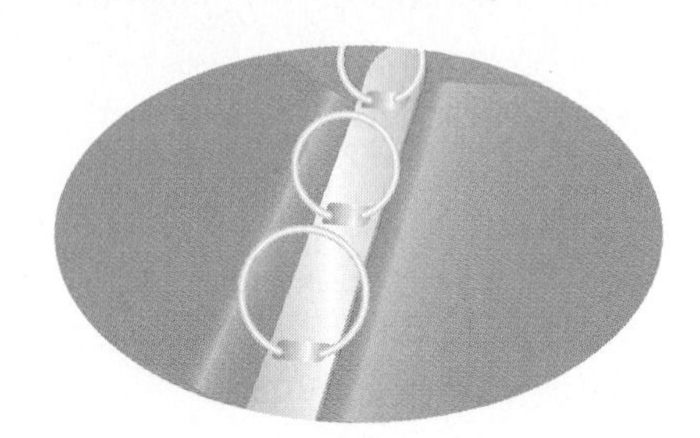

Three-Ring Binding Three-ring, or loose-leaf, binding is another simple method of securing pages, using a folder with clips or rings. Pages can be easily inserted or removed, and many folders have a spine sleeve, which allows for labels or titles. However, because the pages must have holes punched, a larger left margin may be necessary. And binders can become expensive, so three-ring binders are typically used when just a small number of copies are necessary.

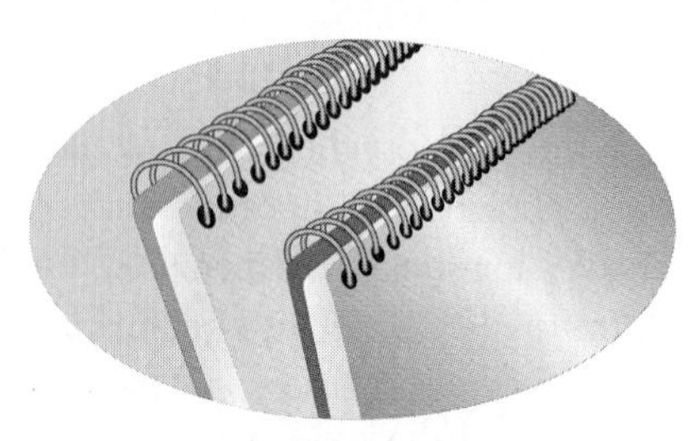

Comb, or Spiral, Binding Printing or copy shops often use comb, or spiral, binding, which keeps pages relatively secure and tidy. One advantage is that documents can be opened flat on a desk or folded back to show just one page. Comb or spiral binding is relatively inexpensive, but it does have two disadvantages: spines cannot generally be printed on, and pages cannot be added or removed after the document is bound.

Saddle Binding Printing professionals typically create saddle-bound documents; this method usually uses fabric stitching or a set of large staples to bind the document. Pages in saddle-*stitched* documents are stitched together in sections known as *signatures*; pages in saddle-*stapled* documents usually have two or three large staples through the spine and into the middle of the whole document. Saddle stapling is less expensive than fabric stitching but is impractical for longer documents. Saddle stitching can be quite expensive and is generally used only for hardback books.

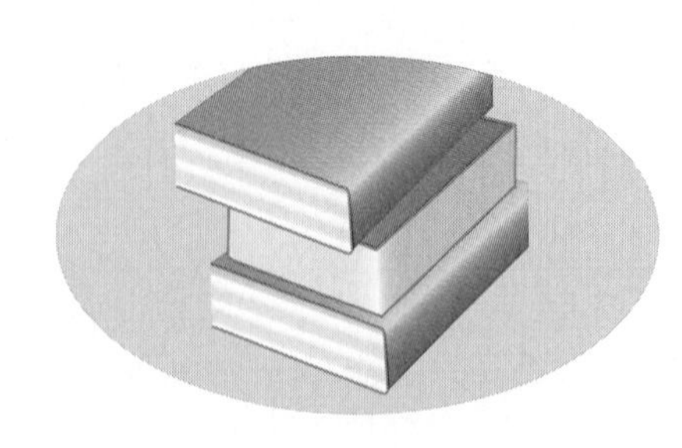

Perfect Binding The standard in the publishing industry, perfect binding, consists of binding single sheets gummed along the spine of the document. This is the method typically used for paperback books and journals. You should consult with printing professionals if you are considering perfect binding; it often requires a wide inside margin so that readers don't have to open the publication flat to read what's printed near that margin. You should also let the printing professionals know how many pages you need to include because some small operations produce covers that will accommodate only a certain number of pages.

ANALYZE THIS

With your class, discuss these various options for binding. Describe some of the documents you've seen recently that have been bound in each particular way. For each example analyze whether that particular binding method was the most appropriate for the document, audience, and occasion. What might readers infer about each document from the way it was bound? Did the authors or publishers choose the most appropriate binding method in each instance?

SUMMARY

- Workplace writers must carefully consider the layout and design of their documents.
- Document architecture refers to the positioning and design of text and visuals for readability. It involves five principles:
    - Balance focuses on how the elements are arranged and aligned to create a unified whole.
    - Connection emphasizes the relationship among elements.
    - Duplication suggests that elements of the same type, importance, or magnitude should use the same basic format, layout, and design.
    - Variation allows contrasts or differences among elements to become apparent.
    - Flow refers to the movement of the reader's eye from one part of a document to another.
- The elements of a document consist of all of the physical or visual components readers see. How writers use them is shaped by space, time, money, equipment, collaborators, and readers.
- Typography refers to the arrangement, style, appearance, and printing of text on a page. Typographical choices include font style, font size, boldface, italics, colored text, and capitalized text.
- Titles and headings forecast and introduce the subject of the text that follows them.
- Captions are descriptive labels used to identify or describe the information in visuals of all types.
- Headers and footers enable readers to easily find specific sections in a document.
- Lists are an effective way to emphasize words, phrases, or groups of items; lists can be ordered or unordered.
- Writers make conscious choices about the length and spacing of lines of text. Line justification refers to the way the words are aligned along the margins of the page.
- White space appears as "empty" areas surrounding text and visuals. Effective use of white space should keep related items together, emphasize important elements, and provide breathing room between blocks of information.
- The physical properties of a document, paper and binding, also present rhetorical choices.

CONCEPT REVIEW

1. Why are layout and design choices important in workplace writing?
2. In what ways does thinking about layout and design affect problem-solving approaches?
3. What role do layout and design play in how you ethically present information to readers?
4. What do functional elements of document architecture contribute to a document?
5. What do aesthetic elements of document architecture contribute to a document?
6. What are some strategies for achieving visual balance in your documents?
7. What role does connection play in designing your documents?
8. What are some strategies for creating visual connections in your documents?

9. In what ways can duplication assist your readers in navigating your document?
10. What are some strategies for employing duplication in your documents?
11. How can varying visual elements help you emphasize or contrast parts of your documents?
12. What strategies might you try in order to use variation in your document design?
13. Why do your documents need to flow visually?
14. What are some strategies for creating documents that flow?
15. What kinds of choices do all workplace writers have to make in terms of typography?
16. How do choices in font styles affect how readers encounter documents?
17. What are some devices to use for emphasis?
18. What role do titles and headings play in document design?
19. What are some strategies for using titles in your documents?
20. What is the relationship between a visual and a caption?
21. When should you use a caption?
22. Why might lists be effective methods for conveying information to readers?
23. What are the visual advantages of using lists?
24. How can line length, line spacing, and justification all contribute to the readability of your documents?
25. How does white space influence how a reader sees a document?
26. What things should you consider in making the best use of white space?
27. What are some of the primary physical properties you need to consider when designing documents?

CASE STUDY 1

Design Patents and Online Booksellers

Layout and design are important in sales documents, and many companies believe that an effectively designed website equates to increased sales. In fact, some online retailers have patented some of their website design features to keep rival companies from employing similar features. In 1999 Amazon.com filed suit against Barnesandnoble.com, claiming it had copied parts of the Amazon.com web page design, including Amazon's patented "1-Click" purchase feature. The 1-Click design allows customers to skip several steps in the checkout process by presetting credit card and shipping information, and Amazon claimed that Barnes and Noble's "Express Checkout" design was a patent violation. According to Amazon CEO Jeff Bezos, "We spent thousands of hours to develop our 1-Click process, and the reason we have a patent system in this country is to encourage people to take these kinds of risks and make these kinds of investments for customers." The infringement suit led to an eventual closed settlement between the two retailers in 2002. Although the details of the settlement have never been released, Barnesandnoble.com no longer uses the Express Checkout feature on its website.

Examine the Amazon.com and the Barnesandnoble.com websites. Do you feel that the 1-Click design gives Amazon an advantage over Barnes and Noble? Do you see other design similarities between the two websites? Do you believe that Barnes and Noble's Express Checkout system was a violation of Amazon's patented design? Of course, when writers borrow content from other documents without permission, we call this plagiarism, but does borrowing design features also constitute plagiarism? In a short letter to your instructor, detail what you would consider an ethical approach to borrowing design strategies from other documents, using these two online retailers as a point of reference.

INSURANCE INSTITUTE FOR HIGHWAY SAFETY

NEWS RELEASE

July 1, 2004

MORE CRASH DEATHS OCCUR ON 4th OF JULY THAN ANY OTHER DAY; JULY 3 IS 2nd WORST

ARLINGTON, VA—The upcoming Independence Day holiday is when the most motor vehicle deaths occur, based on experience during 1986–2002. Each year on this holiday, an average of 161 people die in crashes. This is 12 more deaths than the average on any other single day of the year and about 40 percent more crash deaths than occur on an average day. The second worst day for crash deaths during 1986–2002 was July 3. July 2 also was among the days with the most deaths.

Six of the 10 days with the most deaths were holidays or near holidays. Besides the high toll on July 2–4, there was December 23, January 1, and September 2. The other four days on the "worst" list were in August.

Institute researchers analyzed data from the federal Fatality Analysis Reporting System, an annual census of fatal crashes on U.S. roads. The motor vehicle deaths were sorted by month, day, and hour. T he 17-year span 1986–2002 was chosen to balance the effects of weekend travel. Researchers also gathered information on the characteristics of the people and vehicles involved in the fatal crashes.

Allan Williams, the Institute's chief scientist, points out that "while more deaths do occur on some of the holidays, the toll of fatalities is relentless every day, all year long." The average during 1986–2002 was 117 deaths per day.

Deaths by type of crash: About three of every four crash fatalities are occupants of passenger vehicles. Another 13 percent are pedestrians, and 7 percent are motorcyclists. While July 4 was the day with the highest average number of passenger vehicle occupant and motorcyclist deaths during 1986–2002, January 1 and October 31 (Halloween) were when the most pedestrians were killed.

Alcohol is a factor in a greater proportion of crash deaths on both the 4th of July and New Year's Day. Forty-one percent of the deaths on the 4th and 51 percent on January 1 involved high blood alcohol concentrations. These proportions compare with 33 percent on December 25 and January 8 (days in close proximity that aren't associated with New Year's) and 31 percent on June 27 and July 11.

Deaths by month and day of week: More miles are traveled in August than any other month, and August averaged the most crash deaths per day (132). However, October and December averaged the highest death rate per billion miles traveled (19.1). January and February averaged the fewest miles traveled and the fewest deaths per day (98). The day of the week with the lowest average fatality count was Tuesday (95 deaths), followed by Mondays and Wednesdays. Far more motor vehicle deaths (158) occurred on Saturdays.

Toll of crash deaths doesn't resonate: "An average of 117 deaths per day is the equivalent of a major commercial airline disaster occurring every day of the 6,209 consecutive days of the 17-year span we analyzed," Williams says. "But there's a big difference in how society approaches these losses.

When a plane goes down, it's big news and there's a concentrated effort to find ways to prevent future crashes. But the toll of highway deaths doesn't attract the same attention." Nationwide tallies of crash deaths aren't available until months after the crashes occur. "However you tally the deaths or sort them by contributing factors, the total of 727,438 human lives lost over 17 years represents a huge burden on the public health," Williams concludes.

DAYS WITH THE MOST CRASH DEATHS, 1986–2002

	Total deaths	Avg. per day
July 4	2,743	161
July 3	2,534	149
December 23	2,470	145
August 3	2,413	142
January 1	2,411	142
August 6	2,387	140
August 4	2,365	139
August 12	2,359	139
July 2	2,340	138
September 2	2,336	137
	ALL DAYS	117

PEDESTRIAN DEATHS, 1986–2002

	Total deaths	Avg. per day
January 1	410	24
October 31	401	24
December 23	373	22
December 20	357	21
November 2	351	21
October 26	350	21
November 3	348	20
November 10	344	20
November 1	340	20
December 18	339	20

CRASH DEATHS BY MONTH, 1986–2002

	Total deaths	Avg. per day	Miles traveled (billions)	Deaths per billion miles
January	51,694	98	2,996	17.3
February	47,247	98	2,860	16.5
March	54,645	104	3,328	16.4
April	55,710	109	3,328	16.7
May	62,426	118	3,534	17.7
June	64,152	126	3,526	18.2
July	68,099	129	3,658	18.6
August	69,731	132	3,677	19.0
September	63,965	125	3,366	19.0
October	66,553	126	3,477	19.1
November	61,145	120	3,237	18.9
December	62,071	118	3,258	19.1

CRASH DEATHS BY DAY OF WEEK, 1986–2002

	Avg. per day
Sunday	132
Monday	89
Tuesday	59
Wednesday	89
Thursday	105
Friday	133
Saturday	158

CASE STUDY 2

Increasing Student Awareness, Increasing Highway Safety

The press release on page 265 was issued by the Insurance Institute for Highway Safety on July 1, 2004. The release discusses car crash fatalities during the July 4th weekend. Imagine that your university's student government has hired you to redesign the press release into a document or series of documents to be placed or distributed around your campus. The primary goal of these documents is to increase student awareness of July 4th car crash fatalities and potentially lower the number of accidents and fatalities among students.

Begin by thinking about the type(s) of documents that would be most useful. Might you create a poster? a pamphlet? Would some other format be more useful? Would it be best to use a combination of different types of documents?

Then redesign the press release into the type(s) of documents you've decided on. Pay particular attention to the five principles and the various elements of document design discussed in this chapter. Think carefully about which design features would be most effective in communicating your message.

Finally, share your choices with your classmates, discussing the design principles and features you've incorporated into your document(s) and the reasons you think your choices would be the most effective.

CASES STUDIES ON THE COMPANION WEBSITE

Publicizing Orton Construction's Upcoming Public Service Campaign

Orton Homes donates materials and labor for several Habitat for Humanity projects. As the personnel director, you are responsible for publicizing the event among employees and motivating them to spend several Saturdays building homes for free.

Crafting a Highly Detailed Document: Bruce Smithers and the Winged Elm

Bruce Smithers spends his vacation in a home maintained by your company. You quickly learn that Mr. Smithers likes to micromanage—even the maintenance of the vacation home that he pays you to maintain. He wants you to plant a tree to shade the porch and has some recommendations. He wants to know if any of his ideas are acceptable.

Managing Miscalculated Bills and a Public Relations Crisis at CPW

You are the director of billing and customer service at Greenwood Commission of Public Works. Bills went out last week, and every household in Reedy Creek was charged for approximately 748% more water than its average monthly consumption.

Putting Fish Online

Manny Rodriguez owns a fish store in a university town and caters to many students. However, these students come in to ask for advice too often, slowing down normal business operations. Manny asks you, one of his employees, to create documents to help new customers take care of their fish.

VIDEO CASE STUDY

Agritechno

Information for New Investors: Defining the Terms

Synopsis

Lev Andropov and Teresa Cox, geneticists with Agritechno, are working in Andropov's lab. Jaylen Castillo, part of the Agritechno Product Development program and liaison to Andropov's lab, comes to the lab to tell Lev and Teresa that Agritechno wants to push into a broader fruit and vegetable market and needs some information from them to develop promotional materials.

WRITING SCENARIOS

1. In a small group, locate three or four different kinds of documents. Consider magazine pages, newspapers, brochures, instructions, reports, manuals, e-mails, web pages—any published document will do. Then compare the design elements used in each of the four sample documents. Discuss why the writers of each of those documents employed the design elements they did. Consider whether any of the design elements are somehow characteristic of the genre. For example, do newspapers tend to use a particular kind of font, column structure, type of paper, and so on?

   As you discuss the similarities and differences in design approach, consider why certain design features might work better in one document than another. Then make a chart that lists the four sample documents and the elements (addressed in this chapter) that those documents use. Include in your chart an assessment, based on your collaborative discussions, of the effectiveness of each design element in that particular document and reasons for the assessment.

2. In a small group, locate either a one-page set of instructions or one sample page from a larger set of instructions. Examine the page to determine which visual elements it contains and whether there might be a way to design the instructions to make them more reader-friendly. Then re-create the page, using the strategy your group develops.

   Finally, write a short memo to your instructor, explaining what changes you made to the document and why you decided to redesign the document in that way. Be sure to turn in the original sample page, your revised version, and your memo, so that your teacher can see how you've changed the document. Your instructor may request that you give a short presentation about the document and the changes you made. Consider showing the class both the before and after documents on an overhead projector, LCD projection, or distributed photocopies.

3. What role do you think collaboration plays in document design? In a small group, discuss the ways in which collaborators might work toward a specific design for a document. Write a short list of the things that are crucial in design collaboration; then write a sentence or two explaining why your group decided that each item was important. Have each group report to the class what was decided, and discuss the consistencies and inconsistencies among the groups' conclusions.

4. Because layout and design are essential to workplace writing, a wide variety of manuals and tutorials exist for this activity. Pick a particular kind of workplace document that is discussed in this text, and locate a tutorial or manual that specifically addresses design issues for that kind of document. What kinds of design features are emphasized? Write a memo and distribute copies to each of your classmates, identifying the kind of document you researched and the primary design elements the tutorial or manual highlighted for that kind of document. Be sure to identify the source from which you gathered your information.

Your class may want to coordinate the memos so that everyone has a guide to the layout and design of a variety of genres. If your class decides on a standardized format for the individual memos, the final collection may resemble an organized packet.

5. Many fields maintain industry standards for document design to ensure accessibility, consistency, and legality. Look into whether a field in which you are interested maintains such standards. Then write a summary of the design standards for a given genre in your chosen field.
6. Overdesigned documents can become difficult to read and can leave readers confused, unable to get the information they need. Locate a document that you would consider overdesigned, and write a short memo identifying what makes it overdesigned. Then redesign the document to be more effective in its delivery of information.
7. Find an example of a document or a website that takes a strong stance on a particular issue or topic. Examine and describe the ways in which the authors emphasize or de-emphasize information within their document to strengthen their position or weaken another position. Which particular design principles or elements do they use? Do the authors' layout and design choices help them achieve their persuasive goals? Are their choices ethical or unethical?
8. Sometimes writers adopt particular design elements to give their documents more credibility. For instance, a manufacturer of diet supplements might present an advertisement designed to look like a medical report rather than an advertisement. World Wide Web pop-up ads frequently use such strategies, as do spam e-mails that mimic legitimate business correspondence. Locate a document designed to look like another kind of document in order to gain visual authority with readers. Analyze the document, noting how it works to convince readers that it is something it's not. Then redesign the document, or a page or two of it, in a way that is more appropriate for the content of the document.
9. The term *design* has many definitions and applications that depend on the context in which they are used. For instance, in engineering, *design* can be used as a verb to identify a plan or a process for making something new. It can also be used as a noun to refer to the plan or process that was developed. We have already offered a succinct definition of *design* as it pertains to technical communication, but that definition is by no means all encompassing. Locate information about how other fields and disciplines define *design*. Then write a memo to your instructor that identifies two other definitions of *design* as it relates to those other fields. Be sure to cite the sources you use.
10. Choose a chapter in this textbook, and find in it at least one example of each design principle mentioned in the beginning of this chapter. Describe how the principle is employed in each instance, and determine whether it is effective for the book's audience.
11. Find a recently published journal article in a field in which you are interested. Evaluate the article strictly on the basis of document architecture and design. Which design features described in this chapter do the authors employ? Which features are most effective? Which features are absent or ineffectively executed?
12. Imagine that your current or future employer has asked you to create a top-ten list of things all workplace writers should know about layout and design. Reread this chapter, think about what is most important, and then create the list.

13. Identify the design features that are included in your word processing package. Then create a document that lists the name of the word processor you use, followed by an outline or list of the features that are available on it. (Use the title/heading schemes described in this chapter.) Focus on the general features available, such as the fonts available in the font library, but don't feel obligated to list all of them. Once you've built a thorough list, identify which design features you knew about before writing the list, which ones you discovered while exploring your word processor for this assignment, which of the features you think you'll use most often, and which you don't anticipate using and why.
14. Through your word processor or web-authoring software, locate a template for a document such as a memo, a letter, a résumé, or a web page. Then use the template to create a sample document. Don't worry about the content; just fill in the template with whatever information you see fit. Then analyze the document. Is it effective or ineffective? Why? Would you have designed it differently if you had not used the template? If you think the document would work better with a different design, redesign the document, and then write a short description of the changes you made and the reasons you think they work better. If you think the document is effective, identify the aspects of the template design that you find valuable, and explain why.
15. Locate a web page you find interesting. Then consider how the designers employ visual elements to help you navigate the page. Do they use variation? repetition? balance? flow? white space? After you've analyzed the page, capture a screen shot of it. On most computers you can do this through your word processor application by pressing the Ctrl, Alt, and Print Screen keys at the same time. Then on the same page with the screen shot, write a description of the design elements you identify in it, and offer a short analysis of their effectiveness.

10

Revising, Rewriting, and Editing

CHAPTER LEARNING OUTCOMES

After completing this chapter, you will be able to do the following:

- Know how to **revise**, **rewrite**, **and edit** your documents
- Consider the **ethical** aspects of revising, rewriting, and editing
- Understand strategies for revising documents: for **purpose**, to **avoid excess**, for **clarity**, for **concision**, for **style**, for **tone**, for **visual effectiveness**, and for **accuracy and timeliness**
- Know specific strategies for **rewriting** documents
- Understand strategies for **proofreading** and editing documents: **formatting**, **typography**, **grammatical correctness**, **punctuation**, and **spelling**
- Understand the role **technology** plays in revising, rewriting, and editing

DIGITAL RESOURCES

On the Companion Website www.prenhall.com/dobrin:

- Case 1: Revising as Improvement: Walker Brothers' Acknowledgment of Order
- Case 2: Editing and Revising for Tone and Clarity: Tandem Watercraft Customer Service
- Case 3: Revise to Meet the Needs of Multiple Audiences: Linearization Test Station Upgrade ROI (Return on Investment) Report
- Case 4: Rewriting a Negative Letter: Sleep Hollow Apartments Pet Violation
- Video Case: Editing the Internal Report at DeSoto Global
- PowerPoint Chapter Review, Test-Prep Quiz, Exercises and Activities

REAL PEOPLE, REAL WRITING

MEET ROBERT UY • Human Resource Generalist with Microsoft

What types of writing do you do at Microsoft?

Most of my communication within Microsoft tends to be e-mail. This consists of meeting agendas and scheduling, feedback on proposed actions, responses to longer documents, and so forth. In addition, I regularly write employee relations investigation reports, performance documentation, communications on HR initiatives, analyses of HR data and trending, and internal training documents.

How crucial is revision and editing in how you write those documents?

I typically write, edit, and rewrite anything that will be viewed by others, looking for ambiguous language, typos, or anything else that could distract from or dilute my core message. Even in situations where time is critical, I try to focus on getting things said correctly the first time. In my opinion, taking the additional time up front is better than further correspondence later to clarify the message or fix problems related to poor communication.

Do you rely on others to review your writing before making it public?

HR documents tend to go out to diverse audiences, so as a practice we review and edit each other's communications before finalizing them. Sometimes this involves running my drafts past my manager or colleagues, particularly for communications that are going to executives or the public. Since my colleagues write similar messages, we often collaborate on the style and structure of our documents to remain consistent.

How do you respond to company documents that seem poorly edited or revised? Do you make a connection between how well a document is edited and the capabilities of the writer?

At some level you always make a value judgment linking the quality of someone's communication with that person's competency in the field. If you send out documents or communications with errors, it definitely undermines readers' confidence in your ability. I see many errors that could be easily rectified or avoided through routine use of spell-checkers and other available technology or by simply asking someone else to take a quick look at the work. When I was actively recruiting, I discarded lots of résumés because of typos and grammatical errors.

What advice about workplace writing do you have for readers of this textbook?

I try to review everything I write with the same care, from casual e-mails to more formal executive presentations. You never know when a document might be seen by someone you didn't expect to see it or used differently from its original intent. I hope that doesn't sound like paranoia; I think of it as being consistent about the quality of my writing.

DILBERT by Scott Adams

INTRODUCTION

To solve problems, documents must be clear and accurate. To make them clear and accurate, workplace writers revise, rewrite, and edit their documents before sending them to readers. Experienced writers understand that *first* drafts are not the same as *finished* drafts and that all writing requires revision, rewriting, or editing. Unfortunately, deadlines sometimes force workplace writers to shortchange these aspects of writing, if not forego them altogether. However, neglecting these aspects can have serious consequences, even in short, simple documents.

A Lawrence, Kansas, woman named Kris Bryan found out how serious these consequences could be when a small newspaper typo led to the theft of many of her possessions. In July 2005 Bryan returned home from work to find strangers loading her possessions into a vehicle. When Bryan asked what they were doing, they showed her a public notice printed in the *Lawrence World-Journal* stating that the items at 1319 Tennessee Street (Bryan's home) would be thrown away if they were unclaimed. The newspaper ad should have read 1339 Tennessee Street—a one-digit mistake (see Figure 10.1). The people she confronted returned Bryan's things, but her TV, DVD player, furniture, and seven-week-old kitten had already been taken. In a public statement Bryan said she wants the newspaper to pay $3,500 for the stolen items, but that can't replace her lost kitten. "Sometimes you can't get away with your mistakes, especially when someone gets hurt because of them," Bryan said.

> Items left at 1319 Tennessee by Deborah Elaine Thomas such as furniture, clothing, dishes, etc. will be thrown away by July 28, 2005, if gone unclaimed.

FIGURE 10.1 Newspaper typo with serious consequences

This example does not suggest that revising, rewriting, and editing consist of simply fact-checking or correcting typos. Checking facts and eliminating typos are merely small parts of the problem-solving approach to writing effective documents. However, the example does suggest that writers should give their documents serious consideration before releasing them to an audience because failure to do so can have serious consequences. Revising, rewriting, and editing are important activities in the writing process; but before this chapter addresses them individually, you need to understand the differences among the three activities in order to make effective choices in how to best use each of them in your own writing.

Revising

revising improving the content and design of a document by making substantial changes to drafts of the document

Re-vision means seeing something again. **Revision** refers to changing a document so that it appears different in both content and design. Revising can improve the substance of a document by addressing content, organization, style, and design. In some instances revision also means updating a document by adjusting its content to reflect new information and contexts. Revision can take a great deal of time because it encompasses many aspects of documents. Many writers conflate revision and editing into a single activity. Still more common, many writers revise as they write, integrating revision throughout the PSA.

Rewriting

rewriting altering a document or piece of text to serve another purpose or audience

Many use the terms *revising* and *rewriting* synonymously to describe large-scale changes to a document; some dictionaries likewise equate them. However, for workplace writers **rewriting** refers to a more substantial activity: it means altering a document to serve another purpose or audience. And changes in purpose or audience can radically affect a document's content, length, style, and format. For example, supervisors within a company might produce a number of progress reports detailing the accomplishments of their various divisions over the course of a fiscal year. The CEO of the company, upon reviewing these reports, might rewrite the information contained in them into a new report for the board of directors. This new document would contain elements of all of the progress reports, but the CEO would have rewritten that information to serve a new purpose and a new audience.

Editing

editing changing the contents of documents to meet the requirements of correctness

Editing is necessary to prepare documents for public view; it refers to changing the contents of documents to meet the requirements of correctness. This typically means proofreading and correcting the small stuff—such as spelling, mechanics, grammar, format, and design—to meet the standards and conventions required by the audience. Editing also includes double-checking facts, dates, addresses, and numerical data to ensure that they've been inserted into the document correctly. Editing is generally done after a document has been written, revised, and/or rewritten.

Editing involves attention to the details of writing and might seem like a dull task, but it is crucial, as the introductory example demonstrates. There is an old parable in business about a company that submits a proposal for a multimillion-dollar contract; the proposal is interesting and accurate but is riddled with mechanical, spelling, and grammatical errors. The company reviewing the proposal turns it down, not because of the editorial problems per se, but because of the larger perception that such errors prompt: If the writers are not attentive to the details in simple documents, how can one be certain that the company will pay attention to the details of the multimillion-dollar project? Documents are a reflection of both the writer and the organization that stand behind the documents. A poorly edited document reflects negatively on those whose names are associated with the document. On the other hand, a well-edited document conveys authority and competence.

GUIDELINES FOR REVISING

Revising improves the content, style, organization, and design of documents. It includes rethinking how material is presented and often requires further planning, research, and even drafting. Thus, revision is part of the entire problem-solving approach because you must continuously ask yourself revision questions as you write:

- Does the document serve its purpose and its audience?
- Does the document make sense?

- Does the organization of the document efficiently present the information contained in the document?
- Does the style of the document suit the kind of document?
- Does the document clearly provide the information it must convey?

Some documents—particularly shorter documents like memos or e-mails—may not require much revision, but they will require some.

Workplace documents may be technically correct and may present accurate information, but they must also be appropriate and readable for their audiences. **Readability** refers to how easily an audience can extract information from a document, that is, how understandable and clear that document is. A number of factors contribute to readability:

readability the ease with which audiences can read and extract information from a document

- Style and word choice
- Clarity and concision
- Organization and presentation of information
- Design and layout

Making a document readable ensures that the audience will not be confused or misled by its information.

Revision, then, is an ethical obligation because it enables writers to confirm accuracy, access, and accountability. Consider the two documents in Figure 10.2; each is correct in terms of the technical information it reports, but which is more readable? The second version is clearer, more concise, and more readable, yet it still maintains the accuracy and the completeness of the information. Writers do fine-tune the accuracy of information during revision, but they attend mainly to issues of readability at this stage. Writers should reread documents carefully and repeatedly to identify the need for a wide variety of revision activities.

Version I: Summary of Specifications for One-Way Wireless Handheld Touch-Panel Screen Requirements

The MD-34G touch panel should be required to uphold the following specifications, as determined by the engineers that designed the handheld unit and in doing so,thereby,also defined these specifications. These specifications have also been approved by the Engineering Division of Matson Davis Electronics and the Matson Davis Electronics Board of Supervisors. It is these specifications by which all MD-34G touch panels should be manufactured in accordance with the specifications set by the engineers. These are the specifications: resolution on the touch panel must be at least 320 x 240 pixels in order for the information on the touch panel to be clearly accessed; the engineers recommend that lower resolutions not be produced. Engineers also recommend that the touch panel screens provide at least a 256-color palette and a screen size no smaller than 3.6 inches. The screens, the engineers confirm, should also be a passive matrix LCD.In addition to these specifications for the touch panel screen,the Matson Davis Electronics engineers have provided specifications for the memory size of the handheld wireless touch panel, the battery requirements of the unit, and the transmitting/receiving capabilities of the unit. Those specifications can be found elsewhere in this document.

Version II: Summary of Specifications for One-Way Wireless Handheld Touch-Panel Screen Requirements

The following specifications have been determined for the production of the MD-34G touch-panel screen by the Matson Davis Electronics Engineering Division and have been approved by the MDE Board of Supervisors:

- Screen resolution: 320 x 240 pixels, minimum
- Color: 256-color palette, minimum
- Screen type: passive matrix LCD

Specifications for other components of the MD-34G are listed in subsequent sections.

FIGURE 10.2 Two examples with different readability

Get an Overview

You should read a document once to be certain that it meets the central goals and purposes you set out to accomplish. As you read, ask yourself these questions:

- What is the purpose of the document? Does the document achieve that purpose?
- Where does the document state its purpose? Can the audience easily identify that purpose?
- Does the information in the document support the purpose of the document?
- Is there any critical information missing that is necessary for the document to achieve its purpose?

Trim the Fat

Important: When revising for readability, writers must be certain that their revisions do not alter the information that the document presents.

A common flaw of workplace documents is that they repeat information unnecessarily. When revising, look for words, phrases, or information that is just not necessary. Trim the excess, leaving only necessary information. Think about this as you would think about buying a good steak: you want the butcher to trim away the fat and gristle; you don't want to pay for the weight of the fat, which you don't want to consume. In fact, eating the fat would fill you up but would not provide the nutrition of the meat. The same can be said about workplace documents: readers do not want to fill up on unnecessary words, nor do they have the time.

Cutting out excessive writing is central to creating a readable document. Examine the revisions made in Figure 10.3. The revised version trims away the repetition and redundancy. In addition, the original version uses simple sentences—subject, verb, object—with little variation, which makes the document sound choppy; the revised version varies sentence structure and length.

Original version:
Introduction

This document provides a description of asymmetric digital subscriber line (ADSL). The description is brief. The description is presented in the form of an informal report. The report is about ADSL. ADSL is a relatively new technology. ADSL is a technology that can be used to transmit digital information. The digital information is transmitted at a high bandwidth over existing phone lines. These phone lines may already be present in homes and businesses. ADSL allows people to maintain a consistent connection to the Internet. ADSL uses most of a single channel on the phone line to transmit downstream to the user. ADSL uses less of the channel to receive information from the user. ADSL allows for analog information to travel over the channel at the same time that data travel over the line. Analog information refers to voice transmissions. In order for ADSL to work on your phone line, a filter must be installed. The filter then blocks signals above a determined frequency. The filter blocks all frequencies above 4 KHz. Voices transmit below 4 KHz. The filter blocks all transmissions that are not voice transmissions. The filter blocks data signals from interfering with voice transmissions. The voice transmissions are normal telephone conversations.

Revised version:
Introduction

This document provides a brief description of asymmetric digital subscriber line (ADSL), a relatively new technology that can be used to transmit digital information at a high bandwidth over existing phone lines already present in homes or businesses. ADSL allows people to maintain a consistent connection to the Internet, using most of a single channel on the phone line to transmit downstream to the user and less of the channel to receive information *from* the user. ADSL allows for analog (i.e., voice) information to travel over the channel at the same time that data travel over the line. ADSL works by using a filter to block signals above 4 KHz because voices transmit below 4 KHz. Thus, ADSL blocks data signals from interfering with the voice transmissions of normal telephone conversations.

FIGURE 10.3 An original and a revised/trimmed version of one document

ANALYZE THIS

Here is another document introduction that needs to be revised for readability. Read the paragraph, and then trim the fat from it. Revise the document so that it is more readable but maintains all of its core information.

The Asgaard Boltar Radar: A Description and Overview

Introduction

The following technical description describes the Asgaard Boltar Radar (ABR). The ABR is described in detail in the following description. The ABR is located at 58 49′7″N 158 33′53″W. These coordinates are the GPS coordinates for Tundra Lake near Bristol Bay, Alaska. The ABR is located near Bristol Bay. The ABR is powered by electricity. The electricity is brought in by a 22-kV line. The power line comes in from Bristol Bay. In addition to the 22-kV power line from Bristol Bay, there are step-down transformers. In addition to the step-down transformers and the 22-kV line from Bristol Bay, the ABR has two generators on site. The generators are portable. The ABR is housed in a complex of small buildings. There is a primary operations building close to the radar antenna. There is a transmitter hall in the operations building. There is an instrument room in the operations building. There is a control room in the operations building. There is an office in the operations building. There is a laboratory in the operations building. There is a workshop in the building. There is a storage room in the building. There are personnel facilities in the building. The building is fully shielded. The shielding protects against radio frequency interference. The shielding protects against electromagnetic compatibility. The building is also designed to withstand the climate conditions of Bristol Bay's arctic climate.

Make It Clear

Clarity ensures that documents are not ambiguous and audiences interpret information as the author intends. The following suggestions are ways to ensure clarity in your documents:

Use Pronouns Clearly

Be certain that when you use pronouns, there is no question about who or what those pronouns are referring to. Pronouns replace nouns.

> **Unclear:** The engineers provided models as long as they were available.
>
> **Revision question:** Does *they* refer to the engineers or the models?
>
> **Explanation:** The antecedent of *they* could be either the models or the engineers, giving the sentence very different meanings.
>
> **Revision:** The engineers provided models as long as the models were available. (*or*) As long as the models were available, the engineers provided them.

Use Modifiers Clearly

Modifiers are words, phrases, or clauses that provide additional information about other words, phrases, or clauses. Modifiers can be adjectives or adverbs and should be placed as close as possible to the words they modify. Awkwardly placed modifiers can disrupt the flow and the meaning of a sentence.

> **Unclear:** The lab equipment was returned to the factory where it had been developed three years earlier by parcel post.
>
> **Revision question:** Does "by parcel post" tell about the manner in which the equipment was shipped or developed?
>
> **Explanation:** Because "by parcel post" is not close to what it modifies, the sentence is unclear, and readers may incorrectly interpret it.

Revision: The lab equipment was returned by parcel post to the factory where it had been developed three years earlier.

Unclear: Early in the meeting even the CEO couldn't hear the speaker next to him.

Revision: Early in the meeting the CEO couldn't hear even the speaker next to him.

Unclear: The technician, following an extensive investigation, was cleared of all charges of liability.

Revision: Following an extensive investigation, the technician was cleared of all charges of liability.

Unpack Sentences That Contain Too Much Information

Sentences that contain too much information lack clarity because they try to do too much. Readers have difficulty understanding and remembering when sentences contain too much information.

Unclear: This report provides a critical evaluation and summary of experimental vibrational and electronic energy-level data for neutral and ionic transient molecules and high-temperature species possessing from three to sixteen atoms in cases of isolated molecules of inorganic species, such as heavy-metal oxides, which are important in a wide variety of industrial chemical systems.

Revision: This report provides a critical evaluation and summary of experimental vibrational and electronic energy-level data for neutral and ionic transient molecules and high-temperature species possessing from three to sixteen atoms. The report specifically addresses cases of isolated molecules of inorganic species, such as heavy-metal oxides, which are important in a wide variety of industrial chemical systems.

Clarify Ambiguous Statements

Information that can be interpreted in multiple ways leads readers to misuse and misunderstand information.

Unclear: Recent inaccurate genome mapping data have created product failures. This problem must be remedied.

Revision question: Which problem must be remedied: inaccurate genome mapping data or product failures?

Revision: Recent inaccurate genome mapping data have created product failures, which must be remedied. (*or*) Recent inaccurate genome mapping data have created product failures and must be remedied.

Unclear: Architects require more exact measurements than plumbers.

Revision question: Do architects require more exact measurements or more plumbers? Or do architects or plumbers require more exact measurements?

Revision: Architects require more exact measurements than plumbers do.

Unclear: The lab technician was exceptionally efficient at cleaning the lab equipment each day. She collected numerous urine samples.

Revision question: What is the connection between cleaning the lab equipment and collecting numerous urine samples?

Revision: The lab technician was exceptionally efficient at cleaning the lab equipment. She also collected numerous urine samples each day.

ANALYZE THIS

The sentences that follow might be interpreted differently by different readers. How might you revise each example to more accurately present the information? For each, identify the revision question you would ask, and then provide a revised version.

Unclear: The biotechnician was much more fascinated with nanotechnology than his assistant.

Unclear: Skewed salinity readings led to inaccurate predictions. These will need to be reassessed.

Unclear: The research assistants completed the economic census of the region early last year. They collected copies of death certificates.

Change Punctuation That Causes Confusion

Proofreading for punctuation errors is generally considered an editing activity, but incorrect punctuation can also affect clarity.

Unclear: The office manager files, reports and letters.

Revision: The office manager files reports and letters.

Compress Sentences Carefully

Compression, sometimes called telegraphic writing, is a strategy used to eliminate unnecessary parts of a sentence. Writers may leave out pronouns, articles, conjunctions, or even transitional phrases, striving for concision (which is discussed next). However, eliminating key words can confuse the meaning of some sentences.

Unclear: Report to investors delayed production.

Revision question: Is *report* a verb or a noun here?

Revision: The report to the investors delayed production.

Unclear: To disengage the boom assembly, pull the red lever. Push the blue lever until even with boom assembly.

Revision question: When in the process should the reader push the blue lever?

Revision: To disengage the boom assembly, pull the red lever. Then push the blue lever until it is even with the boom assembly.

Order Words Thoughtfully

The ordering of words is crucial in conveying information clearly; word order can alter the meaning of a sentence and the information that is emphasized.

Unclear: Over the budget looked the IT supervisor.

Revision: The IT supervisor looked over the budget.

Use Active and Passive Voice Appropriately

Writers are often told to use active instead of passive voice in their writing, and generally this is good advice. Active sentences are those in which the subject of the sentence acts through the verb; in passive sentences the subject of the sentence is acted upon. Consider this sentence:

Word processors are used by workplace writers.

In this sentence the subject is *word processors*, and the verb is *are used*. The sentence is in passive voice because the subject is receiving the action of the verb rather than contributing to it. Consider this revision:

Workplace writers use word processors.

Now the subject, *workplace writers*, is doing the action, *use*, to the direct object, *word processors*.

However, workplace writers sometimes find it necessary to use passive voice for clarity, emphasis, or accuracy. Consider this sentence:

> The feasibility study was completed ahead of schedule, allowing construction to begin early.

This sentence is clearly written in the passive voice; the writer wanted to emphasize the study's early completion and the resulting benefits. Of course, the sentence could be revised like this:

> The research team completed the feasibility study early, allowing construction to begin early.

The active voice redirects the emphasis from the completion of the feasibility study to the team that completed the study. Such choices are crucial for writers.

Deciding between active and passive voice can also be important for clarity. Passive voice can sometimes create ambiguity and obscure information.

> **Unclear:** Production costs were greatly underestimated.
>
> **Revision question:** Who underestimated the production costs? Is it important for the reader to know that information?
>
> **Revision:** The development group greatly underestimated production costs.

Choosing between active and passive voice can involve clarity as well as ethics and professionalism. Often the passive voice is used to redirect attention or to veil responsibility, and this use can be entirely ethical. For instance, if a client owes you money, the passive voice can be less accusatory and can maintain goodwill: "This bill has not yet been paid" rather than "You have not paid your bill." Professional writers must consider the full effect of active or passive sentences.

Don't be deceived into thinking that the passive voice makes writing more objective because it avoids using personal nouns and pronouns. These sentences are equally subjective, but the first is more elusive:

> The proposals were never reviewed because the mailing labels were addressed incorrectly.
>
> The committee never reviewed our proposals because Chris incorrectly addressed the mailing labels.

ANALYZE THIS

Here are five sentences written in the passive voice. Consider which of them should be revised for clarity and which could remain as they are in order to emphasize certain information without compromising clarity or professional ethics. For those that need revision, change them into active sentences without altering their essential information. For those that can remain passive, write a short explanation about why the sentences work in passive voice.

1. Mid-year committee reports have been submitted, and budgeting requests are being considered.
2. The geotechnical reconnaissance was botched on-site, so the liquification features resulting from the earthquake were not properly cataloged.
3. The e-construction projects were begun by the IT division.
4. The problems in development were solved just before redesign began.
5. Because semiconductor use has been limited, continued developments in nanotechnology have resulted in submicron lithography.

Eliminate Numerous Modifying Nouns

Writers often use nouns to tell about other nouns. For instance, *budget*, *committee*, or *lab* could describe *report*, as in *budget report*, *committee report*, and *lab report*. And writers sometimes use strings of nouns, such as *annual laboratory committee budget report*, which can become unwieldy and unclear. You should break up noun strings by including prepositional phrases, articles, and active verbs and by varying sentence structures.

> **Unclear:** Consumer product consent forms must be signed in order for claims to be reviewed by adjusters.
>
> **Revision:** In order for adjusters to review claims, consumers must sign a consent form for the product.

Be Concise

concision expressing information clearly and briefly

Workplace writing is best when it is clear and concise. Chapter 1 identifies concision as one of the key objectives in workplace writing because readers don't have the time, or the interest, to read long documents in order to access small amounts of information. **Concision** is the expression of information in as few words as possible. It does not compromise information but cuts out unnecessary words.

There are two primary objectives in revising for concision:

- To eliminate unnecessary writing—wordiness
- To eliminate unnecessary information—excessiveness

Fortunately, revising for concision works hand in hand with revising for clarity; therefore, much of your revising accomplishes both at the same time.

Cut Empty Phrases

Some inexperienced writers try to make their writing sound more professional by using phrases that they think sound elevated, sophisticated, or experienced. Usually, though, this approach just makes the writing wordy. Often, these phrases can be revised into single words that convey meaning more clearly and concisely. Table 10.1 lists commonly used wordy phrases and ways to revise them for clarity and concision. Various other strategies can be helpful, too.

Avoid Unnecessary Modifiers

> **Wordy:** The report doesn't require any particular deadline.
>
> **Revised:** The report doesn't require a deadline.
>
> **Wordy:** This year's budget request must be very accurate.
>
> **Revised:** This year's budget request must be accurate.
>
> **Wordy:** The project analysis contains some sort of mistake.
>
> **Revised:** The project analysis contains a mistake.

Reduce Descriptive Clauses Clauses that begin with words like *who*, *which*, or *that* can often be reduced without compromising the meaning of the sentence.

> **Wordy:** The reaction, which was recently identified, was not the only catalyst.
>
> **Revised:** The recently identified reaction was not the only catalyst.
>
> **Wordy:** The chairman will address only the feasibility study that is most acceptable.
>
> **Revised:** The chairman will address only the most acceptable feasibility study.

Avoid Dummy Subjects Some sentences begin with words like *it* or *there* and provide the real subject of the sentence *after* the verb. These words are referred to as dummy subjects, or expletives. The solution is usually to shift the true subject *before* the verb.

TABLE 10.1
Wordy phrases and revisions

Wordy Phrase	Concise Revision
at the present time	now
aware of the fact	know
for all intents and purposes	actually
the reason, for the reason that, owing or due to the fact that, in light of the fact, considering the fact that, on the grounds that, this is why	because, since, or why
on the occasion of, in a situation in which, under circumstances in which	when
as regards, in reference to, with regard to, concerning the matter of, where ___________ is concerned	about
it is crucial that, it is necessary that, there is a need or necessity for, it is important that, cannot be avoided	must or should
is able to, has the opportunity to, has the capacity for, has the ability to	can
it is possible that, there is a chance that, it could happen that, the possibility exists for	may, might, or could
the majority of	most
prior to	before
readily apparent	obviously

Wordy: It is the decision of the board that the project be terminated.

Revision question: To what does *it* refer?

Revised: The board decided to terminate the project.

Wordy: There are numerous connectors attaching the coupling to the housing.

Revised: Numerous connectors attach the coupling to the housing.

Avoid Strings of Prepositional Phrases Writers often use strings of prepositional phrases to clarify sentences. These strings can often be reduced to one or two stronger modifiers.

Wordy: The condition *of* the patient was documented *in* the patient profile written *by* the nurse *on* duty *during* the after-hours shift.

Revision: The after-hours nurse documented the patient's condition in the patient profile.

Avoid Nominalizations Nominalizations are verbs or other words that have been changed into nouns. Writers employ nominalizations to work with weak verbs or to support unneeded prepositional phrases. Nominalizations are revised for concision and clarity by returning the noun to verb form. The verb derived from the nominalization is likely stronger and better than the nominalization.

Nominalization: The role of the project manager is the supervision of employees.

Revision: The project manager supervises employees.

Avoid Excessive Information Inexperienced writers often include excessive information, especially in introductory clauses. These clauses simply delay readers and add nothing substantive.

Unnecessary information: In response to your query of February 28, 2007, our primary microscopes do have digital imaging capabilities.

Revision: Our primary microscopes do have digital imaging capabilities.

Unnecessary information: In response to your request for applications, I am writing to apply for the position of Civil Engineer.

Revision: I wish to apply for the position of Civil Engineer.

ANALYZE THIS

Many technical and workplace writers consider the shifting of negative wording to positive wording to be an important part of revising. Notice in the preceding section on concision that many of our subtitles appear in negative form, often using the word *avoid.* When might negative forms be useful in workplace writing?

Locate other documents that rely on negatives. Could the negatives in those documents be revised into positives? What effect would those revisions have on the document? Should we have revised our negative subtitles into positives? Why or why not? What revisions might we have made?

Consider Style

All of these writing strategies affect your writing style, that is, the way your writing sounds to your readers. Revising for style refers to re-visioning the way your audience reads your writing. Look at these two examples:

The report was filed by the two attending technicians. The technicians had observed the incident. The report details the observations made by the technicians.

The two attending technicians who had observed the incident filed a report detailing their observations.

The first sentences are technically correct, but the revised version is much more readable. As you revise for style, pay attention to how your document "sounds" to your audience.

Your Own Style

Before thinking about style revision, you should identify the stylistic choices you already make.

- *Sentence length:* Do you tend to write short or long sentences?
- *Word choice:* Do you tend to use short or long words? Do you use many words where one would do? Do you repeat certain words?
- *Paragraph length:* Are your paragraphs long and wandering or short and to the point? Are your paragraphs dominated by a single long sentence, or are they composed of numerous midlength sentences?
- *Location of emphasis:* Where in your sentences do you emphasize points? Are your sentences front-loaded with information, or do they hold information until the very end? Where in your paragraphs do you emphasize information?
- *Sentence consistency:* Are your sentences all written in a similar pattern: subject, verb, object? Or do you vary sentence construction?
- *Use of transitions:* Are your transitions smooth and announced, or are they rough, requiring readers to make leaps between blocks of information?

Short Sentences

As mentioned earlier, combining numerous short sentences into a single, more dynamic and sophisticated sentence can make documents more engaging and readable. Consider these versions:

Short sentences: Allied health workers may work with bone marrow transplant cases. Acute lymphoblastic leukemia is one kind of cancer they may work with. Chemotherapy is used in conjunction with transplants. Transplants are difficult procedures. Locating bone marrow donors is also difficult. Some transplant facilities maintain donor registries.

Revision: Allied health workers may work with bone marrow transplant cases, including patients with acute lymphoblastic leukemia. Because transplants are such difficult procedures, chemotherapy may be used in conjunction with the transplants. And because locating donors can be challenging, many transplant facilities now maintain donor registries.

Long Sentences

Revising long, rambling sentences into shorter, more readable sentences can also improve your writing style.

Long sentence: ASD Technical, an information technology company specializing in providing new companies with database administrators and computer analysts capable of working with software engineers, is looking for two computer programmers to write, test, and maintain programs of various designs in COBOL, Prolog, JAVA, Smalltalk, and C++ for software engineers who have recently completed a new system application for monitoring corporate transactions at the international level.

Revision: ASD Technical is looking for two computer programmers. ASD is an information technology company that specializes in providing new companies with database administrators and computer analysts. The two new programmers would work with ASD engineers to develop software for a new application designed to monitor international corporate transactions. Applicants should be experienced in working with COBOL, Prolog, JAVA, Smalltalk, and C++.

Jargon

jargon specialized or technical language used within a particular group, organization, or field

Jargon is often described pejoratively to mean specialized language that excludes some people from understanding. But jargon can actually be a necessary part of workplace writing. **Jargon** refers to the specialized or technical language of a particular community and is used to communicate complex ideas in simple ways.

For instance, information technology experts might use jargon terms such as *cache* or *cookie* when talking about the World Wide Web. For specialists the terms are simply an efficient way of discussing detailed concepts, even though people not versed in IT terminology might not understand the terms.

A *cache* is space on a web browser that stores information about what websites a user has visited with that browser. When a user returns to a website already visited, the browser can access that site through the cache, rather than going through the server on which the site is located. Sometimes a user may not see the most recently updated version of a page when the browser takes the user to the site stored in the cache.

A *cookie* is a message sent and stored on a browser by a web server. When a user logs onto a web page, the server retrieves the cookie from the browser and is able to customize the web page for the user, based on the information stored in the cookie. This explains why some web pages, particularly business web pages like Amazon.com, appear to remember the user. They simply retrieve information from the cookie on the user's computer and plug it into the page to create the appearance of familiarity.

Now imagine that each time web designers talked about a corporate web page, they had to explain the idea of a cache or a cookie because there were no specific jargon terms to make their conversation more efficient and economical.

Jargon helps clarify and economize communication if all participants are familiar with it. However, jargon can also lead to confusion and frustration if the audience is unfamiliar with the terminology. Some writers use jargon intentionally to obfuscate meaning or to elevate their credibility. Both of these uses are unethical.

When revising, you should follow four primary guidelines about jargon, all of which center on audience concerns:

1. *Use jargon only when you are certain the audience will understand it.* If there is any doubt, avoid the jargon terms.
2. *Use jargon sparingly to convey expertise and to gain audience confidence.* Overuse of jargon usually pushes audiences away rather than drawing them in.
3. *Define jargon terms for audiences.* Sometimes the goal of a workplace document is to inform and instruct an audience. In these instances defined jargon can give audiences access to a term and the authority to use it. Jargon can be defined in the text or in a glossary with a text reference.
4. *Never use jargon to mask information or to confuse your audience.* Readers tend to become suspicious of writing cluttered with jargon.

See chapter 15 for more on definitions.

IN YOUR **EXPERIENCE**

It is impossible to avoid jargon. Every business, academic discipline, sport, hobby, organization, culture, and community has its own jargon. Think for a moment about the groups to which you belong and the jargon those groups use in their communication. Select one of the groups. What are some jargon terms of that group? How did you learn them? How do you use them? When do you use them? When do you avoid them? List ten jargon terms you use in any given community, define each term, and tell how it is used, including who uses it.

Pay Attention to Tone

Think of tone as the writer's attitude projected in a document. A letter of complaint might project anger; a contract proposal might project confidence and enthusiasm; an accident report might project frustration and disappointment; a recall notice might project remorse. Even documents intended to appear impartial project an attitude. All documents have a tone, and it affects how readers react to the documents.

Consider the e-mails in Figures 10.4 and 10.5; think about the tone in both. How are the recipients apt to respond to each? Notice that each e-mail asks for the same thing, but the second does so professionally whereas the first does so aggressively.

Now consider these differences in tone:

Casual: Ed: Here's that report you asked for. I'll call you later to see if you like it.

Formal: Director Deans: Enclosed you will find the requested report. Our office will be in touch to confirm its approval.

Tone has many components: level of formality, gap between writers and readers, emotional connection between writers and readers, level of bias, positive or negative attitude, professionalism, confidence in writers and/or readers.

One way writers can control tone is through word choice. Consider these examples:

Emotional: I'm getting pissed off at the lab technician.

Professional: The lab technician is beginning to frustrate me.

Notice, too, how the shift in the subject of the sentence redirects the focus from the writer to the technician.

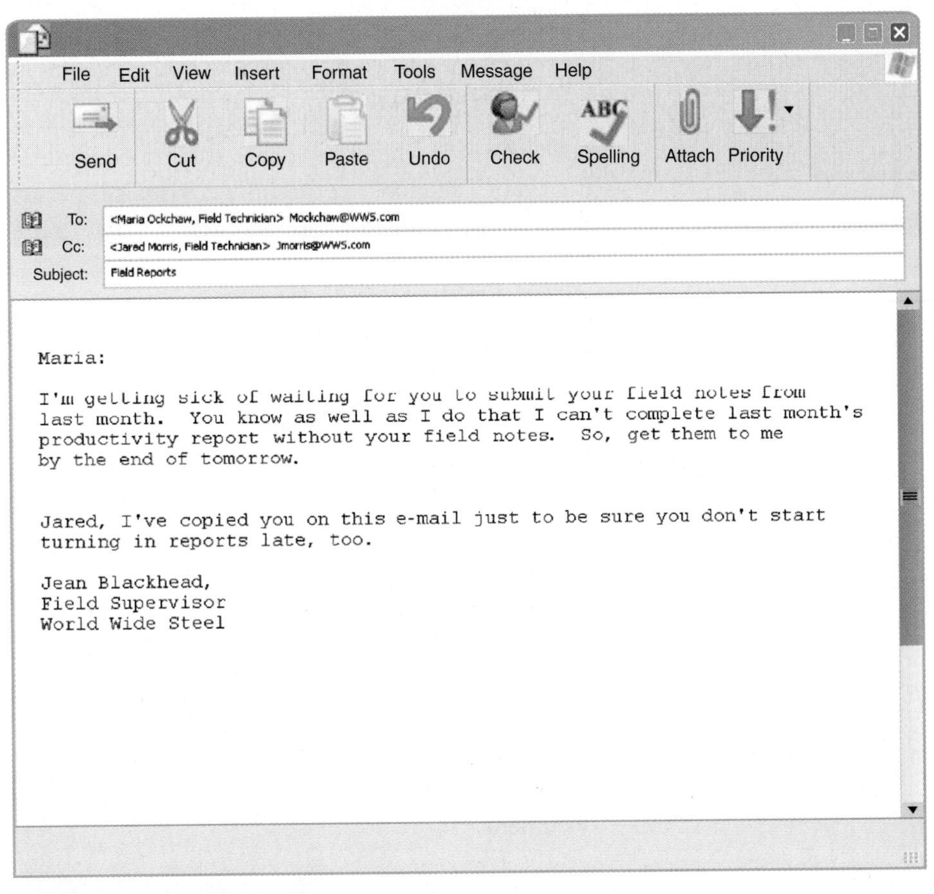

Maria:

I'm getting sick of waiting for you to submit your field notes from last month. You know as well as I do that I can't complete last month's productivity report without your field notes. So, get them to me by the end of tomorrow.

Jared, I've copied you on this e-mail just to be sure you don't start turning in reports late, too.

Jean Blackhead,
Field Supervisor
World Wide Steel

FIGURE 10.4 An e-mail with an aggressive tone

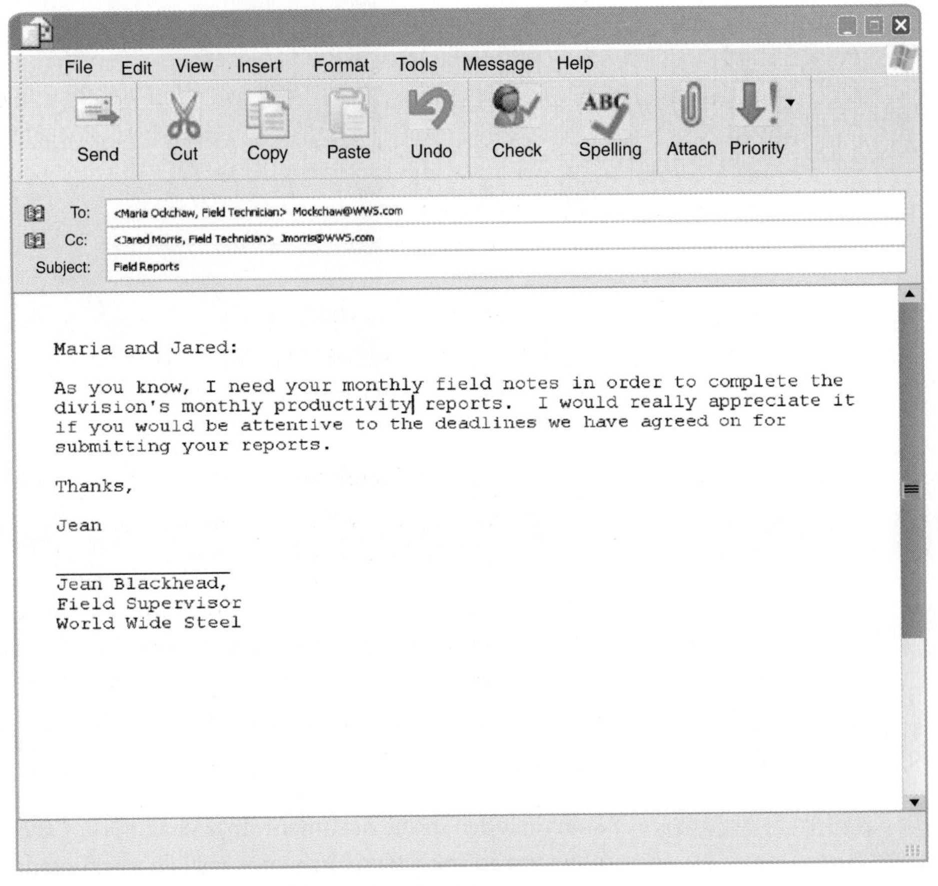

Maria and Jared:

As you know, I need your monthly field notes in order to complete the division's monthly productivity reports. I would really appreciate it if you would be attentive to the deadlines we have agreed on for submitting your reports.

Thanks,

Jean

Jean Blackhead,
Field Supervisor
World Wide Steel

FIGURE 10.5 An e-mail with a professional tone

Look Again at Visuals

See chapter 8 for more on using visuals and chapter 9 for more about design elements.

Sometimes writers neglect to revise visuals because they assume that only words need changing; however, workplace writers need to be as critical of their visuals as they are of their words when they revise their documents. Remember, revision is about "seeing again," and looking carefully at visuals is particularly important. First drafts of visuals can be as unclear and inaccurate as first drafts of writing.

Selection of Visuals

Sometimes the visual you create during the drafting process—for instance, a pie chart or a bar graph—does not fulfill the needs of the final document. During revision you should ask these kinds of questions:

See chapter 8 for more on the uses of exploded views and photographic images.

- Did I use the right kind of visual?
- Would a chart convey the information better than a graph?
- Would a photograph be clearer than a line drawing?
- Would a line drawing be more dynamic than clip art?
- Would an exploded view be more readable than a straightforward diagram?

As an example, examine the differences between the exploded view of a minilathe in Figure 10.6 and the photograph of a minilathe in Figure 10.7. Diagrams with exploded views and photographic images serve different rhetorical purposes, so your visual choices might need to be revised as you complete the final draft of a document.

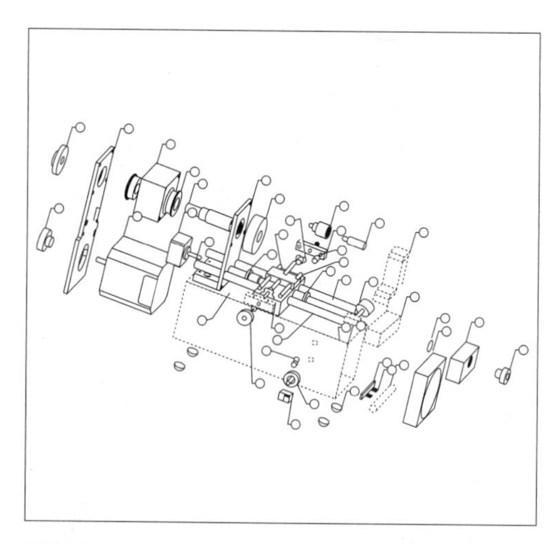

FIGURE 10.6 An exploded view of a minilathe

FIGURE 10.7 A photograph of minilathes

FIGURE 10.8 A sketched visual tagged for revision

Final Versions

When writing longer documents, writers often insert thumbnails or placeholders to identify where a visual will go. Sometimes writers insert a storyboard or a draft version of the visual to identify for themselves or collaborators what kind of visual will appear in the final version. Writers might use a tag such as "Image to be professionally rendered during production." Figure 10.8 shows a rough sketch that is tagged to be revised.

Whatever type of placeholder writers choose to use, they must be certain that the final version of the visual actually appears in the revised document. They should also be sure that any textual references to the visual and any caption describe the final version included in the document. In addition, they should check to see whether there are any gaps in the document where visuals were intended to be; sometimes writers plan to add images during later parts of the writing process but forget to do so.

Placement and Relationship to Text

See chapter 8 for more on relationships between text and visuals.

When revising visuals, writers should also confirm that visuals appear where they are supposed to in the document. Readers get frustrated when texts refer to visuals that do not appear where they should. In addition, readers get confused if no explanation of a visual appears in the document. Writers should always confirm that visuals and text correspond.

Accuracy and Readability

Providing visuals that can be seen by users with disabilities is an ethical issue of accessibility.

See chapters 3 and 8 for more on making visuals readable in terms of accessibility and disability.

Writers should review charts, graphs, tables, and figures to confirm that the information they convey is both accurate and compatible with the information and data provided in the document. In addition, writers should make sure that all visuals can be easily read; they should check specifically for visuals that contain too many details, are difficult to interpret, or are too small to be read without magnification. Resizing, recoloring, realigning, and redrawing are typical revision activities. Writers should pay particular attention to the following features of charts and graphs:

- *Labeling:* Be sure all labels are correctly placed and are easy to read.
- *Lines:* Be sure grid lines are distinct from data lines. Try to eliminate grid lines if possible; but if they are necessary, be sure they are more subtle than the data lines.
- *Distortion:* Eliminate any features or effects that might distort information. Effects like 3D chart renderings can mask information or make it difficult to interpret.
- *Emphasis:* Confirm that the information you need to emphasize is actually emphasized. Check the use of color for emphasis.

Distraction and Excess

Eliminate any distracting visuals that hinder your readers' ability to access information. Some visuals may be more distracting than they are useful. In addition, be sure to cut those visuals that are excessive; use only the visuals you absolutely need. Writers often overpopulate drafts with visuals, but excessive use of visuals can be distracting and can interfere with readers' ability to navigate a document. Excessive visuals can also make a document unnecessarily large. For electronic documents like web pages, this can be inconvenient and can hinder accessibility.

A related concern is visuals that are just there to look good. If you discover that you've included such a visual, you should probably cut it. Decoration doesn't mask or change information, and it can diminish your credibility.

Design, Layout and Color

Examine the layout of each page, checking for page breaks, placement of visuals, font sizes, heading levels, headers, footers, and other design elements. In addition, be sure that the colors you are using are transferable to the medium of the final document. And consider how the colors will appear in the final document—are they effective? are they overused? Finally, look at the overall visual scheme of the document to be sure that visual appeal is consistent throughout the document.

Be Accurate and Timely

When revising documents, you should check for missing or deleted text. For example, look for things like references to appendixes that are not included or paragraphs that were accidentally deleted. In addition, you should verify that multiple references to the same concepts, data, or events are all consistent. And finally, check for timeliness by asking these questions:

- Have other documents been submitted since you began writing this document that might affect the information you are presenting?

- Have other events affected your information?
- Are your data still relevant?
- Have new laws, regulations, or policies gone into effect that might affect your document?

Review Online Documents Regularly

Online documents—such as websites, blogs, and wikis—require regular revision for accuracy and timeliness. Unlike printed documents, which are presumed to be accurate and correct at the time they are published, online documents are expected to be accurate and correct at any time they are publicly available. Consequently, if you publish or maintain a website, you should revise it regularly. Even though some authors choose to update a website's contents at regular intervals, such as weekly or monthly, all authors should update their websites whenever any information in them changes. The authors of blogs and wikis are expected to revise those documents even more frequently—sometimes daily. Many online documents contain information about when they were last updated; this date should be changed whenever an online document is altered, no matter how small the revision.

Aside from adding or changing the content in online documents, writers should make sure that all of the links in their documents are still active and, if not, should change or remove the problem links. Also, if a document links to an image, the author must be sure that the image is still available and has not changed since it was inserted into the document.

See chapters 3 and 17 for more about online documents.

GUIDELINES FOR REWRITING

Rewriting a document involves more substantial changes than those made during revision. This process produces a document altogether different from the original in terms of audience, purpose, or even genre. For instance, a company that manufactures jet-drive propulsion units for recreational watercraft produces an owner's manual for that jet-drive, as well as a service manual for maintenance technicians. Each manual is important for its audience, but the information included in the maintenance manual would have to be rewritten for the owner's manual because of audience differences. This is the crux of rewriting: presenting information in new ways to meet the demands of a different rhetorical situation.

Work from Large to Small

Rewriting complex documents to produce simple documents is easier than rewriting a simple document to produce a more complex one. For instance, in the example of the jet-drive propulsion unit, writers would have difficulty beginning with the owner's manual and rewriting it to produce the maintenance manual because the owner's manual does not contain the complex details required in the maintenance manual. Starting with the maintenance manual would allow the writer to more efficiently extract what is needed for the owner's manual. However, writers must be sure to understand the necessary components of each document; sometimes the new document requires different organization from that of the original.

Consider Audience and Purpose

More often than not, the needs of a new audience necessitate the rewriting. Writers then must consider new ways to convey the information efficiently and appropriately. They should reevaluate word choice, organizational strategies, format, layout, design, content, and, of course, purpose. Generally, a new audience means a new purpose, and the purpose of the new document must drive how the information is rewritten.

Address Ethical Concerns

The rhetorical choices made during rewriting have ethical ramifications. Think back to the example of the jet-drive manuals. The maintenance manual, designed for technicians, contains detailed and complex information. The writers of the owner's manual could decide not to provide some of that information so that owners would have to rely on service technicians to perform the maintenance. That choice to leave out information might be made to protect owners from attempting maintenance procedures that could put them in danger. Or the information might be excluded to ensure that owners must pay specific technicians for the maintenance. Rewriting documents brings ethical considerations, particularly in deciding what information to include or exclude.

GUIDELINES FOR EDITING

Editing plays an important role in how others see and judge you as a professional. Because most writing you produce in the workplace is meant to benefit your employer, it is your ethical and professional responsibility to produce high-caliber documents. Writers should think of editing and proofreading as two parts of a single activity:

- *Proofreading* is a detailed, careful reading of a document, looking for mistakes.
- *Editing* is identifying and making the needed corrections in a document.

In other words, proofreading and editing are inseparable.

Most writers use a system of marks to identify needed corrections in their documents. Often called proofreaders' marks (see appendix A), these marks signal the changes to be made. Writers who proofread their own work might mark corrections as they proofread so that they can make the corrections at a later time, or they might make the changes as they proofread, particularly if they are editing the document in an electronic format. Some writers don't proofread their own documents but have others proofread for them. Using standard proofreaders' marks helps these writers and proofreaders communicate.

Proofreaders read documents carefully and pay attention not only to the overall structure and content of a document, but also to each individual paragraph, sentence, and word. Because you edit documents after you revise or rewrite, the editing phase should come just before the document is distributed.

General Areas of Concern

When you proofread and edit your documents, you should pay special attention to several areas of potential difficulty:

For more on formating and typography, see chapter 9.

- *Facts.* Check to be sure the correct facts, numbers, or data are used in the document.
- *Formatting.* Check to be sure that all parts of the document are included and are in the appropriate order.
- *Typography* (fonts, spacing, margins, etc.). Check for consistency among subheadings, section spacing, font sizes, font types, and use of specialized fonts (boldface, italics, colored fonts, etc.).
- *Grammar.* Check sentence-level grammar especially.
- *Punctuation.*
- *Spelling.* Pay specific attention to commonly misspelled words, names, and discipline-specific terms and jargon.
- *Capitalization.* Check for consistent use of capitalization in names, titles, and other proper nouns.

See appendix A for a detailed list of grammar rules, punctuation rules, commonly misspelled words, capitalization rules, and commonly used abbreviations.

- *Abbreviations.*
- *Electronic addresses.* Check that URLs and e-mail addresses are complete and still current.

Common Errors

Most errors are evident and easily corrected, and often writers repeat the same errors throughout their writing. Many errors are simple oversights. We list here common areas of error but refer you to appendix A for a complete list of the grammatical issues to consider, as well as the individual rules for each.

Familiarize yourself with grammar and punctuation rules to avoid mistakes that will need editing.

Factual trouble spots

- Outdated or incorrect statistics
- Mathematical errors
- Transposed numbers
- Misspelled names or addresses
- Incorrect dates

Grammatical trouble spots

- Sentence fragments
- Run-on sentences
- Dangling and misplaced modifiers
- Pronoun/noun agreement
- Subject/verb agreement

Punctuation trouble spots

- Commas with compound sentences
- Commas after introductory clauses
- Commas after introductory phrases
- Commas with dates
- Semicolons with independent clauses
- Semicolons with series
- Colons introducing lists
- Quotation marks for emphasis
- Punctuation after parentheses
- Hyphens with compound adjectives
- Apostrophes with plural possessive nouns
- Ellipsis points

Capitalization trouble spots

- Company and organizational names
- Titles of objects, products, or services

Abbreviation trouble spots

- States and countries
- Units of measure
- Units of time

Electronic formatting trouble spots

- Precision in all e-mail addresses
- Precision in all URLs

Editing Strategies

Workplace writers should make it a priority to edit their documents or have another reader do so before the document reaches its intended audience. Mindful of the realities of time constraints, we offer some strategies for editing your own documents, as well as editing colleagues' documents. However, whether you are editing your own or someone else's work, it is important that you edit in the best possible conditions because editing requires that you pay attention to details. As much as possible, you should try to edit under these conditions:

- *Be well rested.* It is easy to miss things in your editing if you're tired. Try to do your editing during the time of day when you are most alert.
- *Be nourished.* Hunger can be a distraction; it can also affect your energy level and ability to concentrate. Likewise, editing immediately after a large meal can be difficult if you are lethargic.
- *Focus.* Don't let distractions interrupt you. Give your full attention to the document at hand.
- *Relax.* Don't rush the editing. Editing should be careful and meticulous.

Editing Your Own Documents

One difficulty in editing your own documents is that you are often too familiar with them; familiarity makes it difficult to see mistakes. Especially with documents that you have just finished, you may read what you *expect* to be on the page, not what is really there. For this reason, many workplace writers have others proofread their documents. However, many rhetorical situations in the workplace do not allow for a second reader. Therefore, writers must develop strategies to overcome their own familiarity.

Put the Document Aside for as Long as Possible Taking some time away from a document allows you to lose some familiarity with it. Then, when you return to it, you are better able to see your choices and edit your work. Of course, this strategy is not possible if the document must be distributed soon after it is written. Even when you face deadlines, however, it's a good idea to plan some time for editing and allow at least a little time away from the document before it is distributed.

IN YOUR EXPERIENCE

Have you ever gotten a paper back from a teacher that included notations of incorrect grammar or punctuation for rather obvious mistakes? And have you ever wondered how you missed correcting those mistakes when you proofread the paper? Those kinds of mistakes are usually missed simply because you are too familiar with the paper to accurately read it as an editor. Similarly, have you ever found a paper that you wrote, read it, and not recognized it as yours or not remembered much about writing it? Again, you become less familiar with your own writing over time. If you have a piece of writing available that you wrote some time ago, read through it and think about what changes you would make if you were editing it today. Then write a short explanation about how you would edit that document to improve it.

During the editing stage your focus should be on correctness, not completeness of information, which should have been addressed during the revision stage.

Read the Document Backwards Another strategy to make your documents unfamiliar is to read them in some order other than from beginning to end. Disrupting the flow of your documents will make them seem less familiar because they do not sound the way you intended them to sound, and you will pay more attention to details. Try these backward-reading strategies:

1. Read the document backwards one word at a time, asking yourself these questions:
   - Is each word spelled correctly?
   - Is any word misused?
2. Read the document backwards one sentence at a time, asking yourself these questions about each sentence:
   - Is this a complete sentence?
   - Does this sentence make sense?
   - Do the subject and the verb of the sentence agree?
   - Is the sentence punctuated correctly?

- Have I used these words correctly?
- How does this sentence lead to the sentence I read previously?
- Could this sentence be combined with the sentence I just read to form a more succinct sentence?

3. Read the document backwards one paragraph at a time, asking these questions:
   - Is this paragraph coherent?
   - Is this paragraph grammatically correct?
   - Does this paragraph provide a smooth transition into the paragraph I previously read?

Read the Document Aloud Sometimes simply listening to how a document sounds is an effective way to hear language that is incorrect. Reading aloud also forces writers to slow their reading and articulate each word rather than skim through the familiar document. Reading aloud might not lead to large-scale revisions, but it does facilitate editing, particularly for shorter documents that may face immediate deadlines. Slow, deliberate reading is the key to this strategy.

IN YOUR **EXPERIENCE**

The next time you send an e-mail, try reading it just as you would normally proofread any document. Then read the e-mail out loud to yourself, being careful to read slowly and deliberately. Do you notice a difference in your own attentiveness to the details of the document?

View the Whole Page on Screen If you edit on-screen, try adjusting the view of your document before you edit. Simply shifting the way you see the document on the screen may alter your familiarity with it enough to allow you to read it more attentively. For example, some writers see only a portion of the screen as they compose because they want the screen fonts to be larger and easier to see. Adjusting the screen to show an entire page at a time presents a different vantage point and adds newness to the document. Sometimes workplace writers reduce their pages enough to view the layout of two or more pages on one screen in order to check format. This view reduces the font size so that the pages cannot be read for content or correctness, but sometimes *not* being able to read the words can help you edit for format.

When editing document formats, some writers find it useful to temporarily convert the document into unrecognizable characters (i.e., using the font-selection controls of a word processor). This approach prevents them from reading the document and allows them to pay attention to format. Figure 10.9 shows a regular letter; Figure 10.10 shows that same letter converted to Greek characters.

Print a Hard Copy to Edit After writing a document on-screen, try printing a hard copy to edit. Having the physical document in hand allows you to see how the document actually appears in print. In addition, a hard copy allows you to make proofreading marks and other comments directly on the document. A hard copy also lets you edit from a different perspective because the document is being read in a different environment from the one in which it was orginally written.

Editing Someone Else's Document

Workplace writers often ask coworkers to proofread their documents. Editing someone else's work carries important ethical and professional responsibilities:

- You must proofread or edit in a professional manner, not simply glancing at the document for a few minutes. The other individual is relying on you to offer useful suggestions.

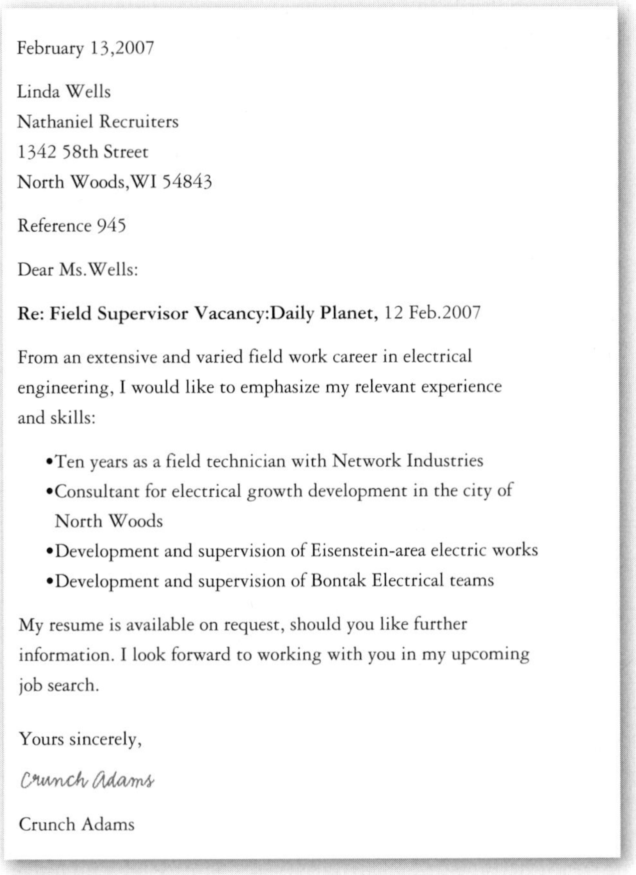

February 13,2007

Linda Wells
Nathaniel Recruiters
1342 58th Street
North Woods,WI 54843

Reference 945

Dear Ms.Wells:

Re: Field Supervisor Vacancy:Daily Planet, 12 Feb.2007

From an extensive and varied field work career in electrical engineering, I would like to emphasize my relevant experience and skills:

- Ten years as a field technician with Network Industries
- Consultant for electrical growth development in the city of North Woods
- Development and supervision of Eisenstein-area electric works
- Development and supervision of Bontak Electrical teams

My resume is available on request, should you like further information. I look forward to working with you in my upcoming job search.

Yours sincerely,

Crunch Adams

Crunch Adams

FIGURE 10.9 A sample letter

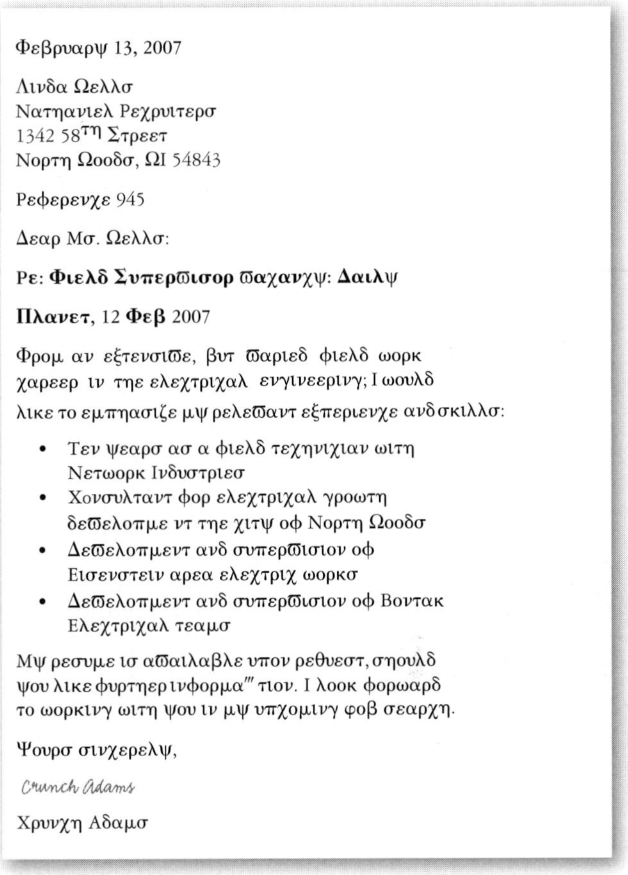

Φεβρυαρψ 13, 2007

Λινδα Ωελλσ
Νατηανιελ Ρεχρυιτερσ
1342 58ΤΗ Στρεετ
Νορτη Ωοοδσ, ΩΙ 54843

Ρεφερενχε 945

Δεαρ Μσ. Ωελλσ:

Ρε: **Φιελδ Συπερϖισορ** ϖαχανχψ: **Δαιλψ**

Πλανετ, 12 **Φεβ** 2007

Φρομ αν εξτενσιϖε, βυτ ϖαριεδ φιελδ ωορκ χαρεερ ιν τηε ελεχτριχαλ ενγινεερινγ; I ωουλδ λικε το εμπηασιζε μψ ρελεϖαντ εξπεριενχε ανδσκιλλσ:

- Τεν ψεαρσ ασ α φιελδ τεχηνιχιαν ωιτη Νετωορκ Ινδυστριεσ
- Χονσυλταντ φορ ελεχτριχαλ γροωτη δεϖελοπμε ντ τηε χιτψ οφ Νορτη Ωοοδσ
- Δεϖελοπμεντ ανδ συπερϖισιον οφ Εισενστειν αρεα ελεχτριχ ωορκσ
- Δεϖελοπμεντ ανδ συπερϖισιον οφ Βοντακ Ελεχτριχαλ τεαμσ

Μψ ρεσυμε ισ αϖαιλαβλε υπον ρεθυεστ, σηουλδ ψου λικε φυρτηερ ινφορμα‴ τιον. I λοοκ φορωαρδ το ωορκινγ ωιτη ψου ιν μψ υπχομινγ φοβ σεαρχη.

Ψουρσ σινχερελψ,

Crunch Adams

Χρυνχη Αδαμσ

FIGURE 10.10 A sample letter converted to a Greek font

- You must proofread or edit in a timely manner.
- Your advice must be useful, correct, and applicable to the document. Vague or empty comments do not help improve the document.

When you agree to edit someone else's document, ask a few questions *before* you begin:

- What is the purpose of the document?
- To whom is the document directed?
- To what should you pay the most attention?
- In what format would the writer like your comments: written directly on the page embedded in an electronic document, written on a separate sheet of paper, e-mailed as an attachment, e-mailed as the body text of an e-mail?
- When does the writer need your comments?

Good feedback usually involves more than just one read. Dedicate one slow, careful reading of the document to revision and another to editing. When proofreading and editing, pay attention to the following facets:

For a detailed list of grammatical and punctuation rules, refer to the handbook in appendix A.

- Does the document make sense?
- Are all of the sentences complete sentences?
- Is each sentence grammatically correct?
- Is each sentence punctuated correctly?
- Does the writer use words and phrases that are clear, or could some words and phrases be changed to improve the document?

- Are all of the words in the document spelled correctly?
- Does the writer use all of the words correctly?
- Has the writer made good choices in verbs?
- Are there any typographical errors?

Be sure to use proofreading symbols and margin comments the writer will understand. If your comments may not be clear, explain your suggestions either face-to-face or in writing. Discussing your suggestions while you both look at the document is the most effective method of ensuring understanding.

TECHNOLOGY

Because you are likely to use a word processor to compose documents, it is crucial to understand all of your word processor's features for drafting and editing documents. Most word processors provide standard features for revising and editing:

- *Cut and paste* allows you to move and replace selected parts of a document, either within one document or between documents.
- *Copy* allows you to insert duplicates of selected text and/or visuals as often as needed.
- *Search and replace* allows you to locate particular parts of a document automatically and replace them automatically.
- *Spell-check* allows you to confirm correct spellings of words in a document.
- *Grammar check* allows you to identify sentences that violate standard grammar rules.
- *Thesaurus* allows you to search for synonyms, antonyms, and other similar words for words you select.
- *Print preview* allows you to view individual pages of a document as they will appear in print.
- *Highlight* allows you to mark words, sentences, or paragraphs with a color notation for emphasis or as a reminder to examine the marked text.
- *Font* allows you to select the font as well as the size and the style of the font in your document.
- *Spacing* allows you to determine how much space will appear between the lines of your document (e.g., single space, double-space).
- *Convert case* allows you to easily adjust from lowercase to capitals or vice versa.
- *Insert date* allows you to easily place the current date in a document. This control function can also be set up to display the current date anytime the document is accessed.
- *Paragraph format* allows you to control the form in which your paragraph will appear, including line spacing, indents, and margins.
- *Merge* allows you to combine information from more than one document, such as merging a file of addresses with a template letter to end up with a separate letter addressed to each of the addresses.
- *Markup features* allow you to track changes made to a document. Many word processors allow you to view the document with or without the changes marked. Markup features also allow you to choose whether to accept the changes or revert to the original presentation.

These features can help you draft documents, but they are also important tools for revision, rewriting, and editing; they can assist you in producing professional-caliber documents. We discuss some of these features here but want to remind you that word processors are no substitute for careful and attentive reading and editing.

Highlighting Sentences

Word processors are great for sentence-level revising and editing because you can highlight single sentences, making them stand out from the rest of the text. Then as you proofread your documents on-screen, the highlight feature can help you remember which sentences you identified as needing attention.

Moving and Removing Text

Word processors allow you to cut, copy, and paste pieces of text within and between documents. This is a useful revising and editing tool because it allows you to move a selected piece of text away from the original document, where you can be especially attentive to it. Word processors also allow you to easily shift text around to reorganize and redesign your documents without rewriting large segments.

Inserting Visuals and Characters

See chapter 8 for more on visuals, including clip art.

Most word processors make it easy to add visuals, such as clip art or images, and special characters, such as accent marks or bullets, without having to reformat a document. This ability is useful when revising and also aids organization, allowing you to insert symbols or visuals to create such things as lists or cover pages. Most word processors, however, are designed primarily for manipulating written text, not visuals; therefore, they are more limited than desktop publishing software and image-manipulation applications.

Using Language Tools

Ode to Spell Checker
Eye halve a spelling chequer
It came with my pea sea
It plainly marcs four my revue
Miss steaks eye kin knot sea.
Eye strike a key and type a world
and weight four it two say
Weather eye am wrong oar write
It shows me strait a weigh.
As soon as a mist ache is maid,
It nose bee four two long and eye
can put the error rite
Its rare lea ever wrong.
Eye have run this poem threw it
I am shore your pleased two no
Its letter purr fact awl the weigh
My chequer tolled me sew
AUTHOR UNKNOWN

A spell-checker, grammar checker, and thesaurus make it easy to check for language problems. However, these tools can check for only certain kinds of problems. For example, spell-checkers look only for words that are *spelled* incorrectly, not *used* incorrectly. The "Ode to Spell Checker" poem makes this point with humor.

Spell-checkers rely on a set database of words—its dictionary. Often discipline-specific jargon won't be found in a spell-checker's database, and the spell-checker then notes those words as spelled incorrectly, even though they are not. Fortunately, most spell-checkers provide a feature that allows new words to be added to the dictionary.

Grammar checkers are equally untrustworthy because they cannot account for all conditions in which a sentence might be constructed. For instance, some situations call for passive verbs, but grammar checkers often flag these as grammatically incorrect. There is no better tool for revising and editing than actually knowing grammar rules. Grammar checkers can highlight particular sentences that you might want to revise or edit, but they should not be trusted completely. In fact, it is a good idea to turn off the grammar-check feature and rely on your own abilities as a writer.

Using a word processor's thesaurus is equally problematic. A thesaurus provides a database of synonyms and antonyms to help you make word choices. However, unless you already know the meanings of the words suggested by the thesaurus, you may inadvertently misuse them because their meanings might not be exactly the same. As much as possible, use words you know and recognize.

EXPLORE

Take a few minutes to use your word processor's thesaurus. Make a list of ten words, and search your thesaurus for each of them. How many synonyms do you find for each? Which seem nearly identical, and which seem only vaguely related? How many antonyms do you find for each? Which seem nearly opposite, and which seem only vaguely opposite?

Designing a Document

Word processors allow the manipulation of text and visuals as you revise and edit your document's layout and design. However, remember that word processors were designed primarily for working with words, rather than implementing design elements. Consequently, many word processors are limited in design abilities or perform design functions in specific ways. Word processors are fine for designing simple documents, such as letters or standard text pages, but many word processors cannot provide the flexibility needed for designing more intricate pages, like those in a set of technical instructions. Desktop publishing applications work with layout and design issues more effectively than word processors. Nonetheless, it is best to produce a document in a word processor first and then transfer it to a desktop publisher, to focus first on the writing and then on the design elements.

SUMMARY

- To ensure clarity and accuracy, writers spend time revising, rewriting, and editing their work before sending it to their readers.
- *Revising* refers to improving a document, making it better.
- *Rewriting* refers to altering a document to serve another purpose or audience.
- *Editing* refers to changing the contents of a document to meet the requirements of correctness.
- *Readability* refers to how easily an audience can extract information from a document.
- To revise, you should get an overview, trim the fat, make the document clear, be concise, consider style, pay attention to tone, look again at visuals, and be accurate and timely.
- When rewriting, you should work from large to small, consider your audience and purpose, and address ethical concerns.
- *Proofreading* refers to a detailed and careful reading of a document, looking for mistakes.
- *Editing* is the activity of making the needed corrections.
- When proofreading, you should consider formatting, typography, grammar, punctuation, spelling, capitalization, abbreviations, and electronic addresses.
- Most word processors provide numerous standard features that are of use in revising and editing.

CONCEPT REVIEW

1. What are the general differences among revising, rewriting, and editing?
2. Generally speaking, when should you edit a document?
3. What are some ethical implications of revision?
4. What are some questions you might ask while revising that will help you keep to the document's purpose?
5. Why would a workplace writer want to trim the fat from a document?
6. What are some approaches to revising for clarity?
7. What is concision, and why do writers need to revise for it?
8. What parts of your writing contribute to the style of your document?
9. What can tone convey in your documents?
10. Why do you need to be attentive to visual elements during revision?
11. Why is timeliness crucial to your documents?
12. What is the difference between proofreading and editing?
13. What should you look for when editing for format?
14. To what does the term *typography* refer?
15. Why is it important to edit for grammatical correctness?
16. When editing for punctuation, what are some key errors to look for?
17. What role can a spell-checker play in editing for spelling?
18. What are some strategies you can use to edit your own work?
19. What are the ethical implications of editing someone else's document?
20. What does it mean to rewrite a document?
21. What role does technology play in revising, rewriting, and editing?

CASE STUDY 1

The Million Dollar Comma

In October 2006, after eighteen months of legal battles, Canadian cable company Rogers Communication had to pay an additional $2.13 million Canadian (about $1,776,000 U.S.) to Aliant Inc. for use of Aliant's telephone poles—all because of a grammar mistake.

Rogers and Aliant had signed an agreement allowing Rogers to string cable lines along thousands of Aliant's utility poles in Canada's Maritime Provinces (New Brunswick, Nova Scotia, and Prince Edward Island) for a cost of $9.60 per pole per year. The contract, Rogers believed, set the price for five years. However, in 2004 Aliant notified Rogers that the price per pole would be increased and that the existing contract had been cancelled. Rogers had believed that the contract was firm and that they would have the option to renew for another five years when the contract was complete.

Citing a single sentence in the fourteen-page contract, Aliant argued that the contract could be canceled with a single year's notice. The sentence in question reads:

> The agreement shall continue in force for a period of five years from the date it is made, and thereafter for successive five year terms, unless and until terminated by one year prior notice in writing by either party.

The Canadian Radio-Television and Telecommunications Commission (CRTC) reviewed the contract and found that Aliant was correct in their understanding and had the right to terminate the contract. The validity of the decision to terminate, the commission decided, was enforced by the placement of the second comma in the sentence. If this comma had not been included, the right to terminate during the first five years would not have applied, and Rogers would have been protected. However, including the comma indicates that termination can occur during the first five years, leaving Rogers open to cancellation. "Based on the rules of punctuation, [the comma in question] allows for the termination of the contract at any time, without cause, upon one-year's written notice," the CRTC indicated. The Rogers lawyers argued that the intent of the contract outweighed the placement of the comma, but their argument did not succeed.

With the CRTC ruling supporting their position, Aliant could raise its rates as high as $28.05 per pole per year. Estimates based on negotiations between Rogers and Aliant suggest that Rogers will pay an estimated additional $213 million to use the poles. In response to the situation, Aliant officials said, "This is a classic case of where the placement of a comma has great importance."

While the Rogers/Aliant case was visible because of the extremely high cost to Rogers, it is not uncommon for disagreements to evolve because of the way sentences are punctuated. Imagine that you are responsible for a corporation that relies not only on contracts with other companies, but on numerous external documents like letters, reports, and proposals for developing and maintaining relationships with other companies. In the wake of the Rogers/Aliant case, you decide to establish rigorous policies regarding procedures for editing all external documents. What might those procedures be? How would you establish editing practices in your company? Using the information found in this chapter, write a document that outlines policies and procedures for careful editing of all external company documents.

CASE STUDY 2

Rewriting an FDA Report Summary

On July 23, 2007, Dr. Andrew C. von Eschenbach, Commissioner of the U.S. Food and Drug Administration (FDA), wrote a memo to the Deputy Commissioner for Policy and the Associate Commissioner for Science of the Department of Health and Human Services, thanking the deputy commissioner for work on a detailed report filed by the FDA Nanotechnology Task Force. The commissioner used the memo as an occasion to officially endorse the report and its recommendations and to authorize the deputy commissioner to "move forward with these recommendations, pursuant to FDA's good guidance practice (GGP) process (21 CFR 0.115), as appropriate."

This endorsement is an important step forward in nanotechnology research, but it is also important because it acknowledges the FDA's responsibility in monitoring nanotechnology components of FDA-regulated foods and drugs. It is an endorsement, then, that affects many people's lives.

Printed here is the executive summary of the "Report of the U.S. Food and Drug Administration Nanotechnology Task Force." Even though the report was written to the director of the FDA to affect policy, as a government document it reaches a larger public audience. You can find the entire report online at http://www.fda.gov/nanotechnology/taskforce/report2007.pdf.

Imagine that you have been assigned the task of rewriting the report with a different audience in mind and a different objective for the report. Imagine that you have been commissioned by the FDA to rewrite the report's executive summary to be accessible to an audience of first-year college students. Consider all of the points of rewriting that must be addressed: language, jargon, clarity, tone, style, accuracy, and medium. Then rewrite the summary that follows to serve as a stand-alone document, not as a summary, to convey to first-year college students an overview of the report.

Executive Summary

As other emerging technologies have in the past, nanotechnology poses questions regarding the adequacy and application of regulatory authorities. The then Acting Commissioner of the Food and Drug Administration (FDA) initiated the Nanotechnology Task Force (Task Force) in 2006 to help assess these questions with respect to FDA's regulatory authorities, in light of the current state of the science for nanotechnology. This report offers the Task Force's initial findings and recommendations to the Commissioner.

The report includes:

- A synopsis of the state of the science for biological interactions of nanoscale materials;
- Analysis and recommendations for science issues; and
- Analysis and recommendations for regulatory policy issues.

The report addresses scientific as distinct from regulatory policy issues in recognition of the important role of the science in developing regulatory policies in this area, rapid growth of the field of nanotechnology, and the evolving state of scientific knowledge relating to this field. Rapid developments in the field mean that attention to the emerging science is needed to enable the agency to predict and prepare for the types of products FDA may see in the near future.

A general finding of the report is that nanoscale materials present regulatory challenges similar to those posed by products using other emerging technologies. However, these challenges may be magnified both because nanotechnology can be used in, or to make, any FDA-regulated product, and because, at this scale, properties of a material relevant to the safety and (as applicable) effectiveness of FDA-regulated products might change repeatedly as size enters into or varies within the nanoscale range. In addition, the emerging and uncertain nature of the science and potential for rapid development of applications for FDA-regulated products highlights the need for timely development of a transparent, consistent, and predictable regulatory pathway.

The Task Force's initial recommendations relating to scientific issues focus on improving scientific knowledge of nanotechnology to help ensure the agency's regulatory effectiveness, particularly with regard to products not subject to premarket authorization requirements. The report also addresses the need to evaluate whether the tools available to describe and evaluate nanoscale materials are sufficient, and the development of additional tools where necessary.

The Task Force also assessed the agency's authorities to meet any unique challenges that may be presented by FDA-regulated products containing nanoscale materials. This assessment focused on such broad questions as whether FDA can identify products containing nanoscale materials, the scope of FDA's authorities to evaluate the safety and effectiveness of such products, whether FDA should require or permit products to be labeled as containing nanoscale materials, and whether the use of nanoscale materials in FDA-regulated products raises any issues under the National Environmental Policy Act.

The Task Force concluded that the agency's authorities are generally comprehensive for products subject to premarker authorization requirements, such as drugs, biological products, devices, and food and color additives, and that these authorities give FDA the ability to obtain detailed scientific information needed to review the safety and, as appropriate, effectiveness of products. For products not subject to premarket authorization requirements, such as dietary supplement, cosmetics, and food ingredients that are generally recognized as safe (GRAS), manufacturers are generally not required to submit data to FDA prior to marketing, and the agency's oversight capacity is less comprehensive.

The Task Force has made various recommendations to address regulatory challenges that may be presented by products that use nanotechnology, especially regarding products not subject to premarket authorization requirements, taking into account the evolving state of the science in this area. A number of recommendations deal with requesting data and other information about effects of nanoscale materials on safety and, as appropriate, effectiveness of products. Other recommendations suggest that FDA provide guidance to manufacturers about when the use of nanoscale ingredients may require submission of additional data, change the product's regulatory status or pathway, or merit taking additional or special steps to address potential safety or product quality issues. The Task Force also recommends seeking public input on the adequacy of FDA's policies and procedures for products that combine drugs, biological products, and/or devices containing nanoscale materials to serve multiple uses, such as both a diagnostic and a therapeutic intended use. The Task Force also recommends encouraging manufactures to communicate with the agency early in the development process for products using nanoscale materials, particularly with regard to such highly integrated combination products.

The guidances the Task Force is recommending would give affected manufacturers and other interested parties timely information about FDA's expectations, so as to foster predictability in the agency's regulatory processes, thereby enabling innovation and enhancing transparency, while protecting the public health.

CASE STUDIES ON THE COMPANION WEBSITE

Revising as Improvement: Walker Brothers' Acknowledgment of Order

Having been in business for thirty years, Walker Brothers Supply bases most of its communications on templates from early years. An owner noticed your technical writing expertise and asked you to revise some of the company's correspondence.

Editing and Revising for Tone and Clarity: Tandem Watercraft Customer Service

A customer complained that his vehicle's engine skips and stalls and contended that Tandem Watercraft should pay for the repairs. You normally write your own customer letters, but when a clerk in the office offered to type the letter for you, you accepted the offer and quickly dictated what you wanted to say.

Revise to Meet the Needs of Multiple Audiences: Linearization Test Station Upgrade ROI (Return on Investment) Report

After working on linearity test systems, you realize it would be more efficient if you replaced the motion control system. To obtain funding for the changes, you draft a preliminary ROI (return on investment) report. After reviewing your first draft, you realize that you must consider multiple audiences.

Rewriting a Negative Letter: Sleep Hollow Apartments Pet Violation

In this case you will assume the position of an assistant property manager for an apartment complex that serves primarily working families and college students. Recently, you have been made aware that tenants in one apartment have adopted a dog and are keeping it in the apartment in violation of the lease.

VIDEO CASE STUDY

DeSoto Global

Editing the Internal Report at DeSoto Global

Synopsis

Jennifer Nichols calls a meeting with Dora Harbin, Bill Kemble, and Tosh Takashi to review a recent internal report written by Bill and Tosh. The report discussed the feasibility of DeSoto Global providing navigation systems for Shobu Automobiles, particularly their luxury line of vehicles. Because Shobu's factories are overseas, the report was meant to discuss how DeSoto could deliver its products, account for quality control, and achieve profits. In addition, the report was supposed to outline any possible difficulties that may arise in working with Shobu. This report was particularly important because Nichols planned on using it as the basis of another report she was working on for DeSoto President Ty Webb. Dora Harbin and Jennifer Nichols read the report prior to the meeting but called the meeting because they had difficulty reading through it.

WRITING SCENARIOS

1. Think about the writing you have produced in the past, either in your workplace or for a class. Write down the process you used to revise and edit documents, focusing on the strategies you used to transform a draft into a final

polished document. Then analyze the strategies you *might* have used to make this process easier or more effective. How did your real revising and editing process differ from an ideal process? Did you use some of the strategies covered in this chapter? Did you use other strategies? What might you have done differently to make the process easier or more effective?

2. Revise, proofread, and edit the letter provided here.

Io Geotechnical Engineering
410 Albermarle Way
Suite 200
Aiken, SC 29802

March 19, 2007

Carmine Chiarelli
Central Engineering
98 W. Palimino Ave
Las Vegas, NV 89044

Dear Miz Chiareli—

Thank you for your recent application? For the position of Environmental Engineer with Io Geotechnical Engineering.

Unfortunatley, do to recent budgte cut, Io Geotech is no longer able to add personell to our engineering staff which is a shame at this time. We were all quiet impressed with you're resume and you're accomplishments with Central Engineering. In any other circumstance, we would have eagerly sought to add you too hour team of professional engineers. It was determined recently that engineers of your caliper will be harder and harder to come by in the future, so it is difficult for us to not accept your application, but we are unable to hire you because of our budget cuts.

We do encourage you to keep an eye out for future Io Geotech job searches as we hope to alleviate the budget crisis and utilize the opportunity to add to our staff.

Again, thanks you for considering us, and we wish you good luck in finding employment some where other than where you work.

Sincerely,

Janine Rowe

Jannie Rowe

3. Develop a working habit of seeking out collaborators to review your writing each time you produce a document and before you release it to its intended audience. Also, always give your full professional attention to documents that

others have asked you to revise and/or proofread. To initiate these two habits, draft a letter to a classmate explaining how you'd like him or her to edit your next graded work in this class.

4. Locate a person who works in a field you are considering pursuing as a career. Ask about the kinds of writing and the kinds of revising, rewriting, and editing the person does on a regular basis. Then write a Real People, Real Writing segment that focuses on the kind of revising, rewriting, and editing that your interviewee describes as part of regular workplace writing.

5. Locate a professional editing service—you should be able to find many online. Find out what kinds of services are offered, and compare those to the guidelines offered in this chapter. Then write a short analysis of whether the editing service would be a worthwhile investment for workplace writers.

6. Bill and Ted are both sales analysts with Bacon, Bacon, and Bacon Food Distribution, the largest wholesale whole-foods distributors in the Southwest. Bill and Ted have a friendly working relationship, but there is a competitive edge between them. Recently, Jill Swanson, the senior sales analyst, has announced her retirement, and both Bill and Ted are hoping to be promoted to her position.

   One Friday afternoon Bill asks Ted to review a report he has just finished. In the past each was comfortable asking the other to review a document before it was sent on. But over the weekend Ted ignores Bill's report until late Sunday night. Then, feeling obligated to look at it, he gives it a quick cursory reading, skipping large parts and offering little for Bill in terms of revision. Consider the ethical issues involved in each of these scenarios:
   a. Ted intentionally did not put much effort into revising the document, hoping that a sloppier document from Bill would influence Jill Swanson's decision about her replacement.
   b. Ted simply shortchanged Bill's document and has consistently done so throughout their working relationship.
   c. In reading Bill's report, Ted noticed glaring errors but decided not to point them out, hoping to better position himself for the promotion.
   d. Ted simply allowed too much time to pass before getting to Bill's document and then was too tired to give it careful attention.

7. Locate a document that contains writing, even just one sentence, that might be considered unethical or deceptive. Then rewrite the document as necessary to more ethically represent the information.

8. Based on what you have read in this chapter, write a short description of what you consider to be your ethical and professional responsibilities toward your readers in terms of revising and editing your documents. Does who your readers are—a supervisor, clients, a general audience—affect your ethical approach to revising and editing?

9. Imagine that you work in the personnel department at North Beach Turtle Rescue and a colleague, Lawrence Rodriguez, has asked you to edit the document that follows before he mails it. Read the letter carefully, and list the corrections (by line of text) that Lawrence should make. When you are finished, examine the list at the end of this chapter to see whether you identified the appropriate corrections.

North Beach Turtle Rescue
1201 Atlantic Blvd.
Melbourne Beach, FL
32951

March 23, 2005

Mary Anne Cavanaugh
2301 Killebrew Drive
Bloomington MN 55425

Dear Ms. Cavanaugh:

Thanks you for sending your resume an cover letter in response too hour advertised position of see turtle biologist. Unfortunately, our Bored of Directors requires that we fill the position with some one who has extensive experience working with sea trutels and who holds a a degree in marine biology. Although we are certain that the two years you have spent caring for your daughter's box turtle has proven to be valuable experience (both for you and for Speedy), we are unable to offer you the position based solely upon that experience.

we do hope, Ms. Cavenough, that you will continue to car for Speedy; and that your enthusiasm for protecting sea turtles continue. We wish you the best of luck, and thank you, again, for submit your rather interesting application.

Respectfully,

Lawrence K. Rodriguez,
Personnel Officer

10. In what ways do design and layout intersect with revision and editing? Write a memo to your instructor detailing the relationship between design and revision.

11. Open the word processor you most frequently use, and look through the tool bars and drop-down menus to see what editing features are available. Make a list of all of those features, identifying whether you have used each in the past and what role each plays in the editing process.

Dinition: Hurricane

A hurricane is a severe type of storm. A hurricane is a type of tropical cyclone. Cyclones refer to low-pressure air systems that form in the tropics. Cyclones are driven by heat, which is generally produced by warm sea water. Hurricanes most often move from east to west across the ocean. Easterly trade winds push hurricanes from the east to the west. Hurricanes can be extremely severe storms. Their winds can be very fast and can cause great damage or create rough seas.

Hurricanes are created by a number of factors. First, hurricanes may be influenced by preexisting weather disturbances. Warm ocean water also fuels hurricanes. Moist air contributes to hurricanes. Light winds above the hurricane may also help a hurricane form. When a number of these conditions occur at the same time or simultaneously for a long period of time, they may cause a hurricane to form.

These kinds of tropical depressions are classified in three categories: tropical depressions, tropical storms, and hurricanes. Atropical depression is an identifiable, organized storm system with sustained winds of no more than 38 miles per hour. A tropical storm is an identifiable and or ganized storm system with winds of 39 to 73 miles per hour. A hurricane is an intense, identifiable, and organized storm system having winds of more than seventy four miles per hour. Hurricanes are then categorized by wind speed. The Saffir-Simpson Hurricane Scale is used to categorize hurricanes. A Category 1 hurricane has winds of 74–95 mph. The Category 2 hurricanes have winds of 96–110 mph. A Category 3 hurricane has winds of 111–130 miles per hour.Category four, 131–155. And finally, a Category five hurricane would have winds that exceed 155 miles per hour. The following charts the storm categories:

Category	Wind Speed
Tropical depression	Up to 38 mph
Tropical storm	39–73 mph
Category 1 hurricane	74–95 mph
Category 2 hurricane	96–110 mph
Category 3 hurricane	111–130 mph
Category 4 hurricane	131–155 mph
Category 5 hurricane	Greater than 155 miles per hour

12. You have just been hired as a researcher/writer for a local news station. As your first duty, you are required to update and revise a number of informative documents that news correspondents might read before delivering broadcasts on those subjects. The first file you plan to revise is the one developed for the station's weather correspondent—who is not a meteorologist but has been broadcasting the weather for the past three years. Because the hurricane season is quickly approaching, you decide to edit the informative document about hurricanes first. The document included here is what you find in the file; revise it accordingly.
13. Some word processors, like Microsoft Office Word, use markup features to keep track of changes you have made to documents as you revise and edit. Why might such a feature be useful to a workplace writer? How might you use such a feature? Write a memo to your instructor explaining how the markup features on your word processor work and how they might be of use.

14. What capabilities does your word processor have to integrate visuals and text? Can you insert visuals? Can you create original visuals? What manipulation tools does your word processor provide, and how might they assist you in revising visuals in your documents? Write a short description of what your word processor provides and an assessment of how those tools might be applied to your revision and editing.

CORRECTIONS FOR WRITING SCENARIO 9

Line 8: Insert a comma after Bloomington.
Line 10: Remove the *s* from *Thanks.*
Line 10: Change *an* to *and.*
Line 11: Change *too* to *to.*
Line 11: Change *hour* to *our.*
Line 11: Change *see* to *sea.*
Line 12: Change *Bored* to *Board.*
Line 13: Change *some one* to *someone*.
Line 14: Change *trutels* to *turtles.*
Line 14: Delete *a.*
Line 19: Capitalize *we.*
Line 19: Change *Cavenough* to *Cavanaugh.*
Line 19: Remove the extra space before *that.*
Line 19: Change *car* to *care.*
Line 20: Delete the semicolon.
Line 21: Add *will* before *continue.*
Line 21: Remove the commas before and after *again.*
Line 22: Change *submit* to *submitting.*
Line 22: Delete *rather.*
Line 24: Delete the comma.

11

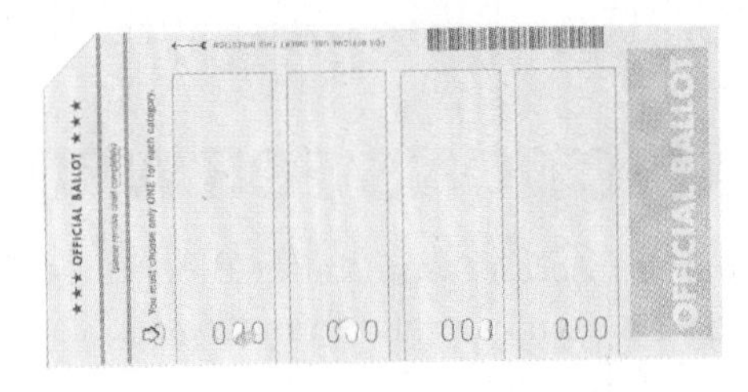

Usability

CHAPTER LEARNING OUTCOMES

After completing this chapter, you will be able to do the following:

- Understand the reasons for **usability testing** on documents and websites
- Know how usability fits into **writing processes** and the **problem-solving approach**
- Know that each usability test must begin with a **planning session**, when test goals are clearly understood and articulated
- Recognize that **participants** in any usability test must be representative of the specified users of the document or website
- Know that participants must engage in **tasks** similar to those a real user would complete
- Know that usability testers must **observe and record** what the participants do and how they react
- Know that usability testers must **collect and analyze the data** received from the tests and change the document or website to make it more usable
- **Collaborate** to conduct a usability test of a document or a website
- Define the **parameters** of a usability test
- Define the **user profile** for a usability test
- Establish the **issues and goals** of a usability test
- Write a valid usability **test plan**
- Know how to **screen and recruit** participants for a usability test
- Report the results and **make recommendations** based on the results of a usability test
- Prepare and conduct usability tests in an **ethical and professional** manner

DIGITAL RESOURCES

On the Companion Website www.prenhall.com/dobrin:

- Case 1: Developing a Usability Test: Bruce Wayne-Edwards and Recruiting Subjects
- Case 2: Reading the Manual: Cory Washington's Technical Documentation
- Case 3: Developing a Usability Test: Nia Barazu and the Game Interface
- Case 4: Making Decisions through Usability Testing: The Tucson Emergency Response System
- Video Case: Usability Testing of the GlobeShare Wireless Customer Support Website
- PowerPoint Chapter Review, Test-Prep Quiz, Exercises and Activities

REAL PEOPLE, REAL WRITING

MEET JEFF RODANSKI • Chief Creative Officer at Slice of Lime

What types of writing do you do in your workplace?

Our company mainly provides design and architecture for corporate websites, although we also help companies develop print, wireless, and illustrated documents. I often work with preexisting websites and deal directly with our clients to figure out who their customers really are and how they are going to use the site. In order to learn more about who these customers or users are, I also send out surveys and forms, as well as create newsletters and promotions to attract new customers.

How do most of your clients envision their users or customers?

Many of our clients have great ideas and the money to spend on new websites but aren't sure how to get started, so we work with them from beginning to end, from initial sketches of websites to full-blown marketing research, which includes analyses of the company's competitors as well as analyses of potential customers or users. In these cases we spend a great deal of time figuring out who these potential users are and how the site can work for them.

Do the end users play an important role in how you design websites?

Absolutely. There are two kinds of designers: those who design for themselves and those who design for the client and the client's customers, which I consider myself.

How do you and your company provide usability testing for the documents you produce for clients?

This is an important part of the design process. We design some software so that potential users can access it online to mimic how it would be used regularly. Since they're using it through this portal, we can track how they use it and record their experiences with it. It is, essentially, an online usability testing area where we'll allow about twenty potential customers to hammer away on our design for a couple of weeks. Afterward we can go back and ask these people questions and have them fill out surveys and questionnaires that help us redesign the product based on their experiences with it. It's crucial to do this kind of testing before the product design is finished and put out there.

What advice about workplace writing and usability do you have for the readers of this textbook?

In document design—particularly with websites—it's absolutely crucial. You just can't assume that users will intuitively figure out how to use the documents you create. Your readers might be surprised to discover how often people misinterpret documents and websites and use them incorrectly. Of course, even if the errors are their fault, users will still attribute the problems to bad document design. Usability testing simply helps make sure that documents are as foolproof as possible because users' inability to navigate documents typically leads to loss of money for and trust in the company.

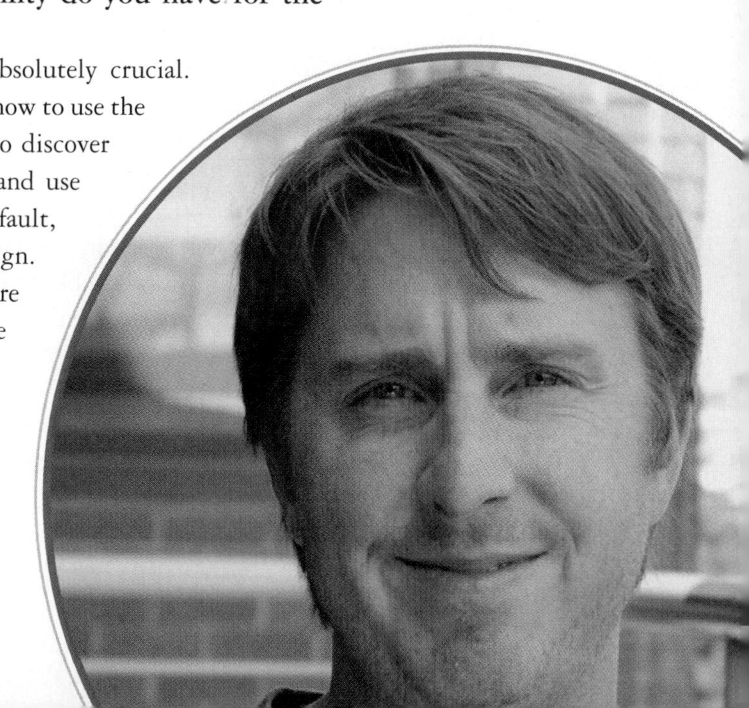

INTRODUCTION

In workplace writing, usability refers to how successfully and satisfactorily a person uses a product, document, or website. Usability refers not only to specified users (i.e., the people most likely to use the product, document, or website) *completing* a task successfully but also achieving their particular goals *effectively* and *efficiently*. Sometimes a person may learn to use a website successfully, but that success is no guarantee that the person's own needs and desires have been satisfied. For instance, a customer who wishes to purchase music online might eventually figure out how to successfully complete the task; however, this success does not mean that this online music store's website is highly usable or satisfying. Customers might be so dissatisfied with their experiences that they vow never to use the website again.

Usability encompasses much more than just success. A product, document, or website should have many other traits:

- *Easy to learn*—Users should be able to quickly start working on a task.
- *Easy to remember*—Users should be able to return to the document after some time without having to learn things all over again.
- *Efficient*—Users who learn how to accomplish a particular task should then be able to maintain high levels of productivity.
- *Satisfying*—Users should like using the product, document, or website.
- *Error free*—Users should be able to accomplish tasks free of errors, and users who do make errors should be able to recover from them easily.

Usability focuses on the user's interaction with a product, document, or website and is tied to the user's wants, desires, and needs. Therefore, usability is always linked to user-centered design. Let's look at a real-life example. The 2000 presidential election was one of the closest elections in American history—and perhaps one of the most controversial as well. On election day, November 8, 2000, national attention was drawn to the state of Florida for several reasons. First, there was a surprisingly large number of votes for conservative candidate Patrick Buchanan in an area of the state known for its Democratic-leaning voters. Second, a large number of Florida voters complained about difficulty using the ballots. The 2000 election serves as an excellent example of problems in usability that affected not just Florida voters but all Americans.

In this particular case the design and usability of ballots caused national debate and furor. In Florida, as in many other states, citizens cast their votes by punching holes in a card with a stylus, a penlike object used to perforate the ballot. However, because of the large number of candidates' names on the ballot and because of the complex layout, the names on the presidential ballot were alternated left and right with a single column of punch-card holes in the middle between the rows of names. Al Gore was second in the left column, whereas Pat Buchanan was first in the right. But the hole for Gore's name was number three, and for Buchanan's name, number two. Thus, although the Democrats were listed second in the left column, punching the second hole would have cast a vote for the Reform party (see Figure 11.1). The arrows that were meant to help were paired with numerals that had no apparent connection to the parties or the holes. And complicating the situation further was the fact that on other parts of the ballot—for senator, for instance—there was only one column, with the first name corresponding with the first hole, the second name with the second hole, and so on. As a result, the ballot card proved not only complicated but also inconsistent.

Such design problems raise numerous questions about the usability of the Florida ballot. A number of other factors could also have contributed to voter mistakes on the ballot:

- Nervousness
- Poor eyesight
- Cultural issues (i.e., not understanding the arrows)
- Spatially challenged voters
- Ballot changes (i.e., the first ballot to use double columns)

But what really is at issue here is the fact that proper usability testing could have prevented at least some of these problems and, perhaps, much of the controversy surrounding the 2000 presidential election.

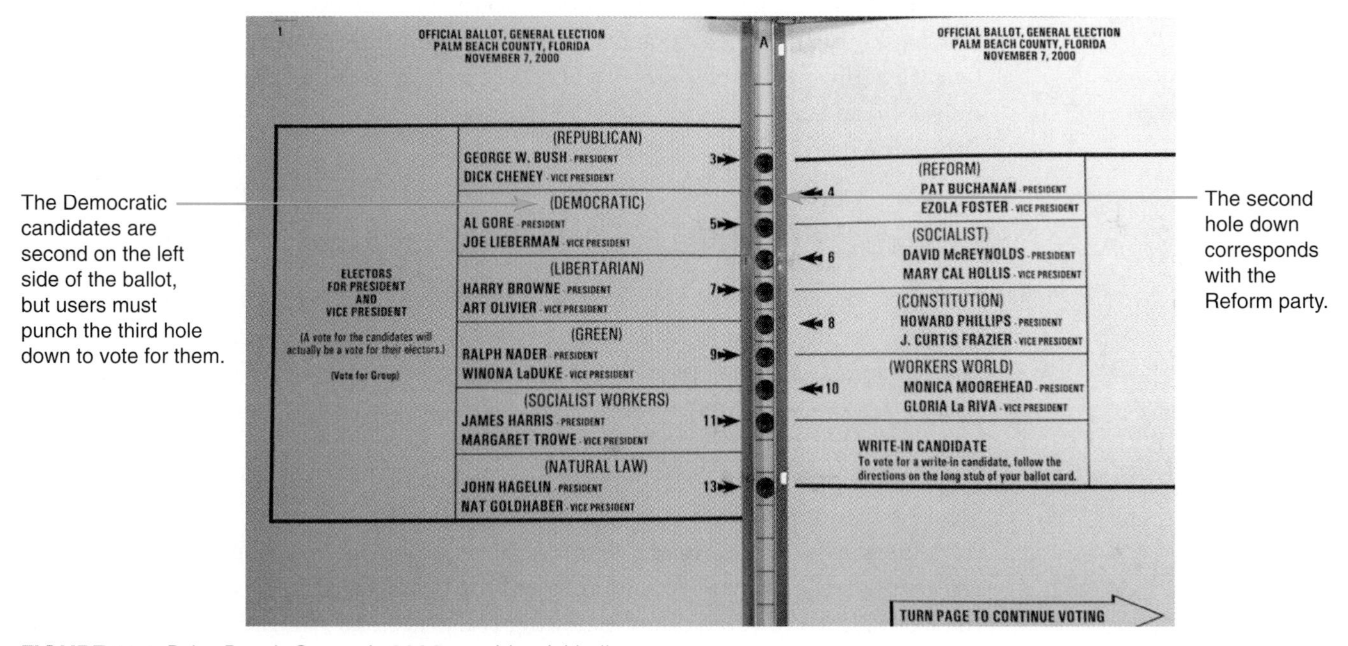

FIGURE 11.1 Palm Beach County's 2000 presidential ballot

ANALYZE THIS

Take another look at the photo of the Florida ballot in Figure 11.1, and consider how it functions as a document with specific kinds of users (i.e., voters). Discuss with your classmates how this document might have been designed differently so that it could have been used more easily and effectively by Florida voters.

With the rise of the Internet and a digital society, usability testing on websites and software programs is increasingly crucial. However, the Internet brings up a host of usability concerns that traditional documents do not.

- *Users may be harder to identify.* Because so many people use the Internet and their numbers are increasing every day, it may be difficult to establish a precise user profile. And since many people surf the Web in ways that are hard to describe, it becomes even more complicated to figure out who the real users of a website are.
- *Websites do not work in a linear fashion.* Traditional documents like books, magazines, and newspapers read from left to right, top to bottom, start to finish. With websites, however, users are given choices about how and where to proceed, and links to different pages are not always apparent or well described.

- *Websites are apt to use diverse color schemes, graphics, and images.* The images may even be moving or animated. As a result, site structures can become obscure and confusing, increasing the likelihood that a visitor's experience will be problematic and frustrating.
- *Websites use untraditional writing styles.* Most traditional documents are written in complete sentences, with the level of formality gauged to the particular audience. Because websites often use phrases or single words, users—particularly transnational users—may misunderstand and misuse the site.

For more about writing for transnational audiences, see chapters 5.

The more that people use and experience websites, the more proficient they become in using them. However, as different websites cater to different audiences, it will be increasingly important that web designers understand these users and design sites that are highly usable for them. Such concerns become even more intense as PDAs, text-messaging cell phones, and increasingly complex e-mail systems permeate our lives. In addition, as web pages—and other kinds of workplace documents—focus on transnational audiences, users and designers will need to attend to issues of usability tied to the needs of transnational users.

USABILITY IN THE WORKPLACE

Usability can affect either external or internal users. That is, a worker can design a product, document, or website for users outside the company or organization (as Jeff Rodanski explains in the Real People, Real Writing section of this chapter) or for users within the company or organization. To make sure that products, documents, or websites are usable, designers and writers engage in usability testing with volunteer participants in either artificial or real-life environments. It might seem that such testing would be required mainly during the development of products such as appliances, computers, toys, entertainment systems, automobiles, and the like. Certainly, in such cases designers need to make sure that specified users can achieve their goals in using these products. However, usability testing is also necessary during the design of many written documents and websites, and these are the main concern of this chapter. Such testing allows writers and designers to catch problems in their documents and websites before the final version is produced.

For instance, an information technology specialist at a large company might be asked to help workers in the finance division use the company's intranet and e-mail systems. She would probably not just send out a manual but would test it with users to see whether it works. It would be a huge waste of time and money to write and print hundreds or even thousands of manuals, only to find that the manuals are not effective for employees.

As you may have noticed, the Review segment of the problem-solving approach includes usability:

Review

- Test the usability of the document
- Solicit feedback and response from peers and colleagues
- Revise or rewrite the document based on feedback
- Edit the document to ensure correctness

Even though usability testing is often part of a Review phase, there are times when it is done *after* a document has been put into use. Sometimes usability problems do not surface until a document has circulated among real users and complaints are registered. In such cases usability testing can help writers and designers create new and effective documents to better suit user needs.

Not every workplace document must be usability tested; it would be a tremendous waste of time and resources to attempt that. However, as you make your way through this chapter and the rest of the text, you will become more aware of which workplace documents need usability testing and which do not.

IN YOUR **EXPERIENCE**

Stop for a moment to think of the products, services, and websites that you use on a daily basis—for instance, computer hardware and software, course registration forms or other documents, campus maps, library services such as online databases or directions to materials on the shelves. Pick one or two of these products or services, and explain how they effectively or ineffectively cater to their specified users.

USER-CENTERED DESIGN

Usability testing is important because not all users are alike; different kinds of people have different feelings about and skills with websites and other documents, particularly those that implement emerging technologies. Usability testing requires putting users at the forefront, and following the PSA is a good way to do so when thinking about developing and testing documents. Usability testing requires a good deal of planning, researching, drafting, and reviewing, but there is no one particular way that all usability testing takes places. Nonetheless, there are a few general characteristics of all usability tests:

- The ultimate goal of the test is to improve the usability of a document or a website for its specified users.
- Each test begins with a planning session, when test goals are clearly understood and articulated.
- Participants in any usability test must be representative of the specified users.
- Participants must engage in real tasks, similar to those a real user would complete.
- A team of usability testers must observe and record what the participants do and how they react.
- This team must analyze the data from the tests and change the document or website accordingly.

To reiterate, usability testing is *not* exclusively meant to make sure a document or a website works properly or to verify that users are able to complete certain tasks. Instead, usability testing assesses whether representative users can smoothly, effectively, and satisfactorily interact with a document or website so that their goals are met.

ANALYZE THIS

Each year google.com is used by millions of people who are searching for various kinds of information on websites around the world. In fact, Google is understood by many experts to be the most efficient, most usable search engine on the Internet. Visit the google.com site. What makes it so usable and perhaps satisfying for both newcomers and experienced users? Is there anything about the google.com website that might be difficult or unclear for users and that would make their use of this search engine problematic and unproductive? Discuss the google.com website for a few moments with classmates.

WHERE USABILITY TESTING TAKES PLACE

Usability testing typically takes place in one of three different areas: inside a lab, outside a lab, or in the field. Where it takes place is typically determined by a number of factors, including what kind of document is being tested, whether the document is already being used, and how much money is available to spend on testing. Good usability testing is generally done early and often in the design and development of a document; good testing does not necessarily require expensive labs to be effective.

In a Lab

The main benefit of usability testing in a lab is that users can be asked to complete certain tasks in a controlled environment. In many cases the lab is made up of two rooms, one in which users engage in tasks that are part of the test and one in which a team of test administrators watches through a glass window or on video monitors and records information about the users' actions. In most cases the participants and administrators have microphones that allow them to interact with one another if necessary.

Labs differ according to the needs of particular tests; however, most labs have video cameras that record the actions of participants in the tests and allow researchers to review those actions. High-tech labs such as those described here are not usually present in or near most workplaces. Instead, they are owned and operated by companies whose sole specialty is testing the usability of products, documents, and websites. Such testing is often—though not always—contracted out to these companies.

EXPLORE

Search on the Web for three companies whose sole specialty is usability testing. Who are these companies, and what specifically do they do? By browsing through their websites, find out what kinds of products, documents, or websites they test, and find out how they actually conduct usability tests.

Outside a Lab

Usability testing inside a lab has its advantages, but labs are not always necessary for quality usability testing and may not even be available. Conference rooms and offices can certainly suffice for many usability tests. One of the authors of this book once was a participant in a usability test that took place in a tent in the middle of a parking lot. In short, usability testing can be accomplished with little more than a room large enough to hold a participant and an observer (or a team of observers) and the equipment needed to conduct the test. To test a software program, for instance, there would need to be enough room for a chair, a desk, and the computer that holds the software; observers would need only pens and paper to record their observations.

In the Field

Usability testing also takes place in the field, which could be a real office, a home, or another workplace environment. Testing in the field takes place after a product, document, or website has been released and is being used in real-life situations. Field testing differs greatly from testing in a lab or other artificial environment because observers cannot control the environment; they cannot remove themselves from the environment by hiding behind a one-way mirror, for instance. In addition, participants can easily fall prey to the typical workplace distractions, and it may be difficult to test more than one or two people because many companies do not want employee time used for usability tests.

However, testing in the field has its advantages as well. Observers can see users working in their actual environments and can consider things like workspace, access to manuals and other documents, kinds of equipment being used, quality of lighting, and so on. Real workplace environments give observers more factors to consider than can be achieved in an artificial lab environment.

Usability testing inside or outside a lab or in the field is expensive in terms of both money and employee time. However, it often makes up for its cost by decreasing the number of help calls from users. Users who are satisfied with a company's products, documents, or website create an advantage for the company and enhance its reputation.

PLANNING USABILITY TESTS

Usability tests require important steps and considerations, much of which takes place long before the actual tests are administered to participants.

Establish a Team

Like other tasks in workplace environments, good usability testing is usually accomplished collaboratively. Because usability testing requires that administrators observe the actions of participants closely and keenly, it makes sense that more eyes can identify more issues and problems to be diagnosed and solved. There is no ideal number of members on a usability team—the size of the project and the location of the facility partly determine how many members are included. A small lab, for example, might not hold more than one or two team members.

What is most important is that the members of the team are all stakeholders in learning more about the usability of the document or the website and that different viewpoints and perspectives about that document or website are represented. For instance, if a company wants to test the usability of a new website that provides online banking for households, the team might best be comprised of not just technical experts but also sales and marketing personnel. This mix provides a well-rounded and dynamic group of individuals who will likely see and understand problems differently.

Define the Parameters of the Test

A usability team wants to learn as much as possible about the usability of a document or a website. However, usability testing can rarely test everything. More often than not, limited amounts of time and a limited budget restrict how much testing gets accomplished. Therefore, usability teams have to decide which particular issues and concerns the test will cover.

When a document or a website is already on the market, usability testers may have a head start in defining the central issues in their testing. For instance, they may have access to surveys, data from focus groups, and information logs from technical support services, all of which may provide some initial insight into user problems, concerns, and preferences.

Because usability testing is not done in a vacuum, real-life constraints on testing help define the parameters of usability testing:

- What is the time frame for the test?
- What is the budget for the testing?
- Where will the testing take place?
- Who is the audience for the test results?
- What other kinds of information are known before testing begins?

Although each of these questions influences how a usability test will be conducted, one of the most important concerns is the audience for the test results. Many times a usability team makes a list of topics to learn about in the usability test, and such a list is usually much longer than can be reasonably addressed in one test. Consequently, the team must trim the list based on the audience for the test results. For instance, a team testing the usability of new features on an e-mail service would focus on one set of concerns if the results were to be reported to marketers and on a different set of concerns if the results were to be reported to computer technicians. People in the marketing department might be most interested in how users arrive at the e-mail's homepage or how they compare the service to that of a competitor, whereas engineers might be more interested in the ease of navigation or the ease with which users recover from errors.

Define the User Profile

Defining the user profile for a particular document or website means identifying the kinds of people who will typically use it, considering demographic information like age, education, gender, or job type. The more specifically and narrowly defined the user profile, the better the results and data pool. However, many products are used by such a wide variety of individuals that there is no easy way to define typical users. In such cases it is important to define different subgroups of users, based on categories such as the frequency with which users access the document or website, the expertise of the users (whether beginners or experienced users), or perhaps the learning styles of the users. Other factors to consider include the physical characteristics of typical users, their attitudes or motivations, their levels of stress, the places where the document or website will most often be used, and the kinds of training users have received before using the product.

IN YOUR **EXPERIENCE**

Think of a website that is central to your life. In what ways might you consider yourself a typical user? If a team wished to set up a usability test for this website, what might an initial user profile look like? Create a detailed list of the characteristics of the typical user.

Establish the Issues and the Goals of the Test

The issues that arise in a usability test are the concerns that team members have about the document or website being tested. However, establishing the main issues is not always easy because usability teams are typically comprised of members with differing concerns. Nonetheless, teams must decide on the central issues that will guide their testing plan. For instance, if a team wants to test an online course registration website for college students, the team might decide on main issues like these:

- Will student users be able to quickly and easily find the online registration site from the college's homepage?
- Will student users be able to easily navigate the site in order to register for upcoming courses?
- Will student users be able to easily and quickly recover if they make an error and register for the wrong course?

Goals are responses to the issues that a team might initially raise; goals are what the usability team hopes and expects a user to be able to accomplish. For example, goals for the online course registration might be these:

- Student users will be able to reach the online course registration quickly, with only one mouse click from the college's homepage.
- Student users will be able to identify the necessary registration tools and register for courses in ten minutes or less.
- Student users will be able to review their course selections multiple times before submitting them. Students can change their course selections as many times as necessary to ensure that errors are easily and quickly rectified.

Goals ultimately shape the usability test and clarify what to test, who to test, and how to test. Goals also provide definitive information about how the results of the test can be measured. For instance, our sample goals do not simply ask whether the online course registration system is easy or user-friendly. Instead, goals provide a concrete and quantifiable means to measure test results: Can students get to the registration page in one mouse click? Can they accomplish their task in ten minutes or less? Can they always make changes to suit their needs and preferences?

Write the Test Plan

Although a usability team may plan for the test long in advance, the test plan itself is the document that puts everything down in writing. Keep in mind that the members of the team are not the only ones who have an investment in the usability test. For instance, managers or other sponsors of the test may not be officially on the team but may wish to see the kinds of ideas and decisions the team has come up with before the actual test takes place. A well-documented test plan allows others to see what the test will accomplish; it is the outline of the test and its procedures. Thus, it clarifies the issues and the goals that make up the test, the process for conducting it, the place where the test will occur, the time frame for the test, the participants, and any resources needed to complete the test and evaluate its results.

There is not just one format for all test plans. The format is contingent on the audiences, the size of the usability team, and the depth and detail of the test itself. Some test plans may be informal e-mail documents that capture only major points of the test, whereas other test plans may be formal reports that document every aspect of the planning process. A small company testing the usability of its employee intranet might require only an informal test plan circulated to a few team members and a manager, whereas a multinational bookstore testing the usability of its online checkout system might require a lengthy, formal test plan circulated to all team members as well as to high-level executives and other company officials.

All test plans, regardless of formality, describe the purposes and processes that will make up the test and include these components:

- *Title page*—lists team members, test sponsor, and relevant dates
- *Table of contents*—provides an easy means for readers to sort through a longer test plan (e.g., one that is more than ten pages)
- *Statement of purpose*—summarizes the test plan and the overall purpose of the usability test
- *Statement of test issues and goals*—lists the problems and concerns the test seeks to address as well as the objectives it hopes to fulfill
- *Description of test environment*—provides a summary of the test location

- *User profile*—describes the specific users of the document or website (and thus the participants in the test)
- *List of participant tasks*—reviews the various test sessions, including the dates, issues covered in the sessions, number of participants, tasks the participants perform, and length of each session
- *Methods of evaluation*—shows types of data to be collected by the team and the methods of collection

The degree to which any usability test succeeds or fails is contingent on the thoroughness of the test plan. Diving too quickly into usability testing without a solid plan is a surefire way to yield few helpful results.

Recruit and Screen Participants for the Test

It is now time to look for participants that fit the user profile(s). The most common way to recruit participants is by contacting them directly or hiring an outside consulting firm to recruit them for you. Recruiting participants directly often involves one or more of the following activities:

- Advertising in a newspaper
- Searching through customer lists from the company's sales and marketing departments
- Contacting professional organizations that focus on the targeted user groups
- Scanning a preformed database of qualified users

Most recruiting efforts require that participants be offered some form of compensation for their time and energy. Depending on the length and depth of a test, compensation might include money or gift certificates or in some cases small gifts or even food.

Recruitment also involves screening participants, usually with a screening questionnaire to ensure that a potential participant fits the user profile. Such a questionnaire might identify demographics (e.g., age, gender, educational level) and might focus on how much skill or proficiency the potential participant has in using the document or website, perhaps how long they have used it and how frequently. Other questions might ask respondents to identify the quality of experiences they have had with the document or website. Figure 11.2 gives an example of a screening questionnaire from a team at Washington State University–Vancouver, which was trying to test and improve the usability of the library's online public access catalog (WebPac).

Questionnaires themselves should not appear complex or intimidating but should seem easy and straightforward. The questions should not allow for any ambiguity but should allow the usability team to quickly and easily assess whether a potential participant fits the user profile. The questionnaire in Figure 11.2, for instance, asks potential participants to answer only ten simple questions. Some questionnaires are much more lengthy and time-consuming, depending on how much information is needed to determine whether a person accurately represents a typical user.

ANALYZE THIS

Look at Figure 11.2 again. Given the purpose of a usability test, why do you believe this questionnaire includes these particular questions? Explain why each question is relevant to the user profile for the library's online catalog.

Testing the Usability of a WebPac: Screening Questionnaire

Thank you for your interest in improving the WSU library's online catalog. In order for us to collect information from a representative group, we ask that you fill out this brief questionnaire. Participants will be chosen from among those who volunteer, but not all volunteers will be chosen to participate.

1. How much experience do you have with a computer?
_____ None
_____ One to two years
_____ More than two years

2. Please indicate your status at WSU:
_____ Undergraduate student
_____ Graduate student
_____ Postgraduate student
_____ Faculty member
_____ Staff member
_____ Other

3. Please indicate your age category:
_____ 18–22
_____ 23–30
_____ Over 30

4. Please indicate your gender:
_____ Female
_____ Male

5. What is your preferred learning style? (You may check more than one.)
_____ Trial and error
_____ Consult with others
_____ Read documentation

6. What is your major (if relevant)? _____
7. Are you employed by the library? Yes _____ No _____
8. Which of the following best describes your WSU library experience:
_____ First-time user
_____ Occasional user (once a month)
_____ Frequent user (at least once a week)
_____ Participant in a library instruction class

9. Which of the following best describes your Internet access/use (you may check more than one):
_____ World Wide Web
_____ Telnet
_____ E-mail

10. How often do you use the Internet?
_____ Never or rarely
_____ Occasionally (once a week)
_____ Frequently (once a day)

FIGURE 11.2 A screening questionnaire

CONDUCTING USABILITY TESTS

In addition to what, who, when, and where to test, usability teams must think carefully about *how* to test. They will be asking participants to complete certain tasks that must reflect the kinds of tasks that typical users are likely to perform. Figure 11.3 gives a partial list of the tasks the usability team from Washington State University–Vancouver put together for its student participants.

IN YOUR EXPERIENCE

Which campus documents or websites have you had difficulty understanding and using in the past? What particular problems did you have, and how would these documents or websites benefit from usability testing?

Testing the Usability of a WebPac:
Tasks List

Does the Washington State University–Vancouver library have a copy of *Gone with the Wind*?

Does the Eastern Washington University library have a copy of *Gone with the Wind*?

Which of the WSU libraries have *Hamlet* by Shakespeare?

Do WSU libraries own the following: Albertson, John, "Supermarkets in the Northwest, "*Advertising Age 68:* 12–25 (1997)?

Find material by IBM in the libraries.

Find books by Laurie Garrett at any WSU location. Look at the detailed information for one of these books.

Find books written about Toni Morrison.

Do WSU–Pullman libraries have a book called *Guns, Germs and Steel* by Diamond?

Do WSU–Pullman libraries have a copy of the periodical *Audubon*?

What years does the Owen library have of this periodical?

Did the Owen library receive the Sep/Oct 1997 issue?

Is volume 94 (1992) of *Audubon* on the shelf at the Owen library?

Do the WSU libraries own the government document *Domestic Price Directory*? Locate the titles of other items shelved close to this government document.

Does the library subscribe to a journal with the international standard serial number (ISSN) of 0002-9114?

Locate books at WSU–Pullman on the use of steroids by athletes.

Look at a detailed record for one of these books.

With only one mouse click use the information on the screen to find more items on the use of steroids by athletes.

Check the Griffin catalog to see if there is an item available with the call number HE 1.2:Ad 7/2.

What material has Professor Christine Oakley put in the library for her Sociology 320 students to use?

Look up the author Evans, Marian. Read the screen and find the books by this author.

What do you do if you have chosen a title search and then discover you really want to do an author search?

FIGURE 11.3 A partial task list

A quick glance at the task list in Figure 11.3 shows that the usability team wants student users to engage in the kinds of tasks that most students would need to perform to complete their course work using the library's online catalog. The task list does not tell students how they should accomplish these tasks—doing so would defeat the purpose of usability testing, which is to observe users interacting with a document or website in the ways that they normally would.

Collect Data from the Test

One of the most important roles of the usability team is to observe participants engaged in tasks and accurately collect the data that derive from the participants' efforts. As discussed earlier, some usability testing involves film recordings of participants engaged in tasks, allowing the team to review the testing scenarios at their convenience. Some testing on websites is conducted with computer programs that trace and collect data about user performance, thereby recording a database that the team can later

Testing the Usability of a WebPac: Data-Collection Form

Participant:___________

Observer(s):_______________

Date:__________

1. Does the Washington State University–Vancouver campus library have a copy of *Gone with the Wind*?

Success: False starts: Begin time: End time:

2. Does the Eastern Washington University library have a copy of *Gone with the Wind*?

Success: False starts: Begin time: End time:

3. Which of the WSU libraries have *Hamlet* by Shakespeare?

Success: False starts: Begin time: End time:

4. None of the WSU libraries have a copy of *Angels and Amazons.* What do you do?

Success: False starts: Begin time: End time:

5. Do WSU libraries own the following: Albertson, John, "Supermarkets in the Northwest," *Advertising Age 68:* 12–25 (1997)?

Success: False starts: Begin time: End time:

6. Find material by IBM in the libraries.

Success: False starts: Begin time: End time:

7. Find books by Laurie Garrett at any WSU location. Look at the detailed information for one of these books.

Success: False starts: Begin time: End time:

8. Find books written about Toni Morrison.

Success: False starts: Begin time: End time:

FIGURE 11.4 A data-collection form

retrieve. Some usability testing relies on observation and note taking during the test itself. Team members who observe and take notes usually use standardized data-collection forms, which allow them to more easily keep track of their observations. Figure 11.4 gives an example of a data-collection form.

The data-collection form in Figure 11.4 provides a straightforward means of taking notes on each participant's actions; it gives space to detail the ways the participant succeeded in completing the task or got off track, as well as the amount of time it took to complete the task. In most cases the amount of time is crucial because most users of a document or a website will not be satisfied with their experience if the task takes a great deal of time, even if they have been successful in completing the task.

Administer Posttest Questionnaires to Participants

Posttest questionnaires are an important part of usability testing because they allow participants, immediately after the testing, to provide feedback about their experiences,

Testing the Usability of a WebPac:
Exit Questionnaire

Date:_______________

Participant #:_______________

Observer(s):______________________________

Your comments are very valuable to us. Please answer the following questions by circling the number that best describes your answer. Also, please give a brief explanation of your answer in the space below each question. Thank you for your time and cooperation!

1. Would you recommend this system to your friends or colleagues?

   Definitely would not — Definitely would

   1 2 3 4 5

2. Were there times when you wanted to use a feature that didn't exist?

   Many such times — Few such times

   1 2 3 4 5

3. Did you find the screen layout easy to understand?

   Little of the time — Most of the time

   1 2 3 4 5

4. Did you understand the terminology used?

   Little of the time — Most of the time

   1 2 3 4 5

5. Did you feel like you knew what was going to happen next?

   Little of the time — Most of the time

   1 2 3 4 5

6. How easy or difficult was it to use the system?

   Very difficult — Very easy

   1 2 3 4 5

7. Are there any other comments you would like to make? (You may use the back of this questionnaire.)

FIGURE 11.5 A posttest questionnaire

while memories are still fresh. Posttest questionnaires usually solicit both qualitative and quantitative responses but should not lead participants into the team's desired responses. To provide a more neutral line of questioning in a posttest questionnaire, the usability team at Washington State University–Vancouver used a Likert scale format, in which respondents tell how much they agree or disagree with a question or statement (see Figure 11.5).

REPORTING USABILITY TESTS

After conducting a usability test, recording the data, and collecting posttest questionnaires, a usability team has a large amount of documented material to study. In fact, the data may have uncovered problems that the team had not previously recognized or foreseen.

Analyze Findings

Large-scale usability tests that involve hundreds of participants provide many challenges in data analysis, some of which are beyond the scope of this particular text. However, several concepts are important in all data analysis of usability testing.

First of all, analysts must distinguish between a problem and the cause of the problem. For instance, recognizing that users are dissatisfied with an online help screen at a corporate website is easy, but determining precisely why this is so may not be so simple. Nonetheless, the data garnered from usability testing are meant to identify the underlying causes of problems and point the way toward solutions.

Before a team can recommend solutions, however, it must determine the severity of each identified problem, that is, its scope and seriousness. In some instances a problem may relate only to a minor part of a document or website; in other instances it may affect the whole document or website. Some problems are large enough to prevent a user from completing a task, whereas other problems may only delay the user. A slight problem might lead to just a cosmetic change, whereas a more serious problem might prompt a complete reworking. Thus, the usability team must use the data to classify the nature of each problem, its cause, and its severity.

Report the Results and Make Recommendations

The format in which the results are reported is determined in large part by the audience—which is not the testing participants, nor is it necessarily the usability team, although in some instances it might be. The report may be fairly informal, perhaps summarized in a memo and a few short supplementary pages. Other, more formal reports may provide the details of the entire usability test, including everything from user profiles to the test methodology to a detailed description of the test location—taking up hundreds of pages. A report's format and formality should correspond to the desires and needs of the audience.

The basic elements of a usability report include these components:

For more about reports and these basic components, see chapters 21 and 22.

- *Cover letter*—Briefly explains the basics of the test, including the reason for the test, the results, and recommendations for improving the usability of the document or website. Cover letters should not be more than one page and can be in a memo format if the audience is internal.
- *Summary*—Reiterates the goals of the test, the nature of the test, test results, recommendations for change, and the reasons for any changes. Because the summary will likely be read by people with different backgrounds, it should be written in jargon-free, nontechnical language so that nonexperts can understand it. It should go into greater depth than the cover letter and should be two to three pages in length.
- *Table of contents*—Guides readers through a report that is lengthy and contains numerous sections.
- *Methodology*—Includes a detailed statement of test goals, the user profile, description of testing location, specific information about the number of participants, the tasks they were asked to do, the means of measuring and evaluating the completion of those tasks, and other factors that influenced the results of the test, such as the test schedule.
- *Results*—Presents the findings of the usability test in terms of the original test goals. If the test produced both qualitative and quantitative information, this section of the report would include a section for each. For instance, one section might detail the findings from direct observation of participants (i.e., quantitative), whereas the second section might present the findings from participant questionnaires or

comments (i.e., qualitative). Each section might include information taken directly from observer notes or participant surveys and could be broken down into subsections that report both positive and negative evaluations.

- *Recommendations*—Lists the recommendations for future action, as determined by the test results. Recommendations usually include changes that should be made to a document or website to make it more usable and may include suggestions for further usability testing. Each recommendation should be discussed in a separate paragraph, labeled with a subheading, and should detail the problem, the recommended solution, and the rationale for the recommendation. In addition, a brief summary of recommendations should follow the final paragraph.
- *Appendixes*—Provides any necessary additional sources of information, such as observer data sheets, screening questionnaires, posttest questionnaires, and any data logs. The amount and kinds of optional information are determined by the formality of the report and the audience. In some instances further reports are composed from the first usability report and would benefit from these kinds of primary information.

Follow Usability Report Guidelines

A few general guidelines pertain to all reports:

- Choose a style and language that are clear, straightforward, and easily comprehended.
- Include charts or diagrams when necessary to clarify.
- Whenever possible, include the actual voices and words of the participants to support your findings and recommendations.

Usability reports are both common and necessary to complete a usability test. It is also commonplace for a report to suggest further usability testing in order to gain better understanding of users and usability. A usability report might be given as an oral presentation to a specific audience of interested parties. Some reports and presentations include a supplementary video documentary, which provides the highlights of the testing process. This approach allows the audience to see the testing site, the participants engaged in their tasks, and other aspects of the testing methodology and procedure.

EXPLORE

Using an Internet search engine, conduct a search for "usability reports." Skim through some of these reports to find out what document, website, or product was tested, and look closely at how the report is structured. Also, pay attention to the user profiles and the tasks participants were asked to perform.

ETHICAL CONSIDERATIONS IN USABILITY TESTING

Because usability testing requires that teams work with other people to evaluate documents and websites and because large sums of money are sometimes invested in such tests, ethical considerations arise when a usability test is being put together.

Briefing Participants about the Test Process

Often a usability testing participant has little or no familiarity with the usability testing process. One of the team members, typically referred to as the facilitator, makes the participant comfortable by providing a briefing session to describe the testing process and procedures in detail. This includes a description of the testing room, any equipment used to record the participant's actions, any other observers, and other information about the document or website being tested and the scenarios involved in the testing.

Although facilitators want participants to be aware of the test procedures and the general reasons for the testing, they must not create a bias in participants that could skew the results. In other words, facilitators should keep a professional demeanor and tone of voice and refrain from encouraging or discouraging the participant in any way. Simply saying something like "Great job!" could affect how a participant responds to later questions; the participant might want to provide responses that would please the observers. Thus, it is important that a facilitator does and says nothing that might lead a participant into an altered response.

Creating Unbiased Questionnaires

Pretest and posttest questionnaires can prompt important ethical concerns. It is crucial to write neutral questions that do not lead a respondent in any direction. In addition, questions must not be racially or gender biased. The ins and outs of preparing unbiased questionnaires are outside the parameters of this chapter, but chapter 6 discusses primary research, which includes surveys and questionnaires. Well-written questionnaires ask straightforward questions, using words and phrases that are easy to understand. Many also utilize a Likert scale.

Using Consent and Nondisclosure Forms

A usability team should get a participant's written consent to be involved in the test. A consent form should describe the procedures of the test, explain how information from the test will be used, and make clear that participants can withdraw from the test for any reason at any time. Some forms may ask for permission to videotape and may ask that the team be able to reveal the names and identities of the participants. All forms require a signature. Figure 11.6 shows an example of a typical consent form.

In addition to the consent form, some usability testing requires participants to read and sign a nondisclosure form. This is necessary when participants are testing documents or websites not yet in the public realm. When documents and websites are under development, it is important that participants not reveal any details that could affect their competitiveness.

Testing in the Field

As discussed earlier, usability teams may go to external workplaces to observe typical users interacting with a document or a website. A field visit requires the team to get permission from workplace authorities in advance, but even with permission a team member may feel out of place or disruptive. Certainly, the presence of an observer can affect the way someone works or performs tasks, so it is important to limit the number of direct interactions with workers on the job. Another ethical concern during a field visit is that a usability team might accidentally observe private or sensitive information that should not be disclosed to anyone else. In addition, if the team is taking pictures or videotaping workers, it is advisable to get specific permission to do so. The team should also explain clearly to company officials what information has been gathered and how it will be used in the future.

Consent Form

Please read and sign this form if you wish to participate in this usability test.

- You will be asked to perform certain tasks on a website.
- We will also conduct an interview with you.
- In addition, you will be asked to fill in a questionnaire.

Participation in this usability study is voluntary; you can withdraw your consent or stop your participation at any time. The results of the test may be used to help improve the website, but all specific information will remain strictly confidential. At no time will your name or any other personal identification be disclosed.

If you have any questions, please contact

_____(name)_____ at _____(phone)_____.

I have read and do understand the information on this form, and I choose to participate in this usability test.

____________________ ____________________

Participant's signature Date

____________________ ____________________

Facilitator's signature Date

FIGURE 11.6 A usability test consent form

SUMMARY

- Usability refers to how successfully and satisfactorily a person uses a product, document, or website.
- Usability encompasses other qualities, such as being easy to learn, easy to remember, efficient, satisfying, and error free.
- Designers and writers engage in usability testing with volunteer participants in either artificial or real-life environments.
- Usability testing may be part of a Review phase or may occur after a document has been put into use.

- To put users at the center of design, workplace writers must recognize the different stages in the development of documents: predesign, usability measurement, and iterative design.
- Usability tests share a few general characteristics:
- The ultimate goal is to improve the usability of a document or a website for its specified users.
    - Each test begins with a planning session, when test goals are clearly understood and articulated.
    - Participants in any usability test must be representative of the specified users.
    - Participants must engage in real tasks, similar to those a real user would complete.
    - A team of usability testers must observe and record what the participants do and how they react.
    - This team must analyze the data and change the document or website accordingly.
- Usability testing typically takes place in one of three different areas: inside a lab, outside a lab, or in the field.
- Planning for usability testing typically includes these steps: establish a team, define the parameters of the test, define the user profile, establish the issues and the goals of the test, write the test plan, and screen and recruit participants for the test.
- Conducting a usability test involves collecting data from the test and administering posttest questionnaires to participants.
- Reporting on usability tests includes analyzing findings, reporting results, and making recommendations.
- Ethical considerations arise when a usability test is put together:
- Briefing participants about the test process
- Creating unbiased questionnaires
- Using consent and nondisclosure forms
- Testing in the field
- Usability testing on websites and software programs is becoming increasingly crucial.

CONCEPT REVIEW

1. Why do workplace writers want to conduct usability testing on documents and websites?
2. How does usability fit into the problem-solving approach?
3. What are the ultimate goals of usability testing?
4. What is a planning session for usability testing?
5. Who should be the participants in any usability test?
6. What kinds of tasks should participants perform in usability tests?
7. In what ways do usability testers observe and record participants' actions?
8. How and why do usability testers analyze the data received from usability tests?
9. How should teams of usability testers be established?
10. What are the parameters of a usability test?
11. What is a user profile, and why is it important for usability testing?
12. What are the issues and goals of a usability test?
13. What is a usability test plan?
14. How do usability testers screen and recruit participants for a usability test?
15. What does it mean to conduct a usability test?

16. What are some of the ways to collect data from a usability test?
17. Why do usability testers create and administer posttest questionnaires?
18. What do usability testers do with the data they get from a usability test?
19. What are the ethical and professional issues that arise when usability tests are conducted?

CASE STUDY 1

Testing Amazon's Website

You and your company, TechnoDart, have been hired as web consultants by Amazon.com, which wants to make its main website more user-friendly. In particular, Amazon.com is most interested in revamping web pages that relate to book sales. To begin work, your company asks you and a team of three others to conduct a full-scale usability test of Amazon.com's website for book sales. To complete this case study, you and your colleagues will need to plan the usability test (including the recruiting of participants), conduct the test, and report your findings. In short, you are to conduct all facets of a complete usability test, including the preparation of all documents necessary to plan, conduct, and report a usability test.

Jakob Nielsen, a computer science expert at the University of Maryland–College Park, once remarked that "it takes only five users to uncover 80 percent high-level usability problems" with websites. With this advice in mind, your usability test should use five test participants as well. Your instructor will assign your teammates for this case study; efficient and effective collaboration will be necessary for you to complete it successfully. Your instructor will also provide additional instructions about whether test participants should come from inside or outside your class.

CASE STUDY 2

TaxTime: Usability and the W-4 Form

Form W-4 of the Internal Revenue Service (IRS) is a document used by all U.S. employees (see Figure 11.7). Imagine that you are an employee of the IRS who has been asked to determine whether the W-4 needs to be revised and updated. Thus, you will need to do a usability test. First, study this document closely. What kinds of users do you think the designers of the W-4 had in mind when they produced this form? Look at the layout of the form, its size, its use of language, the sizes of the fonts, its references to those who can help if the user has questions, and so on. Before you can conduct a usability test, you will need to construct a viable usability test plan, which includes the following: title page, table of contents, statement of purpose, statement of test issues and goals, description of test environment, user profile, list of participant tasks, and methods of evaluation. Be sure to review the section of this chapter that discusses test plans, and then develop a test plan for this project.

Form W-4 (2008)

Purpose. Complete Form W-4 so that your employer can withhold the correct federal income tax from your pay. Consider completing a new Form W-4 each year and when your personal or financial situation changes.

Exemption from withholding. If you are exempt, complete **only** lines 1, 2, 3, 4, and 7 and sign the form to validate it. Your exemption for 2008 expires February 16, 2009. See Pub. 505, Tax Withholding and Estimated Tax.

Note. You cannot claim exemption from withholding if (a) your income exceeds $900 and includes more than $300 of unearned income (for example, interest and dividends) and (b) another person can claim you as a dependent on their tax return.

Basic instructions. If you are not exempt, complete the **Personal Allowances Worksheet** below. The worksheets on page 2 adjust your withholding allowances based on itemized deductions, certain credits, adjustments to income, or two-earner/multiple job situations. Complete all worksheets that apply. However, you may claim fewer (or zero) allowances.

Head of household. Generally, you may claim head of household filing status on your tax return only if you are unmarried and pay more than 50% of the costs of keeping up a home for yourself and your dependent(s) or other qualifying individuals. See Pub. 501, Exemptions, Standard Deduction, and Filing Information, for information.

Tax credits. You can take projected tax credits into account in figuring your allowable number of withholding allowances. Credits for child or dependent care expenses and the child tax credit may be claimed using the **Personal Allowances Worksheet** below. See Pub. 919, How Do I Adjust My Tax Withholding, for information on converting your other credits into withholding allowances.

Nonwage income. If you have a large amount of nonwage income, such as interest or dividends, consider making estimated tax payments using Form 1040-ES, Estimated Tax for Individuals. Otherwise, you may owe additional tax. If you have pension or annuity income, see Pub. 919 to find out if you should adjust your withholding on Form W-4 or W-4P.

Two earners or multiple jobs. If you have a working spouse or more than one job, figure the total number of allowances you are entitled to claim on all jobs using worksheets from only one Form W-4. Your withholding usually will be most accurate when all allowances are claimed on the Form W-4 for the highest paying job and zero allowances are claimed on the others. See Pub. 919 for details.

Nonresident alien. If you are a nonresident alien, see the Instructions for Form 8233 before completing this Form W-4.

Check your withholding. After your Form W-4 takes effect, use Pub. 919 to see how the dollar amount you are having withheld compares to your projected total tax for 2008. See Pub. 919, especially if your earnings exceed $130,000 (Single) or $180,000 (Married).

Personal Allowances Worksheet (Keep for your records.)

A Enter "1" for **yourself** if no one else can claim you as a dependent . . . . **A** ______

B Enter "1" if:
- You are single and have only one job; or
- You are married, have only one job, and your spouse does not work; or
- Your wages from a second job or your spouse's wages (or the total of both) are $1,500 or less.

. . . **B** ______

C Enter "1" for your **spouse.** But, you may choose to enter "-0-" if you are married and have either a working spouse or more than one job. (Entering "-0-" may help you avoid having too little tax withheld.) . . . . **C** ______

D Enter number of **dependents** (other than your spouse or yourself) you will claim on your tax return . . . . . . **D** ______

E Enter "1" if you will file as **head of household** on your tax return (see conditions under **Head of household** above) . **E** ______

F Enter "1" if you have at least $1,500 of **child or dependent care expenses** for which you plan to claim a credit . . **F** ______

(**Note.** Do **not** include child support payments. See Pub. 503, Child and Dependent Care Expenses, for details.)

G **Child Tax Credit** (including additional child tax credit). See Pub. 972, Child Tax Credit, for more information.
- If your total income will be less than $58,000 ($86,000 if married), enter "2" for each eligible child.
- If your total income will be between $58,000 and $84,000 ($86,000 and $119,000 if married), enter "1" for each eligible child plus "1" **additional** if you have 4 or more eligible children. **G** ______

H Add lines A through G and enter total here. (**Note.** This may be different from the number of exemptions you claim on your tax return.) ▶ **H** ______

For accuracy, complete all worksheets that apply.
- If you plan to **itemize or claim adjustments to income** and want to reduce your withholding, see the **Deductions and Adjustments Worksheet** on page 2.
- If you have **more than one job** or are **married and you and your spouse both work** and the combined earnings from all jobs exceed $40,000 ($25,000 if married), see the **Two-Earners/Multiple Jobs Worksheet** on page 2 to avoid having too little tax withheld.
- If **neither** of the above situations applies, **stop here** and enter the number from line H on line 5 of Form W-4 below.

Cut here and give Form W-4 to your employer. Keep the top part for your records.

Form **W-4** | **Employee's Withholding Allowance Certificate** | OMB No. 1545-0074 **2008**

Department of the Treasury
Internal Revenue Service

▶ **Whether you are entitled to claim a certain number of allowances or exemption from withholding is subject to review by the IRS. Your employer may be required to send a copy of this form to the IRS.**

1 Type or print your first name and middle initial. | Last name | 2 Your social security number

Home address (number and street or rural route) | 3 ☐ Single ☐ Married ☐ Married, but withhold at higher Single rate.
Note. If married, but legally separated, or spouse is a nonresident alien, check the "Single" box.

City or town, state, and ZIP code | 4 **If your last name differs from that shown on your social security card, check here. You must call 1-800-772-1213 for a replacement card.** ▶ ☐

5 Total number of allowances you are claiming (from line **H** above **or** from the applicable worksheet on page 2) | 5 |

6 Additional amount, if any, you want withheld from each paycheck . . . . | 6 | $

7 I claim exemption from withholding for 2008, and I certify that I meet **both** of the following conditions for exemption.
- Last year I had a right to a refund of **all** federal income tax withheld because I had **no** tax liability **and**
- This year I expect a refund of **all** federal income tax withheld because I expect to have **no** tax liability.

If you meet both conditions, write "Exempt" here . . . . ▶ | 7 |

Under penalties of perjury, I declare that I have examined this certificate and to the best of my knowledge and belief, it is true, correct, and complete.

Employee's signature
(Form is not valid unless you sign it.) ▶ **Date** ▶

8 Employer's name and address (Employer: Complete lines 8 and 10 only if sending to the IRS.) | 9 Office code (optional) | 10 Employer identification number (EIN)

For Privacy Act and Paperwork Reduction Act Notice, see page 2. Cat. No. 10220Q Form **W-4** (2008)

FIGURE 11.7 IRS Form W-4

Form W-4 (2008) Page 2

Deductions and Adjustments Worksheet

Note. Use this worksheet *only* if you plan to itemize deductions, claim certain credits, or claim adjustments to income on your 2008 tax return.

1 Enter an estimate of your 2008 itemized deductions. These include qualifying home mortgage interest, charitable contributions, state and local taxes, medical expenses in excess of 7.5% of your income, and miscellaneous deductions. (For 2008, you may have to reduce your itemized deductions if your income is over $159,950 ($79,975 if married filing separately). See *Worksheet 2* in Pub. 919 for details.) . . . 1 $

2 Enter: { $10,900 if married filing jointly or qualifying widow(er); $ 8,000 if head of household; $ 5,450 if single or married filing separately } . . . 2 $

3 **Subtract** line 2 from line 1. If zero or less, enter "-0-" . . . 3 $

4 Enter an estimate of your 2008 adjustments to income, including alimony, deductible IRA contributions, and student loan interest 4 $

5 **Add** lines 3 and 4 and enter the total. (Include any amount for credits from *Worksheet 8* in Pub. 919) . 5 $

6 Enter an estimate of your 2008 nonwage income (such as dividends or interest) . . . 6 $

7 **Subtract** line 6 from line 5. If zero or less, enter "-0-" . . . 7 $

8 **Divide** the amount on line 7 by $3,500 and enter the result here. Drop any fraction . . . 8

9 Enter the number from the **Personal Allowances Worksheet,** line H, page 1 . . . 9

10 **Add** lines 8 and 9 and enter the total here. If you plan to use the **Two-Earners/Multiple Jobs Worksheet,** also enter this total on line 1 below. Otherwise, **stop here** and enter this total on Form W-4, line 5, page 1 10

Two-Earners/Multiple Jobs Worksheet (See *Two earners or multiple jobs* on page 1.)

Note. Use this worksheet *only* if the instructions under line H on page 1 direct you here.

1 Enter the number from line H, page 1 (or from line 10 above if you used the **Deductions and Adjustments Worksheet**) 1

2 Find the number in **Table 1** below that applies to the **LOWEST** paying job and enter it here. **However,** if you are married filing jointly and wages from the highest paying job are $50,000 or less, do not enter more than "3." . . . 2

3 If line 1 is **more than or equal to** line 2, subtract line 2 from line 1. Enter the result here (if zero, enter "-0-") and on Form W-4, line 5, page 1. **Do not** use the rest of this worksheet . . . 3

Note. If line 1 is *less than* line 2, enter "-0-" on Form W-4, line 5, page 1. Complete lines 4–9 below to calculate the additional withholding amount necessary to avoid a year-end tax bill.

4 Enter the number from line 2 of this worksheet . . . 4

5 Enter the number from line 1 of this worksheet . . . 5

6 **Subtract** line 5 from line 4 . . . 6

7 Find the amount in **Table 2** below that applies to the **HIGHEST** paying job and enter it here . . . 7 $

8 **Multiply** line 7 by line 6 and enter the result here. This is the additional annual withholding needed . . 8 $

9 Divide line 8 by the number of pay periods remaining in 2008. For example, divide by 26 if you are paid every two weeks and you complete this form in December 2007. Enter the result here and on Form W-4, line 6, page 1. This is the additional amount to be withheld from each paycheck . . . 9 $

Table 1

Married Filing Jointly		All Others	
If wages from **LOWEST** paying job are—	Enter on line 2 above	If wages from **LOWEST** paying job are—	Enter on line 2 above
$0 - $4,500	0	$0 - $6,500	0
4,501 - 10,000	1	6,501 - 12,000	1
10,001 - 18,000	2	12,001 - 20,000	2
18,001 - 22,000	3	20,001 - 27,000	3
22,001 - 27,000	4	27,001 - 35,000	4
27,001 - 33,000	5	35,001 - 50,000	5
33,001 - 40,000	6	50,001 - 65,000	6
40,001 - 50,000	7	65,001 - 80,000	7
50,001 - 55,000	8	80,001 - 95,000	8
55,001 - 60,000	9	95,001 - 120,000	9
60,001 - 65,000	10	120,001 and over	10
65,001 - 75,000	11		
75,001 - 100,000	12		
100,001 - 110,000	13		
110,001 - 120,000	14		
120,001 and over	15		

Table 2

Married Filing Jointly		All Others	
If wages from **HIGHEST** paying job are—	Enter on line 7 above	If wages from **HIGHEST** paying job are—	Enter on line 7 above
$0 - $65,000	$530	$0 - $35,000	$530
65,001 - 120,000	880	35,001 - 80,000	880
120,001 - 180,000	980	80,001 - 150,000	980
180,001 - 310,000	1,160	150,001 - 340,000	1,160
310,001 and over	1,230	340,001 and over	1,230

Privacy Act and Paperwork Reduction Act Notice. We ask for the information on this form to carry out the Internal Revenue laws of the United States. The Internal Revenue Code requires this information under sections 3402(f)(2)(A) and 6109 and their regulations. Failure to provide a properly completed form will result in your being treated as a single person who claims no withholding allowances; providing fraudulent information may also subject you to penalties. Routine uses of this information include giving it to the Department of Justice for civil and criminal litigation, to cities, states, and the District of Columbia for use in administering their tax laws, and using it in the National Directory of New Hires. We may also disclose this information to other countries under a tax treaty, to federal and state agencies to enforce federal nontax criminal laws, or to federal law enforcement and intelligence agencies to combat terrorism.

You are not required to provide the information requested on a form that is subject to the Paperwork Reduction Act unless the form displays a valid OMB control number. Books or records relating to a form or its instructions must be retained as long as their contents may become material in the administration of any Internal Revenue law. Generally, tax returns and return information are confidential, as required by Code section 6103.

The average time and expenses required to complete and file this form will vary depending on individual circumstances. For estimated averages, see the instructions for your income tax return.

If you have suggestions for making this form simpler, we would be happy to hear from you. See the instructions for your income tax return.

FIGURE 11.7 IRS Form W-4 (*continued*)

CASE STUDIES ON THE COMPANION WEBSITE

Developing a Usability Test: Bruce Wayne-Edwards and Recruiting Subjects

Wayne-Edwards Staffing developed an online applicant screening process. You have been asked to test the system and create a detailed proposal to simplify and improve it. You need to develop a prototype, construct a usability test, develop a testing plan, implement the plan, and draft a proposal with your recommendations based on the test.

Reading the Manual: Cory Washington's Technical Documentation

After hearing a number of complaints from customers, Barks Insurance Software Company has hired a group of technical writers and has transferred you to the newly established technical writing department.

Developing a Usability Test: Nia Barazu and the Game Interface

You recently applied for the position of Director of Product Testing at Xtreme Gaming. Because of recurring problems with product testing, the CEO wants all applicants to construct a pre-development usability test for a new product.

Making Decisions through Usability Testing: The Tucson Emergency Response System

You must decide how to develop training materials for the city of Tucson's emergency response system, decide which documents and media forms to use (print, web, or video), provide justification for these decisions, and design appropriate usability testing.

VIDEO CASE STUDY

GlobeShare Wireless

Usability Testing of the GlobeShare Wireless Customer Support Website

Synopsis

Several GlobeShare employees are discussing the customer support section of the GlobeShare Wireless website. A recent questionnaire of GlobeShare customers indicated that the website was difficult to use. Michelle Bankler, a product tester, and Arriana Reggio, a web designer, are assigned to deal with the problem. They decide to conduct a usability test involving a group of potential customers to find out exactly what's wrong with the website.

WRITING SCENARIOS

1. Develop a user profile for each of the following websites: www.cnn.com, www.barnesandnoble.com, and www.espn.com. You will need to look at each website and identify clues about users in the design and message. Which of these sites probably has the most diverse range of users? most narrow range? What evidence do you have to support this conclusion? Be sure to review the

section on user profiles in this chapter and complete your responses in a document type assigned by your instructor.

2. Locate a document or website that you believe needs to go through usability testing. What kind of people would be most appropriate for the usability team? (Remember that usability teams should offer multiple perspectives.) Decide where this document or website should be tested, and explain your responses in detail in a memo to your instructor.
3. Research a profession or career that interests you, and find out more about the kinds of documents produced in those workplaces. List three or four of those documents, and then discuss each in a paragraph or two, addressing the likelihood of its needing usability testing. In as much detail as possible, explain in a memo or letter to your instructor why these specific documents would need to be tested.
4. Companies that maintain websites are curious to find out who visits their sites. Such information is highly relevant to marketers, advertisers, and web designers. But since the Web is an anonymous medium, how do companies find out who visits their sites, and how do they tabulate this information? Write a short, research-based report that explores how companies determine the user profiles of visitors to their websites.
5. Look again at Writing Scenario 4. What kinds of ethical concerns might arise when companies gather information about users of their websites? Explain your responses in a one-page essay.
6. Search online for pictures of the Florida ballots or descriptions of the voting procedures used in the 2004 and 2008 presidential elections. How did the ballots and procedures change from those in 2000 (as discussed at the beginning of this chapter and shown in Figure 11.1)? In what ways did the new ballots and procedures appear more usable for voters? Write your responses in a short memo or other document assigned by your instructor.
7. On the Internet, search for companies that have had usability testing done *in the field*. In a short, informative essay explain which companies you found, what specifically was tested, and what results followed from the tests. Then discuss whether usability testing is more prevalent than you originally thought and whether you foresee any type of usability testing taking place in your future workplace. Prepare your responses in a document type assigned by your instructor.
8. Usability testing is performed on many documents and websites, although certainly not on all of them. In a short essay describe at least three documents or websites that you believe have an ethical obligation to test usability for the sake of their users. What is it about these documents or websites and user needs that calls for usability testing before the documents or websites are released to users? Explain in detail in a document type assigned by your instructor.
9. Find a document at school or at home that you have used in the past—an instruction manual, for instance. In a short essay or memo describe the purpose of this document and its user profile, and discuss the usability issues that this document poses. For instance, what sorts of things make this particular document highly usable or difficult to use?
10. Imagine that you are in charge of conducting a usability test for the employment application form for McDonald's or some other fast-food

company. One of your tasks would be to find participants for the usability test. Design a screening questionnaire of at least five questions that potential participants would fill out to determine whether they are a good match for this usability test. Keep in mind that such a survey would require you first to have a good sense of the user profile.

11. If you were to design the particular questionnaire in Writing Scenario 10, what specific ethical issues would you have to consider to make this survey valid and neutral? Respond in a document type assigned by your instructor.
12. Visit your school's home website. Browse through it to familiarize yourself with the features that students should know how to use. If you were performing a usability test for this website, what particular aspects would you include? Review the sections in this chapter on test goals and task lists, and then create a list of goals and a task list for such a test.
13. Describe your favorite website or one that you visit often. Imagine that you are leading a usability team whose job it is to test this website. In a few paragraphs describe the goals you would set for the test and the kinds of team members who would be best suited to help you with such a test. In your analysis be sure to look at whether the site's technological applications are easy to use and current. Detail your responses in a document type assigned by your instructor.
14. Usability testing can take place in high-tech labs, in conference rooms, or in the field, depending on the documents or websites being tested. Each of these locations has its advantages and disadvantages. Describe two or three kinds of documents or websites that would be best tested in a lab, two or three best tested in a conference room, and two or three best tested in the field. Explain in detail the rationale for your decisions—refer to specific documents and websites if you like.
15. Digital technologies that enhance communication (e.g., websites, PDAs, computer software programs) are becoming more advanced and prolific. Choose one new communication technology that you currently use, and write a short essay that discusses its usability. In particular, recall your first few experiences with it, and discuss whether the technology was easy to learn, easy to remember, efficient, satisfying, and error free. For each of these five aspects, provide a detailed response about your experiences.

12

E-mail, E-Messages, and Memos

CHAPTER LEARNING OUTCOMES

After completing this chapter, you will be able to do the following:

- Distinguish between short **internal and external** documents
- Understand several aspects of those types of documents: **level of formality**, **introductory and concluding information**, and **level of detail**
- Understand why the **problem-solving approach** is important in short documents
- Consider the **privacy rights** you have when sending an e-mail message
- Recognize the basic elements of e-mail messages: **subject lines**, **recipients**, **message content**, **signatures**, and **attachments**
- Consider how **electronic messaging** is used in the workplace
- Distinguish between **instant messaging** and **text messaging**
- Understand the basic elements of memos: **identifying information** and **appropriate content**
- Understand basic memo **formatting** approaches
- Recognize different **situations** requiring a memo

DIGITAL RESOURCES

On the Companion Website www.prenhall.com/dobrin:

- Case 1: Alban Paper: New Philosophies for a New Era
- Case 2: Opening the Lines of Communication: Addressing Concerns about Network Safety
- Case 3: When Does Workplace Communication Become Too Much Communication?
- Case 4: Poor Communication: A Case of Friendship at Capital Oil
- Video Case: Poncho and Lefty to the Rescue: The Case of the Missing Microfiber Filters
- PowerPoint Chapter Review, Test-Prep Quiz, Exercises and Activities

REAL PEOPLE, REAL WRITING

MEET MATT KEFFER • Senior Sales Associate for Epicor Software

What types of writing do you do at Epicor Software?

In a typical week I write approximately fifty to seventy-five messages—many through e-mail or memos. I may respond to prospective clients' questions, create sales and marketing materials, interact with internal consultants and administrative staff, discuss opportunities and industry trends with external consultants, and communicate with trade organizations and publications. My writing can be as simple as a one-line internal e-mail and as complex as a graphic-heavy marketing piece that will be viewed by thousands of organizations.

What do you try to accomplish in the first paragraph of a memo or an internal e-mail?

People have limited time to analyze and interpret information. My writing style is direct, and I try to convey the intent of my memo or e-mail as quickly and efficiently as possible.

How do you determine the level of detail to include in the memos and internal e-mails that you write?

I determine the level based on two criteria: the complexity of the topic and the technical aptitude of my audience. Client requirements can be very intricate and require detailed analysis, whereas setting up an internal sales meeting requires little detail other than the time and place. Also, the interaction that I have with an internal consultant who is a manufacturing or distribution expert will look much different from a request to my administrative assistant.

How significant are the subject lines of the memos and e-mails you send or receive?

Subject lines are critical in today's business environment. I want the recipients of my memo or e-mail to have an idea of its content before they even open it. Subject lines are especially important if the document involves time-sensitive information regarding a prospect or client. An e-mail regarding a key client will take priority over details on the company picnic, so I want readers to know the subject matter by the time they've read the subject line.

How do you determine whether to send a short internal message as an e-mail or a memo?

If the subject matter has a long-term or dramatic impact on a coworker's job function (such as a policy change), I'll create and distribute a memo. Almost all other correspondence will be via e-mail.

What advice about workplace writing do you have for the readers of this textbook?

In the software world we talk about the difference between data and information. In my experience, good writers present information that is focused and helps people make better decisions. Poor writers often fail to think about their readers' needs and present unsorted data that is too broad and contains unnecessary details and confusing facts. My advice is to use a style that is direct and informative, focusing on the specific information readers need.

INTRODUCTION

E-mails, e-messages, and memos are perhaps the most common and widely distributed forms of communication in workplaces today. These three types of documents generally transmit smaller chunks of information, often addressing immediate or short-term topics rather than lengthy, long-range subjects that might be addressed in proposals, reports, or other documents. E-mail is used to communicate with diverse audiences and cannot be defined as strictly internal or external. E-messages, or text messages, began as a method of transmitting short messages informally within an organization, but they have become more widely used internally and externally in both formal and informal contexts. A memo (short for *memorandum*) is almost always an internal document used to communicate information quickly and succinctly to colleagues and coworkers.

All three forms are normally used for nonsensitive information that will not elicit a strong emotional response from the reader. For example, e-mails, e-messages, and memos are often used to transmit specific information about meetings or changes in policy, updates on future events and activities, short introductory or follow-up questions about a product or service, or other current issues pertaining to a workplace problem. In some contexts e-mails and memos appear to take on the same purposes and approaches, particularly when e-mails employ memo formatting to convey information. Similarly, it can be difficult to tell e-mails and e-messages apart other than through the devices on which they are transmitted and received.

This chapter—focusing on e-mail, e-messages, and memos—and the next chapter, focusing on letters, both address relatively short documents. The primary distinction among these document types has to do with the audience for which each is produced. Quite simply, memos are used for shorter written correspondence *within* an organization (i.e., internal correspondence), whereas letters are used for shorter written correspondence to audiences *outside* an author's organization (i.e., external correspondence). E-mail and e-messages can be used for both.

SHORT INTERNAL AND EXTERNAL DOCUMENTS

There are many similarities between internal and external documents, and the boundaries can sometimes overlap and become blurred. However, there are also some general differences in layout, content, and style, primarily because they have different audiences and purposes.

Level of Formality

Because internal documents are written from one employee to another within the same company, they tend to be less formal than documents sent to outside audiences. When individuals know each other well and work together toward common goals, they are often able to forego some of the formality necessary for communication with people outside the organization. Some internal documents—particularly memos and internal e-mails and e-messages—can be more friendly and conversational than outgoing documents.

However, not all internal communication should be informal or casual. Some internal situations—such as those that have serious consequences for you, your readers, your company, or the public—may require more formal communication, which shows that you have carefully considered the situation and take it seriously. Furthermore, when writing for transnational audiences, workplace writers should be attentive to cultural expectations regarding formality. In some cultures an informal or casual tone in professional correspondence could be construed as insulting. In addition, you should

See chapter 5 for more on transnational communication.

not confuse informality with sloppiness. Writing should always be grammatically, mechanically, and factually correct, even in casual correspondence with colleagues.

Introductory and Concluding Information

Short, internal correspondence frequently relinquishes formal greetings and closing statements, which are often meant to establish a bond between the author and the reader. Formal greetings—for example, "Dear Sir"—and formal closing statements—for example, "I look forward to working with you"—are useful ways to make a connection with a new client or to show respect, but with memos, e-mail, and e-messages between coworkers, such connections may already exist.

Similarly, detailed contact information—such as the author's title within the organization, phone number, and mailing address—is often unnecessary in an internal document, because these details are probably known by colleagues and coworkers. However, if an e-mail is intended for a client, customer, or applicant, it may be necessary to include the introductory and concluding information to show respect for the reader, to create goodwill, and to provide information about who the author is.

Level of Detail

Internal correspondence can often omit the comprehensive background details or lengthy descriptions that might be found in external documents. When writing to people you've worked with closely, you may be able to work from a basis of common knowledge about a topic, product, plan, or idea. A memo or an internal e-mail or e-message doesn't need to repeat information that readers are sure to know. Nonetheless, an effective internal message will not overlook any details a reader needs in order to act; even if readers share common knowledge, they may need approval, feedback, or advice before they can take necessary steps.

Letters and external e-mails, on the other hand, should always provide enough detail to ensure that readers understand the message and can act on it accordingly, even if this means providing basic background or primary information. Readers might not be thinking about the subject or know much about it and may need the key facts and contextual information. This is particularly important if you aren't sure who might read your message other than your primary audience. Insufficient detail or background information can be a problem with e-mails that begin as documents intended for one audience but wind up in the inbox of a completely different audience. Remember that electronic messages are easily forwarded to others you may not have initially considered as readers.

IN YOUR **EXPERIENCE**

Consider the last e-mail you wrote to someone about a work or school topic, such as a question about a summer job or a request to meet with an academic advisor or professor. Was the e-mail written in a formal or informal style? Did you include a formal or informal greeting or closing statement? Did you include any specific contact information about yourself? If you were to rewrite that e-mail as a memo, would it be different? Is there anything you might have done more effectively in the e-mail you sent?

PSA

Problem Solving in Short Messages

Clearly, creating an effective short document requires more than simply determining whether the audience is internal or external. Although short documents often require less time and effort than some of the longer documents created in the workplace, you should still use the problem-solving approach to help you gauge what is appropriate in each situation.

- During the Plan phase consider the problem at hand—who is involved and what information is needed—rather than dashing off a hasty response or reply.
- Sometimes research is necessary, even for concise messages; if so, gather the facts and details needed to make your message useful.
- Organization and design are crucial as you draft short documents; you have limited space and are often communicating with busy readers who need information to act.
- Reviewing your documents is important, even if they are just a few sentences long. Solicit peer response whenever possible; in most instances the extra time spent revising will save your readers time in understanding your message.
- How you distribute a short document can be especially important; your decision might determine whether you use e-mail, e-messages, a memo, or a letter, which can influence your reader's ability to receive needed information.

To further understand the importance of the PSA in short messages, consider the e-mail exchange between Michael Brown, former chief of the Federal Emergency Management Agency (FEMA), and Marty Bahamonde, a FEMA employee (see Figure 12.1). The original message was sent by Bahamonde two days after Hurricane Katrina struck New Orleans; Brown's response appears above it. The level of formality and detail in the two messages is quite different. Bahamonde's message begins with a more formal greeting—"Sir"—and uses a more formal tone throughout, which suits the gravity and seriousness of the situation. Bahamonde's message provides short but detailed sentences describing the situation in New Orleans and inviting response.

Brown's reply, on the other hand, is much different. It offers no formal language, using a familiar "thanks" in the introduction, followed by an informal sentence

From: Brown, Michael D
Sent: Wednesday, August 31, 2005 12:24 PM
To: 'Marty.Bahamonde@dhs.gov'
Subject: Re: New orleans

Thanks for update. Anything specific I need to do or tweak?

-----Original Message-----
From: Bahamonde, Marty <Marty.Bahamonde@dhs.gov>
To: 'michael.d.brown@dhs.gov' <Michael.D.Brown@dhs.gov>
Sent: Wed Aug 31 12:20:20 2005
Subject: New orleans

Sir, I know that you know the situation is past critical. Here some things you might not know.
Hotels are kicking people out, thousands gathering in the streets with no food or water.
Hundreds still being rescued from homes.

The dying patients at the DMAT tent being medivac. Estimates are many will die within hours. Evacuation in process. Plans developing for dome evacuation but hotel situation adding to problem. We are out of food and running out of water at the dome, plans in works to address the critical need.

FEMA staff is OK and holding own. DMAT staff working in deplorable conditions. The sooner we can get the medical patients out, the sooner wecan get them out.

Phone connectivity impossible

More later

Sent from my BlackBerry Wireless Handheld

FIGURE 12.1 An e-mail exchange between FEMA employees

fragment. It provides no real detail or instructions in response to the details of the initial message and does not present any solutions, advice, or support to help solve the problems that Bahamonde and others faced. Twelve days after this message was sent, Brown resigned as FEMA director, and this e-mail and others like it were later used as evidence in a U.S. House committee probe of the government's response to Hurricane Katrina.

ANALYZE THIS

It is obvious that Brown's response to Bahamonde required little time and effort to create—and consequently did little to help solve the problems. Reread the e-mail exchange, and review the problem-solving approach. Then consider the ways in which Brown's response might have been more useful had he applied that approach to this situation.

Using the PSA yourself, write a more effective response to Bahamonde's message. You might begin by reflecting on the situation Bahamonde was in and the sort of response that would have been most suitable. You might also do a bit of basic research before you begin writing. You will certainly want to revise the document before you distribute it, given the seriousness of the situation.

Share your responses as a class, and compare them to Brown's response. In what ways are all of your responses different from Brown's? In what ways did you employ the PSA as you wrote your responses?

Privacy Issues with Short Messages

As the FEMA example demonstrates, messages that are intended to be private do not always remain so. This is true of all short messages, particularly when those messages reveal some error, oversight, or problem on the part of the author. Poor or ill-chosen messages often have legs—they seem to wind up going where they were not meant to go. Memos and letters can be easily photocopied or scanned and sent to public audiences. E-mail messages are especially likely to be sent to unintended audiences, since it is easy with little thought or effort to forward a message from one person or group to another. Workplace writers must remember that e-mail is not a private means of communication. In fact, e-mail messages are best considered as public documents, because they always have the potential to become so. E-mail messages, like memos and letters, have been used in countless civil and criminal cases. Consider these recent examples:

- In a Massachusetts class-action suit over the dangers of the diet-drug-combination Phen-Fen, the court allowed this e-mail from a company executive to be admitted: "Do I have to look forward to spending my waning years writing checks to fat people worried about a silly lung problem?"
- Chevron settled a lawsuit for $2.2 million that involved an interoffice e-mail giving "25 reasons why beer is better than women."
- Former star investment banker Frank Quattrone was convicted of obstructing federal investigations into stock offerings at Credit Suisse First Boston. Central to the case was an e-mail Quattrone forwarded, telling employees it was "time to clean up those files" after he learned of the investigation.
- More than five hundred of former West Virginia Governor Bob Wise's intimate e-mails to and from a state employee were obtained under the Freedom of Information Act and made public in 2003. The employee's husband filed for divorce, and Wise didn't seek reelection in 2004.[1]

[1] Examples from Ann O'Neill, "E-mail Can Bounce Back to Hurt You," *CNN.com*, November 7, 2005, http://www.cnn.com/2005/LAW/11/03/email.legal/index.html.

Such examples may cause you to wonder exactly what privacy rights you do have in regard to your e-mail messages. Most e-mail services—including private services like AOL and gmail, as well as university or business accounts that are provided to users—have a specific privacy policy covering a range of issues. Figure 12.2 gives an example of an e-mail privacy policy. Many e-mail users are unaware of their privacy rights: some users never give their rights a thought; others assume, often wrongly, that they "own" their messages. Having a specific name and password can give the illusion of privacy; it is easy to assume that since your password is private, your messages are, too. In reality, privacy rights vary from service to service. Some commercial services work hard to keep your messages private, others have more lenient privacy policies, and some universities and businesses retain the right to access all of the e-mail you send and receive through their service. It is important to know and understand your e-mail privacy rights and to consider them before you send messages that you'd like to keep private.

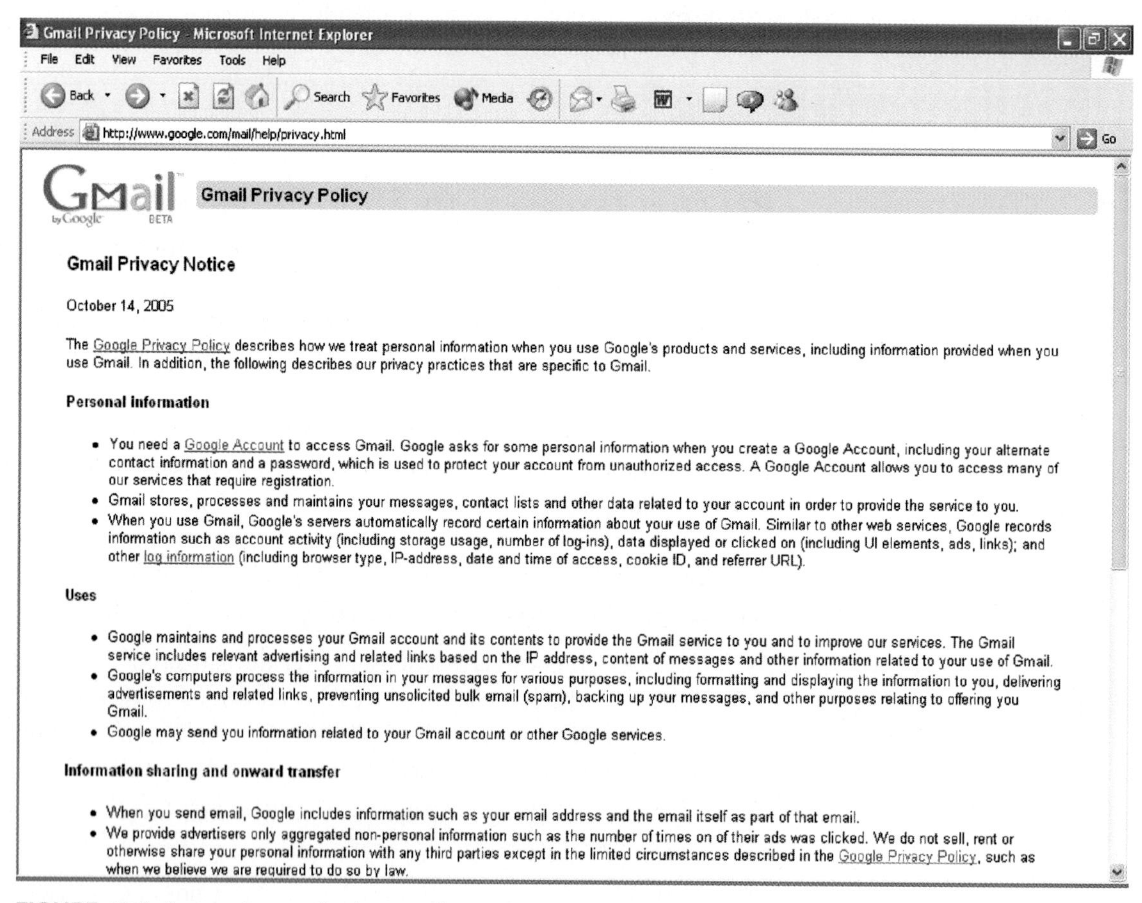
Gmail Privacy Policy - Microsoft Internet Explorer

File Edit View Favorites Tools Help

Address http://www.google.com/mail/help/privacy.html

Gmail by Google BETA — Gmail Privacy Policy

Gmail Privacy Notice

October 14, 2005

The Google Privacy Policy describes how we treat personal information when you use Google's products and services, including information provided when you use Gmail. In addition, the following describes our privacy practices that are specific to Gmail.

Personal information

- You need a Google Account to access Gmail. Google asks for some personal information when you create a Google Account, including your alternate contact information and a password, which is used to protect your account from unauthorized access. A Google Account allows you to access many of our services that require registration.
- Gmail stores, processes and maintains your messages, contact lists and other data related to your account in order to provide the service to you.
- When you use Gmail, Google's servers automatically record certain information about your use of Gmail. Similar to other web services, Google records information such as account activity (including storage usage, number of log-ins), data displayed or clicked on (including UI elements, ads, links); and other log information (including browser type, IP-address, date and time of access, cookie ID, and referrer URL).

Uses

- Google maintains and processes your Gmail account and its contents to provide the Gmail service to you and to improve our services. The Gmail service includes relevant advertising and related links based on the IP address, content of messages and other information related to your use of Gmail.
- Google's computers process the information in your messages for various purposes, including formatting and displaying the information to you, delivering advertisements and related links, preventing unsolicited bulk email (spam), backing up your messages, and other purposes relating to offering you Gmail.
- Google may send you information related to your Gmail account or other Google services.

Information sharing and onward transfer

- When you send email, Google includes information such as your email address and the email itself as part of that email.
- We provide advertisers only aggregated non-personal information such as the number of times on of their ads was clicked. We do not sell, rent or otherwise share your personal information with any third parties except in the limited circumstances described in the Google Privacy Policy, such as when we believe we are required to do so by law.

FIGURE 12.2 A sample e-mail privacy policy

EXPLORE

Find the privacy policy of your e-mail service, and write a short summary of your rights. Print a copy of the privacy policy, as well as your summary of it, and bring both to class. As a class, discuss the similarities and differences among the various privacy policies. Do the policies seem fair and ethical? Are they presented clearly, or do they contain difficult or complex jargon?

IN YOUR EXPERIENCE

Take a moment to think about your own e-mail habits. Are there personal e-mails you have sent or received that you'd be embarrassed to have someone other than your intended audience see? What if your family members had the right to see all of the e-mails you have sent or received?

Consider the e-mail systems you use to send or receive those e-mails. Do you use e-mail systems provided by your school? How would you react if your teachers could see all of the e-mails you have sent or received? Have you ever sent an e-mail to one person, expecting that person to be the only one to read it, and then discovered that your e-mail was forwarded on to others?

As a class, consider the difference between a personal e-mail account and a professional e-mail account. Discuss the professional responsibility of using your school account to conduct your school business and your personal account to conduct your personal business.

BASIC ELEMENTS OF E-MAIL MESSAGES

Chances are good that you use e-mail on a regular basis—95 percent of college students report using e-mail for social communication at least once a week. However, only 5 percent of those same students use e-mail to communicate with work colleagues.[2] As a result, many entry-level employees begin their careers with a clear understanding of how e-mail works but little experience with or rhetorical understanding of e-mail communication in the workplace. Consequently, this section provides specific rhetorical strategies for communicating effectively in the workplace through e-mail.

See chapter 3 for more information about e-mail etiquette.

An internal e-mail takes on many of the characteristics and features of a paper memo, and an external e-mail looks much like a letter. In fact, most of the significant differences between e-mails and memos or letters are a result of how technology shapes the reader's experience. As we point out in chapter 3, computer technology not only affects the ways we produce and disseminate technical documents but also often affects the content of those documents as well.

Subject Lines

In e-mail messages the subject line should be both concise and specific. Many recipients receive dozens, if not hundreds, of e-mails every day and often scan subject lines to decide which e-mails should be read immediately and which can be read later or deleted. If your subject line is something vague like "Your Message" or "I Have a Question," the reader may think it is spam (i.e., unsolicited junk e-mail) and delete it instantly. In addition, because some e-mail programs show only the first few words of the subject line, you should emphasize the important information first and try to limit the length of the subject line. Some e-mail programs now use spam filters to block indiscriminate, unsolicited, or mass e-mails; these filters often identify spam by recognizing common or repetitive e-mail subject lines. Vague subject lines like "Your Message" also risk being blocked by filters.

Although using all capital letters in your workplace communication (or in any written communication) is often seen as rude or overbearing, you may wish to use all caps in the subject line—but only if your message is urgent. This can help to draw

---

[2] Steve Jones, *The Internet Goes to College*, September 15, 2002, http://www.pewinternet.org/pdfs/PIP_College_Report.pdf.

the attention of busy colleagues who receive many messages, alerting them that the information in your e-mail is critical or time sensitive. Consider these examples:

SUBJ: Hiring Committee Meeting

Subject: District 7 Development

Subject: Pharmacology Report

Subject: TAX AUDIT TODAY

One final point: in an e-mail conversation when you and another person are simply replying to each other's e-mails, you may wish to occasionally create a new subject line. Some e-mail programs place an "Re" before each message that is a reply. When your subject line begins to look like this—"Re: Re: Re: Re: Thursday's Meeting"—it is time for a new subject line.

Recipients

E-mail makes it easy—sometimes almost too easy—to send a message to multiple recipients. Think carefully about your intended audience as you compose e-mail messages, and tailor what you have to say to that specific individual or group. Avoid the urge to "Cc" (i.e., carbon copy) multiple groups or individuals with the same e-mail unless it is suitable and written for them. Instead, you may wish to compose several different but similar e-mail messages about the same subject to meet the specific needs of each reader.

Likewise, avoid the urge to "Reply to all" if your response to a message is directed only to the sender. Most people would like to reduce the amount of unnecessary e-mail they receive, so you should send reply messages only to appropriate recipients. And when you do reply to a message, it may be useful to quote just a sentence or two from the original message to put your reply in context; many e-mail programs automatically identify the quoted text by placing a > sign before each line. However, you need not include the entire message. Your reply will be more useful if you include only that part of the sender's e-mail that is important, as in this example:

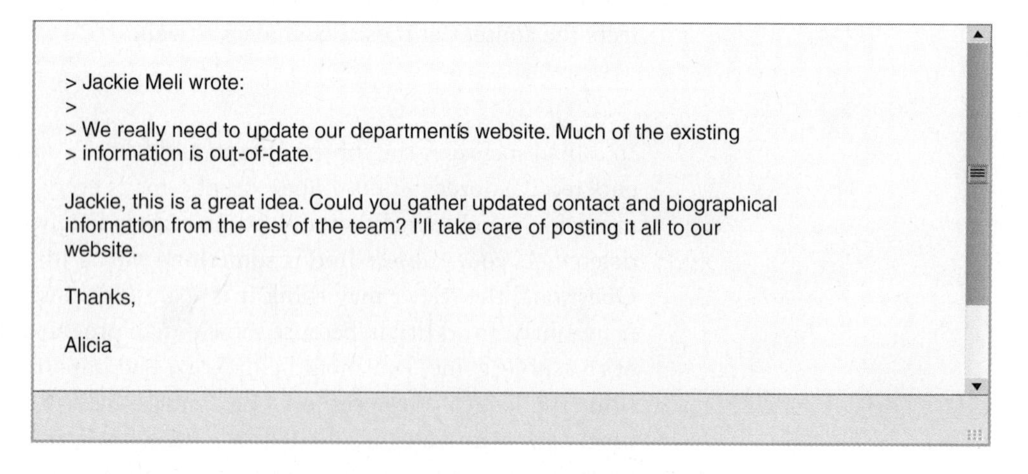
> Jackie Meli wrote:
>
> We really need to update our departmentís website. Much of the existing
> information is out-of-date.

Jackie, this is a great idea. Could you gather updated contact and biographical information from the rest of the team? I'll take care of posting it all to our website.

Thanks,

Alicia

Finally, be considerate when forwarding a message from one person to another. The sender may have intended the message just for you, so don't send it to others without the sender's permission. Deciding when to forward e-mails is as much an ethical decision as it is a matter of being thoughtful.

See chapter 4 for more on ethics.

Message Content

Brevity is the key to an effective e-mail. As previously mentioned, the recipients of workplace e-mails may be very busy and may receive many e-mails every day. Consequently, e-mail messages should provide only the information that is pertinent to the situation. Although transnational readers may have different expectations in the structure and presentation of e-mail messages, most still expect the messages to be concise. In other words, brevity may be the one thing that all effective e-mail messages have in common—regardless of their origin.

See chapter 5 for more on transnational communication.

In most instances you should state your reason for writing within the first sentence or two of the message. In addition, you should try to use topic sentences in any body paragraphs so that readers can easily determine the main point or idea of that section of text. The key is to create a message that readers can skim quickly to find information. You should avoid long, drawn-out introductions, discussions, or analyses in your e-mail messages; most readers expect e-mail messages to be clear, direct, and succinct. Therefore, as you revise e-mail messages before sending them out, you should focus on streamlining the content.

You should also remember that the recipient will be reading your message on a monitor. Whenever possible, create e-mail messages that can be read on a single screen because scrolling through a long, detailed e-mail can be confusing and difficult. If you need to give lots of specific information, you may want to send a printed memo or include a formatted document as an attachment to your e-mail message.

Figures 12.3 and 12.4 provide examples of the same message presented in two different ways. Figure 12.4 is ineffective: the subject line is wordy, the message contains irrelevant information, important details are buried, and the message has not been carefully edited. Figure 12.3 is much more effective: the subject line is concise, the information in the message is succinct and emphasizes important details, and the message is free of typographical errors.

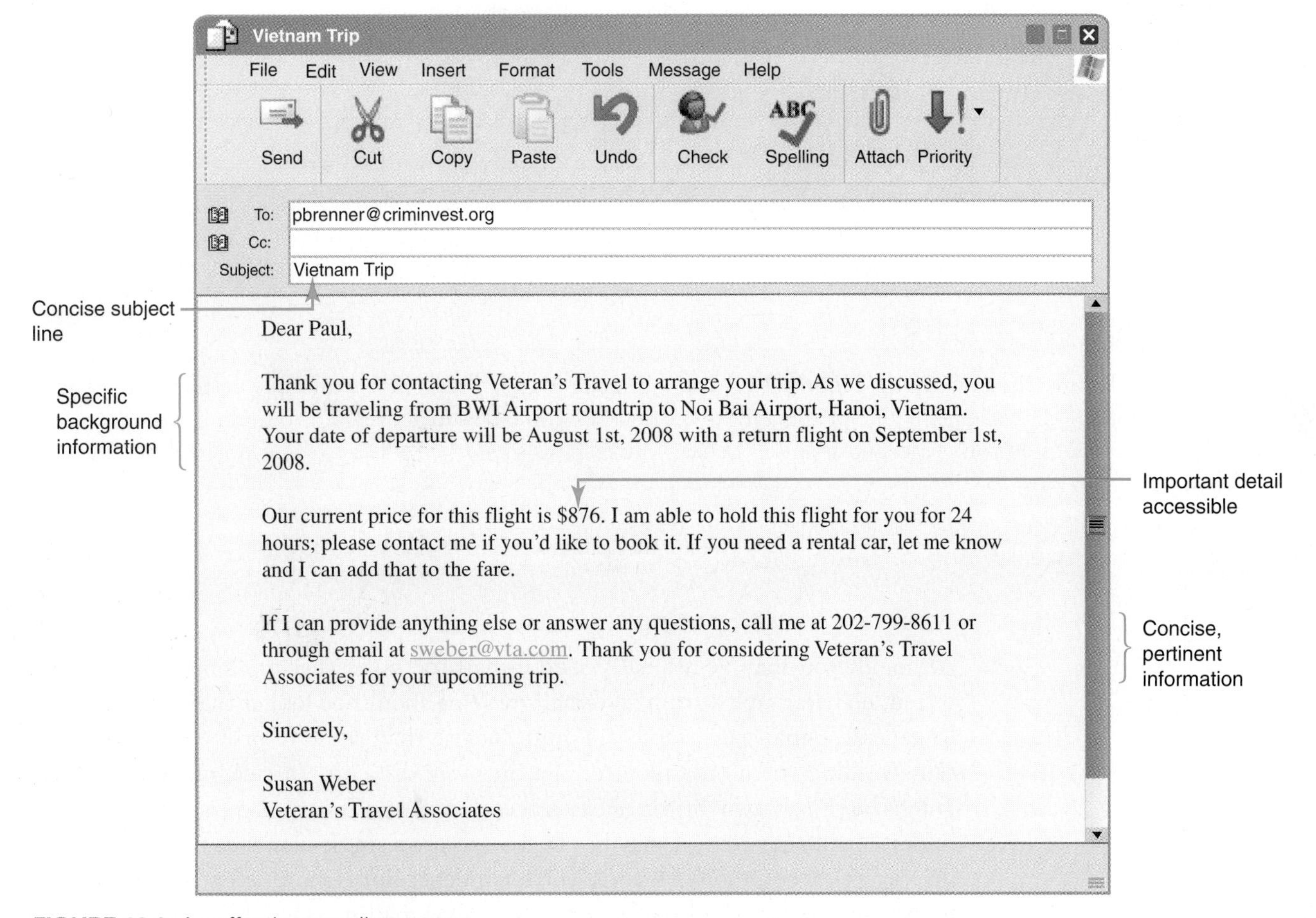

FIGURE 12.3 An effective e-mail message

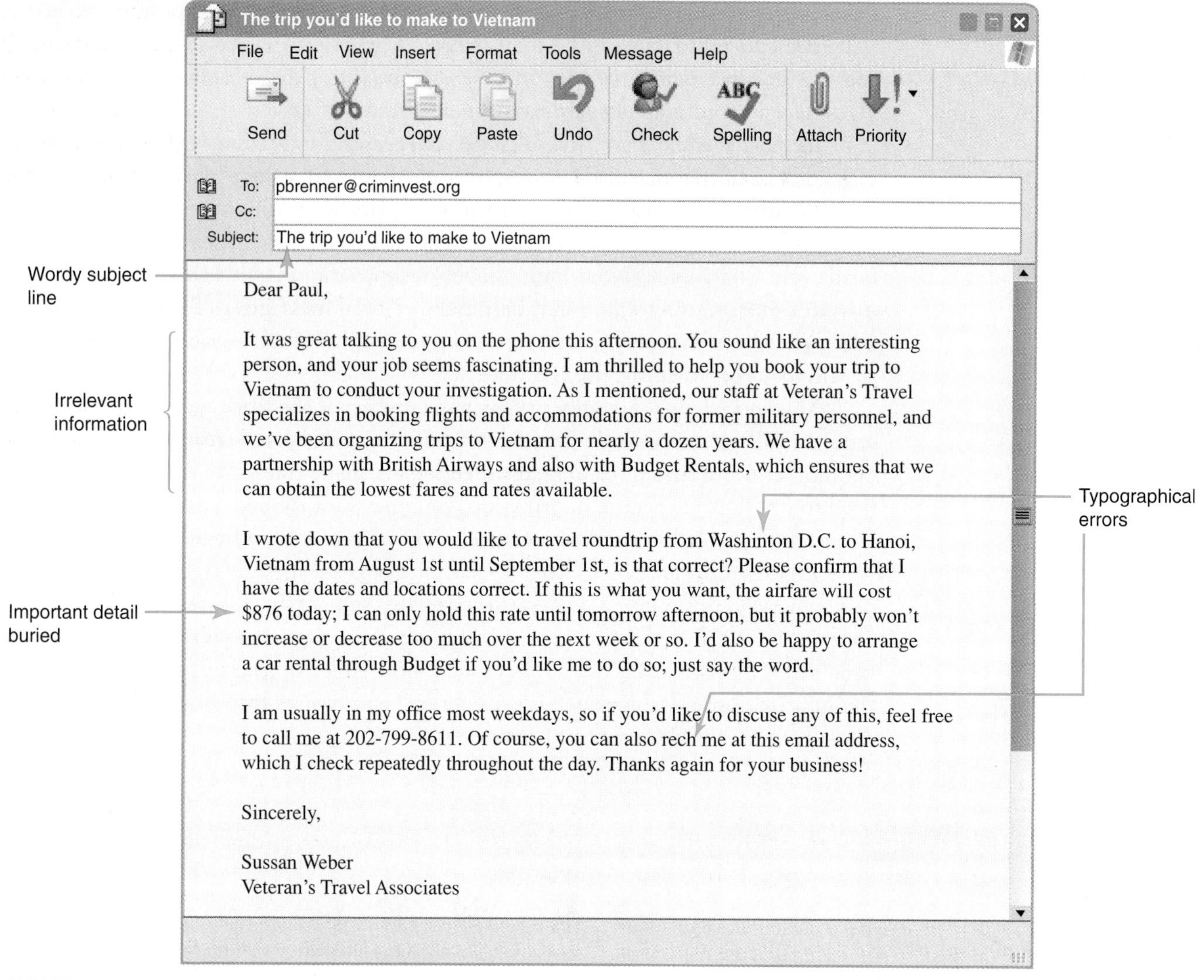

FIGURE 12.4 An ineffective e-mail message

ANALYZE THIS

Spend some time analyzing Figures 12.3 and 12.4. Other than the changes noted in the call outs, what specific revisions do you see in Figure 12.4? What, precisely, makes the second example better? Could the second example be further revised to make it even more effective? How so?

Paragraphs and Spacing

Although you may occasionally send e-mail messages that are longer than a single screen, no paragraph within an e-mail message should be longer than a single screen. In general, e-mail paragraphs are more concise than those in printed documents because reading them on a monitor is a bit more difficult. In fact, many e-mail paragraphs are just two or three sentences long; and when communicating highly important information, it may be effective to use a single-sentence paragraph. Such a practice is less common in printed correspondence but is more acceptable in an e-mail. Of course, the length of the paragraphs in an e-mail message, like that in any other form of written communication, ultimately depends on the rhetorical situation.

You'll typically want to use block format for your paragraphs—that is, no initial indentations and a single line of white space between paragraphs. In fact, most e-mail programs do not allow you to indent paragraphs. You'll also want to be careful when cutting and pasting a message from a word processor into an e-mail because the formatting used in some word processors can create spacing errors in an e-mail that won't be immediately visible to you. When possible, you should simply rewrite the message or include it as an attachment if you are concerned about formatting.

Other Formatting Issues

Italics, underlining, bolded text, bulleted and numbered lists, tables, graphics, and visuals can be quite useful in printed documents, but these features are problematic in e-mail messages. Some e-mail programs simply do not allow you to generate these features in e-mails, and if you are able to produce them, they may appear garbled to readers who use a different software or web-based program to read their e-mails. For example, you may compose a message that looks like this:

Copies of *Technical Communication in the Twenty-First Century* will be distributed as follows:

School	Contact	Date
Michigan State U.	Karla Hayashi	January 10
Ohio State U.	Luke Bailey	January 21
University of Colorado	Kevin Luber	February 9

But your recipients may see a message that looks like this:

Copies of &%Technical Communication in the Twenty-First Century&% will be distributed as follows:

School Contact Date
Michigan State U.Karla Hayashi January 10
Ohio State U.Luke Bailey January 21
University of Colorado Kevin Luber February 9

See chapter 8 for more on using visual elements to emphasize information.

In general, it is best to avoid formatting features like these in e-mail. If you wish to call attention to specific information and must communicate through e-mail, you have a number of options. Many e-mail programs do now include standard word processing features like the ability to add bold, italicized, underscored, size-varied, or even colored text. E-mail clients who have those capabilities should use them to emphasize just as they would in any other document but should do so sparingly, remembering that their audience's e-mail program may not have the same capabilities. Text-based e-mail clients (i.e., those that don't rely on visual cues beyond the basic text) should consider these methods of calling attention to specific information:

1. Use CAPITAL LETTERS to designate a heading.
2. Use a single line of white space between each phrase or word to designate a list.
3. Use an *asterisk* on either side of a word to designate emphasis or italics.

4. Use an_underscore character_at the beginning and ending of a passage to indicate underlining.

Signatures

In some e-mail programs your name will appear alongside your e-mail address in the "From" line, but sometimes only your e-mail address will be listed. In either case it is useful to include your name at the bottom of each e-mail you send. Some people use their e-mail program to create a standard signature that is automatically inserted at the end of each message. Such signatures function like a business card, giving contact information about the sender—for example, job title, phone, fax, web page URL, mailing information. Some people add quotations or **ASCII** text art to their signature to express some aspect of their personality, but these additions should be kept to a minimum because they can slow the transmission of e-mails to people with slower Internet connections. Moreover, longer signatures can become annoying in a series of messages back and forth because recipients may need to scroll past them repetitively to see the entire conversation. Signatures may look like this:

ASCII American Standard Code for Information Interchange, the basic protocol used in most computers to represent letters, numbers, and other characters without formatting

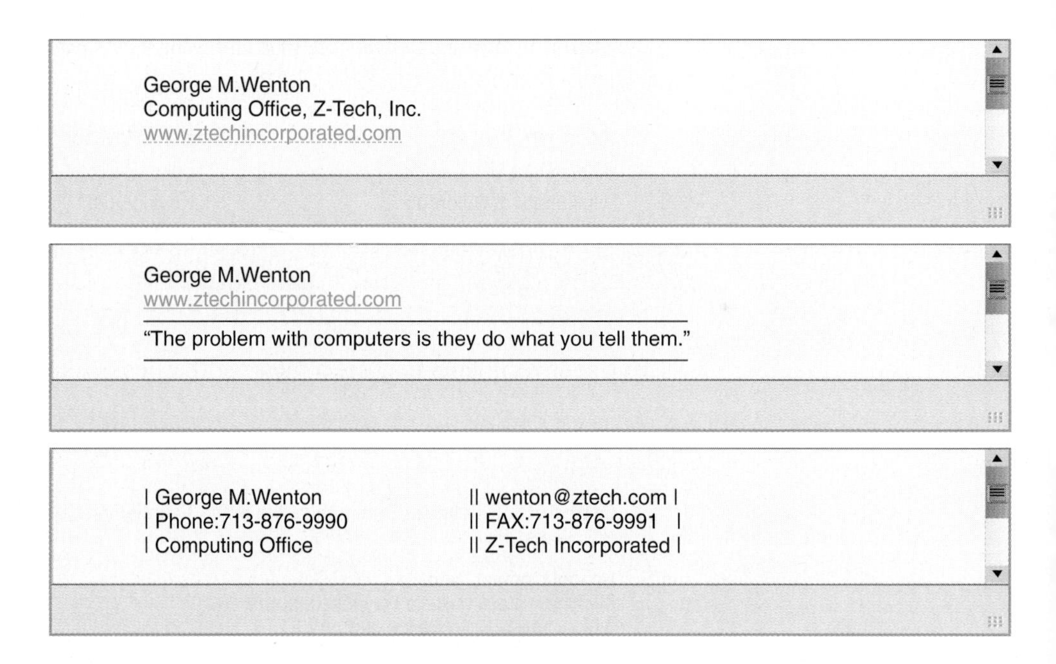

Attachments

E-mail can be a useful way to transmit documents such as reports, proposals, and contracts because it is quick and cheap and transmits an exact duplicate, unlike a fax. In fact, the purpose of many e-mail messages is simply to send a longer formatted document, which is attached to an e-mail. You should send information as an attachment, rather than including it in the body of a message, in the following situations:

- When the message is lengthy
- When formatting is important
- When visuals are an important part of the message
- When the message will be printed and used for a particular purpose

When sending an attachment with an e-mail message, you should make specific reference to the attachment in the body of your e-mail, describing its content, its format (i.e., the program used to create it), and sometimes its size. Without this information recipients may not open the attachment for fear of opening a computer virus. Figure 12.5 shows just the top portion of a lengthy message containing formatting; Figure 12.6 shows the same message transmitted more efficiently as an attachment.

Whenever possible, you should minimize attachments by eliminating unnecessary graphics or visuals. Large attachments can be slow or difficult to download, and some e-mail programs have limits on the file sizes they will accept. In some situations it may be necessary to break a longer document into sections, sending each section as an attachment in a separate e-mail.

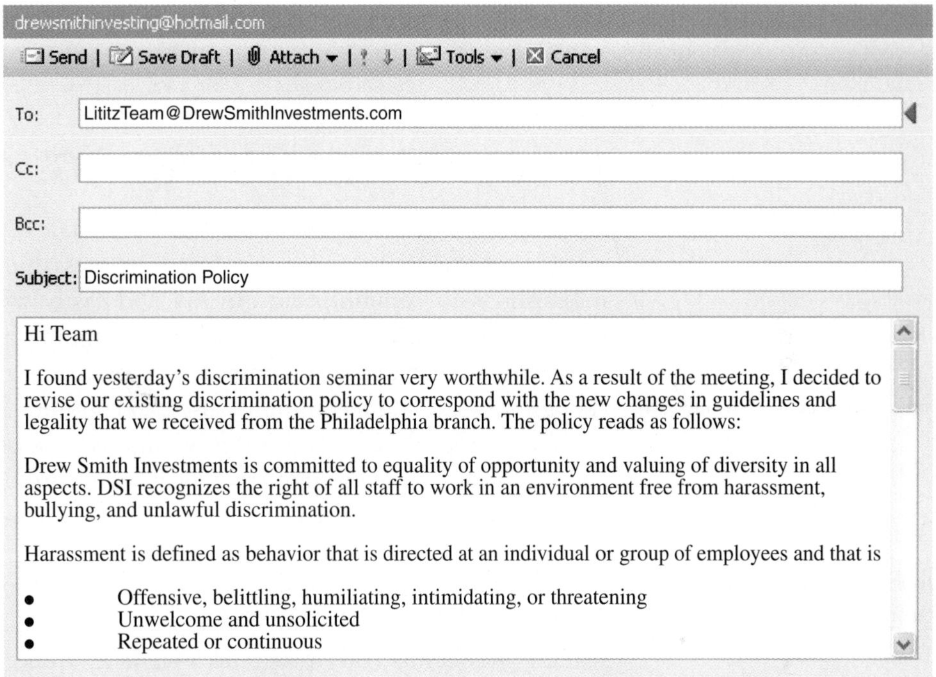

FIGURE 12.5 Top portion of an e-mail message with lengthy content in the body

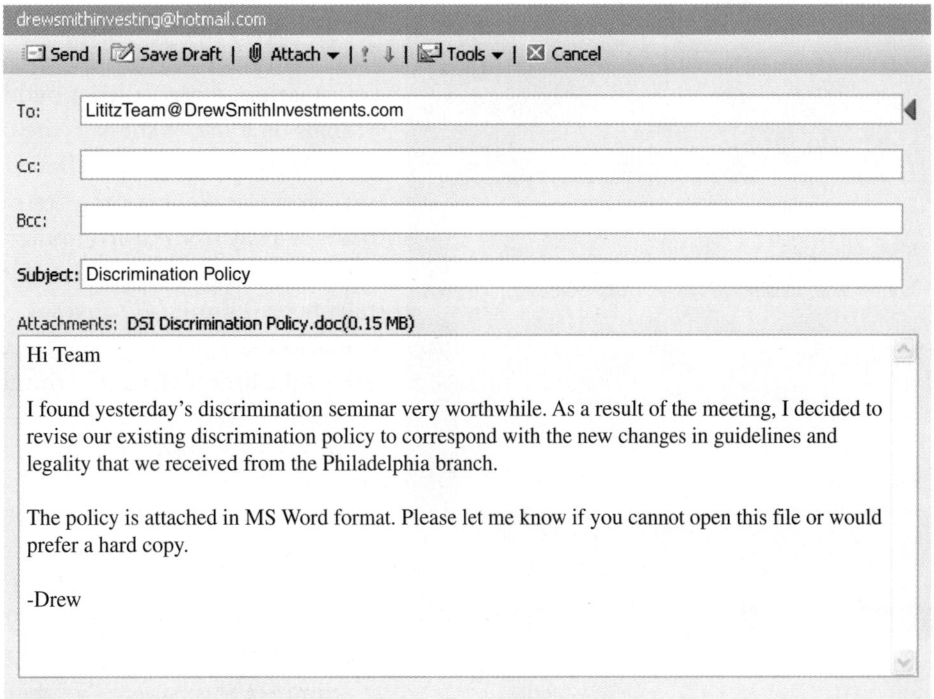

FIGURE 12.6 Complete e-mail message with lengthy content attached

IN YOUR EXPERIENCE

Because e-mail messages are so quickly and easily sent, some writers are more likely to overlook the basic conventions of written communication. However, e-mail messages represent you and your company just as any other form of written communication does. Consequently, you should use complete sentences and proper sentence conventions (e.g., beginning with a capital letter and ending with punctuation), as well as carefully spell-checking and proofreading all of your e-mails before you send them.

You probably have a number of e-mail messages in your account right now. If you do, spend some time looking over them. Do they use proper sentence conventions? Do they appear to be edited and mechanically correct? Of course, such care may not have been necessary if the messages were written for personal reasons, but if these messages had been sent to an employer, customer, or colleague, would they have represented the author in a professional way?

ELECTRONIC MESSAGING

Although e-mail is the primary form of electronic communication in the workplace, many workplace writers rely on electronic messaging—sometimes referred to as e-messaging—to communicate quickly and easily with others. There are two basic types of e-messaging: instant-messaging programs and wireless text-messaging devices. Both can be used to communicate with internal and external audiences, although they are more commonly used to communicate with groups or individuals within a company or organization because the messages themselves are often short and informal. The use of these two forms of electronic messaging continues to grow, keeping pace with the need for ubiquitous, up-to-the-minute communication in the workplace.

synchronous communication communication with the sender and receiver transmitting and receiving at the same time

asynchronous communication communication with the sender and receiver transmitting and receiving at different times

Instant Messaging

Instant messaging (IM) programs—such as those offered by Yahoo!, AOL, and MSN—allow users to communicate in real time (i.e., **synchronously**), whereas e-mail messages are usually stored in a folder that can be read at the recipient's convenience (i.e., **asynchronously**). In addition to simple text communication, most IM programs allow users to send Internet links and files back and forth, and most offer presence information, often called a buddy list, indicating whether individuals on a list of contacts are currently online and available to chat. Most IM programs allow users to set their online status so that group members can determine whether they are available, busy, or away from the computer. As a result, communication via instant messaging can be less intrusive than communication via telephone. Figure 12.7 gives an example of instant messaging in the workplace.

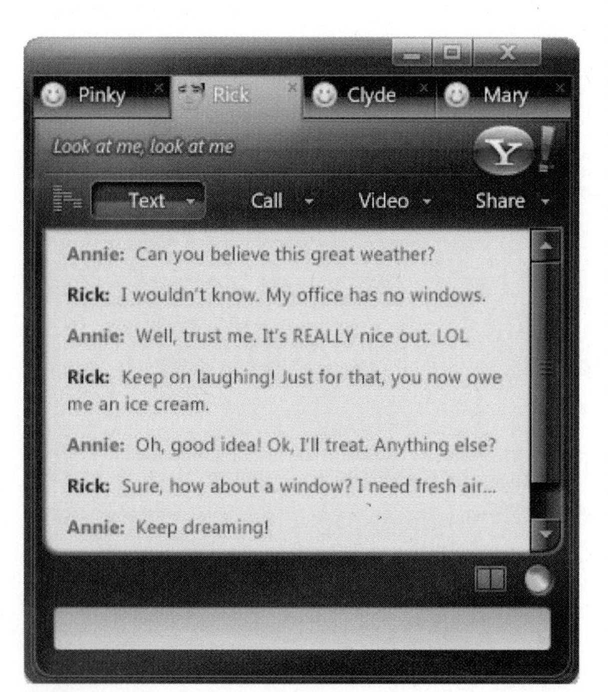

FIGURE 12.7 An example of instant messaging

Like other forms of communication, instant messaging has its own conventions and etiquette. Follow these guidelines in your instant messages:

- Keep messages short and informal to read more like a dialogue than a speech.
- Assess the rhetorical situation to determine which sentence conventions are appropriate. When in doubt, use Standard English, and always pay attention to the basic rules of grammar, mechanics, and spelling, even though some other people may not.

- Avoid using all capital letters unless you need to place great emphasis on something.
- After you have sent an instant message, give the recipient a chance to respond before sending a new message.
- If someone's online status is away, asleep, or otherwise unavailable, do not send continual prompts. You can send a single instant message, asking for a response when the person is online, or you may wish to send an e-mail.

Text Messaging

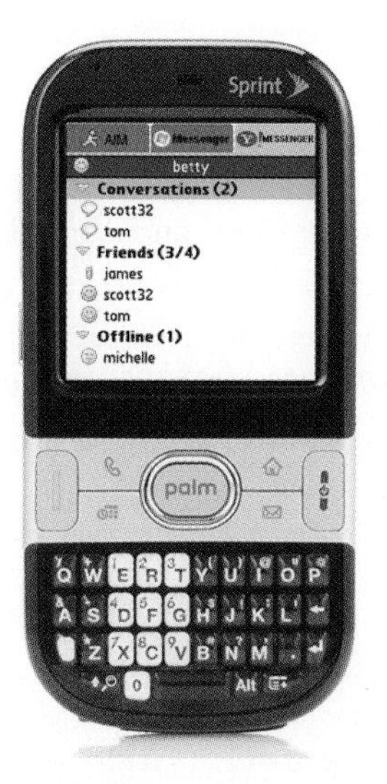

FIGURE 12.8 A personal digital assistant (PDA)

Wireless text-messaging devices, such as smart phones or personal digital assistants (PDAs), are often used to send messages from remote locations. The messages are often short and text based, although sending large or multimedia files wirelessly is becoming easier and less expensive than it has been in the past. Typically, messages sent from a smart phone have a limit of just a few hundred characters, and the sender generally uses the phone buttons to input text into a message. Messages sent from PDAs can be longer, and the user interface generally consists of a small keyboard on the device, which makes typing easier and faster (see Figure 12.8). Nearly *one trillion* text messages were sent in 2007, although most of those were personal rather than workplace messages.

Even though text messaging has been primarily used for personal or informal communication, many businesses and organizations have begun using text messaging as a quick method of conveying up-to-date information to clients. Many airlines, for instance, now send customers text messages about updated flight information, and large organizations like universities now use text messaging to alert students, faculty, and staff about urgent information, such as unexpected school closures during inclement weather or other emergencies. Thus, text messaging is being used to convey information internally and externally, as well as formally and informally.

Text messaging can create interesting rhetorical situations for you in your workplace because you may not be sure how your recipients will receive your message. If you are sure your readers will get the message via another wireless device or a phone, the message should follow the general conventions of instant messaging. However, as the introductory FEMA example demonstrates, your recipients may sometimes receive your wireless text messages via e-mail. In that case a message with short phrases, abbreviated words, or no punctuation may seem abrupt or out of context. Although Marty Bahamonde's message was originally sent from his BlackBerry handheld device, his message is longer and more developed than the typical text message, perhaps because of the situation, perhaps because Bahamonde knew that Brown was likely to read the message through e-mail. Notice that Bahamonde's message contains a closing tag that identifies it as coming from a handheld device (see Figure 12.1). If your device does not automatically include such a tag, consider including one yourself if the method of transmission is likely to affect your recipient's reading.

Because of the limited message length and user interface of most wireless devices, particularly smart phones, many writers use a hybrid language known as texting, which often includes the following features:

- Using abbreviations extensively, especially numbers for words—for example, "4" in place of "for"
- Omitting vowels from words, such as "txt msg"
- Replacing spaces between words with capitalization, such as "ItsTimeToGo"
- Using emoticons to convey a mood or emotion, such as :-) for a smile

There are hundreds of texting abbreviations in use today, including these:

B4	before
BTW	by the way
CUL8R	see you later
F2F	face to face
IOW	in other words
NRN	no reply necessary
OBO	or best offer
RTM	read the manual
TIA	thanks in advance
WYSIWYG	what you see is what you get

Although texting can make your messages easier to compose and shorter in length, you should be careful in professional settings. Because there is no standard texting language, recipients may misinterpret what you have written or fail to understand it. For example, if you sent the message "DoUWnt2Go2NYCL8R?" your colleague might not understand that you're asking, "Do you want to go to New York City later?"

EXPLORE

A number of online sources provide comprehensive dictionaries of texting language. Explore some of these sources, such as netlingo.com or lingo2word.com. Write a short description of one of these sources and its usefulness to your own workplace writing. Do you use or anticipate using texting in your workplace? What terms might be useful for you? Do you believe that most of your coworkers would understand texting language, or would they need an explanation first?

BASIC ELEMENTS OF MEMOS

In the past, memos were the primary means of internal communication in the workplace, but e-mail is now used in place of memos in many instances because it is fast, cheap, easy to use, and easy to store. However, memos are still quite common and will continue to be, because they are the better choice for more detailed communication, for correspondence in which formatting is important, and for more important or confidential correspondence. The self-test at the end of this chapter will help you determine whether an e-mail or a memo is better in a given rhetorical situation. In some highly important internal situations, a letter may be used in place of a memo—for example, in hiring, firing, promoting, or censuring an employee; communicating a formal policy change; or detailing a contract or agreement between various departments or offices.

See chapter 13 for more information about letters in highly important or confidential writing situations.

There are minor variations among organizations in the format of memos. Some use a company logo or letterhead at the top of their memos, others assume that the format itself designates internal communication, and still others indicate that the correspondence is internal with the word *Memo* or *Interoffice* at the top of the document. Some organizations use boldface for identifying information, others use capital letters for the first letter of each major word, some use single spacing, and others use double-spacing (see Figure 12.9). In short, there is no one right way to format a memo, although some formats may be more effective than others. To familiarize yourself with the formatting styles used within your organization, it is best to look at an example of a successful internal memo or ask trustworthy colleagues to explain the formatting styles they use.

Date: April 16, 2009
To: Bill Wright
From: Angela Anderson
Subject: BUDGET REQUEST FOR PORTLAND TRAVEL—MAY 20, 2009

Energy Systems Limited, Eugene, Oregon

Date: April 16, 2009
To: Bill Wright, Payroll Coordinator
From: Angela Anderson, Field Consultant
Subject: Budget Request for Portland Travel—May 20, 2009

Interoffice Memo

To: B. Wright

From: A. Anderson

Subject: Budget Request for Portland Travel—May 20, 2009

Date: April 16, 2009

FIGURE 12.9 Varying memo formats

Identifying Information

Despite formatting variations, most memos begin with four key components, which provide some of the background information readers need to understand the message.

- *To:* Because every memo is written to some individual or group, a line identifying the intended audience is listed in the top section of nearly all memos. In most cases you will list the primary audience for the memo, although you'll want to carefully consider other audiences that might read your document. Your audience may be an individual, a group, or even several individuals or groups.

  Organizations differ in how to address the intended audience: some use the person's full name, some use the first initial and the last name, and others list the person's full name and job title (see Figure 12.9 for varying memo formats). If you work for a large organization, do not know the individual you are addressing, or are giving bad news, it may be appropriate to use the full name and title to maintain some formality. Listing a title can also provide useful context for others in the organization who do not know the recipient or who may read your message at a later date.

- *From:* In most cases you will be the person listed in the From line because you are sending the memo. Sometimes, though, you may wish to list the name of the person taking responsibility for the memo (e.g., if you are composing the message for your supervisor) or list several names (e.g., those on your team or committee) or list the name of a department or division. If you are listing several individual names, you should order them alphabetically or in descending order of rank. Here again you'll have to decide whether to use a full name, a first initial and last name, or a full name and title.

  Memos often include the author's initials, handwritten in blue or black ink, after the typed name to show that he or she has reviewed the memo and accepts responsibility for it. The From and the To lines should always use the same format, spacing, font, and font size.
- *Subject:* Because your readers may receive many memos, you should create a subject line that they can skim quickly before deciding to read the body of the memo. Your goal is to be specific yet concise, giving your readers precise information about the subject of the memo but omitting anything confusing, misleading, or unnecessary. If times, dates, and locations are important to your message, you should try to include them in the subject line. Your subject should be a phrase rather than a complete sentence, should not contain end punctuation, and should fit on one line if possible.
- *Date:* The date a memo was sent can be significant in tracing the history of an issue, idea, or problem within an organization. You should give the date the memo was sent, not the date you began writing the memo. Some organizations prefer standard dating (May 20, 2009), whereas others prefer military style (20 May 2009). You should use whichever dating method is most common in your organization. However, if the memo is likely to be scanned, faxed, or photocopied, military style may be more readable.

Content

Because memos are skimmed for content, the information in them should be easy to find, clear, and concise. The first paragraph of a memo should explain your reason for writing. In fact, many memos begin with a stand-alone statement that identifies the purpose, such as "I am writing to request . . . ," "I want to tell you about a problem . . . ," or even "The purpose of this memo is to inform you . . . ." Other memos include the main point or purpose for writing in a longer first paragraph. A good test is to read aloud to a colleague the first paragraph of your finished memo and ask him or her to identify the basic purpose of the memo. If there is any uncertainty, you may need to revise that introduction.

What you choose to include in each following paragraph depends on your subject, your reason for writing, and the relationship you have with your readers. At the very least you must be sure to include all of the necessary information your readers will need to make an informed decision or to respond appropriately. Sometimes you may want to summarize the activities or events that led up to the memo, such as an earlier meeting or a change in policy. If you refer to an earlier meeting or a conversation you had with your readers, it is usually best to include the specific date and/or location of that meeting.

Organization is important in memos; each paragraph should contain a central idea or focus. If you need to include lots of details in a paragraph, consider using a bulleted or numbered list or a heading to help your readers understand the information. Because memos typically conclude with a restatement of the action you hope your readers will take, you might place this information at the end of your memo for emphasis. The memo guide shown in Figure 12.10 provides further detail about the content, organization, and formatting of a typical memo. This guide should not be used to simply "plug in" information, but it can serve as a general model of what is typically found in a memo.

Letterhead of the Organization

To:	Name and perhaps title of the recipient
From:	Name of author, followed by handwritten initials
Subject:	Or "Re"in reference to an earlier message
Date:	Standard or military format

Main point or thesis
Unless there is some reason to be indirect, a summary of the memo's content and purpose should be given immediately. Some memos begin with one stand-alone sentence identifying the main point; others include it in a longer first paragraph.

Introductory paragraph
In addition to the main point of the memo, the first paragraph may describe the action the writer wants the reader to take or may contain background information about the writer, the subject, or the writer's relationship with the reader. An introductory paragraph in a memo is typically three to six sentences long.

Individual sentences
Each sentence in the memo must be clear, precise, and error free. The writer should spend time revising each sentence to meet these criteria. Sentences in memos are typically five to fifteen words long.

Spacing
Memos typically use single spacing between lines of text in order to include as much information as possible on each page. Most memos also use a block format for paragraphs—that is, a single line of white space between paragraphs and no indentations.

Body paragraphs
Each body paragraph should contain one central idea, usually found in a topic sentence at the beginning or the end of the paragraph. Each body paragraph should follow a logical order, allowing the reader to anticipate the subject or topic of each following paragraph.

Concluding paragraph
The concluding paragraph generally does not include a closing statement or detailed contact information for the writer. Instead, the concluding paragraph often restates the action the reader should take or describes a plan for the future, the benefits or difficulties of the topic just presented, or a time and place to discuss the subject further.

Enclosure, copy, or distribution information
The writer may wish to include additional information with the memo, send a copy to someone else, or distribute the memo to a group. That information is typically placed at the end of the memo, much as it is in a letter (see chapter 13).

FIGURE 12.10 A guide for creating memos

ANALYZE THIS

Most word processing programs contain several different memo templates that allow you to insert the contents of your message into a predetermined format. Choose one word processing program, and analyze the memo templates that are included with it. Do you find these templates to be useful guides in organizing information, or do you feel that they would restrict you in creating memos for specific purposes and audiences? Do these templates lend themselves to a particular type or style of message?

See chapter 21 for more on informal reports.

Length and Formatting

Although some companies and organizations limit the length of memos to one page, you may sometimes need to write memos that are significantly longer. For instance, some informal reports are written as multipage memos. In such cases each page beyond the first should be printed on plain paper (i.e., no letterhead). These additional pages should include, at the top, the name of the person to whom the memo is sent (flush with the left margin), the page number (in the center), and the date (at the right margin). That information will look like this:

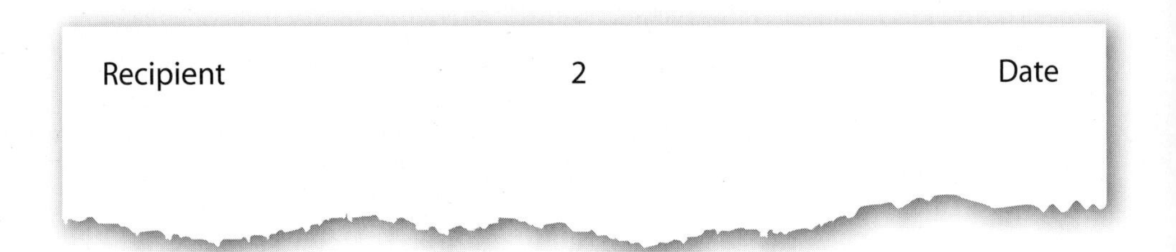

If possible, use the header function of your word processing program to automatically insert this information; this will help you avoid formatting and spacing problems if you revise or edit your memo later.

Regardless of the length of your memo, be sure that the final page has at least five lines of text on it. If it does not, you may wish to manipulate the spacing, font, font size, or margins to make your document appear longer or shorter. Similarly, throughout the memo you should avoid widows (i.e., only one line of a paragraph at the bottom of a page) and orphans (i.e., only one line of a paragraph bumped to a new page). Many word processing programs have commands to address these problems.

Design Features

Because memos are often skimmed quickly, various design features such as headings, lists, and tables may help to highlight and organize information in your memos. Likewise, visuals such as graphs, charts, and photographs can help to explain or identify the textual information in your memo. These design features become increasingly important as the length of your document grows. Many of the examples in this chapter use design features to deliver information more effectively. See chapters 8 and 9 for more information about visual rhetoric and layout and design.

TYPES OF MEMOS

Workplace writers find themselves in a variety of situations requiring memos; the circumstances that call for a short document can be positive, negative, informative, persuasive, or even a combination of these. Some of these situations are so common that the documents relating to them have specific names and functions. Even though you should always focus on your audience and the message of your memo, you may be asked to create or help create a specific type of message, and it can be useful to understand what some of these types are. Each of the following memo types is organized and structured in particular ways.

Directive Memos

Some memos provide information about policies or about tasks the readers should perform. These are called directive memos, or directives, because they direct readers in how to act or proceed in a given situation. Directives should begin with a clear, concise statement of the purpose of the memo. If your audience has a choice in how to act or proceed, you should explain that choice. Figure 12.11 is a directive memo that informs a group of employees about how to purchase wireless devices.

Progress or Status Memos

Memos are often used to supply information about the status of a project or assignment, describing what has been done so far and what remains to be done. Known as progress or status memos, these documents are generally written to

Asimov Robotics, Inc.

Date: December 4, 2008

To: Engineering Division

From: Del Spooner, Payroll Division DS

Subject: Wireless Device Purchases

Statement of purpose →

As of January 1, 2009, all engineers are instructed to carry a wireless device when visiting job sites.

All Engineering Division employees are asked to make their own purchases. Please limit the price of your wireless device to $600. You may choose to purchase a device that is more expensive, but you must pay the balance of the cost.

Choices available to fulfill the directive →

You have two choices for making these purchases:

Numbered list explaining how to proceed

1. You may use our corporate account through www.wirelessdevice.com to order your wireless device. Please select ground shipping and request insurance on the delivery. To obtain the account information, contact Susan Calvin in the Payroll Office at ext. 4554.
2. You may purchase a wireless device at any local electronics store if you need it immediately. If you choose this option, use your own credit card, and turn in your receipt to Susan. You will be reimbursed for the expense within two to four weeks.

As you know, our office also has an account with a wireless service provider. After you have obtained your device, see Kevin Lanning in the IT Department to set up your account.

If you have any questions about this purchase approval or how to get your wireless device, feel free to contact me at ext. 4553 or at spooner@asimov.com.

FIGURE 12.11 A directive memo

supervisors or managers. Figure 12.12 is a progress memo, describing the status of a bridge construction project.

Trip or Field Reports

Workplace writers may need to visit an outside location to gather information, test a product or service, or meet with customers or colleagues. Such visits are generally documented in memo format. Trip reports usually document the specific events, expenditures, and outcomes of business trips. Field reports are used to provide details about off-site visits, inspections, or tests. Figure 12.13 is a trip report describing a seminar attended by a telecommunications employee.

Blossom Hill Construction
218 Convenience Way
Minneapolis , MI 55401

Date: October 3, 2009
To: Blossom Hill Construction Project Leaders
From: Val Rowe, Public Information Coordinator
Subject: Blossom Hill Construction Weekly Update

Overview of the memo

This week, Blossom Hill Construction continues its work on the Light Rail Transit (LRT) Connection in downtown Minneapolis. The following is a summary of this week's construction activities.

Downtown Minneapolis – Light Rail Transit (LRT) Connection

Bulleted lists to help organize the information

- Construction crews are scheduled to begin setting girders and beams for the south half of the 5th Street North bridge over the BNSF railroad late this week. The south half of the bridge is being constructed for Hiawatha LRT use.
- Demolition of the south half of the 5th Street North bridge over I-394 is scheduled to begin later in October. Detailed traffic control plans—including lane closures, temporary I-394 on and off-ramp restrictions, and other detour information—will be distributed in future *Blossom Hill Construction Updates.*
- See photos of LRT Connection construction below.

Images to provide further clarification

South View of 5th Street bridge

East View of LRT Construction

If you have any questions or concerns regarding Blossom Hill Construction or any other aspect of the project, please contact Chris Lindser at 651-326-5540 or by e-mail at clindser@blossomhill.com.

FIGURE 12.12 A progress memo

Specific introductory details

Chronological explanation of what was observed and discussed

Analysis and overview

MEMORANDUM

To: Tyler Durden, Call Center Coordinator
From: Marla Singer, Network Supervisor
Subject: Incoming Calls Management Institute Seminar
Date: November 19, 2007

On November 13, 2007, I attended a seminar in Los Angeles that was sponsored by Incoming Calls Management Institute. The name of the seminar was "Understanding and Applying Today's Call Center Technologies." The seminar leader was Raymond K. Hessel, a consultant with Parker-Morris Corporation.

We talked about strategic alignment—how important it is that our call center technology enable our call center strategy, which supports our business strategy. We looked at the customer contact challenges presented today. We must use multiple media and build relationships with customers—not just service their needs over the telephone.

We learned the why, what, how, and when of a broad array of call center technologies, from basic functions to advanced tools. We started with a discussion about infrastructure—application architectures and network architectures. We talked about some of the emerging architectures that move intelligence outside the switch and leverage Internet Protocol (IP) for voice switching.

We looked at the specific capabilities of each technology, as well as the business challenges it addresses. For each one we discussed the major vendors and how their offerings varied. Our materials packet included a handout listing vendors by category, along with their corporate URLs. There is also a valuable resource list with online learning centers, magazines, conferences, organizations, and other sources to turn to. Updates to this will be available to us ongoing at www.parkermorris.net.

The instructor explained how the various technologies work together and how different configurations can address similar issues. We also talked about the benefits to the company, the call center supervisor or manager, the CSR, and, of course, the customer.

We had group exercises to apply what we were learning and a lot of good discussion among the class members. In addition to the materials already mentioned, the course packet included a number of other useful takeaways, including a glossary, a what-fits-where table to help identify when to use different technologies, Call Center and Voice Response Best Practices and Trends, and a planning and implementation toolkit.

I think this course was worth attending because I learned about the newest strategies in our field. I would recommend that others from our organization attend. I learned a lot about the many call center technologies and have new ideas about why they make sense in our business and what we should do next. I would like to meet with you to discuss how we can use some of the helpful tools to get started on our strategic call center direction here at Paper Street Communications.

FIGURE 12.13 A trip report

Response Memos

The purpose of response memos is to provide audiences with information they have requested. Most response memos begin with a clear, specific answer to a question or request, followed by explanation or details about the subject. If the readers need further information to act, you should include it in the memo. Figure 12.14 is a memo written in response to a request for a groundbreaking ceremony at a newly constructed warehouse.

Lab Reports

The results or outcomes of in-house experiments, procedures, or studies are often described in lab reports. These memos are sometimes written from one researcher or expert to another and may contain mathematical or numerical data about the subject.

Kenneth Fifer Realtors
307 Mesa Ave,
Taos, NM 87571

Date: February 4, 2009
To: Gary Kunkelman, Commercial Realty Department
From: Terry Baker, Advertising Promotions Manager
Subject: Gila Cliff Groundbreaking Event

Clear answer to the request →

Your department's request for a groundbreaking ceremony at the Gila Cliff Warehouse has been approved.

The Advertising and Promotions Department will take care of the following items in preparation for the groundbreaking ceremony on March 17th, 2009:

Bulleted lists to explain the different responsibilities →

- **Advertising the event in *Taos Commerce Magazine***
- **Tent rental**
- **Purchase of ceremonial ribbon, scissors, and shovels**
- **Producing and printing flyers announcing the event**
- **Ordering catering for approximately 100 guests**

We ask that the Commercial Realty Department take care of the details listed below. These Haspects of the ceremony are necessary but are beyond the scope of our department. You should:

- **Deliver copy of what you want in the announcement flyers (text only)**
- **Distribute flyers after we've delivered them to you**
- **Make personal phone calls to special guests**
- **Arrange for guest speakers or advisory board members**
- **Clean up Gila Cliff location in preparation of the event**

Forward-looking statement that builds goodwill →

We are thrilled that we can help with this exciting event, and we know it will be a great success. The Gila Cliff warehouse signals an important development in Fifer Realtors' plan for the future. We will be available to coordinate these activities with you, should you need us. Contact me personally at tlb@fiferrealty.com or call me at extension 8799.

FIGURE 12.14 A response memo

Most lab reports explain the methods or steps taken in the experiment or procedure, followed by the results and analysis. Figure 12.15 is a lab report summarizing a chemistry experiment conducted in a training lab.

Minutes of a Meeting

Formal workplace meetings may be documented in memos. The minutes of a meeting usually contain the time, date, and location of the meeting; a list of who was present; and a chronological summary of what was discussed or accomplished at the meeting. Most meetings end with a call for announcements, which is reflected in the minutes of those meetings. Minutes can be written in various formats and may not

Subject: Determination of the solubility product constant for lead iodide
To: Dr. Stephan Wilson
From: Eugene Smith, Chemistry Intern *ES*
Date: 10 October 2008

This memo summarizes a lab experiment performed in the Redox Corporation Training Lab. The purpose of the experiment was to determine the value of the solubility product constant of lead (II) iodide. An approximate value was obtained at room temperature, and exact values were determined at higher temperatures.

The solubility product constant (Ksp) of a substance numerically represents the position of equilibrium, as the substance dissolves at a certain temperature. At equilibrium the substance is dissolving at the same rate as it is crystallizing. For lead (II) iodide the constant expression is Ksp = $[Pb^{2+}]\,[I^-]^2$.

The following data were obtained and/or calculated during the experiment:

Numerical data

T.T.	Ppt.	Temperature at dissolution (°C)	Solubility (M)	Ktr	Ksp
A	Y	63	0.0050	5.0×10^{-7}	5.0×10^{-7}
B	Y	54	0.0040	2.6×10^{-7}	2.6×10^{-7}
C	Y	44	0.0030	1.1×10^{-7}	1.1×10^{-7}
D	Y	35	0.0020	3.2×10^{-8}	3.2×10^{-8}
E	N			1.4×10^{-8}	$> 1.4 \times 10^{-8}$
F	N			4.0×10^{-9}	$> 4.0 \times 10^{-9}$

Specialized language written specifically for expert audience

At a given temperature the value of the Ksp expression is called a Ktr value. The Ktr value is equal to the Ksp value at the temperature at which a precipitate just begins to form, or just dissolves, assuming the concentrations have remained constant while the temperature is varied. Since at room temperature test tube E did not form a precipitate, it can be assumed that the value of Ksp is greater than the Ktr value of 1.4×10^{-8}. Because test tube D did form a precipitate at room temperature, it can be assumed that the value of Ksp is smaller than its Ktr value of 3.2×10^{-8}. These two values give a range of Ksp from 1.4×10^{-8} to 3.2×10^{-8} for PbI_2 at room temperature. The Ksp values are shown in the table.

From these values it can be seen that as the temperature increases, the value of Ksp also increases, as does the solubility. The relationship between the value of Ksp and the solubility was expected, based on the Ksp expression. The relationship to temperature may be explained by the availability of more energy at higher temperatures, so that the bonds in the solid may be more easily broken.

Results and analysis

The results of this experiment indicate that the solubility of a solid increases at higher temperatures. It also gives a range of Ksp values for PbI_2 at room temperature—1.4×10^{-8} to 3.2×10^{-8}. This is slightly higher than the published value of 8.5×10^{-9}.

FIGURE 12.15 A lab report

follow a standard memo format. Figure 12.16 contains the minutes of a town development board meeting.

Figures 12.11, 12.12, 12.13, 12.14, 12.15, and 12.16 give examples of memo types written for a variety of workplace situations. Spend some time analyzing these memos, and then use them as a reference when you create memos of your own.

ANALYZE THIS

As you can see, the six sample memos in Figures 12.11 to 12.16 are all formatted differently and contain different identifying information and different body layout. Study the design features in each. Do you think that these six memos were consciously designed to be as they are? Does the rhetorical situation for each dictate its layout? In other words, do the subjects, purposes, and audiences of these six memos help to shape the way they are written? Explain. Could the formats used in these memos be used interchangeably? Why or why not?

Details about the meeting

A list of who was present

Chronological summary of discussion and actions

Minutes of September 2, 2008, Meeting

A regular meeting of the Downtown Development Board was held on September 2, 2008, in AD 205.

Present: Presiding Officer Julie Etson, Secretary Carl Webley, Nancy Millard, Chris Strain, Tabitha Stevens, David Ganns, Jun Kim, Susan Bosh, Barry Ford, Raul Cortez, Max Schmidt, Becka Pauls

The meeting was called to order at 3:00 PM.

1. **Minutes:**
   David Ganns moved to accept the minutes of the 8/19/08 meeting. The motion was seconded, and the minutes were approved unanimously.
2. **Presentation—Owens Paving**
   A representative from Owens Paving gave a short presentation about ways to cap sidewalks to prevent cracking. A sign-up sheet was passed around for board members who wished to visit a worksite to see how the capping works.
3. **Banker's Report—Nancy Millard**
   Nancy addressed our upcoming budget, noting that we have a 50K surplus for the current fiscal year. Discussion ensued concerning the use of this surplus. Nancy agreed to chair a committee to explore options; contact her to participate.
4. **Standing Committee Reports:**
   Waste Management—Tabitha Stevens
   A new company (A-Plus Waste) has been contracted for waste management; they will continue the route and schedule we have used in the past.
   Landscaping—Chris Strain
   Chris noted the excellent job our landscapers did in preparation for the bicentennial celebration. He has sent them a formal letter of thanks and asks that we extend their contract an additional five years.
   Technology—Raul Cortez
   Plans are underway to purchase new computers for Development Board members who need them. See Dan to make your request.
   Historical—Becka Pauls
   Several old maps of the area have recently been found and are currently being cleaned and prepared for framing. They will be displayed in the town hall after the work is completed.
   Borough Manager—Barry Ford
   The search for a new borough manager continues until October. At that time an interview committee will be constituted to begin reading applications. Contact Barry if you'd like to participate.
5. **Announcements:**
   - A production of *Cat on a Hot Tin Roof* will be held at Bock Theater next week. Tickets are on sale at the box office.
   - The *Downtown Gazette* is offering 20 percent discounts on advertisements from now until the end of the year in an effort to increase circulation.
   - New stop lights are being installed in the business district; traffic may be rerouted around areas where work is taking place.

The meeting was adjourned at 4:00 PM.

FIGURE 12.16 Minutes of a meeting

EXPLORE

Earlier in this chapter you learned that e-mail messages are not private or confidential. That same reality applies to memos as well; even a memo intended for a private audience can wind up in the hands of unintended readers. Of course, the worst-case scenario occurs when an embarrassing memo becomes available to the public. The website www.internalmemos.com contains hundreds of such memos and e-mails from companies like IBM, SlimFast, Warner Records, and even the Pentagon.

Find one or two interesting memos on this website; some require a paid subscription, but many are free. Bring the memos to class and discuss them, focusing on what makes each message embarrassing or difficult for the author. Then discuss whether the memo might have been written differently had the author known that it would be made public. For more information about the website, read the CNN.com article found at http://www.cnn.com/2002/TECH/internet/08/22/net.internalmemos/.

CHOOSING BETWEEN E-MAIL AND A MEMO

Even though there are many similarities between printed memos and internal e-mail messages, you will probably encounter numerous situations in which one of these two mediums is a better choice. To determine which medium to use in a particular situation, we encourage you to assess the circumstances that appear in the following checklist. If you address these issues in each new situation, you'll probably make the right choices. In the event that a situation meets criteria in both categories, consider sending the message as a memo attached to an e-mail message.

E-mail Memo Checklist

Condition	E-mail	Memo
1. The message will be sent to a **large group**.	✓	
2. The message requires a **rapid response**.	✓	
3. Recipients are **geographically distant**.	✓	
4. Recipients will need to **modify/revise** the message.	✓	
5. Recipients may need to **print and distribute** part of the message.	✓	
6. The message will be **longer than one computer screen**.		✓
7. The message requires **careful formatting**.		✓
8. The message requires **detailed graphics and visuals**.		✓
9. The message contains highly **important information**.		✓
10. The message contains somewhat **sensitive information**.		✓

SUMMARY

- Internal e-mails, e-messages, and memos are used to transmit small chunks of nonsensitive information and communicate with colleagues and coworkers.
- Internal e-mails and memos are different from external e-mails and letters in the following ways:
  - Internal documents tend to be less formal than external documents.
  - Internal documents generally do not include formal greetings and closing statements.
  - E-mails and memos are often less detailed than other types of documents.
- The PSA is as important in short documents as it is in longer documents.
- Workplace writers should consider their privacy rights when sending e-mail messages.
- E-mail messages are usually made up of certain key elements: subject lines, recipients, message content, signatures, and attachments.
- Electronic messaging is becoming increasingly important in workplace communication.
  - Instant messaging allows individuals to communicate in real time, generally from their computers.
  - Text messaging allows individuals to use their smart phones or other wireless devices to communicate from remote locations.
- Memos start with the following identifying information: To, From, Subject, and Date.
- Memos contain several key elements:
  - A main point or thesis at the beginning of the memo
  - An introductory paragraph that provides background, context, or an overview
  - Body paragraphs that contain one central idea each
  - A concluding paragraph that restates the message

- Memos are often one page long but sometimes longer.
- Graphics, visuals, and other design features can help to organize and highlight important information in a memo.
- Memos come in a variety of types, including directive memos, progress or status memos, trip reports, field reports, response memos, lab reports, and minutes of a meeting.

CONCEPT REVIEW

1. What is the difference between an internal and an external document?
2. How should you use the problem-solving approach in short documents?
3. Are e-mail messages and e-messages private?
4. What are the basic elements of an e-mail message?
5. Why is brevity important in an e-mail message?
6. How long should you make most e-mail paragraphs?
7. Why do e-mail messages use a block format?
8. Why is formatting problematic in e-mail messages?
9. When should you use an attachment in your e-mail message?
10. What are the two primary forms of electronic messaging?
11. What is synchronous or asynchronous communication?
12. How is texting different from standard writing?
13. When should you use a memo, and when should you use an e-mail message?
14. What are the basic elements of a memo?
15. How do most memos begin?
16. How long are most memos?
17. What situations might require a memo?

CASE STUDY 1

Communicating Budget Cuts at Lake Clarke National Park

Lake Clarke National Park, located in upstate New York, covers an area of approximately eighty square miles and provides hiking, camping, swimming, fishing, and boating opportunities for nearly 200,000 visitors each year. Approximately two hundred employees work for the Lake Clarke National Park Service (LCNPS). In September 2006 the director of LCNPS issued a series of memos to employees, discussing budget cuts that would affect staffing and hours of operation at the park. Although the memos were intended for employees only, several were published on the Internet, including a September 20, 2006, e-mail message sent to park supervisors (see Figure 12.17). Local media outlets—including newspapers, a regional magazine, and websites—criticized the LCNPS memos, suggesting that they encouraged LCNPS employees to intentionally mislead the public on the impact of the budget cuts. Examine the memo in Figure 12.17, considering these questions:

- In what ways does the memo encourage LCNPS employees to intentionally mislead the public?
- The heading of the memo refers to "operational cutbacks," yet the body of the message refers to "service level adjustments." What are the effects of these two phrases? Do they mean the same thing? Why does the memo place the phrase "service level adjustments" in quotation marks throughout the memo?
- The fifth paragraph of the memo addresses the ways in which each supervisor should communicate with the public about specific plans. Does this paragraph raise any ethical questions concerning accuracy and truthfulness in

communicating with the public? Do you believe the LCNPS has an obligation to be truthful in the information it discloses (or fails to disclose) to the public? Why or why not?

- Aside from any ethical issues, do you feel as if this is a well-written message? Did the LCNPS director of operations communicate clearly and succinctly with her staff? Would the staff members have known what was expected of them?
- Why did LCNPS officials choose to distribute this message through e-mail? Would the message have worked better as an attachment? Would the message be different if it had been distributed through postal mail?

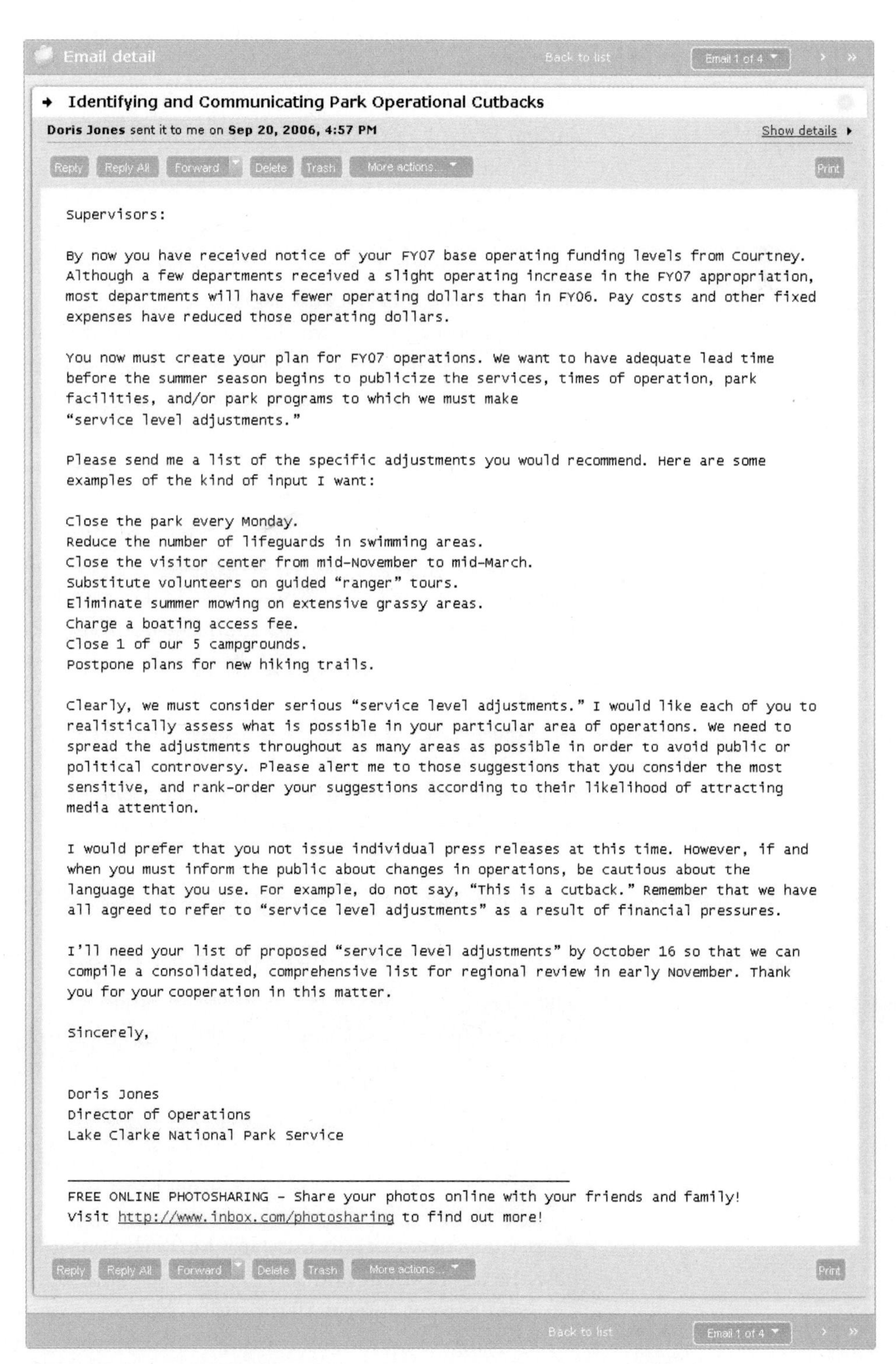

Email detail
Back to list
Email 1 of 4

Identifying and Communicating Park Operational Cutbacks

Doris Jones sent it to me on **Sep 20, 2006, 4:57 PM** Show details

Reply | Reply All | Forward | Delete | Trash | More actions... | Print

Supervisors:

By now you have received notice of your FY07 base operating funding levels from Courtney. Although a few departments received a slight operating increase in the FY07 appropriation, most departments will have fewer operating dollars than in FY06. Pay costs and other fixed expenses have reduced those operating dollars.

You now must create your plan for FY07 operations. We want to have adequate lead time before the summer season begins to publicize the services, times of operation, park facilities, and/or park programs to which we must make "service level adjustments."

Please send me a list of the specific adjustments you would recommend. Here are some examples of the kind of input I want:

Close the park every Monday.
Reduce the number of lifeguards in swimming areas.
Close the visitor center from mid-November to mid-March.
Substitute volunteers on guided "ranger" tours.
Eliminate summer mowing on extensive grassy areas.
Charge a boating access fee.
Close 1 of our 5 campgrounds.
Postpone plans for new hiking trails.

Clearly, we must consider serious "service level adjustments." I would like each of you to realistically assess what is possible in your particular area of operations. We need to spread the adjustments throughout as many areas as possible in order to avoid public or political controversy. Please alert me to those suggestions that you consider the most sensitive, and rank-order your suggestions according to their likelihood of attracting media attention.

I would prefer that you not issue individual press releases at this time. However, if and when you must inform the public about changes in operations, be cautious about the language that you use. For example, do not say, "This is a cutback." Remember that we have all agreed to refer to "service level adjustments" as a result of financial pressures.

I'll need your list of proposed "service level adjustments" by October 16 so that we can compile a consolidated, comprehensive list for regional review in early November. Thank you for your cooperation in this matter.

Sincerely,

Doris Jones
Director of Operations
Lake Clarke National Park Service

Reply | Reply All | Forward | Delete | Trash | More actions... | Print

Back to list
Email 1 of 4

FIGURE 12.17 LCNPS memo

- Do you think that the memo would have been made publicly available on the Internet if it had not already been in electronic format? In other words, what role did the genre of the message (i.e., e-mail) play in the distribution of the message to the public at large?

Your instructor may ask you to revise the e-mail message based on your responses.

CASE STUDY 2

To Text or Not to Text: That Is the Question

In June 2003 the British newspaper *The Telegraph* reported that a thirteen-year-old student had submitted a school essay that started like this: "My SMMR hols wr CWOT. B4, we usd 2g02 NY 2C my bro, his GF & thr3: -kds FTF.ILNY, it's a gr8 plc." According to the article, the entire essay was written in text-messaging language. The student's teacher was mortified by the girl's use of "hieroglyphics" and said she could not translate the paper. *The Telegraph* offered a translation: "My summer holidays were a complete waste of time. Before, we used to go to New York to see my brother, his girlfriend and their three screaming kids face to face. I love New York, it's a great place." The paper then initiated a conversation about whether texting was destroying the English language.

In 2006 a project was launched to translate many of literature's greatest works into text-message language. For instance, a plot summary of *Romeo and Juliet* might read like this: "A feud between two houses—Montague and Capulet. Romeo Montague falls in love with Juliet Capulet, and they marry secretly. But Romeo kills Juliet's cousin and is banished. Juliet fakes her own death. As part of the plan to be with Romeo, she writes him a letter, but it never reaches him. Everyone is confused, and both lovers kill themselves." That summary would be translated like this: "FeudTween2hses—Montague&Capulet. RomeoM falls_<3w/_JulietC@mary Secretly Bt R kils J's Coz&isbanishd. J fakes Death. As Part of Plan2b-w/R Bt_ leter Bt It Nvr Reachs Him. Evry1confuzd—bothLuvrs kil Emselves."

Text-messaging language has become prevalent in many contexts in which it might seem out of place—for example, in school essays or translations of Shakespearean plays. Do some research, and identify other contexts in which text messaging is being used even though it might seem out of place. Then write a short report about text messaging outside of workplace environments; consider whether the use of textmessaging in those contexts might be useful.

CASE STUDIES ON THE COMPANION WEBSITE

Alban Paper: New Philosophies for a New Era

The EPA tested the water in a stream adjacent to Alban Paper and found that it contained more than three times the allowable limit of dioxins. Alban Paper is not looking forward to EPA fines but is more concerned about losing the company's excellent reputation.

Opening the Lines of Communication: Addressing Concerns about Network Safety

As assistant network administrator for PHP, your job is to help employees understand how the applications and services they use interact with the network and how they can keep patient information private. You are often putting out technological fires while trying to communicate to PHP employees about how they can prevent such fires from cropping up.

When Does Workplace Communication Become Too Much Communication?

The structure of Liberty Lighting recently came under review, and the venture capital board advised instituting a more traditional organizational plan for employees. An instant-messaging system now used to communicate between employees and between employees and their families has come into question.

Poor Communication: A Case of Friendship at Capital Oil

As the leader of an oil company's audit team, you face difficult challenges when a coworker embarrasses your department. Your task involves rhetorically navigating both personal and professional relationships while advocating professional intradepartmental communication.

VIDEO CASE STUDY

Agritechno

Poncho and Lefty to the Rescue: The Case of the Missing Microfiber Filters

Synopsis

In this case Agritechno geneticist Lev Andropov is working in his lab with his lab assistants Benjamin Goodman and Adam Simpson, whom Andropov playfully calls Poncho and Lefty. Lev has been looking through a supply cabinet for a box of microfiber filters he needs for the work he is doing, but he is unable to locate any. When he asks his lab assistants where the filters are, they tell him that they are out of them and that they placed a request three weeks ago with the purchasing department to order more, but the purchasing department has not delivered the filters yet.

WRITING SCENARIOS

1. In a small group of three or four, discuss some problems you've encountered on your campus, such as long lines at the financial aid office, a lack of parking spaces, or dangerous walkways in front of the library. Decide on one problem that is most pressing or important, and write a short memo to the dean of students or the university president. Outline the problem, and recommend specific ways to solve it.
2. In a small group of three or four, select one of the six sample memos in this chapter (Figures 12.11–12.16). Study the memo, and then create an e-mail message transmitting the same information. Consider whether the information should go in the body of the e-mail or should be an attachment. If you opt for an attachment, think about how best to describe it in the body of the e-mail. If you have access to a computer, send the re-created message to your instructor through e-mail.
3. Imagine that your class has been invited to attend an employment fair, but it takes place during the time that your technical writing class meets. In a small group of three or four, write a memo to your instructor, requesting that your class be canceled to allow you to attend the fair. Provide specific reasons that the fair would be beneficial, and explain why it would be a worthwhile use of class time. Then compare your memo with others in the class.

4. The Internet contains thousands of examples of memos and e-mail messages. Search for one that you think is especially effective or ineffective, and write a short summary and an analysis of it.
5. The Internet contains many sites that provide advice or basic instruction in writing effective memos. Do a search using a phrase such as "writing effective memos" or "memo writing advice," and select a site that gives such instruction or advice. After you have read through the site's offering, compare it to the advice in this chapter. What advice is similar in the two sources? What advice is different? Which provides better instruction? Why?
6. This chapter introduces a number of technological terms, phrases, or keywords, such as *ASCII*, *synchronous*, and *asynchronous*. Some are defined in the margins, whereas others are not. Select one important keyword or phrase from this chapter, and research it. Then write a comprehensive definition of the word or phrase.
7. The relatively new medium of e-mail creates new rhetorical situations for workplace writers—for example, determining what is fair, considerate, and appropriate when communicating with others through e-mail. In a small group of three of four, develop a short list of guidelines for ethical e-mail communication. If you need help, do an online search for information, perhaps beginning with Virginia Shea's netiquette at http://www.albion.com/netiquette/book/index.html.
8. Make a short list of problems or issues in your local community, town, or city, and then select one issue that seems most important and urgent. Write an e-mail message to your congressperson, describing the problem or issue and what might be done to correct it. Remember that your congressperson receives dozens, perhaps hundreds, of e-mails every day, so you'll have to make your message brief, clear, specific, and memorable. You can find your congressperson's e-mail, phone, and mailing information at the "Write Your Representative" web page at http://www.house.gov/writerep/.
9. Write a brief memo to the instructor of this course, outlining what you've learned so far and what you hope to learn by the end of the term. If there are specific subjects or topics you'd like to see the instructor cover that are not listed on the class syllabus, use this memo as an opportunity to mention them. In addition, you may wish to provide some feedback on those aspects of the class you've found most useful or interesting or those you've found confusing or unclear. Be specific and diplomatic about any positive or negative feedback you provide.
10. Find an example of a memo online, and analyze the layout and design it uses. What particular layout and design choices did the author make? How is this example different from or similar to the examples provided in this chapter?
11. Imagine that your employer has asked you to design and create a standard signature that employees should use in their e-mail messages. What information should it include? In what order should the information appear? Will the signature include anything other than basic text? Design a sample signature to serve as a model for other employees.
12. Imagine that you have just been sent the following internal e-mail message:

    > We need your yearly sales figures by close of business today. If we receive them later than that, we will lose an account. All salespeople should call the office as soon as possible with this information.

    You realize that one of your salespeople is at a seminar and can be contacted only through a text message. Rewrite this message as a text message to be sent via smart phone, keeping it as short as possible while retaining key information. Feel free to use texting abbreviations. Compare your message to that of a classmate. Which uses fewer characters? Are they equally effective?

13. Do some basic research on current trends in wireless handheld devices. In a short memo or e-mail to a fictitious employer, recommend that a particular device or service be purchased for all employees. Discuss the features, compatibility, and price of the device or service that support your recommendation.
14. Imagine that your instructor has agreed to hold virtual office hours two evenings per week, when students can ask questions in an online setting about the readings, assignments, or other aspects of the class. Your instructor wants to use an instant-messaging program to conduct these virtual office hours but has not yet decided on a program and has asked for your input. Do some basic research on the various IM programs available to download. In an e-mail message describe the program you believe to be the best choice, highlighting the features and usefulness that make it the best choice.
15. Chances are good that your college or university provides an e-mail account for all or most students. Research your school's policy regarding student e-mail accounts, and then write a one-page memo to incoming freshmen about those accounts. Be sure to mention how to obtain an account as well as how to use it for courses and educational correspondence. If such a document already exists, read it carefully, and attempt to revise it to make it more clear and effective for new students.

PROBLEM SOLVING IN YOUR WRITING

Select one of the documents that you wrote and designed for this chapter. In a short memo to your instructor, analyze the rhetorical choices you made in the layout, design, and content of the document. As part of your analysis, you may wish to address whether you used any of the models in the chapter as a guide or template. How did you determine which formatting features would be most appropriate for the situation? How did you determine the order, placement, and paragraphing of information in the document? How did you determine an appropriate subject line?

13

Letters

CHAPTER LEARNING OUTCOMES

After completing this chapter, you will be able to do the following:

- Understand why the **problem-solving approach** is important in writing letters
- Consider how letters are used in a **real-world situation**
- Recognize the basic elements of letters: **heading or letterhead**, **date**, **recipient's address**, **salutation or greeting**, **introductory/body/concluding paragraphs**, **closing phrase**, **signature**, and **page headers**
- Consider **additional features** often found in letters
- Identify and select **block style** or **modified block style**
- Understand the basic patterns for **organizing** letters
- Recognize common types of letters:
  - **adjustment letters**
  - **claim letters**
  - **collection letters**
  - **confirmation letters**
  - **inquiry letters**
  - **rejection and refusal letters**
  - **sales letters**
  - **thank you and congratulatory letters**

DIGITAL RESOURCES

On the Companion Website www.prenhall.com/dobrin:

- Case 1: Bad Loans and Bad Vibes: An Assistant Manager Is Placed in a Sticky Situation
- Case 2: Cutbacks and Quality Care at St. Martin's Hospital: A New Medical Professional Faces Ethical Dilemmas in the Workplace
- Case 3: Fletcher Global Medical Manufacturers: An Engineer Confronts Quality-Control Issues and Helps Combat Negative Publicity
- Case 4: The Charles Grove Memorial Scholarship: A Case for Endowment
- Video Case: Testing the GlobeTalk Handheld Devices
- PowerPoint Chapter Review, Test-Prep Quiz, Exercises and Activities

REAL PEOPLE, REAL WRITING

MEET DAVE FERRON • Vice President of SunTrust Bank

What kinds of writing do you do as a commercial banker?

I do a lot of internal writing, such as memoranda, e-mail, and credit underwriting reports, which take the form of a memo and have a specific flow with consistent components. I also correspond by e-mail with clients or prospective clients. In the last several years, more and more customers have felt comfortable using e-mail. It seems that we connect a lot quicker by e-mail than we might otherwise. As a result, I do less of the formal letter writing than I used to, but I still write many letters. The bulk of these are thank-you letters to customers.

Does SunTrust have a structured form for writing letters?

We have a template for commitment letters, but we often deviate from that template. There are specific clauses we have to include, but we can add as much detail as we need. On occasion the reader may happen to be, say, a licensed attorney or the owner of a company and may want the language to change a bit. What I have to do then is understand what the issues are with our specific language, and I have to convey those issues to our in-house legal council to get an agreement to the proposed changes in the language before I can change a standard document.

How would you say you learned to write within your company?

It was trial by fire. When I first started in banking about nine years ago, I attended a training program that included a writing course, which involved some basic instruction. I also used letters that others had written for the company as a guide. I don't follow any standard form now, but when I began, I got examples of different types of letters and used them as models.

When you write letters now, do you involve others in your writing? Do other people see your letters before you send them?

If I involve anyone in my writing, it is typically my assistant. But because we are all so pressed for time, there seem to be only two scenarios for how I get letters written. One is that I do the letter, and it just goes out without anyone else revising or editing it. The better scenario is to send it to my assistant for review. I'd say maybe 80 percent of the time no one reviews my letters, so I need to write, revise, and edit them very carefully.

What advice about workplace writing do you have for the readers of this textbook?

I don't think you can get enough writing classes, honestly. Whether it's creative writing or technical writing, learning about writing is critical. I think that the people who do the best, whether in this industry or in other fields, are those who communicate best both verbally and in writing. Those who communicate well tend to succeed and go to the top. I guess the advice, then, is to work on your writing as much as you can, and that's not a marshmallow answer.

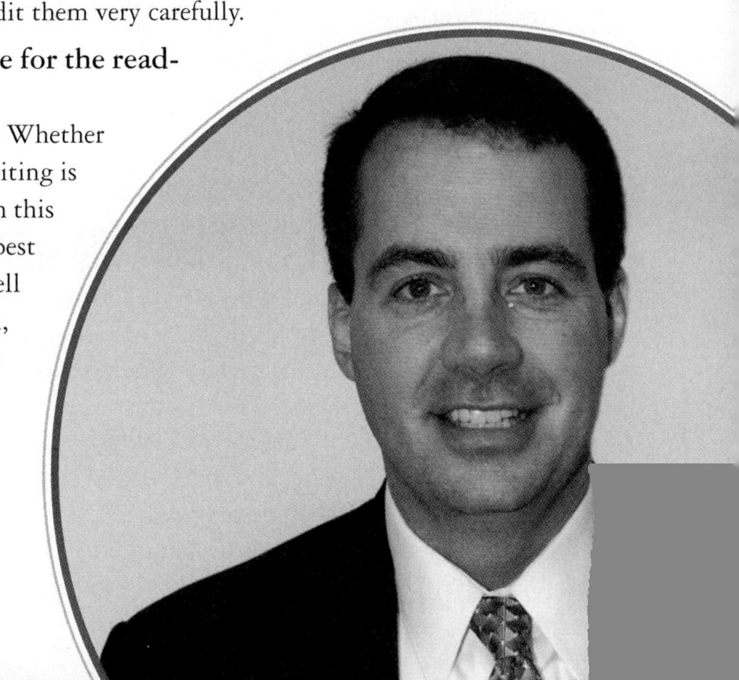

INTRODUCTION

Every day millions of letters are sent and received worldwide. Although e-mail and e-messages are quickly becoming the vehicles for much correspondence in the business and technical world, the letter is still a primary means of solving external problems in business and industry. Although in your professional career you may not write in all of the genres described in this book, you will surely write letters, regardless of your job description. Letters are usually addressed to someone outside your organization, and like e-mails, e-messages, and memos, letters generally transmit smaller chunks of information. This chapter addresses the basic elements found in most letters, provides organizational strategies for communicating positive, negative, and persuasive messages through letters, and discusses common types of letters.

There are distinct differences between internal and external messages. Letters are often more formal than memos because they usually address audiences who are not part of the writer's organization and who may not share the author's goals, perspectives, or language. Letters may also contain more background information and less specialized language than memos because the recipient of a letter may not have the same knowledge and level of familiarity with a product, service, or idea as someone within the writer's organization. In addition, most letters include a more formal greeting (e.g., "Dear Sir") and closing statement (e.g., "I look forward to working with you"), which help to establish a bond between the author and the reader.

PSA

Letters solve problems ranging from formal to informal, specific to general, and individual to public. Regardless of the problem, letters should never be written hastily or without careful attention to detail; the problem-solving approach can be a useful guide in creating letters. Workplace writers must carefully plan how they will respond to situations requiring letters, they must conduct research when necessary to adequately address or solve the given problems, and they must carefully draft and review their letters before they distribute them.

IN YOUR **EXPERIENCE**

Think about the last letter you wrote or received. What was the subject, purpose, and audience of that letter? Was the letter of a personal or a professional nature? How did the subject, purpose, and audience of the letter influence the style, tone, and level of formality used in the letter? What were the external problems for which the letter was written? What rhetorical problems did the letter deal with? In a small group or in class, discuss the subjects, purposes, and audiences of the letters you've written or received, looking for similarities and differences among them.

As just one example of workplace problem solving through letters, Figure 13.1 deals with a highly public problem faced by the U.S. Food and Drug Administration (FDA). In 2004 several clinical studies of Vioxx, an arthritis and pain medication produced by Merck, concluded that the medication increased the risk of heart attack and stroke in some patients. News of these studies became the focus of close media scrutiny, and Merck responded by withdrawing Vioxx from its $2.5 billion market in over eighty countries. Needless to say, the FDA wanted to spread the news of this withdrawal quickly in order to minimize both the health risks to consumers taking Vioxx and the risk of liability for health-care providers prescribing the medication. Consequently, the FDA chose several genres and media to help spread the word about Merck's withdrawal of Vioxx, including a series of letters to distributors, health-care professionals, and patients who were then taking the medication. The FDA's letter to the public is an adjustment letter that describes a change—in this case, a withdrawal—in a product or a service. It also relates a negative message because the information was apt to disappoint, concern, or frustrate patients who were using Vioxx.

FDA employees must have reflected carefully on the rhetorical situation facing them in order to determine which document formats would be most useful in addressing this problem. When a series of letters was decided on as a primary means of communication, it is safe to assume that every aspect of those letters, from layout and design to word choice and phrasing, was carefully analyzed and discussed. In fact, we can guess that a host of FDA employees—scientists, technical writers, lawyers, and administrators—helped to research, write, and revise each letter before it was released. Thus, FDA used letters as a primary method of addressing a serious external problem, just as you might do when attempting to solve a problem in your workplace. Examine the FDA's letter in Figure 13.1, and then address the questions in the Analyze This box, which correspond to it.

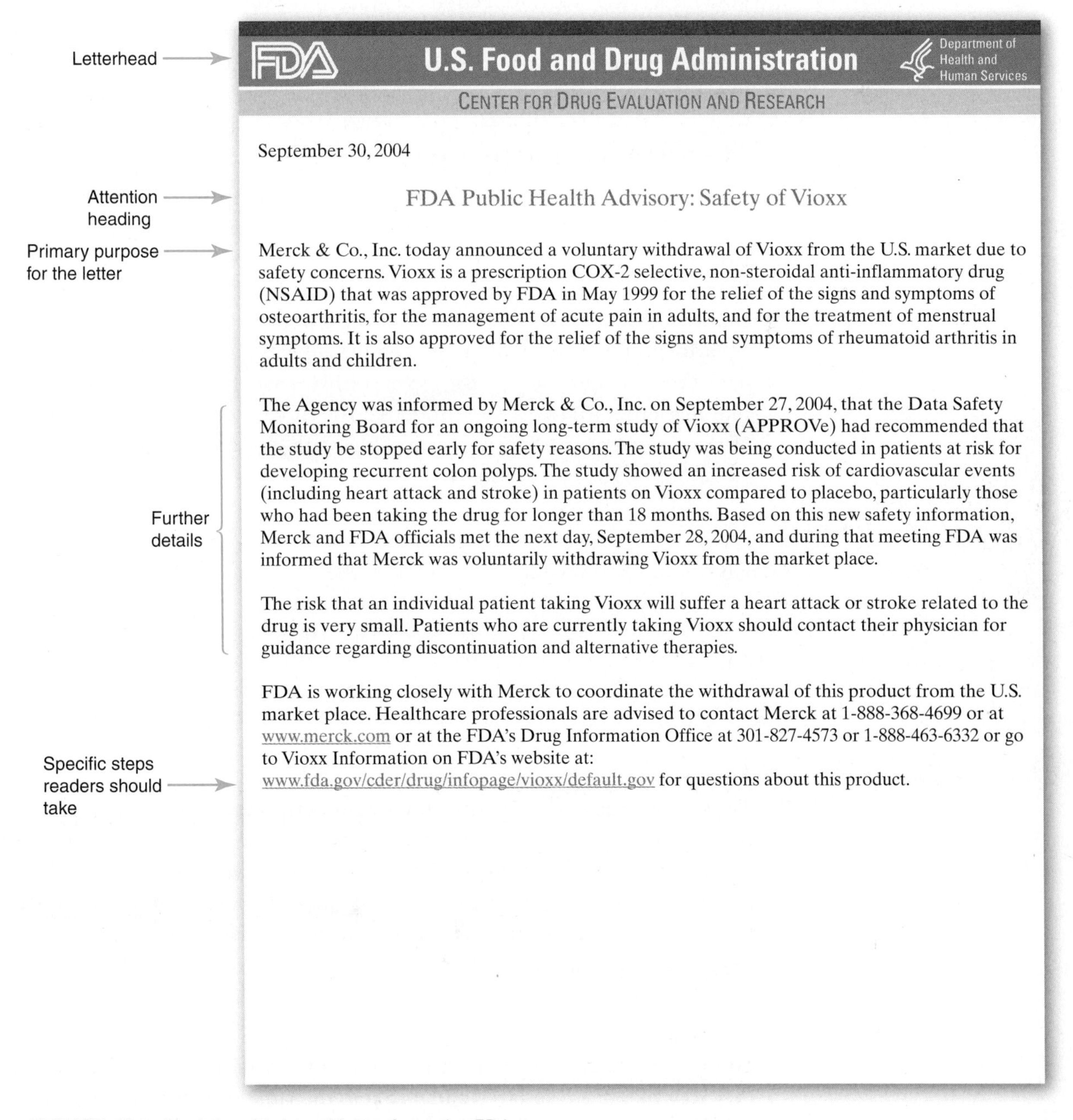

FDA U.S. Food and Drug Administration — Department of Health and Human Services

CENTER FOR DRUG EVALUATION AND RESEARCH

September 30, 2004

FDA Public Health Advisory: Safety of Vioxx

Merck & Co., Inc. today announced a voluntary withdrawal of Vioxx from the U.S. market due to safety concerns. Vioxx is a prescription COX-2 selective, non-steroidal anti-inflammatory drug (NSAID) that was approved by FDA in May 1999 for the relief of the signs and symptoms of osteoarthritis, for the management of acute pain in adults, and for the treatment of menstrual symptoms. It is also approved for the relief of the signs and symptoms of rheumatoid arthritis in adults and children.

The Agency was informed by Merck & Co., Inc. on September 27, 2004, that the Data Safety Monitoring Board for an ongoing long-term study of Vioxx (APPROVe) had recommended that the study be stopped early for safety reasons. The study was being conducted in patients at risk for developing recurrent colon polyps. The study showed an increased risk of cardiovascular events (including heart attack and stroke) in patients on Vioxx compared to placebo, particularly those who had been taking the drug for longer than 18 months. Based on this new safety information, Merck and FDA officials met the next day, September 28, 2004, and during that meeting FDA was informed that Merck was voluntarily withdrawing Vioxx from the market place.

The risk that an individual patient taking Vioxx will suffer a heart attack or stroke related to the drug is very small. Patients who are currently taking Vioxx should contact their physician for guidance regarding discontinuation and alternative therapies.

FDA is working closely with Merck to coordinate the withdrawal of this product from the U.S. market place. Healthcare professionals are advised to contact Merck at 1-888-368-4699 or at www.merck.com or at the FDA's Drug Information Office at 301-827-4573 or 1-888-463-6332 or go to Vioxx Information on FDA's website at:
www.fda.gov/cder/drug/infopage/vioxx/default.gov for questions about this product.

FIGURE 13.1 Product withdrawal letter from the FDA

ANALYZE THIS

Discuss or write about how effectively the letter in Figure 13.1 deals with a major workplace problem. What were the external (i.e., real-world) problems that faced FDA? Does this letter adequately address those problems? Why or why not? What were the various rhetorical problems facing FDA as it attempted to communicate with the public? Why do you think the FDA chose to communicate this information in a letter? Did the letter have an appropriate level of formality and detail for the audience? What information does this letter highlight or downplay? Does the FDA present the information in an ethical manner?

BASIC ELEMENTS OF LETTERS

Letters can take various forms and can include different stylistic and organizational features. These variations are often shaped by the external problems a letter might address, as well as by the rhetorical problems surrounding the writer, the audience, and the purpose for writing. As Dave Ferron suggests in the opening interview, a company or organization sometimes requires its employees to use a template as a guideline for outgoing letters, but experienced writers often adapt a template to suit the occasion. As you compose workplace letters of your own, consider using the primary elements described here.

Heading or Letterhead

If it is available, you should use stationery with company letterhead for all outgoing letters. Such stationery appears more professional than blank paper and allows readers to easily identify the source of the letter before they read it. As Figure 13.1 shows, the FDA used a simple graphic along with the name of the company so that customers could quickly determine the source of the letter.

If you are using blank paper, you should create a heading that contains your address (but not your name) at the top of the first page. If you choose to center the header information, it should come before the date; if you choose to align it on the left side of the document, it should go below the date. Figure 13.2 provides three different variations of this information.

EXPLORE

Letterhead is not only useful in providing contact information about a company, but can also help to define and emphasize the company's image, goals, and beliefs for new readers. Do a web search using a term like "company letterhead" or "letterhead samples," and analyze the letterheads you find. Do these letterheads tell you anything about the company's goals or beliefs? What colors, images, and font choices do they reflect? Do the letterheads contain address, phone, e-mail, or web page information about the company?

Date

The date a letter was sent may be important in tracing the history of an idea, an action, or a problem, particularly if potential legal issues are involved, as is the case with the Vioxx letter. Always include the date your message was sent to your audience, even if you began composing the message earlier. And you should simply list the date—never use a "Date" line in a letter. The date generally appears two lines below the heading, aligned left (see Figure 13.2).

Recipient's Address

You should address letters to a specific individual whenever possible; this helps ensure that your message is received and read. On the same line as the name, you should also include the recipient's job title if possible. Using a title like Mr. or Ms. is optional, although it is generally advisable to include more formal titles like Dr. or Rev. when appropriate. You should use the postal service's two-letter state abbreviations in any addresses you include in your correspondence. The recipient's address is generally placed two lines below the date, aligned left.

Salutation or Greeting

The salutation should begin with the word *Dear*, continue with the recipient's last name, and end with a colon—not a comma. If you don't know the recipient's gender or professional title, use both the first and the last names in the salutation. If you don't know the name of the recipient, refer to the department or use a generic name that

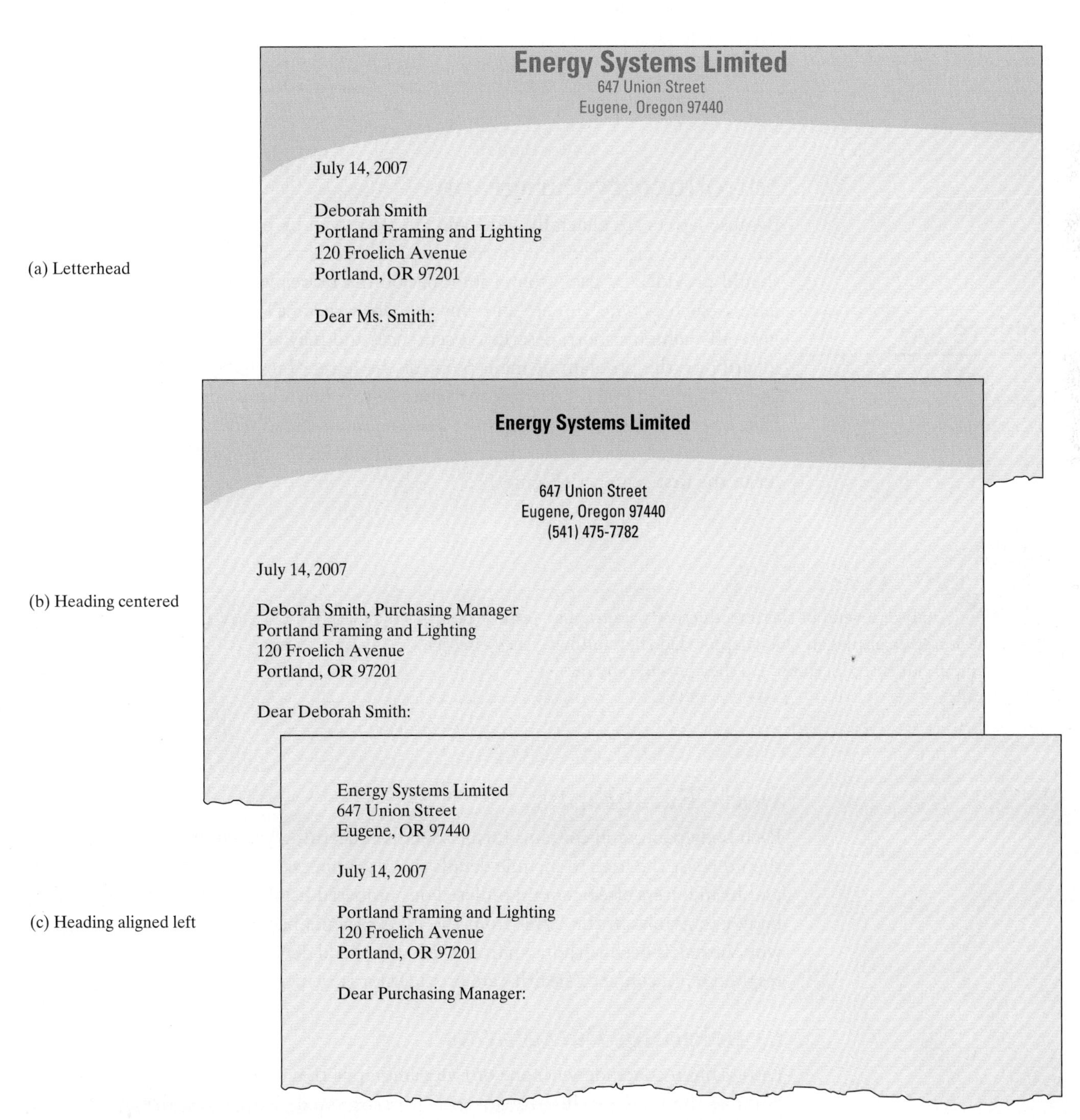

(a) Letterhead

Energy Systems Limited
647 Union Street
Eugene, Oregon 97440

July 14, 2007

Deborah Smith
Portland Framing and Lighting
120 Froelich Avenue
Portland, OR 97201

Dear Ms. Smith:

(b) Heading centered

Energy Systems Limited

647 Union Street
Eugene, Oregon 97440
(541) 475-7782

July 14, 2007

Deborah Smith, Purchasing Manager
Portland Framing and Lighting
120 Froelich Avenue
Portland, OR 97201

Dear Deborah Smith:

(c) Heading aligned left

Energy Systems Limited
647 Union Street
Eugene, OR 97440

July 14, 2007

Portland Framing and Lighting
120 Froelich Avenue
Portland, OR 97201

Dear Purchasing Manager:

FIGURE 13.2 Three heading variations

identifies your recipient (see Figure 13.2). Avoid salutations such as "Dear Sir" or "To Whom It May Concern" because they are vague and impersonal and show a lack of audience awareness. The salutation is generally placed two lines below the recipient's address, aligned left. Remember that the degree of formality in a salutation or greeting is a matter of cultural practice. Be sure to confirm cultural practices when writing letters for transnational audiences.

For more on transnational audiences, see chapter 5.

IN YOUR EXPERIENCE

Although they are short and simple, it is important to use the correct two-letter abbreviations, especially when sending letters out of state. You certainly know your own state's abbreviation, but how many of the other two-letter state abbreviations do you know? Make a list of the ones you know. Then visit the U.S. Postal Service website containing all of the abbreviations, http://www.usps.com/ncsc/lookups/abbr_state.txt. How many did you not know? Did you get any of the others wrong? Why did you know some abbreviations and not others?

Introductory Paragraph

Because letters are generally short and are often read by busy audiences, it is important that they are effective in introducing their subjects. The introductory paragraph should provide any background information the reader will need to understand the letter—such as who you are and why you are writing. The introductory paragraph generally contains short, descriptive sentences and may also include a thesis sentence identifying the problem or main point of the letter. The introductory paragraph is generally placed two lines below the salutation. When addressing a highly important issue or problem, you might consider a single, thesis-based sentence as your first paragraph. Notice how the letter from the FDA identifies the primary purpose for the letter in the first sentence (Figure 13.1).

EXPERTS SAY

"I have made this letter longer, because I have not had the time to make it shorter."

BLAISE PASCAL
1657

ANALYZE THIS

Analyze several of the introductory paragraphs in the letter examples in this chapter. Do they follow the advice about introductions? Do they provide any necessary background details? Do they provide a thesis? Are they effective? Why or why not?

Body Paragraphs

Each body paragraph contains further details concerning your message, although you should resist the urge to include details that are unnecessary or irrelevant. Body paragraphs in letters often contain bulleted or numbered lists to improve readability. The body paragraphs in the letter from the FDA provide background information (e.g., why Vioxx has been withdrawn from the market) and details about how readers should respond (e.g., contact a health-care provider for treatment).

Concluding Paragraphs

Letters often conclude with one or two paragraphs that sum up or reiterate the main point of the message. If there are specific steps you'd like your reader to take, you should mention them in the concluding paragraphs. If you do not wish to encourage specific action or if you have done so earlier, you can use a final paragraph to state something positive about the relationship between you and the recipient. This is particularly important if you have delivered negative information to your reader; the conclusion can help build goodwill and a positive image of you and your company. In some letters you might conclude by providing details about where, when, and how the recipient could contact you for further information or feedback. The FDA letter includes phone numbers and websites for both the FDA and Merck in case readers have additional questions.

Closing Phrase

Many workplace letters contain a closing phrase followed by a comma. Although there are a number of appropriate closing phrases, some are more formal than others. The phrase you choose should reflect the relationship you have with the recipient.

- Closing phrases for highly formal letters:
  Respectfully yours, Yours sincerely, Respectfully,
- Closing phrases for moderately formal letters:
  Sincerely, Cordially, Thank you, Regards,
- Closings for informal letters:
  Best wishes, Kindest regards, Best,

The closing phrase is generally placed two lines below the last paragraph, aligned left. As is the case with salutations, closing phrases should conform to the cultural standards of the audience. Remember to confirm such cultural standards when writing to transnational audiences.

For more on transnational audiences, see chapter 5.

Respectfully yours,	Sincerely,	Best,
Shawn Blake	A. R. Komenaka	Mike
Shawn Blake Acquisitions Editor	A. R. Komenaka Division of Research	Mike Englert

FIGURE 13.3 Closing variations

Signature

Your full name should be typewritten four lines below the closing phrase, aligned left. If you are writing in an official capacity and your job title is not included in the stationery's letterhead, you should include your title on the line directly below your typewritten full name. You should then sign the letter in the space between the closing phrase and your typewritten name. Use blue or black ink for your signature; these colors convey a professional image and will remain legible even when scanned or photocopied (see Figure 13.3).

Page Headers

As with memos, you may sometimes need to write letters that are significantly longer than one page. Each page beyond the first should be printed on plain paper (i.e., no letterhead) of the same quality, weight, and color as the first page. These additional pages should include, at the top, a header containing some identifying information. A common approach is to include the name of the person to whom the letter is sent (flush with the left margin), the page number (in the center), and the date (at the right margin). Additional pages would look like this:

Recipient	2	Date

When possible, you should use the header function of your word processing program to automatically insert this information and avoid formatting and spacing problems if you revise or edit your letter at a later time.

Again, as with memos, you should be sure that the final page of your letter has at least five lines of text on it. If it does not, you may wish to manipulate the spacing, font, font size, or margins. And remember to avoid widows and orphans, too. As mentioned previously, many word processing programs have commands to address these problems.

ADDITIONAL FEATURES

Some letters contain additional features for specialized purposes, such as highlighting information, making note of others who helped create the letter, drawing attention to attached documents, or designating recipients who are not listed in the address line.

Introductory Line

When you are sending information that is highly important, requires an immediate response, or is directed toward a particular individual or division within a large company, it is useful to include an introducing line. As the name suggests, these lines introduce vital information about the subject or the audience of the document; there are three types.

Subject Line

This line is used to announce and highlight the topic of your letter. Subject lines are always used in memos and e-mail messages and are sometimes used in letters as well. You should use all caps for the title (SUBJECT) and then capitalize the first letter of each major word in the line. Use underlining or boldface type to highlight the information contained in the line. A subject line is generally placed two lines below the recipient's address, aligned left.

SUBJECT: Short, Descriptive Phrase Here

Attention Line

This introducing line is used when you wish to address a specific department or position within a company but do not know the recipient's name. You should use the same formatting and positioning features in an attention line that you'd use in a subject line.

ATTENTION: Department or Position of Recipient

Promotional or Attention Heading

When sending a letter to many different readers (i.e., a mass mailing), you might choose to delete the salutation and use either a promotional heading or an attention heading before the body of your message. Promotional headings are often used in sales letters to announce discounts or sales; attention headings are often used to highlight problems, significant changes, or other important information.

It is common to leave two spaces above and below the heading to set it apart from the rest of the text. You should consider using all capital letters or bold or italicized text to draw attention to the heading, as these two examples demonstrate:

ANNUAL CLEARANCE SALE AT TUBERIDER'S SURF SHOP
FDA Public Health Advisory: Safety of Vioxx

Typist's Initials

Although most workplace writers type their own letters, someone else may occasionally type a letter for you. In that case you should place directly below your typed signature (with or without a line space) your capitalized initials followed by the typist's lowercase initials.

YI/ti

Enclosure Line

Letters often introduce or are accompanied by other documents. If you plan to send something else with your letter, you should include an enclosure line directly below the typist's initials, noting the number of enclosures, or naming them if they are highly important or valuable. Here are three examples:

Enclosure: warehouse contract
Encl: 3
Enclosures: monthly rent check, warehouse contract

Distribution Line

You may send copies of your letters to other internal or external recipients and may wish to note those individuals. If you do plan to distribute copies of your letter, directly below the typist's initials or your typed signature if no typist's initials are necessary, you should include a carbon copy notation—CC, Cc, cc, or Copy—followed by the additional names.

Cc: Randy Westgate
Copy: Amy Behr
cc: S. Wile
D. Plastino

If you are simply making a copy of an original document for your own files, you can note this with the word *Copy* placed directly below the typist's initials.

FORMATS FOR LETTERS AND ENVELOPES

Your company or organization may have its own method of formatting letters. If so, follow Dave Ferron's advice, and ask a supervisor or trusted colleague for an example that you can refer to as a model. However, if your company has no particular method for formatting letters, consider using one of the two common formats diagrammed in Figure 13.4: block style and modified block style.

Hector Esquivel
Delta Omega Gamma Fratemity
500 Mercer Way, Coolidge, OK 54534
[One blank space]
April 1, 2007
[One blank space]
Mr. Anthony Buffone
815 Oak Drive
Collegetown, OK 12055
[One blank space]
Dear Mr. Buffone:
[One blank space]
This letter is to inform you that you owe the sum of $500.000 to the College Town University (CTU) chapter of the Delta Omega Gamma Fraternity. This amount consists of the following:
[One blank space]
$200.00 Spring 2005 semester membership dues
$200.00 Fall 2005 semester membership dues
$100.00 Spring 2006 semester membership dues
$500.00 Total Due
[One blank space]
If this debt is not paid within ninety (90) days, the College Town University chapter of Delta Omega Gamma Fraternity reserves the right to pursue the debts through all appropriate means, including, but not limited to
[One blank space]
• Collection agencies
• Small claims court
[One blank space]
Delta Omega Gamma is willing to consider payment arrangements. However, you must pay the debt listed above in full or sign a mutually agreeable payment plan by May 1, 2007, or Delta Omega Gamma will begin to pursue its legal rights at that time.
[One blank space]
Sincerely,

[Three blank spaces with signature in blue or black ink]

Hector Esquivel
Chapter Treasurer

(a) block style

Hector Esquivel
Delta Omega Gamma Fratemity
500 Mercer Way, Coolidge, OK 54534
[One blank space]
April 1, 2007
[One blank space]
Mr. Anthony Buffone
815 Oak Drive
Collegetown, OK 12055
[One blank space]
Dear Mr. Buffone:
[One blank space]
This letter is to inform you that you owe the sum of $500.000 to the College Town University (CTU) chapter of the Delta Omega Gamma Fraternity. This amount consists of the following:
[One blank space]
$200.00 Spring 2005 semester membership dues
$200.00 Fall 2005 semester membership dues
$100.00 Spring 2006 semester membership dues
$500.00 Total Due
[One blank space]
If this debt is not paid within ninety (90) days, the College Town University chapter of Delta Omega Gamma Fraternity reserves the right to pursue the debts through all appropriate means, including, but not limited to
[One blank space]
• Collection agencies
• Small claims court
[One blank space]
Delta Omega Gamma is willing to consider payment arrangements. However, you must pay the debt listed above in full or sign a mutually agreeable payment plan by May 1, 2007, or Delta Omega Gamma will begin to pursue its legal rights at that time.
[One blank space]
Sincerely,

[Three blank spaces with signature in blue or black ink]

Hector Esquivel
Chapter Treasurer

(b) modified block style

FIGURE 13.4 Two common letter formats

Block Style

Block style is the most commonly used format for professional letters; in fact, most of the examples in this chapter use a block format. In essence, a block style aligns everything except the company logo or letterhead along the left margin. It uses single spacing for all text, with a space between paragraphs. This style looks sleek and businesslike and conveys a sense of order and symmetry. Most readers find it helpful because their eyes can skim a document vertically rather than jumping back and forth along the left margin. But a block format is equally useful for the author; the left alignment of all text is easy to remember and duplicate, even without a letter template or model. Figure 13.4a shows a sample letter written in block format (with typical spacing).

Modified Block Style

Modified block style is slightly less formal than block style. Modified block style aligns the return address and the closing phrase and signature along the right margin, with all other text aligned left. Because a modified block style uses a different alignment for the writer's return address, closing phrase, and signature, it can be useful to highlight information about the author. Figure 13.4b shows a sample letter written in modified block format (with typical spacing).

Most letters are written with single spacing, no indentations, and a space between paragraphs. This is the most readable spacing convention and is the most common. Indentations are occasionally used in letters, but you should never use indentations along with a space between paragraphs, since this would result in a document with too much wasted space.

IN YOUR EXPERIENCE

The choices between block and modified block styles are somewhat arbitrary. There is no single right or wrong style to use, other than the style endorsed by your company. If you were asked to select the formatting style for a large corporation for which you worked, which style would you choose? Why? Do these formats lend themselves to particular types of workplace problems? Explain.

Envelope Style

The envelope will often be the first thing your reader sees, and you want to make a good first impression. We recommend using the same quality and type of envelope as the stationery it contains. Typically, business letters are sent in a #10 envelope, although we encourage you to use a letter-size manila envelope for very important letters, those requiring signatures, or documents with several pages. Documents should not be folded in a letter-size envelope and should be inserted in such a way that recipients can pull them out of the envelope and begin reading, without having to reposition the documents.

You should *always* type your mailing and return addresses single spaced and in the same font style you used for the letter. The reader's name and address belong in the center of the envelope, with your return address in the upper left corner. Figure 13.5 illustrates the proper styling of business envelopes.

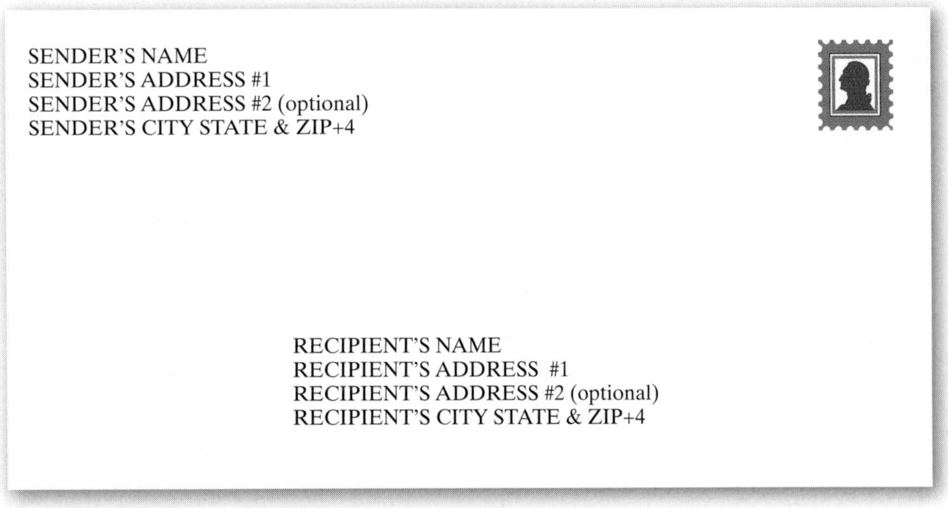

SENDER'S NAME
SENDER'S ADDRESS #1
SENDER'S ADDRESS #2 (optional)
SENDER'S CITY STATE & ZIP+4

RECIPIENT'S NAME
RECIPIENT'S ADDRESS #1
RECIPIENT'S ADDRESS #2 (optional)
RECIPIENT'S CITY STATE & ZIP+4

FIGURE 13.5 Proper styling of a business envelope

PATTERNS FOR ORGANIZING A LETTER

Workplace writers use letters to solve all sorts of problems and to address many unique situations. Some writers find it useful to think about those problems and situations as being of particular types and then try to write a letter that fits the specific type. For example, some organizations have forms for composing adjustment, claim, and inquiry letters, allowing employees to insert the details into the appropriate spaces. We provide examples of some of these common letter types later in this chapter. This approach does take some of the guesswork out of letter writing—it becomes a matter of just plugging information into predetermined places. Unfortunately, real-world writing situations rarely fit perfectly into types, and rhetorical situations don't come with labels identifying the pattern to use. Moreover, using particular types as models shapes both the way a writer communicates the message and the way readers interpret it.

Instead, the purpose for writing should shape the letter, and the form or pattern of the letter should depend on what is included. Therefore, rather than focusing on

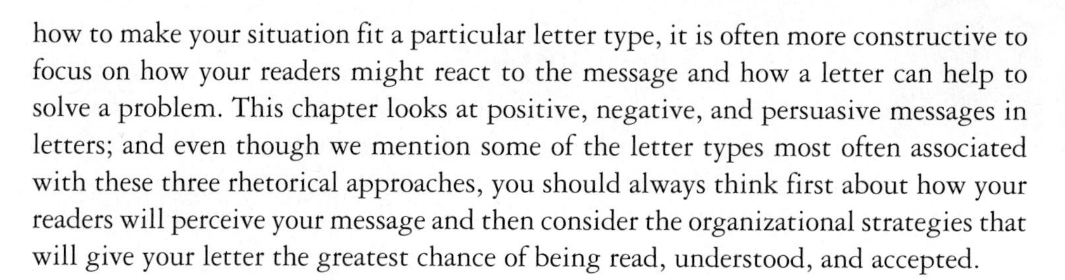

how to make your situation fit a particular letter type, it is often more constructive to focus on how your readers might react to the message and how a letter can help to solve a problem. This chapter looks at positive, negative, and persuasive messages in letters; and even though we mention some of the letter types most often associated with these three rhetorical approaches, you should always think first about how your readers will perceive your message and then consider the organizational strategies that will give your letter the greatest chance of being read, understood, and accepted.

Positive Messages

When you give your readers news that benefits them or their organization, the message is a positive one. Many situations require positive messages: agreeing to a request, accepting a proposal, offering a job, giving a discount, presenting a reimbursement or favorable adjustment, acknowledging that an order has been received, extending a deadline. These are all situations in which your readers' response will likely be positive. In many of these cases, you won't be asking your readers to do anything else; you'll simply be telling them the good news. We hope that most of the letters you write will be to communicate good news!

You should use the following pattern of organization for positive messages:

1. *Give the good news first.* When you have good news to communicate, state it immediately and directly. Any delay can temporarily confuse readers, causing them to wonder whether the letter carries good news or bad. The first sentence of your letter should state the specific benefits, advantages, or other positive information you have to offer, including the most important details of the good news, such as the job title you're offering, the percentage of the discount you can give, the date to which the deadline is extended, or the specific request to which you're agreeing.
2. *Provide the necessary follow-up details.* Nothing can spoil good news more quickly than a message that doesn't contain all of the information readers might need. After stating the good news, be sure to give readers any background details or specific dates, times, places, and prices that will help them fully understand the message. Try to anticipate and answer any questions they might have about your message, and give these further details in order of decreasing importance. If no response or reaction is required, say so.
3. *Be clear about potential negative elements or limits.* Sometimes a positive message comes with limits or potential drawbacks. For example, a discount may have a time limit, a reimbursement may require returning the previously purchased item, or a job offer may not have all of the benefits the reader requested. If your message contains potentially negative elements or limits, be sure your reader understands them, but don't overstate the negative—present negative elements as positively as possible.
4. *Explain the advantages and benefits for the reader.* If there are benefits for both your company and the reader, be sure to focus on the reader. Clearly state the advantages and rewards the *reader* can expect from the good news. And remember that all benefits are not monetary; individuals can benefit from improved service, better products, or more time. However, avoid emphasizing benefits (a) that make the reader sound selfish (e.g., "You'll make more money here than in your current job), (b) that are obvious (e.g., "Having a job will keep you from being poor"), or (c) that are not likely or significant (e.g., "The $1 discount will help you save money for larger investments").
5. *Use a goodwill ending to cement the relationship.* The conclusion of your letter should highlight and solidify the working relationship you have with your reader, emphasizing that the reader is your primary concern. Some writers finish with

the three *p*'s: something *personal* about the relationship, something *positive* about working with the reader, and something *prophetic* about the relationship in the future. But keep in mind that the personal details in the conclusion should focus on the business relationship, not the reader's family, hobbies, or private life.

Figure 13.6 provides an example of a positive message.

Northwestern Technological University
Department of Engineering

April 13, 2007

Lucy Steigenga
1120 Tarpon Drive
Islamorada, FL 33036

Dear Lucy:

Good news →

We are pleased to offer you a graduate administrative assistantship (GAA) in the Department of Engineering for the 2007–2008 academic year. This support package includes the following:

- A stipend in the amount of $5,000 per semester
- A tuition waiver covering full-time enrollment (9–15 hours for graduate study)
- Enrollment in the NTU Student Health Insurance Plan

Follow-up details →

In exchange for this assistantship package, you will be assigned 20 hours of university employment per week. Your assignment may consist of a teaching assistantship, for which you will be assigned to assist a professor in two undergraduate courses, or your assignment may consist of a research assistantship. Most GAA recipients do some teaching and some research in a typical academic year.

Clear description of limits →

Because of the number of applicants seeking support packages, we ask that you inform us **within two weeks from the date of this letter** about your acceptance of this support package. Please sign the enclosed acceptance form, and return it to my office—105 Staten Hall. Information on GAA requirements and conditions is detailed on the attached sheet; acceptance indicates that you have read and agreed to these conditions.

Goodwill ending →

We look forward to welcoming you into our program this fall and hope you will find the coming years at NTU productive and interesting. If you have any questions concerning this support package, please contact Sandy Ogden, senior faculty secretary, at 712-282-4556.

Sincerely,

Chris Barabazon

Chris Barabazon
Director, Graduate Programs in Engineering

Enclosures
cc: [personnel]

FIGURE 13.6 A congratulatory letter

EXPLORE

Goodwill endings are used in many types of letters that cover a variety of positive, negative, and persuasive situations. Find an example of a letter—one you have recently received, one sent to a friend or family member, or one posted on the Internet—and investigate its conclusion. Does it contain a goodwill ending? Does it state something personal, positive, and prophetic about the relationship between the sender and the recipient? Is it an effective ending, or could it be rewritten? Compare your letter with those of your classmates, looking for similarities and differences in the letters' conclusions.

Negative Messages

Along with the many positive letters you'll send as a working professional, you'll also find yourself having to give bad news to your readers—that is, news that will disappoint them, make them unhappy, or cause them difficulty. In those cases the message is negative. Unfortunately, there are *many* situations that readers construe as negative, such as mistakes that you or your company has made, proposals or offers you are rejecting, increases in the cost of a product or service, decreases in coverage or services, or the termination of a contract, agreement, subscription, or order.

Negative letters are perhaps the most difficult to write, but a clearly written and well-organized letter can help minimize a reader's negative feelings toward you. In many situations you will continue to have a working relationship with the reader, and in all cases you want the reader to have a positive, or at least neutral, image of you and your company. An effective and clear negative message also helps reduce or eliminate future correspondence on the subject, which means less work for you. You want your readers to clearly understand the news you're giving them; understand what led up to it; feel that you've treated them fairly, reasonably, and respectfully; and believe that they'd have made the same decision if they had been in your position.

You can use the following pattern of organization for communicating negative messages:

1. *Prepare the reader for the negative message.* It is often useful to begin with a sentence or two, sometimes more, that serve as a transition to the news you must communicate. Introductory statements that delay negative news or prepare the reader for it are called *buffers*. Like other introductions a buffer often provides context or background for the information that follows it. You should choose a buffer that is appropriate for your situation, although it will often take one of the following forms:
   - A chronology of events that led to the negative message
   - General principles or ideas about the subject of your message
   - An expression of thanks for something the reader has done
   - Good news that helps to balance the bad news
   - A reference to attachments or enclosures

   Buffers can often help ease the reader into the message, but they must not imply a positive message. If you do choose to begin with a buffer, keep it short, because long, rambling introductions frustrate readers—especially when they finally come to the negative message.

buffer an introductory statement that delays negative news or prepares a reader for it

2. *Clearly state the negative—once.* Make the negative message clear, specific, and direct. Don't be evasive or vague—there's nothing worse than having to explain a negative letter to a confused reader. When possible, get the message across in one short, direct sentence—avoid long, rambling sentences in which the negative message might be overlooked. If you are able, phrase the message in positive terms, referring to any benefits or possibilities the situation creates. Avoid placing blame on the reader, unless it is essential for an understanding of the message. And after you've clearly stated the negative message, avoid rehashing it again—once is usually enough.

3. *Present alternatives, compromises, solutions, or possible actions if they exist.* Effective communicators think about how their readers will feel or react. When possible, you should suggest alternatives, compromises, or other solutions that your readers might consider. This not only helps them accomplish what they want, but also lets them know that you really care about their needs. However, you should avoid going into too much detail or giving a chronology of what

will happen if they pursue another option—just present the choices, and allow your readers to decide whether they want to try an alternative.

4. *End with a positive, forward-looking statement.* Even the most negative messages can have a positive aspect. If something good can come to the reader as a result of the message, mention it. Often, your conclusion will look ahead to a happier future when your reader's needs can be satisfied, even though you are unable to satisfy them at the present time. If you plan to work with the reader in the future, say that you are looking forward to that interaction. If not, thank the reader for any positive experiences you've had in the past. And if you anticipate further questions that you are unable to address at the moment, give specific contact information in the conclusion.

Figure 13.7 provides an example of a negative message.

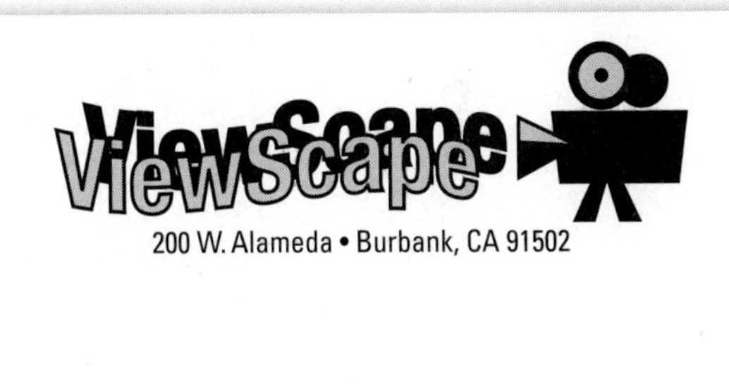

May 29, 2007

AVN Entertainment
Canyon & Patrick Streets
Los Angeles, CA 91554

Dear AVN Production Team:

Thank you for contacting Viewscape about digital video scoring for your upcoming video series to be filmed on June 20, 2007. We have enjoyed working with you over the past few years to create sophisticated DVD and VHS products, and we hope to work with you on other upcoming projects.

Unfortunately, we will not be able to provide MPEG scoring for the June 20 series. As you know from our phone conversation, Viewscape works in DV, DVCam, and DVCPro audio formats. Since you plan to film the June 20 video series entirely in MPEG format, our scoring software would be incompatible with your production.

If you decide to change your production to one of the formats we can accommodate, we would be happy to work with you on the series. In fact, because we value working with AVN, we can offer free use of our editing bay to complete the project if you switch to a format in which we currently work.

If you'd like to discuss this project further, please contact me at 818-220-1010 or through e-mail at amys@viewscape.com. Best of luck with the video series!

Thank you,

Amy Scott

Amy Scott
Viewscape Acquisitions Manager

FIGURE 13.7 A refusal letter (in modified block style)

ANALYZE THIS

Analyze the following sentences, and determine which of the five common types of buffers each most closely resembles.

1. "Thank you for visiting our office last Wednesday to introduce Webcom's new high-speed DSL service." (A letter discontinuing service with an Internet service provider)
2. "Enclosed you will find the annual sales report for Deerborne Construction." (A letter announcing a drop in yearly home sales)
3. "Three years ago we began using employees from WorkStaff Temp Agency. For the first two years this worked well for us, but recently WorkStaff temps have been arriving to work unprepared, late, and sometimes not at all." (A letter canceling service with a temp agency)
4. "Many people purchase computers because of a company's reputation for good customer service. At Computer Systems Limited we take customer service very seriously." (A letter announcing a decrease in customer service hours)
5. "AquaLad Pools has a policy of offering a 10 percent discount to customers whose swimming pool installation date is more than thirty days past our original estimate." (A letter informing a customer that an installation date has been delayed)

Persuasive Messages

Much workplace writing involves persuading others. When you are encouraging others to think or act in a way that benefits you or your company, you are writing a persuasive message. Many of the letters you will send will involve some form of persuasion—selling a product or service, requesting help or materials, recommending strategies or solutions, proposing plans or guidelines, providing positive or negative feedback that you want readers to accept. In fact, many of the positive and negative messages you send will also contain a persuasive element.

To persuade your readers to act in a certain way, you must consider whether they have previous experiences or preconceived notions about the subject of your message. Thus, your first goal should be to overcome any objections that might prevent or delay them from acting. In all cases you will also need to provide enough information that your readers will be able to act. And persuasive messages, like all other pieces of writing, should create or cement a bond between you and your readers and should help to build a good image of you and your organization.

You can use the following guidelines to shape your persuasive message:

1. *Capture your readers' attention.* Most readers today are very busy, so it is important to capture their attention and interest from the outset—particularly if you want them to act in a certain way. An attention getter provides common ground between you and your readers and can take many forms:
   - Asking a question readers might be considering
   - Citing a surprising fact or statistic
   - Describing a new product or service
   - Summarizing a previous event or transaction
   - Challenging a commonly held belief
   - Providing a specific example
   - Offering proof that you can solve readers' problems
2. *Connect readers' interests with your goals.* After you've got your readers' attention, make the transition from the introduction to what you want them to do. In most cases you should show how the desired action is in their best interest. But avoid overstating the benefits; readers respond best when given factual, reliable information.
3. *Provide convincing details.* After you've made your readers aware of the purpose of your message, provide all of the details necessary to convince them.

Be specific about costs, quantities, deadlines, or other important details, but also show that the advantages clearly outweigh any negatives. If possible, give your readers a reason to act promptly; show that acting now will save time or money or that delaying the action will only make things more difficult later. Anticipate any questions that might keep someone from acting immediately, and attempt to answer these questions in your message. Conclude this section with one main point or idea that you want your readers to remember.

4. *Tell readers how to act.* You'll want to spell out the specific steps readers should take. Be clear about the order of the steps if there are several steps involved. If readers need to call, write, or e-mail, give them the specific contact information. Avoid threatening or demanding language; readers respond better when they want to act, not when they feel forced to act. If appropriate, offer to call them, or invite them to call you.

Figure 13.8 provides an example of a persuasive message.

LinkageLongDistance

September 7, 2007

Mike Strube
President
Strube Flight Instruments
2020 Pagan Way
Anthem, AZ 85086

Promotional heading that captures reader's attention →

YOU CAN CUT YOUR COMPANY'S LONG DISTANCE PHONE EXPENSES BY 30%!

Dear Mr. Strube:

Connection with reader's goals →

Thank you for inquiring about Linkage Long Distance services. We offer the same fiber optic telephone lines that your current service offers, except we charge a lot less for our high-quality service. And with Linkage Long Distance, you never have to worry about lost sales due to a power or system failure. In the event of an emergency, our computer will automatically reroute the calls from your 800 system to anther location of your choosing.

We deal exclusively with small businesses like yours. Unlike AT&T and MCI, we don't spend millions of dollars on expensive advertising campaigns. We choose to pass the savings on to you. In fact, most of our new business is generated the old-fashioned way: our customers recommend us to their colleagues.

Our Small Business Plan has been designed to suit your needs:

Convincing details →

- **Delayed payment options during your off-peak months**
- **Guaranteed rates for two years**
- **A FREE month of long distance to new customers**

I've enclosed a brochure that further details our state-of-the-art rerouting system and delayed payment options. **Please be advised the free month of long distance is available only to new customers who sign with us by December 1.** You must act quickly to take advantage of this exceptional savings opportunity.

Telling reader how to act →

I'd like to meet with you to discuss how Linkage Long Distance can immediately begin saving you up to 30% on your monthly long distance expense. I'll contact you next week at your Anthem warehouse to schedule an appointment; if you'd like to meet earlier, call me at 623-285-2290. I look forward to meeting with you soon.

Sincerely,

Raymond E. Kitch
Raymond E. Kitch
Regional Sales Manager, Linkage Long Distance

FIGURE 13.8 A sales letter

IN YOUR EXPERIENCE

Sales letters are among the most common letters sent through the mail today. Chances are good that you have received a sales letter at some time, and chances are also good that you skimmed it quickly and threw it away. Based on your experience with sales letters, why do they often go unread? Does their success have to do with their subject or with the way that subject is presented in letter form? With your class discuss your general experience with sales letters, and reflect on the ways that you could make your own sales letters more effective. Is there anything you could do to make the sales letter in Figure 13.8 more useful or effective?

TYPES OF LETTERS

Because letters are particularly important in conjunction with employment and service applications, we devote more attention to those types of letters in chapter 14.

Letters cover a wide range of topics; each example shown in this chapter depicts a different type of letter. As with memos, some situations requiring a letter are so common that the documents relating to them have specific names and functions. Because you may be asked to create or help create a specific type of letter, it can be useful to understand what some of these types are, even though you should always focus on your audience and the message of your letter. In your profession you may encounter some of the following letter types, each of which is organized and structured in particular ways.

Thank-you and Congratulatory Letters

Letters can be used to offer thanks for a successful transaction or meeting or to recognize an achievement or accomplishment. Both situations are positive, and the letters that address them are often structured in similar ways. Thank-you and congratulatory letters often provide information concerning future actions or correspondence, and they should always strive to develop goodwill between the author and the recipient. Figure 13.6 on page 379 is a congratulatory letter offering a graduate student an assistantship.

Rejections and Refusals

Letters can be used to inform readers that they will not receive what they want. Although rejection and refusal letters give bad news, they often use buffers and present alternatives, both of which can help to minimize negative feelings toward the author. Figure 13.7 on page 381 is a refusal letter informing a video production company that the video scoring service they have requested cannot be granted.

Sales Letters

Although some sales letters are mass-produced with little sense of audience, an effective sales letter can have a significant impact on a reader. Sales letters describe the benefits of a product or a service and are often written to potential or existing customers to stimulate interest. Such letters should provide specific details to meet the readers' needs. Figure 13.8 on page 383 is a sales letter written in response to an inquiry about long distance telephone service.

Claim Letters

Letters are sometimes used to request adjustments or reimbursements for defective goods or services. Customers or consumers usually write these letters. When writing a claim letter, you should provide as many specific details as possible. Including the order number, the exact item or service purchased, and both the order and delivery dates can help to expedite a claim and avoid disagreements. Figure 13.9 requests a reimbursement for damaged lab equipment.

EXPLORE

Find an example of a letter you've received or have found elsewhere. Does it seem to represent one of the common types of letters described here? What makes it that type of letter? If it does not seem to represent one of the common types, how would you classify it? In other words, what type of letter is it? Does it seem to fit more than one category? Is the message primarily positive, negative, or persuasive? With your classmates compare the letters you've found; consider their similarities and differences. You may also wish to tally the numbers of each type of letter. Is one type more common? Why?

Dahl Agriculture
10 Plains Vista
Lincoln, Nebraska 44822

July 21, 2007

Herick Lab Equipment
101 Dartmouth Plaza
Ames, Iowa 50098

Dear Sales Manager:

Pertinent details

On July 17, 2007, we received a UPS shipment of four (4) Herick Homogenizer Silent Crushers. We noticed that the box showed signs of water damage, so we immediately opened it and tested each homogenizer. Upon testing, we found that two of the homogenizers were inoperable.

Consequently, we are returning the two defective homogenizers to Herick Lab Equipment. We have been able to locate two additional homogenizers from our regional warehouse, so we will not need replacements.We paid $1,150 each for the two homogenizers, which should bring our total reimbursement to $2,300. I have enclosed a copy of our bill with this letter—the two items being returned are circled. Please send the reimbursement to the office address listed at the top of this letter.

Contact information

We have been pleased with Herick equipment in the past and anticipate purchasing products from you in the future. If you have any questions about this claim, please call me at 802-445-1195 or e-mail me at billings@dahlag.com.

Sincerely,

Lisa Billings

Lisa Billings
Director, Agriculture Research

Enclosure: receipt for homogenizers

FIGURE 13.9 A claim letter

Adjustment Letters

Letters can also be written in response to claims. Adjustment letters address specific problems and provide answers, solutions, or compensation for the problems. Some adjustment letters are written in response to a specific claim made by an individual or an organization; others deal with adjustments made by a manufacturer or a service provider. Figure 13.10 is an adjustment letter responding to a customer's request for a computer upgrade; notice that the letter does not give the customer what he has requested but still provides a solution.

ComputerNation, Inc.
68 Etching Street • Lubbock, TX 79416

October 10, 2008

Paul Esqueda
500 Tulpehocken Road
Wolfforth, TX 79382

Dear Mr. Esqueda:

Summarizes the recipient's perspective →

Thank you for writing to us about the computer you purchased from our store on September 19th, 2008. I understand how difficult it is to deal with a miscommunication.

Summarizes the author's perspective →

According to your letter, you were expecting a computer equipped to perform advanced engineering design functions. After purchasing the computer, you discovered that it lacked the proper hardware and software to suit your needs. You are now asking our company to upgrade your computer to your specifications free of charge and to send you the needed software as soon as possible.

Says no but provides a solution →

Our advertisement for the computer stated that it came equipped to perform basic word-processing and simple graphics design, and your computer came with what our advertisement promoted. This computer is suitable for many personal computing situations, but it is not designed to do the sort of advanced computing that you require.

The upgrades you requested can be purchased at the Computer Nation store where you purchased your computer. Our sales staff can assist you in finding the components that you need.

Please contact us if we can be of any further assistance.

Sincerely,

Jason Tremblay

Jason Tremblay
Claims Manager
ComputerNation, Inc.

FIGURE 13.10 An adjustment letter

Collection Letters

Collection letters ask readers to pay for goods and services they have already received. These letters typically describe the amount owed and provide a payment history. They should provide due dates for payments as well as penalties or repercussions for missing these payments. Figure 13.11 is a collection letter detailing membership dues owed by a fraternity member.

ΔΩΓ

Delta Omega Gamma Fraternity

500 Mercer Way • Coolidge, OK 54531

February 1, 2007

Mr. Anthony Buffone
815 Oak Drive
Collegetown, OK 12055

Dear Mr. Buffone:

This letter is to inform you that you owe the sum of $500.00 to the College Town University (CTU) chapter of the Delta Omega Gamma Fraternity. This amount consists of the following:

Uses lists to clairfy →

$200.00 Spring 2005 semester membership dues
$200.00 Fall 2005 semester membership dues
$100.00 Spring 2006 semester membership dues

Emphasizes important information →

$500.00 Total Due

Describes what will happen next →

If this debt is not paid within ninety (90) days, the College Town University chapter of Delta Omega Gamma Fraternity reserves the right to pursue the debt through all appropriate means, including, but not limited to

- Collection agencies
- Small claims court

Delta Omega Gamma is willing to consider payment arrangements. However, you must pay the debt listed above in full or sign a mutually agreeable payment plan by May 1, 2007, or Delta Omega Gamma will begin to pursue its legal rights at that time.

Sincerely,

Hector Esquivel

Hector Esquivel
Chapter Treasurer

FIGURE 13.11 A collection letter

Confirmation Letters

Some letters are written to provide a record of previous conversations or agreements. These letters are often sent as a follow-up to a face-to-face or phone conversation; they may serve as an informal contract for what will happen in the future. Confirmation letters are also known as acknowledgment letters. Figure 13.12 is a confirmation letter recognizing a participant's willingness to serve on a steering committee and informs him of when the first meeting will be held.

Pinewoods Preservation Society
http://www.pinewoods.com

March 1, 2007

Kevin Finnerty
2991 Stockton Dive
Asbury Park, NJ 07712

Dear Mr. Finnerty:

Thank you again for agreeing to serve on the Pinewoods Preservation Society (PPS) steering committee. Your participation and service will help to protect New Jersey's pinewoods for generations to come.

As a member of the steering committee, you will be called on to provide leadership, technical assistance, and recommendations for program development. Below are some examples of ways you can contribute to the goals of our PPS:

Lists potential actions →

- Participating in monthly conference calls
- Building PPS membership
- Serving as a resource for the steering committee and other PPS members
- Seeking funding to support PPS activities
- Conducting statewide and regional education events for PPS members
- Supporting outreach efforts to raise awareness about pinewoods preservation

Provides details about future action →

Our first meeting will be held at Crystal Monastery on March 30, 2007, at 7 p.m. After this initial meeting, the steering committee will convene primarily by conference call.

We are sending this preliminary notice to remind you to save the date. Directions to the meeting location, an agenda, and a list of committee members will be provided closer to the meeting date. Please confirm your attendance by contacting Jennifer Melfi at 201-273-9093.

We look forward to meeting with you!

Anthony Sirico

Anthony Sirico
President, Pinewoods Preservation Society

FIGURE 13.12 A confirmation letter

Inquiry Letters

Inquiry letters ask for something or make a request. They should provide adequate background so that readers can understand the inquiry and should also inform readers of what is expected or requested of them. Figure 13.13 on page 389 is an inquiry letter inviting an office developer to submit a proposal.

Deb Webber
Rhombus Accounting
565 N. Broad Street
Tacoma,WA 93499

April 3, 2007

Fred Olson
Space Management Solutions, Inc.
10 Division Street
Yakima,WA 98502

Dear Fred:

Provides context →

As you know from our previous conversations, Rhombus Accounting plans to develop a new accounting office in Yakima this summer. We envision a networked environment consisting of approximately 50 workstations, each of which will need high-speed Internet, phone, and fax capabilities. Because of your excellent work in creating our current office in Tacoma, we are contacting you to discuss developing this new office.

Describes plans →

Several of our team managers will be in Yakima on May 15 to survey the planned office, and we would like to invite you to present a proposal on that date outlining state-of-the-art solutions for our workspace needs. Please let me know if you are interested in delivering such a presentation. If you are, I will e-mail you the office specifications and our budget for the project.

Gives details about future action →

If possible, please contact me by April 15 to confirm this engagement. We have been impressed with your work in the past and hope to continue our productive partnership. Feel free to call me at 854-775-2211 or e-mail me at dwebber1@rhombus.com if you'd like to discuss any of this.

Sincerely,

Deb Webber

Deb Webber
Network Administrator

FIGURE 13.13 An inquiry letter

SUMMARY

- Letters are used to transmit small chunks of information, communicate with individuals outside an organization, and discuss a variety of topics, from simple requests to detailed responses and descriptions.
- Letters (i.e., external documents) are different from memos (i.e., internal documents) in the following ways:
  - Letters tend to be more formal than internal correspondence is.
  - Letters often contain more contextual, background information than memos do.

- Letters generally include formal greetings and closing statements, whereas memos usually do not.
- Letters typically contain the following basic elements: a heading or company letterhead, a date, the recipient's name and address, a salutation, an introductory paragraph, body paragraphs, a concluding paragraph, a closing phrase, a signed and typed signature, and page headers if the letter is longer than one page.
- Letters can also contain additional features: a subject line, an attention line, a promotional heading, typist's initials, an enclosure line, and a distribution line.
- Most letters use one of two common formats: block style or modified block style.
- The structure of a letter can be determined by the nature of the message—positive, negative, or persuasive.
- The pattern of organization for a positive message is to give the good news first, provide the follow-up details, be clear about negative elements, explain advantages and benefits, and use a goodwill ending.
- The pattern of organization for a negative message is to prepare the reader, clearly state the negative once, present alternatives if they exist, and end with a positive statement.
- The pattern of organization for a persuasive message is to capture the reader's attention, connect the reader's interests with the writer's goals, provide convincing details, and tell the reader how to act.
- Specific letter types include thank-you and congratulatory letters, rejections and refusals, sales letters, claim letters, adjustment letters, collection letters, confirmations, and inquiries.

CONCEPT REVIEW

1. Who is the general audience for most letters?
2. What does the Vioxx letter teach about letters?
3. What is usually placed in a heading or letterhead?
4. What date should you put on a letter?
5. Where should you place the recipient's address in most letters?
6. What are common salutations?
7. What should the introductory paragraph contain?
8. What should the body paragraphs contain?
9. What should a concluding paragraph contain?
10. How should you structure additional page headers?
11. What are the additional features you might find in a letter?
12. What is block style?
13. What is modified block style?
14. How should you organize a positive message?
15. How should you organize a negative message?
16. How should you organize a persuasive message?
17. What is a thank-you or congratulatory letter?
18. What is a rejection or refusal letter?
19. What is a sales letter?
20. What is a claim letter?
21. What is an adjustment letter?
22. What is a collection letter?

23. What is a confirmation letter?
24. What is an inquiry letter?
25. What are the differences between letters and memos?

CASE STUDY 1

Responding to Critics of a Park Service Message

The first case study in chapter 12 describes a series of memos written by the director of the Lake Clarke National Park Service (LCNPS). The memos were written to employees about budget cuts that would affect staffing and hours of operation for the park, which is visited by nearly 200,000 hikers, campers, swimmers, fishermen and -women, and boaters each year. Although the memos were intended for employees only, several were published on the Internet, including the September 20, 2006, e-mail message sent to park supervisors (see Figure 12.17).

Two local newspapers and one regional magazine have reprinted that e-mail message and have written negative articles about it. One correspondent called the e-mail "dangerously unethical," another stated that the memo "betrays public trust in our park service," and an editorial called for the resignation of Doris Jones, the director of LCNPS and author of the message.

Director Jones has now asked you, as an employee of LCNPS, to create a response letter to the public that will be sent to these newspapers and the magazine. Your goal is to explain the information found in the e-mail and attempt to dispel any negative feelings toward LCNPS. You'll want to draw from the information found in the memo but rewrite it in the form of a letter to the public. Before you begin, you should reread the memo in Figure 12.17 and consider the following questions:

- Based on the information you have, how do you envision your audiences? Will they be generally receptive or hostile to the information in your letter?
- What tone, style, and level of formality will work best in this response letter?
- Will your letter be apologetic, or will you take the perspective of someone who is correcting a misconception?
- Will you structure your letter as a positive, negative, or persuasive message? Will you use one of the patterns of organization described in this chapter?

CASE STUDY 2

Imagine that, at the end of spring semester, your favorite professor, who recently wrote a letter of recommendation for you, hands you a sealed envelope with your name on it. Inside you find a letter describing a colloquium the professor is organizing for next year's freshman orientation. The colloquium is entitled "What I Like about This School," and the letter invites you to be a guest speaker, taking the stage for about ten minutes. The end of the letter gives the professor's summer address, asking you to mail a response by mid-July.

Decide whether you will speak at this colloquium, and then write a detailed response letter to the professor. As you are deciding, consider whether you have an ethical obligation to this professor in return for the letter of recommendation. If you choose to participate, you might tell the professor what you'd like to discuss. If you choose not to participate, you should offer some explanation. Use one of the three patterns of organization for your letter, based on whether you view this as a positive, negative, or persuasive situation.

CASE STUDIES ON THE COMPANION WEBSITE

Bad Loans and Bad Vibes: An Assistant Manager Is Placed in a Sticky Situation

In this case you are an assistant sales manager at a large financial company and are asked to write two letters that help rectify problematic loan records. You must compose these letters while simultaneously considering various ethical concerns involving stakeholders.

Cutbacks and Quality Care at St. Martin's Hospital: A New Medical Professional Faces Ethical Dilemmas in the Workplace

You work on an emergency medical team at a hospital and must communicate with supervisors via e-mails and letters in order to handle various kinds of discrepancies with your work schedule and assignments during a time of administrative cutbacks.

Fletcher Global Medical Manufacturers: An Engineer Confronts Quality-Control Issues and Helps Combat Negative Publicity

As a quality-control specialist at Fletcher Global Medical Manufacturers, which produces dialysis machines and chemicals, you help investigate the deaths of several people in Europe whose deaths may be linked to your company's machines. In this case you must write several different kinds of letters to different audiences that address the investigations and the ways Fletcher will handle this difficult situation.

The Charles Grove Memorial Scholarship: A Case for Endowment

This case study outlines the difficulties facing a medical consultant trying to start a memorial scholarship. The task involves analyzing and implementing effective rhetorical and organizational skills in the process of writing letters to professional organizations and personal contacts.

VIDEO CASE STUDY

GlobeShare Wireless

Testing the GlobeTalk Handheld Devices

Synopsis

A group of GlobeShare Wireless employees learn that the new GlobeTalk Messaging handheld units are almost ready to be shipped to customers. Michelle Bankler, a product tester for GlobeShare Wireless, convinces her team to do a final field test before the units are sent. The field test leads to a shocking situation.

WRITING SCENARIOS

1. You and a colleague (i.e., a classmate) work as exhibit coordinators for Coriolis Press, a publisher of scientific and technical books. You've just learned that your top-selling author, Dr. Dana Bunker, would like to do a speaking and book signing tour in San Diego. Dr. Bunker is a world-renowned expert on weather patterns and a Nobel Prize finalist. Write a letter to a large bookstore in San Diego, asking whether the store would like to host Dr. Bunker as a speaker. Be specific about who Dr. Bunker is, what he will do, when he will

be available, and what the bookstore will need to provide. Use one of the three patterns of letter organization, based on whether you view this as a positive, negative, or persuasive situation.

2. You work with a small group that manages a commercial landscaping business, and you recently contracted with a graphic design firm to create a series of ten landscape design models to include in your written proposal to landscape a new shopping complex. In your phone and e-mail conversations with the graphic design firm, you were clear that you wanted all ten of the models by May 1, but the firm delivered five of the models to you on May 1 and the other five on May 15. You included the first five in your proposal, and you were given the contract for the shopping complex. Now, the graphic design firm has billed you for all ten of the models they created, even though they did not deliver all of them on time.

   As a group, decide whether you should pay for all of the models, the half that you received on time, or none of them. Then write a letter to the graphic design firm explaining the situation. You may wish to consider the claim, refusal, and confirmation letters as potential models (see Figures 13.7, 13.9, and 13.12), while also considering whether to structure this as a positive, negative, or persuasive rhetorical situation.

3. You work for the graphic design firm that created the ten landscape models, and you have just received the letter written in the previous scenario. Write a response. You may wish to consider the adjustment, collection, and confirmation letters as potential models (see Figures 13.10, 13.11, and 13.12), while also considering whether to structure this as a positive, negative, or persuasive rhetorical situation.

4. Find and analyze a letter written by someone in your future profession. You might ask a professional for a copy of a sample letter, request one from a professor in your major field, or even look online for a letter to analyze. As you evaluate the letter, consider the following questions:
   - What is the subject of the letter?
   - What format does the letter use?
   - Does the letter follow one of the patterns of organization described in this chapter? Would the letter have been more effective if it had?
   - Does the author use jargon specific to the field or profession?
   - How is this letter similar to or different from other letters you've seen?
   - What external problem or issue does the letter address?
   - What rhetorical problems faced the author?

5. Browse through a local, regional, or national news source, and find a workplace issue or problem that is described in an article. Drawing on this news source and any other information you can find about the organization(s) involved, write a letter addressing the problem. You might write the letter from the perspective of the company involved or responsible, or you might write as a customer or a member of the public. Think carefully about the audience you are addressing, and employ the appropriate strategies for the positive, negative, or persuasive rhetorical situation of your letter.

6. The Internet is filled with advice about writing letters. Spend some time browsing through websites devoted to writing workplace letters, and compile a list of suggestions and strategies you believe to be useful. You may wish to structure your list as a top ten list of things workplace writers should know about writing letters.

7. You work as an appraiser for an antiques dealer. One of your biggest clients recently approached you about an antique desk for which she paid $25,000. She has asked you to write a letter to her insurance company, confirming that the desk is worth the $25,000 she paid so that she can insure it for the full amount. After examining the desk and consulting your sources, you find that this type of desk was selling for $25,000 a few years ago but is now selling for only $10,000 to $15,000. Carefully consider the ramifications and the ethics of this situation, and then write the letter to the insurance company.

8. You work as a tutor at your university, providing help to students in your discipline or field. The director of the tutoring center has offered to double your hourly pay if you agree to occasionally tutor students in other disciplines when the tutors for those disciplines are booked. The director assures you that such occasions will be rare and the majority of your tutoring will remain in the subject or discipline with which you are familiar. However, you are concerned about whether you can provide any useful tutoring for the additional students, because you might not know anything about their particular disciplines. Write a letter to the director, explaining why you will or will not take on overflow students from other disciplines.

9. The word processor you use probably contains several templates for letters. Examine these templates, and compare them with the samples offered in this chapter. Are they identical? If not, what are the differences? Write a one- to two-page document analyzing and comparing the various letter templates and examples found in your word processor and in this chapter.

10. Review chapter 12. Then write a one- to two-page document outlining the design similarities and differences between memos and letters, focusing on the rhetorical reasons behind those similarities and differences.

11. Find an example of a letter you've received or have located on the Internet, and write a one- to two-page analysis of the design features it uses. Make specific reference to those features discussed in this chapter, and describe any ways in which the design of the letter could be improved.

12. Using one or more software programs, create a letter template that you can use for your own personal correspondence. The template should contain the key information that your readers might need (e.g., your name, address, phone number, and e-mail address) and should convey a sense of style and character as well. You might include these elements in a letterhead, or you might simply create text and add an icon or symbol to the page.

13. Write a letter to your instructor explaining how to use a particular word processor's letter template, often called a template wizard, to create a workplace letter. Be sure to name the program you are describing, provide step-by-step instructions, and analyze whether the template wizard is easy to use and effective.

14. Write a letter to your college or university president, emphasizing the need for a new technological product or service at your university. You might, for example, make the case for Wi-Fi service for the campus, newer computers in the library, high-speed Internet service for the dorms, or laptops that students can borrow with a student ID. If necessary, do some research into these technologies, but assume that the president knows little about the technology itself.

PROBLEM SOLVING IN YOUR WRITING

After you've completed either of the Case Studies or one of the first two Writing Scenarios in this chapter, think about how you solved the problem that faced you—that is, the need to write a letter in response to a complex rhetorical situation. In a one- to two-page letter to your instructor, describe the ways in which you employed the problem-solving approach, described in chapter 1, to successfully complete that particular exercise. Consider the following questions to guide your analysis:

- In what ways did you plan the writing of this letter?
- Did you do any research to help you complete the assignment?
- Did you employ any specific strategies to help you draft the letter?
- Did you spend much time reviewing the assignment?
- How did you distribute, release, or transmit the assignment to your instructor?

14

Finding and Obtaining Employment

CHAPTER LEARNING OUTCOMES

After completing this chapter, you will be able to do the following:

- Recognize what you should do to **prepare** for your future career
- **Search for a job** using a variety of strategies
- Write **recommendation request letters**
- Write **letters of inquiry**
- Write **chronological, skills, entry-level, online, and scannable résumés**
- Recognize the basic **components** of résumés
- Write a **curriculum vitae**
- Write **job application cover letters**
- Write **follow-up letters**
- Write **job acceptance and rejection letters**
- Prepare for a **job interview**
- Negotiate a **job offer**

DIGITAL RESOURCES

On the Companion Website www.prenhall.com/dobrin:

- Case 1: Tyrus Johnson: Preparing for the Job Search
- Case 2: Elizabeth Walker Searches for a Job
- Case 3: Shakira Solomon Creates a Job Application Packet
- Case 4: Writing a Position Description and Job Ad for a Web Page Designer
- Video Case: Finding Employment at Cole Engineering–Water Quality Division
- PowerPoint Chapter Review, Test-Prep Quiz, Exercises and Activities

REAL PEOPLE, REAL WRITING

MEET STEPHANIE HINSON • Landfill Director with the Pima-Maricopa Indian Community

What kinds of writing do you do as a landfill director?

A great many memos and letters to county residents, consultants, and various county departments are required on a weekly basis. Some of my communication is directed to regulatory agencies as well, so being thorough and precise is extremely important. I also compose and deal with technical specifications for equipment and construction projects.

What do you look for in a résumé?

The content of the résumé should highlight specific information in two major categories: relevant experience and education. In my opinion the ideal length is two pages. Understandably, those who are just out of college will likely have only one page because of a lack of work experience.

What do you look for in a cover letter?

I would advise people not to simply reiterate the contents of the résumé. I view cover letters as an extension of the individual. If an applicant has taken the time to research the company or organization and can relay how he or she could be an asset for reasons *other* than what is spelled out in the résumé, then I view the cover letter as valuable.

How do you react when you find typos, misspelled words, or formatting errors in an application?

Typos and misspelled words are the easiest reason to eliminate an applicant. If individuals don't have the time and talent to prepare a résumé without mistakes, I infer that they won't have the attention to detail and desire to work to the best of their ability as employees.

How should an interviewee prepare for a job interview?

Find out as much as possible about the company, agency, or organization. Most have websites or at least some pertinent information that can be found on the Internet. By demonstrating an interest in the company up front, an applicant can significantly impress an employer during an interview, because it translates into a strong desire to work. For those right out of school and inexperienced in job interviews, practice. Have a friend or family member role-play and ask sample interview questions.

Should an interviewee do anything to follow up after an interview?

Although it seems to be a fading tradition, a prompt thank-you letter after the interview can get you bonus points. Thank the employer for the time spent, note one or two things discussed in the interview that appealed to you, and succinctly restate your desire to work there.

What advice about workplace writing do you have for readers of this textbook?

Once in the workplace, review correspondence from your superiors, and try to imitate the format and style that they use. Pay particular attention to detail, and use whatever tool is available to you, whether it is spell and/or grammar check or a fellow coworker as an editor. Although you should not become dependent on such tools to do the writing for you, they can be useful in fine-tuning your writing before you send it out.

INTRODUCTION

Although it may seem obvious, finding and obtaining employment is a prerequisite to creating nearly all of the other types of documents covered in this book; most workplace documents are created by individuals who work for particular companies and organizations. However, finding professional employment is becoming increasingly difficult in today's marketplace. Job applicants can no longer expect to simply stumble onto the job of their choice—in many cases even entry-level positions are flooded with applicants. As a result, people who do find employment are sometimes forced to work in jobs outside their field, in jobs that do not require a college degree, or in jobs that are unrewarding and thankless.

In 2000 the Internet job service Monster.com ran a series of commercials playing on the public's fears about the job market. The commercials, which cost $4 million and premiered during Superbowl XXXIV, featured black-and-white close-ups of children describing the jobs they want when they grow up. Their descriptions were ironic, focusing on the realities of modern employment rather than the idealistic jobs most children imagine. The children made statements like these: "When I grow up, I want to file all day," "When I grow up, I want to claw my way up to middle management," "I want to have a brown nose," "I want to be a yes-man," "I want to be replaced on a whim," "I want to be forced into early retirement." The commercials closed with the text *What did you want to be?* appearing on a black background.

Those commercials were enormously popular and were voted among the 'top ten commercials of all time' by readers of *USA Today*. Their popularity may lie in the fact that they reveal a truth about the job market: some people are forced to take unsatisfying jobs. Nevertheless, great professional jobs still exist, and following the suggestions in this chapter should help you maximize your chances of getting the job you want. Rather than simply describing how to write a cover letter and a résumé, this chapter addresses the entire process of finding employment, from the things you should be doing now to prepare for a job search, to following up with potential employers after you've applied and interviewed for one or more jobs. The chapter covers these activities: preparing for a job search, creating employment-related documents, preparing for and participating in job interviews, and responding to interviews and job offers.

PREPARING FOR A JOB SEARCH

Successful job applicants are those who begin thinking about their careers long before their first interview. Some college students assume that a job will be waiting for them upon graduation, others just don't consider what they will do after they finish school, and some are so nervous about the whole process that they put it off until their last semester. Even though it may be difficult to think beyond the twenty-page economics paper due next Thursday, there are some things you should do now and in the near future to find the job you want after receiving your degree.

This section prepares you for upcoming job searches, and our recommendations come with a time line of steps to follow—from today until you apply for and find professional employment. Implementing these suggestions will maximize your chances of finding the job you want while minimizing the stress, time, and effort involved in the job application process itself.

Gather Textual Information *Begin Now*

One important thing to do now is to get details about the type of job you are seeking, and a good way to begin is by looking at websites, books, and articles devoted to that particular career. Although these sources may not answer all of your questions, they will give you a general sense of the jobs available in your field and the duties and requirements those jobs entail. Many of these sources also provide information about the demand for new employees in the field, average beginning salaries, and regions of the country most in need of such workers.

You can begin with a basic web search, using the job you have in mind as your keyword, just to get a sense of what is out there. You will probably come across personal web pages created by individuals who hold the job you're researching; websites developed by professional organizations concerned with issues, trends, and opportunities in that specialization; online magazines and journals targeted to professionals in the field; and some web-based directories devoted to the field itself. All of these can be useful in getting a sense of the job you're pursuing and what it entails.

After completing a basic web search, you might want to visit the library or your school's placement center to browse through their collection of career-related publications. These will run the gamut from this month's issue of a magazine devoted to your intended career (e.g., *Mechanical Engineering Magazine* or *Nursing Monthly*) to handbooks and bibliographies that supply information about various occupations or provide the names of other books containing that information (e.g., *The Directory of Career Resources*, *Occupational Outlook Handbook*, or *The Encyclopedia of Careers and Vocational Guidance*). Ask a reference librarian or someone in the placement office to help you locate informative sources, but keep in mind that more and more of these traditionally print-based sources are moving to the Web, so you may be able to find many of them online.

Interview Experts *Begin Now*

Perhaps the best way to get current, straightforward information about a career is to interview an expert. Experts may range from a professor with whom you have studied, to a recent graduate now working in the field, to a manager or supervisor with a lifetime of experience in the profession. Each of these experts has something different to offer: your professor probably has information about employment trends in the field and ways you, as a student, can best prepare for your career; recent graduates can explain how they went about getting their jobs and can give you some specific advice; and a manager or supervisor knows what specializations are most needed and who is currently hiring.

You may be surprised to find that most experts are eager to share their knowledge if you approach them in the right way. Let them know up front that you are not asking for a job interview but are merely soliciting information about the profession itself; this is particularly important when interviewing a manager or supervisor. To interview a working professional, it can be helpful if someone introduces you to that person. Draw on your resources: ask friends and family whether they know someone in this profession, call on your school's placement office to help you set up an interview, or contact your university's alumni association. If these resources fail, call (don't e-mail) a reputable firm in your field to request an informational interview.

After arranging the interview, you need to prepare for it by writing down a list of important questions to ask, leaving space below each question to jot down some notes during the interview. Avoid writing down responses word for word, however—the best interviews proceed like a conversation rather than an inquisition. If you schedule an interview for fifteen to twenty minutes, limit your questions to ten or fewer. Consider the following questions:

- How did you become involved in this profession?
- What course work and training prepared you for your job?

- What should I do to prepare for this career?
- What do you like most/least about your job?
- What do you do in a typical day?
- What is the short- and long-term career outlook for this profession?
- Can you recommend any books, magazines, or websites about this profession?
- What type of personality traits seem to work best in this career?

Although your primary objective is to gather information about the profession, you may also be cultivating a relationship with someone at a company or an organization to which you might apply in the future or from whom you might ask for a letter of recommendation. Consequently, you should dress professionally, be flexible about the length and time of the interview, and send a thank-you letter immediately after the interview.

Interact with Professionals *Begin Now*

Professionals interact with each other in a number of ways, sometimes face-to-face and sometimes through the Internet. Face-to-face interactions may result from membership in formal organizations and associations, which often hold regular conferences and meetings. Online interactions may occur through e-mail lists, bulletin boards, and online networking sites like LinkedIn or even more general social sites like MySpace and Facebook. These points of connection allow individuals to learn more about trends, issues, and new ideas in their profession, and they allow *you* to find out more about the career you have in mind.

As you conduct preliminary research and informative interviews, keep abreast of the organizations in your field and the ways to join them. Joining a professional organization is usually inexpensive and can be a great way to learn more about the field. Many conferences and meetings welcome future members of the profession, and you can learn a great deal through observation. In addition, Internet forums for particular professions are usually free—all you need is an e-mail address and access to a computer.

ANALYZE THIS

Find and join an Internet forum (such as an e-mail Listserv, electronic newsletter, or online networking site) geared toward your profession. Then spend some time reading the messages and, if possible, interacting with the members. You should also access the archive of previous messages, if those messages are available. After you have a clear idea about the group you've been studying, write a short description of this online group. Consider the following as you write your description:

- Why do individuals participate in this Internet group?
- What characteristics do the group members share?
- What important issues or topics are discussed in this forum?
- How often do people participate/contribute?
- What is the general tone, length, and perspective of most messages and postings?

After completing your description, compare it to that of one or more of your classmates, pointing out similarities and differences among various professions.

Apply for Career-Related Training or Volunteer Programs *Begin Now*

Although you may not be prepared for employment in your profession, you can still gain valuable experience through training or volunteer positions related to your field. Most colleges have work-study programs designed to help students pay for tuition—often through on-campus jobs related to their major. For example, many information technology and computer science majors work as part-time attendants in their school's computer labs. In addition, many students participate in supervised internships related to their majors, earning academic credit and sometimes a salary in the service of various firms and organizations. Or you might volunteer to help a nonprofit or community-service organization. Volunteering can offer wonderful work experience, but you must remember your ethical responsibility to those you are helping—you should not volunteer just to develop your résumé. You may also consider military service, where you can get free training in many technical fields while serving your country. All of these positions provide important experience, which can later set you apart from other candidates applying for the same job.

Begin a Working Résumé or a Personal Data File *Begin Now*

Although you may not be sending out your résumé for several years, it is important to write down employment-related names, dates, experiences, and qualifications as soon as possible. Some people choose to record all of these details in a working résumé, whereas others simply jot down pertinent material in a file on a computer disk—often called a personal data file. Regardless of which method you choose, make a point of updating the information and recording details at least once a year.

Creating a working résumé or a personal data file can be useful for a number of reasons. First, it helps you recall specific names, addresses, dates, and details that may be needed on a more formal résumé in the future. Second, seeing the information written down shows gaps in your training or experience that you need to fill in. And third, this information is useful if you need to contact a work-related colleague or boss for a letter of recommendation.

IN YOUR **EXPERIENCE**

The preceding sections describe five things that you should already be doing to prepare for your future job search. Which of these things are you already doing? Which things have you not yet begun to do? What is your plan for the things you have not yet begun? Write a narrative describing the things you've already done to prepare for your job search, and outline how and when you plan to accomplish the other job preparation activities mentioned here. In a small group share your narrative with your classmates, swapping ideas and thoughts about job search preparation.

Request Letters of Recommendation and References *One Year before Graduation*

References are extremely important in landing your first professional job, and you should begin thinking about them as early as possible. References come in two forms: letters of recommendation and verbal references. For you as a student, a primary source of letters of recommendation will be professors with whom you have studied. You usually begin to cultivate relationships with professors in your major by taking one or more courses with them. While enrolled in those courses, you should participate often in class discussions and try to meet with the professor during office hours at least once during the semester, perhaps to discuss an upcoming essay or a reading assignment or even to conduct an informal interview about career opportunities in your field. You might briefly mention your career goals at this time.

During your final year of course work, it will be easier to approach that professor for a letter of recommendation if you have already established a relationship. You should request the letter in person—not through e-mail or phone—and you should ask during an office visit. Don't corner a potential letter writer before or after class, because the professor will probably have other things to think about at that time. Of course, this contact may be easiest to accomplish if you are currently enrolled in the professor's class.

Before requesting a letter, compile a list of things that the person might mention about you. The following details might be useful:

- Courses you've taken with that professor
- Essays you've written
- Grades for particular assignments and courses
- GPA (in major and overall)
- Degree you will receive and date of graduation
- Academic awards
- Volunteer, internship, or work-study experience
- Job experience related to the career you're seeking

Although you cannot tell the letter writer exactly what you would like in the letter, you can provide information to assist that person with important details. One approach is to include the details in a letter to the professor, officially requesting a recommendation; these are called *recommendation request letters*. Even though many letters of recommendation contain glowing adjectives—such as hard working, studious, or intelligent—the most effective letters are filled with details and concrete facts. Thus, providing specific information in a recommendation request letter not only makes the writing easier for the person writing the recommendation, but also improves the quality and the content of the letter. Figure 14.1 on page 408 provides an example of a recommendation request letter.

Because most job applications request two or three references, you may also want to ask the supervisor of a job you held, the director of a community organization or a volunteer project in which you participated, or even an influential family friend to write a letter for you. You'll still want to provide a list of details the person might mention about you, although the details themselves depend on the type of interaction you've had with the letter writer. It is best if you've known or worked with this individual for a year or more because letters of recommendation should come from people who know you well. In some circumstances confidential letters are better; they presumably allow the writer to be more honest in evaluating you since you will not see the letter. If you choose to use confidential letters or the writer requests that option, have the writer seal the letter in an envelope or submit it to a dossier service.

Although letters of recommendation are highly useful, some job listings simply request a list of individuals who serve as references for you. Like letter references, verbal references generally fall into two categories: *character references*, who speak to your integrity, and *professional references*, who speak to your professional abilities. Before listing anyone on your résumé, ask for permission to do so. Not only is it unethical to list someone as a reference without asking first, but it can also backfire if the reference is contacted unexpectedly. If you anticipate that a specific employer will contact your references, let them know ahead of time so that they can think about what to say. Keep track of the following details about your references:

- Full name and professional title
- Work address, e-mail address, and telephone number
- Home address and phone number, not typically shared with an employer
- Length of time and capacity in which you've known the person

Begin Compiling a Dossier or Portfolio *One Year before Graduation*

As you gather knowledge and experience in preparation for your career, you should also begin to collect and organize relevant documents in a dossier. A *dossier* contains the basic credentials and materials needed for most job searches and applications, including college transcripts, letters of recommendation, and any other certificates, letters, or records that chronicle your achievements.

There are several ways to compile a dossier; however, your school's placement center or career services office is often the best place to begin because such a center usually provides an inexpensive, sometimes free, dossier service. You could also use a professional dossier service, which might be a bit more expensive but can sometimes provide quicker turn-around on requests. Many dossier services now place credentials online, where you can read through them (with the exception of confidential letters), update your file from home, and order copies to be sent to prospective employers. We recommend using a service, although some people like to maintain control of their dossiers themselves.

Dossier services provide a number of benefits that make your job search easier. They make copies of your reference letters, even if they are confidential, so that you can spare your references the task of writing the same letter repeatedly. In addition, dossier services maintain your documents in a secure, organized manner, eliminating the danger of your misplacing or damaging important originals. And in most cases one simple request sends all of the documents you ask for, so there is little danger of failing to include an important document in a large file. Furthermore, transmission is usually quick, and you'll often get confirmation that your dossier has been received, which can be very reassuring. Finally, sending a dossier from a service just seems more professional; the dossier usually comes printed on quality paper, bound in a packet with a cover, and packaged in a printed mailer. Because some employers question the integrity and authenticity of self-maintained dossiers, using a service reassures them that your materials are genuine.

If you apply for a job that emphasizes writing, artistic, or creative skills, you might create a *portfolio* of documents and visuals that represent your best work. Portfolios can include any of the following:

- Essays that received high marks
- Published writing, such as articles in a campus newspaper
- Charts, graphs, or posters
- Printed copies of web pages
- Artistic or graphic creations

Of course, you would want to redesign your portfolio to suit the unique nature of each job you seek. And if an employer does not request a portfolio, you might offer in your cover letter to send it, or you might take it to the interview. A portfolio can be a great icebreaker during awkward or uncertain moments in an interview, and it can make you more memorable to employers who might be interviewing dozens of applicants for a position.

Begin Your Job Search *Six Months before Graduation*

As you get closer to graduation, research and apply for jobs in your profession. You will have many avenues for this search, so don't be disappointed if you hit a few dead ends along the way. Explore as many options as possible.

Employers

Through all of the previous research you've done, you probably have developed an interest in several companies in your field. You can find out about jobs within those companies by searching their websites, speaking with the human resources or personnel departments, or contacting the departments most likely to hire someone with your qualifications.

Even if no position is listed, you can send an unsolicited letter to the employer—often called a *letter of inquiry*. Although this may seem pushy, many professionals favor this technique to find or change positions. Many internal positions are unadvertised, and occasionally an unsolicited letter can be the impetus for a company to create a new position for an outstanding applicant. Although applying for a job listed in the newspaper might seem to be the most obvious method of gaining employment, studies show that applying directly to an employer is twice as effective as answering a newspaper ad.[1] Figure 14.2 on page 410 shows a sample letter of inquiry.

Personal Contacts

Another benefit of earlier job inquiries and interviews is that people become aware that you're in the job market. In other words, these basic activities help you develop a network of professionals who can alert you to new jobs in your profession or can put you in touch with employers. Contact your resources: friends, relatives, past employers, professors, and anyone else who can help you find employment. Sometimes who you know is as important as what you know; networking is one of the most successful methods of finding a job in any profession.

Internet

An increasing number of jobs are advertised through the Internet—some exclusively so. If you are searching for a position in a large corporation or in a technology-related field, job listings can almost always be found through the corporations' websites. In fact, a recent survey of Global 500 corporations reveals that 91 percent of these companies use the Web for recruitment—up from only 29 percent four years earlier.[2]

Even if you have not targeted a specific organization for your job search, you can still use the Web to find employment. Internet job services have multiplied in number and in comprehensiveness in the last few years. Although some sites simply list jobs, many offer such services as job counseling and career advice, editing help for job-related documents, articles and statistics about market trends, and detailed lists of resources. Many of these sites also allow you to post résumés to specific employers or to a database of individuals who are seeking employment.

If you plan to use an Internet job service, examine the following sites as a starting point:

America's Job Bank	www.jobbankinfo.org
Monster	www.monster.com
Head Hunters	www.headhunters.com
Career Mosaic	www.careermosaic.org
College Grad	www.collegegrad.com
Job Hunter's Bible	www.jobhuntersbible.com
Career Builder	www.careerbuilder.com
Job Hunt	www.job-hunt.org
Rockport Institute	www.rockportinstitute.com

---

[1]U.S. Department of Labor Employment and Training Administration, "Tips for Finding the Right Job," 1996, http://www.doletagov/uses/tip4jobs.pdf.

[2]Tales Research, "Global 500 Website Recruiting, 2003 Survey," 2003, http://www.taleo.com/research/articles/strategic-article1.php?id=28.

Because these job sites have become so numerous, you can also use a search engine such as Google to search for "employment" or "jobs," combined with the name of your profession. The website dice.com, for example, specializes in jobs in technology, whereas teachwave.com is designed specifically for K–12 educators. Narrowing your search can be helpful, especially if you are overwhelmed by the number of job listings on a larger Internet job service.

In addition, online networking sites are becoming useful in job searches. As mentioned earlier in this chapter, sites such as LinkedIn, Myspace, and Facebook are coming to the attention of employers as they seek and consider potential employees. Some employers advertise positions through these sites; others use the sites to actively seek candidates with particular specializations. Employment recruiters (known as headhunters) also search these sites for potential clients. According to a 2008 survey conducted by the recruitment firm Robert Half International, 67 percent of executives believe professional networking sites will be useful in the search for job candidates in the next three years.[3]

Most of these sites require you to create a user profile, which displays information about you to other viewers. Make sure that your online profile portrays a professional image; don't post anything that you wouldn't want an employer to see. Even if you don't list your online profile on your résumé, employers may search for it to find out more about you.

Newspaper Listings

Newspapers are still a great way to find employment opportunities, especially if you are looking for your first "real" job after graduation. Consider looking at the employment section of more than one source, especially if you live in a heavily populated area with more than one newspaper; some employers may not advertise in all of the newspapers in their area. Scanning the newspaper listings can also alert you to trends in your profession; a company with many listings may be open to hiring someone with your credentials, even if the company doesn't have a listing for someone with your particular training and skills. Some job seekers even run their own ads, which may be read and acted on by an employer who spots it.

Because most newspapers are now on the Web, you should scan their online job listings frequently. In many cases printed ads run on a weekly basis, whereas online listings may be updated every day. Reading a listing online can give you a head start on applicants who wait for the printed listings.

ANALYZE THIS

Find a newspaper in your area or one in a town or city in which you'd like to work. Read through the classified section, and search for jobs in your intended career. Do any such jobs exist? What criteria or qualities do those job listings specify? If a job in your intended career is not advertised, are there jobs in related fields you'd consider? Are there jobs that might serve as a good stepping-stone to your intended career? Don't be discouraged if no listings exist; job listings change frequently, and as we suggest in this chapter, not all jobs can be found in the classified section. Write a short analysis of the job market, based on what you are seeing in the classified section of the newspaper.

Journals and Catalogs

In addition to newspapers, you should explore other print-based publications. Many jobs in technical and professional fields can be found through the major journals in those fields. Job listings also appear in the public relations catalogs of larger

[3]Mike Sachoff, "Employers Going Online to Find Job Seekers," April 21, 2008, http://www.webpronews.com/topnews/2008/04/21/employers-going-online-to-find-job-seekers.

companies, many of which are designed to attract new college graduates. Visit your library to browse through journals in your field, and visit your placement office to explore listings in a public relations catalog.

Placement Office

Your school's placement office provides information about jobs in your field and helps to match students with potential employers. Many placement offices allow students to put a copy of their dossier on file, which is then made available to representatives of businesses and government agencies, who use such files to contact prospective employees. Using the placement office is usually free, and simply dropping off a dossier may place you on a list that might be examined by hundreds of potential employers.

EXPLORE

Spend some time getting to know your university's placement office. You can simply visit the office after class, or perhaps your instructor will invite a placement officer to visit your class to describe the services that are available. When you speak with a placement officer or representative, be sure to ask specific questions about your future job search and intended profession. You may find it useful to jot down several questions ahead of time.

After you've learned as much as you can, write a memo to your instructor outlining the services available through your placement office, the specific resources you plan to use to help with your own job search, and anything the placement office might do differently to help students.

Professional Placement Service

Although professional placement services usually cater to managerial or supervisory professionals, they can be useful for finding entry-level positions as well. In most cases professional services use what is known as a headhunter, an individual who actively searches for a job for you. Placement services can be useful for finding jobs you weren't aware of, but they charge a fee that must be paid by either the employer or the new employee. Many placement services specialize in a particular profession; you can find information about them on the Web.

CREATING EMPLOYMENT-RELATED DOCUMENTS

According to a survey conducted by the National Association of Colleges and Employers (NACE), effective communication is the single most important characteristic employers look for in job candidates, averaging 4.8 on a 5-point scale. NACE Executive Director Marilyn Mackes asserts that "regardless of the job market, employers want to hire those candidates who complement their work-related skills and experience with the interpersonal and communication skills so critical to workplace success."[4] As a result, employers will carefully scrutinize the job-related documents you send to them, and your ability to communicate effectively in these documents may determine whether you are interviewed for a job. In other words, as Stephanie Hinson suggests in her opening interview, employers often assess a potential employee's communication skills based on job application materials, often before meeting the applicant.

Consequently, it is imperative that your job application materials are clearly organized, easy to read, and factually, mechanically, and grammatically correct if you hope to be interviewed for a job. Employing the problem-solving approach can help

---

[4]Marilyn Macker, quoted in "News for Media Professionals," January 15, 2004, http://www.naceweb.org/press/display.asp?year=2004.prid=184.

PSA

you create those materials. Successful applicants carefully plan ahead before creating job-related materials; they consider the persuasion strategies, document formats, and delivery methods that will be the most effective. They also conduct research and gather information about their potential employers, learning as much as possible about company mission, goals, products, and services. Successful applicants then carefully draft their documents in a way that will engage their readers and present information clearly and logically. In addition, these applicants solicit feedback and revise their documents to ensure that they meet all appropriate standards. Finally, these applicants consider the best way to distribute their documents to be sure that they are transmitted and received in a manner that is most suitable for their readers.

Recommendation Request Letters

As mentioned earlier, some jobs require an official letter of recommendation, whereas others simply require the names and contact information of people who can speak on the applicant's behalf to potential employers. Regardless of the form of the reference or recommendation, applicants should write a recommendation request letter to each person solicited for that role. Again, applicants often choose a mixture of individuals who can address work experience, educational achievements, and/or personal integrity. The letter of request should contain the following information:

- *A specific request.* In most cases the letter should begin with some brief information about who you are and what your relationship is with the reader, followed by a specific statement of purpose, such as "I am writing to request a letter of recommendation" or "I am writing to ask whether I can place your name on my list of references." Your reader needs to know exactly what you are requesting. You should also specify whether the requested recommendation is for a specific job or for a broad job search; that information will influence how the letter is written.
- *Pertinent details.* You want to phrase the details you include as a reminder of interactions you have had with this person; you should never tell a potential reference exactly what to say. With a professor you might mention classes you've taken, grades you've received, essays you've written, and other academic information. With a former employer you might mention the job you did, dates of employment, and other appropriate information. You should include any other activities and experiences that you think might be helpful.
- *A time frame.* If you need a letter by a specific time, you should mention the date near the end of the letter. Whenever possible, allow at least three to four weeks for the writer to work on the letter; never ask someone to write a letter that needs to be finished in just a few days. If you haven't already secured the person's agreement to write a letter or serve as a reference, ask for a response in a week or so. After that time you can follow up with a phone call or an e-mail message as a friendly reminder.
- *Contact information.* You should always provide your phone number and e-mail address and perhaps your home or campus mailing address. If a letter is to be written for a specific position, include the appropriate mailing information. If the letter is for a broad search, include the mailing information for your dossier service, or offer to pick up the letter at the writer's convenience, if appropriate.
- *An offer to discuss the request further.* A follow-up discussion may be helpful if you have not had recent contact with the person. In any case do not limit your availability; offer to meet with the person at his or her convenience. Even if there are specific days on which you are not available, it is better to simply offer to meet with the person and then arrange an appropriate time after he or she contacts you.

- *An expression of thanks.* Be sure to thank the individual for *considering* your request; do not assume a positive response. If the individual does agree to write a letter of recommendation or serve as a reference, offer additional thanks for the time and effort expended. No one is obligated to help you.

15578 Whiskey Way
Twin Falls, ID 83301
(208) 734-9102
jdcourt@coled.edu

August 11, 2008

Professor G. A. Onslo
Department of Engineering
Coltran College
Twin Falls, ID 83301

Dear Professor Onslo:

As you know, I was enrolled in your Introduction to Engineering course in fall 2006 and your Theories of Environmental Engineering class this past semester, and I have also served as your research assistant for two years. I am currently putting my career portfolio together and am gathering materials for my dossier. I am writing, per our conversation, to request from you a letter of recommendation. I would appreciate it if your letter could address the following:

Specific request →

- My ability to grasp and implement environmental engineering theories
- My reliability and performance as an engineering student
- My ability to successfully meet assignment deadlines
- My work with you as a research assistant

Pertinent details →

It would also be helpful if your letter could specifically address that you have worked with me as both a student and a research assistant and for how long. Please address your letter generically, because it will be used to apply for a number of jobs.

I would appreciate it if you could have the letter ready in three weeks. Please call or e-mail me to let me know when it might be convenient for me to pick up the letter, or you can send it directly to the Campus Career Resources dossier service in Emerson Hall. Thank you very much for your time and consideration in writing this letter. If I can answer any questions, please do not hesitate to contact me.

Time frame →

Expression of thanks →

Sincerely,

Jamaica DeCourt

Jamaica DeCourt

Encl:Résumé

FIGURE 14.1 A recommendation request letter

IN YOUR **EXPERIENCE**

It is likely that you have asked a person to recommend you for something at some time. It may have been a recommendation for your first job, a school or extracurricular activity, or perhaps admission to where you are now—in college.

Write about the experience of asking for that recommendation. Was it difficult or intimidating? How did you approach the person? Did you write a letter? If so, was it similar to the letter shown in Figure 14.1? What might you have done differently in asking for the recommendation? Do you think that change would have influenced how the person responded?

Letters of Inquiry

If you hope to work for a specific organization that does not currently have a job available in your field, you can send that organization a letter of inquiry, asking for an informal interview. Many companies and organizations welcome such inquiries and interviews even if they are not currently hiring; it gives them a chance to learn about potential employees, promote their company, and add you to a list of future prospects.

Who you send this letter to is important because you want to make contact with someone who actually has an interest in your qualifications and has the ability to hire you. The general rule is to go as high up the corporate ladder as possible but with someone who has the time and specific knowledge of your area to make the interview worthwhile. For larger firms you should contact a division supervisor or a director of human resources; for smaller firms you might write to the president or the owner. If you know someone within the organization, it may be useful to ask that person to refer you to someone within the organization who has the ability to hire you; and if you are able to get a referral, you should consider mentioning at the beginning of your letter of inquiry the person who referred you.

Before writing a letter of inquiry, you should learn as much as you can about the company, using the strategies outlined earlier in this chapter. As you write the letter, you should follow these guidelines:

- *In the opening sentence request an informal interview.* Make it clear to the reader that you are not applying for a specific job listing but are asking for a chance to discuss future openings in the organization.
- *Mention your qualifications for a future position in the firm.* Your goal is to show that you could *potentially* work for the organization in the future. Mention specific training or degrees, as well as years of experience in the field, if applicable, but do not go into detail about what you could bring to the firm.
- *State your reasons for contacting this particular organization.* This is an opportunity for you to show that you've done your homework. If you can show your knowledge of the firm and explain your interest, the reader will be more inclined to offer an informal interview. If you can do so without being too obsequious, try to use the firm's corporate language, slogans, or buzzwords—but be careful here.
- *Include a copy of your résumé, if appropriate.* The reader can skim through your résumé and qualifications before deciding whether to meet with you.
- *Provide contact information.* Include your phone number and e-mail address and your home or campus mailing address, if appropriate.
- *Offer to meet at the reader's convenience.* And don't forget to thank the individual for considering your request for an informal interview. Figure 14.2 gives a sample letter of inquiry.

Résumés

A résumé contains key information about your experiences and training and functions as the main tool in getting a job interview. There are few universal rules for creating an effective résumé, but it must be clearly organized, filled with specific details, and free from all errors. Aside from these basic criteria, there is no single right way to prepare a résumé. However, employers expect résumés to follow a generally accepted format and guidelines.

Remember that a résumé is a reflection of your professionalism; even small errors in your résumé suggest to readers that you are not detail oriented or concerned with quality. As Stephanie Hinson suggests, errors in a résumé are the easiest reason to eliminate a candidate. Even highly qualified candidates may be eliminated from consideration if their résumés or cover letters contain errors.

EXPERTS SAY

"The first thing any employer sees is a snapshot of that person in the form of a résumé, a CV, or a cover letter. Those documents speak volumes about the candidate's ability to communicate effectively."

STORMY BROWN,
Employee Relations Manager with Coca-Cola Enterprises

1300 Quail Cove Road
Matthews, VA 23109
(804) 423-9327
jocox@vaam.edu

July 29, 2008

Ms. L. M. N. O. Pritchard
Director of Personnel
Old Dominion Biotech
876 Sundowner Pkwy.
Chesapeake, VA 23321

Dear Ms. Pritchard:

Request for an informal interview →

I am currently seeking a position as a DNA and protein techniques specialist and am requesting an informal interview with Old Dominion Biotech. From the information provided on your web pages, I am aware that Old Dominion conducts functional analysis of full-length DNAs in its work to develop a system for screening gene functions.

Qualifications →

As you can see from the enclosed résumé, I majored in biology with a primary area of research in genetic screening. In addition, I served as a work-study lab assistant in the genetics lab while I was a senior at Virginia A&M. These experiences have helped prepare me for a career in biotechnology and genetics.

Reasons for contacting this organization →

I appreciate your time in providing me with any information and am available to meet with you at your convenience to discuss the possibility of employment with ODB.

Sincerely,

Joel Cox

Joel Cox

Encl:Résumé

FIGURE 14.2 A letter of inquiry

Chronological Résumés

The most common type of résumé is a chronological résumé, which summarizes your experience and training in the order in which it occurred, beginning with the most recent and going backward in time (i.e., actually *reverse chronology*). This type of résumé highlights the facts, emphasizing degrees, job titles, and dates. You should use a chronological résumé under any of these conditions:

- You have the appropriate or typical experience for the position for which you are applying.
- Your work experience shows increasing promotion or responsibility over time.
- You have impressive qualifications, job titles, or honors.

Figure 14.3 shows a sample chronological résumé.

Emily Sanchez
1482 Purple Lane
Philadelphia, PA 19601 (215) 723-8068
esanchez@temple.edu

CAREER OBJECTIVE
To obtain a position as an accountant in a nationally recognized firm

EDUCATION
Temple University, Philadelphia, PA
Bachelor of Science, May 2008
Major: Accounting, Overall GPA: 3.4/4.0

WORK EXPERIENCE
Accounting Intern, August 2006–present
Grote Construction, Mohnton, PA

- Review and adjust ledgers
- Prepare state and federal tax returns
- Manage accounts payable and receivable

Bookkeeper, May 2004–August 2006
Watkins Associates, Ambler, PA

- Calculated payroll for more than thirty employees
- Trained and supervised assistant bookkeeper

HONORS AND ACTIVITIES
Dean's list, Fall/Sping 2007
Vice President, Temple University Accounting Society, 2005–2007
Volunteer, Tax Assistance Program 2004–2006

REFERENCES

Scott Grote	Barry Golston	Patricia Mizzen
Owner, *Grote Construction*	Professor of Accounting	Manager, *Watkins Associates*
517 Cherry Street	Temple University	815 Sycamore Road
Mohnton, PA 19601	Philadelphia, PA	Ambler, PA
(215) 464-3332	(215) 438-5669	(215) 889-7866

FIGURE 14.3 A chronological résumé

Skills Résumés

Another type of basic résumé is the skills résumé, which emphasizes the experience and skills that qualify the applicant for a position, rather than specific degrees, job titles, and dates. Use a skills résumé under any of these conditions:

- Your experience or training is not typical of most applicants for the job for which you are applying.
- You have extensive experience or training in another field, and you are making a career change.
- You have significant gaps of time between jobs, education, or training.
- You want to highlight the breadth of your experience in different categories, rather than focusing on titles, degrees, or dates.

Although skills résumés can be effective in some situations, there is a stigma against them in certain professions, particularly when a large number of applicants are competing for one position. Even though an applicant may have significant experience in a related field, interviewers are sometimes more likely to focus on those who have taken the usual route to the position for which they are applying. Figure 14.4 gives an example of a skills résumé.

Entry-Level Résumés

Even if you have no professional experience, you can still create a useful résumé. Entry-level positions or internships often require a résumé, and readers of those résumés understand that you may be applying for your first job and may have little or no related experience. In such a case, you should create an entry-level, or no-experience résumé. These résumés follow the same basic format as the chronological and skills résumés, but they serve to highlight who you are as a person rather than your credentials or experience.

Zollie Felder
1440 Centerville Road
Greeley, CO 80631
(970) 555-3888 Zfeld@hotmail.com

CAREER OBJECTIVE
To obtain a managerial position in an information technology firm

SUMMARY OF RELEVANT SKILLS
Business Management and Sales

- Raised $20,000 in venture capital to develop medical industry website
- Designed and managed online sales distribution processes
- Successfully acquired over 50 product dealers throughout the United States, Europe, and Japan

Graphics and Design

- Designed and engineered cutting-edge web portal in medical sales
- Conceived and designed all advertising promotions for Medi-Net Inc.

Hardware and Software

- Proficient running and operating web servers and Internet applications including HTML and XHTML
- Experience with graphics software, including all Adobe products
- Experience with word processing and spreadsheet software

EMPLOYMENT HISTORY
2005–2008 Owner and Manager, *Medi-Net Incorporated*, Greeley, Colorado
1999–2002 Medical Supplies Specialist, *United States Air Force*

EDUCATION
1998 Diploma, Edmonton High School, Edmonton, Oklahoma

REFERENCES
Available upon request

FIGURE 14.4 A skills résumé

Most people have some skills or experiences they can include in an entry-level résumé, even if those skills or experiences do not demonstrate their ability to do the particular job for which they are applying. In these situations you should select details that show you as trustworthy, compassionate, intelligent, and interesting. As Figure 14.5 demonstrates, you can produce an acceptable entry-level résumé from your basic involvement in everyday activities. Most entry-level résumés are no longer than one page, although you should try to fill the page if you can.

Basic Elements of Résumés

As Figures 14.3, 14.4, and 14.5 demonstrate, the types of résumés differ in what information is included, as well as in how that information is organized, but they do share some general types of information: contact information, career objective, education, experience, honors and awards, activities, and references.

Jim Rennix
300 Sugar Lane, Baton Rouge, LA 70801
home: 225-889-3874
cell: 225-686-3210
jer5@1su.edu

EDUCATION

Port Allen High School, Baton Rouge, LA
2002–2006

Baton Rouge Community College, Baton Rouge, LA
2008–present

WORK EXPERIENCE

Pet Sitter, 2004–present

- Provide pet sitting services including dog walking, feeding and yard care.

Child Care, 2002–2005

- Provide child care for several families after school, weekends and during school vacations.

ACHIEVEMENTS

- Governor's Volunteer Award, 2005
- Neighborhood Watchdog Award, 2006

VOLUNTEER EXPERIENCE

- Big Brothers/Big Sisters, 2004
- Baton Rouge Literacy Program, 2005
- Habitats for Humanity, 2007

COMPUTER EXPERIENCE

- Proficient with Microsoft Word, PowerPoint, and Internet

INTERESTS/ACTIVITIES

- Member of Port Allen Soccer Team
- Active flats and saltwater fisherman
- Proficient in bass guitar

FIGURE 14.5 An entry-level résumé

Of course, your résumé may not contain all of these categories; no two résumés ever contain the same details.

You should experiment with different approaches to categorizing your information: combine two or more short categories; add new categories not listed here; try rephrasing the categories to better suit your information. If you are particularly strong in one area, try placing it ahead of other areas. Then, after you have experimented with several versions, show two or more of them to a friend, classmate, family member, or professor to get some feedback on which résumé best emphasizes and organizes your information. And remember that a good résumé is not static—it is constantly being revised and tweaked as you apply for new jobs and gain new information to include in the résumé.

Contact Information. All résumés contain the individual's contact information, clearly displayed at the top of the first page. Many résumés include the person's full name in boldface, centered at the top of the page and followed by the person's address directly below it. There should be a comma after the city name and a two-letter postal abbreviation for the state, like this:

Columbia, PA 17512

In some cases you may wish to include both your home address and a school or work address. If so, align these on the left and right margins of the page, balanced evenly. Below your name and address, you'll want to include your complete phone number, including the area code. List the area code in parentheses, like this:

(614) 477-8176

You should also include your professional e-mail address directly below your phone number. If you do not have an e-mail address, you should obtain one soon. If you have a professional-quality homepage that you would like potential employers to see, list the URL below your e-mail address. If possible, list your e-mail and homepage addresses as hyperlinks within your document to highlight them.

Having an e-mail account is a prerequisite for many professions. See chapter 12 for more information about e-mail.

Career Objective. Some individuals beginning a career or switching to a new one provide a career-objective statement to identify their goals for the next five to ten years. If you choose to include a separate category for your career objective, keep the description short (i.e., one to two sentences) and realistic. Because a good career objective can be difficult to write, some people avoid them altogether or simply list under their name the career they desire, like this:

Jordan Goldenberg	Jill Hershey	Todd Kriner
Creative Director	Legal Consultant	Technical Writer

If you do include a career objective, be sure it is specifically written for the job for which you are applying. Use the specific language of the job description if possible, and mention only goals that you might reasonably accomplish in that particular position.

Generally speaking, your career-objective statement simply says that you wish to attain a position in your field, although the statements can take a number of forms. Here are some sample statements for various fields:

- *Accounting:*
  To attain a position as an accountant in a nationally recognized firm.
  Seeking employment as an entry-level staff accountant.
  To attain a position in public or private accounting, with a long-term goal of becoming a certified public accountant.
  Seeking a position as an accountant in a large organization. Long-term plans include advancing to a managerial position within the financial branch of the company.

- *Biology:*

  Recent biology graduate seeking practical position in biotechnology, medical, or pharmaceutical industry.

  To acquire an internship in medical research.

  Seeking position with biological research organization.

  To attain employment as an entry-level biochemical laboratory technician in an organization that analyzes blood and other bodily fluids for chemical influence.

- *Computer technology:*

  A position in the computer technology field so that I can contribute to the organization's advancement with the skills that I have acquired.

  Objective: Information systems administration. CAD/CAM system maintenance. Computer-aided design and detailing.

  I seek a position to best employ my troubleshooting skills, my technical knowledge, and my leadership abilities and to best use my knowledge of operating systems, hardware, software, data communications, and networks.

  To acquire a position supervising and maintaining systems using the Unix operating system, Linux, TCP/IP, HTML, JavaScript, and the Internet.

- *Engineering:*

  Seeking employment as a structural engineer to work with stress analysis or civil engineering.

  Seeking a position as a software engineer to work with object-oriented design and analysis in software development.

  Seeking a position in mechanical design with a focus in robotics.

  To attain a full-time engineering position in civil engineering, transportation engineering, or construction management.

- *Management:*

  Objective: Entry-level human resources position with a specialization in HR management and a unique perspective on the latest HR techniques.

  Seeking to provide top-quality HR support to ensure a competitive advantage in the global marketplace.

  Management position within a company that uses my skills in sales, marketing, and administration.

  Seeking management position for a comprehensive wholesale medical supply company.

- *Criminal justice:*

  Objective: A position in law enforcement, security, or investigation.

  To attain a position as a parole officer.

  Seeking employment as a law enforcement officer.

  To obtain a position in law enforcement in order to promote justice and equality in a community atmosphere.

Education. If the degree you have or are seeking qualifies you for the position you hope to attain, you should list your education as the first major category on your résumé. You can list the information in any order, but be consistent with the order you choose. Include the degree you received or will receive, the date you graduated or will graduate, the complete name of the school, and the city and state in which the school is located. Be sure to list all schools from which you have earned degrees: community colleges, four-year colleges, and universities. For most professional jobs you should not list your high school.

After providing this basic information, you can include further details; use your judgment about what is helpful. Grades are worth mentioning only if they are at or above a 3.0 on a 4.0 scale. Course work can be useful if it applies directly to the desired job or if you need to add to a sparse résumé, but do not list courses that are irrelevant to the position. You might include some or all of the following details below each degree:

- Overall GPA
- GPA in major if higher than overall
- Specific courses related to the position
- Credit hours in a major related to the position
- Title of a thesis you wrote, a directed study you completed, or a specific academic honor

Experience. The experience section lists any employment experience since high school, including part-time jobs, summer employment, military training or experience, and work-study, internship, or volunteer positions. Include work experience that precedes college only if it pertains to the job for which you're applying or if you have minimal recent experience. Some people with a great deal of work experience place this section before the education section, particularly if the work experience is more pertinent than the education. For example, a job candidate applying for a position as a retail manager may have majored in English but has had seven years' experience in various retail stores.

If possible, try to follow the same format and sequencing in the experience section that you use in the education section. Include the job title in boldface, the name of the employer, the city and state in which the employer is located, and the dates you worked there. For summer jobs you can simply say "Summer" and give the year.

Below this basic information, include some details to describe each position you've held. Avoid using long, complete sentences; most résumés include two to three bulleted, single-line descriptions of positive, results-oriented activities completed in that position. Bulleted lists are typically indented. As you write your descriptions, choose action verbs from the categories listed in Figure 14.6. For example, details about a sales job might read as follows:

- Supervised twenty sales employees
- Scheduled monthly sales meetings
- Wrote detailed marketing reports

IN YOUR **EXPERIENCE**

According to the U.S. Department of Labor, between the ages of eighteen and thirty-five, Americans hold an average of ten different jobs. Although a long list of short-term jobs is not going to look good on your résumé, it is OK to "try on" various occupations as you prepare for your career. And since you can never be sure which of these jobs might provide the best experience for a future career—most Americans average four careers during their lifetime—it is best to keep track of them all in a working résumé or a personal data file. A job you hold now may provide experience you can use in a career later in life.

Take a few minutes right now to jot down the jobs you've held, the dates and duration of those jobs, and the activities and skills that were part of the jobs. Then talk with one or two of your classmates about which attributes of your former jobs might be useful in your future profession.

Honors and Awards. Hard work and dedication have likely earned you some recognition for your accomplishments in school, work, or service over the past few years,

Management	Communication	Research	Technical	Teaching
assigned	authored	collected	analyzed	advised
contracted	communicated	conceived	built	coached
directed	drafted	disproved	devised	coordinated
executed	edited	evaluated	fabricated	developed
implemented	formulated	examined	inspected	encouraged
launched	moderated	identified	operated	facilitated
managed	negotiated	interpreted	overhauled	initiated
organized	persuaded	investigated	programmed	instructed
oversaw	publicized	reviewed	remodeled	presented
supervised	reported	surveyed	upgraded	taught
Financial	**Creative**	**Helping**	**Clerical**	**Organizational**
administered	conceived	advised	assembled	arranged
allocated	created	aided	classified	distributed
analyzed	designed	counseled	compiled	maintained
appraised	established	demonstrated	generated	ordered
audited	formed	diagnosed	maintained	purchased
balanced	illustrated	expedited	monitored	recorded
budgeted	instituted	facilitated	operated	reserved
calculated	introduced	inspired	prepared	scheduled
estimated	originated	referred	processed	updated
planned	produced	supported	recorded	verified

FIGURE 14.6 Commonly used action verbs for résumés

and this section is the place to emphasize that. Feel free to change the wording of the section heading to fit with whatever you list below it. Include the following kinds of information in this category, but use discretion; some honors and awards may not be relevant for the position for which you are applying:

- Scholarships and fellowships
- Academic or service awards
- Nominations for awards
- Organizational positions, such as campus vice president or treasurer of the local PTA
- Publications
- Academic honor societies
- Sports achievements, such as varsity letters, leadership roles, and special accomplishments, such as state finalist

Activities. Employers often like to know what activities applicants have participated in on a regular basis, especially if applicants are recent college graduates. You can include activities that showcase skills and ability to work with others, such as the following:

- Membership in professional organizations
- Volunteer work
- Membership in campus clubs, fraternities, or sororities

- Activities that highlight special skills, abilities, or responsibility, such as school band or scout leader
- Membership in community organizations

Keep in mind, though, that certain readers may have biases or prejudices against particular organizations. Think carefully about how potential employers might perceive your activities. For example, some might see a fraternity affiliation as a means of gaining leadership and teamwork experience, whereas others might see it as evidence of beer-drinking elitism.

References. References can play an important role in obtaining employment. However, experts disagree as to how you should include them in your résumé. Some argue that you should include names and contact information on your résumé to distinguish yourself from other candidates and to provide your readers with easily accessible information. Others suggest that the phrase "References available upon request" lets your readers know that you have references ready, should they want them. Still others suggest that you should not include any information about references or that you should include them on a separate page.

You'll need to decide what works best in each situation and what impact your references might have on potential employers. If you do decide to include references on your résumé, it is most common to list three references and their contact information. The best way to list them is to create a table with invisible borders, placing each reference in one of three equally spaced cells, like this:

Professor Kennedy Collier Greenville University 501 University Park Greenville, OH 45321 (475) 707-6283 kcollier@greenville.edu	Chris Hansel Director Maytown Animal Shelter 647 Union Street Maytown, VT 17601 (803) 285-5007 chansel@mas.com	Crystal Smith Personnel Manager Speedy Auto Lube 102 N. 2nd Street Hoganville, PA 21612 (690) 440-1220

Include a full name and title for each reference. If the title does not describe the profession, include it below the name. Also include a full mailing address, a work phone number, and an e-mail address if the reference provides it. As previously suggested, you should ask these individuals whether you can use them as references and should let them know whether to anticipate a call, letter, or e-mail anytime soon.

If you have compiled a portfolio, consider adding the phrase "Portfolio available on request" directly below your list of references, or you can include a hyperlinked URL to the portfolio if it is available online.

Because design is crucial, we encourage you to read chapters 8 and 9 as you create or revise your résumé.

Design Elements. The layout and design of a résumé are important because résumés do not follow the typical sentence-and-paragraph format of most documents and are usually skimmed quickly by busy readers. The central goal is to make the information clear and easy to locate. Readers should never have to search to find particular details, and information should be arranged to emphasize strengths.

As you begin to write your résumé, experiment with different font styles to find one that reflects your own personality while fitting the overall image of the company to which you're applying. One strategy is to use the same font you find on the company's website or promotional materials. Or you can simply choose a conservative, professional font style, such as Times New Roman or Arial. Use the same font style throughout the résumé and the cover letter, perhaps varying the font sizes to emphasize your name and section headings. We recommend using 14-point font for headings and 12-point font for text, although you may wish to try a slightly smaller font size if it allows you to fit all of your information on one page.

To position your résumé information in a reader-friendly way, make use of white space, bullets, and indentations. White space helps readers find the information they need. You should use enough of it so that your information does not look crowded together, but not so much that the résumé looks sparse or empty. Use an extra line of white space between sections to set them apart. Bullets are effective for listing details that follow a title or heading. Indenting details can create a sense of flow through your document, highlighting major headings while also making the details that follow them accessible.

You should use bold type to emphasize your name at the top of the résumé; the names of sections within your document; the titles of jobs you've held; and particular skills, credentials, or abilities to which you'd like to draw attention. If you plan to list a publication, use italics. Avoid underlining text in a résumé; it can be difficult to read, does not scan or print well, and can give your résumé a cluttered appearance.

Résumé FAQs

How long should my résumé be?
Most résumés are one to two pages long. Be sure you fill at least one full page. If you can, make your résumé fill at least 1½ pages. Studies show that individuals are more likely to interview a candidate with a longer résumé, even if they say they prefer shorter résumés.

What should my résumé emphasize?
Unless you are sending out dozens of résumés, you should tailor each one to fit the particular job listing. In general, emphasize things you've done that (a) are relevant to the position, (b) highlight your communication and people skills, (c) show your superiority to other applicants, (d) show your consistency and dependability, and (e) are recent.

Should I include specific details in my résumé?
Because job recruiters and interviewers look for details that separate you from other applicants, you should include details that describe and provide evidence of your accomplishments. Omit details that are redundant, that don't strengthen your case, or that are irrelevant to the position for which you are applying.

Should I use complete sentences in my résumé?
In most cases you should use phrases and sentence fragments, with no punctuation. Use complete sentences only if they are the briefest way to present the information.

Should I refer to myself specifically in the résumé?
Avoid using *I, me* or *my* if possible, because these words can be interpreted as being arrogant or self-centered. Use them only if they are unavoidable or if they are the briefest way to say something. You can use first person in your cover letter.

How should I describe things I've done?
When possible, use gerunds (the *–ing* form of verbs) to describe your actions, whether past or present. Gerunds sound more dynamic and energetic than nouns. So rather than saying you did "website design" (noun) or you "designed websites" (past tense verb), use a gerund and say that your responsibilities included "designing the company website." If using a gerund makes the description too long, use present or past tense verbs.

Should I stick with white paper?
Paper choice contributes to the overall style of your document. For professional jobs white or off-white paper is usually best. For jobs in which creativity is important, you can experiment with different light-colored papers until you find one that conveys the style you want.

Can I omit details that portray me in a negative light?
Lying in a résumé can get you fired, sued, or even arrested—so don't do it. However, it is perfectly fine to omit details that are not to your credit. For example, you might choose to omit jobs that you held for a very short time, memberships in organizations that might seem unprofessional or radical to potential employers, or your GPA if it is low.

Online Résumés

For more on web page design, see chapter 17.

Many job seekers create online versions of their résumés, and we encourage you to do so. Although simply posting your existing résumé to the Web is a good start, we also encourage you to use the features of the Web to create a version that gives a high-tech impression of you and your information. Most word processors such as Word or WordPerfect give you the ability to easily create useful and interesting online résumés. As Figures 14.7 and 14.8 demonstrate, online résumés can come in all shapes, sizes, colors, and designs. One piece of advice, though—if you are applying for a job that requires the ability to code web pages, do not create your online résumé using a word processor, because it creates messy codes that most technology experts try to avoid. Consider these suggestions as you design an online version of your résumé:

- Include an e-mail link at the top of your résumé and perhaps at the bottom as well.
- Be sure to use keywords in your résumé; employers often search for specific terms and phrases when doing online and database searches for potential employees.
- Omit your mailing address and phone number; your online version allows readers to contact you more easily through the e-mail link.
- Omit page numbers and headers from your résumé if you are converting from print to electronic format.
- If your online résumé is long, consider using internal links (i.e., bookmarks) to allow readers to easily navigate from one part of the résumé to another.
- If you have a great deal of information to include on individual pages, create a front page with links to various sections of your résumé.
- If you have other online documents or websites that you have created, include hyperlinked URLs in the résumé so that readers can access them.
- If you think readers may want to print your résumé, include links to downloadable, print-based versions, such as a Word or ASCII version (scannable résumés are discussed next).
- Use a few simple and professional graphics or perhaps a photo of yourself, but avoid animated graphics and elaborate backgrounds; for employers who may wish to print your résumé directly from the Web, these extras don't translate well to print.
- Consider creating a border around your online résumé, to give it the effect of a printed page.
- Consider including a short quote (no more than one sentence) about your abilities or credentials from a reference with a prestigious or noteworthy title; list the person's name and title below the quote.

EXPLORE

Online résumés open up even more possibilities for the layout, design, and content of your information. Visit some professional websites, or do a web search, to view examples of online résumés. Analyze which features work well and which seem ineffective or unprofessional. Then create a list of the features you'd like to incorporate in your own online résumé.

Remember that your résumé should fit with the image of your profession and the company to which you are applying. Cutting-edge images and formatting might be persuasive to an advertising or graphic design firm but might be a turnoff to a conservative banking firm.

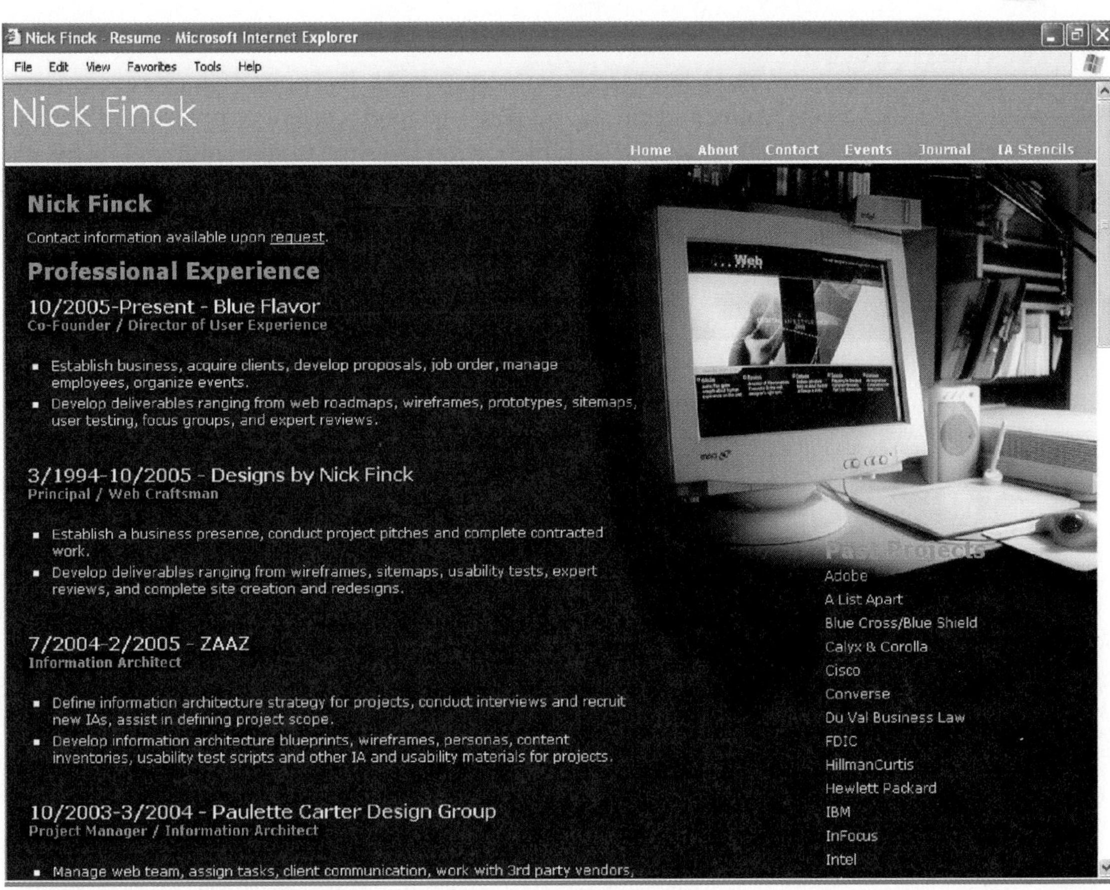

FIGURE 14.7 An online résumé

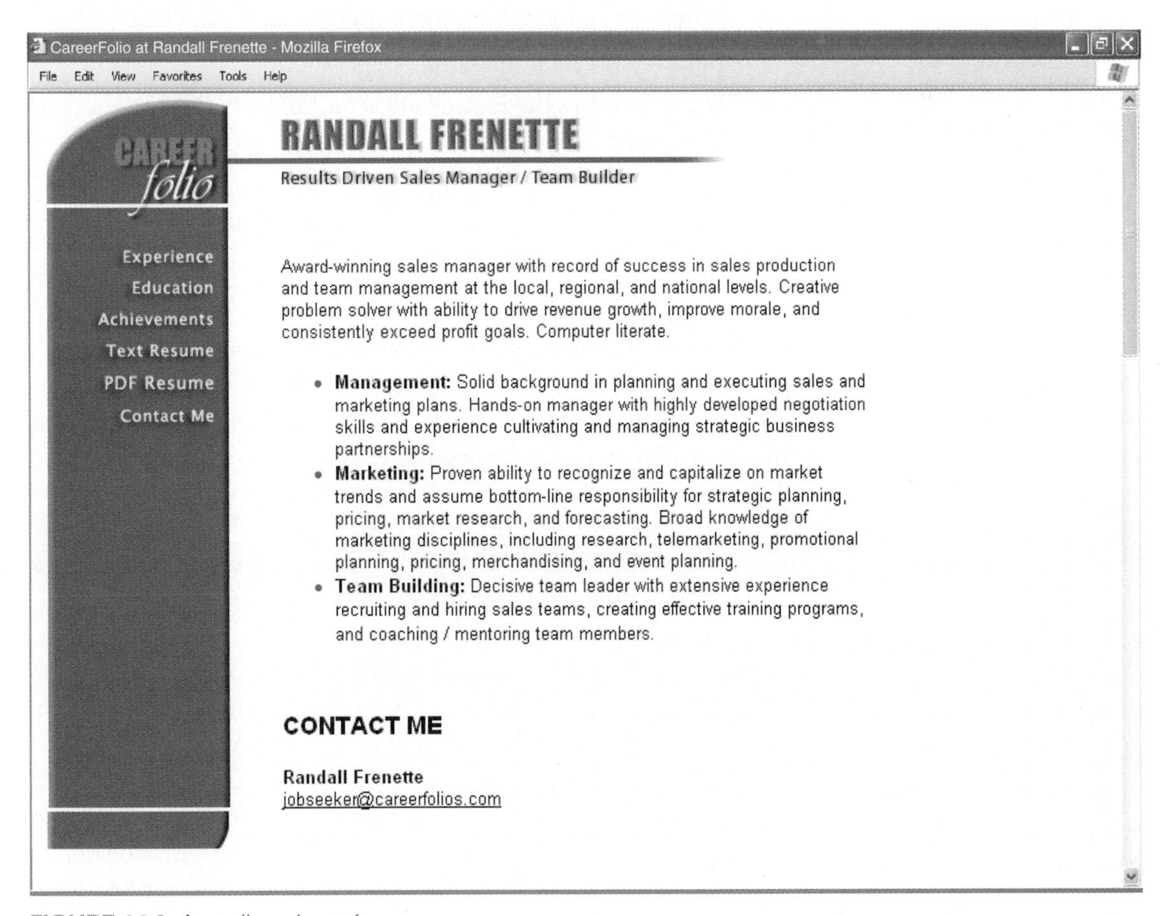

FIGURE 14.8 An online résumé

Scannable Résumés

Many large corporations automate the application process, entering résumés directly into a database of potential and current employees. Because corporations like Microsoft, General Electric, and Verizon Communications receive thousands of applications every day, they often scan résumés into computerized systems that categorize them according to specialization, geographic region, keywords, and more. In fact, the entire process is automated in some instances, with special data systems managing interview schedules, results of tests, and interview notes; sending out job-offer letters; and entering into payroll the file of an applicant who has been officially hired.

As software and hardware capabilities increase, more companies are using databases to manage job applications and résumés. Consequently, if you plan to upload a résumé to an electronic database, it is best to use a format and style conducive to the automated search and retrieval systems that these databases use. You do not need to completely rewrite your résumé, but you should create a scannable, plain-text version, following these guidelines:

- Use a standard typeface that will not jumble together when scanned (e.g., Times New Roman), and use 12- or 14-point font.
- Avoid underlining, italics, and boldface text, which scanners can misread; if you need to emphasize something, use all caps—sparingly.
- Use specific keywords and terms common in your profession; a large company may simply search its database for applicants whose résumés contain those terms.
- Do not indent text or use columns; most scanners cannot format them correctly.
- Use a ragged right margin, not full justification; scanners misinterpret the gaps in spacing that occur when a document is fully justified.
- Eliminate any graphics, including borders, lines, and bullets.
- Put each phone number on a separate line.
- When sending the résumé electronically, save it as plain text or ASCII to avoid formatting problems.
- If you are mailing a résumé that will be scanned, send a high-quality laser copy on white paper; do not fold or staple the paper.

Figure 14.9 shows a scannable résumé.

Curriculum Vitae

A curriculum vitae is a particular type of résumé that outlines your credentials for an academic position, graduate school application, fellowship, or grant; it contains much more detail than a résumé does about your academic and professional achievements. The term comes from the Latin *curriculum* ("course") and *vitae* ("of life"). Also known as a *vita* or *CV*, it contain lists of data such as educational institutions attended, degrees received, positions held, presentations delivered, publications authored, professional affiliations, and languages spoken.

Because of these details, CVs are usually two pages at least and can be many pages in length, some upwards of twenty pages for advanced academic positions. Each field has a different standard for how CVs are structured; you should consult with experts or professors in your particular field or contact professional associations for guidelines and examples. Be aware that the term *curriculum vitae* has a particular meaning for U.S. readers; in European Union countries the CV is a standard document in the *Europass*, a collection of documents designed to facilitate the migration of professionals between member countries. The Companion Website for this text provides an example of a curriculum vitae for a graduate school application.

Vince Metzger, MCSE
vmetz@mcse.com

MICROSOFT-CERTIFIED SYSTEM ENGINEER with broad knowledge of information systems and qualifications in programming, maintenance, systems analysis, and troubleshooting.

* Five years'experience in user support and training. Able to explain technical concepts and communicate with users and system administrators at all levels—from novices to MIS managers.

* Database specialist with three years'experience in applications development and support.

* Effective teamworker with ability to lead technical teams, manage changing priorities, and resolve problems in complex, multiplatform environments.

***Technical Skills

Operating Systems: DOS, Windows NT/XP/Vista, Mac OS X, Linux, Java OS
Internet Protocols: TCP/IP, FTP, HTTP
Programming Languages : Java, C, C++, PHP, JavaScript
Applications: Microsoft Office, Outlook, Adobe Creative Suite, Apple iWeb

***Experience

HALF-PLUS INTERNATIONAL, INC.–St.Louis, Missouri

Contract Computer Technician (2005–Present)

Design and set up Microsoft Access databases at job sites nationwide. Track provider information, copyright applications, and depreciation schedules for 3,000 health-care providers.

SYSTEMGUARD CORPORATION–Portland, Oregon

Database Manager (2002–2005)

Developed applications based on Microsoft Access. Prepared weekly reports detailing network and data access activity.Worked closely with MIS team to coordinate installation of workstations, printers, scanners, and servers. Researched new software, prepared feasibility reports, and coordinated installation projects companywide. Trained users and provided technical troubleshooting.

DAILY COMPUTER SOLUTIONS–Frampton, North Dakota

Computer Technician (2001–2002)

Used MS Access (2.0 and 7.0) to develop database applications for inventory management and auditing. Trained employees in software, Internet, and Windows NT use. Configured hardware for PCs and provided user support.

***Education

CHICAGO COMMUNITY COLLEGE–Chicago, Illinois

Microsoft-Certified System Engineer: MCSE, MCP+I, MCP (May 2003)

FIGURE 14.9 A scannable résumé

Transnational Résumés

For more on writing for transnational audiences, see chapter 5.

It is important to note that résumés may contain different components and may be designed differently if they are written for jobs outside the United States. Some employers outside the United States may expect much more detail in a résumé, including personal details such as your age and marital status. In countries like Germany and India, for example, it is common for résumés to contain a picture of the applicant at the top of the résumé. In addition, some credentials may not translate to audiences in other countries, which may have different educational systems and may not understand references to grade or university levels or types of degrees. Be sure to modify or explain any credentials your readers may not understand. Similarly, when referring to experience or expertise, be sure to use terminology your readers will understand; avoid using terminology that is specific to your organization or to U.S. audiences. If you are submitting your résumé in English, find out if the recipient uses British English or American English, and modify your résumé accordingly. Some transnational audiences also expect more details about the positions you've held, complete sentences in descriptions of those positions, and a greater emphasis on yourself and your personal goals.

As with CVs, you should consult with experts when creating a résumé for a transnational audience. Some websites offer examples of résumés written for particular audiences outside the United States. Figure 14.10 gives an example of a résumé written for a transnational audience.

Jim Nance
555 Minetta Lane
Duncan, New York, USA 11020
212-942-5221
jimnance@hotmail.com

Personal details →

Age: 25
Nationality: American
Objective: A marketing position with a fast growing consumer products company in Latin America

Complete sentences describing positions held →

WORK EXPERIENCE

Assistant Brand Manager, 2005–2007
Homegoods Incorporated, New York, New York, USA
During my two years working for the Healthcare Products Department (annual turnover $1.4 billion USD) I worked on various assignments of increasing responsibility, including an analysis of the Brazilian and Columbian markets for possible launch of new products. During this project I coordinated and worked closely with our Latin American offices. Other activities included the successful launch of a new line of organic shampoos and a specialty display initiative within client stores that resulted in a doubling of the number of displays over a 10-month period.

Marketing and Communications Intern, 2003–04
Homegoods Incorporated, New York, New York, USA
During the two summers that I interned at this company, I learned the basics of international marketing communication, including how to conduct focus groups in different countries and the basics of merchandizing in Latin America.

EDUCATION

St. John's University, Brooklyn, New York, USA

Bachelor of Arts Degree, May 2004

Explanation of terminology →

Major Fields of Study: Political Science, Spanish Language and Literature

GPA 3.55 on a 4.0 scale, Dean's List (top 20% of graduating class)

Senior Thesis: "The Role of the Rising Middle Class in the Democracy Movement in Brazil"

LANGUAGES

English: Native fluency
Spanish: Fluent
Portuguese: Proficient

INTERESTS

- Latin American literature
- Volleyball and baseball
- Jazz saxophone
- Hiking and backpacking

FIGURE 14.10 A résumé written for a transnational audience

Ethics and Résumés

As we mentioned earlier, your résumé probably affects your workplace writing life more than any other document because it is the document by which you gain employment. Without a strong résumé to earn you the employment you seek, you have less opportunity to write other professional and technical documents in the workplace. The résumé, then, must be a document that convinces readers without question; it must persuade readers to interview and hire you. But when you do not have the experience or education necessary to get a particular job, you must make some important ethical decisions.

How you represent yourself in your résumé is a matter not only of rhetorical choice, but of ethical choice as well. For instance, claiming to have supervised a staff of employees is much different from identifying that you reported to a supervisor. Exaggerating or fabricating résumé information is a serious ethical and professional breach. As a professional, you should opt to represent yourself, your experience, and your potential in the most realistic and honest manner. And if the basic notion of honesty does not convince you, consider these realities:

- Potential employers often confirm information on résumés by checking with previous or current employers.
- Switching jobs does not necessarily mean that you are switching communities, which may be more interconnected than you think. For example, civil engineers in one specialty area know other civil engineers in that same area; they attend the same conventions, workshops, and organizational meetings. Thus, the chances are good that if you fabricate your work experience, someone will know what your actual experience has been.
- Exaggerating your skills, language ability, software proficiency, or other abilities may come back to haunt you if you are called on to perform those skills.
- False information about your educational background can be easily spotted; many employers check educational credentials with the schools cited. And an employer may well know another alumnus from your school, who may share conflicting information.

Ultimately, it is in your own best interest to represent your experience and training accurately. Of course, all résumés do emphasize the strengths of the individuals that they represent, while also downplaying or omitting shortcomings or negative aspects. There's nothing wrong with presenting your best self in your résumé. However, there is a big difference between accentuating your best features and fabricating information.

Job Application Cover Letters

When sending a résumé to an employer, you should include a cover letter with it. The cover letter addresses the specific job for which you are applying and emphasizes the most important qualities that should make you attractive to the employer, highlighting qualifications and experiences that set you apart from other candidates. Cover letters are particularly important when an employer must decide between several equally qualified applicants. Consequently, you should use the cover letter to provide interesting, useful details about yourself. Employers often skim a cover letter just before an interview and may use it to guide the interview. You should anticipate questions about anything that is mentioned in either your letter or your résumé.

You should compose the version of the résumé you plan to send *before* writing the cover letter, even though the cover letter will appear first. After you complete your résumé, skim over it for details to emphasize; find the three or four most important points of the résumé, and then be sure to include them in your letter. However, even though the cover letter allows you to elaborate, avoid unnecessary or irrelevant information. Because readers often look at stacks of résumés and cover letters, the information you include in your cover letter should be easy to find, relevant, and specific.

In most cases cover letters should be one complete page, although they can occasionally extend to a second page if you have extensive experience or if longer cover letters are common in that field. As always, check the conventions of your field and the type of job for which you are applying in order to determine the standard. Be sure to use the same type of paper and font style that you used in your résumé, and use a laser printer to ensure a professional appearance. The cover letter follows the same basic format as other types of letters.

See chapter 13 for more on letter formats.

Introductory Paragraph

The introductory paragraph begins by identifying the specific job for which you are applying. If a job listing includes a reference number, include it here. You may also wish to refer to your source of information about the job, whether a newspaper or online job listing, a friend or coworker, or some other source. The introductory paragraph concludes with a brief summary of how you meet the primary requirements for the position. Keep the introductory paragraph short—three to five sentences should cover it.

Body Paragraphs

Body paragraphs provide further details about your qualifications for the position. Include specific facts, figures, and details about your qualifications here. Organize each body paragraph around a type of qualification; for example, the first body paragraph might describe your work experience, the second your education, and the third your community service and other activities. Use a bulleted or numbered list if it helps you emphasize details. Make sure no body paragraph, or any other paragraph in the letter, is longer than six lines because busy readers prefer short chunks of information.

Concluding Paragraph

The concluding paragraph refers to the résumé and any other documents included in the application. You might tie the letter and the résumé together with one main point of emphasis in the conclusion. If possible, the concluding paragraph should present the reader with a specific opportunity to meet with you, such as an upcoming visit to the reader's location or a conference or meeting that you both will attend. Try to avoid routine and presumptuous conclusions, such as "I look forward to your response" or "Contact me at your earliest convenience." Figure 14.11 shows a sample cover letter.

ANALYZE THIS

Successful applicants—like all good workplace communicators—are able to consider things from their readers' perspective, anticipating not only what those readers will want to read but how and where they will read it. Think about the mindset and characteristics of a typical employer in your intended profession. Then write a short description of how you envision that employer considering your application for an interview. Consider these questions about employers in your profession:

- What is the employer's initial impression of your application?
- What sorts of details will the employer be looking for in your letter and résumé?
- How much time will the employer devote to your application?
- Will the employer jump from section to section or read the documents from start to finish?
- What might impress, captivate, or interest the employer?
- What might separate your application from others?

After completing your description, compare it to those of other classmates, pointing out similarities and differences among them.

213 Cherry Street
Davidson, VT 02445
(732) 898-5059

November 23, 2008

Ms. Julia Bowers
Ephrata Middle School
Ephrata, VT 02480

Dear Ms. Bowers:

I am writing to apply for the position of Art Director at Ephrata Middle School (position #3345), as advertised in the November 20, 2008, issue of the *Ephrata Times*. I received my MA in Art History from the University of Vermont in 2000, and I have been an art teacher at Davidson High School for the past four years. I am confident that my education and experience make me a good candidate for the position at Ephrata Middle School.

During my graduate training at the University of Vermont, I took a number of courses that have prepared me to lead an art program like the one at Ephrata Middle School. I earned an A in my Art Administration course and an A- in Art Assessment, both of which provided useful instruction in how to develop and maintain art programs in K–12 education. In addition to my course work, I organized a graduate student conference in art program development in 1999, which was attended by more than fifty graduate students and faculty statewide.

Along with my teaching duties at Davidson High School, I also developed a student art auction, served on the Davidson High School Program Board, and managed and produced *Davidson Memories,* our school yearbook. These experiences could serve me well as the Art Director at Ephrata Middle School.

I have enclosed a copy of my résumé with this letter, and I have also completed the Ephrata Middle School Employment Form online. I will be in your area on Friday, December 14, and would appreciate an opportunity to discuss my qualifications in greater detail. I will try to contact you on Monday, December 10, to talk about a possible meeting time.

Sincerely,

Eric Rhodes

Eric Rhodes

Encl: Résumé

FIGURE 14.11 A job application cover letter

PREPARING FOR JOB INTERVIEWS

If you've written an effective résumé and cover letter and have submitted them to positions for which you're reasonably qualified, you will probably begin hearing from potential employers relatively quickly. They may contact you through e-mail but more likely will call you; be sure that your answering machine or voice mail message is brief and professional. When you speak with the employer directly, keep the conversation focused on when and where you will interview; save your questions for the interview itself. Accommodate the employer's schedule as much as possible, and express your eagerness to meet at the interview.

Once the interview is arranged, begin preparing for it. Good applicants spend lots of time prepping for each interview in which they participate, which gives them a definite edge over other applicants.

Gather Information

You should gather information about the company and the individual(s) with whom you will interview. Use the same resources described earlier to learn even more about the company, but take your search one step further, searching local newspapers, professional journals, and the Internet for the most current information available. If possible, find out about the interviewers as well, such as their official titles and the names of the departments they oversee.

Anticipate Questions

Anticipate questions the interviewer(s) will ask you, and write out specific answers for each. The questions will test your knowledge of the profession as well as your ability to fit into the organizational structure. Expect some inquiries about you and your specific qualifications; rehearse how you will answer questions about your education, previous employment, and anything else mentioned in your cover letter and résumé. The employer may also ask you to address a work-related problem or issue, and your ability to answer confidently and promptly will be as important as the answer you give. In addition, you may be asked about your desired salary, so research the current market value of individuals with your qualifications. Finally prepare several questions of your own to ask the interviewer(s); most interviews conclude with some time for your questions.

Conduct Mock Interviews

Practice with a mock interview in which you address the questions that may arise in the interview. These work best if the person conducting the mock interview knows something about the position for which you're applying; your university's career center can help you arrange a mock interview. However, even if you give the mock interviewer the questions you would like to rehearse, you should encourage him or her to ask some surprise or follow-up questions to keep you on your toes. Consider videotaping the mock interview to analyze your performance.

Prepare Physically

Prepare yourself physically during the time leading up to the interview. Make copies of documents that might be useful, including your cover letter and résumé, and place these in a professional-looking briefcase or folder to take with you. Also, take care of grooming and clothing needs the day before the interview to avoid feeling rushed. Think carefully about the image you want to convey when you walk into the interview, and choose your clothing according to the formality or informality of the organization. Then get a good night's sleep, and eat a small, healthy meal an hour or two before the interview. If you can, clear your schedule for a few hours before the interview to avoid feeling pressured or distracted. Finally, try to take a short walk immediately before the interview; it will dispel nervous energy and help you focus.

PARTICIPATING IN JOB INTERVIEWS

Although we focus here on face-to-face interviews, an increasing number of employers now conduct interviews by phone, through video, or even online, and many of the same guidelines apply to these other methods of interviewing. Regardless of the format, most employers interview more than one person for a position because top applicants often look equally qualified, based on their letters and résumés. The interview is the employer's opportunity to discover important differences, and it's your opportunity to prove yourself the best candidate for the position. Remember that you are being evaluated from the moment you walk into the interview until the moment you leave. Your goal should be to portray yourself as a confident, energetic, and intelligent person, and that image should be conveyed in everything you say and do.

In a face-to-face interview you should enter with a confident smile, making eye contact with everyone in the room; sometimes a committee will be conducting the interview. As the interviewers introduce themselves, repeat their names as you say hello, giving each a firm handshake. When they invite you to sit down, maintain good posture, leaning slightly forward. Avoid nervous gestures, such as crossing your arms or tapping your heel, which can suggest that you are resistant or uneasy. Even if only one member of the group asks the questions, be sure to make eye contact with everyone throughout the interview. Good interviews should seem more like a lively conversation than an inquiry, so try to interact with the interviewers rather than simply answering their questions.

Beginning of Interview

The beginning of the interview is usually designed to break the ice between you and the interviewers, who may tell you a bit about the company, as well as the surrounding area if you have traveled for the interview. The interviewers may also ask you some informal questions, such as where you are from and what your interests and hobbies are if you've mentioned them in your résumé. This part of the interview is usually short and may seem unimportant, but avoid being too flippant or overly casual. Interviewers draw conclusions quickly and may form a positive or negative impression of you from the very beginning of the interview.

Middle of Interview

The middle of the interview is much longer; it can last from ten minutes to an hour, depending on the position for which you've applied. Interviewers typically ask the questions during this part of the interview, but you can use the questions as an opportunity to highlight your strengths and achievements. For example, an interviewer might ask you, "How many employees did you supervise in your last job?" Rather than simply replying "Ten," you might describe what their various duties were, what you did to keep them motivated and focused, how you juggled different personalities and obstacles, and what you learned from the experience. As mentioned earlier, you should also feel free to ask questions of your own during the interview, but do so only after you have responded to the question you were asked. In the previous example you might have followed your response to the question with an inquiry about the individuals you would be supervising in this position.

Be specific with your responses; interviewers remember those who gave details about themselves more than those who spoke in vague generalities. If you are asked a question for which you have no answer, find a way to direct it to something positive about which you can talk. Speaking slightly off topic is better than not saying anything. And do remain calm and cool, even if asked questions that seem unfair, confusing, or unanswerable. Some employers like to test potential employees' ability to handle difficult situations.

EXPERTS SAY

"Whenever you are asked if you can do a job, tell 'em, 'Certainly I can!' Then get busy and find out how to do it."

THEODORE ROOSEVELT

Conclusion of Interview

The conclusion of the interview is usually short, but it is very important because it gives you a chance to lead the conversation. The interviewers may conclude by describing their hiring process and their time line for making a decision. Then they will probably ask you if you have any questions. Develop two or three questions that demonstrate your ability to do the job well and your genuine interest in working for the company. Avoid asking questions about the salary or benefits; these may make you look self-centered. Instead, focus on questions about the responsibilities you would have, the long-term goals of the company, and other subjects important to your success in working there. If you have any documents or materials you want those to see, distribute those at this time, followed by brief comments. Thank them for considering you for the position, and conclude with a positive, forward-looking statement about your potential role in their company.

EXPLORE

The Internet is filled with information and strategies for participating in job interviews. Investigate some of the sites devoted to job interviewing, and assemble a top ten list of strategies you think are essential to interviewing success. Compare your list with those of your classmates to determine which strategies you have in common and which are different.

Follow-up Letters

Within twenty-four hours of your interview, send a follow-up response to the individual who led the interview. Doing so reinforces positives from the interview, makes you appear highly interested in the job and motivated to get it, and encourages the interviewer to remember you when making the hiring decision. You should respond by postal mail or e-mail, depending on the company and the position. In general, if the company or your position focuses on technology in some way, you should reply by e-mail. If you think the interviewer does not read e-mail or values formality and tradition, send a letter.

The letter you write should follow the basic letter format discussed in chapter 13. The letter should be short—no more than one page in length—and structured as a thank-you letter:

- The introductory paragraph should remind the interviewer of the job for which you applied and the date you were interviewed and should thank him or her for meeting with you.
- One or two short body paragraphs should briefly highlight your qualifications for the job, emphasizing things to which the interviewer responded favorably during the interview. If possible, use company language or jargon that came up during the interview.
- A short concluding paragraph should restate your interest in the job and refer to the next step, whether you will be waiting to hear from the company or will call to learn about the status of your application.

Figure 14.12 shows a sample follow-up letter.

NEGOTIATING JOB OFFERS

Waiting to hear whether you have been offered a job can be excruciating—you may jump every time the phone rings. You can put that nervous energy to good use by planning your response and ways to negotiate for the best employment possible. Consider these strategies:

- *Focus on the positive* when you are offered a job. Express genuine appreciation for the offer, and emphasize the things you like about the company, its management, and its employees. Avoid saying anything negative during this initial conversation.

302 Maple Street
Atlanta, GA 34060

October 11, 2006

Ms. Hillary Pruett
Human Resources Manager
Highmark Fashion, Inc.
5353 Skyline Drive
Duluth, GA 32030

Dear Ms. Pruett:

Reference to the specific job →

I enjoyed interviewing with you during your recruiting visit to the University of Georgia on October 10. The management trainee program you outlined sounds both challenging and rewarding, and I look forward to your decision concerning an on-site visit.

Qualifications →

As mentioned during the interview, I will be graduating in December with a bachelor's degree in Fashion Merchandising. Through my education and experience I've gained many skills, as well as an understanding of retailing concepts and dealing with the general public. I have worked for seven years in the retail industry in various positions from sales associate to assistant department manager. I think my education and work experience would complement Highmark's management trainee program. I have enclosed a copy of my college transcript and a list of references that you requested.

Restatement of interest →

Thank you again for the opportunity to interview with Highmark Fashion. The interview reinforced my strong interest in becoming a part of your management team. I can be reached at (802) 458-1123 or by e-mail at escobar@georgia.edu should you need additional information.

Sincerely,

Ariana Escobar

Ariana Escobar

FIGURE 14.12 A follow-up letter

- *Ask for twenty-four hours to decide* whether to take the job. Most employers will grant you this time, especially if you are currently employed and need to resign from your present position. Some individuals ask for time to talk with family members or colleagues before making such a big decision.
- *Be patient* if negotiations take some time. Carefully considering what is being offered will keep you from making hasty emotional decisions you might later regret. If the employer pushes you to make a quick decision about anything, explain that you plan on staying in your next job for a long time and that careful negotiations will ensure a better long-term relationship between you and the company.

- *Be firm but flexible* when negotiating salary and benefits. If you are offered a salary that you believe to be low, you have several options: (a) ask how the salary figure was determined and whether it is negotiable; (b) ask for a higher salary after a trial period; (c) ask for other benefits that are important to you, such as a sales commission, computer hardware or software that would enable you to better perform your job, a bigger benefits package, or a larger office. Never demand to have your desired salary and benefits met, but do try to politely negotiate for a fair package.
- *Call other companies* with which you recently interviewed if you feel they may be considering you for employment. Do not try to pressure them into making you an offer, but do let them know that you have been offered another position. If they are interested in you, they will let you know. If you receive two or more offers at the same time, you can use them as negotiating points, but you must be careful not to push too hard and create bad feelings.
- *Look at the big picture.* If you think you would be genuinely unhappy working for a company that offered you a job, can you afford to reject the offer and keep looking? If you are offered two jobs, which would be the more satisfying in the long term? Avoid getting caught up in the heat of the moment and making a decision you might later regret.
- *Don't haggle over minor details* during or after the negotiation. Be willing to concede some small issues in the interest of starting the job on a positive note. And after you officially accept a job, do not attempt to immediately reopen negotiations for more benefits or a higher salary because you would appear indecisive or scheming.
- *Always send a letter of response* to any employer who offers you a job, regardless of whether you accept or reject the job offer.

ANALYZE THIS

After carefully considering the negotiation strategies described in this section, do some freewriting (see chapter 7) about the things you would say if you were offered a job in your profession. After you write your response and think about it, have a classmate or a friend conduct a mock phone call offering you the job and trying hard to persuade you to accept it immediately. You should diplomatically but firmly request some time to consider the position.

After the mock job offer, discuss it with your classmate or friend, soliciting input about specific aspects of the conversation. You might also ask other friends or classmates to observe the conversation and offer their suggestions as well.

Acceptance Letters

A job acceptance letter may be the easiest letter you'll ever write. The letter should begin by stating your acceptance of the job offer in the first sentence. The letter should also express your appreciation for and enthusiasm about the job. Because the acceptance letter sometimes serves as a contractual agreement between you and the employer, you may wish to briefly describe the terms of your employment if they were

1123 Cedar Avenue
Denver, CO 85202
(706) 452-6505
bchen@hotmail.com

January 8, 2008

Mr. Tom Morey, Director of Personnel
InfoSys Analysis
1002 Technology Park Drive
Boulder, CO 85304

Dear Mr. Morey:

Acceptance of the job offer →

Terms of employment →

As I mentioned in our telephone conversation on January 3, 2008, I am pleased to accept the position as Radar Systems Analyst working with Ms. Helen Hoover in the Navigations and Electronics Division of InfoSys Analysis. I am also pleased with the conditions of employment, including the yearly salary of $46,000 and the Employment Benefits Program specified in your letter of January 4, confirming my acceptance.

Thank you for offering to supply me with information about housing and for sending the detailed Moving and Living Guide. I'm sure it will answer any questions I might have concerning travel, moving, and reimbursement.

The enclosed questionnaire gives my reactions to the recruitment program of InfoSys Analysis, which played an important part in my acceptance of employment with you.

I will attend training and orientation on February 1 at your Personnel Office and will bring with me the required documents. I am looking forward to joining InfoSys Analysis in Boulder.

Sincerely,

Bill Chen

Bill Chen

Encl: Questionnaire

FIGURE 14.13 A job acceptance letter

negotiated by phone. Your acceptance letter should generally be only one page, unless you need to cover details about your employment that might be forgotten or misinterpreted. The employer usually sends an offer letter at the same time, sometimes with a contract. Figure 14.13 shows a sample acceptance letter.

Rejection Letters

A rejection letter should be brief and clear, declining the offer in the first paragraph. If it is appropriate, provide one or two sentences explaining why you are declining the offer. Avoid saying anything negative about the company, however, because you may want to work for them in the future. Conclude the letter by thanking the employer for the offer. Figure 14.14 shows a sample rejection letter.

120 Long Lane
Beston, IL 60943
(236) 752-8954

April 20, 2007

Representative Lisa Ranieri
Lincoln Office Complex
Washington, DC 20515

Dear Representative Ranieri:

Thank you for your employment offer for the position of Legal Researcher. I am unable to accept the offer because I have been accepted to the University of Chicago Law School and plan to begin school this August.

I want to express my sincerest appreciation for the opportunity to interview with you. Becoming involved in the political process in Washington, DC is very appealing, but I believe that my long-term goals are better served by continuing my education. I am therefore respectfully withdrawing myself from consideration.

Yours truly,

Karen Fisher

Karen Fisher

FIGURE 14.14 A job rejection letter

extrinsic motivators rewards that come from without, such as money, titles, and awards

intrinsic motivators rewards that come from within, such as contributing to a common good or producing something of high quality

IN YOUR **EXPERIENCE**

Before you begin negotiating the terms of your employment, you should consider what will motivate you and make you happy or unhappy in a particular job. Psychologists typically describe two types of motivators—*extrinsic* motivators, which are seen as rewards or incentives that are given by someone in power for a job well done, and *intrinsic* motivators, which are positive attributes of the activity itself. Even though you may be concerned with the tangible, extrinsic benefits of the employment—such as the salary, bonuses, the number of vacation days, and the size of your office—you should also consider the intrinsic benefits of the position, such as the amount of respect and responsibility you'd have, the degree to which you'd work with people you find interesting, and the contribution of the job to society. Even though a decent paycheck may be your primary concern right now, studies show that intrinsic motivators are more important in determining the long-term happiness of most employees.[5]

Think and then write about your own motivations in seeking employment. What things are most important to you? What things are not very important? Are you more motivated by the intrinsic or the extrinsic aspects of your career? After you've written down your thoughts, compare your motivations with those of your classmates.

[5]Kimberly Elsbach, "Rewards for Professionals: A Social Identity Perspective," *Innovator* (fall 1998), http://www.gsm.ucdavis.edu/innovator/fall1998/rewards.html.

SUMMARY

- You can begin preparing for your future career in these ways:
  - Gathering information through library, Internet, or placement center research; through interviews; and through professional organizations, meetings, e-mail lists, electronic bulletin boards, and online networking forums
  - Applying for work-study, internship, or volunteer programs
  - Beginning a working résumé or a personal data file
  - Requesting letters of recommendation and references
  - Beginning to compile a dossier or portfolio
  - Beginning a job search
- A job search can involve these components: employers, networks, Internet, newspaper listings, journals and catalogs, placement office, and professional placement service.
- Employment-related documents include recommendation request letters, letters of inquiry, résumés, job application cover letters, follow-up letters, and acceptance and rejection letters.
- You can prepare for a job interview by gathering information about the company, anticipating questions you might be asked, completing a mock interview, and preparing yourself physically.
- When negotiating a job offer, you should follow these suggestions: focus on the positive, ask for time to decide, be patient during negotiations, be firm but flexible, call other companies with which you've interviewed, look at the big picture, don't haggle over small details, and send an acceptance or rejection letter immediately after deciding.

CONCEPT REVIEW

1. How can you use a library to help you prepare for a job search?
2. How can you use the Internet to help you prepare for a job search?
3. How can you use your school's placement center to help you prepare for a job search?
4. What can an interview with an expert do to help you prepare for a job search?
5. What types of questions might you ask an expert during an interview?
6. What textual or electronic forums do professionals use to communicate and interact with each other?
7. How can work-study, internship, or volunteer programs help you prepare for a job search?
8. Why should you begin a résumé before you are ready for a job search?
9. What is the most appropriate method of requesting a letter of recommendation?
10. What should a dossier contain?
11. What options do you have for maintaining and sending your dossier?
12. What Internet job services might be useful in your job search?
13. How can the problem-solving approach help you create job application materials?
14. When should you write a recommendation request letter, and what should it include?
15. When should you write a letter of inquiry, and what should it include?
16. What is a chronological résumé?
17. What is a skills résumé?
18. What types of information should you include in a résumé?
19. Why are layout and design important in a résumé?
20. What are the benefits of creating an online résumé?
21. When might you need a scannable résumé?

22. What is a curriculum vitae?
23. How are transnational résumés different from those written for U.S. readers?
24. What ethical issues must you consider when creating a résumé?
25. What should you include in a job application cover letter?
26. How should you prepare for a job interview?
27. What should you do or not do during a job interview?
28. What is a follow-up letter, and when should you write one?
29. What negotiation strategies should you consider when offered a job?

CASE STUDY 1

Responding to a Salary Discrepancy

Trisha Ackerman recently earned her degree in chemistry from a large state university. She did an exhaustive search of employers and interviewed with a number of pharmaceutical development companies, including Ambichem, Inc. Trisha liked the level of responsibility she would have in the position with Ambichem and felt she could work well with Angie Winman, the supervisor for the position and the person with whom she had interviewed.

Consequently, Trisha was thrilled when Angie called to offer her the job. Although Angie was hesitant to increase the starting salary, Trisha successfully negotiated a 5 percent increase during their phone conversation. Trisha expected to begin working for Ambichem in approximately two weeks. However, as she began packing for her move across the state, Trisha received the official offer letter and contract from Ambichem. The letter specified the initial salary offer made by Angie, not the 5 percent higher salary they had negotiated.

In a small group, discuss the situation and craft a response from Trisha to Angie. Think carefully about the best way to handle the situation, and consider not only what Trisha should say, but what medium she should use to communicate. Should Trisha communicate through phone, e-mail, or a letter? What specific language should she use to rectify the situation?

CASE STUDY 2

Transnational Résumés in a Warehouse Construction Project

You work in the human resources department of Construx, Inc., a midsize engineering firm based in San Antonio, Texas. Your firm has recently merged with a smaller engineering firm in Chihauhua, Mexico, named Forma Perfecto. Because of the merger, your firm has decided to employ five engineers from the new Chihuahua branch to help develop a new warehouse facility near the San Antonio branch. It is anticipated that the work will take approximately one year to finish, and Construx is offering housing and a completion bonus to employees from the Chihuahua branch who are selected for the job.

Because Forma Perfecto specialized in warehouse construction and because of the completion bonus, twenty of the thirty engineers from the Chihuahua branch have sent résumés. You are responsible for sorting through their résumés and selecting the candidates who seem to have the best experience, training, and abilities to do the job successfully. However, after you begin sorting, you find that many of the résumés are structured differently than typical U.S résumés. You don't understand the educational details; the ways in which the candidates' qualifications are structured, organized, and discussed; and in some cases the language and word choices used in the résumés.

You discuss this situation with Vice President Cheryl Hudson of Construx, who suggests that you write a letter to the applicants from the Chihuahua branch, providing suggestions about how to write a résumé for U.S. audiences and inviting the applicants to resubmit their résumés. Write this letter to the applicants from the Chihuahua branch.

CASE STUDIES ON THE COMPANION WEBSITE

Tyrus Johnson: Preparing for the Job Search

This case asks you to put yourself in the position of a university student who is seeking a job in environmental engineering and who must contact two individuals for letters of recommendation. In particular, you must closely consider the contexts in which to make these particular requests and then produce the requests themselves.

Elizabeth Walker Searches for a Job

In this case you will analyze the particular issues and constraints facing a university student seeking a job as a radiographer. You will need to examine how to go about searching for specific kinds of positions as well as trying to find a position in a particular region of the country.

Shakira Solomon Creates a Job Application Packet

Shakira Solomon is a senior at Michigan State University who will soon be in the job market for a position in accounting. This case asks you to look closely at her academic, work-related, and extracurricular accomplishments during the past few years. In doing so, you will discuss her strengths and weaknesses as a job candidate as well as her future career field. In addition, you are asked to compose relevant documents related to her job search.

Writing a Position Description and Job Ad for a Web Page Designer

Bad Dog/Big Dog Clothing produces three lines of specialty shirts. As the hiring manager, you have been asked to hire a web page designer who can create and manage a promotional site for your company. Your company prefers to hire an experienced web designer who is both fluent with the software needed to run a commercial website and familiar with clothing sales.

VIDEO CASE STUDY

Cole Engineering–Water Quality Division

Finding Employment at Cole Engineering–Water Quality Division

Synopsis

Because of their increased success in a number of areas, Cole Engineering has recently begun hiring a number of new engineers to contribute to various divisions within the company. The Water Quality Division places a job ad and begins to interview possible additions to its team. This case shows the interview of Patrick Little with Asher Harris, Director of the Water Quality Division, and Cynthia Moore, Simulations Lab team leader. The interview is difficult, and the questions, at times, identify flaws in Patrick's application materials.

WRITING SCENARIOS

1. In a small group, create a top ten list of things all students should do before graduation to prepare for a future job search. Then as a class discuss the list each group formed. What similarities do the lists contain? What differences? Are most of the items on your group's list taken from this chapter? What other items did your group think should be included?
2. Conduct a mock job offer with one other person. The employer should write down some fictitious details about the job being offered. The potential employee should write down some possible responses or strategies for negotiation. After you've both prepared, conduct the mock job offer, imagining that you are speaking by phone, and negotiate any salary or benefits. Provide feedback on both roles, and then reverse the roles.
3. Locate and examine one or two Internet job services; then analyze them based on the quality and comprehensiveness of the information they contain. Give each a grade and justify that grade.
4. Interview a professional in your intended occupation, using the questions listed in this chapter and any other questions that come to mind. Then summarize your findings in a memo to your instructor.
5. Using one of the resources mentioned in this chapter, find a job advertisement that matches the qualifications you will have upon graduation. Write a résumé and cover letter that respond to the ad, and submit these to your instructor, along with a photocopy of the actual job listing. You may invent some of the qualifications you will have at graduation, but be realistic about what you will be able to include in your résumé at that time.
6. After you have written a résumé and a cover letter for the previous Writing Scenario, exchange these materials, including the job advertisement, with a classmate. Then write down ten or fifteen questions an employer might ask about your classmate's qualifications and ability to do the job. Conduct mock interviews for each other, and provide feedback on both the interviewing skills and the answers provided.
7. Like all other situations involving communication, the job search process requires ethical behavior. What ethical decisions might you encounter as you create your résumé and cover letter? What ethical decisions might arise as you interview for one or more positions? As a class or individually, write a short code of ethics for job seekers.
8. During a job interview, employers are not generally allowed by law to ask personal questions, such as health history, political or religious beliefs, sexual orientation, or marital status. However, such topics do come up in some interviews, and you should be prepared to address them. Imagine that a potential employer asks you several personal questions during a job interview. Spend some time freewriting responses to those personal questions. Think about the importance of getting the job offer as well as your own ethical feelings about discussing personal details. Which questions would you answer? Which questions would you refuse to answer, even if it meant sacrificing the job offer?
9. Employers have ethical responsibilities, too. Which aspects of a job should an employer disclose in a job listing or an interview? Does an employer have an ethical responsibility to describe any negative aspects of a job? Which negative aspects is an employer *not* ethically bound to reveal to applicants? Consider such things as workload, colleagues, job duration or schedule, workplace safety

or comfort. List the negative things you think an employer should reveal to a potential applicant. Even though you may not be in a position to hire anyone soon, this exercise can help you ask better questions as you interview

10. Consider the layout and design choices in Figures 14.3 (a chronological résumé) and Figure 14.4 (a skills résumé). Which design features do they have in common? Which are different? In a letter to your instructor, describe the specific similarities and differences between these two types of résumés, focusing on how information can shape design.
11. Although this chapter offers several design models for résumés, there is no single "right way" to design or construct one. Find several examples of print-based résumés—in books, in your placement office, or online—and compare them. In a letter to your instructor, describe the design elements you see in each one, emphasizing those you find most effective.
12. Find several examples of online résumés that contain visuals, and analyze those résumés, focusing specifically on their use of visuals. Which of them use visuals effectively? Which are ineffective, confusing, or unprofessional? Write a letter to your instructor describing the use of visuals in those résumés and addressing when and how visuals can be useful in online résumés.
13. Find examples of job listings and/or résumés in your intended profession, and analyze which technological skills, certifications, or abilities are necessary or helpful in that profession. Then do some freewriting on your own abilities. Are you technologically prepared to enter your profession? What skills or abilities do members of your profession have that you do not? How can you go about getting the training or experience you will need?
14. If you've already created a résumé, convert it to either an online or a scannable version. Consider the ways in which technologies (specifically, the Internet and the fax machine) influence your readers' experiences and your own content and design choices.
15. In a memo to your instructor, outline the major advantages and disadvantages of conducting a job search through the Internet. What do you gain by using the Internet for your job search? What is difficult or problematic about conducting a job search through the Internet? What sources might you use? Could you conduct a thorough job search in your profession exclusively through the Internet?

PROBLEM SOLVING IN YOUR WRITING

Select one of the documents you wrote and designed for this chapter, such as a résumé, a cover letter, or some other job-related document. Then consider how you employed the problem-solving approach in creating that document. Write a letter to your instructor identifying how you planned your response, conducted research to gather information, drafted the document, reviewed it, and finally distributed (or plan to distribute) the job-related document.

15

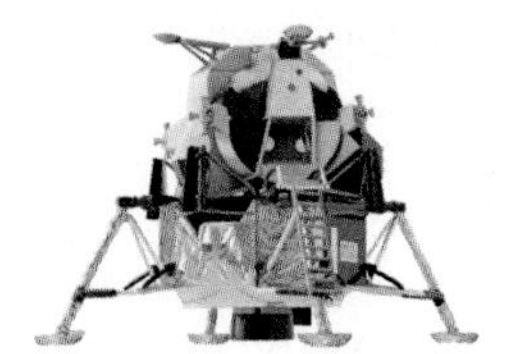

Technical Definitions

CHAPTER LEARNING OUTCOMES

After completing this chapter, you will be able to do the following:

- Understand that writing definitions is a **common activity** in workplaces
- Recognize that **workplace definitions** usually differ from **dictionary definitions**
- Understand how definitions can **solve problems** for audiences
- Understand the many **different kinds** of definitions and their purposes
- Compose definitions that **describe**
- Compose definitions that **compare and contrast**
- Compose definitions that **classify**
- Compose definitions that **provide examples**
- Compose definitions that **illustrate with visuals**
- Compose definitions that **combine types** of definitions
- Understand how definitions are woven into various workplace writing **genres**
- Recognize the most appropriate ways to **place definitions** in documents
- Consider the **ethical dilemmas** inherent in writing workplace definitions
- Know the **common mistakes** writers make when composing definitions

DIGITAL RESOURCES

On the Companion Website www.prenhall.com/dobrin:

- Case 1: American Dream Mortgage Consulting and the Confusing 80-20 Loan
- Case 2: Argon Gas: Effective Insulator or Insidious Carcinogen?
- Case 3: Introducing Portland Public Schools to New Computer Equipment
- Case 4: Clarifying Accounting Terms for a New Client
- Video Case: Information for New Investors: Defining the Terms
- PowerPoint Chapter Review, Test-Prep Quiz, Exercises and Activities

REAL PEOPLE, REAL WRITING

MEET ORLANDO LAMAS • Project Architect with Telesco Associates

What types of writing do you do in your workplace?

As an architect, I spend time in my office, clients' offices, and construction sites. My most important writing projects are proposals, which involve getting interviewed by clients, taking notes, identifying clients' needs, writing bid documents, and putting everything together in a formal proposal. These proposals showcase our company, our work, and how we'll complete a job. They often determine whether we land the job.

As an architect, do you need to define technical terms for your different audiences?

This is a big, complicated part of my job. On a daily basis I communicate with four or five different groups: perhaps clients, engineers, construction workers and managers, contractors, suppliers, or owners of buildings and properties. The difficulty is that these different groups don't always share the same knowledge and terms; they may have difficulty communicating with each other. I have to define technical terms differently for several different audiences.

Is there a particular way you write these technical definitions?

Usually it involves research into a certain discipline or area. I have to understand how engineers or contractors think—for instance, how they define things and understand their roles in construction—in order to communicate with them. To define a technical term for these different audiences, I have to look for some common ground among these groups and use that as a way to define a term for all of them. And sometimes this might even involve using a simile or metaphor to get my point across.

What are the most important considerations for you as you write technical definitions?

Precision and clarity. The more legal weight a document carries, the more technical it is going to be, so it's crucial that the definitions in documents are precise and clear. If they aren't and a mistake happens, we're the ones liable—legally and financially.

What advice about workplace writing do you have for readers of this textbook?

People think that careers in architecture involve only drawings and visual design. But there's a strong link between visual design and writing. Some people understand things better with visuals, some with writing, but there's always a connection between the two. And in both cases non-clarity hurts design.

INTRODUCTION

In September 2006 scientists officially introduced a new planet named Eris as part of our solar system. Eris had been first detected in 2003 and was initally called 2003 UB313, but its status as a planet was uncertain for three years. The reason for this uncertainty: scientists had to rethink the definition of *planet* altogether before deciding how to classify this celestial body. Figure 15.1 shows a recent diagram of the solar system that includes 2003 UB313 (Eris).

The debate about the definition of the term *planet* had actually begun in 1999 when a number of planetlike objects were discovered at the farthest reaches of the solar system—beyond Pluto. These objects were round and orbited the sun like other planets, but they were much smaller than the other planets, and they had wild and elongated orbital patterns that stretched well above and below the orbital planes of the other planets in our solar system. In short, because these objects shared some but not all characteristics with the planets Mercury, Venus, Earth, Mars, Jupiter, Saturn, Uranus, and Neptune, there was uncertainty among scientists about whether they should be called planets. Perhaps surprisingly, astronomers had not previously developed a definition of planets that accounted for these new celestial objects, nor had they developed a different term for them. Consequently, when Eris was discovered, scientists engaged in vigorous debates about its planetary status.

Amid the controvery, the International Astronomical Union (IAU) proposed in August 2006 a way to define planets and other bodies in the solar system:

> (1) A planet is a celestial body that (a) is in orbit around the Sun, (b) has sufficient mass for its self-gravity to overcome rigid body forces so that it assumes a hydrostatic equilibrium (nearly round) shape, and (c) has cleared the neighbourhood around its orbit.
>
> (2) A dwarf planet is a celestial body that (a) is in orbit around the Sun, (b) has sufficient mass for its self-gravity to overcome rigid body forces so that it assumes a hydrostatic equilibrium (nearly round) shape, (c) has not cleared the neighbourhood around its orbit, and (d) is not a satellite.
>
> (3) All other objects orbiting the Sun shall be referred to collectively as "Small Solar System Bodies."

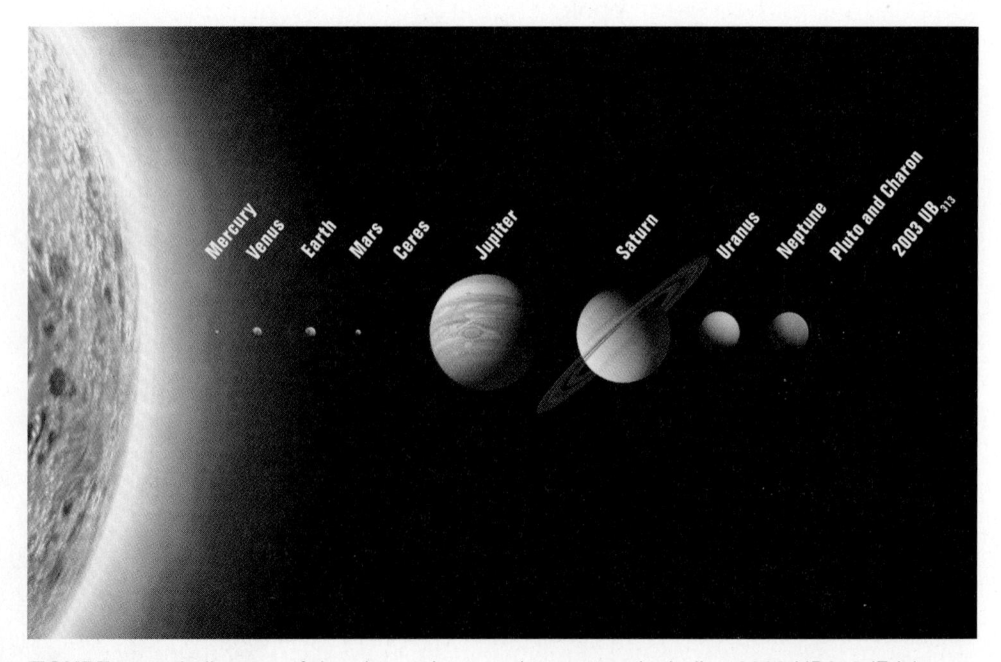

FIGURE 15.1 A diagram of the planets in our solar system, including 2003 UB313 (Eris)

The IAU also created a resolution that both Pluto and Eris would be considered dwarf planets.

By 2008, however, many scientists again began arguing that these definitions were not adequate, and a Great Planet Debate was set to commence at Johns Hopkins University in August 2008 to further discuss, among other things, how to define the term *planet* and other celestial bodies in the solar system. Such issues of definition are likely still taking place as you read this chapter.

Most people understand clearly that science is a process of discovery; scientific knowledge is constantly changing because scientists are always learning new things about the worlds we live in. However, since most of us first learned about the planets in third or fourth grade, we assume that the definition of a planet is such a basic idea that it must have been settled long ago. Obviously it has not. And although this may seem like a relatively small issue in the larger scope of scientific inquiry, it is relevant nonetheless. How scientists define things like planets does carry consequences: on one level, by changing the definition of *planet*, our understanding of our solar system and the universe must change, if only slightly. On a different level, think of all the books and textbooks that must be revised if the IAU no longer considers Pluto a planet, or if Eris must be added as a kind of planet. How many millions of dollars have been spent on science textbooks that now include outdated information about the planets?

This example of redefining *planet* also ties into problem solving. In recent years scientists discovered a variety of new bodies in the solar system that did not fit well into previously established definitions, resulting in a good deal of confusion about these celestial objects. In an attempt to remedy this problem, the IAU decided to reconsider the definitions they had previously relied on; even scientific fields such as astronomy must pay attention to how they use language and discourse to describe and define their objects of study. What they write and how they write may be as important as what they see and what they study; and writing and revising technical definitions may often be a crucial part of their writing.

Those in other professional fields are also regularly involved in arguments and debates over ways to define terms important to the operation of their organization, institution, or company. However, not all debates and discussions of definitions revolve around previously defined terms like *planet*. Some workplace writers must be involved in creating definitions of new things that did not previously exist. These are called **neologisms**, which are, essentially, new words or new usages of words.

neologism a new word, expression, or usage of a word

Because we live in a world with ever-increasing technologies, for example, there is often a need to introduce new terms and definitions to various audiences. The word *Internet*, for instance, probably crosses our path in one form or another every day; however, this was not always the case. If someone had spoken the word *Internet* thirty or forty years ago, no one would have understood its meaning. At some point that term was brand new to most audiences and had to be introduced to them in clear and effective ways to make people comfortable with the concept as we know it today. Many workplace writers do not simply develop new products and services but also develop the language used to describe and define these products and services for their intended audiences.

EXPLORE

Locate two or three definitions of *planet*. How do these definitions differ from the IAU's definition of *planet* (see page 442)? Where did you find the definitions, who are their audiences, and what are their purposes? Do you find them easier or harder to understand than the IAU's? Explain.

Dictionary Definitions

There is a good possibility that when you think about definitions, you associate them with dictionaries. This chapter, however, teaches you to expand your understanding of definitions beyond dictionaries, to look at how various kinds of technical and professional documents use definitions in more complex ways. Look, for example, at an entry for the word *phenylephrine*, found in the *American Heritage High School Dictionary*:

> **Phenylephrine** *N.* An adrenergic drug, $C_9H_{13}NO_2$, that is a powerful vasoconstrictor and is used to relieve nasal congestion, dilate the pupils, and maintain blood pressure during anesthesia.

Perhaps you already knew the meaning of the word *phenylephrine*, or perhaps you had some general idea of what it means but did not necessarily know the details. Or maybe, like many of us, you had no idea what the word means, and even *after* reading this definition, you are still confused about what it is and what it is used for.

For the medical expert this definition may be highly suitable. For the nonexpert, however, no matter how thorough or impressive this definition sounds, it has shortcomings. For example, the definition uses other jargon words like *adrenergic* and *vasoconstrictor* and throws a chemical compound at us. If readers don't know what these other terms and symbols mean, they will likely not understand the definition itself and will need to look up other words to make sense of phenylephrine. In addition, the definition provides little contextual information. We are not told when the anesthesia is used or for what purpose, nor do we know whether phenylephrine is always used when anesthesia is administered or only sometimes. Furthermore, we might wonder whether phenylephrine is used only on humans.

The limitations of this definition might be due in part to the fact that it is written in a high school dictionary. Maybe the writer of the definition assumed that high school students would need only very basic information about phenyl-ephrine. In addition, high school dictionaries are quite short in comparison to collegiate and other professional dictionaries; maybe the definition is severely constrained by a need to keep it short. Look at another definition of *phenylephrine*, found in a medical dictionary (see Figure 15.2). It is clearly not written for a high school audience.

Definitions in most dictionaries provide explanatory information but lack contextual information, in part because different dictionaries cater to different audiences. This limitation doesn't mean that dictionary definitions are irrelevant and useless—most writers and professionals keep a trusty dictionary by their sides. However, dictionary definitions are not the only kinds of definitions found in professional and technical documents.

In many cases your writing will avoid the typical dictionary or textbook definition; instead, workplace writing often includes definitions that exist in the context of a larger document and are geared to specific audience members. This chapter discusses different kinds of definitions, the ways they fit into technical and professional documents, and the ways they cater to particular readers inside and outside the workplace.

Definitions are a necessary part of professional and technical communication because they do the following:

- Describe and clarify information for readers
- Allow experts and specialists to communicate with nonexperts and nonspecialists
- Work to help solve external and rhetorical problems

Phenylephrine Hydrochloride

Benzenemethanol, 3-hydroxy-α-[(methylamino)methyl]-, hydrochloride; (*Various Mfrs*)

OH

C—CH_2NHCH_3 • HCl

H

OH

(−)-*m*-Hydroxy-α-[(methylamino)methyl]benzyl alcohol hydrochloride [61-76-7] $C_9H_{13}NO_2$.HCl (203.67).

Preparation—*m*-Hydroxyphenacyl bromide is condensed with methylamine and the carbonyl group then is reduced to carbinol via catalytic hydrogenation. The phenylephrine so formed is dissolved in a suitable solvent and neutralized with HCl.

Description—White or nearly white crystals; odorless; bitter taste; melts between 140° and 145°.

Solubility—Freely soluble in water or alcohol.

Uses—A *direct-acting sympathomimetic* with strong α-agonist and negligible β-agonist and CNS activity. See the general statement for the actions, uses, adverse effects, precautions and drug interactions of direct α-agonists. It is used in the treatment of *paroxysmal supraventricular tachycardia* and to support *blood pressure*. It can be used in the presence of heart-sensitizing anesthetics because of its lack of significant beta-adrenergic cardiac stimulant actions. It also is used as a nasal, scleroconjunctival and uveal *decongestant*, as a *mydriatic* and to decrease aqueous humor formation in *open-angle glaucoma*. It is included in some local anesthetic preparations to prolong local and regional anesthesia and in combination with inhaled bronchodilators.

It is absorbed orally and, since it is not attacked by MAO, it is effective by mouth for orthostatic hypotension. By the intravenous route the duration of action is about 15 to 20 min and by the intramuscular route 30 to 120 min.

Dose—*Intramuscular or subcutaneous, adults*, for *mild to moderate hypotension*, *initially* **2** to **5 mg**, repeated at intervals no less than 10 to 15 min, *or*, for *prevention of hypotension during spinal anesthesia*, **2** to **3 mg** 4 min before anesthetic; *children, to treat hypotension*, **100 μg/kg** (or **3 mg/m²** of body surface), repeated in 1 to 2 hr, as needed, *or, prior to spinal anesthesia*, **44** to **88 μg/kg**. *Intravenous injection, adults*, for *mild to moderate hypotension*, **500 μg** (*slowly*!), repeated at intervals no less than 10 to 15 min *or*, for *severe hypotension* during spinal anesthesia, *initially* **200 μg**, with subsequent increments no greater than 200 μg, up to a total of 500 μg/dose, **or** for *paroxysmal supraventricular tachycardia, initially* **up to 500/μg** given over 30 sec, with subsequent increments of 100 to 200 μg, up to a total of 1 mg/dose. *Continuous intravenous infusion, adults, initially* **100** to **180 μg/min**, usually reduced to **40** to **60 μg/min** for *maintenance*. To *prolong spinal* and *local anesthesia*, **2** to **5 mg** added to a *spinal unesthetic* solution *or* **1 mg** (as **0.005%** solution) to 20 mL of *regional anesthetic* solution. *Topical, intranasal, adults*, **2 or 3 drops** or **1 or 2 sprays** of **0.2** to **1.0**% solution, or a **small amount** of **0.5** % jelly into each nostril every 3 to 4 hr; *children, all age groups*, **2** or **3 drops** of solutions as follows: *up to 2 yr*, **0.125**% every 2 to 4 hr; *2 to 6 yr*, **0.125**% every 2 to 4 hr or **0.16**% every 4 hr; *6 to 12 yr*, **0.25**% solution every 3 to 4 hr. *Topical*, to the *eye*, **1 drop** of **2.5** to **10**% solution into the conjunctival sac, repeated once in 5 min, if ncessary, for *opthalmoscopy*, or, 2 to 3 times a day for *chronic mydriasis*, or, **1 drop** of **0.12**% solution 2 or 3 times a day for *scleroconjunctival decongestion* and *blepharospasm*, or, **1 drop** of a **2.5%** solution, repeated at 15-min intervals for **4** doses for *preoperative mydriasis*, or, **1 drop** of **10**% solution once a day for 3 days, to be alternated with miotics, for *postiridectomy mydriasis*, or, **1 drop** of **10**% solution to the upper surface of the cornea for *glaucoma*; *children*, as adults, except 10% solution is not to be used.

Dosage Forms—Injection: 10 mg/1 mL; Ophthalmic Solution: 0.120, 1, 2.5 and 10%; Nasal Jelly, 0.5%; Nasal Solution: 0.125, 0.160, 0.2, 0.25, 0.5 and 1%.

FIGURE 15.2 Definition of *phenylephrine hydrochloride* in a pharmaceutical guidebook

Knowing how to write definitions is an important part of workplace writing. And because dictionary definitions don't always offer a comprehensive description of terms, you should be prepared to develop your own definitions in the workplace. Your audience, the particular problem, and the rhetorical situation at hand will all influence how you define a word or concept, where the definition fits into your writing, and what kind of definition you use.

ANALYZE THIS

Consider the two different definitions of *phenylephrine* discussed in this section (the first from the *American Heritage High School Dictionary* and the second from the medical dictionary). What specifically are the differences between these two definitions? What are the differences in the information presented and the way it is presented? What does the medical definition tell you about the audience? What kinds of problems do you envision each of these definitions helping to solve?

SUBJECTIVITY OF DEFINITIONS

This chapter makes the point that definitions can help solve problems for audiences; however, definitions themselves can sometimes *cause* problems and difficulties because definitions are often written subjectively rather than objectively, thus causing ambiguity and debate. The Introduction of this chapter provided an example of how the IAU's definition of *planet* included some celestial bodies but not others, therein causing confusion among astronomers about how to categorize 2003 UB313 when it was first discovered. That confusion still exists today. It is not always easy to alter people's understanding of definitions once they have been set.

Nonetheless, in the IAU case, readers at least know that a definition is being proposed. In some cases, definitions are implied in documents and are not pointed out to readers directly. For example, the National Trust for Historic Preservation is an organization that works to save buildings and landmarks that represent important aspects of American heritage. In one of the National Trust's projects, called Dozen Distinctive Destinations, the organization recommends to audiences "distinctive" places to visit. The National Trust for Historic Preservation describes the project as shown in Figure 15.3.

In this document the National Trust discusses its project to recommend twelve distinctive destinations, but it does not explicitly state for audiences how it is defining *distinctive*; that is, there is no obvious definition of *distinctive*. Instead, audiences must infer throughout the passage that distinctive destinations are those that offer visitors an "authentic" experience of "dynamic downtowns, cultural diversity, attractive architecture, [and] cultural landscapes." In addition, distinctive destinations "represent the richness and diversity of America's cultural heritage," and they serve as models to other towns and cities by "preserving their historic fabric and spirit of place." Thus, the National Trust asks readers to put together a definition of *distinctive* based on a variety of other key terms, which are not themselves necessarily clear-cut. What does a dynamic downtown look like? What characterizes cultural diversity? Who is to say what makes for attractive architecture? How do we recognize a cultural landscape?

We do not mean to suggest that the National Trust should have presented an obvious and explicit definition of the term *distinctive* or any other term on their website. However, it is important to recognize that definitions often run through documents without any direct reference to themselves *as* definitions. This situation can cause difficulties for readers, especially when important terms are implicitly defined with other words and phrases that can only be subjectively understood. In our example, by not defining its terms, the National Trust leaves its selections for the Dozen Distinctive Destinations open to debate and contention among those who may define these terms differently. In some instances, leaving key terms undefined, or loosely defined, is acceptable; writers must understand audience needs and know what is acceptable.

KINDS OF DEFINITIONS

There are a number of ways to describe the different kinds of definitions, but this section discusses the different kinds in terms of what they do.

Each year since 2000, the National Trust for Historic Preservation has selected 12 vacation destinations across the United States that offer an authentic visitor experience by combining dynamic downtowns, cultural diversity, attractive architecture, cultural landscapes and a strong commitment to historic preservation and revitalization.

The destinations selected in 2008 range from a French colonial village along the banks of the Mississippi River that captures the pioneer spirit of the early settlers, to a small Texas town that serves as a gateway to the unspoiled terrain of the 19th century western frontier, to a gorgeous Southern city with roots three centuries deep, and a coastal town renowned for its seafood and historic buildings.

"These twelve communities represent the richness and diversity of America's cultural heritage," says Richard Moe, president of the National Trust for Historic Preservation, "and in preserving their historic fabric and spirit of place are models for other towns and cities."

This is the ninth time the National Trust for Historic Preservation has announced a list of *Dozen Distinctive Destinations*. To date, there are 108 *Distinctive Destinations* located in 42 states throughout the country. In each community, residents have taken forceful action to protect their town's character and sense of place. Whether by enacting a local preservation law to protect historic buildings against demolition, rewriting zoning codes to prevent commercial sprawl, removing regulatory barriers to downtown housing, making downtown areas more walkable, enacting design standards, or taking some other major step that demonstrates a strong commitment to their town, residents have worked hard to preserve the historic and scenic assets of their communities, with rewards that transcend town limits.

FIGURE 15.3 Dozen Distinctive Destinations of the National Trust for Historic Preservation

ANALYZE THIS

Why do you think the National Trust for Historic Preservation leaves so many important terms undefined, given the fact that these terms are subjective and open to debate? In light of the mission of the National Trust and its audiences, does it matter? Do you think the organization should compose the document in Figure 15.3 any differently? Explain your answers in writing or in class discussion.

Definitions That Describe

In one way or another, all definitions attempt to describe the thing or concept being defined, but here we mean more specifically definitions that describe what something looks like, feels like, sounds like, smells like, or tastes like. For instance, a writer putting together a training manual for new pilots might begin to define *aileron* through description:

> **Ailerons** are small, slender, hinged sections on the outer portion of a wing.

Or a writer composing a manual for inexperienced PC users might suggest this:

> A **window** is a large rectangular portion of your computer screen.

In each of these definitions that describe, the writers have tried to appeal to readers' visual senses by describing the shapes and sizes of ailerons and windows.

Because purely physical descriptions may be inadequate to enable inexperienced, unfamiliar readers to understand the given terms, definitions that describe also often include what the thing does, how it functions, and for what purpose. Each of the above definitions would likely be expanded for clarification:

> **Aileron** A small, slender, hinged section on the outer portion of a wing, used to control an aircraft's rolling and banking movements. Ailerons usually work in opposition: as the left aileron is deflected upward, the right is deflected downward, and vice versa.
>
> **Window** A large rectangular portion of a computer screen, which displays an open program or the contents of a file independently of the other areas of the screen.

Certainly these two definitions leave much to be desired for someone completely unfamiliar with the terms, but they do provide a basis for understanding by describing both *appearance* and *use*. The amount of description that a definition provides depends on the audience and the needs of the audience.

Many important terms do not have physical or tangible qualities that writers can easily describe. In such cases a definition might focus on what the thing does, how it works, or where it comes from. Consider a term like *computer virus*. Such a thing seems quite tangible in its ability to wreak havoc on our computers and operating systems, but the computer virus itself cannot be described with sensory details like sight, touch, taste, sound, or smell. Instead, a computer virus would need to be described on the basis of what it does, how it works, and/or where it comes from:

OTHER PROFESSIONAL DEFINITIONS IN THE WORKPLACE

In the following definition, notice how fuel cells are defined according to descriptions of what they do and how they are used:

Dow and General Motors (GM) Corporation are inaugurating the world's largest fuel cell project to-date for power generation. The project is based at Dow's Freeport, Texas, site, where hydrogen is created as a coproduct. GM's fuel cells will convert hydrogen to electricity. When complete, the fuel cells will provide a portion of the power used at the plant.

Hydrogen fuel cells offer a cleaner, more efficient option for power generation than coal, natural gas, or other fossil fuels. Both greenhouse gas and air pollution are reduced. Fuel cells could also reduce need for foreign oil and provide an option to costly natural gas.

If the project proceeds as planned, up to 400 fuel cell units could be producing up to 35 megawatts of power at Freeport. That is enough energy to power 25,000 homes for a year. And, it is more than 15 times greater in scale than any other fuel cell system to date. Although the fuel cells will provide only two percent of the total power the Texas plant uses, the project is a first step in developing fuel cells into a major power source. Dow and GM are discussing future use of fuel cells at other Dow sites in both the U.S. and Europe.*

*"The Dow Global Public Report 2003: Case Studies," http://www.dow.com/publicreport/2003/studies/fuelcell.htm. Courtesy of The Dow Chemical Company.

> A kind of computer program that hides itself within another program in order to produce copies of itself and insert them into other programs or files, usually to perform some kind of malicious action, such as destroying data or rendering a program inoperable.

A person might have to actually experience the effects of a virus to truly understand what it can do and how much frustration and mayhem it can cause, but this definition provides a basic explanation of how viruses work and what effects they create.

Definitions That Compare and Contrast

Professional and technical definitions are sometimes effective when they compare or contrast the term at hand with something else that the audience is already familiar with. If, for instance, you returned to the United States from a visit to England and wished to explain the British game of cricket to a friend, it is likely that you'd begin this definition by saying that cricket is similar to baseball. Although you would need to provide more details because there are a number of differences between the two, this initial comparison at least gives a frame of reference for your friend.

Or let's say you needed to write a computer user's manual for an audience of non-experts and one section of the manual needed to explain flash drives; it would probably be beneficial to compare a flash drive to the older 3.5-inch floppy disk. Doing so would allow your audience to immediately recognize it as a kind of storage device for programs and files. More explanation would follow, but this comparison would allow the audience to begin to understand its general function:

> A small, lightweight, portable Universal Serial Bus (USB) device that allows users to store, carry, and transfer files, much like the older 3.5-inch floppy disk.

Such a definition gets at the heart of the flash drive, but to provide a thorough definition, you would need to contrast a flash drive and the older 3.5-inch floppies, perhaps detailing its smaller size and lack of moving parts:

> A small, lightweight, portable Universal Serial Bus (USB) device that allows users to store, carry, and transfer programs and files, much like the older 3.5-inch floppy disk. The flash drive, however, differs from the older floppy disks; it is much smaller (only 70 mm long, 30 mm wide, and about 11 mm in height) and more durable and contains no internal moving parts. In addition, flash drives hold a great deal more data than floppy disks. A typical 128-megabyte flash drive holds as much storage space as eighty-eight 3.5-inch floppy disks.

OTHER PROFESSIONAL DEFINITIONS IN THE WORKPLACE

In the following definition, notice how protected membrane roofs (PMRs) are defined in part by being compared and contrasted to traditional roofs:

A Protected Membrane Roof (PMR) system using STYROFOAM insulation shields and protects a roof's waterproof membrane unlike a traditional roof system that does not cover the membrane. On a PMR roof, the membrane is placed beneath STYROFOAM insulation. The PMR roof protects the membrane against the most common causes of failure—sunlight, extreme heat, extreme cold, weather, and foot traffic. Traditional roofs are typically replaced every 7 to 10 years. PMR roofs have lasted over 30 years. The cost savings to building owners are obvious. However, the benefits do not end there.

Currently, three to four percent of all waste in U.S. landfills comes from old roofs. By more than tripling the time a roof lasts, landfill space is conserved. In addition, STYROFOAM insulation resists water so well it can be reused if the roof is ever replaced, which saves even more landfill space.

PMR roofs help the environment in another way. Ever notice how large cities tend to be hotter than surrounding areas? This is because the many buildings and paved surfaces create an effect called a "heat island." PMR roofs stay cooler than black membrane surfaced roofs, helping to reduce this heat island effect.*

* "The Dow Global Public Report 2003: Case Studies," http://www.dow.com/publicreport/2003/studies/roofs.htm. Courtesy of The Dow Chemical Company.

Definitions That Classify

We often make sense of the things around us by putting them into particular categories. Many definitions—technical and otherwise—work by classifying or categorizing a term within a larger group in order to provide a context or an association. For example, someone writing a field guide to birds and animals in Texas could safely assume that many people would not know what an ocelot is. However, if the writer begins a definition that tells the audience that an ocelot is a kind of wildcat, much of the mystery of the term quickly disappears because most, if not all, readers can identify a wildcat, at least to some extent.

OTHER PROFESSIONAL DEFINITIONS IN THE WORKPLACE

In the following definition, notice how the term *dioxin* is defined by classification; it is placed in a larger category of chemical compounds:

> The term "dioxin" refers to a family of chemical compounds that are unintentional by-products of certain industrial, non-industrial, and natural processes, often involving combustion. The concern about dioxin is that it persists in the environment and can accumulate in the body through the consumption of some foods containing fat.*

*"The Dow Global Public Report 2003: Environmental Stewardship," http://www.dow.com/publicreport/2003/stewardship/dioxin.htm. Courtesy of The Dow Chemical Company.

Definitions That Provide Examples

Some definitions work well when a writer provides a clear-cut and familiar example of the thing or concept being defined. A government agency that develops a website to give consumers information about the fuel efficiency of all vehicles sold in the United States would probably provide different categories of vehicles in order to make the website more user-friendly: compact cars, sedans, light trucks, sport-utility vehicles, and sports cars, among others. However, the website would also need to provide readers with definitions of these different categories. For example, *sports cars* might be defined as vehicles with sleek body styles, superior acceleration, little cargo space, and so forth. Although such a definition would be useful, any definition of *sports cars* would benefit by providing specific examples: perhaps Chevrolet Corvette, Dodge Viper, or Porsche 911. Many readers can recognize these cars by sight and can use their experience with these examples to make sense of the ways the vehicles are being defined and discussed on the website. However, definitions that use examples are not always as straightforward as this; writers cannot always point to an object that provides a recognizable, clear example that audiences will immediately grasp.

In an essay about nature and biology, E. O. Wilson discusses a concept that he calls "the insurance principle of biodiversity"—not exactly a phrase that makes much sense at first glance. He writes,

> Ecosystems are kept stable by the insurance principle of biodiversity. If a species disappears from a community, its niche will be more quickly and effectively filled by another species if there are many candidates for the role instead of a few.

This helps clarify Wilson's meaning of the insurance principle of biodiversity, but Wilson knows that such a definition is not entirely complete, that a good example would help readers understand his writing. Thus, he continues with the following, even using the word *example* to alert his readers:

> Example: A ground fire sweeps through a pine forest, killing many of the understory plants and animals. If the forest is biodiverse, it recovers its original

composition and production of plants and animals more quickly. The larger pines escape with some scorching of their lower bark and continue to grow and cast shade as before. A few kinds of shrubs and herbaceous plants also hang on and resume regeneration immediately. In some pine forests subject to frequent fires, the heat of the fire itself triggers the germination of dormant seeds genetically adapted to respond to heat, speeding the regrowth of forest vegetation still more.

Wilson then goes on to provide a second, lengthier example of his insurance principle in order to clearly define this concept. Examples in this case help make an abstract term or phrase easier for audiences to understand. In most cases it is abstract terms that most need specific, detailed examples in order for a definition to register completely with readers. And the world of workplace and technical writing is filled with such abstractions in need of clarity.

OTHER PROFESSIONAL DEFINITIONS IN THE WORKPLACE

In the following definition, notice how the company's values are defined largely through examples. Thereafter, the document goes on to list other values, such as agility, respect, and innovation.

> Our mission will be accomplished by living according to values that speak to the economic, social, and environmental responsibilities of business and society:
>
> **Integrity**—We believe our promise is our most vital product—our word is our bond. The relationships that are critical to our success depend entirely on maintaining the highest ethical and moral standards around the world. As a vital measure of integrity, we will ensure the health and safety of our communities, and protect the environment in all we do.
>
> **Unity**—We are one company, one team. We believe that succeeding as one enterprise is as important as succeeding independently. Balancing empowerment and interdependence makes us strong. As one company, Dow's impact on the world is far greater than the impact of any one of its parts. We will work together, building relationships to create ever-greater value for the customers and consumers we serve.*

*"Dow's Essential Elements: Values—Who We Are," http://www.dow.com/about/aboutdow/vision.htm. Courtesy of The Dow Chemical Company.

Definitions That Use Visuals

Using visuals in definitions is a strategy that writers in the workplace sometimes use to help readers understand their meaning. Rather than additional sentences and words, visuals can give readers quick and clear understanding. This chapter earlier mentioned that an ocelot is a kind of wildcat and that readers would be better able to understand this by using a categorical definition. However, readers would likely benefit even more from seeing an image of an ocelot in addition to a written definition (see Figure 15.4). For this reason many documents provide visuals along with definitions in order to help create clarity for readers. Figures 15.5 and 15.6 show the effect of illustrations on two other definitions discussed earlier.

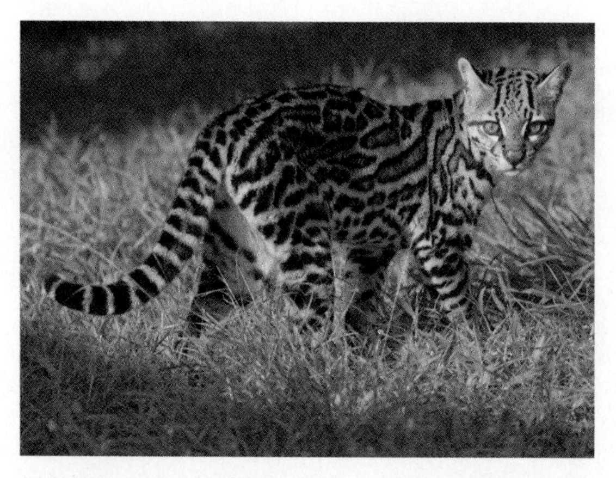

FIGURE 15.4 Ocelot definition with a visual

Visuals are also necessary to help explain concepts that cannot be seen with the naked eye. For instance, in 2006, *Wired* magazine did a story about the Zobot computer virus, which crashed the Department of Homeland Security's US-VISIT screening system that was used by the Immigration and Customs Enforcement Bureau. The article defines and describes the Zobot virus, telling how it worked and what problems it caused. However, in order to help readers understand the terminology used in defining the virus, the article also provides a graphic. Figure 15.7 shows the diagram that allows readers to follow the virus's path and comprehend how it managed to breach the government's screening system.

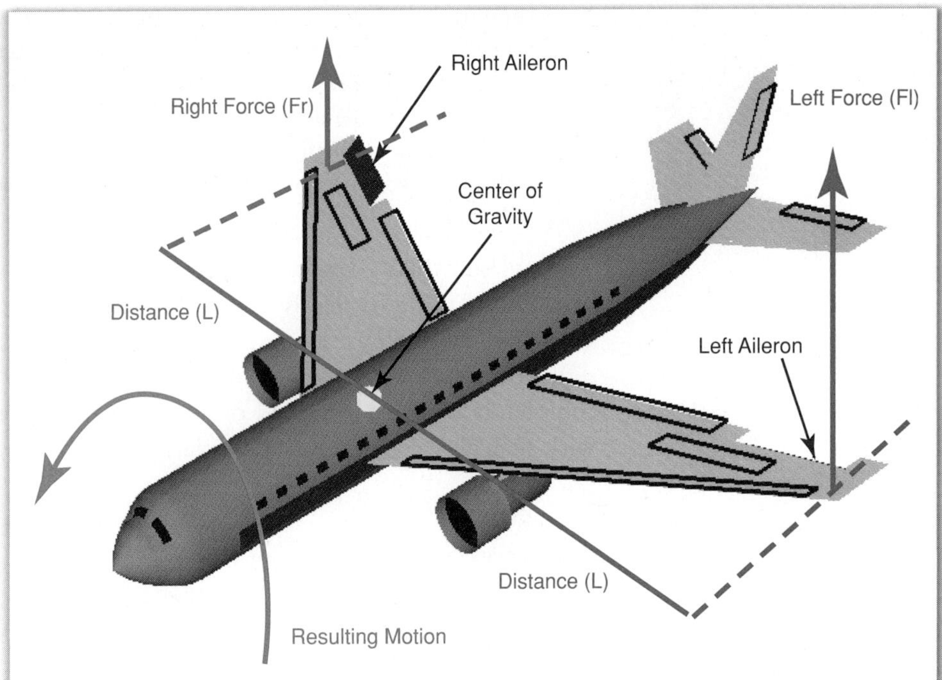

Ailerons are small, hinged sections on the outboard portion of a wing, used to generate a rolling motion for an aircraft. Ailerons usually work in opposition: as the right aileron is deflected upward, the left is deflected downward, and vice versa. The illustration shows what happens if a pilot deflects the right aileron upward and the left aileron downward.

FIGURE 15.5 Aileron definition with a visual

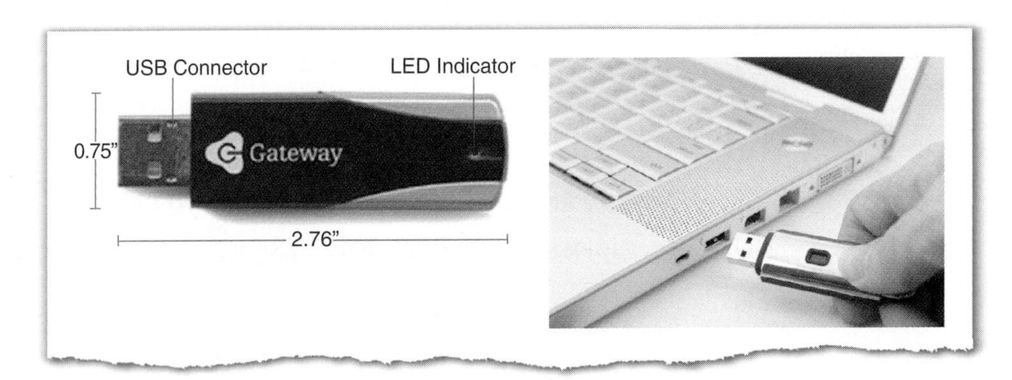

FIGURE 15.6 Flash drive definition with a visual

For more on creating and implementing visuals, see chapter 8.

Because of space limitations in many technical and professional documents, writers cannot provide visuals for every term that needs explanation. Consequently, writers must decide which visuals are most appropriate for their audiences and the purposes at hand.

EXPLORE

Locate a copy of a college-level dictionary, and flip through the pages. Does the dictionary provide any illustrations? If so, list three terms that include illustrations. Why do you think the editors of the dictionary chose to include illustrations for those particular terms and not others? If the dictionary does not have any illustrations, find three words that you believe would benefit from illustrations. Explain why you have identified these three as needing illustrations.

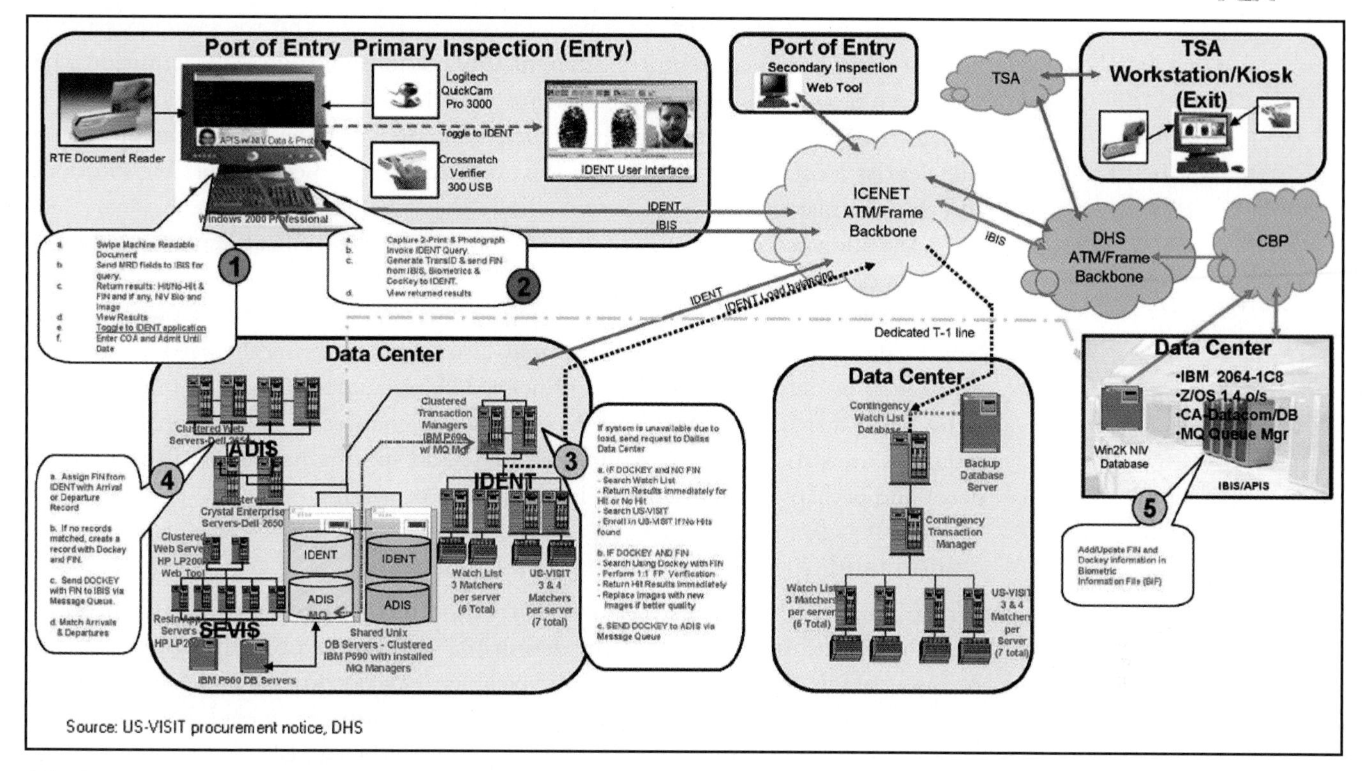

FIGURE 15.7 Diagram of the Zobot virus path

WRITING DEFINITIONS WITH THE PROBLEM-SOLVING APPROACH

Workplace writers must know how and when to use these different kinds of definitions in their writing. Using the problem-solving approach can help because it asks writers to consider technical definitions in terms of audience, purpose, genre, and strategies. Review the steps of the PSA, printed on the inside back cover of this textbook. Try to remember several key points about the PSA:

- It helps writers produce documents that solve workplace problems.
- It helps writers address the rhetorical problems that arise as they compose a document.
- It is not a linear method.

Planning Definitions

To produce a definition in your writing, you should first address several questions: What is the purpose of this definition (i.e., what problems will it solve)? Who are the audience and the stakeholders? And how much do the audience and the stakeholders already know about this term or concept? These questions encapsulate the heart of the Plan phase when composing definitions.

Knowing the Audience

A dentist who is writing or speaking to another dentist would likely not need to define *root canal*. This same dentist, however, would probably need to spend a good deal of time explaining the procedure to the individual who will experience the root canal. When speaking to the patient, the dentist encounters a problem and has a purpose—explaining the procedure to the patient in a way that is accurate and honest yet also comforting and reassuring. If the patient is a child, the dentist might have multiple audience members, or stakeholders—the patient and the patient's parents. Each of these audiences requires different explanations and kinds of definitions, because each faces a different set of problems that the dentist must address: the patient may fear the procedure itself whereas the parent may fear the cost of the procedure. The dentist must review the bases of the problems, the audiences, possible solutions, and the information needed to help facilitate the solutions.

The audience determines the amount and kind of technical language or jargon used in a definition. A writer must be able to gauge how much technical vocabulary the audience will comprehend because that assessment determines the structure, length, and type of definition used. It may also help determine the need to integrate visual aids. A definition is further influenced by an audience that does not speak or write English as its first language. As discussed in chapter 5, transnational audiences require writers to discover how much detailed and specific information the audience needs and perhaps to adjust the style and format of writing.

Understanding the Problem and the Purpose

Writing clear definitions also requires that writers recognize what their audiences will *do* with the definitions and how the definitions will be used to help solve problems. As discussed in chapter 2, readers must be able to *use* documents and solve problems with them. Definitions are no exception; readers do not usually read definitions simply to enhance their knowledge but rather to accomplish a task.

For example, the U.S. Army Corps of Engineers and the Environmental Protection Agency define *wetlands* like this:

> Those areas that are inundated or saturated by surface or ground water at a frequency and duration sufficient to support, and that under normal circumstances do support, a prevalence of vegetation typically adapted for life in saturated soil conditions. Wetlands generally include swamps, marshes, bogs, and similar areas.

But these agencies do not define *wetlands* as an end in itself. They define the term so that readers will know how to identify, use, or protect wetlands as the law necessitates. Definitions, then, *do* things; they often allow readers to act and make decisions in appropriate legal and ethical ways. In terms of purpose, most definitions fall within one or more of the following three categories.

Explanatory Definitions. Explanatory definitions attempt to explain a concept, thing, or idea to readers—to give them the meaning of a term so that they can understand it and, in some cases, act accordingly. An astronomer, for instance, who writes a scientific article for her peers that discusses the discovery of a celestial body such as 2003 UB313 would name, define, and describe this new body in a way that explains it clearly to her audience. Such an explanation might lead her audience of fellow astronomers to rethink how they understand celestial bodies, planets, and the universe.

Operational Definitions. Operational definitions enable audiences to engage in some kind of activity or operation. Operational definitions take many forms, such as a description of a tool or a process. Most of us, for instance, would probably be a bit confused the first time we read a recipe that asked us to whip, blanche, or carmelize something. We would have to know what these terms meant before we could do them. Workplace writing often requires individuals to work with new terms, concepts, and procedures that must be defined for audiences who need to know how to do something in a particular manner.

No doubt as a student, you have been given course syllabi that included a variety of operational definitions—for example, late-work policies, attendance policies, essay assignments, and so forth. Your instructors probably defined the meaning of late work, attendance requirements, and the parameters of essay assignments. And such definitions probably gave you a sense of how the course would operate, as well as how you would engage in course activities. Similarly, a variety of workplace documents—such as manuals, handbooks, memos, and e-mails—provide definitions that allow audiences to take part in activities in certain designated ways.

Deliberative Definitions. Deliberative definitions clarify the meanings of terms, ideas, and concepts so that audiences can decide on a future course of action. Often we need to know the meaning of certain things in order to make decisions or to deliberate on the best way to proceed. For example, an inexperienced investor would need to know the meanings of such things as stocks, bonds, certificates of deposit, and money markets. Figure 15.8 shows just one example of information that might confuse an uninformed investor; imagine receiving a response like that after inquiring about investment options. If you were unfamiliar with the bulleted terms, you would need specific and detailed definitions before you could make any decisions about what investments to pursue.

As another example, the E. O. Wilson essay mentioned earlier spent a good deal of time defining the insurance principle of biodiversity in order to help readers make a decision—to work to ensure biodiversity in the world's natural environments. Wilson's work contributes to long-running debates about the best possible course of action for humans to take to protect the planet.

IN YOUR **EXPERIENCE**

What is the job you now have or most recently had? What terms or ideas did you not understand at first that had to be defined for you? Write about how this rhetorical situation played out. What problems did you face by not knowing these definitions? Or what problems were solved by knowing the definitions? How did you learn the definitions? What purpose did the definitions serve? Did the person who gave you the definitions, either in oral or written discourse, seem to adequately review the rhetorical situation? Explain.

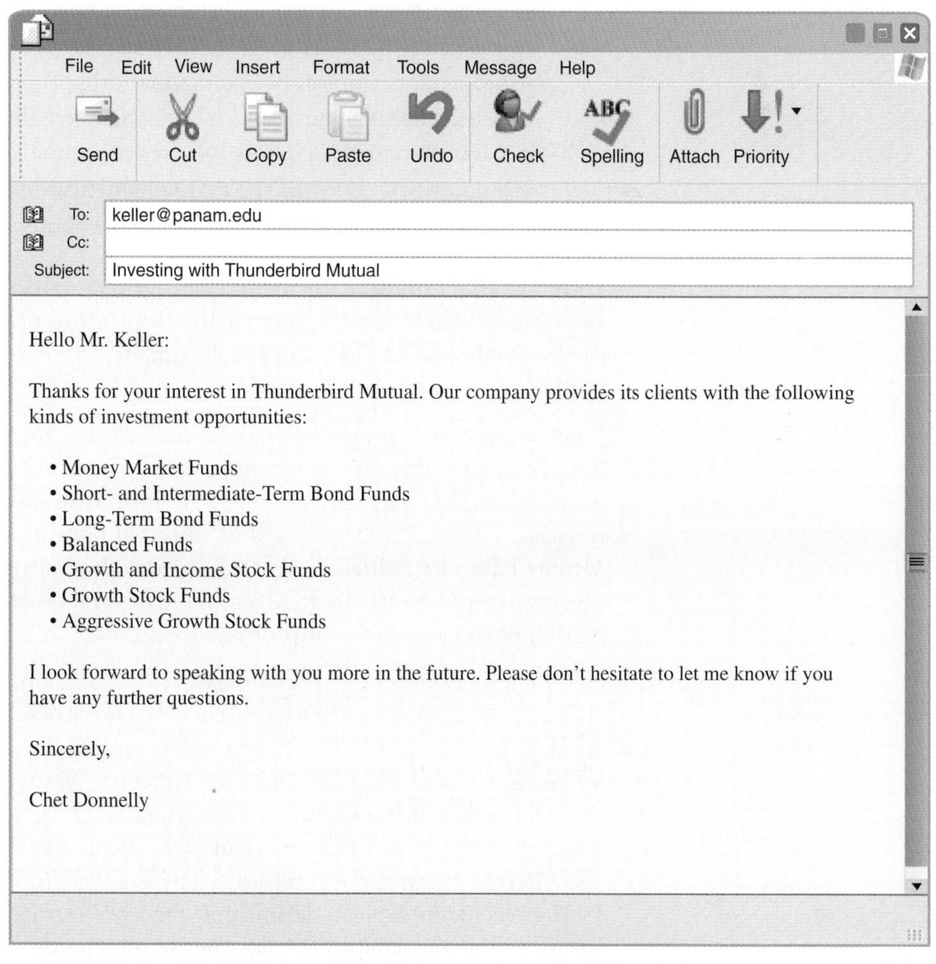

Hello Mr. Keller:

Thanks for your interest in Thunderbird Mutual. Our company provides its clients with the following kinds of investment opportunities:

- Money Market Funds
- Short- and Intermediate-Term Bond Funds
- Long-Term Bond Funds
- Balanced Funds
- Growth and Income Stock Funds
- Growth Stock Funds
- Aggressive Growth Stock Funds

I look forward to speaking with you more in the future. Please don't hesitate to let me know if you have any further questions.

Sincerely,

Chet Donnelly

FIGURE 15.8 Investment broker's e-mail with undefined terms

Considering Length and Placement

When workplace writers create definitions, rarely are they writing "just" a definition, disconnected from a larger text or context or from an audience that needs the definition. In other words, writers do not often create definitions that stand alone; their definitions exist within a larger written document or website. An instruction manual may define *computer monitor*, *CPU*, *printer*, *mouse*, *keyboard*, and *USB cord*, but defining is not the manual's larger purpose. The definitions are meant to contribute to the larger goal of enabling the user to put the computer together and operate it correctly.

Because definitions exist to support the meaning and the purpose of a document, writers must consider the appropriate length and placement of a definition in the document. In some circumstances definitions need to be extensive and thorough, and in other cases short definitions may do the job. Such choices are determined in large part by the genre of the document. Consider these two examples:

- *Memos*—Memos are often written for internal audiences likely to be familiar with the terms being discussed. Thus, definitions in memos usually do not need to be detailed or lengthy, in part because memos do not typically communicate long, detailed information. If memo writers believe that readers will not understand the meaning of a particular term or concept, they might direct readers to another document where they can find more detailed definitions. See chapter 12 for more on memos.
- *Manuals*—Manuals often inform readers about policies and procedures, explaining how to complete certain processes like conducting experiments, building and fixing machines, operating equipment, or filling out forms and documents. Good definitions in manuals serve many important purposes, safety being one of them. Writers of manuals, therefore, must determine the technical knowledge of their readers in order to produce definitions that ensure readers can follow the manual properly. Manual writers are less concerned about length and more concerned about whether the definition does its job correctly. Those who write manuals often include definitions throughout the body of the manual and also include a glossary at the end of the manual for easy accessibility. See chapter 19 for more on manuals.

Internal Placement. In many cases writers need to define terms and concepts in the midst of their texts, or internally. Depending on the term to be defined, the audience needs, and the genre of the document, an internal definition may be short, parenthetical, or extended.

Short Definitions. Short internal definitions are typically a sentence to a paragraph in length and are usually categorical, placing the term in a category of similar terms. Such definitions may also provide descriptions of use or other distinguishing characteristics. Figure 15.9 shows a short internal definition from Major League Baseball's Mitchell Report, which sought to explore the ramifications of professional baseball players' use of performance enhancing drugs. Pay particular attention to the ways this section of the report defines *steroids.*

Parenthetical Definitions. A parenthetical definition is a type of short definition that is separated from the rest of a sentence by parentheses or a pair of dashes. Parentheses allow writers to define a term or concept quickly and efficiently, often as a reminder for an audience already familiar with the term. Writers would not want to put a complex definition in parentheses if their audiences are unfamiliar with the term or it needs to be explained in more than eight or nine words. Parenthetical definitions are also effective with foreign phrases. The excerpt from the Mitchell Report in Figure 15.9 uses parenthetical definitions for two terms: *anabolic process* and *catabolism.* Parenthetical definitions are not meant to provide readers with lengthy or detailed definitions but can help clarify jargon that could potentially cause confusion.

The term "steroids," when used in the context of athletic performance enhancing drugs, refers to a class of drugs more precisely known as anabolic androgenic steroids. Anabolic steroids are natural or synthetic versions of testosterone, the primary male sex hormone. Steroids foster the anabolic process (muscle growth and the increase of muscle mass) and also limit catabolism (the breakdown of protein in muscle cells). As a result, steroid users can increase the muscle gain resulting from strenuous exercise and maximize the impact of a high protein diet. In addition, because of their anti-catabolic effect, steroids reduce the soreness that normally results from strenuous exercise, which allows an athlete using steroids to exercise more frequently, even daily. (pp. 5–6)

FIGURE 15.9 A section of the Mitchell Report with a short internal definition of anabolic steroids

Extended Definitions. Sometimes audiences need more information than a short definition can provide. For example, a transnational audience, an audience whose first language is not English, or an audience not technically proficient in a document's material may require longer, extended definitions that provide much detail about the term. Such definitions need more than a paragraph or two to provide the necessary information. There is no set length for extended definitions; writers provide as much detail as they believe their audiences need to understand or use the definition.

See chapter 5 for more on transnational audiences.

Extended definitions can use any of the different kinds of definitions outlined earlier. The example in Figure 15.10 is an extended definition written by New Technologies, Inc. for potential clients. The company's "focus is on the highly technical issues which are involved in the processing of computer-related evidence and in the identification and elimination of computer security risks." To assist their clients in trial preparation, they provide a series of extended definitions, one of which is excerpted in the figure. Notice

The term "Computer Forensics" was coined back in 1991 in the first training session held by the International Association of Computer Specialists (IACIS) in Portland, Oregon. Since then, computer forensics has become a popular topic in computer security circles and in the legal community. Like any other forensic science, computer forensics deals with the application of law to a science. In this case, the science involved is computer science and some refer to it as Forensic Computer Science. Computer forensics has also been described as the autopsy of a computer hard disk drive because specialized software tools and techniques are required to analyze the various levels at which computer data is stored after the fact.

Computer Forensics deals with the preservation, identification, extraction and documentation of computer evidence. The field is relatively new to the private sector but it has been the mainstay of technology-related investigations and intelligence gathering in law enforcement and military agencies since the mid-1980's. Like any other forensic science, computer forensics involves the use of sophisticated technology tools and procedures which must be followed to guarantee the accuracy of the preservation of evidence and the accuracy of results concerning computer evidence processing. Typically, computer forensic tools exist in the form of computer software. Computer forensic specialists guarantee accuracy of evidence processing results through the use of time tested evidence-processing procedures and through the use of multiple software tools, developed by separate and independent developers. The use of different tools which have been developed independently to validate results is important to avoid inaccuracies introduced by potential software design flaws and software bugs. NTI's computer evidence-processing tools were intentionally developed by separate in-house software developers to deal with these potential problems because the accuracy of the results is extremely important. It is a serious mistake for a computer forensics specialist to put "all of their eggs in one basket" by using just one tool to preserve, identify, extract and validate the computer evidence. Cross-validation through the use of multiple tools and techniques is standard in all forensic sciences. When this procedure is not used, it creates advantages for defense lawyers who may challenge the accuracy of the software tool used and thus the integrity of the results. Validation through the user of multiple software tools, computer specialists and procedures eliminates the potential.

FIGURE 15.10 Extended definition of *computer forensics* for potential clients of New Technologies, Inc.

how this extended definition explains not just what computer forensics is but when it started, how it works, and what purposes it serves. The definition provides a great deal of information for audiences who are likely not experts in this area of forensics.

External Placement. External definitions are those that are not placed inside a particular text. Instead, they supplement texts such as manuals or reports and are placed, usually, in separate sections. A glossary is an easy way for writers to codify all terms and concepts that might give readers some difficulty, and it allows readers to find all terms in alphabetical order in one specific location—usually at the end of a document. Glossaries are not practical, however, for shorter workplace documents. And their use in longer documents is contingent on whether the audience needs a glossary to make sense of the document.

ANALYZE THIS

The following is a short internal definition found at the beginning of "Understanding Colic in Horses," a technical report written by equine specialists at Oklahoma State University:

> Most horse owners have dealt with colic. Rather than a disease, colic is a condition of pain. Specifically, colic refers to abdominal pain most often originating from the digestive tract. Because it is a condition of pain rather than a specific disease, causes are numerous and sometimes difficult to diagnose accurately.

Decide whether this definition works as an explanatory, operational, or deliberative definition. Why would or wouldn't this definition work as a parenthetical definition? What kind of information would need to be added to make it an extended definition? How would it need to be rewritten to function as a dictionary or glossary entry?

Researching Definitions

Many definitions in workplace writing are not the result of individuals simply sitting down and writing them; they are written only after considerable research and collaboration. Because workplace writers often have to construct definitions that deal with unfamiliar ideas and concepts, additional research may be necessary. For instance, a lawyer may be involved in a case or a lawsuit that involves jet engines. Much of the case may rest on the technical aspects of the engine, and the lawyer would need not only to understand some of these technical terms but also to define them for other lawyers and jurors. Research can be a complex but necessary activity to produce accurate and effective definitions.

Drafting Definitions

Like writing any other kind of text, writing definitions requires drafting, designing, and, in some cases, using visual aids. Drafting includes putting together an initial definition and placing it within the context of the document. The planning decisions made earlier regarding the audience, purpose, and length and placement of the definition will shape the entire drafting process. Part of document design will be fitting the definition into the scope of the document. In addition, some definitions should be supplemented with visuals and diagrams to help audiences understand them. Because definitions exist within the context of larger and more complex documents, writers must concern themselves not just with what a definition says or means but also with how it fits into and contributes to the document as a whole.

Reviewing Definitions

When writing documents that include definitions, workplace writers often test how well their definitions work—how well they do or do not fit into and contribute to the document in which they are placed. Much of this testing leads to revision of the definition and its larger document.

For more on usability testing, see chapter 11.

Feedback about definitions can come in many forms. Sometimes it derives from coworkers and other collaborators who have some stake in making sure the definitions are successful. At other times feedback can come about through usability testing—testing the definitions and the document on individuals who represent the intended audience. Both kinds of feedback lead to comments and suggestions about how definitions can be revised and improved.

IN YOUR **EXPERIENCE**

What symbols or signs have you encountered recently—at school, work, or home—that sought to define important information for their audiences? What were these, where were they, what was their purpose, and how did they work? Did you find them effective or not? Explain.

ETHICS AND TECHNICAL DEFINITIONS

Like all other aspects of writing in the workplace, writing definitions involves a series of ethical and responsible decisions—part of the rhetorical thinking that takes place throughout the entire process. For instance, thinking about the purpose of a definition is closely linked to ethics because definitions can potentially mislead and confuse or, even worse, cause danger or harm for audiences. Individuals who work with dangerous objects and products—beginners and specialists alike—must fully understand these objects and products in order to use them safely: what they are, what they do, how they work, and what the consequences are for misuse, for example. Thus, it is necessary for writers of definitions to make sure that readers have all the information necessary to use products safely.

However, clear and thorough definitions do not involve ethical thinking *only* when they refer to dangerous products like chainsaws, chemicals, or firearms. Those who write definitions in workplace documents face ethical considerations when readers are relying on those definitions for correct, unbiased information to help them make decisions. If you are applying for a loan or a mortgage, for example, you would certainly hope that all the terms and procedures of the process are defined clearly and truthfully for you and that nothing is hidden.

Full Disclosure

Full disclosure means, essentially, that writers are not leaving out any information that an audience would need to comprehend the term being defined. A good many of us, no doubt, have felt cheated at some time because we didn't understand the meaning of terms in a contract or application, for instance. In some cases the lack of clear definitions may have been purposeful, at other times not. Regardless, professional and technical communication must include definitions that provide readers with all the information they need for the purposes at hand.

Appropriate Language and Style

Writers should compose their definitions in a language and writing style that readers will clearly grasp. Most of us have seen definitions that are not understandable simply because of the language and writing style used. Certainly we know that much legal writing is problematically unclear for some audiences. Look at the way two fairly common terms are defined in U.S. copyright law:

> "Copies" are material objects, other than phonorecords, in which a work is fixed by any method now known or later developed, and from which the work can be perceived, reproduced, or otherwise communicated, either directly or with the aid of a machine or device. The term "copies" includes the material object, other than a phonorecord, in which the work is first fixed.

> For purposes of section 513, a "proprietor" is an individual, corporation, partnership, or other entity, as the case may be, that owns an establishment or a food service or drinking establishment, except that no owner or operator of a radio or television station licensed by the Federal Communications Commission, cable system or satellite carrier, cable or satellite carrier service or programmer, provider of online services or network access or the operator of facilities therefore, telecommunications company, or any other such audio or audiovisual service or programmer now known or as may be developed in the future, commercial subscription music service, or owner or operator of any other transmission service, shall under any circumstances be deemed to be a proprietor.[1]

Although this kind of writing is certainly appropriate for some audiences, mainly lawyers, it is not so for all audiences. Many readers would find these definitions inaccessible and meaningless. In some cases, definitions can induce people to make poor or unsafe decisions simply because they were unable to fully comprehend the meaning of terms and concepts. Knowing the audience and writing definitions in a language and style that audience members will fully understand are not just good rhetorical moves but are ethical ones as well. The two are intertwined tightly.

EXPLORE

Locate a definition of a term important to your future career or current academic major. Where did you find the definition? Who wrote it? What is its purpose? Who is the audience for this definition? As far as you can tell, does the definition disclose what it needs to in order to be ethical? Is the style appropriate for its audience and purpose? Write about your findings, or discuss them with classmates.

COMMON MISTAKES IN COMPOSING DEFINITIONS

It is important to note a few errors that writers tend to make when constructing definitions. These are problems and hazards to look out for when you craft your definitions:

Circular Definitions

Circular definitions use the term being defined as part of the definition itself and are therefore usually uninformative. For instance, if you define *web browser* as "a computer program that allows users to visit and navigate websites," you have a circular definition; the definition begins with a reference to the Web and ends with a reference to the same thing. The definition is faulty because a reader who does not understand what the World Wide Web is will find little value in the explanation.

Synonyms in Definitions

Synonyms in definitions can often confuse readers because they receive no distinguishing explanation of the term. If, for instance, you define *philanthopy* as "a kind of charity," the reader is given no specific information, just a word with a similar meaning. Definitions need to provide more specific characteristics. A more accurate definition would assert that philanthropy is an activity intended to promote the welfare of humankind.

Definitions That Are Not User-Friendly

Definitions that are not user-friendly do not take the user's or audience's knowledge, background, and needs into account. Technical definitions, for example, written at the level of an inexperienced user, might not draw the sophisticated distinctions that a proficient user

---

[1] "Subject Matter and Scope of Copyright," *Copyright Law of the United States of America*, http://www.copyright.gov/title17/92chap1.html.

needs; on the other hand, a highly technical, specialized definition might not suit the needs of inexperienced users. To draw on an earlier example, defining the term *flaperon* as "a control surface combining the functions of a flap and an aileron" hinges on the reader's understanding of the terms *flap* and *aileron*. Consequently, this definition would not be appropriate for an inexperienced user unfamiliar with such terms but would be appropriate for someone more comfortable with airplane mechanics. The audience is key.

SUMMARY

- Many debates inside and outside the workplace revolve around disagreement over the definition of ideas and terms.
- Writing definitions is a common activity in most workplace settings and involves problem-solving strategies.
- Workplace definitions differ a great deal from dictionary definitions.
- Definitions describe and clarify information for readers and allow experts and specialists to communicate with nonexperts and nonspecialists.
- Definitions in workplace writing usually do one or more of the following: describe, compare and contrast, classify, provide examples, or illustrate with visuals.
- Definitions in workplace writing usually have particular purposes that can be described as explanatory, operational, or deliberative.
- Workplace definitions are usually integrated into longer documents, such as memos, reports, and proposals.
- Definitions in workplace documents can be placed internally—for example, short, parenthetical, or extended definitions—or externally, such as glossary definitions.
- Ethical definitions provide audiences with full disclosure and an appropriate language and writing style.
- Common mistakes in composing definitions for workplace documents include circular definitions, use of synonyms in definitions, and definitions that are not user oriented.

CONCEPT REVIEW

1. In what ways can definitions be a source of controversy inside and outside the workplace?
2. How do workplace definitions typically differ from dictionary definitions?
3. Why is knowing the audience important in writing definitions?
4. What does it mean for a definition to describe?
5. What does it mean for a definition to compare and contrast?
6. What does it mean for a definition to classify?
7. What does it mean for a definition to provide examples?
8. What does it mean for a definition to illustrate with visuals?
9. What does it mean for a definition to have an explanatory purpose?
10. What does it mean for a definition to have an operational purpose?
11. What does it mean for a definition to have a deliberative purpose?
12. Why are definitions integrated into longer documents?
13. What are internal definitions?
14. What are parenthetical definitions?
15. What are external definitions?
16. What does it mean to provide full disclosure in definitions?
17. What does it mean to write definitions in an appropriate style?
18. What is a circular definition?
19. Why is it problematic to use synonyms in definitions?
20. What is a user-oriented definition, and why is it important?

CASE STUDY 1

Bioterrorism and the Lone Star State: Writing Definitions for Multiple Audiences

As a recent college graduate with a degree in biochemistry, you land your first job at the Texas Department of State Health Services. Your job is to assess potentially dangerous biochemical health risks that may affect the public. Another part of your job description, however, is to work closely with public relations writers who put together press releases, pamphlets, reports, and websites that inform the public about important health issues. One of your first tasks on the job is to help the writers devise an extended definition of *bioterrorism* for the department website. After an initial meeting and an hour of work, you and the writers come up with the following definition:

> **What is "bioterrorism"?**
>
> Bioterrorism, or a biological attack, is the intentional release of germs or other biological substances, such as toxins and poisons, that can cause illness and death among people. The possibility also exists for a terrorist to use new, genetically engineered agents that are harder to treat.

Your supervisors agree that this definition will work for the time being but that you and the writers need to elaborate on this definition and compose an extended definition that will go on the website in two weeks. You and the writers, therefore, must get to work assessing the definition and expanding on it to make it more clear, effective, and relevant to the website's audience.

Given what you have learned in this chapter, consider the following: Who is the audience? What is the definition's purpose? How might the genre (i.e., government-sponsored website) affect how you present an extended definition? What kinds of definitions will be most effective? What is done well in the current definition, and what needs to be revised and expanded? In a collection of informal notes, begin to describe how this short definition could best be revised to make it better suited to its audience and audience needs. You will need to discuss what kinds of ethical ramifications are involved in writing a definition of *bioterrorism*.

You might browse the Texas Department of State Health Services' website (http://www.dshs.state.tx.us/preparedness/defaults.shtm) to get a feel for the site and the way it communicates information in general and definitions in particular. Depending on your instructor's directions, you can either write out an extended definition of *bioterrorism* for the website or simply discuss how the definition should be revised. Your instructor may also suggest that you work on this case study in a small group. *Hint*: remember that this is for a website; thus, you should consider which portions of text to link to other pages.

CASE STUDY 2

Soaring Gas Prices and Hybrid Car Sales: Composing a Technical Definition of Hybrid Car Technology

The rise in gasoline prices has led many consumers to purchase more fuel-efficient hybrid automobiles. Sarah Schroeder, the editor of a daily newspaper in Oklahoma City, wants to run a story about the increase in hybrid automobile sales; however, she wants it to go beyond simple statements about sales figures and actually explain to readers some of the technologies behind hybrid vehicles. Because none of the journalists at the newspaper have the technical expertise to define and discuss hybrid vehicles, she hires a mechanical engineer, Elizabeth Keene, as a technical consultant to help develop the story. Elizabeth is not asked to actually write the newspaper story but to develop an extended, explanatory definition of hybrid vehicles, with the particular purpose of showing readers how hybrid technologies work. The definition is to include at least one visual to help audience comprehension. The journalist working on the story will then work the definition and visual into the article.

You are to take on the role of Elizabeth Keene, developing the extended, explanatory definition of a hybrid vehicle and submitting your work in memo format to Sarah Schroeder (see chapter 12 for more on memos). You may use any combination of the definition types discussed in this chapter, and you may either develop your own visual or use a preexisting one, so long as it is correctly cited. Keep in mind that your definition is ultimately going to play a role in a newspaper story; most readers are not apt to be technical experts. And when conducting research for your definition, remember to evaluate the credibility of your sources so that your information is accurate and correct. Finally, be prepared to discuss with your instructor and classmates the rhetorical choices you made and the reasons you made them.

CASE STUDIES ON THE COMPANION WEBSITE

American Dream Mortgage Consulting and the 80-20 Loan

You serve as a counselor for first-time home buyers who are seeking mortgages. During part of your counseling activities, you realize clients are increasingly confused by the definition of a particular kind of mortgage (the 80-20), and you must consider ways to address this problematic definition.

Argon Gas: Effective Insulator or Insidious Carcinogen?

In this case you assume the role of a lead sales representative at Industrial Glass, which serves the needs of residential and commerical customers. One of your potential clients has misunderstood the meaning of a particular term related to one of your company's products, and you must remedy this situation with correct definitions.

Introducing Portland Public Schools to New Computer Equipment

You are to assume the role of IT director for a school district in Portland, Oregon. You and an outside consulting firm have formulated a plan for systemwide upgrades of computers in the district's classrooms. Part of your job is to inform district administrators and teachers about these upgrades, so you must think about ways to define new computers, systems, and software to audiences that do not have much technical knowledge of these subjects.

Clarifying Accounting Terms for a New Client

Your accounting firm has recently acquired a new client who wants to open a sporting goods store. However, he has very little knowledge of business or accounting terminology, so you must determine a way to communicate clearly with him without using too much jargon.

VIDEO CASE STUDY

Agritechno

Information for New Investors: Defining the Terms

Synopsis

Lev Andropov and Teresa Cox, geneticists with Agritechno, are working in Andropov's lab. Jaylen Castillo, part of the Agritechno product development program and liaison to Andropov's lab, comes to the lab to tell Lev and Teresa that Agritechno wants to push into a broader fruit and vegetable market and needs some information from them to develop promotional materials.

WRITING SCENARIOS

1. Compose a short definition (two to four sentences) of three of the following terms for an audience unfamiliar with them: *technical communication*, *rhetoric*, *ethics*, *memo*, *usability testing*. Use three different kinds of definitions discussed in this chapter.
2. Locate a manual—print or online—and find an example of a technical definition that uses a visual (image, graph, chart, or illustration). Discuss how the visual works: who is the audience, what purpose does the visual serve, and is it effective or ineffective? Why does the writer(s) of this manual include a visual for this particular definition? Explain in a document type assigned by your instructor.
3. Compose a list of five terms (preferably technical ones from academic courses or professional environments) that you do not currently know or fully understand. Then locate three different definitions for each of these five terms—from articles, reports, dictionaries, or textbooks—and explain how and why the definitions differ for each of the terms. Provide your analyses in a document type assigned by your instructor.
4. Online or in the library, locate the main sources of definitions in your future career area. Is there a particular dictionary, handbook, or guidebook for those who work in this field? What is it, and how can individuals access it? What do you find interesting about this source? Explain your findings in a short memo to your instructor.
5. Find a copy of the dictionary, handbook, or guidebook you identified in the previous exercise, and choose three terms from it to analyze. What kinds of definitions are they? What are their purposes (i.e., explanatory, operational, deliberative)? How are the definitions placed in the source? Explain your findings in a short memo or other document type assigned by your instructor.
6. Research any field of study that interests you, and make a list of two or three terms that are currently debated in this profession or field. In a memo to your instructor, explain what these debates are and how definitions are central to them. What in particular is being debated? What different sides are taking part in the debates? Why do you think the debates will or will not be resolved in the near future?
7. Make a list of governmental organizations and agencies that often define terms for Americans (e.g., the Environmental Protection Agency, the Federal Bureau of Investigation, and the Internal Revenue Service). Choose one of these organizations, and locate its website. Using one of the note-taking strategies described in chapter 6, describe how the website provides definitions to its viewers. What are these definitions? What specific ethical obligations did the writer(s) have when composing these definitions? Do the definitions seem to disclose all they should? Are they written in appropriate styles?
8. Recall a time in your life when you were given information that contained a poorly written definition. What was the definition? Where was it located? What was its purpose? Did it keep you from completing a task? What made it a poorly written definition, and how might it have been more effective? Write your responses in a document type assigned by your instructor, and be prepared to discuss your responses in class.
9. Choose a term that you are familiar with in your field of study but one that is highly specialized and complex, a term that is used mainly by experts in the

field. Compose an extended definition of that term (one to two pages) for a nonspecialized audience that is not familiar with it. Then describe in a letter to your instructor how you wrote the definition. What type of definition was it? What was its purpose? What specifically did you do to ensure that this definition was an ethical one?

10. Compose a short letter or e-mail to students who will take this course in the future. Give them information about the course: what it is like, what they should expect, and what kind of work the course entails. Discuss what you believe are the three most important terms or concepts they should understand in the course, and include short definitions of these. After writing the letter, provide a few paragraphs of reflection on your definitions: what kinds of definitions they were and what purposes they served for this audience.
11. Write an extended definition of your current college major, assuming an audience that is only slightly familiar with it. Remember that extended definitions are usually multiple paragraphs and often use more than one kind of definition. (If you want to but cannot include an illustration, you might discuss the visual that you think would be effective.) Assume that the purpose of this extended definition is deliberative, helping students decide whether they would like to major in this area.
12. Write a two-page letter to a prospective employer in which you explain why you are a good candidate for employment. Describe and define yourself, your accomplishments, and your abilities. Try to use at least two or three different kinds of definitions, and be aware of how you place those definitions in the letter. Remember your audience and your purpose in writing.
13. In what ways have new media technologies made locating definitions easier than in the past? In what ways has our access to more definitions increased or decreased our tendency to agree with the definitions that we do find? Take notes on your ideas, and be prepared to discuss them in class.
14. If you were to design a brochure for prospective students at your university, what definitions do you think would need to be included? What kind would they be? Where might they be placed? How might these definitions differ if they are to be presented on a website instead of in a printed handbook? How would web technology influence how these definitions would be written and presented? Explain your responses in either memo or letter format to classmates.

PROBLEM SOLVING IN YOUR WRITING

After you've completed either of the case study exercises or one of the writing scenarios in this chapter, think about how you solved the problem that faced you—writing or using one or more definitions. In a one- to two-page letter or memo to your instructor, describe the ways you employed the problem-solving approach to successfully complete the case study or the writing scenario. Let the following questions guide your thinking:

In what ways did you plan before and/or during writing?

Did you do any research to help you complete the assignment?

Did you employ any specific strategies to help you draft the definition(s)?

Did you spend much time reviewing the assignment?

How did you distribute or transmit the assignment to your instructor, if it was turned in?

16

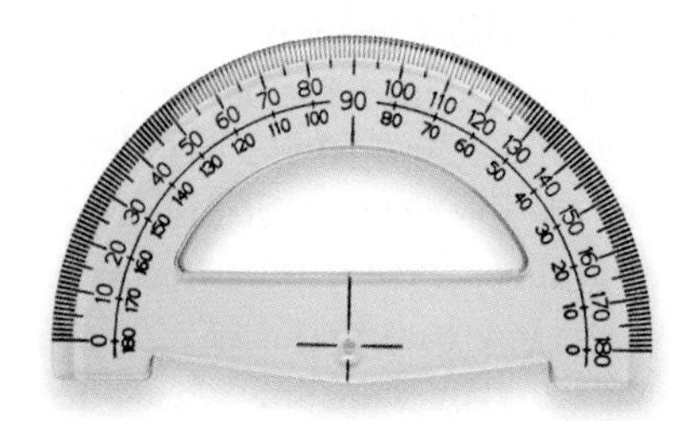

Technical Descriptions and Specifications

CHAPTER LEARNING OUTCOMES

After completing this chapter, you will be able to do the following:

- Know the purpose of **technical descriptions and technical specifications**
- Know the difference between **technical descriptions and definitions**
- Recognize how technical descriptions **solve problems and serve audiences**
- **Incorporate** technical descriptions and specifications into other documents
- Write **introductions, backgrounds, and parts and characteristics** sections of technical descriptions
- Recognize the importance of **visuals** to descriptions and specifications
- Write descriptions using **sensory detail** appropriate to audience needs
- Understand appropriate **organizational patterns** of technical descriptions
- Recognize writers' **ethical obligations** when producing technical descriptions
- Write both **short and long** technical descriptions for various audiences and purposes

DIGITAL RESOURCES

On the Companion Website www.prenhall.com/dobrin:

- Case 1: 10-Megapixel SLR Cameras for Interstate Insurance Adjustors
- Case 2: Hampton Place: New Electronics for Elderly Residents
- Case 3: Mid-Texas Power and the Search for Fuel-Efficient Automobiles
- Case 4: Choosing the Right Bike for the Job
- Video Case: Writing a Description of the GlobeTalk Messaging System
- PowerPoint Chapter Review, Test-Prep Quiz, Exercises and Activities

REAL PEOPLE, REAL WRITING

MEET CLINT HOCKING • Game Designer and Creative Director at Ubisoft Entertainment

What types of writing do you do as a game designer?

In the conception stage of creating a video game, I work on concept documents. In preproduction I e-mail a lot in support of those documents, as well as contribute to design specification writing. In full production I write evaluations of the implemented design for content creators or systems programmers.

Who are the typical readers of your workplace documents?

One audience is the Ubisoft editorial team members, who use these documents to evaluate the direction of the project to ensure that it meets the company's goals. Once the concept meets their approval, shorter documents are created for individual team members, which are supported with constant communication.

What role do descriptions play in the writing you do?

Descriptions are a critical component of the documentation I write and of the field of game design in general. We are a young industry, and game designers have only recently begun to formalize their concepts; in general, we largely reinvent our vocabulary every time we make a game. This is a major problem for our profession that I work hard to solve through accurate, detailed descriptions and definitions.

Can you describe how you used descriptions in a document you recently created at work?

Immediately following the green-lighting of our most recent Splinter Cell game, I put together a creative brief of the project for every member of the editorial team. The document was light on text and full of diagrams and images that present the themes of the game and its core systems, as well as the art direction and visual presentation. Its purpose was to ensure that everyone shared the same general vision and had a reference, enabling them to make quick and accurate decisions about the creative vision for the project.

What role does ethics play in the writing you do?

Video games are embedded in popular culture and have the capability to make statements about the human condition. We are at a stage where it is easy for the uninformed to misinterpret the *context* of a game as being the *content* of the statements that the game is making. A game that uses the context of urban crime to explore freedom is easily mistaken for a game that promotes criminal behavior. In that sense, game designers have an ethical responsibility to make games that say important things and that protect the medium itself from the ethical (and legislative) backlash that can arise from a lack of understanding of our medium.

What advice about workplace writing do you have for the readers of this textbook?

We need the written word—but not as much as we use it. If you can talk to someone in person—do. Workplace writing should always be the fallback solution to a communication need that cannot be handled in person.

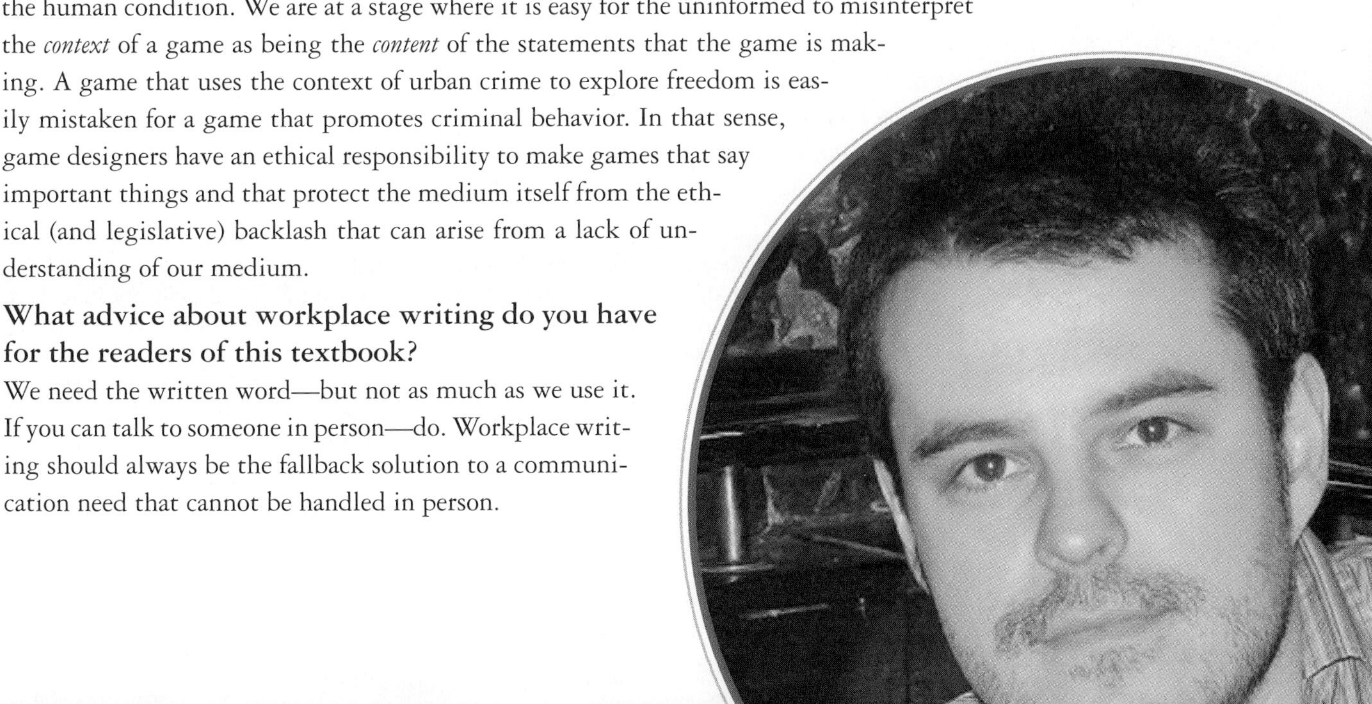

INTRODUCTION

A technical description provides concrete details, precise words, and visuals to show readers what an object, mechanism, or product looks like, how it is put together, and/or how it works. Some technical descriptions detail processes instead of objects—for example, how a customer's credit card is processed online by a hotel's website. Composing technical descriptions, however, requires writers to recognize audience needs and expectations as well as pay careful attention to language use. In most cases audiences need to know answers to one or more of the following questions:

- What is it?
- What does it look like?
- What does it do?
- What is it made of?
- How does it work?
- How has it been put together?

Many professionals rely on technical descriptions to learn and perform their workplace tasks—making products, using products, fixing products, selling products, and sometimes simply understanding products.

It may seem at first glance that technical descriptions and specifications are used only for highly technical objects and processes and that technical descriptions and specifications are not particularly relevant to most everyday workplaces. It is true that not every workplace writer will compose technical descriptions and specifications on a regular basis; however, many will. In addition, it is important to note how descriptions and specifications surround so much that happens in all workplaces.

For example, in 1949 the U.S. government created the General Services Administration (GSA) as an agency to help manage and support the basic functioning of all other federal agencies. In particular, the GSA recommends and supplies products, vehicles, office space, and other services for federal agencies; and it attempts to do so in ways that minimize costs. The GSA deems itself "the federal government's premier acquisition agency" and oversees about $66 billion of spending each year. Interestingly, one of the tasks of the GSA is to develop standards for all products and services that can be purchased for use in federal agencies. Even for products such as furniture, the GSA provides detailed descriptions and specifications of all acceptable types. It is not difficult to realize that all federal agencies need things like filing cabinets; however, it is probably somewhat surprising to learn that the GSA provides a thirteen-page technical description and specification of the types of filing cabinets that are acceptable for use in federal agencies. Here is an excerpt from one section of the technical description, which focuses on two characteristics of approved filing cabinets:

2.5.1 Vertical dividers for drawers and shelves. The material used in the dividers shall be of steel formed or reinforced so that the finished item will have stiffness at least equal to that of an uncoated flat sheet of steel. All exposed surfaces shall be smooth and free of burrs and rough edges. The dividers shall retain both legal and letter size documents adequately without excess overlap and shall be designed to snap securely in the drawers and shelves without the use of tools or attaching devices. When painted a color other than the color specified for the unit, they may be used provided color harmony is maintained. Each cabinet shall have two dividers per drawer or shelf.

2.5.2 Out stops and bumpers. The cabinet drawers and shelves shall have two noise reducing out-stops (one on each side) that prevent drawers and shelves from falling out of the cabinet when they are fully extended. All stops and bumpers shall be installed in a manner to withstand a normal rebound without damage. The use of adhesive as the only method of securing the stops or the noise absorbing material is not permitted.*

*Commercial item description, www.gsa.gov.

The GSA provides other technical descriptions and specifications of stools, chairs (folding style, rotary, and straight), desks, credenzas, and tables, among other types of furniture. However, this is not even the tip of the iceberg; the agency handles everything from paint to copy machines to fitness equipment to vehicles to buildings. In other words, just about anything that federal employees (hundreds of thousands of them) touch, use, work with, or work in has probably been described and specified by the GSA.

Although this may all seem a bit odd, most companies and organizations probably have similar technical documents that describe the characteristics and specifications of products and services that have been approved for use in the workplace. It is not by accident, for example, that hamburgers from McDonald's and Burger King taste pretty much the same regardless of where you buy them; these companies have developed descriptions and specifications that standardize their food in all locations.

Descriptions, then, are a normal part of workplaces. Even though many workers may not directly recognize how their work is affected by technical descriptions and specifications, there are instances when technical descriptions play a more obvious and crucial role in helping workers accomplish their jobs successfully. Consider the following workplace situations that require technical descriptions:

- *Manufacturing:* A lumber company provides customers (i.e., manufacturers of wood flooring) with technical descriptions of all its different woods for sale, including information about their properties, toxicities, and shrinkage rates. These technical descriptions help manufacturing companies efficiently, effectively, and safely make floors from the lumber.
- *Engineering:* A civil engineer samples and tests the soil where a new office building is going to be constructed. The technical details of the soil must be described so that engineers and builders know whether it is safe and feasible to construct a building on that specific site.
- *Computer science:* An electronics firm that develops fuel-cell-powered mobile phones must provide details and descriptions of how the phones work and how they use methanol for power in order to get government approval to produce them.
- *Medicine:* Pharmaceutical companies provide technical descriptions of drugs to both pharmacists and consumers, identifying chemical makeup, appearance, recommended dosage, proper usage, and side effects.
- *Environmental studies:* A research scientist who studies how traffic in large metropolitan cities contributes to global warming writes a technical description explaining the detailed processes by which this interaction takes place, including a visual diagram of the cause-and-effect relationships between traffic patterns and global warming.

In some respects technical descriptions are types of technical definitions, which are discussed in chapter 15. Like definitions, technical descriptions are not always stand-alone documents but are often parts or sections of other documents, such as memos, letters, instructions, manuals, proposals, reports, websites, and presentations. However, unlike many definitions, technical descriptions require writers to describe products, mechanisms, or processes with tremendous attention to physical details—parts, pieces, functions, and features, for example.

It is easy to mistakenly believe that definitions and descriptions are the same thing. For instance, the following might seem like a description of a spark plug:

> A device inserted in the head of an internal-combustion-engine cylinder that ignites the fuel mixture by means of an electric spark.

This tells us a good bit about spark plugs but would make a poor technical description because it discusses only what the spark plug does. It does not provide a description of the spark plug itself, what it looks like, what materials it is made of, what parts make it up, and so on. Such sharp attention to physical details and properties is the reason that technical descriptions are discussed on their own and are treated separately from technical definitions.

This attention to detail requires writers to think carefully about audience needs and the kinds of problems the technical description helps solve. Figure 16.1 is a technical description of a theater at Harvard University; the description derives from a more complex website about all the various theaters on the campus.

This technical description of the Loeb Theatre is for audiences considering the theater for a particular theatrical production. The technical information about capacity, lighting, and sound lets readers know whether this theater is the right place to hold their productions. It even includes a visual diagram of the theater's layout to help audiences visualize the space.

Loeb Experimental Theatre

General Information

- Space Contact(s): Loeb Technical Director, (xxx) 555-1212
- Overall Capacity (fire code): 120
- Recommended Seating Capacity: 80
- Overall Dimensions: 38′6″ by 51′0″(see plans)
- Loeb Technical Director maintains up-to-date space inventory and technical information at his own page.

Set/Props Information

- Largest Door Opening: airlock double doors 12′w × 8′h
- Ceiling: Full grid can support pretty much anything. Height is 16′.
- Floor: Replaceable black masonite. Can be screwed and painted.
- Walls: Black painted concrete/mesh panels. Solid walls can be painted; acoustical surfaces cannot.

Shows in the Loeb Ex have access to the Loeb shop for construction of their sets and props. The grid can be used to suspend set pieces and drops, and permanent movable flats are already in place (16′ × 4′). See plans for further details.

The space is a neutral black box with a narrow balcony and an overhead grid. The hanging hard flats allow you to break up the space or even create a wall across it. The 3′× 8′ audience seating risers in 8″, 16″, 24″, 32″ heights can be arranged in any desired fashion (while allowing for proper fire exits).

Lighting Information

- In-house Dimming: 12 × 2.4 kW
- In-house Circuits: One cable for each dimmer on 3 sides of the grid. Each cable is 50′, enough to reach most points in the grid and on the ground (see plans). All circuits have 3-pin twist-lock (L5-20) with nub in.
- In-house Control: ETC Express 48/96 board w/ color monitor. Houselights run off board, permanently wired to dimmers 91 through 96.
- In-house Instruments/Accessories: 25 × (4.5 × 6), 25 × (6 × 9), 3 × (6 × 12), 20 × (6″ Fresnel). 4 × (6′3 circuit PAR 38 striplights) (all twist-lock)
  2 × 30′cables, 6-20′cables, 6-15′cables, 12 two-fers, various SP-twist and Edison-twist adapters. 10 floor mount stands, various gel frames and gels from HRDC stock, templates and holders from Loeb Technical Director on request, other accessories from ART stock by permission of ART master electrician.
- Hanging Locations: The Ex has a full overhead grid consisting of steel mesh with lighting pipes spaced at 4′ intervals running E-W. In addition, there is a pipe on the S wall, and plugs can be made for the grid track to hang units below the grid. Floor mounts and booms are also possible.

The space itself is a very neutral black box; any set must be created essentially from scratch. Set designers tend to end up hanging lots of things from the grid, so it's worth checking with them to make sure there won't be any interference with lights.

The lightboard is on a custom table with the monitor. This is on a long cable and can be moved around on the balcony for running the show.

continued

FIGURE 16.1 Technical description of the Loeb Experimental Theatre

Sound Information

- Amplifier
- Speaker
- Sound board
- Sources: CD, Minidisc
- Subwoofer and amp

The Ex has a full in-house sound installed in a rolling rack which can be placed anywhere in the balcony. Speakers can be installed anywhere in the space.

Plans

Ground plan, sections and grid plan available in Vector works, DW6, PDF, or printed hard copy from Loeb Techical Director and Loeb website.

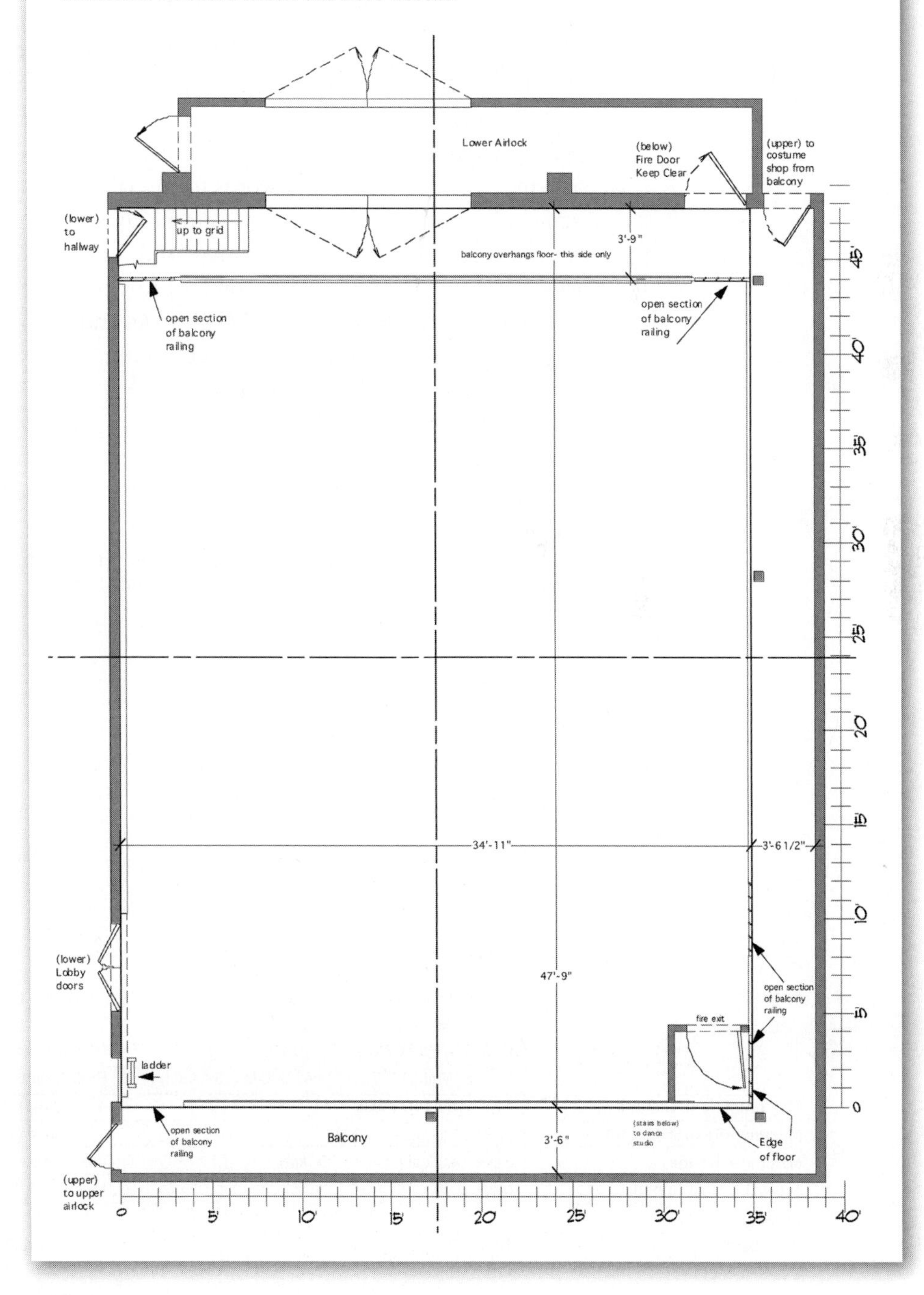

FIGURE 16.1 *Continued*

ANALYZE THIS

Look through Figure 16.1 again. If you were using this document to determine whether the Loeb Experimental Theatre would be appropriate for your theatrical production, which particular pieces of technical information would help you? Explain specifically how and why the information presented in this description would be beneficial. In what ways would it not be helpful?

Technical Specifications

Technical specifications are similar to technical descriptions in many ways, but they do have some distinct differences. Technical specifications typically do not use traditional forms of writing, such as complete sentences and paragraphs. Rather, technical specifications usually rely heavily on visuals—graphs, charts, tables, or illustrations—and include short phrases or words to describe parts or procedures. Like descriptions, however, technical specifications commonly serve as smaller sections of larger documents, such as instructions, manuals, and reports. Figure 16.2 shows technical specifications in the form of a table providing technical information about the characteristics of the Boeing 757 (200 series and 300 series). Notice how the table uses words, phrases, and numbers without relying on complete sentences, paragraphs, and other forms of traditional writing.

Technical specifications are relevant not only for those communicating in workplaces; many are written for audiences needing technical information about products they have purchased. These technical specifications are usually included in manuals and instruction booklets. Figure 16.3 shows the technical specifications of a remote control for a DVD player; it is part of the larger manual that came with the product.

See chapter 19 for more on manuals.

757 technical specification

	757-300 Technical Characteristics	**757-200 Technical Characteristics**
Passenger Seating Configurations		
Typical 2-class	243	200
Typical 1-class	280	228
Cargo	2,370 cu ft (67.1 cu m)	1,670 cu ft (43.3 cu m)
Engines	Rolls-Royce RB211-535E4B - 43,500 lb (193.5 kN)	Rolls-Royce RB211-535E4 - 40,200 lb (179 kN)
	Pratt & Whitney PW2037 - 36,600 lb (162.8 kN)	Rolls-Royce RB211-535E4B - 43,500 lb (193.5 kN)
	Pratt & Whitney PW2040 - 40,100 lb (178.4 kN)	Pratt & Whitney PW2037 - 36,600 lb (162.8 kN)
	Pratt & Whitney PW2043 - 42,600 lb (189.4 kN)	Pratt & Whitney PW2040 - 40,100 lb (178.4 kN)
Maximum Fuel Capacity	11,466 gal (43,400 l)	11,489 gal (43,490 l)
Maximum Takeoff Weight	272,500 lb (123,600 kg)	255,000 lb (115,680 kg)
Maximum Range	3,395 nautical miles (6,287 km)	3,900 nautical miles (7,222 km)
Typical Cruise Speed	Mach 0.80	Mach 0.80
Basic Dimensions		
Wing Span	124 ft 10 in (38.05 m)	124 ft 10 in (38.05 m)
Overall Length	178 ft 7 in (54.5 m)	155 ft 3 in (47.32 m)
Tail Height	44 ft 6 in (13.6 m)	44 ft 6 in (13.6 m)
Interior Cabin Width	11 ft 7 in (3.5 m)	11 ft 7 in (3.5 m)
Body Exterior Width	12 ft 4 in (3.7 m)	12 ft 4 in (3.7 m)

FIGURE 16.2 Technical specifications for the Boeing 757 in a table format

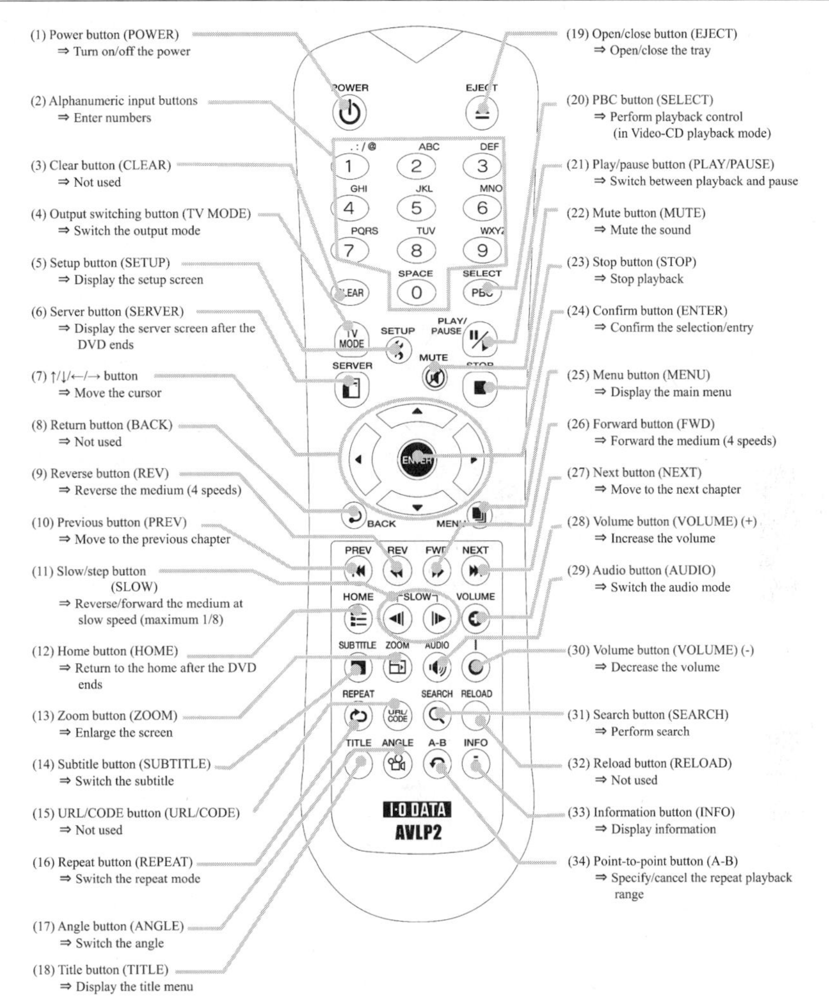

FIGURE 16.3 Technical specifications for a remote control

The technical specifications for the Boeing 757 and the JVC remote control are similar in that they provide a great deal of technical information in a small amount of space. This is one benefit of technical specifications: short phrases and visuals give readers information quickly and efficiently. However, because technical specifications do not use complete sentences and paragraphs, writers must pay close attention to the layout and design of these documents—the key to the documents' success. Word choice is important, but the visual aspects must render the information clearly to readers.

See chapter 9 for more on layout and design.

EXPLORE

Find the technical specifications online for a product that you use regularly, such as a car, computer, phone, or MP3 player. Remember that the specifications might be a section of a longer document such as a manual. After locating the technical specifications, note the way they are put together visually. Is there a table, chart, graph, or illustration? Given the document and its purpose, why do you believe the writer(s) chose to present the information in this visual form? Write your responses down, and be prepared to share them in class.

COMPONENTS OF TECHNICAL DESCRIPTIONS

PSA

The main components found in technical descriptions are not all found in every description, nor are they always found in the order presented here. Workplace writers must know their audiences and their purposes in writing to determine which sections are necessary and which are not. This determination this should take place in the Plan phase of writing.

Introduction

The purpose of the introduction to a technical description is to do one or more of the following:

- Identify the object, product, mechanism, or process to be described
- Discuss what background information the audience needs to know
- Give a general, brief description of the object, product, mechanism, or process
- Provide an overview of the rest of the technical description (if the description is longer than ten pages and has multiple sections, writers should use a table of contents)

In all cases the introduction should be short and to the point.

The Congressional Research Service sent a report to Congress that discusses ricin and functions in large part as a technical description. The introduction to that document (see Figure 16.4) is succinct but lets readers know what is described, provides background information, and forecasts the rest of the document all in just a few sentences.

Introduction

On February 2, 2004, the deadly toxin ricin was detected in the Dirksen Senate Office Building. As of this writing, early indications suggest that the toxin may have been mailed to Senator Frist's offices. All mail sent to government offices on Capitol Hill is sterilized by irradiation. However, this procedure was designed to kill bacteria, such as anthrax, not to inactivate preformed toxins such as ricin. Ricin is often mentioned as a potential bioterror weapon. This report describes what ricin is, how it is made, it's effects, a brief history of it's use, it's potential for use as a bioterror weapon, and how it is currently regulated.

FIGURE 16.4 Introduction to a technical description

Background

Sometimes the thing writers describe is not entirely familiar to audiences, or there might be contextual information that readers need to know to fully understand the description. In such cases writers provide detailed background sections for their audiences. In the article "Knee Brace Use in the Young Athlete," Thomas J. Martin, MD, composes a document that serves as a

> revision of a previous statement on prophylactic knee bracing and provides information for pediatricians regarding the use of various types of knee braces, indications for the use of knee braces, and the background knowledge necessary to prescribe the use of knee braces for children.[1]

---

[1] Thomas.J. Martin, "Technical Report: Knee Brace Use in the Young Athlete," *Pediatrics* 108, no. 2 (2001).

That document has a distinct purpose: to help pediatricians prescribe the correct kind of knee brace for young athletes. To accomplish this task, Martin includes a variety of technical descriptions of knee braces. In addition, even though his audience is highly educated, Martin provides background information (see Figure 16.5) to help his readers make sense of the descriptions and to ensure that the document serves its purpose. A reading of the excerpt in the figure makes it clear that Dr. Martin's document is written for medical experts, yet even experts may still need background information. Don't assume that background information is only for nonexperts.

Background

Pediatricians are appropriately becoming more involved in the care of young athletes. The knee is one of the most commonly injured joints in athletes. The correct care of knee injuries is an important part of any sports medicine or general pediatrics practice and may include the use of braces. Therefore, the pediatrician should be knowledgeable about knee bracing. This statement is an update of a previous statement on prophylactic knee bracing and includes information for pediatricians regarding the use of various types of knee braces, indications for the use of knee braces, and the background knowledge necessary to prescribe the use of knee braces for children. Acute and over use injuries to the knee are seen as a result of participation in virtually all athletic activities. Injuries to the ligamentous structures of the knee in the young athlete are becoming more common. The medial collateral and anterior cruciate ligaments are prime stabilizers of the knee and can be injured when direct or indirect forces are applied to the knee. In a growing child, the distal femoral physis is subject to these same forces and may also be injured. In the skeletally immature child, acute trauma to the knee is most likely to cause injury to these 2 ligaments and/or to the distal femoral physis. Patella subluxation, dislocation, or tracking abnormalities can occur as a result of mechanical predisposition as well as direct or indirect stress to the knee. Cumulative microtrauma or over use can lead to patellofemoral disorders or apophysitis of the tibial tuberosity (Osgood-Schlatter disease), which are common in adolescents.

FIGURE 16.5 Background section of a technical description

Parts and Characteristics

The main section of a technical description is the discussion of parts and characteristics. *Parts* refers to the physical, tangible portions of the thing itself. For example, a desktop computer has the following parts: a monitor, a CPU, a keyboard, a mouse, and a printer. *Characteristics*, on the other hand, refer to describable qualities that are not parts. For example, a computer monitor might be 17 inches wide, and the CPU might weigh 8.7 pounds. Inches and pounds are not physical parts of the object but are characteristics of those parts.

Writers identify and discuss parts and characteristics according to audience needs. A manual that serves as an introduction for new automobile mechanics, for example, might divide an engine into its basic parts: engine block, cylinders, pistons, radiator, spark plugs, and so one. A manual written for advanced mechanics might go into much greater detail, perhaps dividing the radiator itself into its many constituent parts. Writers also describe those parts and characteristics at a level of detail appropriate for audience needs.

In addition, technical descriptions sometimes describe the functions of those parts and characteristics—how they work and/or how they are put together. Audiences sometimes can make better sense of the description if writers explain what the part does as well as what it looks like. Technical descriptions commonly describe physical characteristics, explain functions, define terms, and discuss relationships among the parts, all within a single description. These discussions of parts and characteristics use phrases, sentences, and/or paragraphs to describe the subject in as much detail as the audience needs. Visuals such as tables, charts, graphs, and illustrations are usually included to help readers.

In "Knee Brace Use in the Young Athlete," Dr. Martin describes the parts and characteristics of many different kinds of knee braces, including one in particular called a knee sleeve. Figure 16.6 gives the parts and characteristics section that describes the knee sleeve. Notice that the document includes descriptions of the possible parts of the knee sleeve: its nylon cover, movable straps, lateral hinge, infrapatellar band. And the passage describes some characteristics of a knee sleeve, such as being expandable, made of neoprene, and perhaps having a buttress of varied shapes. In addition, this section discusses the purpose and function of a knee sleeve—increasing warmth, providing compression, and so on. The description tells the audience what the knee sleeve does and why it is used.

Knee Sleeves

Knee sleeves are expandable, slip-on devices usually made of neoprene with a nylon cover. They increase warmth, provide even compression, and may enhance proprioception. Knee sleeves may provide a feeling of support to the knee. Plain knee sleeves may be used to treat postoperative knee effusions and patellofemoral syndrome. Used in this capacity, the purpose of a knee sleeve is to decrease knee pain. When a knee pad is added, it provides protective cushioning to the patella and anterior knee.

The knee sleeve may be modified to include an opening for the patella, or more movable straps, or a buttress. The buttress may be circular, C-shaped, J-shaped, or H-shaped. With these modifications, the knee sleeve is often referred to as an extensor mechanism counterforce brace and is used to treat patellofemoral joint disorders, including patella subluxation, patella dislocation, and patellofemoral syndrome, all of which are very common in athletes. The pathophysiology of patellofemoral syndrome is unclear, but it has been postulated to occur as a result of abnormal tracking of the patella on the femoral trochlear groove. The knee sleeve helps compress the tissue and limits patella movement. The extensor mechanism braces are designed to apply a medially directed force to the lateral patella, thereby improving patellofemoral tracking and decreasing the likelihood of lateral patella subluxation or dislocation. Used in this capacity, they may be of benefit in the athlete with an unstable patella. These braces may also contain a lateral hinge that incorporates an extension stop.

When a strap is placed inferior to the patella, it may be used to treat Osgood-Schlatter disease and patellar tendonitis. This infrapatellar band is used to decrease the traction forces at the tibia tuberosity for patients with Osgood-Schlatter disease and on the patellar tendon for patients with patellar tendonitis.

FIGURE 16.6 A portion of a parts and characteristics section of technical description

Visuals

Despite all that a good description can do for its intended audience, it may also need visuals to help readers understand more fully and clearly. Technical descriptions, like other workplace documents, incorporate visuals for many reasons, as discussed fully in chapter 8.

In the case of the knee sleeve description, the writer is providing readers with information about an object that might be hard to envision, as evidenced in the passage in Figure 16.6. Providing an illustration of knee sleeves not only provides an example of written information but also clarifies a difficult description and helps the audience retain an understanding of it. Figure 16.7 shows a table with illustrations that let readers see the different kinds of knee sleeves.

Most technical descriptions, and all technical specifications, require visuals of one kind or another, even if it is simply a chart, table, or graph to organize information. The kind of visuals, their complexity, and the number included depend on audience needs and the purpose of the document itself.

Brace Category	Indication	Comments
Knee Sleeves†		
Plain sleeve	Postoperative knee effusions; patellofemoral syndrome	Insufficient for treatment of an unstable knee. Should only be worn during sports activities if swelling occurs. Simple to fit and inexpensive.
Sleeve with knee pad	Protection and padding of the anterior knee	
Sleeve with buttress	Patella subluxation, patella dislocation; patellofemoral syndrome	Improves patellofemoral tracking.
Sleeve with strap	Osgood-Schlatter disease; patella tendonitis	Decreases traction forces on the tibia tuberosity and patella tendon.

FIGURE 16.7 Visuals used to clarify a technical description

ANALYZE THIS

Return to Figures 16.5, 16.6, and 16.7. Compare your understanding of the knee braces in the textual descriptions with your understanding of them when visuals were provided. In what ways did the visuals enhance your understanding? Do you think the intended audience for the document could have gotten the necessary meaning from just the text? In what ways are the visuals unhelpful, or in what ways might the visuals be rendered in more effective ways? Explain your answers in as much detail as possible.

COMPOSING TECHNICAL DESCRIPTIONS

Workers and consumers rely on technical descriptions to learn and perform tasks, but the purpose of technical descriptions typically is not to instruct someone to do a task. Technical descriptions can help readers by describing the parts and characteristics of a product, for example, but that is not the same thing as explaining to readers how they themselves can assemble or use the product.

Composing technical descriptions requires writers to recognize that these documents help audiences solve problems, thus making the problem-solving approach quite relevant. Writers must plan to meet the needs of the audiences who will read their technical descriptions; in particular, writers must know the level of technical sophistication the audiences hold. For example, general readers of a technical description of a small lawn mower require easy-to-read and uncomplicated data, quite different from the technical description of a large dam written by one engineer for another. Writers must also remember the purpose of the description and the ways audiences will use it. Some descriptions that are part of owner's manuals, for example, are used by readers to understand products such as DVDs and computers. Other descriptions that are part of a technical report may be used by audiences to simply understand a process, such as how and why earthquakes happen in California.

In addition, writers of technical descriptions often must conduct research to complete their documents. For instance, a physician writing technical descriptions of different knee braces for young athletes might offer his expert opinions on the subject matter, but he must also show how these opinions correlate with beliefs and ideas in the wider medical community. Thus, his technical descriptions and specifications will result in part from research about knee braces. And like other workplace writing, technical descriptions benefit from the same PSA focus on drafting, reviewing, and distributing. Some specific suggestions for writing technical descriptions follow.

Audience Analysis

This chapter and others in the book have continually asked you to consider audience needs when writing any workplace document; we tell you to always know your audience and, in the case of technical descriptions, provide the descriptive detail that the audience requires. Nonetheless, some writers mistakenly think that complex objects or processes necessarily require more detailed description than simple ones require. In fact, a seemingly simple product might require a tremendous amount of description, whereas an extremely complex mechanism might require only a short, general description. It all depends on what the audience needs.

For example, a governmental agency like the Food and Drug Administration (FDA), which inspects and approves new food products, might require lengthy and complex descriptions (e.g., hundreds of pages) of a restaurant chain's new line of organic menu choices; consumers, on the other hand, might want only short paragraph-long descriptions. The amount of detail a writer uses in technical descriptions is not determined by the complexity of the object or process; it is determined by what the audience needs.

Descriptive Detail

Writers of technical descriptions must determine the specific descriptive detail to provide for readers. Appealing to their five senses usually means describing according to characteristics. The following is a list of characteristics to consider when writing technical descriptions:

color
height
width
length
depth
weight

shape
texture
location
pattern
ingredients
materials

age
finish
temperature
moisture content
smell
sound

This list is certainly not exhaustive, nor is it arranged in any order of importance. And of course, few descriptions would use all of these characteristics. Instead, writers choose characteristics based on what they are describing and which categories best help readers understand. In most instances readers need descriptions that appeal to their sense of sight and touch, although there are times when taste, smell, and hearing are important. For example, technical descriptions of chemicals might include information about smell.

Well-written technical descriptions provide sensory detail for audiences but also include visuals to help audiences further understand the information. No matter how much descriptive detail a writer uses, visuals can always help ensure clarity. Figure 16.8 is a passage from a field guide about requiem sharks, and Figure 16.9 is a visual from the same field guide. Notice that the passage is full of details about this shark species, yet the inclusion of the visual helps readers better understand the information by letting them see illustrations of the shark and its parts.

> Members of Carcharhinidae [a species of shark] are variously distinguished by the presence of precaudal pits; lack of spiracles (present on tiger sharks and occurring rarely on lemon sharks; Compagno, 1988); bladelike teeth with single cusps; first dorsal fin origin usually above pectoral fin or slightly posterior to pectoral fin inner corner (except on the blue shark with the dorsal fin base midpoint closer to pelvic fin origin than pectoral fin axil); second dorsal fin smaller than first dorsal fin and above anal fin (second dorsal fin and first dorsal fin almost equal size on lemon sharks); fifth gill slit over or posterior to pectoral fin origin; no fleshy keels along sides of caudal peduncle (except on tiger sharks and blue sharks); well-developed nictitating membrane along eye socket lower margin.
>
> Freshly caught carcharhinids have a variety of body colors that are often muted color mixed with gray tones dorsally, and pale yellow or white laterally or ventrally; the exception is the blue shark which is counter shaded brilliant blue above and white below. Juvenile tiger sharks have distinct mottling and vertical bars that often fade with age; Atlantic sharpnose sharks (*Rhizoprinodon terraenovae*) often have light spots along the body. Juveniles for most species have fins tipped dusky or black. Adults generally have dusky fin tips; however, in the western North Atlantic Ocean blacktip sharks (*Carcharhinus limbatus*) greater than 80.0 cm TL have black fin tips except for a pale or white anal fin (Branstetter, 1982); all fin tips are black on spinner sharks (*Carcharhinus brevipinna*) greater than 80 cm TL (Branstetter, 1982) ; oceanic whitetip sharks have white fin tips on the first dorsal fin, pectoral fin, and caudal fin.
>
> Sharks that may possibly be misidentified as belonging to Carcharhinidae include species of Odontaspididae (sand tiger sharks), and Lamnidae (mackerel sharks). Members of these families have all five gill slits anterior to the pectoral fin origin; sand tiger sharks have short pectoral fins and large pelvic fins; mackerel sharks have lunate caudal fins (upper and lower caudal fin lobes of almost equal length) and well-developed caudal keels that extend past the caudal fin origin.

FIGURE 16.8 A technical description of requiem sharks

ANALYZE THIS

Choose an object that is located near you at the moment: a pencil, a cell phone, a book, or a piece of electronic equipment. Using sensory detail, write a complete description of this object. Focus only on its parts and its characteristics, rather than on what the thing does or is used for. You might consult the list of characteristics in this section to help you decide which aspects are most important.

Organization

Like other workplace documents, technical descriptions use an organizational pattern suitable for the subject matter and audience needs. Chapter 7 covers many organizational patterns in detail, but technical descriptions all tend to use versions of the division strategy. As you may recall, this strategy is based on the idea that some things can best be understood as a series of smaller parts; the division strategy allows a writer to divide and even subdivide an object into its various parts. There are three variations of division patterns that workplace writers use for technical descriptions.

See chapter 7 for more about organizational strategies.

General to Specific

This organizational strategy can be used in two ways: the first provides descriptions that progress from general to specific information; the second moves from specific to more general. Purpose and audience help determine which is more appropriate. A technical

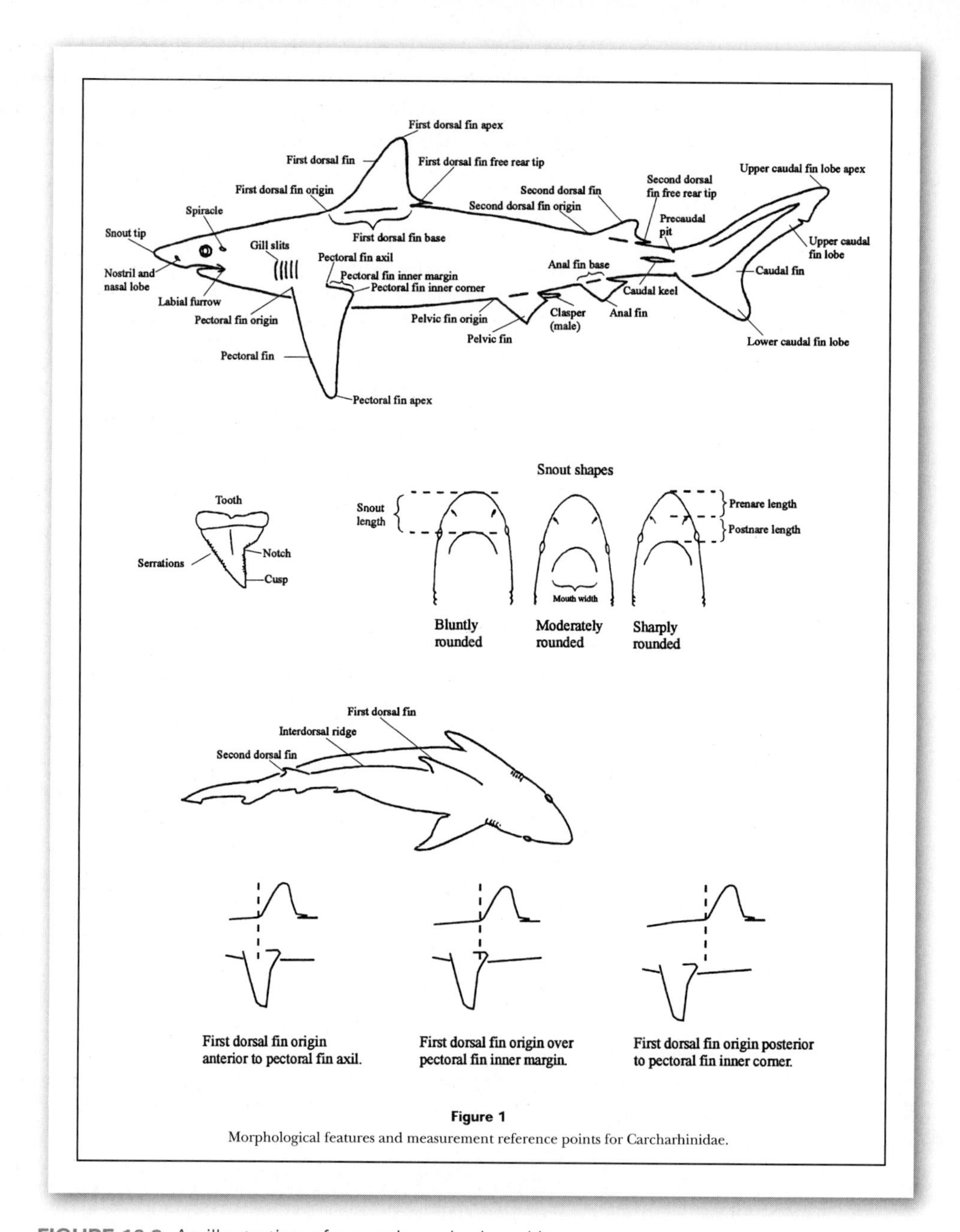

FIGURE 16.9 An illustration of a requiem shark and its parts

description of a complex telescope, for example, might first provide a general overview of the object, then move into detailed descriptions of its specific parts, and finally provide descriptions of smaller pieces that make up those parts, all the way down to individual screws. On the other hand, a technical description can work in the opposite order. For example, a proposal to build a new shopping center might begin by describing the immediate site the shopping center would occupy. Writers might then describe outlying areas and finally the region in general to further explain their reasons for the site selection.

Spatial

The spatial strategy helps readers navigate information pertaining to physical objects or places; it is useful in technical descriptions that detail relatively large objects or places. For example, a technical description of an airplane might detail the different parts by moving the reader from the front of the plane to the back, or from the bottom to the top. Or a technical description of a building might detail the different rooms or areas by moving from the front of the building to the back, the bottom to the top, or the left to the right. In short, spatial organization is appropriate when writers must discuss parts and pieces as they relate to each other in space. The diagram of the requiem shark in Figure 16.9 relies mainly on spatial organization.

Chronological

A chronological pattern guides readers through a sequence in time, moving from start to finish. Some technical descriptions describe processes—why events happen and/or how things work over time. For example, technical descriptions using a chronological pattern might detail how a complex mechanism like a dam or a motor works or how and why a natural process like photosynthesis or a thunderstorm happens. Figure 16.10 is a technical description of how electricity is produced. Notice how the description takes readers through a step-by-step chronological organization.

How Electricity Is Produced

It's hard to imagine our homes without electricity. There would be no TV, computer, or video games. You'd have to do your homework by candlelight or oil lamps. You wouldn't be able to listen to your favourite bands on the radio or CD player—instead you'd have to make your own music!

But what is electricity?

Electricity is a form of energy that starts with atoms. You can't see atoms because they're too small, but they make up everything around us. There are three parts to an atom: protons, neutrons and electrons. Electricity is created when electrons move from atom to atom. There are a number of ways to make electrons move, but most electricity is produced at power plants.

How do power plants work?

It all starts with a source of power. At Yukon Energy, we use water to create most of our electricity. That's why most of our plants are called hydroelectric facilities: hydro means water. Power plants that use water to make electricity are built near rivers. Our Whitehorse plant, for instance, is on the Yukon River. Dams are built across rivers to hold back the water. The water is then directed through big pipes and it falls against the blades of giant turbines. The turbines have blades on them that turn when the water hits them, just like the blades of a pinwheel turn when you blow on them. Once the water hits the blades, it returns to the river. The turbine blades are attached to a big metal rod, and at the end of that rod are large magnets. When the blades turn, they make the rod and the magnets spin very fast. The magnet end is surrounded by heavy coils of copper wire, and the spinning magnets cause electrons in the wire to begin to move, creating electricity.

What happens to the electricity after that?

It moves through wires into what's called a power transformer. The electrical voltage (the strength at which electricity flows) is fairly high, and the transformer makes it even higher to help it flow through wires called transmission lines. Those wires are attached to wooden or metal poles that you see along roads and throughout communities. All the wires are made of metal—usually aluminum or copper. That's because metal is a good conductor—electricity travels through it easily. By the way, water is also a good conductor, and because our bodies are mostly made of water, electricity can travel through us easily. That's not something we want to happen though, because if we have electricity going through us we'll likely be seriously hurt or even killed. That's why grown-ups warn you to stay away from high voltage sites and not to stick your fingers in a wall plug. Electricity travels fast—about 310,000 kilometers per second! If you moved that fast, you could probably make several trips around the world in the time it takes to turn on a light! Sometimes, when electricity has to travel a long way, it gets a little weaker as it moves along the lines. It needs a boost (like you need food to replace the energy you've burned after playing outside all day). That's where substations help. Substations are large box-like power transformers that sit in fenced-in areas. You'll see signs on the fences that say high voltage—stay away, and it's really important that you obey those signs (remember what you read about electricity being able to travel easily through your body).

How does electricity get into my house?

When wires reach your house, another transformer on the power pole makes the electricity just the right voltage so you can use it safely. The wire is connected to a meter box that keeps track of how much electricity is being used. There are wires in your house connected to plugs, also called outlets. These outlets let you plug in your boom box, television set, or anything else electrical. What an amazing journey electricity takes to get to your home!

FIGURE 16.10 A technical description using chronological organization

ANALYZE THIS

Compare the styles and language used in Figures 16.6 (about knee sleeves), 16.8 (about requiem sharks), and 16.10 (about electricity). Who are the audiences for each of these three documents? From what you can tell, what is the purpose of each? How does the writing style of each of these three contribute to your understanding of the audience and the purpose of each? Discuss your answers to these questions with a small group of classmates.

Headings

Technical descriptions often use headings and subheadings to designate individual sections; for example, introductions, background sections, and body paragraphs might require headings. In many cases technical descriptions themselves are sections in longer documents, such as reports or proposals, which may have sections in their bodies titled, for instance, "Description of Building Site" or "Technical Description of X-578A Laser Printer." No matter where writers place their technical descriptions, they usually describe each part or piece of the object, product, mechanism, or process in a separate paragraph or section, and those paragraphs or sections are usually marked with a descriptive heading.

Parts Lists

Parts lists are sometimes used in technical descriptions to itemize the parts of an object, product, mechanism, or process. Writers whose technical descriptions detail many different parts (ten or more) may want to include a list in their introduction to let readers know what lies ahead. Readers probably would like to know ahead of time if the description they are reading is going to take them through the details of many parts.

ETHICS AND TECHNICAL DESCRIPTIONS

Even though technical descriptions are usually smaller segments of larger documents, writing technical descriptions entails a particular set of ethical obligations and responsibilities.

Objectivity

Ethical technical descriptions require objectivity. Those who write technical descriptions should not try to persuade audiences that the object, product, mechanism, or process being described is amazing, indispensable, or intimidating. In short, writers should not make value judgments about the thing being described. Don't try to sell it, and don't try to degrade it. Describe it in as much detail as the audience needs, but do so in objective, unbiased ways. For example, in describing a suspension bridge, you should not say that it is a dramatic and inspirational architectural achievement, but you might mention that the bridge spans a 200-foot-long gorge by utilizing two 77-foot-high arches that connect cardinal red cables to the roadway. Let readers decide whether the bridge is dramatic and inspirational.

Alerts

As mentioned earlier, technical descriptions simply describe and should not provide readers with any kind of instructions or procedures to follow. Consequently, audiences should not be at risk of damaging anything, including themselves. In some cases, however, writers may decide to include special notices about the things they are describing if those things are potentially dangerous. Such alerts could be either words and phrases or visuals, such as warning signs. Alerts in technical descriptions are not nearly as crucial as they are in instruction booklets and manuals, but extremely hazardous objects or processes may warrant special notices, at the writer's discretion.

Usability Testing

Some technical descriptions require usability testing to ensure accuracy. Because technical descriptions range from one-page memos to several-hundred-page reports, the extent of usability testing must depend on the length of the technical description, its purpose, and its audience. A technical description of a large corporation's computer network, which is sent out to thousands of employees, might necessitate usability testing to ensure that audiences understand the information. Because technical descriptions are so often part of other documents, it is more typical that the entire document is usability tested, rather than just the technical description itself. Nonetheless, writers are ethically responsible to determine whether usability testing of their technical descriptions is necessary to ensure that readers can use the descriptions safely and effectively.

See chapter 11 for more on usability testing.

IN YOUR **EXPERIENCE**

Recall three or four documents—either print or electronic—that you have come across in the past and that describe products. What are the products, what are the purposes of the descriptions, and who are the audiences? For each of the products, decide whether the description is objective. What in particular seems to make it objective or unobjective? Explain.

SUMMARY

- A technical description provides concrete details, precise words, and visuals to describe objects, products, mechanisms, or processes to particular audiences.
- Technical descriptions answer questions such as these: What is it? What does it look like? What does it do? What is it made of? How does it work? How has it been put together?
- Technical descriptions are often parts or sections of other documents.
- Technical specifications usually rely heavily on visuals—graphs, charts, tables, or illustrations—and use words or short phrases to describe parts of the product or process.
- Technical descriptions help audiences solve problems.
- The introduction to a technical description attempts to do one or more of the following:
    - Identify the object, product, mechanism, or process to be described
    - Discuss what background information the audience needs to know
    - Give a general and brief description of the object, product, mechanism, or process
    - Provide an overview of the rest of the technical description
- Writers provide detailed background sections when the subject is not entirely familiar to audiences or they might need contextual information to fully understand the description.
- Parts are the physical, tangible portions of the thing itself; characteristics are the describable qualities that are not parts.
- Technical descriptions incorporate visuals for many reasons.
- Writers of technical descriptions provide descriptive detail that appeals to readers' five senses.
- Technical descriptions use a division organizational pattern, as well as others that suit the subject matter and audience needs, including general to specific, spatial, and chronological.
- Writing technical descriptions entails a particular set of ethical obligations and responsibilities.

CONCEPT REVIEW

1. What are the purposes of technical descriptions and technical specifications?
2. What are the differences between technical descriptions and technical definitions?
3. Why is attention to audience and purpose so important in writing descriptions?
4. Why would writers incorporate technical descriptions and specifications into other documents, such as proposals, instructions, manuals, and reports?
5. What purposes do introductions to technical descriptions serve?
6. What purposes do background sections of technical descriptions serve?
7. What is a parts and characteristics section of a technical description?
8. What are the differences between parts and characteristics?
9. Why are visuals necessary in most technical descriptions and specifications?
10. What is sensory detail, and how much is necessary in descriptions?
11. What are some typical organizational patterns used in writing technical descriptions?
12. How and when do writers use headings and parts lists in technical descriptions?
13. What are the ethical concerns of those who write technical descriptions?

CASE STUDY 1

Researching and Writing Vehicle Descriptions and Specifications for the General Services Administration

You were recently hired in an entry-level position at the General Services Administration (GSA) to help collect data and determine standards for federal vehicles. Your supervisor informs you that the GSA is looking into purchasing a fleet of new vehicles for rangers in the national parks. Even though the terrain in the various parks is highly diverse, the GSA would like to standardize these vehicles as much as possible. To begin the analysis, your supervisor requests that you develop both technical descriptions and specifications for the most recent versions of the following vehicles: Chevy Tahoe, Ford Explorer, GMC Yukon, Mercury Mariner, and Jeep Wrangler. Your supervisor tells you that you need to focus only on those aspects of the vehicles that are most relevant to a ranger's work. Thus, things like power, safety, and ground clearance are most important; you should think carefully about a ranger's work in order to determine other facets of the vehicles that would be relevant.

You will need to conduct research on these vehicles in order to compose your descriptions and specifications, keeping in mind that your immediate audience is your supervisor, who will include your findings in the body of a larger report that will go to the decision maker who will decide which vehicles are to be purchased. Remember to include relevant visuals in your work and to cite properly any sources that you have used in developing your descriptions and specifications. Also remember that descriptions and specifications need to be objective and unbiased. Because the websites and brochures that advertise vehicles are usually anything but objective and unbiased, you must be careful to avoid the language of the automobile makers.

CASE STUDY 2

Descriptions and Specifications of Beef Quality at Constellation Foods

You have recently been hired in the public relations department at Constellation Foods, one of the nation's largest producers of steaks and other beef products for restaurants and grocery stores. The company produces everything from fresh steaks to frozen hamburgers to canned beef products. At this point in time, Constellation Foods wants to provide its customers with more information about its products and plans to develop a website that discloses that information. You are part of a team that will work to do this. Your supervisor gives you the initial task of researching information about the eight different quality grades of beef established by the United States Department of Agriculture (USDA). The long-term goal is to show how all of Constellation's products fit within these different quality grades, but for now you are simply to collect information about these different grades and compose descriptions and specifications for the beef that falls within each.

In order to complete this work assignment, you'll need to research beef quality grades, get a sense of the terminology used in describing beef, and develop language that will make your descriptions clear and understandable. In addition, you will develop specifications for each quality grade that make use of relevant visuals. Your supervisor reminds you that this information will ultimately be part of a website that is geared to all customers, so it should be written in language that is clear and concise for nonexperts; all jargon and technical terms should be defined for the audience. You are to submit your descriptions and specifications using layout and design principles that make the information clear and readable. Your work will then be reviewed by your supervisor and integrated with other information on the website.

CASE STUDIES ON THE COMPANION WEBSITE

10-Megapixel SLR Cameras for Interstate Insurance Adjustors

In this case you serve as a regional claims director for an insurance company. You are given approval to purchase new cameras for insurance adjustors in your office (with a $66,000 budget). This case asks you to choose the best camera for your staff's needs and inform staff members about the technical aspects of this camera.

Hampton Place: New Electronics for Elderly Residents

This case revolves around an apartment manager's need to explain a new security system to elderly residents. In particular, this case asks you to analyze this audience, comprised mainly of the elderly, in order to provide them with technical descriptions that they can easily and accurately use.

Mid-Texas Power and the Search for Fuel-Efficient Automobiles

As the director of transportation for Mid-Texas Power, you are asked by the company president to investigate the purchase of new vehicles for company employees, although the president gives you few parameters to help guide your investigation. You are asked to consider many different types of vehicles that would suit employee needs, look closely at the rhetorical situation surrounding this particular problem, and draft technical descriptions that would help you and the company make decisions about which vehicles to purchase.

Choosing the Right Bike for the Job

You have recently been hired by a bicycle rickshaw company. However, there is a small problem: the company doesn't own any bicycles yet. You must help your boss research and decide which bicycle best suits the company's needs.

VIDEO CASE STUDY

GlobeShare Wireless

Writing a Description of the GlobeTalk Messaging System

Synopsis

In this scenario Sheryl Malnip, GlobeShare's financial manager, explains that a group of investors are interested in the GlobeTalk Messaging System. Before the investors agree to provide monetary support to further develop and market the system, they want a written description of its functions and uses. After some discussion Don assigns Mike and Amber to write a description of GlobeTalk aimed at these potential investors.

WRITING SCENARIOS

1. Choose an object or a product to write a technical description about. This can be an everyday object such as a pen, a cell phone, or a laptop computer, but it should be an object that you have access to. Compose a one- to two-page technical description of the object. When you are finished, get together with two or three classmates, and share your description with them, as well as the actual object itself. Discuss the strengths and weaknesses of the description, make any necessary revisions, and then submit your revised description to your instructor.
2. If you were to create a visual(s) for the object or product you wrote about in the previous scenario, what kinds of visuals would be most appropriate for an audience with little knowledge or background information about this object or product—tables, graphs, charts, or illustrations? Provide details about what appropriate visuals would look like and what information they would need to contain.
3. Make a list of ten specific kinds of documents that you have encountered in the past that included technical descriptions and/or technical specifications. What were these documents? What were their purposes? Who were their audiences? Why did these particular documents need technical descriptions and/or specifications? Provide your responses in a document type assigned by your instructor.
4. Through Internet search engines or library databases, do some historical research on technical descriptions. What were some of the first examples of technical descriptions? Who wrote them and why? Were the examples relatively recent or very old? In what ways do these first technical descriptions differ from those written today? How are they similar? Provide your responses in a document type assigned by your instructor.
5. Find at least three different sources that discuss how the water purification process reverse osmosis works, and write your own technical description of the process. Assume that your audience is nonscientific and knows nothing about this process.
6. Think of three technical descriptions that would warrant special notices to audiences. What are these technical examples, and why would writers have an ethical obligation to include a special notice? Either in class discussion or in a letter to your instructor, explain your answers.
7. Review the sections of the report on "Knee Brace Use in the Young Athlete" (Figures 16.5 and 16.6). Given the purpose and the audience of this particular document, what kinds of ethical responsibilities did this writer need to consider? Remember that chapter 4 is devoted entirely to ethics.
8. Review the different sources you found for reverse osmosis (in Writing Scenario 5), or locate three new sources. Which of the three sources seems most ethical? Which seems least ethical? Explain your answers. Consider the purpose and the audience of each of the sources, and decide whether these factors play into your assessment of the ethics of each source. Provide your responses in a document type assigned by your instructor.

9. Compose a technical description of the classroom in which your technical communication class meets (or whatever class requires you to use this textbook). Include both parts and characteristics of the classroom, as well as any visuals—tables, charts, graphs, or illustrations—to aid in the description. Compose this description for an audience unfamiliar with the classroom.
10. Compose technical specifications for the classroom in which your technical communication class meets (or whatever class requires you to use this textbook). Before starting this assignment, be sure to review the differences between technical descriptions and technical specifications.
11. Choose any nonelectronic object or product that you are familiar with: perhaps a golf club, a fishing rod, a stapler, a mechanical pencil. Compose a technical description of the object or product, and include the description in a memo to your instructor. Be sure to think about the different sections of technical descriptions, their organizational patterns, and their use of descriptive detail. Your instructor may already know a bit about the object or product, so be sure to provide detail appropriate for the audience.
12. Note the exact make and model of the computer you regularly use, either at home or at school. Conduct an Internet search to find information about this particular computer; visit at least four or five websites that contain information. Detail your findings in a letter to a classmate. Did you find any technical descriptions or technical specifications for the computer? What in particular made the information you found a technical description or technical specifications? If you did not find a technical description or specifications, what did you find? Why is or is not the Internet a good and reliable place to find technical descriptions or specifications of personal computers? Explain your answers in detail.
13. What two or three technological devices do you currently own (or want to own) that require technical descriptions and specifications? What is it about these particular technological devices that necessitates technical descriptions and specifications for owners? In other words, why would you need such information? Be prepared to explain your answers during a class discussion.
14. Think about the technological product that you most frequently use: a phone, a video game system, a PDA, a computer, an MP3 player, or anything else that you use daily and know how to use well. Without looking at already-written technical descriptions, write a thorough technical description of this product. Then in a short memo or letter to your instructor, explain why you could or could not write a thorough technical description of the product. What about the product and your understanding of technical descriptions would allow you to write it or would constrain you from doing so? What steps would you need to take in order to start the description? What would be easiest and most difficult about writing the description of the product?

PROBLEM SOLVING IN YOUR WRITING

Think back to the case study or any of the writing scenarios you were asked to complete for this chapter. Each one asked that you solve or think through a particular kind of problem. Which particular steps in the problem-solving approach did you find yourself using? During your work on the case study or the writing scenario, were you consciously using the PSA or not? As well as you can, trace the steps you took to complete the assignment, and explain how they did or did not correspond to those in the PSA.

17

Websites and Online Environments

CHAPTER LEARNING OUTCOMES

After completing this chapter, you will be able to do the following:

- Understand the **function of websites** as workplace documents
- Understand the differences between print documents and **web documents**
- Understand the differences between **intranet and Internet** web pages
- Understand and use basic **terminology and concepts** associated with writing web pages
- Recognize various **web technologies** and their applications
- Consider the ethical implications of designing web pages, particularly issues of **accessibility**
- Use standard **guidelines** for writing web pages
- **Plan and organize** a professional web page

DIGITAL RESOURCES

On the Companion Website www.prenhall.com/dobrin:

- Case 1: On the Intranet: Mildred Jackson and the Internal Job Postings
- Case 2: Playing Well with Others: Cole Blackburn and Creating a Children's Website
- Case 3: Website Comparisons: Ben Bryson and the Competition's Website
- Case 4: Optimizing a Website for Search Engines
- Video Case: Revising the Cole Engineering Website
- PowerPoint Chapter Review, Test-Prep Quiz, Exercises and Activities

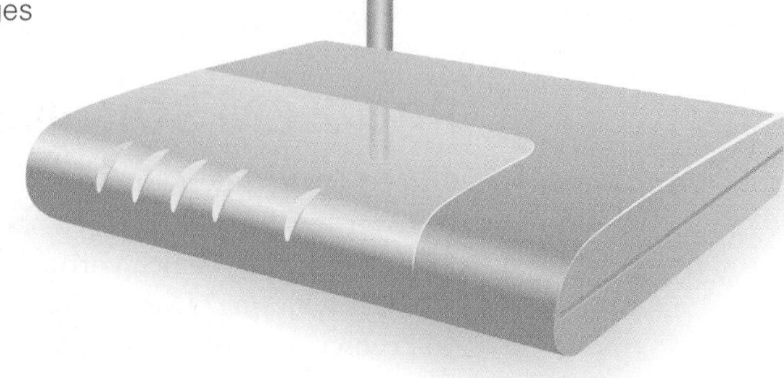

REAL PEOPLE, REAL WRITING

MEET JEN DEHANN • Instructional Designer for Adobe Systems, Inc.

What types of writing do you do in your workplace?

I've been writing technical documentation and online articles about Adobe's web and video products for over two years. I'm involved in creating online and in-product tutorials, articles, and samples for a variety of products, making sure our customers have accurate and appropriate instructional resources and our instructional strategy is consistent across products.

Can you describe an important document you recently created at work?

I recently created an online beginner's tutorial on Flash for the Adobe Developer Center. The goal was to make the tutorial as simple as possible, covering what the user absolutely needs to know to start using the software and product documentation, while not assuming any prior knowledge.

Who are the typical readers of your workplace documents, and how do different audiences influence the style and content of your writing?

I have primarily written for beginner ActionScript coders and new users of Flash. It's important to understand what your users do and don't know since it's often difficult for them to find missing information because of a lack of tool vocabulary and community experience. It's also important to link related information.

What advice do you have about organizing various pages on websites or online documents?

Writers need to think about what experience users have when they first use a website or software and how they find information to solve a problem. From that starting point, try to create a good learning path through the documentation, website, or tutorials. Good cross-references, tables of contents, subject organization and hierarchy, and searchable terminology are important.

How important is collaboration in the websites you create?

All of the work I have done has involved collaboration. Articles, tutorials, and documentation sections are reviewed, edited, and proofed by several individuals. Engineers review the documentation technically and sometimes online articles as well. Feedback from outside the company is very important. I have a blog where I sometimes run surveys or ask for feedback on articles or parts of the documentation, and several members of our team attend industry conferences to speak with users and get feedback on what we produce.

What advice do you have about writing for the Web for readers of this textbook?

The most important thing is to interact with your audience, online or in person. Similarly, go to where the users are. One of the most useful tools I have is my forums, which are focused on the topics I write about (web and video). Reading questions and seeing what common issues the users have in their day-to-day work is incredibly valuable.

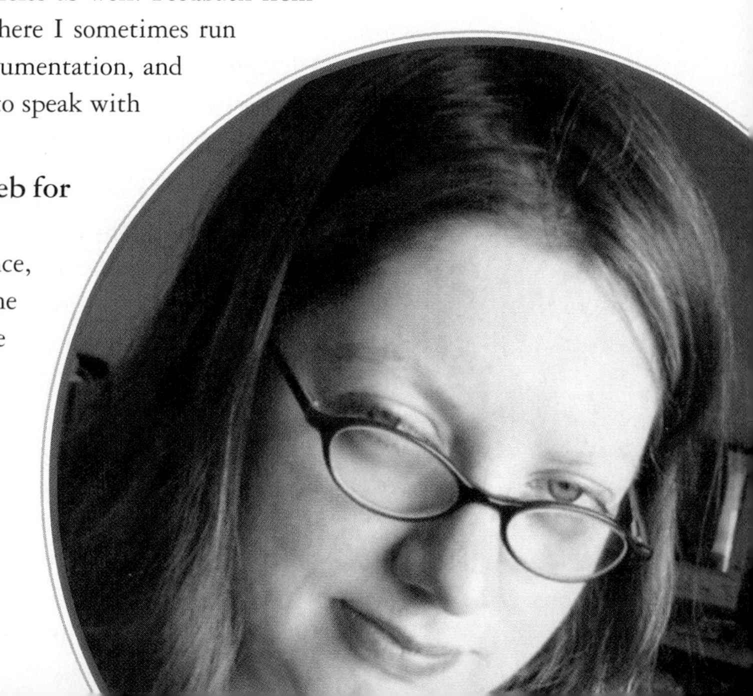

INTRODUCTION

In the past, creating websites in technical and professional workplaces was usually left to the techies—the individuals in information technology (IT) departments or at a company that specialized in website creation and maintenance. Today, the ability to create and maintain websites is increasingly important for professionals of all sorts. In fact, in some technical fields it is expected that new employees will know how to write for the Web when they are hired. Even those who take no part in the actual creation of a workplace website may be asked to determine what sort of information is contained in a site, how it should be organized, and even what types of visuals and design elements best represent the company. Knowing how to create a website—or at least understanding the basic principles involved—is becoming an essential part of solving workplace problems. This chapter explains the basic elements of websites and shows that website creation is within your capacity as a workplace writer.

Websites, like other genres, are documents; they rely on words and visuals to convey ideas and information to readers. In fact, referring to web "pages" is related to our thinking about them like print pages. Many of the strategies of effective textual and visual communication discussed throughout this book are equally applicable to either print or electronic-based documents; in fact, in today's workplace it can be difficult (and often pointless) to distinguish between print-based documents and those written for the Web. However, despite the similarities there are some differences you will need to take into consideration, and we discuss those here.

See chapters 8 and 9 for more on design elements that are applicable to web pages.

Because websites are documents, the process of creating them is similar to the process of creating other types of workplace writing. We encourage you to use the problem-solving approach to create websites, just as you would use it to create other forms of written communication.

- *Plan* for a website by determining your goal or purpose in creating it as well as the audience you'll address.
- *Research* the problem or topic to determine the information you'll include in the website.
- *Draft* and design the website, organizing the pages, text, and graphics to communicate the message effectively.
- *Review* and edit the website after soliciting feedback from peers, colleagues, or other interested individuals.
- *Distribute* the website by publishing it online and then notifying the intended audience through appropriate channels. Thereafter you'll need to continually maintain, refine, and update all aspects of the website.

To see the problem-solving approach in action, consider the website created by Kasun Development Corporation, a small industrial-warehouse developer located in south-central Pennsylvania. The corporation recognized the need to market its warehouses through a website to an increasingly networked group of industries that might be interested in leasing or purchasing warehouses. Kasun's typical customers require high-ceiling warehouse space that can be customized to meet their needs. After some discussion, the Kasun team decided on a homepage containing a brief explanation of what the company does, as well as aerial photographs of a few of the company's properties. The corporation determined that additional pages on the website should focus on the properties available, provide some basic corporate information, and include contact information for potential clients and customers. To create a professional site, the team worked closely with a web design firm that incorporated all of their ideas into a website, which was then presented to and evaluated by a group of current and potential clients before being placed online (see Figure 17.1). The Kasun team employed all of the components of the PSA in the creation of their website, and they continue to revise and update the page to meet the needs of their customers.

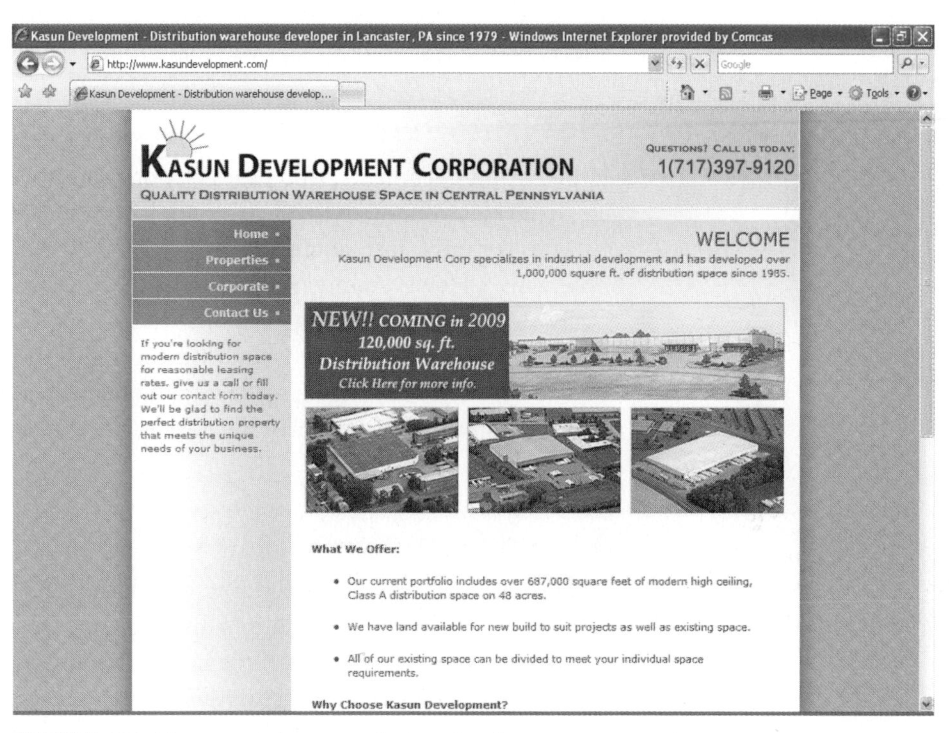

FIGURE 17.1 Homepage for a warehouse development corporation

ANALYZE THIS

Analyze the homepage of the Kasun Development Corporation website. Based on what you know about the company and its potential customers, does this website appear to be effective? Why or why not? What do you think about the choices in page layout, color schemes, images, and organization? Does the site appear to provide adequate information? Is there anything missing or irrelevant? You might also visit the corporation's current website at www.kasundevelopment.com to view the entire site. Has the site changed since it was published in this textbook? How and why? Write a memo to your instructor, analyzing the effectiveness of this website.

BASIC DIFFERENCES BETWEEN A WEBSITE AND A PRINT DOCUMENT

Although there are many similarities in creating a website and a print-based document, there are also some key differences.

Size and Dimension

Print documents work within a fixed two-dimensional space and rely on a static relationship between textual and visual elements. Print documents are also limited in size by the paper on which they are printed. A newspaper, for instance, can include more information on a single page than a report can on standard 8 1/2" × 11" paper. Web pages, on the other hand, can be viewed only according to the size of the monitor on which a reader accesses the page, and readers can often see only a small part of the page at a given time. Thus, to read a web page, a reader must scroll through it to get all of the information it contains. Because web pages are partially controlled by the technology available to readers, creating two-dimensional relationships between text and visual elements is more complicated—you can't be certain that what you display on the page will be exactly what readers see. For instance, your text may reference a visual, like a chart or a table, but readers who access the page on a small monitor may not see the chart or table until they scroll down to it. This physical aspect of the page affects how you should present the content and visual elements of a web page.

Navigational Features

Unlike print documents, web pages rely on their hypertextual qualities to allow readers to navigate among multiple pages to locate information. In print documents, navigation is really just about turning pages; it is, for the most part, linear and progressive. However, much of the web experience is grounded in the ability to move freely through linked web pages in various ways. This characteristic greatly affects how you should organize web pages within your website. In print documents it's fairly easy to identify what page a reader is on; on a web page a reader needs specific navigational information—identifying where the reader currently is and where various links lead.

Visual Components

Print documents are often superior to web documents in presenting high-quality visual elements. Most computer monitors display visuals of a lower resolution, whereas print documents can display visuals of the highest resolution, resulting in stark, clean images. For example, most of the images in this textbook are of a higher quality than their counterparts on the Companion Website. In addition, as the resolution of web-based visuals increases, the file size increases as well, slowing the time to download a file. And because of the limited screen size, web-based visuals must be small enough to fit on the screen, whereas large-document production can accommodate larger visuals.

See chapter 8 for more on resolution and visual file size.

Multimedia

Print may have an advantage in being able to display high-resolution visuals, but web pages can include multimedia tie-ins. Web pages can easily include video, animation, pop-ups, and sound features. Such features can increase the time needed to access a web page and may not be accessible by all readers, but they can be added efficiently and can increase the reader's interaction with the page. Such features may also be used to *increase* accessibility. For instance, some sound features may provide access to visually impaired audiences.

Accessing Speed

Print-based documents are typically easier and quicker for readers to access than web pages area. A reader can flip between pages of a document as quickly as necessary, instantly viewing the information on each, whether textual or visual. In contrast, a number of factors affect the speed with which a reader can access a web page—for example, the size of the file, the bandwidth of the reader's system, the size of the reader's screen, and the number of users accessing a website at a given time. Access speed is important to consider in the design of web pages, as some audiences may be unable or unwilling to visit sites that take them too long to access. This is particularly true of websites that may be geared toward transnational audiences, since some geographic areas may not have high-speed Internet connections or users may have limited access on shared workplace terminals. You should carefully assess your audience before uploading content that some readers may have difficulty accessing or viewing.

See chapter 5 for more information about transnational communication.

Because of these differences, website designers should not simply use the same criteria they would have used to create a print document. Even though some software applications allow writers to directly convert a print document into a web page, doing so does not account for the differences between print and web documents. Rhetorical problem solving should be used to address those differences.

INTRANET AND INTERNET WEB PAGES

As you begin to think about designing and writing web pages, it is important to note the two distinct kinds of web pages: those made for intranet systems and those made for the Internet. Knowing your audience is crucial.

Intranet Web Pages

Intranet web pages serve internal network systems and are designed to be seen by members of the writer's company or organization. Intranet web pages have these characteristics:

- Their internal audiences are often relatively small and easily defined.
- Intranet websites generally aren't designed to sell products.
- Intranet web pages tend to use fewer graphics and are more information-heavy.
- Intranet web pages often rely on the repetition of a standard format. A Cascading Style Sheet (covered later in this chapter) allows easy repetition of design styles.
- Many intranet systems establish templates for common design and informational elements, such as titles, dates, headers, and footers.

Internet Web Pages

Internet web pages should be thought of as public documents; audiences other than the ones you anticipate may access your web page. Internet web pages have these characteristics:

- Internet homepages often contain introductory or contextual information about the company or organization.
- Internet web pages are likely to contain splash pages to introduce the site or company—more on this later.
- Internet websites are likely to use design variations to create reader interest.
- Internet web pages may contain a great number of textual elements, visuals, color, sound, animation, and/or video.
- Many Internet websites include branding or marketing information to identify the company to potential customers.
- Most Internet web pages provide clear navigational features because diverse readers may need navigational guidance.

WEB TERMINOLOGY

Web pages and web technology are talked about using a specialized language. The following miniglossary is offered so that we can efficiently address key components of web page creation within the jargon of the World Wide Web. We are offering only basic definitions here; most of these concepts are taken up in greater detail throughout the chapter.

Web Page

A *web page* is an online document that is made available through the Internet or an intranet. Web pages are made available to public audiences via a server that gives readers the ability to access those pages.

Website

A *website* is a collection of related web pages. Other information—such as images, sound, or video files—may be part of the larger website.

Homepage

Homepage refers to the main page that users see when they access a website. You should think of a website's homepage as its front or cover page, containing key introductory information for readers. A homepage provides information about what readers will find in the pages to which it is linked. It also offers navigational aides and direction to readers.

Web Browser

A *web browser* is a software application that allows users to locate, read, and interact with pages on the Web. Web browsers operate by communicating with servers on which web pages are stored. Web browsers communicate by using a hypertext transfer protocol (HTTP), which allows users to access information from a server as well as send that server information. This is what allows web pages to be interactive. Some of the most popular browsers include Microsoft Internet Explorer, Mozilla Firefox, Netscape, and Apple Safari.

URL

A *URL* is the Web address for the location of a web page. URL is an abbreviation of "uniform resource locator," which suggests that a URL provides access to a single (i.e., uniform) location of a resource. Postal mailing addresses begin with specific information and end with more general information. URLs provide information in the reverse order, from general to specific. They are constructed like this:

protocol://authority/directory/filename.type

The URL for this book's web page is http://www.prenhall.com/dobrin/home.html.

- The *protocol* refers to the set of standards or system used to read and access the web page. The standard web protocol is http://, which indicates a hypertext document.
- The *authority* refers to the server on which the web page is stored. For instance, the web pages for this text are stored on the Prentice Hall server.
- The *directory* refers to the larger file in which all of the web pages associated with that page are stored. The directory is also sometimes called the path. Our web pages are stored in the "dobrin" directory.
- The *filename* refers to the specific file for which the URL is an address.
- The *type* refers to the kind of file that the page is. Most web files are written in hypertext markup language (HTML) and are written as either .html or .htm file types (more on this difference later).

When creating URLs or entering URLs into a browser, keep these factors in mind:

- URLs do not use back slashes (\); they use only forward slashes (/). In talking about URLs, you need only say "slash," not "forward slash."
- URLs are case sensitive to information in the directory and filename. Consequently, http://www.prenhall.com/dobrin is a different address from http://www.prenhall.com/DOBRIN.
- Because URLs are case sensitive, it's a good idea to use only lowercase letters in the URL so that your readers won't confuse the use of capitalization.
- URLs convey meaning to readers just as any other part of the document does, so you don't want your URLs to represent your company or organization in a negative manner. For instance, Prentice Hall would probably not approve of our assigning a URL to a textbook web page about bad editing, such as http://www.prenhall.com/TCTC/crappyediting.html.
- Because many websites contain numerous, perhaps even hundreds, of pages within a given directory, it is important to develop a strategy for keeping track of how you assign URLs.

Hypertext

Hypertext is any text that is found on a web page and is linked (i.e., connected) to other text, visuals, multimedia, or pages. The World Wide Web relies on hypertext to form its "web"; hypertext can be thought of as a system for moving from page to page on the Web.

Link

A *link* (also known as a hyperlink) is a connection between an element of a web page—text, picture, button, and so on—and another web page or another part of a web page. Users activate links by placing their cursor over a link and clicking their mouse. Links are often designated by a difference in color or the appearance of an icon, although it is also common for links to be hidden, or not readily identified. The word *link* can also be used as a verb, meaning to link or connect one page to another in a larger website.

Interface

Interface refers to the connections between a user's computer and the interconnected servers of the Web. The interface is both the point at which the user's computer and the network intersect and exchange information and also the name given to what happens during that intersection and exchange. Thus, the term can be used as a noun or a verb. For example, one might say, "My computer's interface with the World Wide Web allows me to obtain information," as well as "My computer is equipped to interface with the World Wide Web to obtain information."

Navigation

Navigation refers to the way in which a reader is directed to move through various web pages. Navigational tools are the rhetorical elements writers place in web pages to guide their readers. These tools can appear as text or as visuals, such as a link to a homepage using the word *home* or using a graphic representation of a home. Some websites use a navigation bar, where all of the navigational links are placed together at the top, bottom, or side of the web page.

Search Engine

Search engines are software applications that allow users to locate specific information on the Web; they locate and list web pages containing information relevant to a reader's search parameters. Search engines work by storing information about a large number of web pages, which are then accessed by a web "crawler," or "spider," that locates specific information. The search engine looks for key terms in places like titles, headings, and metatags (more on this later) and then indexes the information it finds based on relevance to the search. Some search engines, like Google, store key terms they have accessed in the past in order to search more quickly, whereas other search engines, like AltaVista, store and search every word of every page they store.

Web page writers must be sure that key terms can be found by search engines when a user searches for a specific term. There is no point in designing a web page that can't be seen by the right audiences, most of which locate web pages by way of search engines.

Site Architecture

Site architecture refers to the way in which web page writers arrange, build, and link the various pages on a website. It refers to the structure of the entire site, not just a given page within that site. Much like document architecture (discussed in chapter 9), site architecture refers to the way all related pages in a group are created in relation to one another.

Site Map

A *site map* is a web page that describes the architecture, or organizational structure, of the entire website. Some site maps contain a textual description of the site's architecture, whereas others contain graphical representations of the pages arranged spatially to show their relationships.

Server Space

Server space is the physical space where web page information is housed. A server is a particular kind of computer designed to store information and provide service to other computers, including access to the information it stores. For instance, one of the key steps in constructing a website is to upload pages to a server, where the pages can be stored and accessed by others.

Cookies

A *cookie* is a small data file that is written into a computer's hard drive while the user is visiting a web page. Cookies store data like user IDs and passwords, preferences, archived shopping cart information, and other personal information. Cookies then share that information with the website, most often to provide access to or to create a personalized page about the user.

WEB TECHNOLOGIES

In addition to the key terms just addressed, it's important to understand some of the basic technologies associated with writing web pages. The discussions that follow are intended to be introductory, to provide you with some basic understanding so that you can think about writing web pages. We will address many of these technologies in greater detail later in this chapter.

Always check the W3C web pages for standards on coding or any other issue that comes up as you create your web pages.

HTML

HTML is the basic language of web page writing. Again, HTML stands for hypertext markup language and is based on the standard generalized markup language (SGML), which in turn was based on IBM's generalized markup language (GML). HTML works by providing codes that computers read in order to understand how to format a web page. HTML is considered the international standard markup language for the World Wide Web and has been approved by the International Organization for Standardization as well as the International Electrotechnical Commission. More recent updates to HTML are monitored by the World Wide Web Consortium (W3C), an international organization headed by Tim Berners-Lee, who invented HTML and the World Wide Web. The W3C is the primary organization for maintaining standards for the Web.

EXPLORE

Explore the W3C web pages. Specifically, look up coding and HTML to see what the W3C says about coding your web pages. Jot down their definitions of coding and HTML, and compare them with the definitions in this chapter. How are they similar or different?

HTML codes tell a computer what to do with the text, visuals, and layout on a web page. The codes provide three basic types of information:

Structural Structural codes tell the computer what level of text will follow. For instance, a code might indicate that the following title should be read as a level 2 heading. This kind of coding does not tell the computer what to do with the text, other than to use its settings for a level 2 heading.

Presentational Presentational codes tell the computer how the text should look. These codes indicate textual adjustments like boldface or italic. In the most recent versions of HTML, these kinds of codes have been updated to a more generic code system; for instance, codes like *strong* or *emphasis* are used in place of *bold* or *italic*. The older codes have been deprecated, meaning that they still work on many computers and browsers but are being phased out.

Hypertextual Hypertextual codes tell the computer what portions of the text are to serve as links, either within a web page or to other web pages. These codes indicate which words or images on a web page can be clicked to lead readers to other pages.

HTML codes are based on English and work as mnemonic abbreviations, which allow web page writers to recall and use the codes as they write HTML. There are three primary kinds of codes:

Tags These codes determine the elements of a web page: paragraphs, inset quotes, and so on. Tags sometimes require end or container tags.

Elements These codes are the parts of the web page that a reader sees. Anything that is included as an element will appear on the web page after it is posted.

Attributes These codes allow writers to control aspects of web page presentation, such as centering or heading level. They control the placement or alignment of the elements on a web page.

Although this chapter does not teach HTML or coding, it is useful to see some of the more common HTML codes, which you can use to begin writing basic web pages. You do not need any specialized software to do so, just a basic text editor, such as Notepad. Tables 17.1 and 17.2 display some of the more common HTML codes.

TABLE 17.1
Basic Tags Used in All Websites

Tag	Function
<HTML>	Tells the Internet browser that the page is written in HTML
<HEAD>	Specifies the head section of the site, which is not shown in the browser window but contains information about the site, such as the title
<TITLE>	Specifies the name of the website, which appears in the blue bar at the top of the browser window
</TITLE>	Ends the title section
</HEAD>	Ends the head section
<BODY>	Specifies all of the information that appears in the body of the website and is shown in the browser window
</BODY>	Ends the body section
</HTML>	Ends the page, after which nothing is recognized by the browser

TABLE 17.2
Additional Tags

Tag	Function
<P></P>	Inserts a paragraph break
 	Inserts a line break
<FONT SIZE = "1">	Inserts a font size ranging from 1 (smallest) to 7 (largest)
<FONT FACE = "Arial">	Inserts a font face
<CENTER></CENTER>	Centers the text or image
<I></I>	Italicizes the text
<B></B>	Applies boldface to the text
<OL></OL>	Inserts an ordered (numbered) list
<UL></UL>	Inserts an unordered (bulleted) list
<LI></LI>	Specifies a list item
<H1>	Inserts a heading size ranging from 1 (largest) to 6 (smallest)
<BODY BGCOLOR = "GRAY">	Specifies the background color (can also use a hexadecimal number)
<BODY TEXT = "WHITE">	Specifies a default text color
<FONT COLOR = "BLUE">	Changes the color of a specific section of text
<IMG SRC = "image.jpeg">	Inserts an image or graphic
<A HREF = "URL">	Inserts an absolute link to an external web page
<A HREF = "page.html">	Inserts a relative link to an internal web page
<A HREF = "mailto: email address">	Inserts an e-mail link
	Ends clickable text in a link

For more on coding web pages, see the Web Page Tutorial on the Companion Website at http://www.prenhall.com/dobrin.

Use the View Source feature of your web browser to see the coding of web pages. Just remember that taking code without giving credit can be unethical.

All HTML documents should be constructed using the basic HTML document structure, sometimes called the HTML skeleton. The HTML document structure provides four primary parts of information, which equate to the basic tags shown in Table 17.1:

- The document type declaration (DOCTYPE) tells a browser which version of HTML is used in the page and whether there is a need to refer to a Cascading Style Sheet (CSS), which we will discuss shortly.
- The head contains the title and other technical information about the web page.
- The title identifies the title of the web page.
- The body contains all of the information and structure of the web page.

Coded, the HTML document structure looks like this:

```
<HTML>
<HEAD>
<TITLE>The name of the web page</TITLE>
</HEAD>
<BODY>
The content of the web page
</BODY>
</HTML>
```

As you can see, HTML codes are presented in <brackets> to distinguish them from the web page text. A <bracket> without a slash indicates the start of a code, and a </bracket> with a slash indicates that the code should stop. The information that appears inside the brackets does not appear on the actual website after it is posted to a server. In fact, you can access the HTML code of most web pages by viewing the source of the web page, which is easily accomplished through most web browsers. Figure 17.2 shows the source of the Kasun Development Corporation page shown in Figure 17.1. One of the best ways to learn about making web pages is to look at the code other web designers have used.

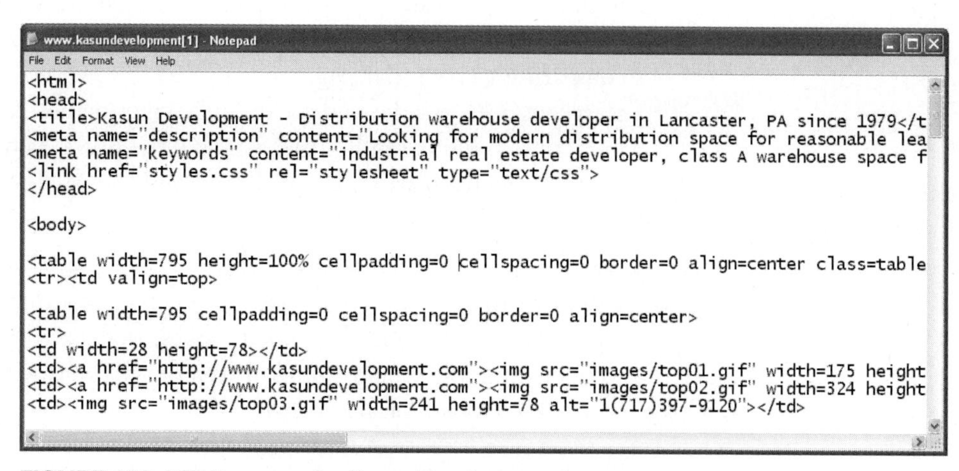

```
www.kasundevelopment[1] - Notepad
File Edit Format View Help
<html>
<head>
<title>Kasun Development - Distribution warehouse developer in Lancaster, PA since 1979</t
<meta name="description" content="Looking for modern distribution space for reasonable lea
<meta name="keywords" content="industrial real estate developer, class A warehouse space f
<link href="styles.css" rel="stylesheet" type="text/css">
</head>

<body>

<table width=795 height=100% cellpadding=0 cellspacing=0 border=0 align=center class=table
<tr><td valign=top>

<table width=795 cellpadding=0 cellspacing=0 border=0 align=center>
<tr>
<td width=28 height=78></td>
<td><a href="http://www.kasundevelopment.com"><img src="images/top01.gif" width=175 height
<td><a href="http://www.kasundevelopment.com"><img src="images/top02.gif" width=324 height
<td><img src="images/top03.gif" width=241 height=78 alt="1(717)397-9120"></td>
```

FIGURE 17.2 HTML source for Kasun Development homepage

XHTML

HTML is the most basic form of hypertext markup, but there is a new version of markup language that is rapidly becoming the standard for web pages—XHTML. XHTML is an *extensible* hypertext markup language, which means it is designed so that users or programmers can expand or add to its capabilities. XHTML provides writers with the same possibilities for their web pages but uses a stricter syntactical set of codes than HTML does. Therefore, a computer has less room to interpret what the web page should do, leading to fewer errors in presentation. XHTML can be thought of as the newest version of HTML; it was developed to be more accessible by a wider range of computers and browsers. However, not all HTML browsers can read XHTML documents, even though all XHTML browsers can read HTML web pages. Many versions of Microsoft Explorer, in fact, are not XHTML compatible. We do not address XHTML coding here, but the differences between XHTML and HTML are minimal.

Tables and Frames

Tables are probably the most versatile and commonly used HTML code feature. Tables place text and visuals on a page by using rows and columns of cells into which the information is placed. Tables are also widely used to specify the width of web pages. And of course, tables are great for their original use: the presentation of tabular data. In most instances the tables on a website are invisible unless the writer wants readers to see an actual table of data or figures. A simple table might look like this:

Name	Home Phone	Cell Phone
Bill Gates	555 777 8554	555 777 8555

Although tables can be useful, they can create problems for viewers using a larger display, forcing them to scroll horizontally to view a table's contents. Tables can also pose problems for documents that will be printed from the Web and can make it difficult to maintain consistency throughout an extensive website. In addition, pages with extensive tables often require more bandwidth and loading time.

Frames serve a similar function in manipulating information on a browser, but they allow the writer to display two or more separate data sources, such as web pages and multimedia files, in a seamless single page. Essentially, frames divide web pages into different parts, or boxes, each of which can act autonomously or in conjunction with another because each is treated as a separate windowlike element. As a result, frames can be useful for presenting an online document with a table of contents or for comparing two similar pages side by side. However, because frames are difficult for search engines to index and can create navigation problems for users, they are not commonly used. Notice in Figure 17.3 how the NIST website uses frames to distinguish between the navigation bar on the left and the page content on the right.

CSS

CSS refers to Cascading Style Sheets, which are markup language documents that give a particular style to multiple web pages; that is, they "cascade" through numerous web pages, giving them common styles or features. CSS files are used to define presentation elements of a web page—like color, layout, and design—and then repeat those same elements in multiple pages. Many web-editing software applications, like Dreamweaver, now automatically convert coding into CSS files so that the writer does not have to code separate style files.

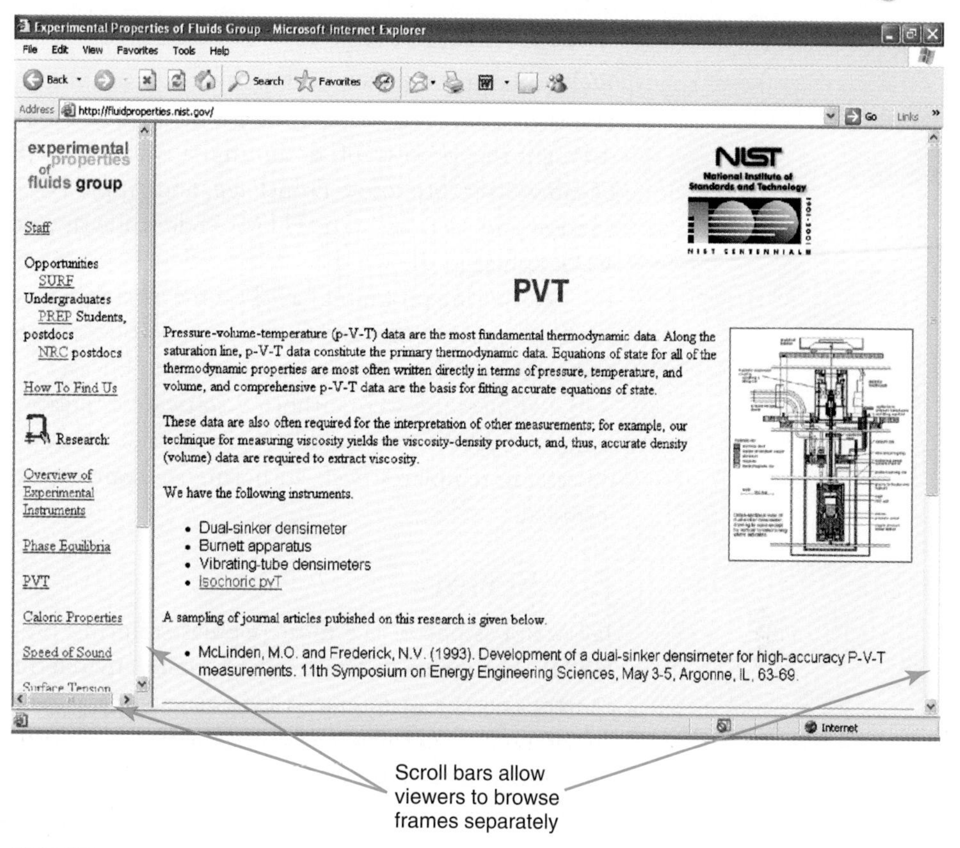

FIGURE 17.3 A website using frames to present a table of contents

CSS coding allows writers to keep style files separate from text files and then apply style to text. In this separation, markup language is used to establish the content and structure of a web page, but not the visual components of the page—its style. Separate CSS-coded pages establish the style. This separation has four primary benefits:

Speed Because a computer needs to download a style sheet only once and can store it in its cache, the resulting web pages can be downloaded more rapidly.

Accessibility A CSS-coded page that works with either HTML or XHTML can be more easily read by a variety of browsers. And browsers that do not support the visual components of web pages—like Lynx—or browsers that don't support CSS can simply ignore the CSS code, rather than presenting garbled code or indecipherable text.

Customization It is fairly easy for a web page writer to adjust a style sheet and apply it to a page of content already in place without having to rewrite the content as well. In addition, a writer can apply that single style sheet to multiple content pages to maintain conformity among pages.

Maintenance If there is a need to make a style change throughout a website, writers can change just the CSS file and apply that revised file to the entire website.

Style sheets can be applied by a reader as well. For example, someone viewing a website could choose a larger font or could specify that a certain language be applied. These features of CSS make it useful in terms of accessibility, a concept we discuss in detail later in the chapter.

Web-Authoring Software

Web-authoring software is designed to assist you in making web pages. Most programs are simple to use and can guide even the most inexperienced web page writer through the process of designing a web page. Programs such as Macromedia Dreamweaver, Microsoft FrontPage, and Apple iWeb create a facsimile of a browser window and then write the HTML code while the user adds and manipulates the text and graphics in the window.

Authoring programs have become increasingly sophisticated, allowing web page writers to create tables and frames, use templates to repeat styles on consecutive pages, and even include animations, blogs, and podcasts in the website. Many professional web designers use web-authoring software to create extensive websites or to include complex features that would otherwise take a long time to code. However, like word processing templates, web-authoring software can sometimes create pages that look formulaic.

JavaScript

JavaScript is one of the most popular scripting programs that can be inserted into HTML documents to create what's commonly referred to as *dynamic content*. Dynamic content consists of interactive elements such as text or images that change when a cursor moves over them (these are called *rollovers*); drop-down menus; and pop-up windows that appear when an item is selected. Figure 17.4 gives an example of a rollover navigation bar that changes color when a cursor moves over it.

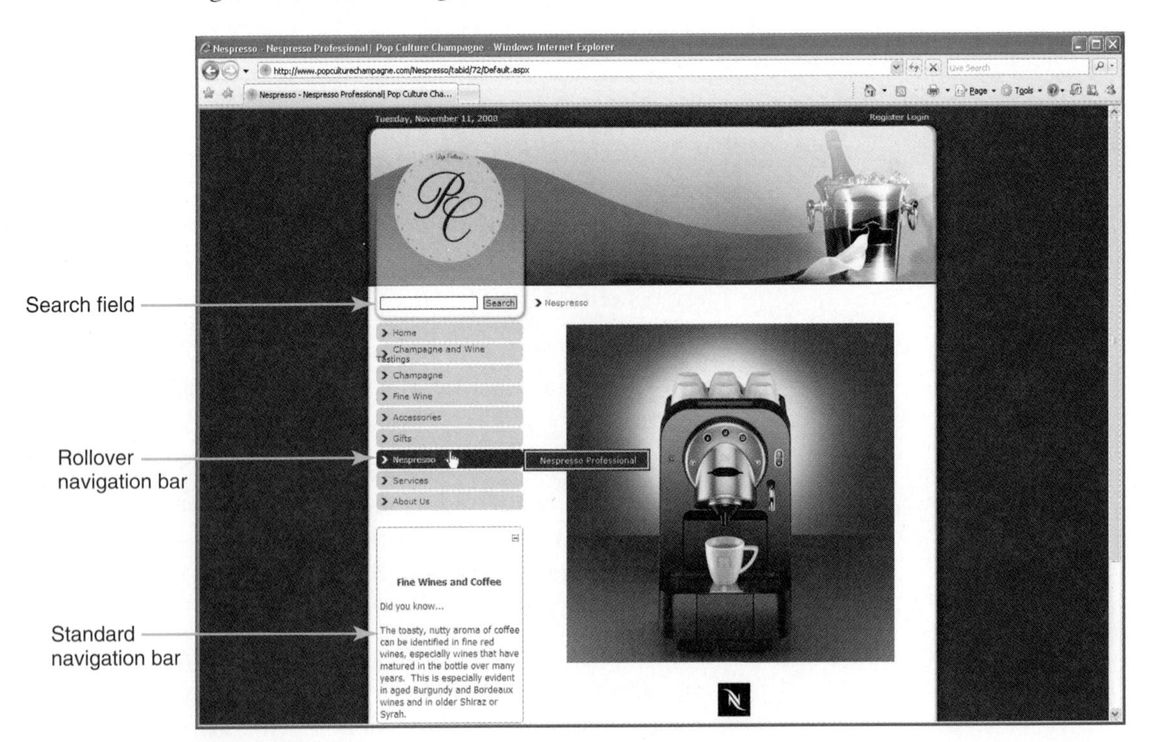

FIGURE 17.4 Rollover created with JavaScript

JavaScript can also be used to react to events—for example, loading an image after a specific amount of time. In addition, JavaScript is sometimes used to detect or save information about the readers of a website—perhaps determining which browser they're using or storing information about them through cookies that are referenced when they revisit the website. When you revisit a site like Amazon.com and it welcomes you by name, you are seeing an example of a script in action. Notice how this example of JavaScript identifies the returning user by name and provides specific recommendations based on previous selections:

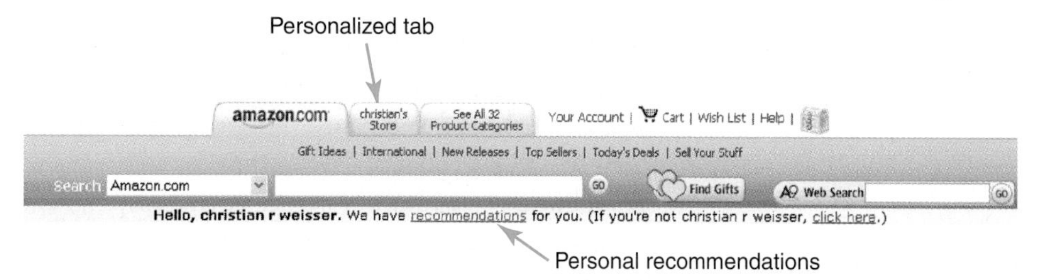

Other, more specific applications of scripting languages can require advanced programming ability. JavaScripts can be fairly simple, such as allowing the text color of a link to change with a rollover, yet they can also offer dynamic ways of presenting and receiving information from readers. Numerous free-use scripts can be found on the Web.

CGI

CGI stands for Common Gateway Interface. Even though CGI programs are advanced web technologies, it is important to be aware of what they do, because many web pages rely on CGI technology to gather information from their clients and/or employees. Whereas standard HTML documents are static and can only deliver information, a CGI program is executed in real time so that it can both transmit and receive dynamic information. A CGI is basically a software application that converts information that is inputted from an outside source, like a web browser, into a form that the server can recognize, like HTML. For instance, a company web page might request customer feedback about product reliability and then publish that information on the web page. The company's CGI would accept the reviews from customers and translate those reviews into HTML code so that they could be posted on the company web page.

CGI is used in many business websites today; it allows designers to create shopping carts, page counters, guest books, order and complaint forms, and response sheets. Data from these sources can be delivered as an e-mail message or as some other document or file. Complex CGI programming can even create personalized pages for individual users and store those pages on the company's server. The U.S. Securities and Exchange Commission's Investor Complaint Form, found in Figure 17.5, is an example of how CGI can be used to transmit information through a website.

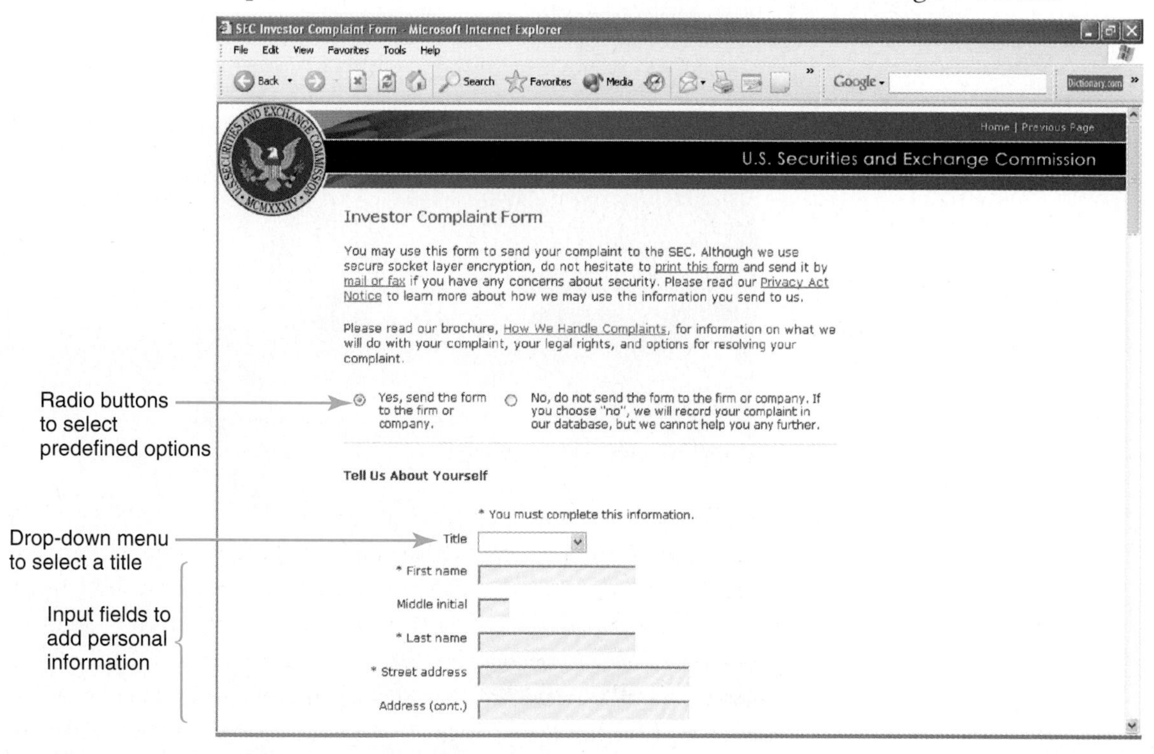

FIGURE 17.5 An example of CGI technology

CMS

CMS refers to content management system, which is a program that allows for collaborative writing of web pages. Sometimes CMS programs are themselves web applications that allow collaborative authors to work on a web page through the World Wide Web, but most often they are independent programs that require particular software applications. CMS is used to create virtual meeting spaces, image galleries, and, more recently, blogs (short for *weblogs*, which are frequent, chronological publications on the Web, often in journal or diary form) and podcasts (audio or video broadcasting through the Web). Free CMS programs are available online, and some newer web-authoring programs allow users to create their own blogs, podcasts, and other related features. CMS programs are particularly useful as systems for managing single-source documents, such as manuals.

Plug-ins

A plug-in is a small computer program that interacts with a web browser to allow the browser to display certain types of documents, show interactive images, play music, or play video. Typically, plug-ins add multimedia functions, which enable users to receive information in more dynamic ways than through normal text and images. Although plug-ins aren't technically part of the web pages you will design, you should be familiar with what they are and how they function. If you plan to include an animation, audio, or video file on your website, you'll need to determine which plug-ins your readers will have access to so that they can open your files. You may also wish to make a specific recommendation about the program they'll need to install.

These programs are called plug-ins because they plug in to a browser. If you access a website that requires a plug-in, your browser will automatically recognize whether you have the plug-in. If you do not have it, the browser will ask whether you want to download the plug-in (see Figure 17.6). If available, plug-ins automatically

FIGURE 17.6 A web page requiring a plug-in

start whenever they are required by a website. Some of the most common plug-ins are Adobe Acrobat, which is used to read PDF files; Windows Media Player, RealPlayer, and Apple Quicktime Player, which are used for audio and video; Flash and Shockwave to play animations and games; and iPix for viewing 3-D images and virtual tours. Most plug-ins are free and can be easily downloaded.

STANDARD WEB PAGE COMPONENTS

As we have noted, this chapter doesn't teach you how to write the code for a web page, but it does offer some basic rhetorical understanding about web pages and how they are written. The sections that follow include suggestions to guide your thinking about writing web pages.

Continuity and Branding

Web pages that are designed to represent companies and organizations must be written a bit differently from the way you might write a personal homepage. In designing professional web pages, writers must attend to issues of continuity and branding.

Continuity

Continuity has two important meanings for web page writers. The first refers to a design that is consistent throughout a website. A continuous design allows readers to become familiar with the same navigation and other design features repeated from page to page and thus easily navigate the site. Second, continuity refers to an uninterrupted flow between the pages within a website. Web pages that contain dead links clearly do not have continuity.

Branding

Branding is a traditional marketing mechanism to ensure that customers have heard of and remember a product or company; the goal is to create a unique or recognizable image for the product or service. In web pages, branding is used to promote the company or organization that sponsors the web page. Branding can be achieved in a number of ways, including repetition of icons and logos, eye-catching graphics or animations, and repeated textual references to the company or product. Icons can be particularly effective for branding. For example, when we see the Nike swoosh, we think of the Nike brand of sportswear and any associated textual references or phrases, such as the Just Do It campaign. Web page writers must be attentive to branding so that readers relate the pages to the company or product the pages support.

See chapter 8 for more on designing icons.

IN YOUR EXPERIENCE

All colleges and universities devote time and money to branding in an effort to distinguish themselves from other academic institutions. Think about how your school brands itself. Can you identify five mechanisms your school uses for branding?

Navigation

Web page writers must provide navigation cues so that readers can find the information they want. Because readers cannot hold an entire website in their hands and see all of the pages that comprise it, as they could do with a printed document, each web page within the site should include information about where readers are currently within the site and where they can go. Web pages may include hyperlinks, navigation bars, a home link, tabs, and search fields to aid navigation.

Hyperlinks

As we mentioned earlier in this chapter, the simplest navigation feature is the link or *hyperlink*, which is a word, phrase, or image that, when clicked, takes readers to another page or section. Hyperlinks should provide a clear indication of where they will lead, and they should be consistent throughout the website. A consistent style for hyperlinks contributes to the overall consistency and continuity of your pages. You can keep hyperlinks consistent by using similar word types or phrases. A repeated icon to indicate a hyperlink can also help readers identify where hyperlinks appear. Visuals, particularly icons, can be highly effective as hyperlinks, because they show readers a subject rather than telling them about it. However, be sure to include a textual reference for each visual hyperlink because some readers view websites with images turned off or with text-only browsers. Textual references are also important for vision-impaired readers who use special browsers to access websites.

Navigation Bars

Navigation bars are hyperlinked lists of the major pages that branch off from your homepage. Navigation bars typically appear at the top or along the left side of a web page. Sometimes they are repeated in textual form (often using smaller, more subtle text) at the bottom of a lengthy page, much like a footer. Navigation bars should be clear and simple to use; they work best when repeated consistently throughout a site. Notice the clearly visible blue navigation bar on the Smithsonian homepage in Figure 17.7. Notice both the clarity of the text and the visual contrast of the blue bar against the black header, which makes the bar easy to see.

Home Link

Websites can contain many pages, making it easy for users to get lost as they navigate through the pages. Websites that have numerous and detailed pages should provide a home link on every page. Home links allow users to return to the homepage of the website with one click. Home links are usually located in the navigation bar (see again Figure 17.7) or in a navigation aid at the bottom of the web page. In addition, many websites include a company logo or icon that serves as a link back to the homepage, thus contributing to branding and continuity as well as navigation.

FIGURE 17.7 A navigation bar

Tabs

Tabs, which are designed to look like tabbed file folders for print documents, can also be used like navigation bars. Tabs are generally used on web pages that have fewer primary links. Notice the red and blue tabs on the U.S. government's web portal, shown in Figure 17.8. Tabs work like a navigation bar, linking readers to specific pages or places in the website. Tabs can also initiate a drop-down menu, a series of sublinks that opens when a tab is clicked. Most web-authoring programs allow you to design tabbed pages or to use plug-ins that will accommodate drop-downs. This method of providing navigation conserves space and looks very sleek and professional, although it may create problems for readers with older browsers that are not enabled to display such applications. Multimedia applications such as JavaScript and Active X allow you to create sophisticated navigation features that can become a significant aspect of the design of a website. However, be careful that such features do not become too elaborate or convoluted; if readers cannot easily use and understand your navigation, it is worthless.

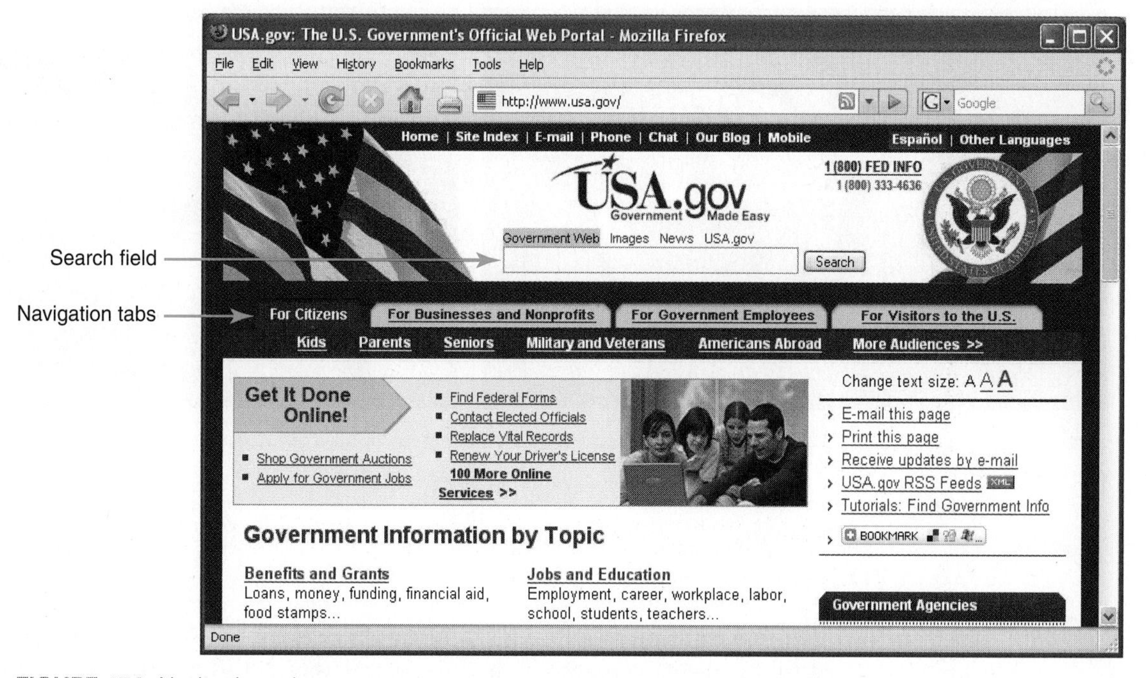

FIGURE 17.8 Navigation tabs

Search Fields

If your website contains a great deal of information that individual readers might need to explore for different reasons, you may wish to include an internal *search field*. Search fields can be useful for finding specific information in text-based documents within a site, a directory of employees or contacts, or a product list. Search fields usually consist of a prominently displayed box on the homepage that allows readers to type in keywords or phrases, which then lead to a list of all of the pages on the site that contain that word or phrase. Although many search engines are available online and in some advanced web-authoring programs, they can be difficult to integrate and manipulate within a website. Notice that Figures 17.7 and 17.8 both contain prominent search fields, which can guide readers to different pages within these two large websites.

Splash Page

Splash pages should be thought of as prehomepages. They usually do not contain any pertinent information other than, perhaps, system requirements to view the web pages that follow. Splash pages are often visually heavy and may contain short animation sequences. In addition, splash pages often include a button to bypass the splash page and move directly to the homepage, and they may include various routes into the page, such as entering via an HTML-only route or a route capable of supporting Flash animation sequences.

Splash pages are the ultimate attention getters and are a good way to show off the creative elements of your organization. However, you need to include a way for readers to skip the splash page; they may have visited the page before or may find splash pages a waste of time. Some splash pages, like the one found in Figure 17.9, automatically open into a homepage once the splash animation or sequence has been completed. Splash pages are used primarily for branding.

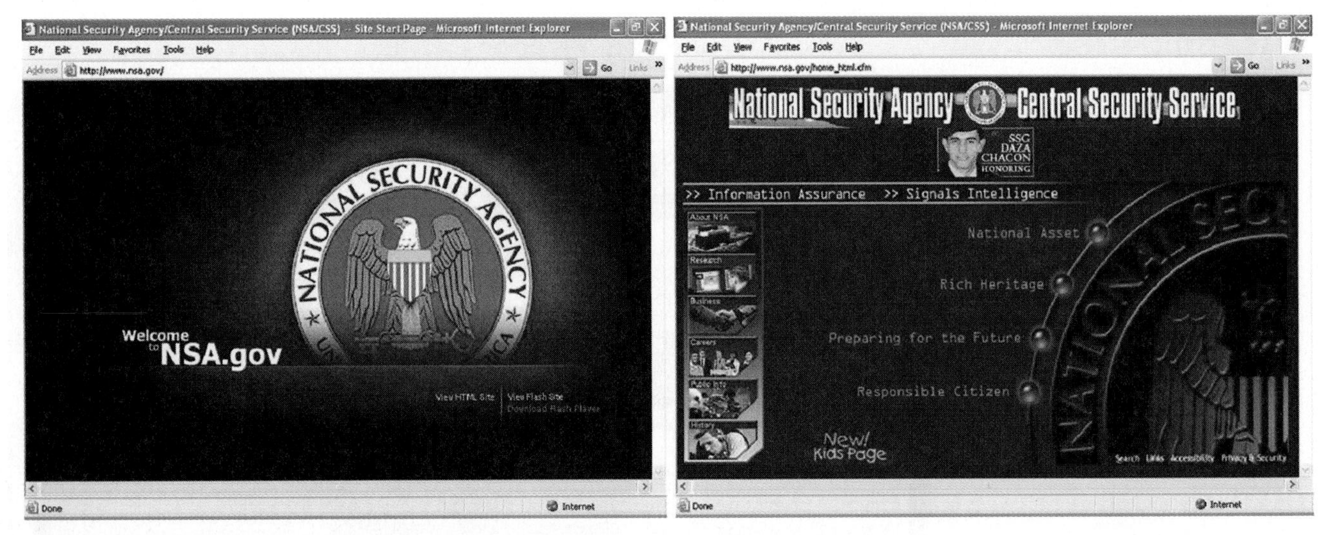

FIGURE 17.9 A splash page that opens automatically into a homepage

ANALYZE THIS

There are pros and cons to using splash pages; some web designers swear by them, whereas others believe they are never appropriate. Look at various splash pages on the Internet, and try to find one effective splash page and one that is ineffective or inappropriate. Then analyze why each splash page is effective or ineffective. What role does the subject, purpose, or audience of the website play in the overall effectiveness? How does the design of the splash page factor in? Does the splash page give the reader any options? Does the length or size of the splash page play a role in its effectiveness?

Homepage

A homepage is often the first page readers access. Nearly all websites have a homepage of some sort, which serves a similar function to that of an introduction in a report or a proposal. Typically, the homepage offers an overview of the site and its offerings. Like the optional splash page, the homepage should explicitly identify the person or organization responsible for the site and should project an appropriate image of that person or organization. You should avoid the urge to include everything on the homepage but should envision it as a gateway to the more specific information that your website contains.

Your homepage should have these characteristics:

- *Clear*—A homepage should convey to the reader exactly what the purpose of the website is and what information the reader can find on the homepage and other linked web pages.
- *Clean*—A homepage should not be cluttered with lots of visuals that distract the reader from the key information.
- *Navigable*—A homepage should be easy to navigate and should make explicit how readers can navigate other pages in the site. Homepages establish and forecast for readers the methods the entire site uses for navigation. Homepages are the first stage in creating continuity.
- *Substantive*—A homepage must provide readers with some information of value. A homepage without any important information does not encourage readers to move deeper into the site.
- *Efficient*—A homepage should open quickly. Large image files or large quantities of data should not appear on the homepage. Readers should be able to get to the first page quickly so that they can begin locating the information they need.

Notice the homepage in Figure 17.10—a clean design providing readers with enough information to begin to locate needed pages within the site. Notice, too, that navigation and style features established in the homepage continue into the next page.

Nodes

Node pages introduce readers to specific topics within a website and can be understood as a second level within websites. In complex websites node pages serve as introductions to content areas within the site, establishing categories of information with a general description and links to pages with more specific information. In effect, such node pages serve as minihomepages because they include contextual information about that part of the website and direct readers to more detailed material. Node pages at the same level often have the same design, which provides a subtle indication of how the pages are ordered. In simpler websites the node pages may be the deepest level of material and may provide the specific details and content that readers are looking for. It is not necessary to make an overt distinction between node pages and the deeper internal pages of a website, although you should follow the general rule of providing increasing detail and specificity as a website progresses.

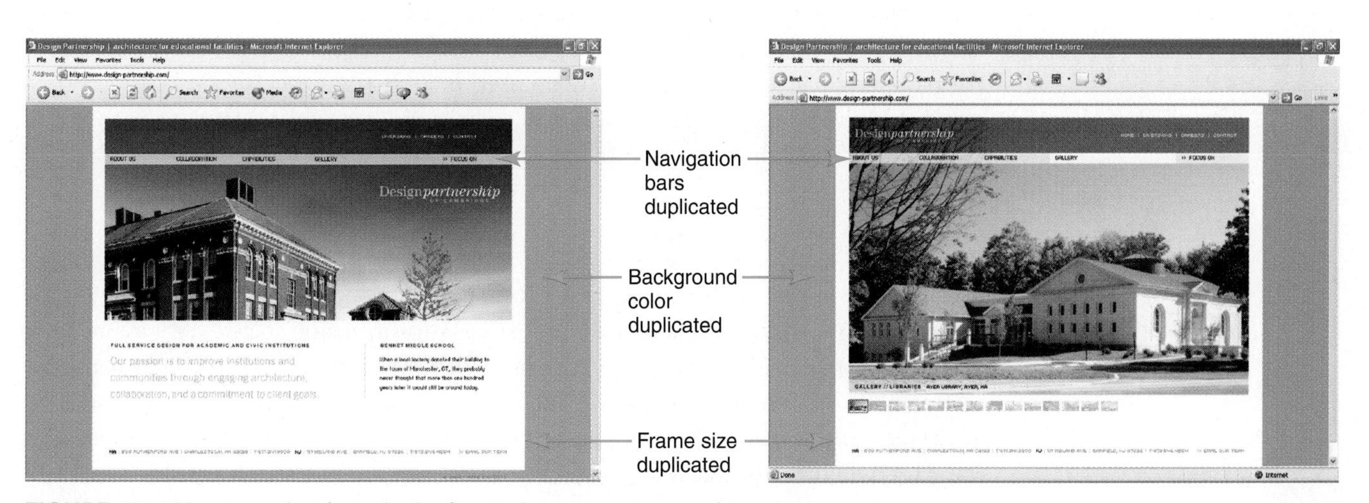

FIGURE 17.10 An example of continuity from a homepage to a node page

Subpage

Subpages are the individual pages of information within a series of web pages. They can be seen as the third level within websites and often contain more specific, detailed information. Subpages may lead to other subpages, and so on, and so on.

Search Optimization and Metatags

Like all documents, web pages are written to reach audiences. Web pages, however, often reach audiences as a result of readers being able to locate them through search engines. It is important, then, for web writers to ensure that their web pages can be easily located by search engines, which locate web pages in two primary ways. In pay-per-click approaches, search engines like Yahoo! charge companies to be sure that their web pages are listed in directories under particular search terms. In organic searches, search engines like Google locate web pages through their content and tags. Most web pages rely on organic searches because pay-per-click can be costly.

Search Optimizers

Search optimizers are methods that web writers use to make sure that their websites can be found through organic searches. Although there are some standard approaches to optimizing search potential, none of the methods have ever been proven completely effective. These are the kinds of questions that web designers ask to determine whether search engines will effectively locate their pages:

- Has the website been up long enough for search engines to have indexed it?
- How often is the content changed or updated?
- How much textual content is included per page?
- Does the content contain common terms that search engines will recognize?
- How many and what kinds of links does the page include?
- How many pages does the server/domain host?
- Does the page have unsafe or illegal content?
- Is the page coded correctly?
- Is the page accessed frequently?
- Are the URLs clean (or are they difficult to locate because of length, capitalizations, or mislabeling)?

In order to have your web page recognized by search engines in organic searches, you do not need to alert the search engine. Based on the content and the tags you use, most spiders will find your page within a few days of its being available on the Web. For pay-per-click search engines, you do need to alert the search service. Most corporate web writers report that the best way to have readers find a web page is to register it with Yahoo! and also to provide content that allows Google's spiders to easily locate your page.

Metatags

Metatags are special HTML tags that provide information about the web page. Unlike other tags, though, metatags do not supply information to browsers about how to read the page. Instead, they provide information about the *page* (*meta* means "referring to" or "about"). Metatags include information like the title, author, date of design, and latest update of the web page—information that won't appear on the page when readers access it. Metatags are hidden codes that can be seen only when someone looks at the code for the page. Remember that one of the best ways to learn how pages are coded is to look at page source information through your browser.

Metatags are useful in improving the likelihood of search engines finding a site because search engine spiders also read and index some metatags. There are three metatags that are key: a title tag, a description tag, and a keyword tag. The first two are the most useful in search optimization because some search engines ignore the keyword tag. For those search engines that do use them, keyword tags allow a writer to place multiple words in a search engine index so that the writer's web page will be identified whenever a reader searches for any of those words. Figure 17.11 shows an example of metatags within a web page code.

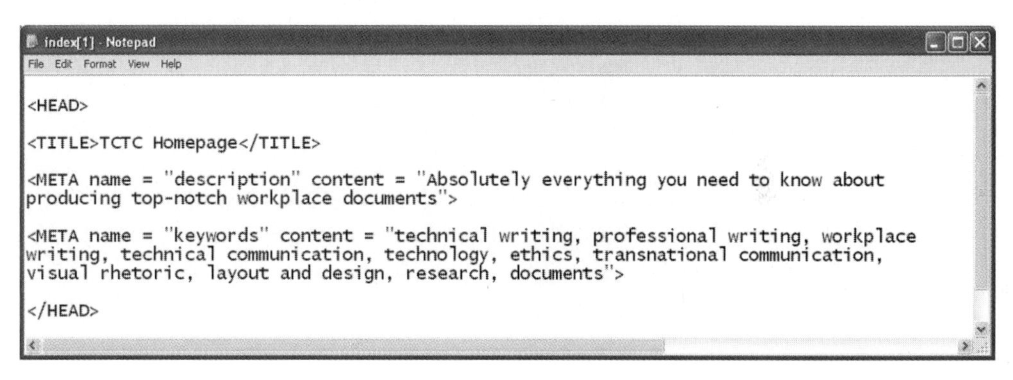

```
<HEAD>

<TITLE>TCTC Homepage</TITLE>

<META name = "description" content = "Absolutely everything you need to know about
producing top-notch workplace documents">

<META name = "keywords" content = "technical writing, professional writing, workplace
writing, technical communication, technology, ethics, transnational communication,
visual rhetoric, layout and design, research, documents">

</HEAD>
```

FIGURE 17.11 Metatags within a web page code

IN YOUR **EXPERIENCE**

Using your web browser, go to a professional homepage with which you are familiar. This can be a page you see regularly or one you have visited for a specific purpose. How and why did you first access this page? Did you find it using a browser? Using the View Source option on your browser, locate the metatag information for the web page. Write a letter to your instructor explaining what metatags the pages use and how those metatags might enhance the chances of a search engine locating that web page.

FAQ

Some websites also contain an FAQ page, which provides answers to frequently asked questions about an organization, its services, or the website. In most cases an FAQ page is a single page within a website, usually structured in a question-and-answer format. However, some pages use internal linking to provide all of the questions with links to the answers below. If you have an extensive FAQ page, it may be necessary to categorize and group the questions into topics or sections. The FAQ page for the Modern Language Association provides answers to common questions about the *MLA Style Manual* (see Figure 17.12).

Site Map

If your website is extensive or multilayered, you might want to include a site map, which is a visual diagram of all of the pages within a website. In most instances a site map is a separate page and should be linked to your homepage. Similar to a book's table of contents, a site map makes it easier for a user to find information without having to navigate through the site's many pages. Site maps can also be important sources of links for search engines to follow. Although many site maps consist of simple text-based links on a page, they can incorporate graphics and visuals to create a more spatial sense of the website's organization. The site map for the PEEL Development Commission provides a graphical map showing different node and sub-pages within the website (see Figure 17.13).

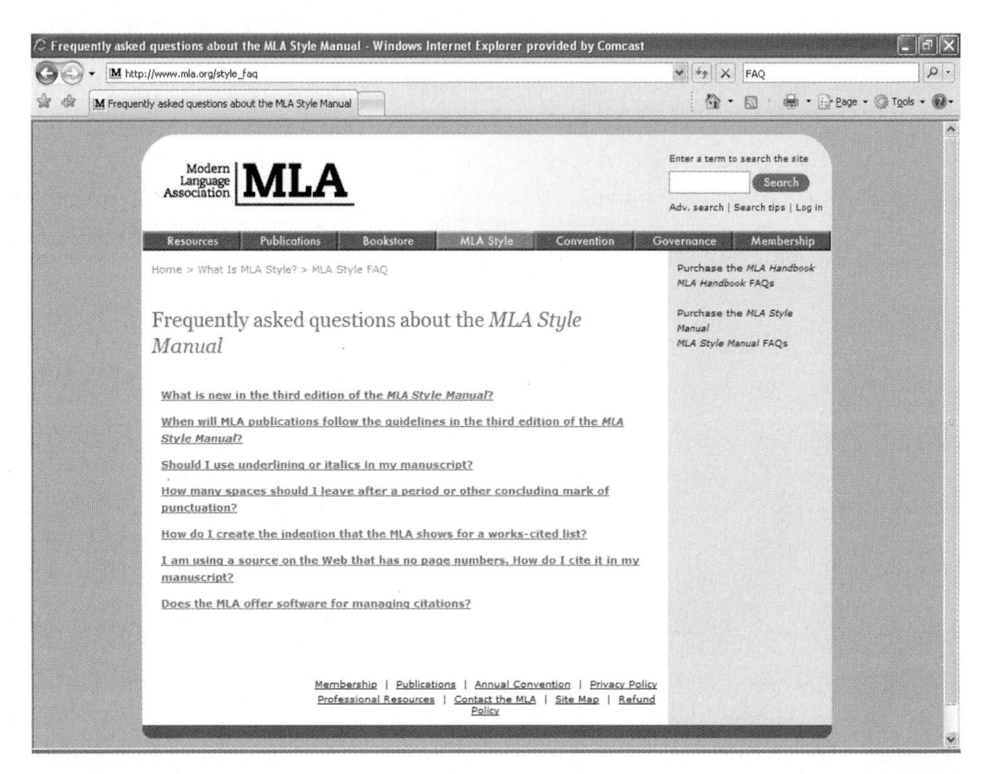

FIGURE 17.12 An FAQ page

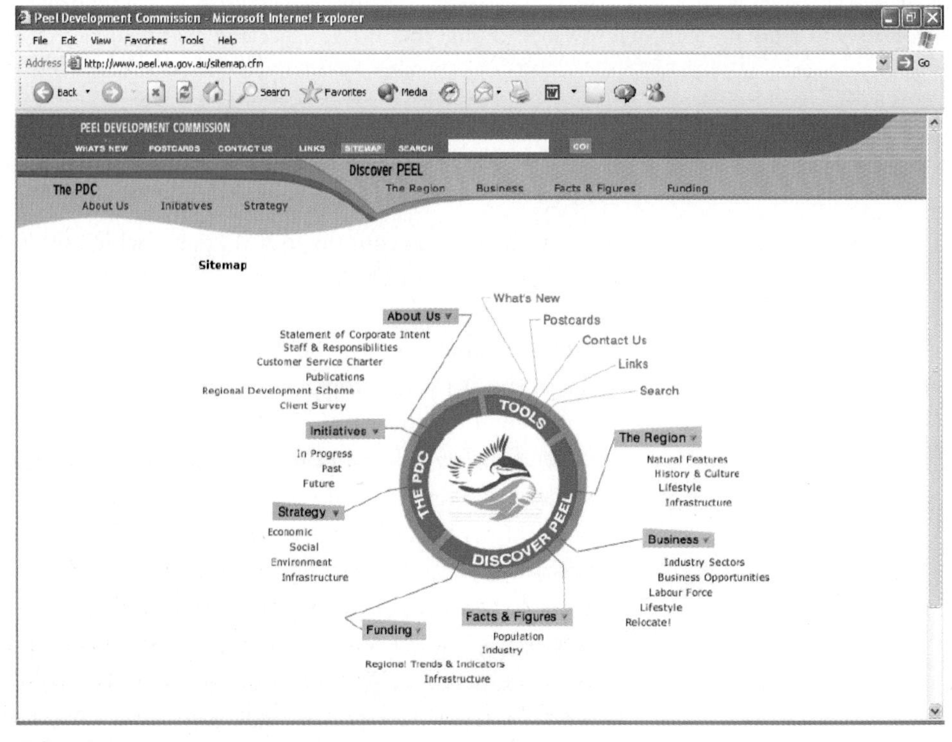

FIGURE 17.13 A site map example

Color

How you use color in a web page can be important for a number of reasons:

- Color can affect accessibility. Be sure to use contrasting colors, and avoid using light colors that might be difficult for visually impaired readers to see.
- Color can emphasize information.
- Color can be used to indicate hyperlinks.
- Color can be used to establish continuity.

See chapter 8 for more on using colors and RGB color schemes.

Remember that web pages, like other electronic documents, create color by emitting light. Therefore, you should use an RGB color scheme, which indicates colors with a hexadecimal code that identifies various arrangements of red, green, and blue. All colors that are viewable on the Web are assigned a six-figure code consisting of numbers or letters. The first two digits determine the amount of red in the color, the middle two digits determine the amount of green, and the last two digits determine the amount of blue. When combined, these form a specific color. A few of the colors and their hexadecimal codes look like this:

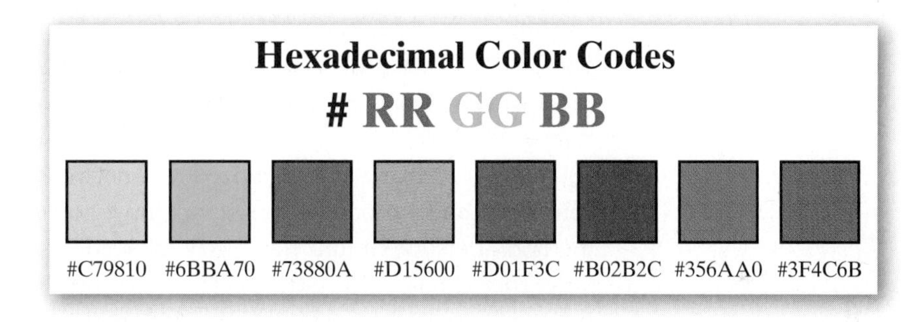

IN YOUR EXPERIENCE

Several of the website examples used in this chapter have received Webby Awards, which recognize excellence in web design, creativity, usability, and functionality. Webby Awards are conferred annually by the International Academy of Digital Arts and Sciences and can be found at http://www.webbyawards.com.

As a class or in a small group, examine some of the current year's nominees and winners. Which of the websites do you think are the best? What makes them the best? Do you agree with the winners chosen in each category? Which of the websites on the list have you visited before? What particular features of these websites would you like to incorporate into your own website?

ORGANIZING A WEBSITE

One of the key differences between print documents and web documents is organization. Print-based documents are often linear; that is, they are organized in a way that encourages readers to move from one idea or concept to another, along a linear path from the beginning to the end. Even documents with information that is not necessarily read in order are basically linear, because page numbers and chapters encourage an ordered progression. Websites, on the other hand, can accommodate more flexible, spatially organized schemes. Websites often allow readers to choose the information that they need and to move through a document in a variety of orders.

This nonsequential, linked method of displaying and reading documents emphasizes multiple connections among topics or ideas, downplaying the strict hierarchy of print. Some theorists suggest that the medium of hypertext is more reader centered, because it allows the reader to decide how and what is read and in what order. Language theorist Jay Bolter suggests,

> The connections of a hypertext are organized into paths that make operational sense to author and reader. In print, only a few paths can be suggested or followed. Hypertext can express cyclic relationships among topics that can never be hierarchical.[1]

See chapter 9 for more on groupings and connections in document design.

Linear organization can work in some online situations, particularly if a series of pages should be viewed in order or they need a strict classification (see Figure 17.14). For example, a set of online instructions, a bibliography, or a phonebook can be organized in a linear fashion if the information in them is of the same kind. However, linear organization doesn't take advantage of the benefits that hypertext offers in terms of linking, and it doesn't allow readers much control over the way they move through information.

Many websites use a hierarchical organization, in which pages and information are grouped together according to various levels and sublevels of information, yet no formal ranking or linear structure within the levels is suggested. This method guides readers with groupings and connections while allowing them freedom and mobility in what and where they choose to browse within the site. The organizational pattern for a hierarchical website might resemble a pyramid (see Figure 17.15).

Other websites are organized as a network, in which all of the pages are interconnected and no one page is more important or central (see Figure 17.16). This method offers maximum flexibility for readers—they can browse the website in any order they choose. However, a network organization requires careful linking to ensure that readers can find the information that they want and do not hit any dead ends as they browse the website. Another option is a modified network with a structure resembling a wheel: the homepage is the organizing point, and various node pages branch off from it.

Because hypertext offers choices in organization, it may be useful to map out the overall organization of your website, including what will appear on the homepage, what will be included in the various node pages and subpages, and how they will all be linked together. Many professional website developers begin by sketching the layout and organization of their website in graphic form on a large sheet of paper, chalkboard, or whiteboard. It is generally best to begin with the basic purpose of your website in the middle of the sketch and then identify the main sections the site will contain. You can use lines and circles to show what these sections are and how they will connect to the homepage.

Another strategy is to write the main ideas or components of the website on individual pieces of paper, like sticky notes, and then try various arrangements to find the best fit. The benefit of this method is that the notes can be easily moved, created, or eliminated. But regardless of how you map out your website, you should include other people in the process. If you are designing a website for a large company or organization, it might be useful to include one member from each of the organization's departments. If you are designing the website for or by yourself, ask members of your intended audience for their input on the map or sketch you've created.

---

[1] Jay Bolter, *Writing Space: Computers, Hypertext, and the Remediation of Print* (Mahwah, NJ: Erlbaum, 2001), 25.

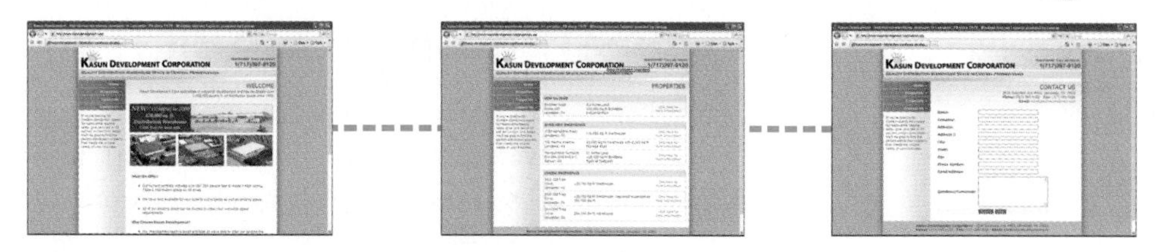

FIGURE 17.14 Linear organization of a website

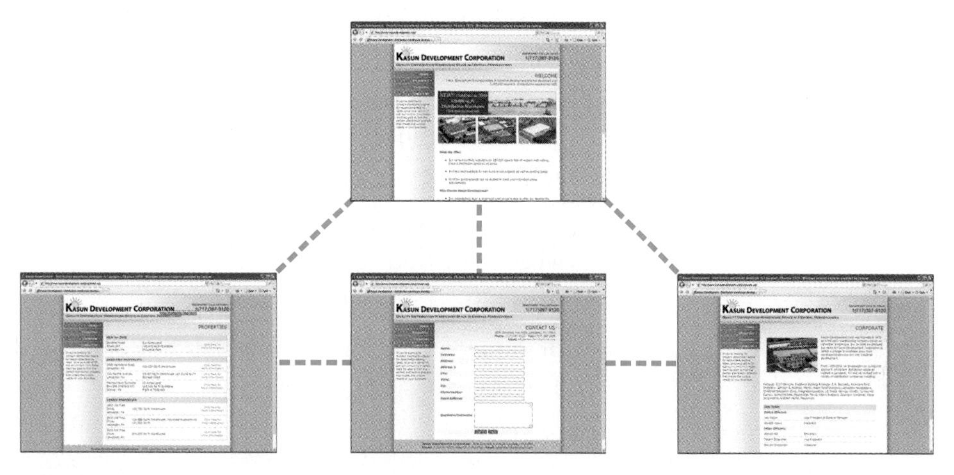

FIGURE 17.15 Hierarchical organization of a website

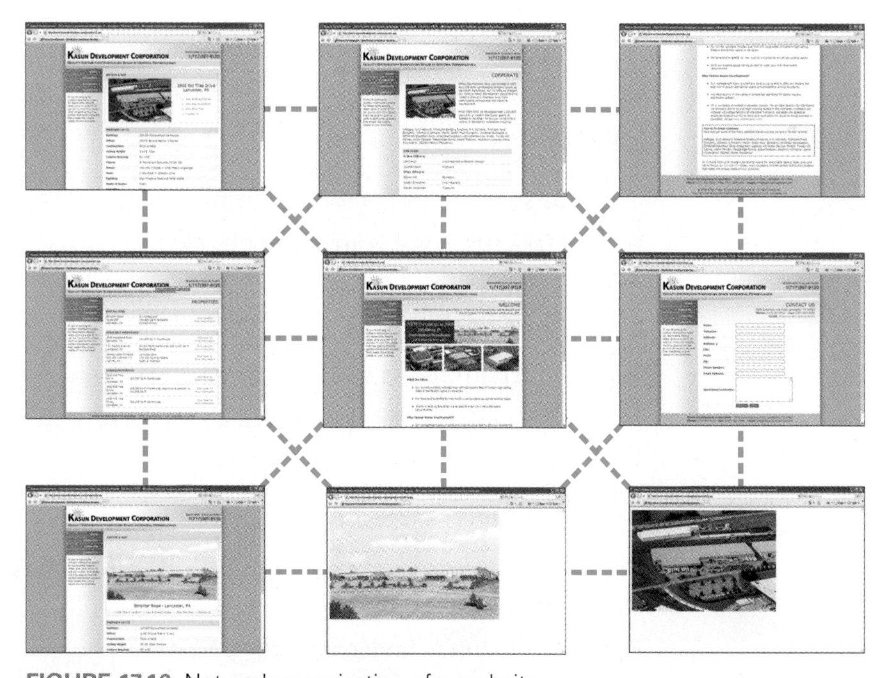

FIGURE 17.16 Network organization of a website

ANALYZE THIS

Take a close look at your university's website, examining it for the components and features we've covered in this chapter. Is there continuity between the homepage and the node pages? Does the website contain the appropriate amount of detail at each level? What sort of navigation does it use? Is the navigation easy to follow and effective?

Write a letter to the designers of the website (or perhaps to the university president), complimenting the aspects of the website that work well and giving specific suggestions for improvements.

WEBSITES AND USABILITY

Like all of the other documents addressed in this text, web pages should be tested for usability. However, testing websites requires attention to some aspects of design that might not be considered with other types of workplace documents.

Navigation

- Test the navigation system to ensure that it is clearly understood and easy to follow.
- Be sure that readers/users always know where they are within the site, where they have been, and what options they have to move on through the site.
- Be certain that navigation markers are consistent throughout the site.
- Make sure that links are clearly identified as links, and that they indicate where they lead.
- Be certain that readers can always quickly return to the homepage.
- Be sure that any internal searches link readers to the correct pages.

Content

- Be sure that content is clear and easy to understand.
- Be sure that readers can easily find important information.
- Be sure that you haven't included too much content on a single page.

Visuals

- Be sure visuals don't overwhelm readers.
- Make sure visuals are clearly labeled.
- Ensure that icons are easily interpreted.
- Check to be sure the file size of visuals doesn't cause slow download times.
- Be certain to use clear, contrasting colors, particularly in text.
- When using copyrighted visuals, make sure you include citation information.

Other

- If the website promotes branding, be sure it does so ethically and professionally.
- Be sure users understand the purpose of the site.
- Ask users whether they enjoy using the site and find it useful.

EXPLORE

A number of websites offer suggestions for testing web pages for usability. Locate two or three such pages. What advice do they offer that might be useful for workplace web writers? Make a list of the usability suggestions you find on these pages.

WEBSITE ETHICS

Workplace writers who create websites face all of the same ethical issues regarding audience and information dissemination that we've discussed throughout this text. Web writers must consider the legal implications of the information they post, they must present honest information, and they must consider issues of confidentiality in what should be posted and what should not. In addition, workplace writers who create websites must consider three other key issues.

Ensuring Accessibility

According to the W3C, "Web accessibility means that people with disabilities can perceive, understand, navigate, and interact with the Web, and that they can contribute to the Web."[2] In addition, Section 508 of the U.S. Rehabilitation Act requires that all federal websites be accessible to employees with disabilities. A recent study by the Disability Rights Commission, however, revealed that 81 percent of public websites failed to meet the minimum standards for web accessibility as established by the W3C.[3]

To make your website accessible, you must first recognize the four different types of impairment that might limit your readers' access—vision, hearing, motor, and cognitive. Your primary goal in accommodating readers with disabilities should be to provide alternative approaches to features that are intended to be seen, heard, acted on, or completed. As you design your website, consider the following strategies:

- Avoid small fonts and light-colored text.
- Be sure colors offer distinct contrast between background and other colors.
- Provide detailed captions for visuals.
- Provide transcripts and descriptions of animation and video.
- Use XHTML and CSS whenever possible to ensure that more browsers can open a page.
- Use headings and lists that show clear differences in levels or items.
- Make hypertext links obvious and logical.
- Limit individual page size so that readers with low bandwidth can open the page efficiently.
- Describe the content of charts and tables.
- Provide alternate content in case plug-ins are not available.
- Provide alternate access, such as keyboard access for users who cannot use a mouse.
- Do not limit the time a reader can access a page or an option.
- Validate your work to ensure that you are passing along correct information.

EXPERTS SAY

"The Internet is an essential tool. And, literally, a lifeline for many disabled people. I have Dragon Dictate. And while I was in rehab, I learned to operate it by voice. And I have enjoyed corresponding with friends and strangers with that system. Many disabled people have to spend long hours alone. Voice-activated computers are a means of communication that can prevent a sense of isolation."

CHRISTOPHER REEVE

EXPLORE

There has been a trend toward legal enforcement of web accessibility in the business sector as well. In August 2004 Ramada.com and Priceline.com settled with the state of New York to make their websites more accessible to the visually impaired, agreeing to pay $40,000 and $37,000 respectively for investigation costs (see http://www.lctjournal.washington.edu/Vol2/a006Fkiaras.html). Conduct a web search to locate other cases in which companies have been found legally required to make their web pages more accessible.

[2] W3C, "Introduction to Web Accessibility," http://www.w3.org/WAI/intro/accessibility.php.

[3] Disability Rights Commission, "The Web: Access and Inclusion for Disabled People," 2004, http://www.dvc-gb.org/Docs/theweb1.rtf.

Presenting a Fair Image

Remember that the web page you design will promote a particular image of the company it represents. It is your ethical and professional responsibility to ensure that the content and design of your web page represent your company positively and fairly.

Providing Accurate Information

Because Internet web pages are designed for public consumption, web writers have an ethical responsibility to provide accurate information. This means not only that immediate content must be accurate, but also that design and style must link readers to the proper information and allow them to navigate freely, rather than diverting them from any information.

ANALYZE THIS

Locate a company web page—any company will do. How does that web page work to convey an image of the company it represents? Write a short analysis of the web page, focusing on the accuracy of the image and the information it portrays.

SUMMARY

- Websites are documents that rely on words and images to convey ideas to readers.
- There are some key differences in writing web pages and print documents—for example, size, dimension, navigation, access, visuals, and multimedia.
- There are two distinct kinds of web pages: those made for the Internet and those made for intranet systems.
- Important web terminology includes *web page*, *web browser*, *URL*, *homepage*, *hypertext*, *link*, *interface*, *navigation*, *search engine*, *site architecture*, *site map*, *server space*, and *cookie.*
- Web technologies include HTML, XHTML, tables and frames, Cascading Style Sheets, web-authoring software, JavaScript, CGI, CMS, and plug-ins.
- Continuity for web page writers refers to a continuous design and an uninterrupted flow between pages.
- Writers must provide navigation cues so that readers can find the information they want: hyperlinks, navigation bars, home links, tabs, and search fields.
- Splash pages, homepages, node pages, and subpages are all part of websites.
- Search optimization and metatags help ensure that search engines find a web page.
- Some websites also contain an FAQ page and a site map.
- Websites can accommodate more flexible, spatially organized schemes, including hierarchical and network patterns.
- Web page writers must ensure accessibility, present their companies fairly, and provide accurate information.

CONCEPT REVIEW

1. What are the basic differences between a print document and a web document?
2. What are the basic differences between an intranet web page and an Internet web page?
3. What is a web browser?
4. What is a URL, and how is it constructed?
5. What is hypertext?
6. What is a search engine?
7. What are cookies?
8. What is HTML?
9. What is XHTML?
10. How do tables and frames function on a web page?
11. What is CSS, and what are some advantages to using it?
12. What are plug-ins?
13. What is continuity in web pages?
14. What is branding, and what are some strategies for branding when designing web pages?
15. What are some good navigation tools to use when designing a web page?
16. What is a splash page?
17. What are some essential characteristics of a good homepage?
18. What are metatags?
19. What is an FAQ page?
20. How does hierarchical organization work when designing a web page?
21. What are the key considerations in addressing the accessibility of a web page?

CASE STUDY 1

Sports-U Redesigns its Website

You work as a technical writer for Sports-U, a sporting-goods manufacturer that specializes in college- and university-related sporting goods, such as college sweatshirts and jackets, balls, hats, and Frisbees. Most of your products are designed using college colors, mascots, and symbols, and many are sold directly from your website, although some are sold in college and university bookstores.

Your supervisor, Alicia Cuttrell, has informed you that you'll be part of the team that will redesign the company's website, which now consists of a single web page on a plain white background with black text describing the equipment Sports-U makes and distributes. Your first responsibility, Alicia tells you, will be to conduct some market research about what viewers want in a website. You've already identified college and university students as your primary audience; now you need to find out what they want in a website marketing university-related sporting goods.

Create and conduct a survey of ten to fifteen questions to determine the elements and features that most viewers would want in Sports-U's website. Draw your survey questions from this chapter as much as possible. For example, you might ask respondents whether they prefer a linear, hierarchical, or hypertextual organization in the websites they visit or whether they prefer simple, text-based websites as opposed to those with lots of color and visuals. Try to survey at least ten people, and then compile your results in a memo written to your instructor.

CASE STUDY 2

Designing a Website for a Nonprofit Organization

Few nonprofit organizations have the resources or support to build a sophisticated website. In fact, many nonprofit websites are poorly designed, out-of-date, or even nonexistent. Find a nonprofit organization that has a very basic website or one that has no website at all. This could be an organization in your own community, a regional group to which you belong, or even a national association with limited funding. Contact the leaders of the organization, and ask whether they would consider a website that you would create for them. If they agree to consider it, do some extensive research on the organization, gather all of the appropriate materials and images from a contact person within the group, design the website, and then post it on a private server. Finally, present the website to the organization's leaders (and perhaps to your instructor), and ask for feedback, offering to revise or redesign the site as they see fit.

CASE STUDIES ON THE COMPANION WEBSITE

On the Intranet: Mildred Jackson and the Internal Job Postings

In this case you serve as the head of corporate training for Twin Sisters Electronics and are asked by company executives to develop new training programs for employees, particularly programs that relate to proper Internet and intranet usage as well as to the different kinds of terminology used when working with these technologies.

Playing Well with Others: Cole Blackburn and Creating a Children's Website

You have recently been hired as a web consultant at a company that produces websites for other companies that market to children. In this case you will look closely at problems that surround the updating of the companies' websites, including ways to make these sites compliant with governmental regulations and ways to make them easier to locate through search engines.

Website Comparisons: Ben Bryson and the Competition's Website

In this case you assume the role of marketing director for *Grow* magazine, which is dedicated to issues related to organic and sustainable farming. You are charged with the development of a new website for your company and are asked to ensure that it follows all applicable web standards, entails a new overall design and marketing strategy, aids in supporting the company's mission and sustainability projects, and helps differentiate *Grow* from its competitors. In addition, you are asked to address an ethical breach committed by one of your company's competitors.

Optimizing a Website for Search Engines

You design and maintain the website for a property management company and use the site for all of your marketing needs as well as for a place of information about the company's apartments. However, your apartment units are failing to lease because of low search engine results. You must figure out a way to optimize your page before students rent from other companies.

VIDEO CASE STUDY

Cole Engineering—Water Quality Division

Revising the Cole Engineering Website

Synopsis

Cole Engineering has a website that is over two years old. Many of its applications are outdated and do not present the cutting-edge feel that the company wishes to portray to clients and customers. In addition, Cole Engineering has expanded its services and areas of research and development. Because the Water Quality Division has grown more than any other division since the website was first designed, this portion of the website is most in need of change and revision. As a web designer, Dennis Blount can give the Water Quality Division a number of suggestions and ideas about the best way to present information through a website; however, he does not have the technical expertise in hydrology and other water-quality issues that would allow him to revise the website all by himself. He needs their input and expertise to make sure that everything presented on that page is technically accurate.

WRITING SCENARIOS

1. In a small group, examine a sophisticated website for a large company or organization. Write a report to your instructor in which you analyze the website, based on the various elements of website design addressed in this chapter, as well as the elements of visual rhetoric and layout and design covered in chapters 8 and 9.
2. In a small group, conduct research and then create a short set of instructions to teach a beginner to create a basic web page. Be sure to conduct some usability testing to ensure that your instructions work. Compare your group's instructions with those of other groups in the class, or compare them with a set of instructions provided by your instructor.
3. If you have created a personal website of your own, save it to a storage device, and trade that storage device with one of your classmates who also has a personal website saved. Then create an alternative or revised version of the other person's website, while that person creates an alternative version for you. Share and discuss these new versions, describing each of the changes or modifications made to the current sites. If you decide to use some or all of the elements from the new version on your actual website, be sure to credit the source.
4. Creating and designing professional websites has become a big business over the past decade or so. Research three web design companies, and write an analysis of the services they offer. If you were planning to outsource the creation and design of a small business website, which of the three companies would you choose? Explain your reasons.
5. Find a current, comprehensive guide to website design, either online or in a bookstore or library. Spend a few hours studying it, and then write a short review of its contents. What subjects does it address that this chapter does not? What subjects are covered in both texts, and are they addressed in the same fashion or differently? Does the more comprehensive guide seem worthwhile and useful for your current or future needs, or does this chapter provide you with all you need to know about creating websites?

6. Find a website for an organization or group whose beliefs, goals, or perspectives differ from your own. For example, you might examine the website of an opposing political or social group or a competing company. Using the elements of website design and creation covered in this book, analyze the website. Does it offer an accurate portrayal of the group? How does the website seek to persuade viewers? Does it present information about the organization or group ethically and fairly?
7. Locate a web page that does not meet the W3C standards for accessibility. How would you redesign the page? Write a short explanation about what needs to be done to the web page to make it more accessible. Be sure to identify the URL so that your teacher can look at the original page. You might consider including screen shots of the page to clarify points of your critique.
8. A number of different websites offer codes of ethics for website design (use search terms like "website ethics" or "web design ethics" to find them). Examine several of these codes, and then create your own top ten list of ethical responsibilities and practices that website designers should follow. If possible, rank them in order of importance.
9. Find the websites of three companies in your intended or current profession. Study the organizational and structural elements of each site, and then write a memo to your instructor analyzing and comparing the three sites. Your analysis should focus exclusively on the pattern of organization used, the introductory and node pages, and the various navigational and directional features. Are all three sites effectively organized? Can you rank them based on the comparison?
10. Using the same three websites you examined in the previous scenario, apply the information about visual rhetoric and images (see chapter 8) and layout and design (see chapter 9), and reevaluate the three sites. Write a more lengthy analysis and comparison of the three sites. Does this expanded analysis influence your opinion of the effectiveness of the three sites? Does it change your ranking of them?
11. Analyze your college's or university's website. How is it organized? What impression does it give you about the school? How does it brand your school? How does it use colors and images to persuade or influence various viewers? Who do you think is the intended audience for the website? In a memo to your instructor, address these questions.
12. This chapter mentions several of the more popular website design software packages used by web designers today. Obtain a copy of one of these software programs (perhaps a trial version if you'd rather not make a purchase). After you've learned its features and capabilities, write a memo to your instructor describing what can be done with the software and explaining whether you'd recommend it to your classmates.
13. Many websites today include sophisticated visuals and multimedia, because most people have access to DSL and cable connections when they access the Internet. Imagine that high-speed connections suddenly disappeared and all web surfers were forced to use slower, dial-up connections to view websites. What impact would this have on the workplace? What problems or opportunities would it create? How would businesses and industries adapt to this change? Write a narrative describing what you think this would be like.

14. In a memo to your instructor, outline the major advantages and disadvantages of using web design software to create a basic website versus using a text editor and coding the site yourself. Conclude by recommending the approach you believe to be better for students in your class.

PROBLEM SOLVING IN YOUR WRITING

Select one of the documents or websites you designed and created for this chapter. Then consider how you employed the problem-solving approach in creating that document. Write a letter to your instructor identifying how you planned and conducted research to gather information, drafted the document or website, reviewed it, and finally distributed (or plan to distribute) the document or website.

18

Technical Instructions

CHAPTER LEARNING OUTCOMES

After completing this chapter, you will be able to do the following:

- Understand how **technical instructions** solve problems
- Consider the role **ethics** plays in writing and designing technical instructions
- Apply some **general guidelines** to writing technical instructions
- Decide which of **many parts** need to be included in the instructions you write
- Understand the role **visuals** play in instructions
- Understand the purpose of a **quick reference card**
- Design useful **help pages**
- Develop and execute detailed **usability testing** for your instructions
- Address key elements of **design** when producing instructions

DIGITAL RESOURCES

On the Companion Website www.prenhall.com/dobrin:

- Case 1: Integrating Pictorial and Textual Instructions: Mathias Contracting of Boise and IKEA
- Case 2: Technical Instructions: Records Archiving at Ross Realty of Houston, Texas
- Case 3: Technical Instructions: Chemical Safety at Roxey Manufacturing
- Case 4: Shutting Down at the End of the Day
- Video Case: Testing the GlobeTalk Handheld Devices
- PowerPoint Chapter Review, Test-Prep Quiz, Exercises and Activities

REAL PEOPLE, REAL WRITING

MEET DIANE STIELSTRA • User Assistance Lead for Microsoft Corporation

What types of writing do you do in your workplace?

I help to produce the online help systems for the Windows operating system and for MSN Messenger. If you are using Windows, any words you see on your computer screen are things that I may have written or helped to write. For instance, when you get error messages, when you look at wizards, when you look at the various commands and instructions, those are all the jobs of the user assistance writer. I also produce a lot of internal writing within Microsoft. In fact, sometimes I think I do more of that than actual external writing because a great deal of work goes into planning how our external writings are going to be positioned. There's a lot of internal writing that goes on around an external piece of writing.

What are your goals in writing instructions?

We want readers to be able to get what they need as efficiently, as quickly, and as easily as possible. That is not easy. It takes time and real ingenuity to figure out how to write things in a really concise, direct, and precise way.

How do you approach writing technical instructions?

First of all, we do a lot of audience research. One of the keys to doing good writing of any kind is being incredibly sensitive to and aware of your audience. To understand what an audience wants, you have to do a lot of research. If we really want to write for users in languages they are going to understand, we need to pay attention to the languages they use.

You must spend a lot of time revising based on that research.

Yes. Nothing ever goes out without having been read by a number of people. Everything is always open to revision for a number of reasons: getting a better fix on the terminology and on the phrasing, for instance. But also, our software is always changing, so we must always revise to accurately reflect the changes in the software.

Is collaboration an important part of your writing?

Yes. We work very closely with our editors. Our writers are generally subject matter experts, and we are kind of equal partners with our editors in owning the whole process. The editors also work a lot with terminology and tone. Those are both crucial areas.

What advice about workplace writing do you have for the readers of this textbook?

The key to being a good writer—and I just can't emphasize it enough—is really knowing your audience and writing directly to them.

INTRODUCTION

"When all else fails, read the instructions." Why is that such a familiar saying? Because so many instruction manuals are so poorly written that we don't trust them, don't want to be confused or bored by them, and certainly don't want to wade through their baffling writing. Do you remember being a kid and getting a new game—like Monopoly or Risk—and opening the instructions, seeing how much you'd have to read, then scrapping the instructions, and making up the rules the way you wanted them? We learn early that reading instructions can be tedious and confusing—not to mention boring. This chapter focuses on writing concise, easy-to-read, ethical, professional instructions that are interesting and useful to readers.

In January 2008 a California tourist visiting the Hudson River Valley in New York drove onto a railroad track after following the directions provided by the on-board GPS system in his rental car. The GPS step-by-step directions—a specific kind of technical instruction—had offered incorrect information that resulted in the rental car getting stuck on the tracks and being destroyed by a train. The driver, who was not injured in the wreck, has been held responsible for the accident and the damages to the rental car, the train, and over two hundred feet of electrified train track. But there is now speculation that the GPS company could be liable in the case. The question is, What happens when inaccurate instructions lead to accidents?

Situations like this raise interesting questions about the responsibility not only of GPS companies but of all those who write instructions to provide accurate and understandable information to readers. We must also question any meaningful difference between inaccurate instructions and imprecise instructions or even incomprehensible instructions. Even though substanderd instructions don't always result in dramatic accidents, understanding the ramifications of poorly written, inaccurate, or incomprehensible instructions must be central to workplace writers who write technical instructions.

Of all workplace writing genres, you are probably most familiar with instructions. In fact, it is nearly impossible to go through a day without encountering some form of instructions. For example, how did you get to your classroom today? Chances are good that you read some street signs, which provided you with specific instructions.

Like many other kinds of instructions, signs solve problems. As readers we understand not just the words on the signs, but also the visual elements—colors and shapes—that are part of the command of that instruction. Even the traffic sign shown above that contains no words still conveys a particular instruction and solves a particular problem—that is, how traffic should proceed when two lanes come together. And other informational signs function similarly, providing the necessary information for readers to make decisions and perform tasks.

No instructions are universal, however, even instructions as simple and direct as traffic signs, because they don't convey accurate information to everyone in all situations. Instead, they address particular audiences in particular rhetorical situations, and

See chapter 5 for more on transnational communication and chapter 8 for more on using visuals:

their contexts are often cultural. Shapes, colors, words, and images—like those found on traffic signs—might be easily interpreted by some readers but would be completely unfamiliar to others. Being attuned to how transnational audiences respond to various visual cues and instructions is important, but more important is to remember that no visual approach will accurately instruct all audiences.

Signs are actually only a small portion of the instructions you encounter daily. Think about how often you read directions on your computer when you encounter a problem: Why won't my CD drive burn a disk? Why doesn't my file save? Why can't I print my document? Why won't my dial-up connect? How can I clean my computer? Instructions are texts that teach readers to solve a particular problem, perform a task, put something together, or repair a product or mechanism.

Hey! This book itself is a technical instruction manual! It helps you solve workplace and rhetorical problems.

Think about a recipe for a moment. A recipe solves a specific kind of problem: how to prepare a particular dish without having to remember all the ingredients and steps. Recipes include the necessary components of good instructions: a title, often a visual of the food, a detailed parts list, easy-to-follow steps (again, some with visuals), cautions and warnings, and other secondary information.

Even solving a simple problem like giving street directions requires an understanding of the basic principles of writing instructions and good document architecture. Consider how basic map and directional resources like Mapquest offer step-by-step directions for getting somewhere by adhering to a standard approach for delivering directions. Take a moment to look at Figure 18.1, which offers directions for getting from the Occupational Safety and Health Administration offices in Washington, DC, to the Library of Congress, also in Washington. Notice the simplicity of the steps, the use of visuals, the shifts in font styles, the necessary background information, and even the links to associated maps.

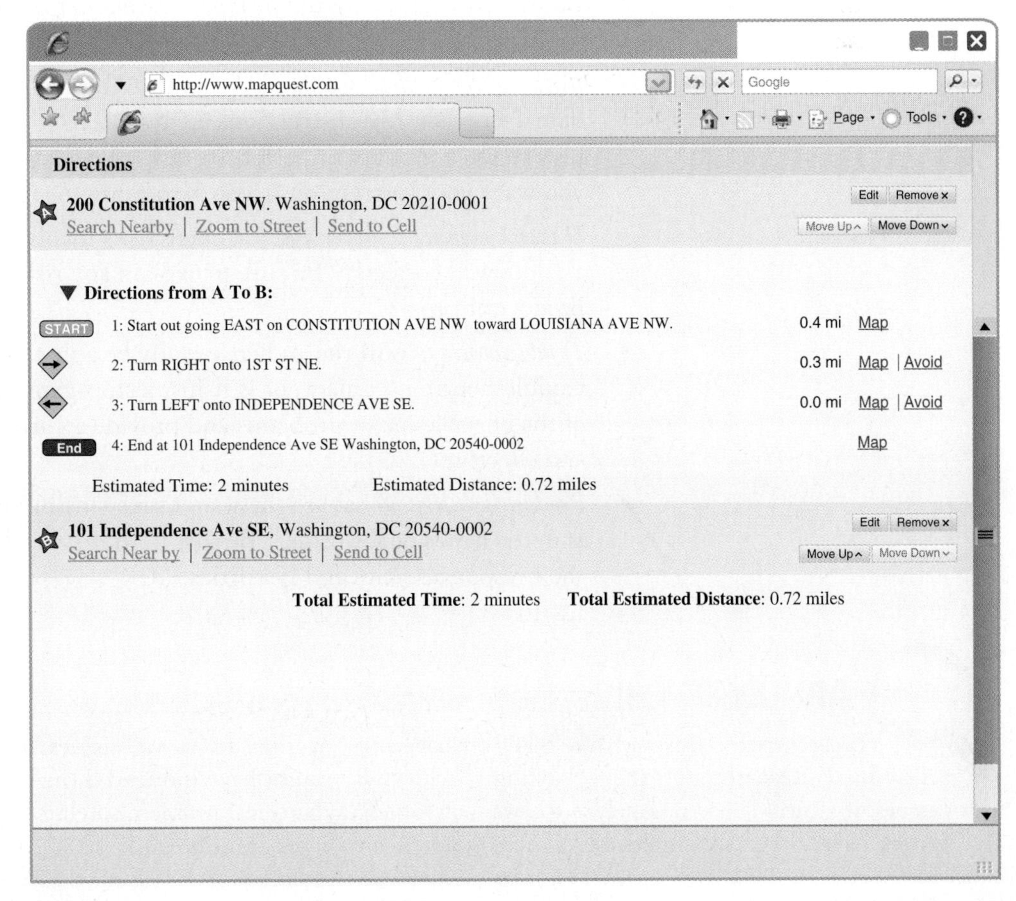

FIGURE 18.1 A simple set of instructions

Many other scenarios prompt workers to write a set of instructions to show readers how to perform a task or solve a problem:

- Putting something together (like a prefabricated bookcase)
- Repairing an object (like troubleshooting a computer software installation)
- Operating a device (like a tractor or a scanner)
- Installing a device (like a new faucet or a DVD player)
- Making a product (like knitting a scarf or making a pinhole camera)
- Maintaining an object (like cleaning a DVD player)
- Packing or unpacking an object (like shipping a printer cartridge back to the manufacturer)
- Working through a process or procedure (like filing a lawsuit)
- Providing documentation (like filling out a student loan application)

In addition, some instructions are written in conjunction with other technical documents.

ETHICS AND TECHNICAL INSTRUCTIONS

As noted in chapter 4, ethics is central to all workplace documents. When drafting technical instructions, writers must consider the following ethical issues:

- *Functionality:* Do the instructions actually teach the audience to solve the problem at hand? Be certain that the instructions allow the reader to successfully complete the task.
- *Safety:* Can someone get hurt because of the negligence of these instructions? Be sure that following the instructions doesn't place the reader in harm's way.
- *Misuse of the product:* Can a reader misuse the product and cause harm or damage because the instructions are not clear? The reader must be shown how to use a product or perform a task, as well as how *not* to. The last thing you want your instructions to do is to cause the reader more problems.
- *Hazards even with proper use:* Are there risks involved even when instructions are followed correctly? Be sure to explain any risks that are inherent in the product or process.
- *Troubleshooting:* Will the audience really be able to solve a problem using the troubleshooting segment, or is it just a stop-gap measure? Be sure to cover all of the possible major problems and provide solutions that are feasible, safe, and effective.
- *Product liability:* What are the writer's responsibilities for this product once it is in the hands of the reader? Be sure both the writer and the reader understand their responsibilities and liabilities.

ANALYZE THIS

Locate a set of instructions that you have recently used. Read through them, looking specifically for issues of ethical concern. Analyze the ways in which the writers have either accounted for or have ignored or overlooked ethical issues. Consider what legal concerns might have influenced the instructions. Consider, too, how rhetorical problem solving plays into the ways ethical concerns are addressed.

KEY ELEMENTS

Not all technical instructions are designed the same, but most share similar key components. Knowing when to use these elements is a crucial part of rhetorical problem solving because most technical instructions don't require all of them. For instance, instructions for making macaroni and cheese might not require any special considerations, but instructions for installing a new motherboard might.

Title/Title Image

Even though some instructions do not need titles, many do. An instruction manual, for example, often begins with a title and a title page. In many cases writers also include an image to show the audience the product being discussed or the problem being solved. Writers can think of the title as a thesis statement, a goals statement, or a solution statement; it tells the audience what the instructions accomplish and what problem they solve. Without a title or a title image, how would you know if the instructions would solve the problem you were facing? Titles should be clear and concise and should let readers know where they will end up.

Byline

Like all technical documents, technical instructions serve as part of an ongoing record. Therefore, audiences must know who produced the document so that they can contact the appropriate person or company if the instructions fail. The byline does not need to include the name of an actual author but can include information about the company responsible for the instructions.

Date

For two primary reasons it is important to include the date the instructions were produced: First, technical instructions may be updated regularly as technologies or procedures change. Think, for instance, about the rapid updating of software components. In some types of instructions, in fact, it may be crucial to also identify a product revision level, which indicates the version of instructions being used. Second, like any other technical document, instructions are part of the documentation history of a problem or a product, and dates track the relationships among different documents produced at different times.

Dates on technical instructions are not always obvious. Often the date appears on the last page of an instruction manual, in a small, inconspicuous font. Where the date is placed in technical instructions is determined by the role the date plays in the instructions. If, for instance, the instructions address a product that is not date sensitive—say, the assembly of a bookcase—then the date can be placed in a less obvious position. However, if the instructions *are* date sensitive—say, instructions for shutting down a store's computers while new software is being updated—then the date should be more prominent. Notice in Figure 18.2 the prominence of the date on the title page of this instruction book.

IN YOUR **EXPERIENCE**

Look at some of the instructions you may have kept for software or electronic devices that you own. Do those instructions identify version numbers or publication dates? If not, why do you suppose they don't? If so, where on the instructions does that information appear?

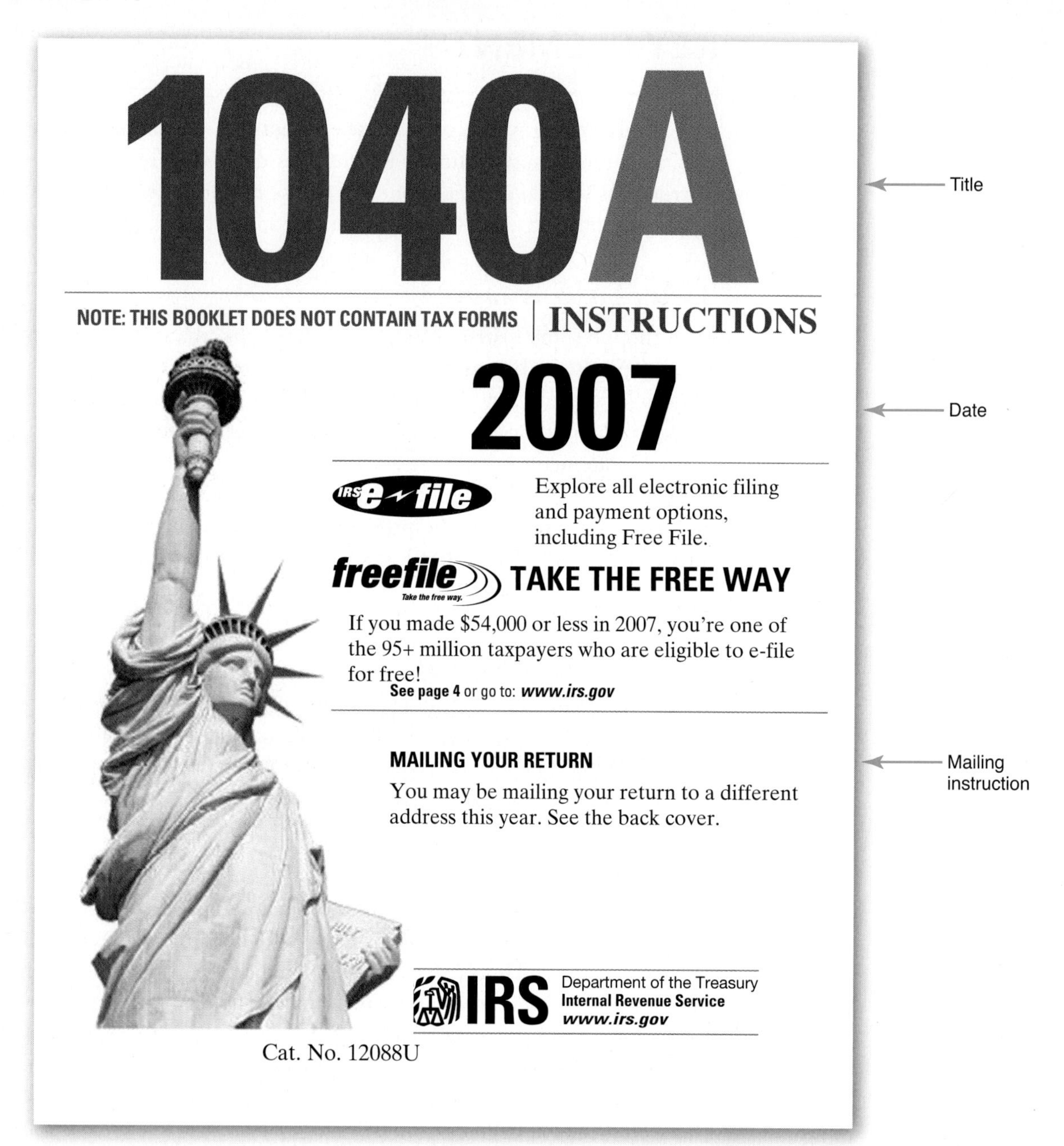

FIGURE 18.2 Instructions showing a date

Introduction

Introductions to technical instructions do not necessarily have to be introductory paragraphs or even be written in narrative form, although in some cases they may be. A technical instruction introduction tells the audience the aim of the instructions and the problem the instructions should help solve. Included in the introduction are the outcomes of the instructions. Because some technical instructions are only part of an ongoing, larger series of instructions that assist readers in multiple stages of problem solving, the introduction should let readers know what will be accomplished when they have finished all the requirements of these instructions. For example, a set of instructions might guide readers in building a small airplane engine, but another set of instructions might be required to install that engine in an airplane. Thus, introductions detail the

DeSoto Global
Tracker1•Owner's Manual

Introduction

Welcome to Tracker1

Congratulations on your purchase of a Tracker1® Global Positioning Satellite system. The DeSoto Global Tracker1® system is a unique system that integrates an automotive navigational GPS system with a tracking and recovering system to give you a reliable positioning instrument. Though easy to use, the Tracker1® is a sophisticated system, and we urge you to read this manual thoroughly before using your new system. This manual is designed to provide you with information for setup, basic operations, and advanced navigation and tracking operations.

Description

The Tracker1® Global Positioning Satellite System integrates a self-contained automotive navigational GPS with a universal tracking chip. The GPS features a detailed road atlas stored in the unit, and offers a 528 kb expandable memory to allow for map downloads from your computer. The GPS system also includes a self-contained antenna, as well as an antenna extension for mounting on your vehicle to improve satellite reception. The Tracker1® GPS operates using two possible power sources: a 12 volt auto plug or six standard AA batteries (not included). When using batteries, the Tracker1® will operate continuously for 36 hours. The Tracker1® also features an adjustable, backlit display panel for maximum readability in any lighting condition.

FIGURE 18.3 An introduction to an instruction manual

scope of the instructions. Figure 18.3 provides an example of a clear introduction to a set of instructions.

Introductions can also identify the organizational structure of the instructions and may include a table of contents, as the example in Figure 18.4 shows. In addition, the introduction can present any terminology or special conventions used in the instruction manual. For instance, instructions for beginners to use sophisticated digital camera equipment might identify terms specific to digital photography. Figure 18.5 provides an example of an introduction that presents key terms. Finally, introductions may include additional information about how readers can use the instructions—for example, reading from beginning to end or skipping around through the instructions to complete particular tasks. See Figure 18.6 for an introduction that provides additional information.

Department of the Treasury
Internal Revenue Service

Table of Contents

FIGURE 18.4 A table of contents in an instruction manual introduction

Parts Nomenclature

Standardized Parts Naming and Numbering System

Parts are identified by plate addresses that include plate numbers, rows, and columns. Plate addresses are prefixd by a catalog abbreviation. Standardized parts names should be used when ordering any part from TelTech.

FIGURE 18.5 An introduction that identifies key terms

Additional Information

For more information about filing a complaint, refer to Gelcone's web page to understand who can file a complaint, when employees can issue a complaint, what information must be provided in a complaint.

FIGURE 18.6 Additional information found in an introduction

Alerts

For ethical and professional reasons, writers use alerts to inform audiences about safety considerations.

Types

- *Danger* alerts indicate the possibility of serious injury or death to the reader. These should be handled with great seriousness.

⚠ DANGER Do not insert arms or other body parts into chopping unit while chopper is engaged. Serious injury or death may result!

- *Warning* alerts indicate the potential for minor or moderate injury to the person or damage to the product or equipment being used. Warnings, too, must be handled seriously.

⚠ WARNING Do not ingest or inhale fule tank cleaner. Ingestion may result in illness.

- *Caution* alerts indicate a range of possible problems. Cautions address the potential for minor injuries to the reader or slight damage to the equipment. They show that the results of the instructions can vary if the instructions are not followed precisely.

⚠ CAUTION Toner cartridge must be inserted as shown, or printer will not close properly, possibly resulting in damage to the printer.

⚠ CAUTION Do not move on to step 16 until all adhesives have completely dried. Wet adhesives will not offer the strength or support needed to install the dowel rod xx in step 16.

⚠ CAUTION Be sure all toggle switches are in the off position before continuing; otherwise you may experience a power surge that may damage the main unit.

- *Notes* provide special suggestions or tips for the reader that make completing the instructions more efficient.

⚠ If you have trouble fitting the machine screw into the predrilled hole, try sanding the hole opening with a small piece of fine grain sand paper.

These examples use different icons to indicate the different alerts. Some technical writing books and manuals suggest using a single icon with changing words. However, we highly recommend that you use visually distinct icons that both attract attention and distinguish the different kinds of alerts. Notice the difference in the two sets of icons that follow this paragraph. Keep in mind, too, that transnational audiences may not recognize red as inherently signaling danger or yellow as signaling caution. When writing for transnational audiences, you must be sure to learn about color meanings for the specific regions for which you are writing.

See chapter 5 for more on transnational communication.

Standards

Some organizations, including government organizations, set standards for how alerts are identified within a given field, industry, organization, or location. Such standards ensure that readers and writers share common understandings. It will be critical for you to familiarize yourself with the standards of your industry.

Many fields, industries, and organizations comply with alert standards set by the American National Standards Institute (ANSI). ANSI has established a series of standards for product safety signs and labels (also known as the ANSI Z535.4 standard), which defines what alert labels are to be used in various conditions. According to ANSI, "A product safety sign or label should alert persons to a specific hazard, the degree or level of hazard seriousness, the probable consequence of involvement with the hazard, and how the hazard can be avoided."[1]

The ANSI Z535.4 standard states that signal words identify the degree of hazard seriousness. The standard defines these signal words:

- *Danger* indicates an imminently hazardous situation that, if not avoided, will result in death or serious injury. This signal word is to be limited to the most extreme situations.

- *Warning* indicates a potentially hazardous situation that, if not avoided, *could* result in death or serious injury.

- *Caution* indicates a potentially hazardous situation that, if not avoided, could result in minor or moderate injury. It can also be used with unsafe practices.[2]

EXPLORE

Look through the ANSI web pages (www.ansi.org). What are some of the standard safety signs and labels that you find? Make a list of six of them. As a class, discuss the various signs and symbols recommended by ANSI.

---

[1] *American National Standard: Product Safety Signs and Labels* (Rosslyn, VA: National Electrical Manufactures Association, 1998).

[2] Ibid.

Corrosive hazard

Electrical hazard

Explosive hazard

Fire hazard

Flammable gas

Flammable liquid

Flammable solid

Non-flammable gas

Irritant

Poison

Radioactive hazard

Biohazard

FIGURE 18.7 Standard alert icons

Attracting Attention

Alerts must stand out from texts visually so that readers are drawn to them. Because readers do not always read instruction manuals with sustained attention to details, alerts need to be visually conspicuous, which can be accomplished in four primary ways. First, graphics and icons can be used that stand out from written text. Countless clip art and downloadable icons are available online, giving writers many options for highly visible, identifiable icons that match alerts and make sense to readers. Figure 18.7 includes some of the standard alert icons, notice their use of shape and color.

See chapter 8 for more on clip art, icons, and visuals in general.

A second approach is to make the text stand out. You can use large, bold lettering for crucial alerts like dangers or warnings, but you shouldn't clutter the document by changing fonts too much.

See chapters 8 and 9 for more on typography.

WARNING: A CHANGE IN FONT AND SIZE HAS OCCURRED!

A third option is to use color. Because most instruction manuals can now be printed in color, you should use this technology to your advantage. Certain colors carry certain understood meanings: red means warning, and yellow means caution, for example

For more on the use of color, see chapters 8 and 9.

WARNING:
Changing text color may attract your reader's attention!

And fourth, positioning on a page can distinguish an alert from other text and can affect whether readers see it or not. For instance, putting the message "Warning: It is important that your text stand out!" right here in this paragraph provides no visual differentiation from the rest of the text. However, different positioning can make the text stand out.

See chapter 9 for more on visual positioning of text.

Warning: It is important that your text stand out!

In addition to attracting attention, writers must be sure to include an explanation of what might result if a reader does not heed an alert. These explanations can take the form of if-then statements: *If* the reader does not do this, *then* this may result, or *IF* the reader does do this, *then* this may result.

⚠WARNING When charging the power unit, be sure the positive and negative connectors correspond before turning on the power source. *If* the connections are reversed, *then* a power surge may cause damage to the unit.

Placement and Repetition

Many alerts should be provided to readers *before* they use instructions, and writers must ensure that readers will not overlook the alerts. Sometimes it may be necessary to warn a reader up front about a general concern and then include a separate, specific alert in the text near the item of concern. In the example in Figure 18.8, from Britax Child Safety's instruction manual for a car seat, warnings and safety information are provided early in the manual (pages 2 and 5 are shown) and then are emphasized in specific contexts (pages 13 and 19 are shown).

In addition to alerts within instructions, some alerts should also be included on the object or product itself to remind users of those alerts. Notice in Figure 18.9 that the alerts noted in the instructions in Figure 18.8 are repeated and emphasized on the product itself. Note, too, that some parts of the instructions can be placed on the product to repeat and emphasize crucial steps.

It's better to alert early and often than late and infrequently. Don't worry about repetition, but don't overdo it either. You won't need to remind readers of basic alerts at every step, and instructions shouldn't be cluttered with alerts. Deciding how and where to place alerts is a critical part of rhetorical problem solving and ethical decision making. You must stay focused on the safety of the reader.

Equipment Needed

Many technical instructions require readers to access and use particular materials to complete the instructions. In many instances those materials are included in the package with the instructions, but frequently readers are required to supply the materials and tools. Readers must be informed about these requirements prior to beginning the first step of the instructions. They want to know up front what tools they will need so that they can work through the instructions without interruption.

In some cases readers also need to know how to use that equipment. This doesn't necessitate a separate set of instructions but may require a brief description. Part of your task as a rhetorical problem solver is determining how extensive your instructions must be to provide readers with details about the tools they need to use to complete the instructions.

Usually, instructions include a list of required tools and equipment and may also include a corresponding list of images, showing readers what the equipment looks like. Deciding how to provide a tools list and images is another part of rhetorical problem solving; working through the PSA can help you decide what kind of presentation to use.

Whatever decisions you make, be sure that images are clearly presented and clearly identified. We have seen instruction manuals that included only images of required tools,which readers could not identify. One such set of technical instructions listed this required tool:

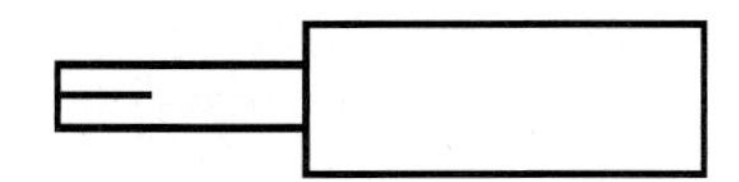

Readers could not figure out what this tool was, leaving them with a new problem, rather than solving the problem they started with. A toll-free troubleshooting line later identified the tool as a Phillips head screwdriver.

WARNING

DEATH or SERIOUS INJURY can occur:

- Use only in a rear facing position when using this restraint in the vehicle. This infant restraint cannot be used forward facing in a vehicle.
- Use only with children who weigh between 4 and 22 pounds (2 and 10 kg) and whose height is 30 inches (76 cm) or less.
- Snugly adjust the belts provided with this restraint around your child. A snug strap should not allow any slack. It lies in a relatively straight line without sagging. It does not press on the child's flesh or push the child's body into an unnatural position.
- An optional Low Birth Weight Foam insert may be used to help obtain a snug fit for infants whose weight is less than 7 pounds (3 kg.). To order optional foam, call Britax at 1-888-4BRITAX.
- Secure this child restraint with the vehicle's child restraint LATCH anchorage system if available or with a vehicle belt.
- Follow all instructions labeled on the child restraint and in these written instructions.
- According to accident statistics, the National Highway Traffic Safety Administration recommends that parents select the rear seat as the safest location for a properly installed child restraint. Please study the sections on Vehicle Safety Belts and Vehicle Seating Positions in this booklet to ensure the child's safety. If in doubt about installing the child restraint, consult the vehicle owner's manual.
- Secure this child restraint even when it is not occupied. In a crash, an unsecured child restraint may injure vehicle occupants.
- Consult your child's doctor before using this child restraint. Some infants must be secured with a device that allows them to lie flat.

2

Safety Information

This child restraint system conforms to all applicable Federal motor vehicle safety standards. This restraint is certified for use in motor vehicles and aircraft.

Fill out the registration card and mail it in today!

Child restraints could be recalled for safety reasons. You must register this restraint to be reached in a recall. Send your name, address, email address if available, and the restraint's model number and manufacturing date to Britax Child Safety, Inc. 13501 South Ridge Drive Charlotte, NC 28273 or call 1-888-4BRITAX or register online at www.BritaxUSA.com/registration. For recall information, call the U.S. Government's Vehicle Safety Hotline at 1-888-327-4236 (TTY: 1-800-424-9153) or go to http://www.NHTSA.gov.

WARNING! DO NOT place a child restraint rear-facing in the front seat of a vehicle with a passenger air bag. DEATH or SERIOUS INJURY can occur. The back seat is the safest place for children 12 and under.

5

Using the Restraint

Recline Adjustment

To adjust recline, release tension of LATCH or vehicle belt then turn knob on front of base until desired angle is achieved (Fig. D).

- *When correctly reclined, the red line on base will be parallel to level ground.*
- *If there is a problem stabilizing the base, place a rolled towel in the crease of the vehicle seat to help level the base.*

D

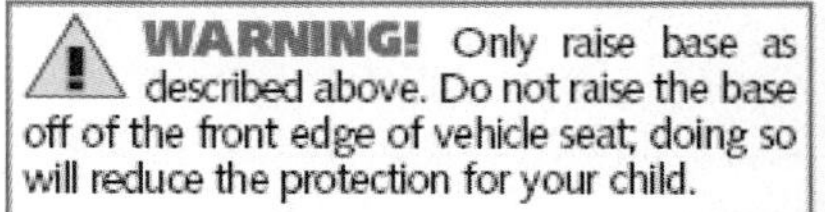

WARNING! Only raise base as described above. Do not raise the base off of the front edge of vehicle seat; doing so will reduce the protection for your child.

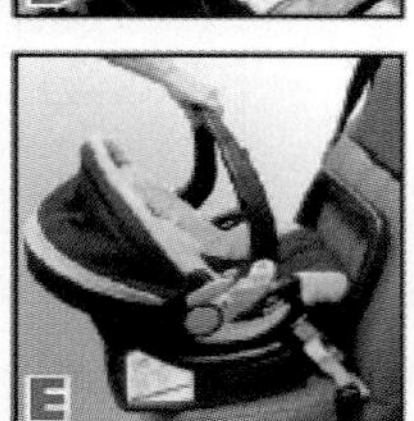

Attaching Restraint to Base

To install, align restraint rear facing with slots in base, then push down until it clicks into base (Fig. E). Pull up from child restraint handle to be sure it is properly secured. Rotate handle to "in vehicle position".

WARNING! Never attach restraint to base unless the base is secured in a vehicle as described on pages 20-25.

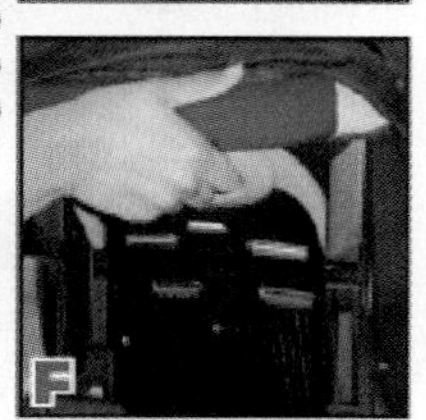

Removing Restraint from Base

To remove child restraint from base, squeeze and hold release handle on the rear of the restraint, then lift restraint by handle (Fig. F).

13

Vehicle Seating Positions

 **WARNING!**

- Forward facing vehicle seats MUST be used with this child restraint. Side facing or rear facing seats CANNOT be used.
- DO NOT place a child restraint rear-facing in the front seat of a vehicle with a passenger air bag unless the air bag is turned off. DEATH or SERIOUS INJURY can occur. The back seat is the safest place for children under 12.

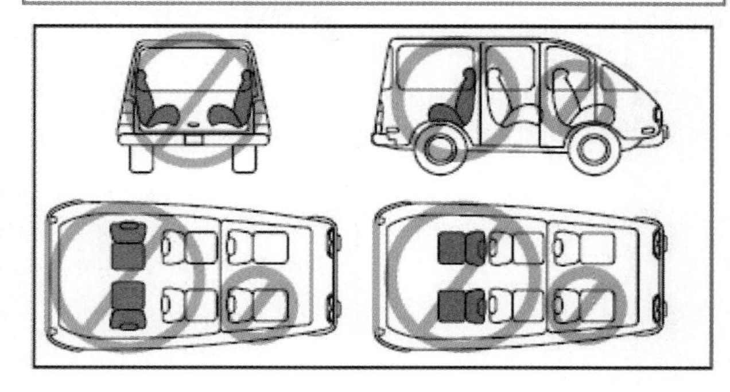

FIGURE 18.8 Repeated alerts in an instruction manual

FIGURE 18.9 Alerts repeated on the product itself

Be sure to include all tools and equipment needed, no matter how minor a piece or how minimal a part it plays in the instructions.

A good tools or equipment list must clearly identify whatever the reader will need to complete the instructions. Notice the correspondence between the images and the text in Figure 18.10. Imagine the uncertainty a reader might experience if the tools list was presented with only the text or only the images (see Figures 18.11 and 18.12). Notice, too, in Figure 18.11 how the centered text makes reading the list even more difficult. Remember that text is a graphic element, and deciding how it will be formatted and presented is part of the problem-solving approach.

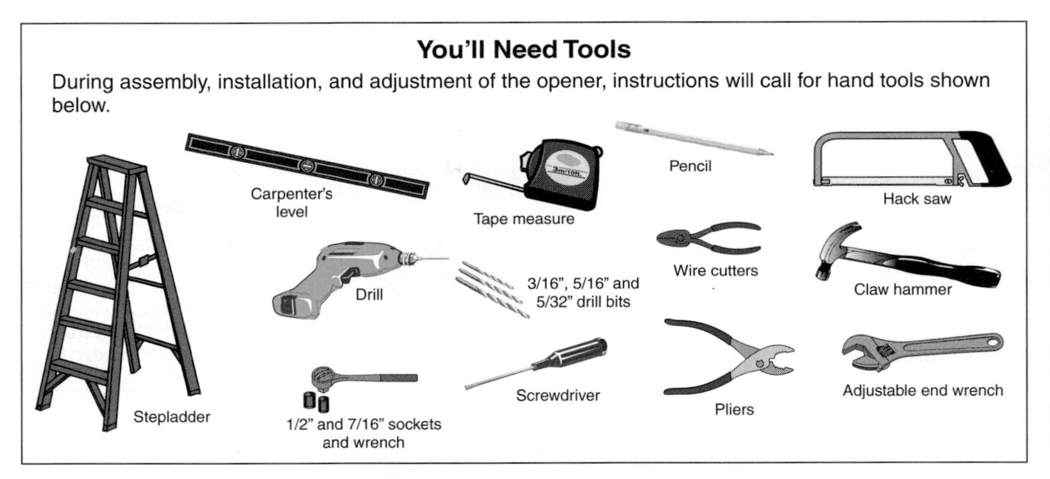

FIGURE 18.10 A tools list with corresponding text and images

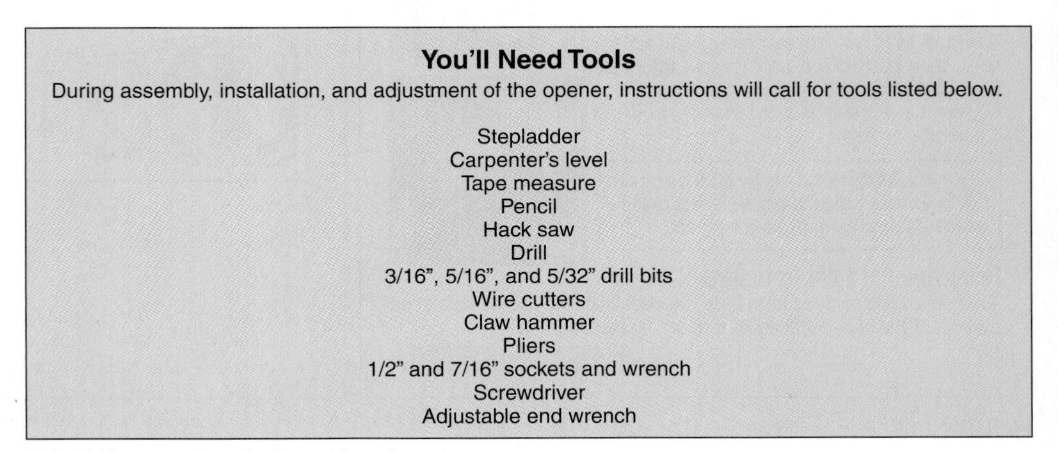

You'll Need Tools

During assembly, installation, and adjustment of the opener, instructions will call for tools listed below.

Stepladder
Carpenter's level
Tape measure
Pencil
Hack saw
Drill
3/16", 5/16", and 5/32" drill bits
Wire cutters
Claw hammer
Pliers
1/2" and 7/16" sockets and wrench
Screwdriver
Adjustable end wrench

FIGURE 18.11 A tools list with only written text

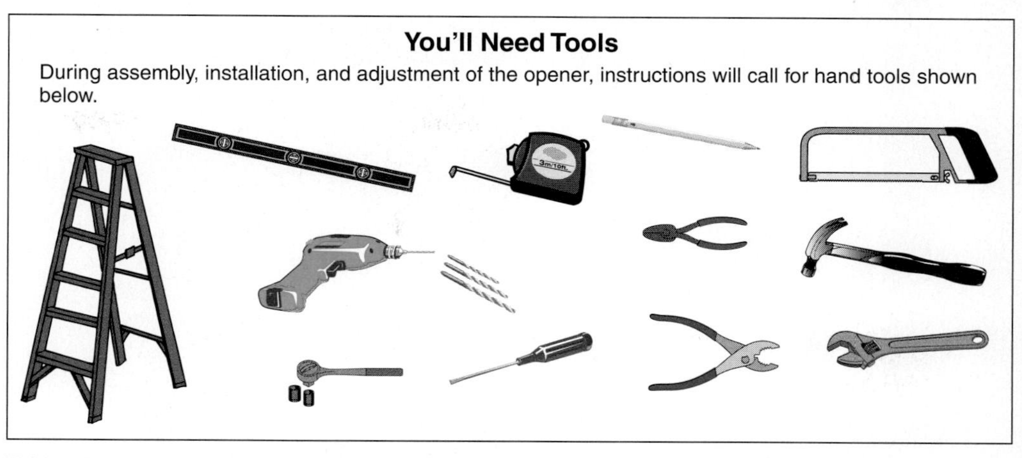

FIGURE 18.12 A tools list with only images

Parts List

A parts list functions much like a tools or equipment list because it provides a detailed list of materials. However, a parts list includes all the parts of the product itself that the reader will need to work with to complete the instructions. The parts list serves two functions:

- To identify what each part looks like and what it is called so that readers can easily identify it when needed
- To serve as a checklist so that readers can confirm that all parts have been included

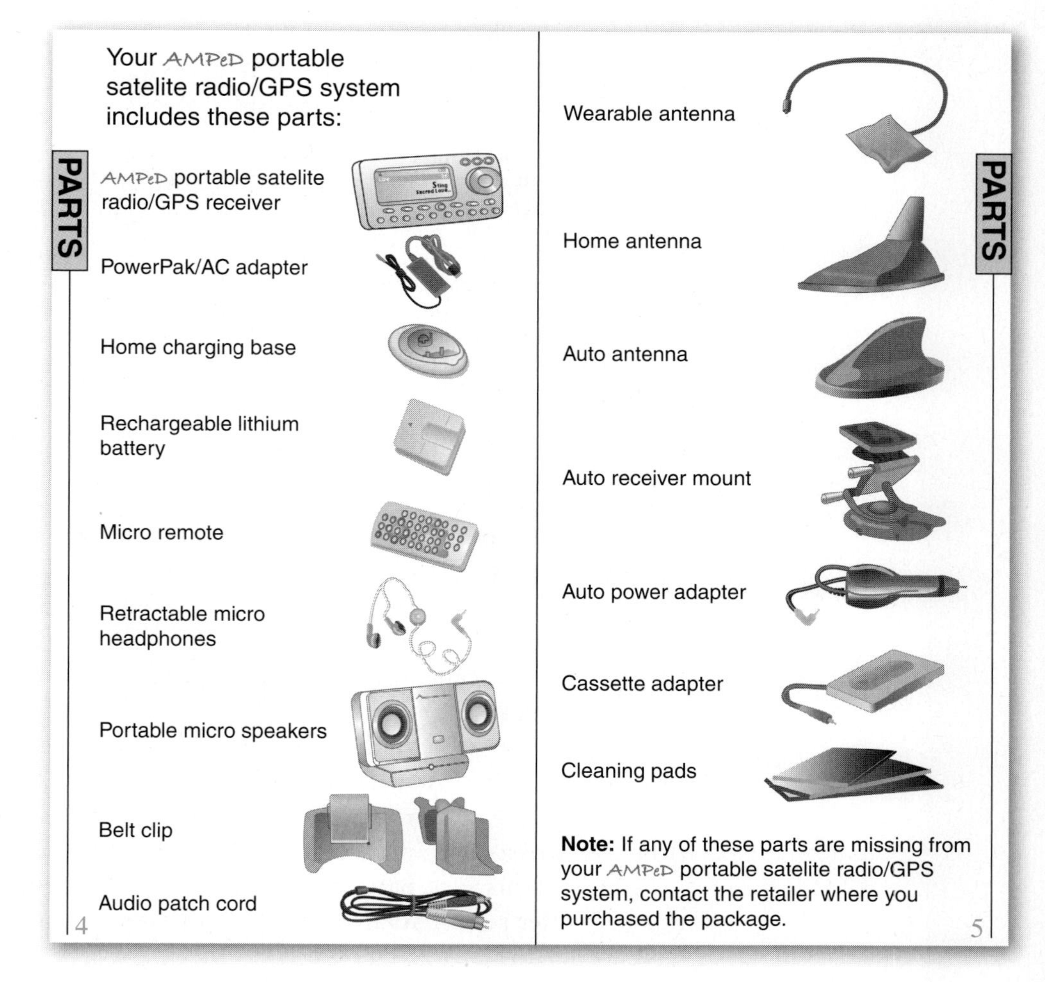

FIGURE 18.13 A parts list

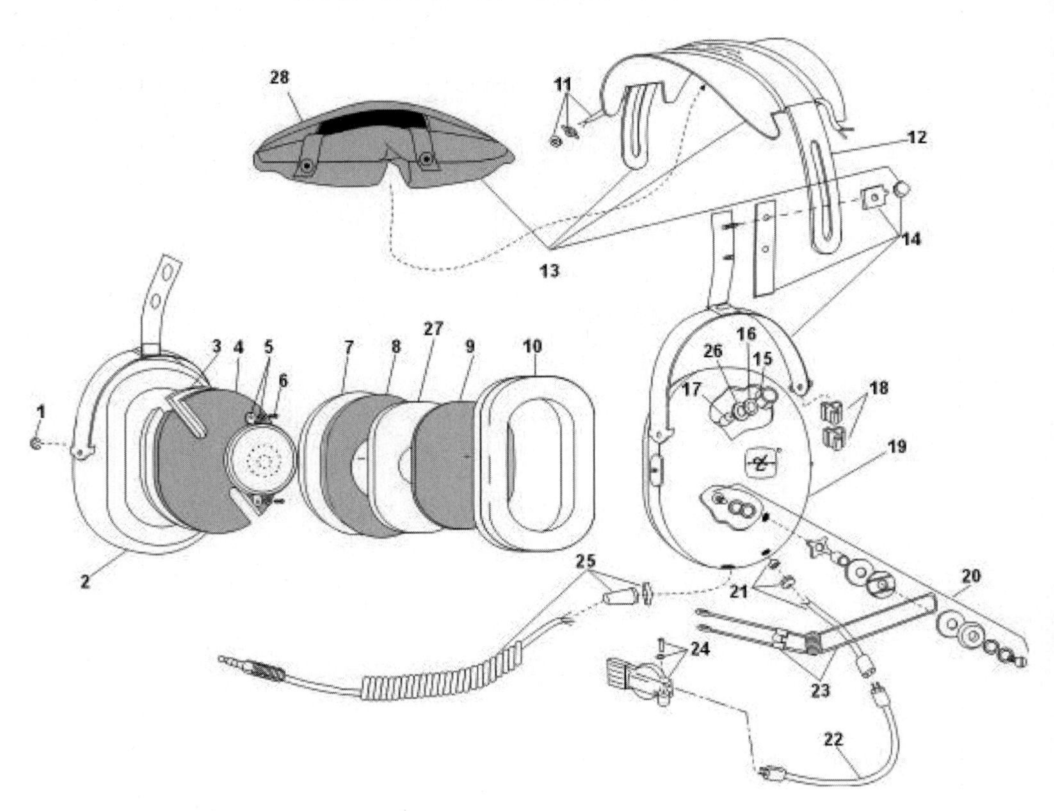

No.	Description	Part No.	No.	Description	Part No.
1.	Retaining Ring (10/pk)	00009M-03	15.	Volume Control Knob	09347P-22
2.	Right Dome	11696P-12	16.	Washer, Neoprene (10/pk)	40688G-21
3.	Filter, Yellow	26475P-10	17.	Volume Control 100 OHM	09349P-28
4.	Filter, Charcoal	26475P-11	18.	Clip, Cord (4/pk)	15125P-02
5.	Earphone Assembly	10376G-30	19.	Left Dome (Mic Side)	11696P-28
6.	Screw (10/pk)	13180P-06	20.	Boom Guide Kit	12840G-11
7.	Filter, Yellow	25629P-09	21.	Mic Jack Cord Kit	10405G-05
8.	Filter, Charcoal	25629P-10	22.	Mic Cord Assy	10516G-14
9.	Filter	14096P-09	23.	Microphone Boom Assy	12765G-04
10.	Undercut Gel Earseal (pair)	40863G-02	24.	M-87/AIC Microphone	09168P-18
11.	Overhead Cord Kit	22607G-01	25.	Comm Cord Kit	18028G-26
12.	Headband Spring	15093P-01	26.	Lock Washer (10/pk)	40688G-09
13.	Headband/Headpad Kit	40688G-36	27.	Filter, Yellow	14096P-18
14.	Stirrup & Clamp Kit	22378G-08	28.	Headpad Restraint, Foam	40867G-01

FIGURE 18.14 Parts list using an exploded view and corresponding numeric list

Have you ever been working your way through a set of instructions only to find that a single part is missing? Frustrating, isn't it? If you had known that the part was missing *before* you began, you could have saved yourself a lot of frustration and time. Or have you ever come to a point in a set of instructions when you couldn't identify a part that was called for? A well-designed, clearly written parts list can alleviate such problems.

Sometimes a parts list identifies the parts a reader needs to gather together to complete the instructions. In those cases it is up to the reader to get all the required parts, but it is up to the writer to provide the details needed to secure the *correct* parts. Recipes, for instance, offer at the beginning of the recipe a list of the ingredients (i.e., parts) the reader needs in order to complete the dish. Recipes usually offer parts lists in a list form, as do many other kinds of instructions (see Figure 18.13).

See chapter 8 for more on exploded-view drawings and line drawings.

There are many ways to present parts lists. Many technical instructions rely on exploded-view parts lists to demonstrate how parts fit together as a whole (see Figure 18.14); they may use a corresponding numeric list or a direct label approach. Exploded-view diagrams are most often rendered as line drawings. Some parts lists identify different parts of a single item by showing users the locations of parts they will need (see Figure 18.15).

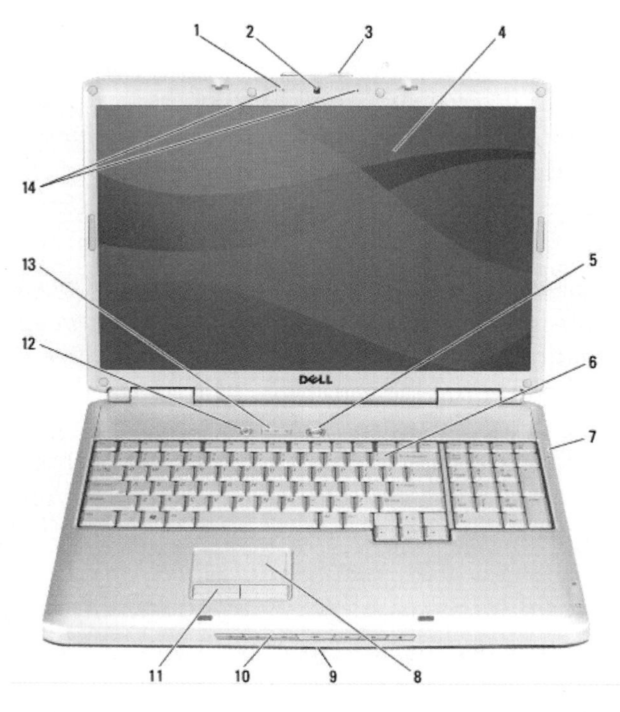

1. camera indicator (optional)
2. camera (optional)
3. display latch
4. display
5. power button
6. keyboard
7. device status lights
8. touch pad
9. consumer IR
10. media control buttons
11. touch pad buttons
12. Dell™ MediaDirect™ buttons
13. keyboard status lights
14. digital array microphones

FIGURE 18.15 A parts list identifying the location of parts of a single item

IN YOUR **EXPERIENCE**

Have you ever worked with a set of instructions and found that you were missing parts listed in the parts list or that you had parts that weren't listed in the parts list? How did that affect your response as a reader/user? What action did you take if and when you couldn't complete the task because you were missing a part(s)? Were you able to resolve the problem?

Steps

Steps are central to instructions. They tell readers what to do and provide the correct sequence of tasks. Determining the correct sequence of tasks from beginning to end depends on the scope of the instructions as well as the audience. Deciding on the best place to begin is part of the Plan phase. Notice in the *Fox Trot* comic strip the discrepancy between the steps of the instructions and the knowledge of the reader. Deciding where users begin the steps and in what order they proceed is also a key component of the Draft phase of the problem-solving approach. Using the PSA helps writers determine the order, number, and detail of the steps included in instructions. In addition, specific guidelines can help writers of technical instructions present the necessary steps in a clear and logical manner.

Number Each Step

Numbered lists allow readers to keep track of where they are in the steps. Without numbers readers must continuously read back through steps to relocate their position. In addition, numbered steps can be extended endlessly; you can't run out of numbers. On the other hand, if you choose to use a letter scheme, you limit yourself to twenty-six steps before you have to devise a secondary organization, such as Aa, Ab, Ac. And numbered steps allow you to create substeps, such as 1.1, 1.2, 1.3. Subset organization is useful for larger steps that require multiple actions before the reader moves to the next major step.

Perform All Steps As You Draft Them

As writers develop the order and details of each step, they should actually work through the process themselves, taking notes on each action performed. Doing the activity multiple times and writing down exactly what actions are performed can help writers develop accurate written instructions.

Begin Each Step with a Verb

Each step should read like a command, using a direct address to audiences. This approach makes clear to readers what they should do. Notice the difference between the first group of instructions that follow and the second:

> As you wait for the epoxy to dry, you should attach shelving unit D to bracket L.
>
> At this point the operator needs to shut down all power to the unit.
>
> The user can now enter the required data.
>
> You can now repeat steps 9 through 11.

> Attach shelving unit D to bracket L as the epoxy dries.
>
> Shut down all power to the unit.
>
> Enter the required data.
>
> Repeat steps 9 through 11.

The first four instructions don't begin with a verb; the second four do.

As always, context dictates the applicability of this guideline. In some cultures such a direct command format may be considered aggressive or even offensive. In such cases you should work with someone familiar with local expectations when you draft your steps. Similarly, when you are producing documents that may be translated into

For more on writing for transnational audiences, see chapter 5.

other languages or that have been translated *from* other languages, your translations must be accurate and culturally appropriate. Really knowing your audience is especially important when working with transnational or transcultural audiences.

Use Positive Commands Whenever Possible

As with other forms of communication, readers respond better to positive language in instructions. Examine the differences in these two examples:

> *Negative:* Do not leave the power on when you are finished with this step.
>
> *Positive:* Turn the power off when you are finished with this step.

> *Negative:* Do not leave rough spots on the platform.
>
> *Positive:* Sand the platform smooth.

Avoid Highly Condensed Language

As writers of instructions try to make individual steps as concise as possible, they leave out words that are not absolutely crucial. However, this effort can be taken too far; removing too many words can confuse readers. A common approach is the telegraphic style, the kind of writing once used in telegrams. This style emphasizes brevity because senders of telegrams used to pay by the word. Notice the difference between these two sentences:

> *Telegraphic style:* Pump #3 corroding.
>
> *Revised for clarity:* Pump #3 *is* corroding.

The easiest way to avoid a telegraphic style is to examine whether you have written the statement the same way you would say it. Also, pay attention to the kinds of words most commonly left out in telegraphic style: articles, linking verbs, pronouns, and prepositions.

Include Only One Action per Step

To make steps clear and concise, you should be as simple and direct as possible. Readers have an easier time working through short, concise steps rather than longer, more intricate steps. Single-action steps avoid confusion. Consider the differences between the instructions for popping corn in Figures 18.16 and 18.17.

Group Similar Steps Together

Readers should be encouraged to perform prerequisite steps in close sequence. It also benefits readers to perform a series of similar steps together rather than performing one task, then another kind of task, and then the original task again. Examine the differences between the instructions in Figures 18.18 and 18.19.

Grouping similar steps together results in a better flow of action for the reader. However, even though the general guideline suggests that you group like activities together, the context may require that you leave like activities scattered throughout the instructions. For instance, in the pallet example, preparing pallet N may require that steps 4 and 5 be performed first, or perhaps the resin on pallet M must have dried first. Like all rules of technical communication, context will dictate how you apply them.

Visually Separate Steps

Usually readers must glance away from technical instructions to perform the required tasks. When they return to the instructions, the steps should be visually demarcated so that readers can easily find their place. The example in Figure 18.17 leaves a space between steps, keeping each step independent and easy to identify. Figure 18.20 shows what that example would look like without those spaces.

Notice the greater difficulty in accessing the steps, particularly if you glance away and return to them. This is a relatively simple set of instructions, but you can imagine the effect if the instructions are dozens or even hundreds of steps long. Cramming

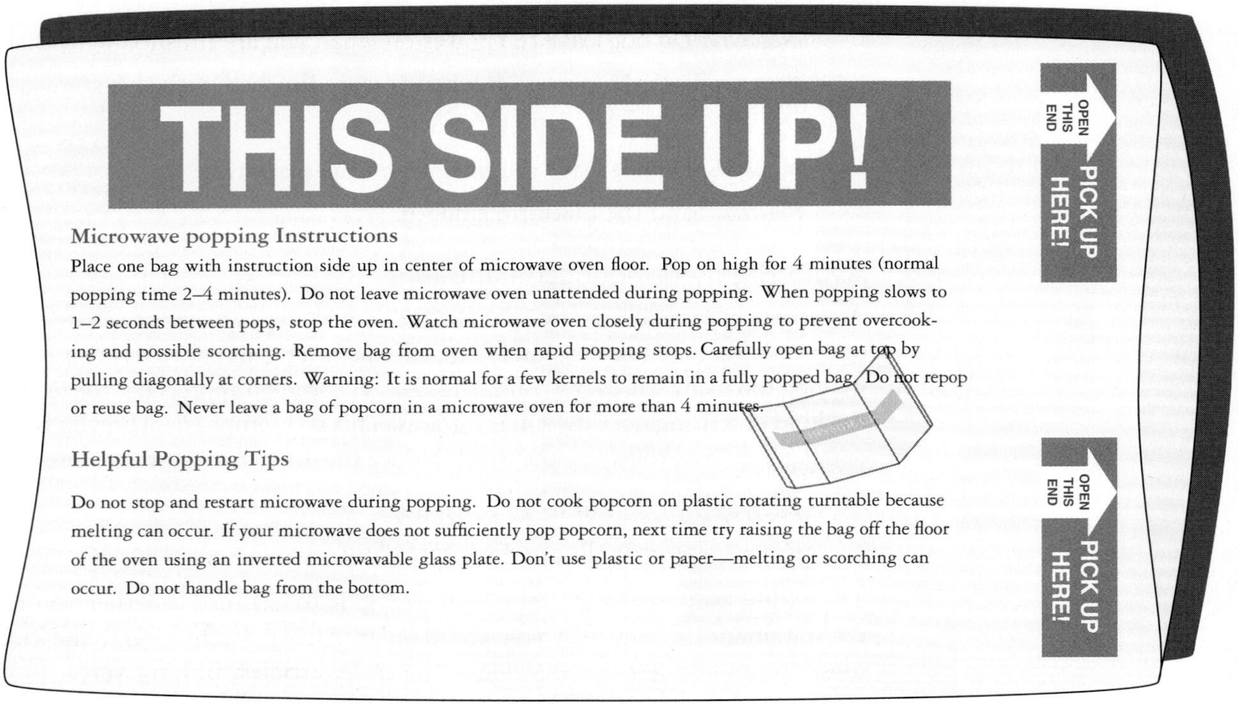

THIS SIDE UP!

OPEN THIS END

PICK UP HERE!

Microwave popping Instructions

Place one bag with instruction side up in center of microwave oven floor. Pop on high for 4 minutes (normal popping time 2–4 minutes). Do not leave microwave oven unattended during popping. When popping slows to 1–2 seconds between pops, stop the oven. Watch microwave oven closely during popping to prevent overcooking and possible scorching. Remove bag from oven when rapid popping stops. Carefully open bag at top by pulling diagonally at corners. Warning: It is normal for a few kernels to remain in a fully popped bag. Do not repop or reuse bag. Never leave a bag of popcorn in a microwave oven for more than 4 minutes.

Helpful Popping Tips

Do not stop and restart microwave during popping. Do not cook popcorn on plastic rotating turntable because melting can occur. If your microwave does not sufficiently pop popcorn, next time try raising the bag off the floor of the oven using an inverted microwavable glass plate. Don't use plastic or paper—melting or scorching can occur. Do not handle bag from the bottom.

OPEN THIS END

PICK UP HERE!

FIGURE 18.16 Instructions with multiple actions per step

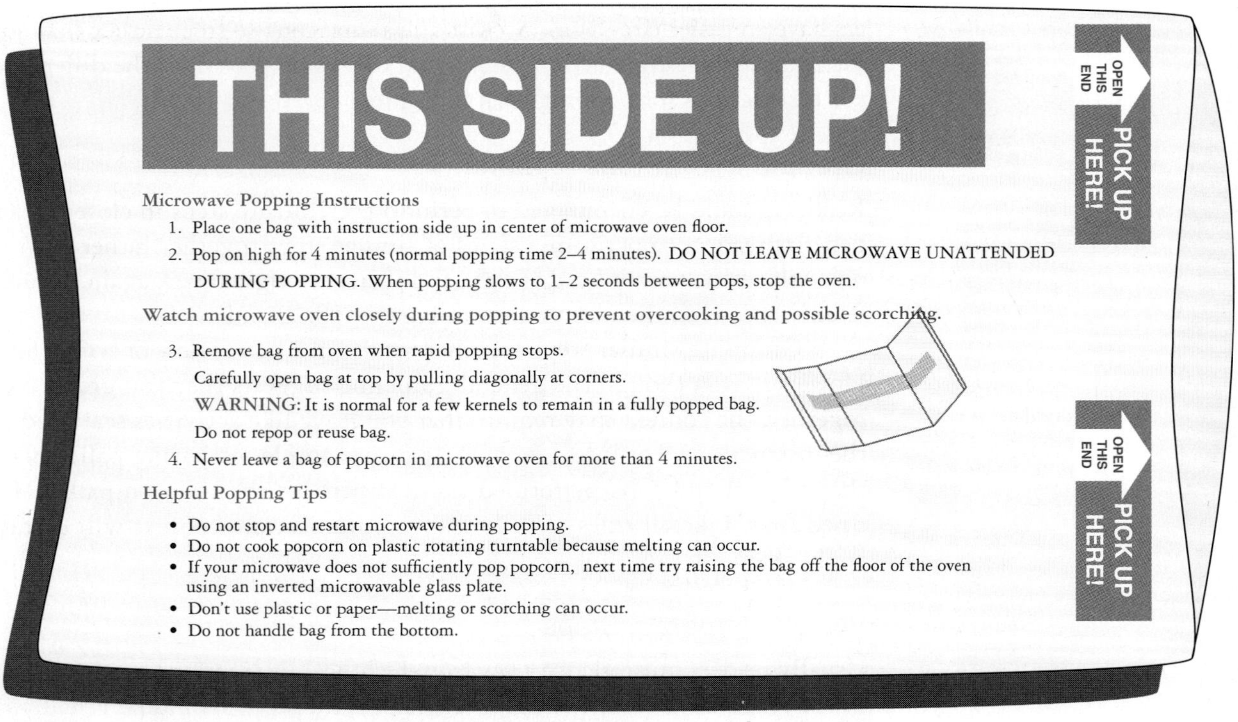

THIS SIDE UP!

OPEN THIS END

PICK UP HERE!

Microwave Popping Instructions

1. Place one bag with instruction side up in center of microwave oven floor.
2. Pop on high for 4 minutes (normal popping time 2–4 minutes). DO NOT LEAVE MICROWAVE UNATTENDED DURING POPPING. When popping slows to 1–2 seconds between pops, stop the oven.

Watch microwave oven closely during popping to prevent overcooking and possible scorching.

3. Remove bag from oven when rapid popping stops.
   Carefully open bag at top by pulling diagonally at corners.
   WARNING: It is normal for a few kernels to remain in a fully popped bag.
   Do not repop or reuse bag.
4. Never leave a bag of popcorn in microwave oven for more than 4 minutes.

Helpful Popping Tips

- Do not stop and restart microwave during popping.
- Do not cook popcorn on plastic rotating turntable because melting can occur.
- If your microwave does not sufficiently pop popcorn, next time try raising the bag off the floor of the oven using an inverted microwavable glass plate.
- Don't use plastic or paper—melting or scorching can occur.
- Do not handle bag from the bottom.

OPEN THIS END

PICK UP HERE!

FIGURE 18.17 Instructions with single actions per step

Step 3: Prepare pallet M by lightly coating the unfinished side of pallet M with a coating of resin provided in canister WR.

⚠WARNING Use resin only in a well-ventilated area. Respiratory damage may occur without ventilation.

Step 4: Assemble light mechanism by installing bulb R into fixture T.
Step 5: Attach light mechanism to power source. Use coupling brackets to ensure connection.
Step 6: Prepare Pallet N by lightly coating the unfinished side of pallet N with a coating of resin provided in canister WR, as you did for pallet M in step 6.

⚠WARNING Use resin only in a well-ventilated area. Respiratory damage may occur without ventilation.

FIGURE 18.18 Instructions with similar tasks not grouped together

Step 3: Prepare pallet M by lightly coating the unfinished side of pallet M with a coating of resin provided in canister WR.
Step 4: Repeat step 3 for pallet N.

⚠WARNING Use resin only in a well-ventilated area. Respiratory damage may occur without ventilation.

Step 5: Assemble light mechanism by installing bulb R into fixture T.
Step 6: Attach light mechanism to power source. Use coupling brackets to ensure connection.

FIGURE 18.19 Instructions with similar tasks grouped together

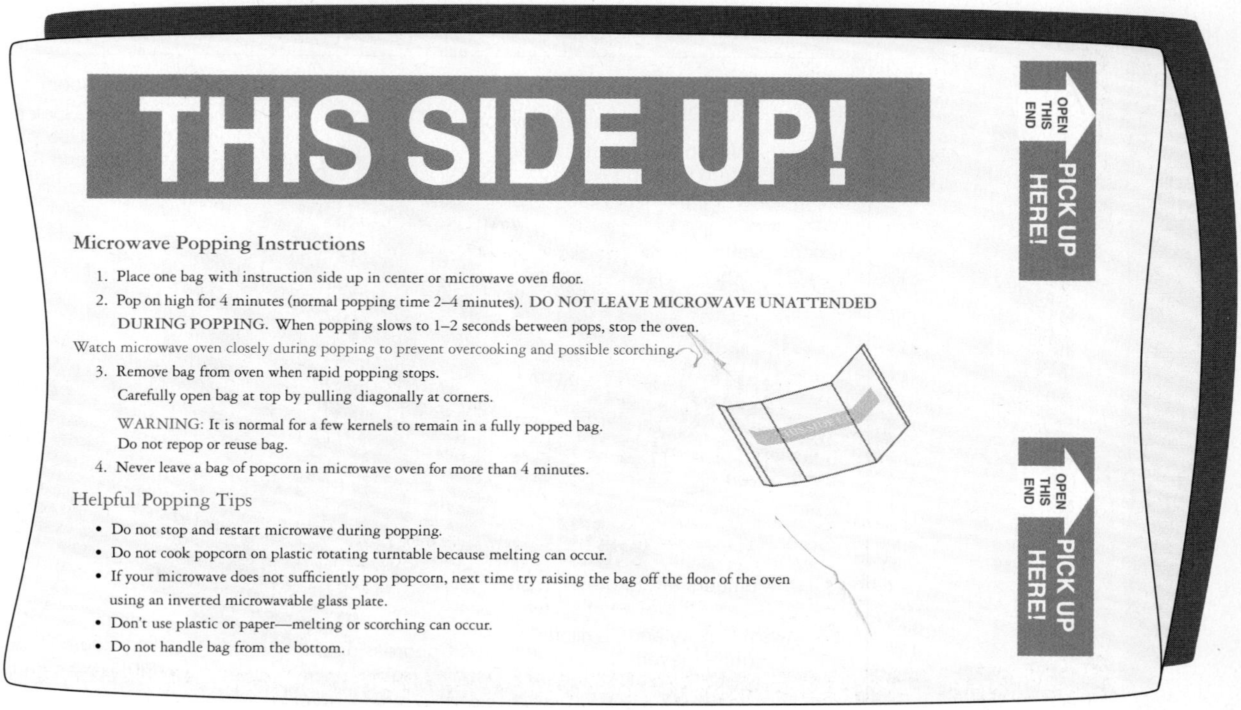

THIS SIDE UP!

Microwave Popping Instructions

1. Place one bag with instruction side up in center or microwave oven floor.
2. Pop on high for 4 minutes (normal popping time 2–4 minutes). DO NOT LEAVE MICROWAVE UNATTENDED DURING POPPING. When popping slows to 1–2 seconds between pops, stop the oven.

Watch microwave oven closely during popping to prevent overcooking and possible scorching.

3. Remove bag from oven when rapid popping stops.
   Carefully open bag at top by pulling diagonally at corners.
   WARNING: It is normal for a few kernels to remain in a fully popped bag. Do not repop or reuse bag.
4. Never leave a bag of popcorn in microwave oven for more than 4 minutes.

Helpful Popping Tips

- Do not stop and restart microwave during popping.
- Do not cook popcorn on plastic rotating turntable because melting can occur.
- If your microwave does not sufficiently pop popcorn, next time try raising the bag off the floor of the oven using an inverted microwavable glass plate.
- Don't use plastic or paper—melting or scorching can occur.
- Do not handle bag from the bottom.

FIGURE 18.20 Instructions without spacing between steps.

text together can create problems for readers—burned popcorn or worse—which can be avoided by simply employing a better spacing scheme.

Clarify Steps with Visuals

For more on using visuals, see chapter 8.

As often as possible, use visuals to clarify steps. Clearly telling a reader how to do something is effective, but showing a reader how to do something is even more effective. Visuals within steps can be used for a number of reasons:

- To indicate location (e.g., where a part fits, where a button is located)
- To indicate action (e.g., how two pieces should be fit together)
- To indicate results (e.g., what the parts should look like when the step is completed correctly, much like a title picture; or screen shots to show what your computer screen should look like when you use or install software)

Pay attention to how visuals interact with your text. Keep the visual near the step to which it is attached.

Sometimes, however, instructions do not need visuals; for example, instructions for filling out forms or for working through a process or procedure might not have steps that can be documented with visuals. You will need to make rhetorical choices about whether your documents need visuals. Figures 18.21 and 18.22 show examples of effective visuals.

EXPLORE

Search your school's web pages for a set of instructions that does not use visuals to guide readers through the steps of a process. Are the instructions clear without the use of visuals? Would visuals enhance the instructions? Why or why not?

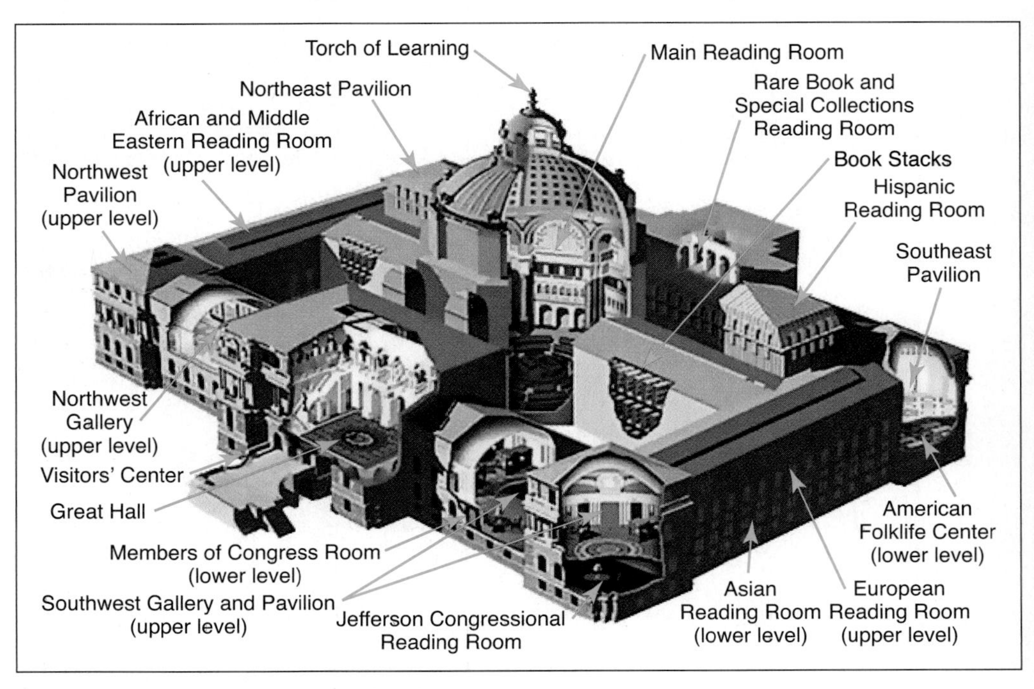

FIGURE 18.21 A visual showing locations

THE PATHOLOGY OF PITCHING

Researchers at the American Sports Medicine Institute broke down the fastball pitches of healthy, elite pitchers frame-by-frame to study the effect of pitching on the human arm. Their conclusion: Unless it's done very carefully, a major league fastball will eventually spell major league arm trouble.

Critical instant Arm acceleration

As the pitcher "cocks" his arm to throw, his arm is rotating into a position for which it wasn't designed.

Ulna bone
Ulnar collateral ligament
Humerus bone
Radius bone

Centrifugal force is literally trying to pull the pitcher's arm out of its socket. The pitcher's muscles respond by pulling back with all their strength.

The total pulling force can add up to as much as 125% of the pitcher's body weight, or 200 pounds or more of pull on the arm.

Critical instant Arm deceleration

The outer ligaments of the elbow absorb much of the energy as the arm decelerates from the throwing motion. The shoulder takes punishment here too, as the head of the humerus — the "ball" in the shoulder's ball-and-socket joint — slams forward and twists.

Annular ligament
Radial collateral ligament

Both the elbow and the shoulder twist backward about as far as they can go, and the momentum tries to push them even farther.

The inside ligament of the elbow and the tendons and cartilage of the shoulder bear the brunt of the pitcher's effort as he begins to whip his arm forward.

Acromion bone
Subdeltoid bursa (cushioning sac of fluid)
Supraspinatus tendon
Deltoid muscle
TWISTING MOTION
HEAD OF HUMERUS
PULL/PUSH
Supraspinatus muscle (one of the rotator cuff muscles)
Scapula (shoulder blade)
Labrum
Cutaway diagram of shoulder

The sharp twisting of the humerus combined with the pushing and pulling of the arm and shoulder muscles can grind the bone surfaces and stretch the tendons of the shoulder.

Ways to minimize the damage include:

Improving overall muscle tone;

Using proper pitching mechanics that reduce the stress on the most vulnerable parts of the joints;

Limiting the number of pitches thrown per game.

The six stages of the pitching motion

The windup
Stress to the body is minimal as the pitcher coils his body, developing potential energy to propel the ball.

Joint under stress

The stride
As the legs spread wide, placement and rotation of the feet is critical; too "open" a stance can put undue stress on the arm.

Arm cocking
The arm assumes a 90-degree angle from the trunk; the elbow cocks to about 90 degrees as the trunk begins to rotate forward and the shoulder begins to rotate backward.

Feet should fall within about 10 degrees of a direct line toward home plate.

Foot contact

Arm acceleration
Huge torque forces build up on first the elbow, and then the shoulder as the pitcher whips his arm forward to fire the ball toward the plate.

Arm deceleration
Stretching, compression and torque hammer the shoulder and the outer ligament of the elbow as the whipping arm reaches the end of its forward motion.

Follow-through
The trunk moves forward and down to help dissipate the tremendous throwing energy.

Maximum inward rotation of the throwing arm occurs here.

LOW STRESS | STRESSES BUILDING | MAXIMUM STRESS

SOURCE: American Sports Medicine Institute

WILLIAM NEFF | THE PLAIN DEALER

FIGURE 18.22 Visual depicting action

The kinds of visuals you may choose to use will be dependent on the needs of the document, the audience, the resources available to you, and even the financial support for production of the instructions. In some cases simple line drawings will suffice; in others, more complex or even photographic images may be required. Because digital photography has become so readily accessible and manipulation of digital images is so convenient, more and more producers of technical instructions have begun to use photographic imaging in written instructions.

Conclusion

After all steps have been presented, writers of instructions usually offer three kinds of final information.

Troubleshooting

Troubleshooting guides provide solutions to common problems that might arise after the reader completes all the steps. When you are designing troubleshooting guides, consider what could go wrong when a reader completes or does not complete all the steps. Remember that troubleshooting guides are written for the reader and must, therefore, focus on the user's perception of the activity or process, as well as the user's skills, abilities, and knowledge of the product or procedure. Troubleshooting guides are generally designed during and after usability testing. These guides also provide contact information, such as a URL or phone number, so that readers can contact technical support if they have further problems that the troubleshooting guide cannot solve.

Expert Systems Expert systems are specialized types of computer programs designed to help users solve problems through a series of yes or no questions. The programs use detailed algorithms that ask questions based on responses to previous questions. With the user's responses to questions, the expert system is able to narrow down the user's problem until it can offer a specific answer to the problem. Expert systems can be thought of as interactive troubleshooting segments for computer programs.

Maintenance Information

To prevent readers and users from running into problems *after* they have completed the instructions, you should provide them with information and tips to properly maintain the product or to follow up on the results of their actions.

Additional Alerts

It is also a good idea to provide readers with additional alerts that become important after they have completed the instructions. For instance, you might alert readers to possible dangers in using the product just described.

QUICK REFERENCE CARD

When writing a detailed or complex set of instructions, workplace writers sometimes produce an abbreviated set of instructions that identifies basic operations. These instructions draw from the larger, central set of instructions and provide the fundamental tasks the reader needs to perform. These quick references give readers speedy access to information about primary operations or functions.

Like the more complete instructions, quick references should be written in clear, concise steps and should consider audience needs. Good use of visuals, attention to language, and consideration of accessibility are all necessary. Figure 18.23 shows an example of a quick reference card for a dash-mounted GPS and depth finder. Because users of the device will likely be piloting a boat and will probably not have the ability to also read through the full instructions, the quick reference card offers easily accessible basic information about the unit. In addition, because reading an instruction manual on a boat will likely be hampered by wind and water conditions, the quick reference card is produced on a piece of waterproof plastic. When you are designing quick reference cards, you should pay close attention to the settings where they will likely be used.

HELP PAGES

Help pages are specialized technical instructions designed to assist people who need help with computer applications. Help pages are usually files embedded in a particular program, or they are online resources for users. Either way, they allow users to search for a specific problem within the pages, and they offer readers step-by-step

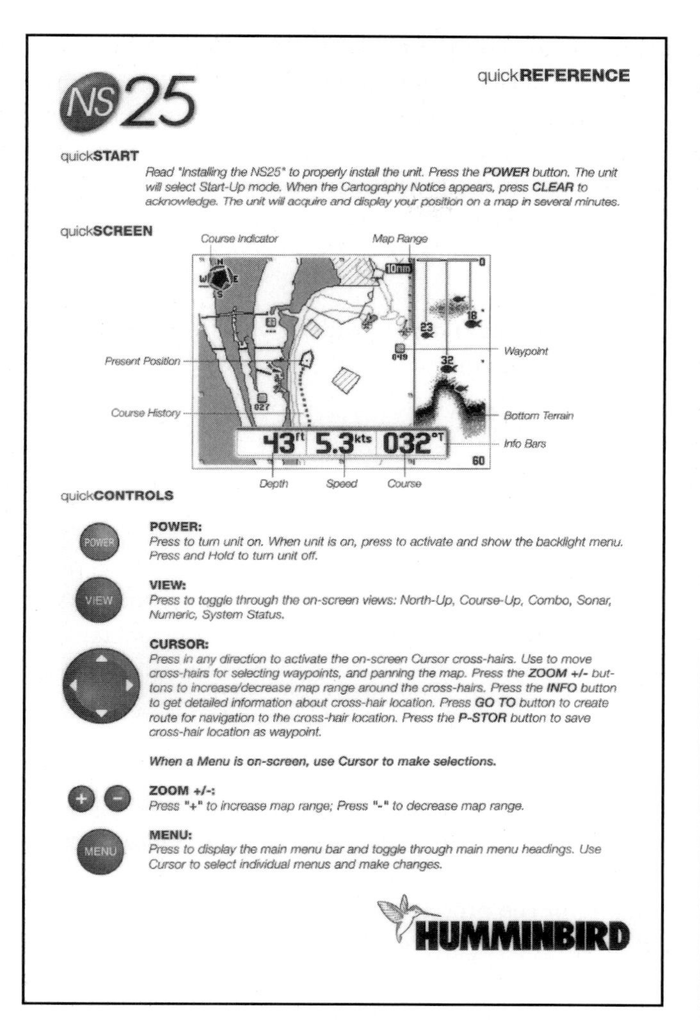
NS25 quickREFERENCE

quickSTART
Read "Installing the NS25" to properly install the unit. Press the **POWER** button. The unit will select Start-Up mode. When the Cartography Notice appears, press **CLEAR** to acknowledge. The unit will acquire and display your position on a map in several minutes.

quickSCREEN

quickCONTROLS

POWER:
Press to turn unit on. When unit is on, press to activate and show the backlight menu. Press and Hold to turn unit off.

VIEW:
Press to toggle through the on-screen views: North-Up, Course-Up, Combo, Sonar, Numeric, System Status.

CURSOR:
Press in any direction to activate the on-screen Cursor cross-hairs. Use to move cross-hairs for selecting waypoints, and panning the map. Press the **ZOOM +/-** buttons to increase/decrease map range around the cross-hairs. Press the **INFO** button to get detailed information about cross-hair location. Press **GO TO** button to create route for navigation to the cross-hair location. Press the **P-STOR** button to save cross-hair location as waypoint.

When a Menu is on-screen, use Cursor to make selections.

ZOOM +/-:
Press "+" to increase map range; Press "-" to decrease map range.

MENU:
Press to display the main menu bar and toggle through main menu headings. Use Cursor to select individual menus and make changes.

HUMMINBIRD

Front

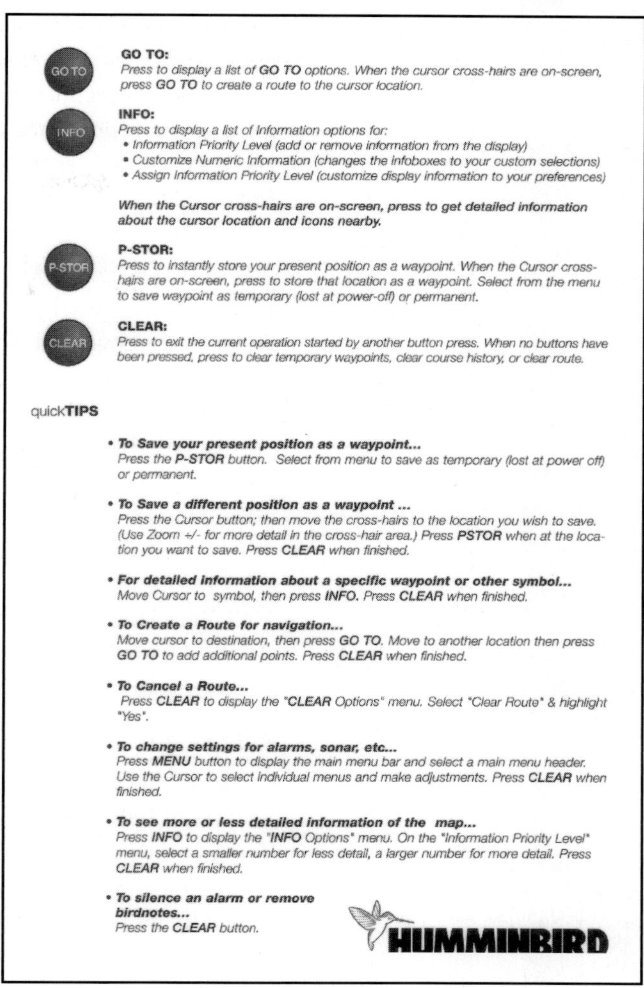
GO TO:
Press to display a list of **GO TO** options. When the cursor cross-hairs are on-screen, press **GO TO** to create a route to the cursor location.

INFO:
Press to display a list of Information options for:
- Information Priority Level (add or remove information from the display)
- Customize Numeric Information (changes the infoboxes to your custom selections)
- Assign Information Priority Level (customize display information to your preferences)

When the Cursor cross-hairs are on-screen, press to get detailed information about the cursor location and icons nearby.

P-STOR:
Press to instantly store your present position as a waypoint. When the Cursor cross-hairs are on-screen, press to store that location as a waypoint. Select from the menu to save waypoint as temporary (lost at power-off) or permanent.

CLEAR:
Press to exit the current operation started by another button press. When no buttons have been pressed, press to clear temporary waypoints, clear course history, or clear route.

quickTIPS

- ***To Save your present position as a waypoint...***
  Press the **P-STOR** button. Select from menu to save as temporary (lost at power off) or permanent.
- ***To Save a different position as a waypoint ...***
  Press the Cursor button; then move the cross-hairs to the location you wish to save. (Use Zoom +/- for more detail in the cross-hair area.) Press **PSTOR** when at the location you want to save. Press **CLEAR** when finished.
- ***For detailed information about a specific waypoint or other symbol...***
  Move Cursor to symbol, then press **INFO.** Press **CLEAR** when finished.
- ***To Create a Route for navigation...***
  Move cursor to destination, then press **GO TO.** Move to another location then press **GO TO** to add additional points. Press **CLEAR** when finished.
- ***To Cancel a Route...***
  Press **CLEAR** to display the "**CLEAR** Options" menu. Select "Clear Route" & highlight "Yes".
- ***To change settings for alarms, sonar, etc...***
  Press **MENU** button to display the main menu bar and select a main menu header. Use the Cursor to select individual menus and make adjustments. Press **CLEAR** when finished.
- ***To see more or less detailed information of the map...***
  Press **INFO** to display the "**INFO** Options" menu. On the "Information Priority Level" menu, select a smaller number for less detail, a larger number for more detail. Press **CLEAR** when finished.
- ***To silence an alarm or remove birdnotes...***
  Press the **CLEAR** button.

HUMMINBIRD

Back

FIGURE 18.23 A quick reference card

solutions to the problem. For instance, your school might provide specific help pages to describe the process of using a campus e-mail account. Those pages probably include information about establishing an e-mail account, creating a password, changing a password, receiving e-mail messages, sending e-mail messages, saving e-mail messages, and so on. If you forgot your e-mail password, you could access the help pages about changing or recovering a password, a far more efficient process than searching through all the pages or sifting through a printed instruction manual.

Many useful help-authoring tools (HATs) provide templates for creating help pages. However, because help pages are written for specific software and application contexts, different organizations and companies write help pages differently. Nonetheless, some general guidelines for writing help pages do exist.

See chapter 3 for more on HATs.

Be Clear, Concise, and Thorough

First of all, be sure that your writing is clear and unambiguous. Remember that these are *help* pages and should not confuse readers. Second, be aware that most people do not like to read extensive information online. In fact, studies show that many people simply print help pages and work from hard copies rather than reading the same information on-screen. Concise pages will keep readers' attention throughout the process and will provide them with efficient information.

However, concise does not mean incomplete. Because help pages are online documents or are embedded in software, they are not limited by the space of printed pages and can cover a larger scope. As a result, help pages can and should offer readers as much information as necessary to account for a variety of problems.

Be Direct but Friendly

Like all other instructions, help pages should tell readers what to do at each step. Again, action verbs lead to direct instructions:

- *Insert* CD into drive.
- *Click* the Edit menu.
- *Install* the driver.

At the same time, though, help pages are assisting customers and users with problem solving. To promote your company's or organization's client relations, you must keep your help pages friendly. Poorly written help pages may fail to help readers, but unfriendly help pages can alienate them.

Provide Examples

Because most problems that help pages can solve are computer-related problems, writers of help pages should show readers visual examples by providing screen shots. These show readers where buttons and menus are located, as well as what should appear on-screen at various stages of the process—including what the screen should look like when the process is complete and the problem is solved.

See chapter 8 for more on screen shots, video, and animation.

Many help pages now incorporate video clips or animations to direct readers through a process. Video or animation produced in Flash or QuickTime can be useful, but you must remember that such formats require more space and are less accessible to readers using slower bandwidths.

See chapter 3 for more on bandwidth, restrictions.

Provide Links

Be sure to provide internal links within help pages to direct readers to other information that is either directly or indirectly related to their problems. Sometimes solving one problem requires that another be solved, perhaps one that the reader hadn't recognized. Be sure that readers can easily navigate through the help pages to solve their problems. Links between pages serve a secondary function as well, making it easier for writers to revise and change individual segments.

Include a Search Feature

Be sure that your help pages are created with a keyword search feature so that readers can quickly access the appropriate pages for their specific problems. If readers must wade through unnecessary information to locate what they need, the help pages become much less helpful. However, a search feature that quickly directs a reader to necessary information is a much-appreciated tool.

Consider Context

As with all other instructions, writers must determine the approaches that best suit the needs of their specific audiences in their specific contexts. For instance, some company help pages use overly simplistic language to facilitate autotranslation, such as that provided by the autotranslator embedded in Google. In such cases the company is more interested in having many readers locate and access the page. And some companies do not use a multipage linked approach, preferring the benefit of having fewer pages to maintain. You must consider these kinds of contexts when you write your help pages.

See chapter 5 for more on bandwidth restrictions.

USABILITY OF INSTRUCTIONS

PSA

Usability testing is especially important for technical instructions. It lets writers determine which parts of the instructions are effective or ineffective, accurate or inaccurate. Are the steps presented logically? Are the instructions understood by the audience? Can readers easily access the information? Usability testing is a crucial part of the Review segment of your problem-solving approach, and it is a great method of solving problems before readers run into them. Frequently writers of technical instructions find that the instructions make perfectly good sense to them but not to their readers. We urge you always to perform usability tests on your technical instructions.

See chapter 11 for more on usability.

We also recommend developing a heuristic—a series of questions—to guide your assessment of the efficiency and effectiveness of the technical instructions. We include here a simplified version of a heuristic, but you will want to expand and customize it to suit the specific audiences and contexts for which you write your technical instructions.

TECHNICAL INSTRUCTIONS USABILITY HEURISTIC

- Do the title and title visual convey an accurate understanding of the goal of the instructions?
- Does the introduction accurately and clearly define the scope of the instructions?
- Are the instructions written with the intended audience in mind?
- Are all alerts easily identified and understood?
- Are the visuals clear and easily understood? Do the visuals offer readily identifiable likenesses of the objects they represent?
- Do visuals like icons work in conjunction with the written text?
- Are visuals in useful proximity to the corresponding text, especially in the case of parts lists, tools lists, and steps?
- Are the instructions—particularly the steps—visually accessible, or is it difficult to rapidly return to one's place in the instructions?
- Does each step focus on only one action or task? Does each step begin with an action verb?
- Do troubleshooting solutions really work?
- Are the user test subjects able to complete the instruction process without any snags?

IN YOUR **EXPERIENCE**

Have you ever worked with instructions that were inconveniently designed? Perhaps they were too small, printed on a single sheet of paper with too much information crammed onto a single page. Or perhaps they were too large and unmanageable as you performed the tasks they described. Maybe they were too long and poorly organized, leaving you to search through pages of information to find what you needed. You may have used instructions that would not stay open because of their tight binding. Maybe the instructions were folded, and the crease of the fold made it difficult to read some of the information.

There are countless ways that instructions might be poorly designed. Think about a set of instructions you've worked with that might have been better formatted. Then identify how the poor formatting affected you and how the writers might have made better rhetorical choices in designing the instructions. Be sure to consider the issues addressed in chapter 9 as you apply the problem-solving approach to this particular design.

DESIGN CONCERNS

PSA

Remember that technical instructions can be presented in any format, ranging from a series of small cards to huge foldouts. The format of instructions is determined by the needs they fulfill, the problems they solve, and the ways the audiences are likely to use them. You should consider the format of the instructions during the Plan phase of the problem-solving approach, when you determine context and audience needs. The following are the kinds of questions to ask yourself when you are deciding on the physical form and design of your instructions:

See chapter 9 for more on document design.

- Will your readers need to open the instructions while working at a desk or in another confined space? Can they spread the instructions out?
- Will your images be too cluttered if you force them onto relatively small pages? Do you need a large space to make visuals readable?
- What materials are available to you for printing?
- What costs limit the printing?
- What technology do you have available to create your technical instructions?
- Can the instructions appear online or in hypertextual format? Must they appear in hard copy?

SUMMARY

- Technical instructions direct audiences in solving problems such as putting something together, repairing an object, operating or installing a device, or maintaining a product.
- Technical instructions include numerous elements.
- The title and title image should give a clear understanding of the goal of the instructions.
- The byline identifies who produced the instructions.
- The date tells when the instructions were produced.
- The introduction should provide information about the aim, outcome, scope, and organization of the instructions.
- Alerts identify dangers, warnings, cautions, and notes related to the instructions.
- An equipment or tools list should identify what the reader needs to complete the steps. Visuals clarify and support a written list.
- A parts list identifies what each part looks like and what it is called.
- The steps should adhere to these guidelines: number each step, perform all steps as you draft them, begin each step with a verb, use positive commands, avoid highly condensed language, include only one action per step, group similar steps together, visually separate steps, and clarify steps with visuals.
- The conclusion offers troubleshooting, maintenance information, and additional alerts.
- Help pages are specialized instructions to help with computer applications.
- Writers should test technical instructions for usability.
- The design of technical instructions should suit the audience and the context.

CONCEPT REVIEW

1. What kinds of problems can instructions solve?
2. What ethical considerations should writers address in writing instructions?
3. What are the components that can be included in a set of instructions?
4. What role does a visual play on the title page?
5. Why should instructions include dates and/or version numbers?
6. What information should be included in the introduction of a set of instructions?
7. What are the four primary kinds of alerts that might be included in a set of instructions?
8. What strategies might you use to emphasize an alert to readers?
9. What is the difference between a tools list and a parts list?
10. What key factors should you consider when writing the steps to your instructions?
11. What is the purpose of a troubleshooting section?
12. What is the purpose of a quick reference card?
13. What are help pages?
14. What factors should you take into consideration when testing the usability of a set of instructions?
15. What kinds of things will affect the design you choose for technical instructions?

CASE STUDY 1

Applying for Aid from the Federal Emergency Management Agency

After Hurricane Katrina a number of questions were raised about the Federal Emergency Management Agency (FEMA) and its response to the storm. One of the recurring complaints involved the difficulty victims had in understanding exactly how to apply for federal assistance. The FEMA web pages (http://www.fema.gov/assistance/index.shtm) offer a user-friendly, three-step process for applying for assistance from federal programs. The page also offers links to application instructions for "public assistance"; for aid to local, tribal, and state governments; and for aid to nonprofit organizations.

Take some time to become familiar with one of these two sets of aid-application instructions. As you analyze the instructions, consider how you are responding to those instructions and how a reader who just experienced a disaster might respond to the instructions. Taking into account the differences in how you and a disaster survivor might read the instructions, write a set of instructions for the disaster survivor, guiding that person through the web-access application. Your instructions should not merely mimic or repeat the FEMA instructions, but should serve instead as a guide to how someone would access and use the FEMA application process.

CASE STUDY 2

Instructions for the Internal Revenue Service

On their web page at http://www.irs.gov/instructions/index.html, the Internal Revenue Service (IRS) makes available more than 170 sets of instructions for various tax forms. This site is designed to help users locate forms and to provide detailed instructions for completing those forms. However, despite IRS efforts to make the process easier, many users have problems understanding how to use the web page.

Review the IRS web page, and consider the rhetorical problems the web page authors might have faced in designing the pages. Pay particular attention to issues of audience, design, and accessibility. Consider, too, what role ethics plays in the way the instruction pages are presented. Then consider what types of problems—in this case, external problems—users might face when accessing these pages. What problems might have brought users to these pages, and what problems might develop for them as they seek solutions through these pages?

Next, consider how you might write and design a set of instructions to teach a first-time user how to use the IRS instruction pages. Develop solutions to both types of external problems users might have. Address all of the rhetorical issues of audience, design, accuracy, access and distribution, and ethics. Then write a set of user instructions to teach a first-time user how best to use the IRS instruction web pages to solve problems.

CASE STUDIES ON THE COMPANION WEBSITE

Integrating Pictorial and Textual Instructions: Mathias Contracting of Boise and IKEA

In this case you oversee the installation of cabinets in a new apartment complex and discover that your workers have difficulty with the installation instructions. Your purpose is to study these instructions closely and decide ways to best solve this problem.

Technical Instructions: Records Archiving at Ross Realty of Houston, Texas

As an employee at Ross Realty, you are asked to oversee the storage of thirty years' worth of paper documents. After hiring temporary employees to sort and store these documents, you must develop a list of instructions that explains to these temporary employees exactly how they must go about this work.

Technical Instructions: Chemical Safety at Roxey Manufacturing

You are a chemical engineer working for Roxey Manufacturing in Houston, Texas. Workers in the plant must handle certain potentially hazardous chemicals in the manufacturing process. Following guidelines produced by the Occupational Safety and Health Administration (OSHA), you are to review the plant's safety instructions for employee handling of these chemicals, and you must also update the posted instructions on how to respond in case of chemical mishaps because employees have had difficulty understanding previous instructions.

Shutting Down at the End of the Day

Eileen Burgess runs a medical research laboratory in a university hospital. One day she realizes that too much equipment is being left powered on when researchers go home, and she wants to write a set of instructions for closing the lab.

VIDEO CASE STUDY

GlobeShare Wireless

Testing the GlobeTalk Handheld Devices

Synopsis

A group of GlobeShare Wireless employees learn that the new GlobeTalk Messaging handheld units are almost ready to be shipped to customers. Michelle Bankler, a product testing employee for GlobeShare Wireless, convinces her team to do a final field test before the units are sent. The field test leads to a shocking situation.

WRITING SCENARIOS

1. Locate a set of less-than-perfect instructions to analyze. Make a list of the common mistakes in the instructions, and offer suggestions for revising and improving each problem.
2. Select a location other than the classroom where you take this class. Then write a set of instructions (i.e., a detailed set of directions) explaining how to get from your classroom to that place. Assume that the audience for these instructions is not familiar with the area at all.
3. The Occupational Safety and Health Administration (OSHA) publishes web pages specifically about warning labels and other alerts. Visit the OSHA website, and search its pages for information about danger, warning, and caution labels. Then write a letter to your instructor, explaining how OSHA addresses alert labels.
4. Go to the web pages of a major technology company—for instance, Apple, IBM, Microsoft, Sony, Hitachi, Dell, Gateway, or Palm—and locate an available help page. Familiarize yourself with the design and approach the company takes in providing help pages. How are the pages searchable? What role do visuals play on the pages? Write a set of technical instructions to teach readers how to locate the help pages within the company's web pages and also how to use those help pages.
5. Chapter 4 addresses the ethics of using visuals that report information. Think about the role of ethics in creating visuals specifically for instructions, visuals that show how to do something. How do visuals used to instruct differ from visuals used to report? How are they similar? Write a set of general instructions (or guidelines) for designing ethically sound visuals for inclusion in the steps of a set of instructions.
6. Visit the websites of three different Internet service providers (ISPs)—AOL, Earthlink, and Roadrunner, for example. On each of your chosen websites, locate two sets of instructions: the instructions for subscribing to the service and the instructions for canceling the service. In a letter to your instructor, detail the similarities and differences between these two sets of instructions, as provided by the three ISPs. Are they equally easy to find on the website? Are they equally detailed and usable? Are they both ethical? What are the different ethical responsibilities for each kind of instructions? Of the three websites is one clearly more ethically written than the others? Explain your responses with detailed information from the websites.

7. Locate any construction modeling set, such as Legos, Tinker Toys, Erector Set, even models for building chemical compounds. Avoid any set that requires gluing or permanently affixing parts; you'll want to be able to take your object apart and put it together again numerous times.

    Using the construction set, build something, anything—a house, a spaceship, a race car, a septic tank, a double helix genetic code for an aardvark. Try to avoid building the designated product found on the box or on the instructions that come with the product. Be creative, and have some fun. Build something new—it doesn't need to fit the conventional form of anything recognizable.

    Once you have settled on a construction, write a set of instructions to guide someone else through the process of building the item. Remember that a good set of instructions teaches the audience how to accomplish the task at hand. Think about how to use the technological tools available to you to really show your audience how to put the thing together. Consider the layout and visual design of the instructions especially and the ethical ramifications of creating such a set of instructions.

    Once you've completed the instructions, dismantle your construction, and place all of the parts in a small bag. Give your instructions and the bag of parts to a friend or classmate, and ask the person to use your instructions to build the designated object. Observe the entire process. That observation should give you ideas about how to revise the instructions; you can also ask for specific comments about the instructions. In other words, conduct a usability test (you may want to look again at chapter 11), and then revise your instructions accordingly.

8. Make a paper airplane of at least eight folds—as simple or as fancy as you wish. Then write and design a set of instructions for building that plane. Be innovative both in your airplane design and in the design of the document. Consider all of the possible ways you might present the instructions. Pay particular attention to the rhetorical problems you will need to solve. What might be the rhetorical situation that requires written instructions for building a paper airplane? Who might be the audience for such a set of instructions? What do you want the instructions to accomplish? What information do you have, and what information do you lack in writing such instructions? How will you organize the information in the instructions? What format and design suit your set of instructions best? What role will visuals play in the instructions? Does technology affect how you will write and design your instructions? How will you test the effectiveness of your instructions? How will you disseminate the instructions to your audience? Once you have completed your instructions, be sure to test their usability before producing your final version.

9. Imagine that you work for a company that produces both small and large power tools, everything from small power drills for casual users to jack hammers for heavy duty demolition. You and a team of other writers are asked to produce instruction manuals for three devices—a power screwdriver for homeowners, a riding lawnmower, and an industrial-strength nail gun for professional carpenters working on large building products. In a memo to the lead writer on the team, tell what size and physical design (e.g., booklet or foldout) are required for each of the three manuals.

10. Locate a set of online instructions for downloading or installing software. Read the instructions, and analyze them to see how they compare with the

guidelines presented in this chapter. Then write an analysis of the instructions, identifying whether you think they are effective according to the parameters set by this chapter, and explain why or why not.

11. Many of us today are adept at ripping and burning CDs, creating electronic photo albums, or even editing home video. The chances are good that the computer you use includes at least a basic application for ripping and burning music CDs and for creating photo albums on CDs. Write a set of instructions that details either of these processes.
12. Select any technology-driven tool that you use regularly: your computer, CD player, DVD player, MP3 player, cell phone, digital camera. Then choose a single function of that device: playing a CD, placing a long distance call, saving a word processing file. Write a set of instructions for someone who has never performed that task, detailing the process using your specific piece of equipment.

PROBLEM SOLVING IN YOUR WRITING

PSA

Look back at one of the sets of instructions you wrote in response to a Writing Scenario prompt: the Legos instructions, the paper airplane instructions, or the CD burning instructions, for instance. As you read back through the instructions, think about how you solved the rhetorical problems you faced in writing those instructions. Using the problem-solving approach as a heuristic, write a short summary of how you solved the rhetorical problems of each of the PSA phases.

19

Manuals

CHAPTER LEARNING OUTCOMES

After completing this chapter, you will be able to do the following:

- Identify the various kinds of **manuals** and their purposes
- Determine whether a **print manual** or an **e-manual** best serves your needs
- Employ a number of **guidelines** in writing manuals
- Identify and select the parts of a manual that best serve your **rhetorical problem-solving** needs
- Address issues of **usability** when writing manuals
- Account for issues of **updating** manuals
- Understand the role of **single sourcing** in contemporary manual and other document production
- Consider the role manuals play in **marketing and public relations**
- Address **ethical** considerations in writing manuals
- Consider the role of **design** in producing your manuals
- Produce a **variety** of different kinds of manuals
- Anticipate writing manuals for **transnational audiences**

DIGITAL RESOURCES

On the Companion Website www.prenhall.com/dobrin:

- Case 1: Cultural Communication Problems in the State Attorney's Office
- Case 2: Malfunction of the Mass Spectrometer: Malfunction of the Manual
- Case 3: Out at Sea: Environmental Difficulties of Writing a Usable Manual
- Case 4: Surviving and Thriving at College: Designing a Manual for First-Year University Students
- Video Case: Out of Kindness, I Suppose: Poncho and Lefty Write Manuals
- PowerPoint Chapter Review, Test-Prep Quiz, Exercises and Activities

REAL PEOPLE, REAL WRITING

MEET STORMY BROWN • Employee Relations Manager at Coca-Cola Enterprises

What types of writing do you do as a human resources manager with Coca-Cola Enterprises?

I write every day, including things like e-mail, internal communications, and announcements. But I also do job descriptions, performance evaluations, training manuals, process manuals, and advertisements, to name a few.

What kinds of manuals do you create at Coca-Cola?

We create process manuals when we come upon a new program or a new function within our division of the corporation. We need to put together a manual to inform our client group—whether that's internal or external—how to go forward with the new procedures. We also create training manuals for our employees when we are introducing either a new internal system or a new employee training program.

Does Coca-Cola Enterprises have a standardized design format for how those manuals are written?

Actually, no. We do have certain expectations of the professional level of the manuals and of the professional skills the manual writers will have when they come onboard in professional roles. The writers usually partner with a team that is a subject matter expert for that function. For example, a recruitment team would be involved with writing the manual about our new recruitment process. We don't have a specified manual program, but we do expect that manual writing is a skill our employees will have or develop as they come into the company.

What format do your manuals generally take when you produce them?

We usually provide both print copy and e-manuals. E-manuals are generally going to be manuals that are a more current format. But we have several instances, especially in production, where our employees don't have access to e-manuals. So for instance, if the manuals are being produced on the manufacturing side, we are more likely to produce hard-copy manuals.

How important is collaboration in how you produce manuals?

It's a huge piece of what goes on. Writing a manual usually ends up touching several different departments but often centralizes within a team that creates it. It is not likely that an individual would write a manual.

What advice about workplace writing do you have for the readers of this textbook?

No matter what, their jobs are going to require writing at many levels. The successful outside job candidate needs to understand that the first thing an employer sees is a snapshot of that person in the form of a résumé, a CV, or a cover letter. Those documents speak volumes about the candidate's ability to communicate effectively. That's the first picture we see, but beyond that they will be routinely communicating on different levels, both formally and informally. Every document that a worker produces, even every e-mail that they send, is going to be evaluated.

NON SEQUITUR ©1996 Wiley Miller. Dist. By UNIVERSAL PRESS SYNDICATE. Reprinted with permission. All rights reserved.

manual 1: Small book generally used for the purposes of reference, containing a listing of information, instructions, or other related data. 2: Physical labor performed by hand.[1]

INTRODUCTION

Simply put, a **manual** is a document that provides information that a reader can reference to solve a specific workplace problem. Manuals instruct readers to complete a task, learn a new skill, operate a piece of equipment, service or maintain a product, or even conduct themselves according to company policy. Manuals range in length and detail from short, quick-reference cards to hypertext e-manuals requiring massive amounts of storage space. Like all other workplace documents, manuals are written according to specific purposes and audience needs and expectations.

Manuals are not more difficult to write than other workplace documents, nor do they require special writing skills unique to manuals. However, because manuals reflect tasks and procedures that are likely to change frequently, manuals sometimes require frequent updating, which demands careful attention because information can change even as the document is being written. Manual writers, therefore, may start by revising previously written manuals, essentially beginning the problem-solving approach in the Review phase, and then will revisit other phases of the PSA to address changes that arise.

Imagine a large engineering firm that works closely with city planners in a large midwestern city. Within that company, engineers, lawyers, and support staff produce countless pages of documents in any given year; and those documents are disseminated among employees of the company and, in some cases, among staff members in the city planners' office. The company uses the most up-to-date document-sharing network and programs available, which require the company to maintain strict policies about how to use the system and which documents can and can't be shared through it. The firm's lawyers have also established policies about what kinds of documents can be shared publicly and what copyright restrictions apply. In addition, the information technology department that maintains the network and the software has its own policies to ensure that the whole system is used properly. All of these policies require the company to develop at least one manual to help solve problems and answer questions. But each time policies change because of shifts in laws, changes and upgrades to the network, or company interaction with the city planners, the manual must also change. And each time a new problem arises, the manual must be brought up-to-date.

Numerous kinds of manuals solve countless kinds of problems. However, even though manuals may share similar elements and their writers may face similar kinds of problems, each manual has a unique purpose and its own rhetorical challenges. This chapter introduces you to the common types of manuals and their primary parts. It also teaches you to consider problem-solving strategies, ethics, technology, and professionalism as you write and design different manuals.

---

[1] Jack P. Friedman, *Dictionary of Business Terms*, 3rd ed. (Hauppauge, NY: Barron's Educational Series, 2000), 406.

IN YOUR **EXPERIENCE**

What kinds of manuals have you used before? In what situations? When have you needed manuals? What problems have you been able to solve because you have had access to manuals? As a class, discuss common and uncommon experiences with manuals. What generalizations can you and your classmates make about how people use and think about manuals?

TYPES OF MANUALS

It is important to know what kinds of problems different manuals solve so that during your rhetorical problem solving, you can select the manual type that suits your purpose and your audience's needs.

Policy Manuals

Policy manuals generally inform employees about information that is crucial to their employment: employee guidelines, company rules, company expectations, business policies, codes of ethics (see chapter 4), corporate standards, and more. Many companies and organizations today have developed *e-policies*, which specifically detail procedures and policies for using Internet and web resources while in the workplace. Simply put, policy manuals convey to readers the *why* and *what* of company policies and explain the goals and approaches of an organization. These manuals are crucial for companies because understanding a company's policies is a starting point for understanding how a company makes decisions. In addition, policy manuals help maintain authority within a company.

Policies are generally implemented from within companies, but they may also be foisted on companies from outside. In either case policies are generally directed from the top down to ensure that employees follow the goals of the company. Policy manuals help solve problems in two ways: first, by informing readers about policies so that they can act accordingly and avoid problems, and second, by providing a resource that employees can access when conflicts or questions arise. Most businesses consider policy manuals to be valuable documents that protect employers and employees in the event of policy disputes.

Procedures Manuals

Procedures manuals are often confused with policy manuals and with instruction manuals (see chapter 18) because they share some of the same parts and goals. However, unlike policy manuals, which convey detailed information about company policies, procedures manuals tell readers how to perform a task or, in some cases, how policies are enacted and enforced. That is, procedures manuals tell readers how to solve problems. Procedures manuals also differ from technical instructions because they typically do not convey how to perform a physical task, such as popping corn, installing software, or building a bookcase; rather, they might direct readers through a bureaucratic problem, such as filing for workers' compensation, applying for a grant, or scheduling a meeting in a particular room. In some instances procedures manuals are combined with policy manuals to create a standard document known as a policy and procedures manual (PPM). Procedures manuals can also be referred to as standard operating procedures (SOP) or a handbook of operating procedures (HOP).

EXPLORE

Search the U.S. government web pages for procedures manuals. What kinds of manuals did your search turn up? Make a list of those manuals, and describe the procedures they contain.

Operations Manuals

Operations manuals, like procedures manuals, describe a particular process. Most often, operations manuals explain the procedures for day-to-day operations within an organization or a company. Like most of the manuals discussed here, operations manuals are prescriptive, clearly detailing the steps required to complete a given process so that users/readers can easily solve problems that might arise in the daily operations of an organization. For instance, an operations manual might walk a new night manager through the steps needed to shut down a computer inventory system at the end of a business day. Often operations manuals are closely linked with policy manuals because particular operations, such as accurately logging and depositing a day's receipts, are company policies as well.

EXPLORE

The Institute of Electrical and Electronics Engineers (IEEE) maintains an operations manual online to describe the procedures through which standards for the IEEE can be developed. Locate this manual online, and read through it. In what ways is it an operations manual as the title claims, and in what ways is it a procedures manual?

Operator's Manuals

Operator's manuals are designed for individuals operating equipment who need to solve any problems they encounter while operating that equipment. Like user's manuals and owner's manuals, operator's manuals can be thought of as detailed how-to guides that assist readers in successfully operating equipment. Their design and purpose are also quite similar to those of instruction manuals (see chapter 18). These manuals are generally written for operators already familiar with the equipment, and the manuals themselves are frequently designed to be used on-site. Consequently, operator's manuals are often concise, containing only the most essential operating instructions and safety warnings.

Operator's manuals are frequently written for various kinds of equipment—military, construction, biomedical, computer, assembly, or manufacturing. In the past many companies and organizations did not rely on specific operator's manuals, but instead used only the information provided by equipment manufacturers. Today, companies frequently develop their own operator's manuals to better suit the needs of those using the equipment. As machinery and equipment have become more complicated, requiring more skilled operators, both operators and manufacturers have often demanded accurate, accessible manuals designed specifically for particular equipment.

ANALYZE THIS

Locate an operator's manual, and read through it, listing the key components in the manual and identifying the organizational approach the writers use. Then write a short assessment of the manual, describing what is effective and ineffective in the manual from the perspective of rhetorical problem solving. What strategies do the writers use to help convey information to the manual's readers?

Owner's or User's Manuals

Owner's or user's manuals are reference documents designed for use with individual pieces of equipment. They often contain elements of several manual types—including instructions, operations, and maintenance—and they are usually designed for both expert and lay audiences. Because they address a wide audience and incorporate several kinds of manuals within a single document, they provide information for solving many kinds of external problems. However, this broader coverage creates a number of rhetorical problems for the writers of owner's manuals, who must address several purposes in one document. Although they are often geared for less sophisticated equipment—such as power tools, stereos, or camping equipment—the most common owner's manuals are found with new vehicles and describe the basic operation of the vehicle, its various features, and even basic driving information. A complete user's manual for a radio receiver is shown in Figure 19.6, near the end of this chapter.

Service and Maintenance Manuals

Service and maintenance manuals often relate directly to operator's manuals. Just as operator's manuals provide information about how to use a piece of equipment, a service or maintenance manual provides information about how to care for and repair that same equipment. Because equipment operators are sometimes responsible for the care of their equipment, service manuals are frequently designed with operators as the intended audience. However, service and maintenance are also frequently performed by service technicians—for example, computer technicians or auto mechanics. Service and maintenance manuals also vary according to the equipment involved. For example, a complex aircraft like a Boeing 777 would not have just one service manual; it requires volumes of manuals for each of its various working parts. On the other hand, a standard lawnmower probably has only a single service manual. Service manuals can be thought of as detailed, or expanded, troubleshooting guides, shorter versions of which may appear in user's or operator's manuals.

ANALYZE THIS

Conduct an online search for maintenance manuals. Locate a complete manual that is available free online. Read through the manual, examining it for features that make it useful or not useful as an online manual. Is the manual designed specifically as an online manual, or is it a print manual that has been presented in PDF format or simply converted to HTML? Think about the rhetorical problems the person who posted the manual might have had to solve in order to make the manual available online. Write a short description of why the manual is or is not useful as an online manual.

Training Manuals

Training manuals prepare readers to perform some kind of task, follow some kind of procedure, or operate some kind of equipment. Most teach new employees how to successfully complete the tasks associated with their positions and may accompany employee handbooks, which provide policies and procedures. Training manuals solve two primary kinds of problems: they prepare individuals to perform a task or use a piece of equipment, or they prepare them to solve problems themselves.

Training manuals differ from previously discussed manuals in being written specifically for audiences unfamiliar with the activity or equipment described in the manual. Thus, training manuals are both prescriptive and instructive, and writers of training manuals must be aware that their audiences are usually unfamiliar with the subject at hand. Training manuals cover a broad range of topics—instructing readers to use a company e-mail system, operate a laser, conduct a seminar, or interview potential employees.

Training videos have become popular forms of training manuals, incorporating training manual methodologies and often working in conjunction with printed training manuals. Numerous companies now produce training manuals and videos and offer training seminars for businesses and organizations. However, to be as up-to-date and specific as possible, companies generally prefer to produce their own training manuals and resources. Many software companies also use training videos—often referred to as tutorials—to teach their customers how to use software applications. For instance, Adobe—the distributor of Photoshop, Acrobat Reader, InDesign, and ImageReady—provides users with detailed video tutorials embedded in their software packages. In addition, the company regularly offers streaming video classes and tutorials through their web pages.

Field Manuals

Field manuals are generally used when workers must work away from the office or base of operations—that is, in the field. Unlike other kinds of manuals, which are designed in conjunction with other documentation and resources, field manuals are self-contained. They must include all the information a reader needs to solve a problem because it is unlikely the user has access to other manuals and resources. However, field manuals are generally concise because those working in the field usually do not want to carry large volumes of printed materials. For example, an engineer who works in telecommunications likely uses field manuals to work with, say, fiber-optics installations or router configurations. Many of these workers do not sit behind desks but move through varying terrains that require them to carry only what is necessary.

The advent of wireless technology has changed the way field manuals are designed because some workers in the field now carry portable computers or PDAs that can download e-manuals on site. However, e-field manuals have not replaced printed field manuals, so the field manual writer must consider both versions.

Lab Manuals

Lab manuals instruct laboratory users in how to operate specific lab equipment, and they provide protocols for working in the laboratory. Lab manuals may cover a single piece of equipment, thus working much like an operator's manual, a user's manual, or instructions. But lab manuals can also cover multiple pieces of equipment housed in a particular laboratory. The beginning of a lab manual usually describes its scope so that readers understand whether they need additional manuals to solve the problems that might arise.

Lab manuals offer specific guidelines for both operations and safety, and they tell users how to follow lab procedures. Many large labs now use online manuals because they can more easily cross-reference information and cover more protocols within a single manual. For instance, a manual that describes the protocols for a cell biology lab might contain information about a range of equipment and processes: microscopes, units and measures, statistics, graphs, computers, preserving membranes, mixing enzymes, cell cycles, cell structures, cell cultures, handling nucleic acids, counting chromosomes, photosynthesis, lasers, centrifuge machines, and so on.

Lab manuals can apply to science-based laboratories, but they can also be used in other kinds of laboratories, such as computer labs, engineering labs, or even usability labs. Laboratories used for teaching—such as the labs you might encounter in a biology or chemistry class—often have detailed manuals to both guide students through processes and experiments and provide policies and procedures to keep students safe in the laboratory environment.

ANALYZE THIS

Search your college's web pages to see whether any lab manuals for students are available online. If they are, skim through them. If they aren't, do a quick search for lab manuals at other colleges and universities. Select one manual, and note its key features and objectives. Does the manual address policies? procedures? equipment? How is it organized? Once you have analyzed the manual, consider what workplace problems the manual seeks to solve. Write a short document that describes those problems and explains how the manual attempts to solve them.

TRANSNATIONAL MANUALS

Perhaps more than any other type of workplace writing, manuals are likely to reach transnational audiences. Because manuals—including instruction manuals (see chapter 18)—are widely distributed with exported products, it is likely that many writers who develop manuals will have to account for transnational needs. Likewise, internal manuals within global corporations may have to convey information to employees from an array of transnational backgrounds. All manuals require consistency in delivering core information, but these may also require attention to localization in order to ensure that content is applicable to various transnational contexts. In anticipation of transnational needs, workplace writers should pay specific attention to the following issues as they begin to plan and design manuals:

See chapter 5 for more about writing for transnational audiences.

Translation—remember that in order for translators to accurately translate documents, the original documents must be clear and concise. Pay particular attention to use of specialized terminology, and be sure to provide translators with a glossary of that terminology. Because many manuals include glossaries of specialized terminology as a standard component, it is a good idea to plan on including a glossary as you plan the manual. Also, you should collaborate with the translator to ensure the most accurate translation. In addition, consider whether a single manual will provide multiple linguistic approaches (as is the case in Figure 19.6) or whether there will be independent manuals for each language into which the manual is translated.

See chapter 5 for more about translation.

Localization—Issues of liability and legal compliance are particularly important when writing manuals. Be sure the manual you are writing complies with all the local laws and governmental policies of the target countries in which it will be distributed. Confirm local patterns of learning, too, since many manuals present step-by-step information. In some cultures the logic with which U.S. writers approach the organization of steps may not make sense. Also, be sure to account for local approaches to numeric representations, like times and dates.

See chapter 5 for more about localization.

Design—Don't forget that most languages take more space to convey information than English does. Keep in mind, too, that some languages don't follow the left-to-right reading patterns of English; use clear spacing and directional information to guide readers. Consider, too, the material form the manual will take and how that form is likely to be received by various cultures.

See chapter 9 for more about design.

Visuals—Even though visuals may seem more "universal," they still require close attention when designed for transnational audiences. Confirm that the visuals you want to use are not offensive, and check that the sequence of visuals follows local cultural reading patterns. Also, be aware of local interpretations of color. Icons can be especially tricky for transnational audiences, so consult a cultural expert.

See chapter 8 for more about using visuals in transnational documents.

Technological access and literacy—Not all target countries have the same access to computer or Internet technologies that you might be accustomed to. Because computers are often shared among coworkers in some countries, time available to read manuals online may be limited, as may be the resources to print hard copies to read off line. Furthermore, not all countries use the same platforms that we might use to read certain kinds of documents. Be sure e-manuals are rendered in formats that can be opened by local applications, or include downloads of required applications. Remember, too, that computer familiarity, training, and literacy may differ in target countries. Make navigation evident and simple.

MANUAL STANDARDS

Like all other documents discussed in this text, manuals are written differently depending on the purpose of the manual, the intended audience, and the standards imposed both by particular disciplines and by individual companies and organizations. Many organizations specifically explain the standard manual formats and approaches for their organizations in order to assure consistency. Standardization does prevent problems of inconsistency and allows more efficient access to and more even production of manuals. As a result, you will likely find that the company or organization you work for sets its own standards for workplace documents, and those standards may deviate from what you learned in your course work. The models provided in this chapter cannot replicate every workplace situation: some manuals will require all of the parts we show you here, whereas others will not. Determining which parts to use will be a central part of your rhetorical problem solving. Remember that technical documents are not formulaic; you can't simply plug information into a set format and expect the document to be perfect. As we have pointed out before, context must drive workplace writing.

PSA

GENERAL GUIDELINES FOR WRITING MANUALS

Many manuals contain elements from more than one type of manual; that is, they may be a mixture of several kinds of manuals. Consequently, as you work through the early phases of the problem-solving approach, you will need to decide which parts must be included. However, certain basic guidelines should govern your writing.

Consider Users First

Because the function of manuals is to instruct readers to solve problems, the construction, content, and layout of manuals are audience centered. Chapter 2 addresses ways to analyze audiences as part of rhetorical problem solving, and a writer's first step is to understand as much as possible about the audience. For manual writers the planning phase is critical, and learning more about the audience may require some research about those who will read and use the manual. Keep in mind, too, that many manuals reach diverse transnational audiences; be sure to consider how your manual will account for linguistic and cultural differences among target audiences.

See chapter 5 for more on writing for transnational audiences.

Understand the Purpose and Function of the Manual

Because manuals often serve multiple purposes, workplace writers must be clear about their goals in order to best convey information to the readers. First, writers must define the purpose(s) of the manual for themselves—a crucial facet of the Plan phase. Second, writers must clearly explain that purpose to the audience. For example, it is not enough to acknowledge that a company needs a training manual; workplace writers must understand what particular training the manual will address and then explain that to the readers. Many manuals contain a specific section called Scope to define exactly what is included.

Because manuals are often written in conjunction with other documents, workplace writers must consider how the manuals interact with them. That is, other documents may affect the function of a manual. For example, a policy manual might address particular policies regarding insurance or other benefits; however, a document explaining those benefits might be needed to fully understand the policy manual.

Develop Detailed Outlines and Overviews before Writing

Because manuals often contain large amounts of information, manual writers may need to develop detailed outlines and overviews of the manual before they begin writing, to ensure that nothing is left out, de-emphasized, overemphasized, or irrelevant. Manual writers focus on the organization of the manual as they enter into the Draft phase of the problem-solving approach. Good outlines allow workplace writers to see the plan of the text and adjust it before writing the document. In conjunction with the description of scope, a solid outline helps writers draft an efficient manual.

Understand Information Sources

Manuals usually require input from several sources as well as multiple writers. For instance, a manual designed to show a service inspector how to maintain a turbojet probably requires several people to contribute: designers, test pilots, the manufacturer, technicians, and so on. In some instances hundreds of people might be used as sources of information; other manuals might require only the writer as a source. Thus, manual writers must determine the appropriate sources of information, compile the necessary information, and confirm its validity.

Agree on Divisions of Labor

Because manuals are rarely written by single writers, those contributing to the production of the manual must understand the range of contributors and the way the work will be divided. The work might be divided by sections, or several contributors might work on a single section. In some instances experts in different areas contribute to different parts of the manual.

Consider the Format and Design

Because manuals can be presented in many different formats, from printed volumes to e-manuals, writers must consider the best manner in which to present the manual, which will affect how the manual is written. For instance, an online manual with hypertext links between sections will be written differently than a manual presented as a booklet, which will be different from a manual presented as a quick-reference card. In most cases writers select the format before the actual writing begins because the format dictates the amount of space devoted to each section, as well as the use of images, graphics, and hyperlinks. To determine which design best suits a given manual, workplace writers should develop a series of questions as they work through the problem-solving approach—questions like these:

- Where will the audience use this manual—in the field? at a desk?
- What size pages will make the information readable and accessible?
- What materials are available for printing?

- What costs limit the printing?
- What technology is available to create the manual?
- Will an e-manual better serve the purpose than a printed manual?
- Will the manual be part of a collection of manuals and other documents, or will it stand alone?

IN YOUR **EXPERIENCE**

Consider the manuals you have used in the past: policy manuals, owner's manuals, maintenance manuals, and so on. What formats and designs for those manuals have you found useful or efficient? Which manuals have been formatted poorly?

Test Manual Usability

Workplace writers should conduct usability tests to ensure that audiences find a manual useful. If the purpose of a manual is to help readers solve a problem, writers want to be sure that it does just that. Thus, the Review phase of the problem-solving approach is as important as any other phase for manual writers; by accepting input from test audiences, writers can make their manuals more efficient and satisfying for readers. Usability testing also allows writers to determine whether the content of the manual is effective and accurate. Are the steps presented logically? Are the policies, definitions, or procedures understood by the test audience? Is the page design reader-friendly? Can readers easily access the information? As suggested in chapter 18 about technical instructions, writers should develop a series of such questions to ask each time they compose a manual.

See chapter 11 for more about usability testing.

Plan for Revision

Unlike many other documents, manuals often require frequent updating as policies, procedures, or problems change. When workplace writers must expect future revisions, they should use several helpful strategies:

- Keep accessible versions of document drafts in the word processing program being used. Revising is more efficient when writers do not have to rewrite all sections of the manual—only those that need to be revised.
- Save drafts of documents with titles that identify which version of the manual they support. Revising is easier when updates can be added to the most recent version.
- Save files in a format compatible with the software used by other workers in the company. By the time it is necessary to revise the manual, someone else may be writing the revisions.
- Alert manual users when the manual is updated. Revisions do little good if potential readers are unaware of the changes.

MANUAL ORGANIZATION

Most manuals use organizational patterns that are similar to those of instruction manuals, described in the previous chapter. But not all manuals direct readers through an active process, such as a set of instructions. The following description of components should be understood as a template that is adjusted to meet the needs of a specific manual. Writers must make rhetorical decisions about which segments and approaches are most efficient for the kind of manual they are writing and which best assist readers in solving the problems they face.

Front Matter

Cover (Title and Visual)

Like any good written document, a manual begins with a title. In many cases writers include a visual with that title to show audiences what the manual focuses on or which company or organization has produced the manual. Writers often think about the manual title as a thesis statement or a goals statement; it tells the audience what the manual accomplishes. Titles should be clear and concise, showing audiences the overall goal, purpose, or product being addressed.

Like all other technical documents, manuals serve as part of an ongoing record. Therefore, it is important that audiences know who produced the document and when. Citing a responsible party also allows readers to know whom to contact if the manual succeeds or fails. The byline does not necessarily include the name of an actual author but might indicate which company produced the manual. In many instances it would be impossible to list all of the contributing writers.

Writers also include the date the manual was produced to enable readers to identify whether the manual is recent or outdated. In addition, manuals need to be recorded as part of the documentation history of a problem, product, process, procedure, or policy; dating the document allows the history to be tracked. Manuals that are updated regularly should include version numbers as well.

Scope

The scope section of a manual details what is covered in the pages of the manual. Although sometimes implied rather than specifically stated, the scope also indicates what is not covered in the manual, directing readers to look elsewhere for that information. Thus, the scope section lets readers know what problems the manual can and cannot solve.

Developing the scope of a manual early on is crucial; it gives writers the opportunity to identify what they need to cover, what research they need to conduct, and what purpose the manual serves. Scope sections can be extremely brief—for example, "To describe the appropriate operating instructions to perform on-site field analysis." Or they may be several pages long. Generally, though, the scope specifically states the breadth of coverage of the manual and does so in short statements. Figure 19.1 gives a sample scope section.

Scope

This manual contains management directives and guidelines that convey and implement Jefferson Lab's policy and commitment to environmental protection, health, and safety. It shall be consistent with applicable federal, state, and local laws and regulations and SURA contractual obligations. The manual shall address identified hazard issues that have been deemed applicable to Jefferson Lab through the Work Smart Standards process (see *Appendix 2410-T1 Jefferson Lab Hazard Issue List*). Its guidance is applicable to all organizational functions and shall be implemented effectively throughout the laboratory.

For clarity and consistency, the EH&S Manual shall be comprehensive. Drawing on diverse resources within and outside Jefferson Lab, the manual shall contain both a description of established EH&S processes and the appropriate job-related guidance and procedures commonly needed to implement work practices that are sound with respect to safety, health, and environmental protection concerns.

FIGURE 19.1 A scope section

Introduction

Introductions to manuals are not necessarily introductory paragraphs, nor are they necessarily written in narrative form, although in some cases they may be. A manual's introduction tells the audience the aim of the manual; that is, it introduces the purpose—for example, "This manual provides operator and technician information for the Yellowfin Petrographic Microscope." Introductions also identify the organizational structure of the manual, usually through a table of contents.

In addition, the introduction includes any terminology or special conventions used in the manual. Depending on the intended audience, writers might need to explain special cases, exemptions, particular details, or case-specific language used in the manual. Writers might also include additional information about how the audience can use the manual. For example, some manuals suggest that audiences first read the entire manual before attempting the steps or following the directions. Other manuals require readers to begin at particular points other than the first page. Figure 19.2 shows the introduction to an online manual.

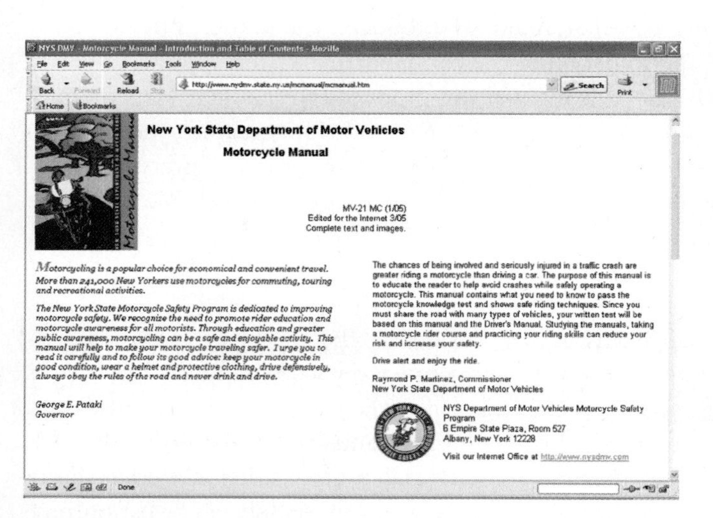

NYS DMV - Motorcycle Manual - Introduction and Table of Contents - Mozilla

http://www.nydmv.state.ny.us/mcmanual/mcmanual.htm

New York State Department of Motor Vehicles

Motorcycle Manual

MV-21 MC (1/05)
Edited for the Internet 3/05
Complete text and images.

Motorcycling is a popular choice for economical and convenient travel. More than 241,000 New Yorkers use motorcycles for commuting, touring and recreational activities.

The New York State Motorcycle Safety Program is dedicated to improving motorcycle safety. We recognize the need to promote rider education and motorcycle awareness for all motorists. Through education and greater public awareness, motorcycling can be a safe and enjoyable activity. This manual will help to make your motorcycle traveling safer. I urge you to read it carefully and to follow its good advice: keep your motorcycle in good condition, wear a helmet and protective clothing, drive defensively, always obey the rules of the road and never drink and drive.

George E. Pataki
Governor

The chances of being involved and seriously injured in a traffic crash are greater riding a motorcycle than driving a car. The purpose of this manual is to educate the reader to help avoid crashes while safely operating a motorcycle. This manual contains what you need to know to pass the motorcycle knowledge test and shows safe riding techniques. Since you must share the road with many types of vehicles, your written test will be based on this manual and the Driver's Manual. Studying the manuals, taking a motorcycle rider course and practicing your riding skills can reduce your risk and increase your safety.

Drive alert and enjoy the ride.

Raymond P. Martinez, Commissioner
New York State Department of Motor Vehicles

NYS Department of Motor Vehicles Motorcycle Safety Program
6 Empire State Plaza, Room 527
Albany, New York 12228

Visit our Internet Office at http://www.nysdmv.com

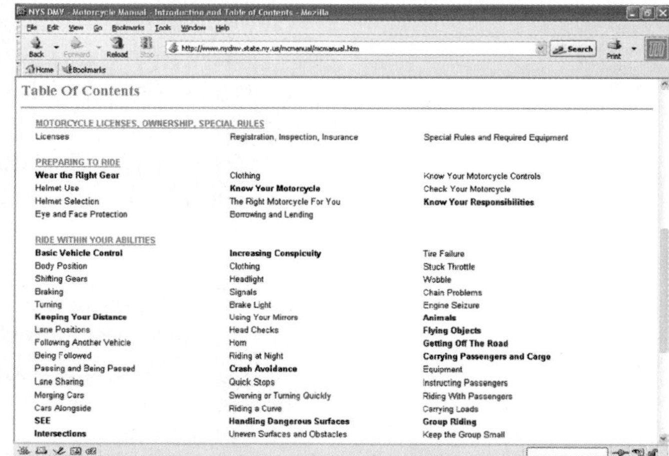

NYS DMV - Motorcycle Manual - Introduction and Table of Contents - Mozilla

http://www.nydmv.state.ny.us/mcmanual/mcmanual.htm

Table Of Contents

MOTORCYCLE LICENSES, OWNERSHIP, SPECIAL RULES		
Licenses	Registration, Inspection, Insurance	Special Rules and Required Equipment
PREPARING TO RIDE		
Wear the Right Gear	Clothing	Know Your Motorcycle Controls
Helmet Use	**Know Your Motorcycle**	Check Your Motorcycle
Helmet Selection	The Right Motorcycle For You	**Know Your Responsibilities**
Eye and Face Protection	Borrowing and Lending	
RIDE WITHIN YOUR ABILITIES		
Basic Vehicle Control	**Increasing Conspicuity**	Tire Failure
Body Position	Clothing	Stuck Throttle
Shifting Gears	Headlight	Wobble
Braking	Signals	Chain Problems
Turning	Brake Light	Engine Seizure
Keeping Your Distance	Using Your Mirrors	**Animals**
Lane Positions	Head Checks	**Flying Objects**
Following Another Vehicle	Horn	**Getting Off The Road**
Being Followed	Riding at Night	**Carrying Passengers and Cargo**
Passing and Being Passed	**Crash Avoidance**	Equipment
Lane Sharing	Quick Stops	Instructing Passengers
Merging Cars	Swerving or Turning Quickly	Riding With Passengers
Cars Alongside	Riding a Curve	Carrying Loads
SEE	**Handling Dangerous Surfaces**	**Group Riding**
Intersections	Uneven Surfaces and Obstacles	Keep the Group Small

FIGURE 19.2 The introduction and table of contents of an online manual

Definitions

Some manuals require readers to understand specific terminology to use the manual properly. And some manuals, like training manuals, introduce readers to new processes and ideas. In such cases part of the introduction may include a focus on terminology applicable to the process. A definitions section provides the meanings of unfamiliar terms used in the manual in order to ensure understanding; if readers misunderstand a particularly important word, they may not be able to use the manual properly. Many manuals do not require definitions pages, but workplace writers include them if there is any thought that the audience might misunderstand key terms. Figure 19.3 shows an online definitions page of a policy manual.

See chapter 15 for more about writing definitions and chapter 16 for more about writing technical descriptions.

Alerts and Special Considerations

As in the case of technical instructions, manuals also require writers to inform audiences about safety concerns; some problems can be dangerous to readers and/or to the product or mechanism covered by the manual. Writers place particular alerts and considerations throughout a manual, but crucial alerts and considerations must be presented *before* the audience reads through the manual.

See chapter 18 for more about alerts and warnings and the ways to use them.

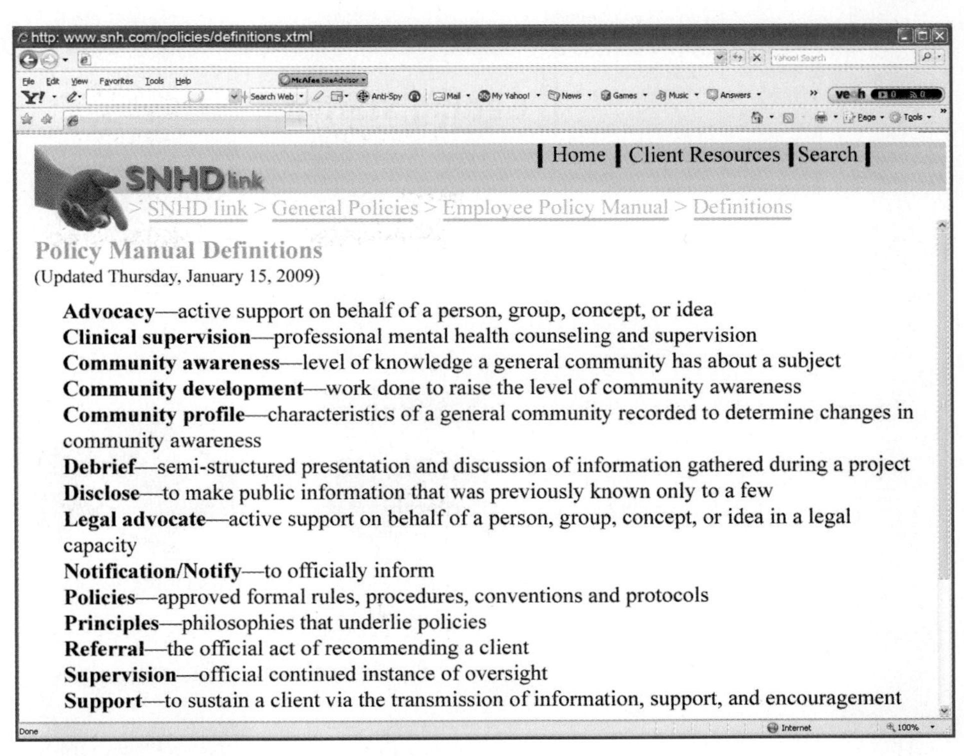

Policy Manual Definitions
(Updated Thursday, January 15, 2009)

Advocacy—active support on behalf of a person, group, concept, or idea
Clinical supervision—professional mental health counseling and supervision
Community awareness—level of knowledge a general community has about a subject
Community development—work done to raise the level of community awareness
Community profile—characteristics of a general community recorded to determine changes in community awareness
Debrief—semi-structured presentation and discussion of information gathered during a project
Disclose—to make public information that was previously known only to a few
Legal advocate—active support on behalf of a person, group, concept, or idea in a legal capacity
Notification/Notify—to officially inform
Policies—approved formal rules, procedures, conventions and protocols
Principles—philosophies that underlie policies
Referral—the official act of recommending a client
Supervision—official continued instance of oversight
Support—to sustain a client via the transmission of information, support, and encouragement

FIGURE 19.3 An online definitions page of a policy manual

Equipment and Tools List

Many technical manuals—particularly lab manuals, operator's manuals, and field manuals—require readers to use particular materials and tools to complete tasks. Often those materials are packaged with the manual, which then includes a section that describes the equipment and/or tools and their proper use. Readers must receive this information before beginning the processes for which the materials are needed.

See chapter 18 for more about equipment, tools, and parts lists.

Parts List

A parts list functions much like an equipment or tools list because it details needed materials. However, a parts list includes all the parts of the product the reader needs in order to complete the tasks described in the manual. In many instances manuals are included with a package(s) of parts. The parts list, then, serves two functions:

- To identify what each part looks like so that readers can easily identify it when needed
- To serve as a checklist so that readers can confirm that all parts have been included

In other cases the parts list identifies what parts a reader needs to gather to complete the tasks described in the manual; in those cases it is up to the reader to get the required parts, but it is up to the writer to provide the details the reader needs to secure the correct parts. Figure 19.4 provides an example of a parts list.

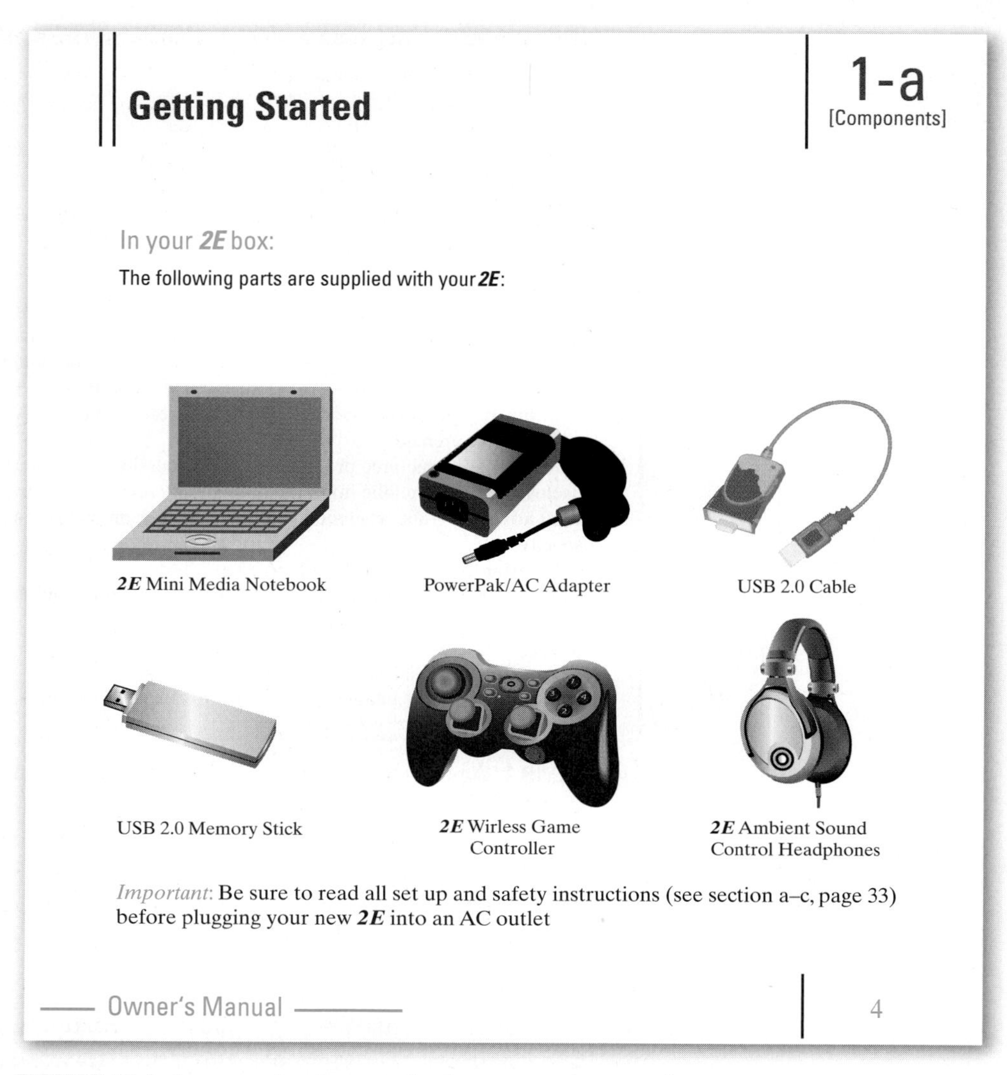

FIGURE 19.4 An example of a parts list in an owner's manual

Body

The steps in a manual are the core of the body section; they guide readers through the problem, product, process, procedure, or policy. When you are considering how to develop the steps in a manual, you should use the ideas found in chapter 18, which provides details about how to develop and write the steps for technical instructions. The basic concepts are the same:

- Number each step.
- Perform all steps as you draft them.
- Begin each step with a verb.
- Use positive commands whenever possible.
- Avoid highly condensed language.
- Include only one action per step.
- Group similar steps together.
- Visually separate steps.
- Clarify steps with visuals.

In some manuals, such as policy manuals, the steps are not written so that readers follow a linear process. Instead, the body may be a list of information; the steps might simply be a list of the policies that the manual is conveying to readers.

Conclusion and End Matter

After all steps are presented, writers offer final information, which is usually of three kinds depending on the type of manual. Again, each of these is covered in detail in chapter 18:

- Troubleshooting
- Maintenance information (unless the entire manual is a maintenance manual)
- Additional alerts

Index

In addition to the three listed types of end matter, writers sometimes include an index, which helps readers locate specific information within a manual. Because some manuals are lengthy, indexes can be useful tools that allow readers to access information easily. In addition, manuals are not always read in linear fashion from beginning to end; some readers skim through them to find specific information, and indexes allow them to locate what they need more quickly. Indexes generally list words and concepts alphabetically, identifying where readers can reference information in the manual. Figure 19.5 gives an example of such an index.

Some manuals place the index after the table of contents and the table of illustrations, because readers often use those resources together to locate information.

Because writing indexes can be an incredibly tedious process, many companies hire freelance index writers to compose the indexes for their manuals. And many workplace writers take advantage of the indexing software available, including Authex Plus, CINDEX, Macrex, and SKY Index. There is also web-indexing software available, such as HTML Indexer, as well as indexers for other utilities and add-ons.

Headings and Subheadings Some indexes in manuals provide more than just corresponding page numbers; they also list headings and subheadings under which the indexed information might be found. This approach places terms within their larger contexts.

Headings would include only recognizable keywords that directly link to materials in the manual. They should anticipate the kinds of subjects the reader might want to look up, and they should include all key terms found in the manual. Subheadings are used if there are multiple page references to a single heading. The general rule is that writers need a subheading if there are more than five references for a heading. However, if there are multiple subheadings (again, more than five), writers should consider reorganizing the manual and the index to convert some of the subheadings into headings, to allow readers more accessibility.

Cross-References Index entries can also include cross-references to other words in the index that are closely related. Cross-references are generally identified with the word *see* and then direct readers to similar, although not identical, information. For example, a manual that discusses the World Wide Web might include an entry that says, "See Internet." Although these are not the same, the writer realizes that readers might see them synonymously.

ANALYZE THIS

Turn to the back of this text, and look closely at the index we've written for this book, a detailed writing manual. What system of organization did we use to write the index? Why did we choose to place the index in the back rather than the front of the book? What characteristics of the index stand out for you?

INDEX

A

B

Index p. 194

FIGURE 19.5 An example of an index in a manual

SINGLE SOURCING

Single sourcing refers to making information available online in such a way that workplace writers can produce many different documents from a single source of information. For example, you might have recently composed a policy and procedures manual for your company. Then the human resources (HR) division asks that you develop a training manual to convey those policies and procedures to new employees. The HR division would also like to develop a training workshop that teaches new employees about those procedures and has asked you to provide materials for the workshop. In addition, because the new procedures and policies affect how business is conducted at your company, the marketing division has asked for documents to help them create marketing strategies that reflect the new policies. And to make matters even more complex, the company's web designers want to incorporate the procedures into the company's website. Thus, your manual has become the single source on which numerous documents are to be based.

The main challenge surrounding single sourcing is ensuring that all of the related documents are accurate and accessible to others. Simply making copies for everyone who needs the original can be time-consuming and expensive. But making the information available online to your workplace community is much more feasible.

Single sourcing is a useful approach for companies that produce a variety of documents, but it can challenge workplace writers who must take into account the needs of all those who will use a single-source document. Writers need to be sensitive to issues of audience, design, access, and word choice; the problem-solving approach can help alleviate some of the snags that might arise. Ultimately, single sourcing is a problem-solving strategy that solves economic problems, saving time and money, and it solves rhetorical problems, finding more efficient ways to disseminate information. Single sourcing can also be useful for companies that disseminate documents to transnational audiences because single-source documents can contribute to localization, internationalization, and standardization of documents. Table 19.1 identifies some of the specific rhetorical problems that single sourcing helps to solve.

See chapter 5 for more about writing for transnational audiences.

TABLE 19.1
Some of the benefits of single sourcing

Rhetorical Problem	Single-Source Solution
Documents written for hard-copy distribution may contain more information than is needed for online resources; appendixes or other extraneous information may not translate easily into online materials.	Content can be easily modified electronically to meet the specific need of the new document.
Cross-referencing in hard copy can be difficult to track and can be difficult to access as a reader.	Cross-references are automatically converted to hypertext links in online single sourcing.
Transferring documents from word processors or desktop publishing to hard copy may adjust the formatting and may reduce visuals or text to smaller, unreadable sizes.	Designs are easier to manipulate in an electronic environment, and design standards can be set and met for different output requirements.
If information is copied from hard copy to a new document, incorrect transcription and the introduction of inaccuracies are possible.	Information in single sourcing is simply transferred, not rewritten, so that inconsistencies don't enter the text.
Manuals and other documents must be revised regularly.	Single sourcing makes changing format and content efficient and accurate.

To assist writers in developing single-source documents, a number of single-source software packages exist. These programs help writers develop formats and approaches that build databases and access points for transmitting single-source documents in various formats. Although these software packages do assist in producing single-source documents, writers should be certain that the programs they use are compatible with the kinds of documents they produce.

MANUALS AS MARKETING AND PUBLIC RELATIONS TOOLS

Because so many manuals are addressed to readers who are not part of the organization that produced them, manuals also function indirectly as marketing and public relations documents. Well-written, user-friendly manuals evoke confidence in readers, both in the task outlined in the manual and in the company that produced the manual. Manuals are often the only contact a reader has with a company, and the reader will attribute problems with a manual to the inferiority of the company.

Workplace writers should always remember that manuals are most frequently consulted when readers are facing problems that they need to solve. If a manual presents more problems, frustration is likely to grow and confidence diminish. However, a manual that actually assists readers will boost their confidence in the company. Effective *internal* manuals function similarly to boost employee confidence, whereas inefficient and ineffective manuals frustrate employees and make their jobs more difficult. In short, *all* manuals should be recognized as valuable products and services that a company provides, and effort and care should be put into their production. Public relations is important, both inside and outside an organization.

ETHICAL CONCERNS

As emphasized in chapter 4 and throughout this text, workplace writers must make ethics a central concern. Because there are so many kinds of manuals, it is crucial that workplace writers seriously consider all of the ethical ramifications related to each type. However, these ethical considerations apply generally to the writing of manuals:

- Safety
- Functionality
- Accurate information
- Tested processes
- Up-to-date policies
- Hazards with proper use
- Troubleshooting
- Liability

Legal concerns usually necessitate that manuals accurately and safely present the needed information, but a professional considers each of these ethical considerations as more than a legal concern. Ethical workplace writers see these issues as personal and professional responsibilities.

SAMPLE USER'S MANUAL

Figure 19.6 includes a complete user's manual for an XM satellite radio receiver and identifies key organizational elements in the manual. This particular manual is intended for transnational audiences.

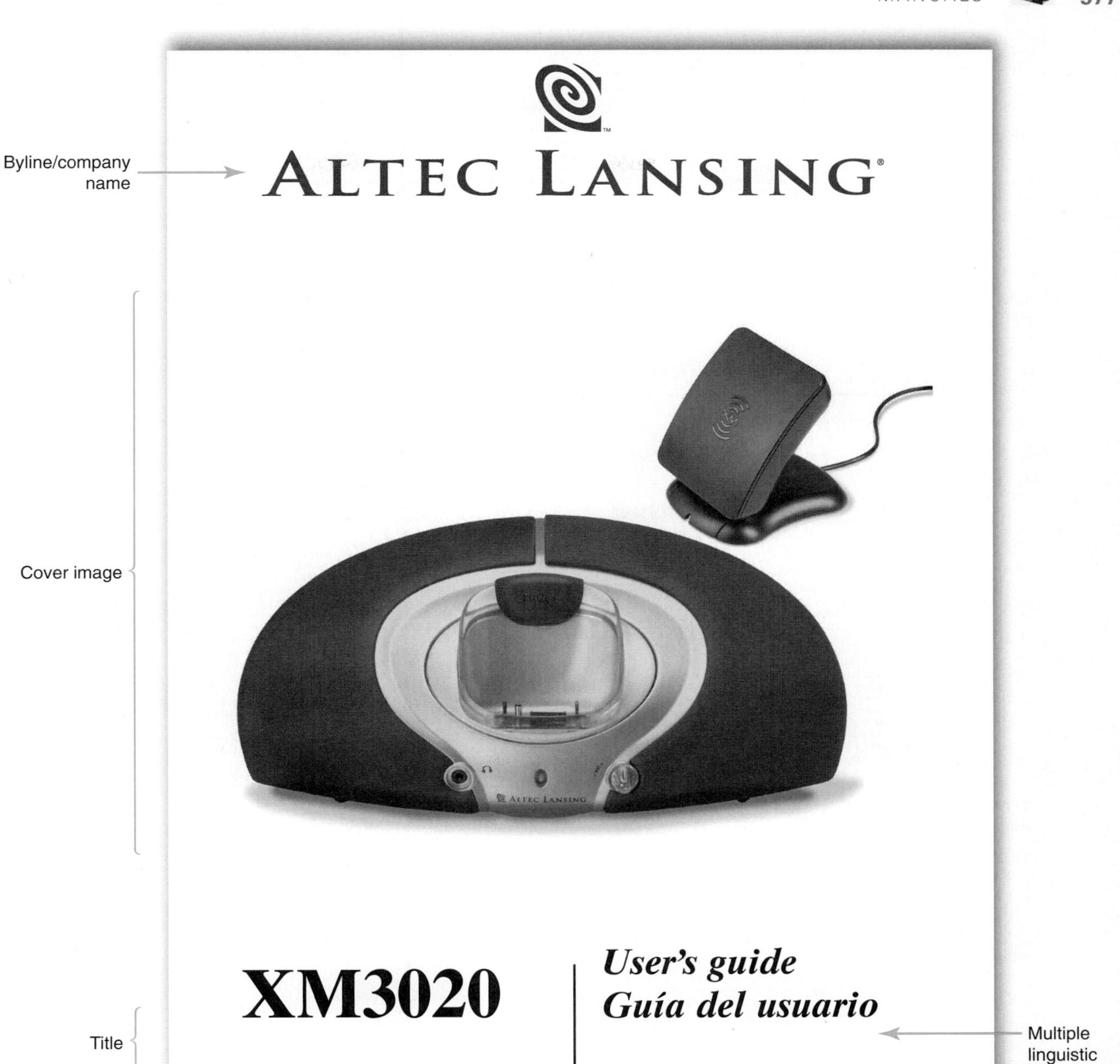

FIGURE 19.6A A user's manual for transnational audiences

Definitions

The lightning flash with arrowhead, within an equilateral triangle, is intended to alert the user to the presence of uninsulated "dangerous voltage" within the product's enclosure that may be of sufficient magnitude to constitute a risk of electric shock to persons.

Caution: To prevent the risk of electric shock, do not remove cover (or back). No user- serviceable parts inside. Refer servicing to qualified service personnel.

The exclamation point within an equilateral triangle is intended to alert the user to the presence of important operating and maintenance (servicing) instructions in the literature accompanying the appliance.

Initial alerts

Caution

To prevent electric shock do not use this (polarized) plug with an extension cord, receptacle or other outlet unless the blades can be fully inserted to prevent blade exposure.

Warning

TO REDUCE THE RISK OF FIRE OR ELECTRIC SHOCK, DO NOT EXPOSE THIS APPLIANCE TO RAIN OR MOISTURE.

Connecting the Power Cord (AC Wall Socket)

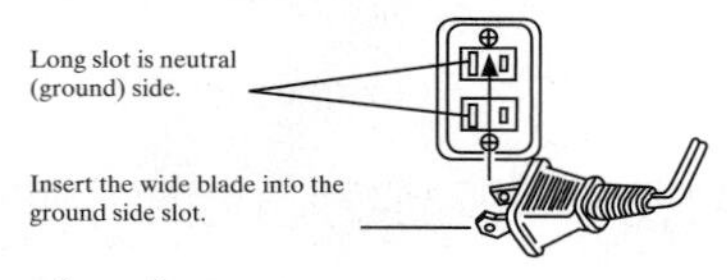

Initial alerts

Important Safety Instructions

1. Read these instructions.
2. Keep these instructions.
3. Heed all warnings.
4. Follow all instructions.
5. Do not use this apparatus near water — This apparatus shall not be exposed to dripping or splashing, and no objects filled with liquids, such as vases, shall be placed on the apparatus.
6. Clean only with dry cloth.
7. Do not block any ventilation openings. Install in accordance with the manufacturer's instructions. The apparatus should not be situated on a bed, sofa, rug or similar surface that may block the ventilation openings. The apparatus must not be placed in a built-in installation, such as a closed bookcase or cabinet that may impede the flow of air through the ventilation openings. Ventilation should not be impeded by covering the openings with items such as newspapers, tablecloths, curtains, etc.
8. Do not install near any heat sources such as radiators, heat registers, stoves, or other apparatus (including amplifiers) that produce heat.
9. No naked flame sources, such as lighted candles, should be placed on the apparatus.
10. Do not defeat the safety purpose of the polarized or grounding-type plug. A polarized plug has two blades with one wider than the other. A grounding type plug has two blades and a third grounding prong. The wide blade or the third prong are provided for your safety. If the provided plug does not fit into your outlet, consult an electrician for the replacement of the obsolete outlet.
11. Protect the power cord from being walked on or pinched — particularly at plugs, convenience receptacles, and the point where they exit from the apparatus.
12. Do not install in an area which impedes the access to the power plug. Disconnecting the power plug is the only way to completely remove power to the product and must be readily accessible at all times.
13. Power source — Use only power supplies of the type specified in the operating instructions or as marked on the appliance. If a separate power supply is included with this apparatus, do not substitute with any other power supply — use only manufacturer-provided power supplies.
14. Unplug this apparatus during lightning storms or when unused for long periods of time
15. Refer all servicing to qualified service personnel. Servicing is required when the apparatus has been damaged in any way, such as power-supply cord or plug is damaged, liquid has been spilled or objects have fallen into the apparatus, the apparatus has been exposed to rain or moisture, does not operate normally, or has been dropped. For service, refer to qualified service personnel, return to the dealer, or call the Altec Lansing service line for assistance.
16. For products which incorporate batteries, please refer to local regulations for proper battery disposal.

Special considerations

Altec Lansing Technologies, Inc. One Year Limited Warranty

What Does The Warranty Cover?: Altec Lansing warrants that its products shall be free from defects in materials or workmanship, with the exceptions stated below.

What Is The Period Of Coverage?: This warranty runs for one year from the date of purchase. The term of any warranties implied by law shall expire when your limited warranty expires. Some states do not allow limitations on how long an implied warranty lasts, so the above limitation may not apply to you.

What Does The Warranty Not Cover?: This warranty does not cover any defect, malfunction or failure that occurs as a result of: improper installation; misuse or failure to follow the product directions; abuse; or use with improper, unintended or faulty equipment. (For information on proper installation, operation and use consult the manual supplied with the product. If you require a replacement manual, you may **www.alteclansing.com**.) Also, consequential and incidental damages are not recoverable under this warranty. Some states do not allow the exclusion or limitation of incidental or consequential damages, so the above limitation or exclusion may not apply to you.

What Will Altec Lansing Do To Correct The Problem?: Altec Lansing will, at its option, repair or replace any product that proves to be defective in material or workmanship. If your product is no longer being manufactured, or is out of stock, at its option, Altec Lansing may replace your product with a similar or better Altec Lansing product.

How To Get Warranty Service: To get a covered product repaired or replaced, you must contact Altec Lansing during the warranty period by email (csupport@alteclansing.com). You must include in your email, your name, address, email address, telephone number, date of purchase and a complete description of the problem you are experiencing. In the United States, you may alternatively contact Altec Lansing by telephone at 1-800-ALTEC88 —please be prepared to provide the same information. If the problem appears to be related to a defect in material or workmanship, Altec Lansing will provide you a return authorization and instructions for return shipment. Return shipments shall be at the customer's expense, and the return must be accompanied by the original proof of purchase. You should insure the shipment as appropriate because you are responsible for the product until it arrives at Altec Lansing.

How Will State Law Affect Your Rights?: This warranty gives you specific legal rights, and you may also have other rights which vary from state to state.

The above limited warranties and remedies are sole warranties and remedies available to purchaser, if, and to the extent, valid and enforceable under the applicable law.

Customer Service

The answers to most setup and performance questions can be found in the Troubleshooting guide. You can also consult the FAQs in the customer support section of our Web site at **www.alteclansing.com**. If you still can't find the information you need, please call our customer service team for assistance before returning the speakers to your retailer under their return policy.

Tel: 800-258-3288
Email: csupport@alteclansing.com
For the most up-to-date information, be sure to check our Web site at **www.alteclansing.com**.

FCC Note

This device complies with Part 15 of the FCC Rules. Operation is subject to the following two conditions:
1. This device may not cause harmful interference.
2. This device must accept any interference received, including interference that may cause undesired operation.

FIGURE 19.6B

Introduction

Parts list

Equipment list

Body (including steps, procedures, and technical information)

XM3020
Powered Audio System

Thank you for purchasing this Altec Lansing product. For generations, Altec Lansing has been first in audio innovation. Today, our line of powered speakers has received more performance awards than any other brand. In all kinds of desktop environments, in every price range, Altec Lansing offers sound of distinction — giving even the most demanding customers the audio enjoyment they seek. Just listen to this!

BOX CONTENTS

- XM3020 Speaker System
- XM Radio home antenna with 20' cord
- Power supply
- Three interchangeable XM2go receiver docking adapters
- Rubber spacer
- User's guide and quick connect instructions

COMPATIBLE XM2GO RECEIVERS:

Works with:
- Delphi MyFi XM2go receiver
- Pioneer Airware XM2go receiver
- Tao XM2go receiver

PLACING THE XM3020 SYSTEM

1. Connect and position the XM Radio home antenna by inserting the antenna connector into the antenna input on the rear of the XM3020 system. Make sure the antenna points south outdoors through a window.

2. Position the XM3020 system on a level surface, preferably close to you (within three feet) and within 20 feet of the XM Radio home antenna.

MAKING CONNECTIONS

To ensure first-time operation as expected, please carefully follow the connection sequences described below in the order indicated.

1. Place the appropriate XM2go docking adapter inside the docking bay. Docking adapters are labeled for XM2go receiver identification.

2. Place your XM2go receiver inside the docking adapter. In most cases, the black rubber spacer included with the XM3020 system should be attached to the rear of the docking adapter in order to fit the XM2go receiver snugly inside the docking adapter. If your XM2go receiver is equipped with a high-capacity battery, do not use the rubber spacer.

3. Insert the barrel connector from the power supply into the DC connector on the rear of the XM3020 system. After this connection is made, insert the power supply into a wall outlet.

4. The XM3020 system is now on stand-by mode and a red LED should be lit.

START PLAYING

Turn on your XM2go receiver. The XM3020 system is now ready to play and a green LED should be lit.

VOLUME

The VOLUME knob on the XM3020 receiver controls the master volume. Turn the knob clockwise to increase the volume and counterclockwise to decrease the volume. Master volume can also be controlled using your XM2go receiver's wireless remote, if available.

WIRELESS REMOTE-COMPATIBLE

The XM3020 system is compatible with the wireless remote that came with your XM2go receiver. Simply point your wireless remote toward the IR window located directly above the Altec Lansing logo on the front of the XM3020 system to control volume and other XM2go functions.

Body (including steps, procedures, and technical information)

SYSTEM SPECIFICATIONS

Altec Lansing's superior sound comes from our proprietary technology, which utilizes custom-built, high-fidelity drivers, state-of-the-art equalization circuitry, and a harmonious mix of the following specifications:

Sound Pressure Level (SPL):	92 dB	System Response:	80 Hz – 20 kHz (-10 dB)
Total Continuous Power:	6 Watts RMS • 3 Watts/channel @ 6 ohms @ 10% THD	Signal to Noise Ratio @ 1 kHz input:	> 65 dB

FIGURE 19.6C

Troubleshooting (use of table format)

TROUBLESHOOTING

SYMPTOM	POSSIBLE PROBLEM	SOLUTION
No Power LED lit	Power supply cord isn't connected to a wall outlet and/or the power supply connector is not plugged into the power supply jack on the back of the XM3020 system.	Plug the power supply into a wall outlet and plug the power supply connector into the power supply jack.
	Surge protector (if used) isn't powered on.	If the power supply cord is plugged into a surge protector, make sure the surge protector is switched on.
	Wall outlet is not functioning.	Plug another device into a wall outlet (same outlet) to confirm the outlet is working.
No sound from speakers.	The XM Radio home antenna is not getting a signal.	Make sure the antenna points south outdoors or through a window. Specific instructions and tips on how to improve signal reception can be found in your XM2go receiver user's guide.
	Volume is set too low.	The VOLUME knob on the XM3020 receiver controls the master volume. Turn the knob clock wise to increase the volume.
	The XM2go receiver is not correctly seated into the docking station.	Remove XM2go receiver and re-seat it in the docking station.
	The XM2go receiver is properly docked and the XM3020 system's green LED is lit, but the XM2go receiver is in the "mute" mode.	Turn off the "mute" function on the XM2go device by pressing the mute button on the receiver or your remote control.
Sound is distorted.	Volume level on XM3020 system is set too high.	The VOLUME knob on the XM3020 receiver controls the master volume. Turn the knob counter-clockwise to decrease the volume.

FIGURE 19.6D

Definiciones

La figura de relámpago que termina en punta de flecha y se encuentra dentro de un triángulo equilátero, tiene por finalidad alertar al usuario de la presencia de "voltaje peligroso" sin aislamiento en el interior del producto que podría tener potencia suficiente para constituir riesgo de choque eléctrico para las personas.

Precaución: Para evitar el riesgo de choque eléctrico, no retire la cubierta (o parte posterior). En su interior hay piezas que no debe manipular el usuario. El servicio debe realizarlo personal de servicio calificado.

El signo de exclamación que se encuentra dentro de un triángulo equilátero tiene por finalidad alertar al usuario de la presencia de importantes instrucciones de operación y mantenimiento (servicio) en la literatura que viene incluida con el artefacto.

Avisos iniciales

PRECAUCIÓN

PARA EVITAR CHOQUES ELÉCTRICOS NO UTILICE ESTE ENCHUFE (POLARIZADO) CON UNA EXTENSIÓN, RECEPTÁCULO U OTRA TOMA DE CORRIENTE A MENOS QUE SE PUEDA INSERTAR COMPLETAMENTE LAS CUCHILLAS PARA EVITAR QUE ESTAS QUEDEN EXPUESTAS.

ADVERTENCIA

PARA REDUCIR EL RIESGO DE INCENDIO O CHOQUE ELÉCTRICO, NO EXPONGA EL ARTEFACTO A LA LLUVIA O HUMEDAD.

CONEXIÓN DEL CORDÓN DE ALIMENTACIÓN (TOMA DE CA DE PARED)

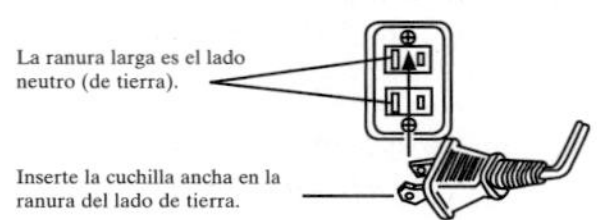

Manual repeated in Spanish for varied target audience

Avisos iniciales

INSTRUCCIONES IMPORTANTES DE SEGURIDAD

1. Lea estas instrucciones.
2. Guarde estas instrucciones.
3. Preste atención a todas las advertencias.
4. Siga todas las instrucciones.
5. No use este aparato cerca del agua —El aparato no debe estar expuesto a goteo o salpicaduras, y no se deben colocar encima objetos llenos de agua, como jarrones.
6. Limpie únicamente con un paño seco.
7. No bloquee ninguna abertura de ventilación. Efectúe la instalación según las instrucciones del fabricante. No coloque el aparato sobre una cama, sofá, alfombra o superficie similar que pueda bloquear las aberturas de ventilación. El aparato no se debe colocar en una instalación empotrada, como un estante o armario cerrado que pueda impedir el flujo de aire a través de las aberturas de ventilación. No impida la ventilación cubriendo las aberturas con artículos como periódico, manteles, cortinas, etc.
8. No instale el aparato cerca de fuentes de calor como radiadores, termorregistradores, hornos u otros aparatos (incluyendo amplificadores) que produzcan calor.
9. Las fuentes de llama descubierta, como velas encendidas, no se deben colocar sobre el aparato.
10. No anule el propósito de seguridad del enchufe tipo conector a tierra o polarizado. Un enchufe polarizado tiene dos puntas, una más ancha que la otra. Un enchufe de tipo conexión a tierra tiene dos cuchillas y una tercera cuchilla que se conecta a tierra. La cuchilla ancha o tercera cuchilla se proporciona para su seguridad. Si el enchufe proporcionado no encaja en su tomacorriente, consulte con un electricista para reemplazar el tomacorriente obsoleto.
11. Proteja el cordón de alimentación para que no lo pisen o prensen — especialmente en los puntos de enchufes, receptáculos, y el lugar donde salen del aparato.
12. No lo instale en un área que impida el acceso al enchufe de alimentación. El único medio de impedir completamente el ingreso de suministro eléctrico al producto es desconectar el enchufe de alimentación que debe estar accesible en todo momento.
13. Fuente de alimentación —Use únicamente fuentes de alimentación del tipo especificado en las instrucciones de operación o como se indique en el artefacto. Si se incluye una fuente de alimentación separada con este aparato no la sustituya con ninguna otra fuente de alimentación —use únicamente fuentes de alimentación suministradas por el fabricante.
14. Desenchufe el aparato durante las tormentas eléctricas o cuando permanezca sin uso por largos periodos de tiempo.
15. Para realizar el servicio técnico acuda al personal de servicio calificado. Es necesario que se efectúe el servicio técnico cuando el aparato se haya dañado de alguna forma, como por ejemplo, cuando el cordón o enchufe de alimentación esté dañado o cuando se haya derramado líquido o se hayan caído objetos dentro del aparato, o que el mismo haya estado expuesto a la lluvia o humedad, no funcione normalmente o se haya dejado caer. Para efectuar el servicio técnico, envíe el artefacto al personal de servicio calificado, devuélvalo al distribuidor o llame a la línea de servicio de Altec Lansing para solicitar ayuda.
16. Para los productos que usan baterías, refiérase a las normas locales para disponer de ellas de manera adecuada.

Consideraciones especiales

ALTEC LANSING TECHNOLOGIES, INC. GARANTÍA LIMITADA DE UN AÑO

¿Qué cubre la garantía?: Altec Lansing garantiza que sus productos no tendrán defectos de material o de mano de obra, con las excepciones que se indican a continuación.

¿Cuál es el periodo de cobertura? Para las unidades adquiridas en la Unión Europea o Asia, la garantía es de dos años, contados a partir de la fecha de compra. Para las unidades no adquiridas en la Unión Europea o Asia, la garantía es de un año, contado a partir de la fecha de compra. El término de cualquiera de las garantías de acuerdo a ley deberá expirar al vencimiento de la garantía limitada. Algunos estados y/o Estados Miembros de la Unión Europea no permiten limitaciones en el periodo de vigencia de la garantía, por lo tanto, las limitaciones antes mencionadas pueden no aplicarse a su caso.

¿Qué es lo que no cubre la garantía? La presente garantía no cubre cualquier defecto, mal funcionamiento o falla que resulte de: instalación inadecuada, mal uso o incumplimiento de las instrucciones del producto; abuso o uso con equipo inadecuado, no correspondiente o defectuoso. (Para obtener información sobre la instalación, operación y uso adecuado consulte el manual proporcionado con el producto.
Si necesita un manual de reemplazo, puede descargar un manual visitando www.alteclansing.com). Asimismo, los daños incidentales e indirectos no son recuperables de acuerdo con esta garantía. Algunos estados no permiten exclusiones o limitaciones por daños incidentales o indirectos, por lo tanto, dicha limitación o exclusión es posible que no sea aplicable a su caso.

¿Qué hará Altec Lansing para corregir el problema? Altec Lansing, de acuerdo a su criterio, reparará o reemplazará cualquier producto que presente defectos en el material o en la mano de obra. Si su producto ya no se fabrica más, o está agotado, de acuerdo a su criterio, Altec Lansing puede reemplazarlo con otro producto similar o mejor de Altec Lansing.

Cómo obtener servicio de garantía: Para obtener el reemplazo o reparación de un producto en garantía, debe ponerse en contacto con Altec Lansing durante el periodo de garantía vía correo electrónico a (csupport@alteclansing.com). En su correo electrónico debe colocar su nombre, dirección, dirección de correo electrónico, número de teléfono, fecha de compra y una descripción completa del problema experimentado. En los Estados Unidos, también puede comunicarse con Altec Lansing llamando al teléfono 1-800-ALTEC88 — por favor esté listo para proporcionar la misma información. Si el problema aparenta ser un defecto en el material o mano de obra, Altec Lansing le proporcionará una autorización de devolución y las instrucciones para el envío de la misma. Los envíos de devolución deberán ser pagados por el cliente, y el envío deberá incluir el comprobante de compra original. Deberá asegurar el envío en forma adecuada puesto que usted es el responsable del producto hasta que éste llegue a Altec Lansing.

¿Cómo afectará la legislación estatal sus derechos? Esta garantía le brinda derechos legales específicos, y además puede contar con otros derechos que pueden variar de estado a estado.
Las garantías y recursos limitados antes mencionados constituyen las garantías y recursos exclusivos disponibles para el comprador, siempre y cuando, y en la medida en que tengan validez y sean exigibles bajo la ley aplicable.

SERVICIO AL CLIENTE

En la Guía de solución de problemas encontrará respuestas a la mayoría de las preguntas sobre configuración y rendimiento. De igual manera, puede consultar las FAQ (Preguntas y respuestas frecuentes) de la sección de soporte al cliente en nuestro sitio Web **www.alteclansing.com**. Si todavía no puede encontrar la información que necesita, comuníquese con nuestro equipo de servicio al cliente para que le proporcione ayuda antes de devolver los parlantes a su distribuidor en virtud de su política de devolución.

Teléfono: 800-258-3288
Correo electrónico: csupport@alteclansing.com
Para obtener la información más actualizada, asegúrese de visitar nuestro sitio Web en **www.alteclansing.com.**

FIGURE 19.6E

Introducción

Lista de piezas

Lista de equipo

Cuerpo (incluyendo pasos (medidas), procedimientos, e información técnica)

XM3020
SISTEMA DE AUDIO AMPLIFICADO

Gracias por comprar este producto de Altec Lansing. Por generaciones, Altec Lansing ha ocupado el primer puesto en innovación de audio. Hoy día, nuestra línea de parlantes amplificados ha recibido más premios por rendimiento que cualquier otra marca en el mercado. En todos los ámbitos de escritorio, en cualquier rango de precio, Altec Lansing ofrece el sonido de distinción — brindando aun a los clientes más exigentes el placer de audio que ellos buscan. ¡Sólo escuche esto!

CONTENIDO DE LA CAJA

- Sistema de parlantes XM3020
- Antena para el hogar XM Radio con cable de 20 pies
- Suministro de potencia
- Tres adaptadores de acople de receptor intercambiables XM2go
- Espaciador de caucho
- Guía del usuario e instrucciones de conexión rápida

RECEPTORES COMPATIBLES XM2GO:

Funciona con:
- Receptor Delphi MyFi XM2go
- Receptor Pioneer Airware XM2go
- Receptor Tao XM2go

COLOCACIÓN DEL SISTEMA XM3020

1. Conecte y coloque la antena para el hogar XM Radio insertando el conector de la antena en la entrada de la antena ubicada en la parte posterior del sistema XM3020. Asegúrese de que la antena señale hacia el exterior al sur a través de una ventana.
2. Coloque el sistema XM3020 sobre una superficie plana si es posible cerca de usted (alrededor de tres pies) y a una distancia máxima de 20 pies de la antena para el hogar XM Radio.

CÓMO HACER LAS CONEXIONES

Para asegurar que la primera operación se lleve a cabo de la manera esperada, siga cuidadosamente las secuencias de conexión que se describen a continuación en el orden indicado.

1. Coloque el adaptador de acople XM2go apropiado dentro del módulo de acople. Los adaptadores de acople presentan una etiqueta que los identifica como receptores XM2go.

2. Coloque el receptor XM2go en el adaptador del acople. En la mayoría de los casos, se debe conectar el espaciador de caucho de color negro que viene con el sistema XM3020 a la parte posterior del adaptador del acople para poder encajar bien el receptor XM2go dentro del mismo. Si su receptor XM2go está equipado con una batería de alta capacidad, no use el espaciador de caucho.

3. Inserte el conector cilíndrico de la fuente de suministro en el conector de CC ubicado en la parte posterior del sistema XM3020. Después de hacer esta conexión, inserte la fuente de suministro en un tomacorriente de pared.

4. El sistema XM3020 se encuentra ahora en modo en espera y un indicador LED rojo debe estar encendido.

INICIE LA REPRODUCCIÓN

Encienda su receptor XM2go. El sistema XM3020 ahora está listo para iniciar la reproducción y un indicador LED verde debe estar encendido.

VOLUMEN

La perilla de VOLUMEN ubicada en el receptor XM3020 controla el volumen maestro. Gire la perilla hacia la derecha para aumentar el volumen y hacia la izquierda para disminuirlo. El volumen maestro también se puede controlar usando el control remoto inalámbrico del receptor XM2go, si es que está disponible.

COMPATIBLE CON CONTROL REMOTO INALÁMBRICO

El sistema XM3020 es compatible con el control remoto inalámbrico que viene con el receptor XM2go. Simplemente dirija su control remoto directamente hacia la ventana IR ubicada sobre el logotipo Altec Lansing en la parte superior del sistema XM3020 para controlar el volumen y otras funciones XM2go.

Ficha técnica

El sonido superior de Altec Lansing proviene de nuestra tecnología patentada que utiliza amplificadores a medida de alta fidelidad, circuitería de ecualización de última generación y una mezcla armoniosa de las siguientes especificaciones técnicas:

Nivel de presión acústica (SPL):	92 dB	Respuesta de sistema:	80 Hz – 20 kHz (-10 dB)
Potencia total continua:	6 Watts RMS • 3 Watts/canal @ 6 ohms @ 10% THD	Proporción señal/ruido @ entrada de1 kHz :	> 65 dB

Cuerpo (incluyendo pasos (medidas), procedimientos, e información técnica)

FIGURE 19.6F

Solución de Problemas[a]

SOLUCIÓN DE PROBLEMAS

SÍNTOMA	POSIBLE PROBLEMA	SOLUCIÓN
El indicador LED no está encendicio.	El cordón de la fuente de suministro no está conectado a un tomacorriente de la pared y/o el conector de la fuente suministro no está enchufado en la clavija de la fuente de suministro ubicada en la parte posterior del sistema XM3020.	Enchufe la fuente de suministro en un tomacorriente de pared y el concector de la fuente de suministro en la clavija de suministro.
	El supresor de picos(si se utiliza) no está encendido.	Si el cordón de la fuente de alimentación está conectado a un supresor de picos, asegúrese que el supresor de picos se encuentre encendido.
Los parlantes no emiten sonido alguno.	El tomacorriente de pared no funciona.	Enchufe otro dispositive en un tomacorriente de pared (la misma toma) para confirmar que esté funcionando.
	La antena para el hoga XM Radio no proporciona señal alguna.	Asegúrese de que la antena para el hogar de XM Radio señale hacia el sur de los exteriores a través de una ventana. Se puede encontrar instrucciones y consejos específicos de cómo mejorar la recepción de la señal en la guía del usuario del receptor XM2go.
	El volumen está fijado demasiado bajo.	La perilla de VOLUMEN ubicada en el receptor XM3020 controla elvolumen maestro. Gire la perilla hacia la derecha para aumentar el volumen.
	El receptor XM2go no está colocado correctamente en la estacion de acople.	Retire el receptor XM2go y velva a colocarlo en la estación de acople.
	El receptor XM2go no está acoplado de manera correcta y el INDICADOR verde del sistema MX3020 está encendido, pero el receptor XM2go se encuentra en modo "mute"(silencio).	Apague la funcion "mute" (silencio) en el dispositivo XM2go pulsando el botón de silencio en el receptor o su control remoto.
El sonido está dis-forsionado.	El nivel de volumen del sistema XM3020 está fijado demasiado alto.	La perilla de VOLUMEN ubicada en el receptor XM3020 controla el volumen maestro, Gire la perilla hacia la izquierda para disminuir el volumen.

[a] **The word *troubleshooting* is idiomatic and does not translate exactly; *Solución de problemas* means "problem solving."**

FIGURE 19.6G

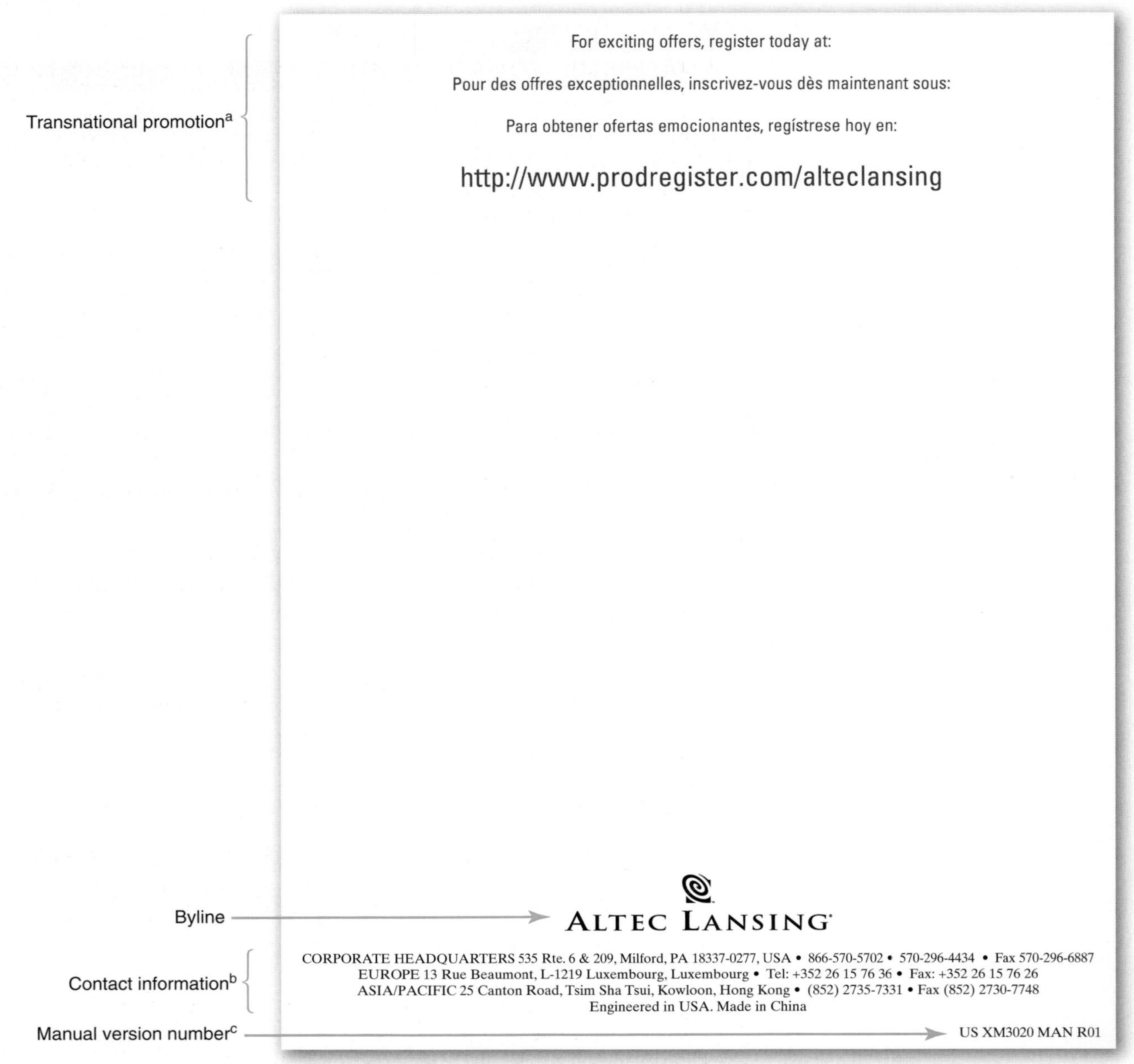

[a] Even though the manual is presented in English and Spanish, this promotional contact information is provided in English, Spanish, and French.

[b] Offices in the Unites States, Luxembourg, and Hong Kong suggest transnational audiences within the company.

[c] This version number indicates that the manual was intended for U.S. distribution.

FIGURE 19.6H

SUMMARY

- A manual provides instruction and other information necessary to accomplish something—that is, to solve a problem.
- The numerous kinds of manuals include these: policy, procedures, operations, operator's, owner's or user's, service and maintenance, training, field, and lab manuals.
- Manual writers should consider readers first.
- Writers should understand the purpose and function of the manual and should develop detailed outlines and overviews before writing it.
- Writers should understand their sources of information and should agree on divisions of labor.
- Writers should consider the format and design of the manual.
- Writers should test their manuals and should plan for revision.
- Manuals include varying components of front matter, body, and conclusion or end matter.
- Single sourcing allows many documents to be created from a single database of information.
- Manuals can also serve as marketing or public relations devices.
- Ethical considerations include safety, functionality, accurate information, tested processes, up-to-date policies, hazards with proper use, troubleshooting, and liability.
- Writers must account for transnational audiences from the earliest phases of writing and pay particular attention to issues of translation, localization, design, visuals, and technological access and literacy.

CONCEPT REVIEW

1. What are the most common types of manuals, and what purpose does each serve?
2. Why do some organizations develop standards for producing manuals?
3. What general guidelines should writers follow when writing manuals?
4. What are the standard parts of a manual?
5. What role does the scope play in the manual?
6. Why might a writer opt to produce a manual as an e-manual?
7. What is the difference between an equipment list and a parts list?
8. What are some basic guidelines for writing indexes?
9. Why is usability testing important for manuals?
10. Why might a manual require updating?
11. What are some benefits and some difficulties of using single sourcing?
12. In what ways do manuals—both internal and external—serve as marketing or public relations tools?
13. What are eight key ethical factors that writers of manuals should consider?
14. What key questions should writers ask in determining how to design and format a manual?
15. What five key issues should writers take into consideration when writing and designing manuals for transnational audiences?

CASE STUDY 1

Writing Standard Operating Procedures for the State of Alaska

A number of helpful resources for writing manuals exist within company web pages. One of the most thorough is the "Standards for Preparing Standard Operating Procedures" of the state of Alaska, Department of Administration, Division of Personnel/OEEO. You can find a PDF version at http://dop.state.ak.us/docpool/pdf/sop/Ch11DivisionProcedures/Sec2PreparingSOP.pdf. The state of Alaska also publishes a number of standard operating procedures (SOPs) on its web pages. One of the more interesting is the policies and procedures of the Division of Personnel; a PDF version is available at http://dop.state.ak.us/ppdb/index.cfm.

The Division of Personnel Policies and Procedures homepage includes a list of new policies and procedures and an outline of older policies and procedures. From either of these lists, select one of the policies and procedures. Then, using that segment of the state of Alaska's SOP, write a new SOP manual that stands alone, not as part of the larger SOP. Use any format you find appropriate, choosing from the possible sections described either in this chapter or in the state of Alaska's SOP for producing SOPs.

CASE STUDY 2

College Success: A Manual

According to the U.S. Department of Education National Center for Education Statistics (USDENCES), in 2004 the graduation rate for students in four-year colleges and universities was 55 percent. During that same year, the graduation rate for two-year colleges was 33 percent. These numbers show that substantial numbers of college students don't complete their degrees. What these statistics *don't* show, however, are the numerous *reasons* that students don't finish—such as financial pressures, familial responsibilities, other opportunities, unmet needs. And unfortunately, some students don't finish college because they simply don't understand what is expected of them until it's too late.

The USDENCES website at http://nces.ed.gov/ipeds/ offers statistical information about many individual colleges and universities; explore those pages to learn what you can about your school. Then explore your college's or university's web pages, and look for information about the school, about graduation rates, and about the things successful students do to graduate. Look specifically at how graduation requirements are publicized.

Then write a manual for students new to your school, telling them the things they must do to complete their degrees. The manual should take into account both official requirements and intangible concerns; it should combine elements of policy manuals, procedures manuals, owner's manuals, and even some instructions. Use the information you find on the USDENCES web pages and your school's web pages as both introductory material and support for the body. Be sure to take into account how your audience might best access the manual, that is, what format of manual might best reach the intended audience. Consider, too, the population and cultural makeup of your audience. If you need to account for transnational audiences, consider carefully how you will address their needs.

CASE STUDIES ON THE COMPANION WEBSITE

Cultural Communication Problems in the State Attorney's Office

As a recent graduate from law school who has not yet passed the bar, you are hired by the state attorney's office in Florida. Your first task is to review office procedural manuals, which were written with a great deal of legal jargon, and develop ideas for revising these manuals for nonspecialists.

Malfunction of the Mass Spectrometer: Malfunction of the Manual

You work part-time as a laboratory technician for a researcher in anesthesiology who owns a very expensive piece of medical equipment. This researcher allows others to use the equipment for a fee and asks you to develop a procedures manual for scheduling use of the equipment as well as handling whatever problems might occur with the equipment.

Out at Sea: Environmental Difficulties of Writing a Usable Manual

In this case you are asked to study a complex rhetorical situation in which a young marine biologist with a great deal of field experience is asked to produce a manual for employees who will assume a similar position to his. After producing a first draft of the manual, the marine biologist is asked by the company to make revisions to it. In addition to researching the problems this employee faces, you are asked in this case to respond to the company that commissioned the employee manual in the first place.

Surviving and Thriving at College: Designing a Manual for First-Year University Students

To encourage students to live on campus, your university has recently begun a program to renovate and expand many of its dorm facilities. The committee in charge of this program, the On-Campus Living Committee (OCLC), has commissioned your group to design and write a manual for first-year students.

VIDEO CASE STUDY

Agritechno

Out of Kindness, I Suppose: Poncho and Lefty Write Manuals

Synopsis

Agritechno has decided to initiate research into developing transgenic strains of apples. Since she has worked alongside Dr. Lev Andropov, an Agritechno geneticist, and because she has expressed interest in the apple project, Agritechno has named geneticist Teresa Cox to head up the new apple division, with her own lab and her own research team. Upon accepting the position, Teresa has requested that lab technicians Benjamin Goodman and Adam Simpson be assigned to her new lab. Teresa tells Lev about her move and about her request to take Ben and Adam—whom Lev has nicknamed Poncho and Lefty—with her.

WRITING SCENARIOS

1. Determine what you would consider to be the best approach to producing a manual—pick any kind of manual. What key factors would you want to require in a set of standards for manual writing? Develop and write a standards guideline for producing the kind of manual you have selected. Be sure to write more than just a list of standards; explain to your audience why you have written each standard as you have. Design the document as though it was directed to an audience that might produce the kind of manual you are targeting.
2. What kinds of manuals do you encounter in your daily life? When do you use manuals? When do you wish you had access to manuals? What might be the situations in which you expect manuals to be important in your career? Write a short assessment of how important manuals are in your life and how important they might be in your future career.

3. Identify and locate your school's policies regarding use of the Internet and the World Wide Web. Those policies are likely to be available in your school's undergraduate catalog and/or on school web pages, perhaps even on the web pages of your school's technology office. What might you change about those policies? How are those policies disseminated (format, access, etc.)? Write a policy manual—one to be rendered in hardcopy—that presents the policies as you'd like them to be. Explain and justify any policies that you change or add.
4. The U.S. Government Printing Office web page provides a link to a search engine for locating information within the U.S Government Manual. You can search the manual by a specific office, such as Homeland Security or EPA, or by subject, such as "engineering" or "electronic." Explore the website (http://www.gpoaccess.gov/gmanual/index.html), select a subject, and then identify what kinds of information the manual provides, as well as what strategies the manual uses to convey that information. Finally, write a letter to your instructor reporting what you found through your research.
5. Many companies now provide manuals for their products and services online. Imagine that you have recently purchased a Yamaha audio or video product—for instance, a set of speakers for your computer—but when the product was delivered, there was no manual included. Do some online research to locate at least one manual for your product. Then write a procedures manual that details how to locate and obtain that manual.
6. The U.S. Department of Commerce International Trade Administration publishes a manual titled "Business Ethics: A Manual for Managing a Responsible Business Enterprise in Emerging Market Economies," which is available online. The manual specifically addresses the need for standards and procedures in responsible business enterprises. Locate and read through this manual, noting specifically the segment on the need for standards and procedures. Then write a letter to your instructor that identifies what kind of manual this is and which specific qualities of the manual support your analysis.
7. Owner's manuals can be useful to readers who own things other than products; for instance, an owner's manual for a first-time pet owner might be quite useful. Caring for a pet involves some ethical considerations, regardless of whether the pet is a dog, cat, goldfish, hamster, python, ferret, tarantula, goat, parrot, hermit crab, or wombat (yes, we know someone with a pet wombat). Select a pet you have, have had, or wish to have, and write an owner's manual for that pet, paying particular attention to the ethics of pet ownership.
8. Locate a manual that contains alerts: dangers, warnings, or cautions. Examine how these alerts are identified in the manual. What do the alerts address? What kind of coverage do they provide? Do they use visuals? Write a letter to your instructor, explaining how the manual uses alerts and what the ethical implications are of the use of alerts.
9. Locate a manual for a piece of electronic equipment: perhaps a DVD player, PDA, MP3 player, cell phone, digital camera, computer. Analyze the manual to see how it addresses these ethical concerns: safety, functionality, accuracy of information, hazards with proper use, troubleshooting, and liability. Make a table that lists these ethical concerns and the manner in which the selected manual addresses them. Be sure to include specifics from the manual itself.
10. You are a member of numerous communities and perhaps organizations—this class, the campus community, your home community, an athletic team, an organized club, a church group, a political organization. Select one group, and consider what its policies are. Those policies may be unwritten—for example,

an athletic team may have an unwritten policy about drinking alcohol the day before a game. Design and write a policy manual for your chosen group. If the policies are too numerous to address in a single manual, narrow your focus to an area of related policies within the overall policy scheme.

11. Locate a manual on the Internet about a subject in which you are interested. Consider how the manual is specifically designed for publication on the Internet. Identify the key components of its accessibility, and think about why the writers may have designed the manual in this way. Write a short assessment of the components that are unique to that document as an e-manual.
12. Locate a manual that you have recently used: for a job, for a new product you've purchased, for your university experience. Work through the manual, identifying the ways it serves you as a reader. Consider the manual's organization, its design and format, and its role in connection with other documents or as an independent document. Write a short assessment of the manual as an effective or ineffective document, detailing your reasons and keeping in mind the manual's target audience.
13. Search the Web for companies that sell single-source software. Closely examine five or more single-source software systems, and consider what each offers. Then compare the approaches: what do the companies do similarly, and what are some of their differences? Write a user's manual for a generic single-source program.
14. Many companies sell or provide templates for writing manuals. Locate some of these templates, and consider what they offer writers. How do they compare with what this chapter suggests that manuals should include? What advantages and disadvantages do you see in using such templates? Create a table that identifies the benefits and disadvantages in using templates to create manuals.
15. Locate a training manual that uses video, whether embedded in the manual or exclusively video based. Watch the video, and consider how it emulates the strategies and organizational pattern of a print manual or an e-manual. Write a short document that explains—using specific examples—how that training video qualifies as a manual under the definitions and characteristics provided in this chapter.

PROBLEM SOLVING IN YOUR WRITING

Look back at Writing Scenario 10, in which you were asked to write a policy manual for a group to which you belong. Consider the role of problem solving in this assignment. What external problems did your manual seek to solve? How did you solve the rhetorical problems that presented themselves as you wrote the manual? Write a description of how you solved these problems and what led you to the rhetorical decisions you made.

20

Proposals and Requests for Proposals

CHAPTER LEARNING OUTCOMES

After completing this chapter, you will be able to do the following:

- Distinguish the different needs of **internal and external proposals**
- Understand the **persuasive qualities** of proposals
- Understand proposals as **problem-solving** devices
- Understand the differences between **solicited and unsolicited** proposals
- Identify the purpose of **sole-source contracts**
- Read and analyze **requests for proposals (RFPs)**
- Write an **effective RFP**
- Identify what information an RFP **must provide**
- Identify what information RFPs **must request**
- Understand the **ethical implications** of issuing and responding to RFPs
- Consider what roles **technology** plays in disseminating RFPs
- Understand the **relationships** between RFPs and other documents
- Distinguish the different rhetorical needs of **sales proposals** and **research or grant proposals**
- Write an **effective proposal**
- Understand and employ the various **components** of a proposal
- Consider the effect **minor errors** can have on proposal writing
- Understand the **relationships** between proposals and other documents
- Consider the role **technology** plays in writing proposals

DIGITAL RESOURCES

On the Companion Website www.prenhall.com/dobrin:

- Case 1: Proposing New Computers for Barksdale Accounting
- Case 2: Banking on Productivity: Seeking Food Service Bids for a Corporate Cafeteria
- Case 3: JMG Construction and Lifeline Safety: Writing a Proposal That Can Save Lives
- Case 4: Testing the Waters: Establishing an Environmental Ethos
- Video Case: Collaboration and Writing a Request for Proposals
- PowerPoint Chapter Review, Test-Prep Quiz, Exercises and Activities

REAL PEOPLE, REAL WRITING

MEET DEVIN MAXEY-BILLINGS • Automated Systems Engineer for Cambridge Technology Partners

What kind of writing do you do as an automated systems engineer?

I write proposals, technical manuals and user manuals, and other technical documentation of automated systems.

When you trained as an engineer, did you know that your work would include this much writing?

I took one technical writing class, and it turns out that I spend half of my time designing large technical systems and about the same amount of time documenting them. It is a competitive advantage for an engineer to be a good technical writer.

Could you talk a bit about the kinds of proposals you've written in the workplace?

The first company I worked for was a firm that completed automated test-system designs for huge manufacturers. An initial proposal was written by a team and delivered to a potential client. If the design was approved, the proposal was used as the initial contract and guided the design from start to finish.

It sounds as if the company you work for now deals primarily with internal proposals.

Internal proposals are generally much shorter and less detailed. Once you've been engineering within a company, your credentials are trusted, and you don't have to include so much technical content in the proposal. You basically identify a problem and write the proposal offering a solution. The two most important parts are the executive summary and the return on investment. If you've got those two sections nailed down, the project will generally be approved.

When you are writing proposals, what role does ethics play?

Most of the time we are in competition with someone else. Our audience might have two proposals from different companies that offer different ways to solve the same problem. The ethical concern is in providing realistic quotes about how long it's going to take to solve the problem. Because networking and reputation are a huge part of how we gain contracts, making good ethical decisions is key.

What advice about workplace writing do you have for the readers of this textbook?

Know your audience. As an engineer you need to keep in mind whether you're writing for another engineer or a salesperson or a financial person. Different parts of a proposal, for instance, might be geared toward different types of people. If the primary reader who is going to review the proposal is a financial person, you should really focus on the budget and the time line. Also, you should make your sentences as simple as possible. English may not be the first language of the people reading your writing.

INTRODUCTION

Workplace writers use proposals to solve problems. The scope, complexity, and subject of proposals can vary greatly, but the primary goal is to offer solutions to readers. Proposals put forward answers, alternatives, or new ways of looking at a subject and attempt to convince readers to solve their problems using the methods, steps, or strategies the proposals describe. Thus, proposals explain both problems and solutions. But as we've pointed out before, workplace problems aren't always negative; in many instances problems are the vehicles used to make changes, maximize opportunities, or increase the success of a particular product, activity, person, or organization.

To propose is to put forward an offer, to make a suggestion that is open for discussion and consideration. A proposal addresses all facets of a given problem and all the details of a potential opportunity—for example, methods, costs, time, personnel and resources, qualifications, and more. As a result, multiple proposals let organizations consider multiple solutions to a given problem and select the best—the most feasible, the most cost-effective, the most efficient, the most timely.

The U.S. federal government relies on both solicited and unsolicited proposals (discussed later in the chapter) to help determine budget expenditures for a range of departments and divisions. In fact, proposals are such a crucial part of how the federal government does business that a simple search of government web pages reveals over one hundred pages dedicated just to guidelines for submitting solicited and unsolicited proposals to various government agencies. Numerous companies and organizations rely on government contracts, all of which they secure through proposals to the government. As just one example, Zorin Industries is a technology developer located in Alexandria, Virginia, that provides electronic benefit transfer (EBT) technologies. EBT systems allow the transfer of government benefits like heath insurance or savings plans by way of a card, much like a debit card, that can be inserted into a special machine. In 2005 Zorin's written proposal earned the company the right to sell EBT card readers across the United States, propelling Zorin to the top of the country's electronic card-reading industry.

A proposal may not offer the only solution, but it should be designed to persuade readers that it offers the best solution. Because it works to persuade readers, a proposal must be written so that all readers understand what is being proposed. As Donald Helgeson puts it in *Handbook for Writing Technical Proposals That Win Contracts*, the golden rule for proposal writers is to "make it as easy as possible for the reader to understand and evaluate your proposal."[1] This chapter helps you do just that.

Because proposals are persuasive, writers must pay particular attention to the rhetorical and ethical choices they make in attempting to convince their audiences. Proposals are often focused on gaining contracts and entering into professional, sometimes profitable, relationships with other people and organizations. Not surprisingly, then, some writers are tempted to provide inaccurate or misleading information in order to secure those contracts and relationships. This chapter also addresses both the rhetorical and ethical choices writers make when writing proposals.

IN YOUR **EXPERIENCE**

Think about the last time someone asked you to help solve a problem. What did you do to offer a solution? Did the situation require you to learn more before you could offer a solution? Or did you know immediately what you would suggest? How did you present your solution? Were you convincing? With your class discuss the ways in which we are all asked to propose solutions to various problems and the ways we make our choices about those solutions.

---

[1] Donald V. Helgeson, *Handbook for Writing Technical Proposals That Win Contracts* (Englewood Cliffs, NJ: Prentice Hall, 1985), 2.

TYPES OF PROPOSALS

Proposals can be divided into two primary categories: internal and external. Like other internal or external documents, internal proposals are written to offer solutions to problems within a company or organization, and external proposals are written to convince readers outside an organization to opt for a given solution. Determining whether a proposal is internal or external is fairly straightforward, but such a determination is essential to making choices about how to write the proposal.

The audience for a proposal is what determines whether the proposal is internal or external. For example, if your company was downsizing its workforce as a result of lower sales, this would be considered an internal situation. However, if your company proposed a joint venture with a competitor as a result of the downsizing, the proposal to that other company would be external because it would be written to someone outside your organization.

Situations and problems are not always *fixed* as internal or external; real-world situations often require a variety of proposals that employ a range of internal and external documents. The previous example might begin with an internal proposal to contact a competitor, followed by an external proposal written to the competitor. Although we employ the designations *internal* and *external* throughout this chapter and the entire book, it is good to remember that documents don't always fit neatly within these categories.

Internal Proposals

Internal proposals can offer solutions to problems that an organization is aware of or problems that the organization does not yet recognize. For instance, suppose an engineering firm employs several dozen engineers working on various projects and makes available to each engineer networked versions of AutoCAD, the premier design software package used by engineers, designers, architects, and other professionals. For most in the firm, AutoCAD serves all of their needs, but several engineers find that they need to integrate their designs into other publication software that does not support an AutoCAD bridge. These engineers often have to re-create their designs in other drafting software in order to incorporate those designs in company documents, such as detailed proposals and reports, as well as publicity materials. Consequently, these engineers draft a proposal to their superiors, identifying why it would be beneficial to purchase other software to bridge design programs and publication programs. In writing the proposal, they are making their superiors aware of the problem while simultaneously offering solutions.

Internal proposals work toward the following goals:

- *To persuade an audience that a problem exists.* In the case of the engineering firm just mentioned, the proposal would identify the problem of integrating several necessary activities.
- *To persuade an audience that the problem is understandable.* If readers know why a problem happens, they are more likely to understand how to solve that problem and are then more likely to accept a proposal that offers a reasonable solution. In the case of the engineers, the intended audience needs to understand why AutoCAD does not work in conjunction with the publication software, so that a solution to the problem makes sense.
- *To persuade an audience that the problem should be fixed.* An internal proposal must detail the extent of the problem: How many people are affected? Is the problem costing the company? If so, how much? In the case of the engineering firm, the engineers would want to document how much time is being spent re-creating the design elements and thus how much money is being wasted. They should indicate how often they must redesign images, as well as how many employees face this problem.

- *To persuade an audience that the problem will escalate if not addressed.* A problem left unsolved may continue to cost a company money, may evolve into a more serious problem, or may be compounded by new and different problems. In the engineering example, the engineers may fall behind schedule if they must spend excess time re-creating design images.
- *To persuade an audience that the suggested solution makes sense.* Convincing readers that there is a problem that needs to be solved is only part of the writer's objective. The proposal must ultimately persuade the audience that the proposed solution should be adopted—that is, that the additional software should be purchased for the engineers.

In most businesses and corporations, managers would prefer not to learn about problems unless the messenger can also provide effective, feasible solutions. And in many cases managers may decide not to address a problem unless it is absolutely crucial because solutions usually require spending money, and spending money affects profitability. Thus, internal proposals must convince readers that solving the given problem is necessary and feasible and makes good business sense. Often, persuasion involves showing readers the long-term benefits of addressing the problem.

Everyday Proposals

Everyday proposals are fairly informal. They often take the form of memos or e-mails and offer solutions to routine problems that do not require great effort to identify, define, or solve. Everyday proposals are usually written for audiences that are already aware of the problem; they may be written as a request, asking permission to solve a problem in a given way. For instance, the example in Figure 20.1 is an e-mail request to order a dozen more training manuals—a simple request to solve a problem in a particular way.

See chapter 12 for more about e-mails and memos.

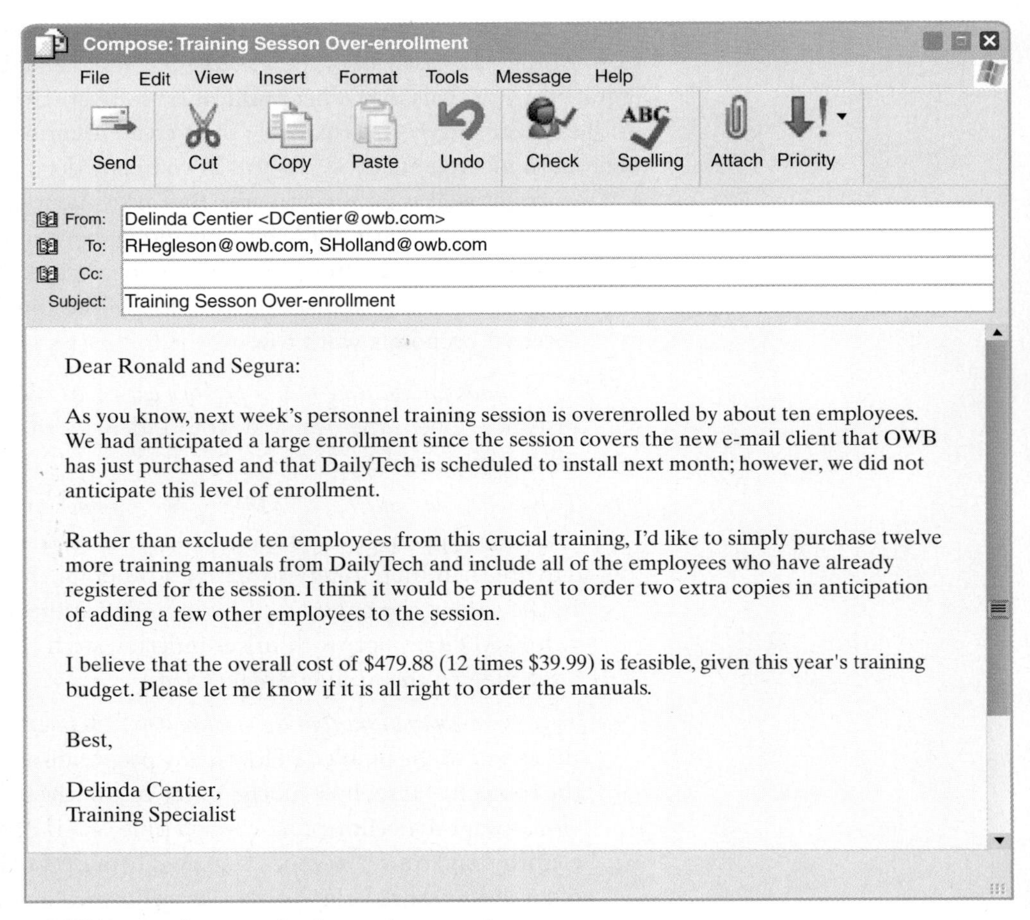

From: Delinda Centier <DCentier@owb.com>
To: RHegleson@owb.com, SHolland@owb.com
Cc:
Subject: Training Sesson Over-enrollment

Dear Ronald and Segura:

As you know, next week's personnel training session is overenrolled by about ten employees. We had anticipated a large enrollment since the session covers the new e-mail client that OWB has just purchased and that DailyTech is scheduled to install next month; however, we did not anticipate this level of enrollment.

Rather than exclude ten employees from this crucial training, I'd like to simply purchase twelve more training manuals from DailyTech and include all of the employees who have already registered for the session. I think it would be prudent to order two extra copies in anticipation of adding a few other employees to the session.

I believe that the overall cost of $479.88 (12 times $39.99) is feasible, given this year's training budget. Please let me know if it is all right to order the manuals.

Best,

Delinda Centier,
Training Specialist

FIGURE 20.1 An everyday internal proposal

Formal Proposals

Formal internal proposals can take many forms but usually contain more information and address more complex problems than informal proposals do. Formal documents may propose solutions to multifaceted or long-term problems and may offer solutions that require greater monetary commitment. Therefore, they may be addressed to multiple audiences and may be read by important people within the organization. They take a more formal tone and often include more substantial details to explain the problem(s) and solution(s). Formal internal proposals often include the standard components of a proposal—all of which are discussed later in this chapter.

External Proposals

An external proposal attempts to persuade readers outside the writer's organization to accept the solutions the proposal offers. Organizations request external proposals because few of them find it cost-effective, or even possible, to provide solutions for all of the problems they face. For example, a company that distributes research material for medicinal chemistry is not able to produce and install the electronic inventory systems needed to monitor and maintain its warehouse. So the company requests proposals from other companies that do produce and install inventory hardware and software. These external proposals—proposals from outside the organization—would be reviewed to determine which proposed solution best solves the problem. The organization might consider timeliness, cost, efficiency, effectiveness, dependability, or other factors that the proposal writer may never have suspected.

In many cases the organizations that solicit proposals are legally, as well as ethically, bound to actually employ those who developed the chosen solution. In other words, an organization can't "mine" proposals for good solutions and then hire someone else to do the work. Many businesses operate solely on their ability to write successful proposals; a construction firm, for example, might obtain all of its work by writing proposals outlining how they would construct buildings, bridges, or roads. Thus, many organizations' profitability is determined by the effectiveness of their proposals.

Sales Proposals

Sales proposals attempt to persuade another organization that the writer's company can best supply a needed service or product. Sales proposals might offer to install computer networks, deliver office supplies, provide transportation, build new buildings, remodel offices, deliver documents, supply parts, manufacture components, and so on. Sales proposals might be written to gain new clients, persuade long-time clients to establish a new contract, or renew and maintain a preexisting relationship. Because sales proposals can be written in many contexts, they can take many forms—from short, individually written proposals to lengthy, collaboratively written documents.

Sales proposals sometimes try to persuade readers that there is a need for additional products or services. For example, the Santa Pilon County corrections facility might fully recognize that it has a need for hash browns but may not recognize that it also needs sausages. A vendor writing a proposal, solicited or unsolicited, to supply the corrections facility with sausages would first need to convince a decision maker that there is a need for the product. This is referred to as *third-party expertise*, which suggests that an outside organization holds a clearer view of a problem and can therefore offer a better solution. Of course, in the process of persuading, sales proposals must also show that the proposed solution is feasible in terms of money, reliability, and performance.

Like other proposals, sales proposals work in relationship with other documents. Sales proposals often lead to contracts, technical documents that frequently reflect what the proposals offered. In addition, sales proposals often lead to progress reports, sales reports, and usability tests. In other words, sales proposals rarely work independently of other professional documents.

Research and Grant Proposals

Research and grant proposals attempt to persuade a sponsoring organization to fund an individual's or a group's research or project. In some cases the organization has sought research projects to fund. For instance, a large pharmaceutical company might offer annual grants for graduate students to conduct pharmaceutical-related research. Other organizations might seek researchers to help solve a specific problem. Research and grant proposals must persuade proposal reviewers that what is to be funded will be important and productive. Although many of us think of research as an academic endeavor, most companies and organizations conduct research to help develop products and services or to expand their abilities to provide products and services. Consequently, research proposals are quite common in the workplace.

Solicited and Unsolicited Proposals

Proposals can also be categorized according to whether or not they were requested. Organizations often request proposals in order to choose a solution that best serves their needs. Proposals written in response to such requests are considered *solicited proposals*. External proposals are generally solicited through requests for proposals **(RFPs)**, information for bids **(IFBs)**, or even requests for quotations **(RFQs)**. Both state and federal governments, for example, publish numerous RFPs each year, many of which are available on government web pages.

RFP request for proposals
IFB information for bids
RFQ requests for quotations

RFPs are formal appeals put out by organizations requesting that interested vendors—that is, organizations that provide goods or services—submit formal proposals in response to a particular set of needs. Vendors respond by proposing the goods, services, techniques, and methods with which they could solve the organization's problem. Organizations can also issue requests for information (RFIs), which generally solicit information before formal RFPs are issued. Some organizations issue IFBs to solicit information from bidding organizations. IFBs are often issued by government agencies and are generally more specific than RFPs. For instance, a company might issue an IFB asking multiple vendors to identify what they charge for a single product. The organization can then review the responses and determine which price and product is most appealing. The Santa Pilon County web page in Figure 20.2 shows an RFP for bids to supply hash browns to the local correctional facility.

In other instances proposals might be offered by organizations hoping to provide a product or service to other organizations that might not realize such a product or service exists. Proposals written without a request are considered *unsolicited proposals*; they hope to bring to light solutions to problems readers may not have known they had. For example, a software company might propose a new program for a mortgage company's website to help increase productivity. The mortgage company may not have been aware of the software nor of the possibility that productivity could be increased. Thus, the unsolicited proposal might be offering a solution to a problem that the mortgage company had not previously recognized.

EXPLORE

A number of government agencies—local, state, and federal—have web pages that make RFPs available. Look through the government web pages for your city, county, and state to learn where to locate posted RFPs. Create a list of at least two RFPs (and the problems for which each seeks proposals) from each of the web pages you explore. Then with your class compare the web pages where you located RFP postings and the kinds of RFPs you found.

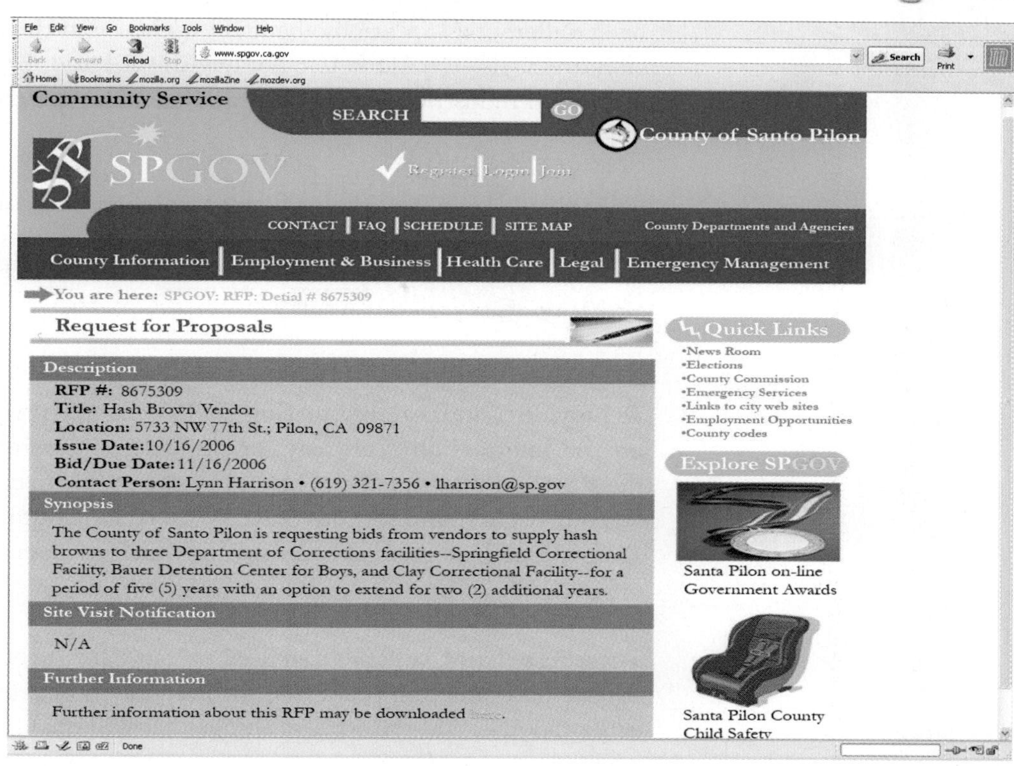

FIGURE 20.2 A request for proposals from Santa Pilon County

WRITING REQUESTS FOR PROPOSALS

Workplace writers produce RFPs as well as proposals. Learning to write an RFP is important in two ways. First, RFPs are fairly common workplace documents, and it is likely that you will compose them during your career. Second, understanding the form and function of RFPs assists you as a proposal writer. In other words, learning how to *write* RFPs helps you better *interpret* RFPs and, in turn, write better proposals to respond to them.

Even though RFPs are not all alike, there are standard items that most RFPs include; some companies even sell templates for creating RFPs. The key elements that appear in most RFPs can be divided into two main categories: the information that RFPs *provide* and the information that RFPs *request*.

Information That RFPs Provide

Title and/or Brief Description

Because some organizations produce large numbers of RFPs, readers need to know up front whether the RFP pertains to them. A title and brief description allow readers to quickly identify the RFPs that interest them and avoid wasting time on RFPs that do not concern them.

Contact Information

RFPs should clearly tell readers where to send their proposals. This contact information also provides readers with information about the audience. For example, if the contact information in an RFP specifies a company name—Biotech Industries, for instance—the proposal writer might assume a general audience of anyone affiliated with the company. But if the contact information identifies a particular department—say, housekeeping and maintenance at Biotech Industries—the proposal writer will make different assumptions about the audience. If the contact information identifies an individual—Dr. Maria Carillo, Divisional Head, Housekeeping and Maintenance at Biotech Industries—the writer has even more audience information to consider.

Proposal Schedule

RFPs should give the deadlines for submitting proposals, as well as the schedule for reviewing proposals and making decisions. Because this information is highly important, RFP writers generally bold or italicize dates. When a series of dates is required, a table or chart may be useful.

Background Information

To provide contextual information, RFPs generally include a detailed section of background information. This may contain a description of the organization submitting the RFP and may detail the history of the problem, as well as previous attempts to solve it. Background sections of RFPs can be long but provide important information about the audience and the contexts in which proposals will be considered.

Proposal Selection Process

Those responding to RFPs also need to know about the selection process: the specific criteria for evaluating proposals, the individuals or groups who will review and evaluate proposals, and the types of things that will qualify or disqualify proposals in the selection process. It may also be necessary to inform readers about how many proposals will be selected because sometimes organizations seek multiple vendors.

Project Summary

So that readers can recognize quickly what is being solicited, RFPs should provide a summary of the project goals and the problem at hand. They should also identify the time line for solving the problem—whether it requires immediate action or a long-term commitment.

Project Details

Readers who determine that an RFP fits their specialty and interest need more details about the project in order to construct the proposal.

Goals RFPs should offer as much detail as possible about the project goals, both primary and secondary goals.

Budget RFPs should identify any budget committed to the project or any budget restrictions.

General Scope In some cases RFPs provide the scope of the project so that readers understand exactly what parameters to work within.

Project Personnel Some projects require a selected proposal to be implemented through specific groups or individuals within the organization that posted the RFP. In these cases the RFP must identify the personnel who would be collaborating with the winning proposal team.

Timeline RFPs should provide a schedule that includes project details, such as a time frame for the project, interim deadlines, and the date the project must be complete, as well as relevant scheduling information.

Requirements Often RFPs identify requirements that proposing organizations need to meet before their proposals can be considered, such as experience or certification. The RFP can also list other requirements that proposal writers need to know, like business, design, function, development, content, technical, or even environmental requirements.

Relationships and Support RFPs usually identify preferred relationships the soliciting organization wishes to have with the organization whose proposal it accepts. The soliciting organization might require the selected vendor to meet with company personnel once a week to discuss progress on the project. Or the soliciting organization might wish to remain detached from the vendor altogether and might then indicate in the RFP that acceptance of the proposal does not indicate any formal connection between the two organizations or any guarantee of future cooperation. Many problems that RFPs address do not go away quickly or easily; therefore, some RFPs indicate that continuing support is expected from vendors once their work is completed. If continued support is expected, the RFP must identify it.

Information That RFPs Request

Effective RFPs request specific information that should be provided in submitted proposals. Later segments of this chapter discuss the proposal structure from the perspective of the proposal writer; here, however, the perspective is that of the RFP writer.

Company/Organization Information

Just as RFPs provide information about the soliciting organization, they also ask about responding organizations. RFPs ask for information about company history, experience, and other background information. They typically ask that proposals identify the equipment, services, or capabilities the vendor has that are relevant to the organization's problem-solving needs. In some instances RFPs ask that proposals include information about personnel, such as how many people work for the company and how many would work on the proposed project. RFPs might even ask that proposals include short biographical accounts or abbreviated résumés of key personnel. And, of course, RFPs ask that proposals include accurate contact information.

Proposed Solution

Because RFPs seek solutions to problems, they request many types of related information.

Summary of Solution RFPs ask for summaries so that proposal reviewers can quickly know whether a proposal interests them. This is a time-saving device.

Proposed Process RFPs request that proposal writers explain *how* they will solve the problem by including a description of the methods they intend to employ.

Task List An RFP asks that proposals include a list of all tasks the vendor will perform to solve the problem.

Time Line RFPs ask that proposals indicate a specific time line that vendors will follow to solve the problem; most problems need to be solved in a timely manner.

Details of the Proposed Solution

RFPs work from general to specific information in order to get the big picture first. They then ask for specific, detailed information in a number of categories.

Special Features What makes the solution particularly appealing? Are there any bonus features?

Functionality Will the solution really work? Why and how?

Options If the proposal is selected, what options or input will the organization have in the final implementation of the solution? Does the solution provide multiple approaches or possibilities?

Add-ons Has the RFP missed anything about the problem? If so, what else should be mentioned?

Originality What makes this solution different from other solutions? Will a tried-and-true solution be the only option? Or are there new ways of thinking about the problem that are more efficient?

Scale Can the solution be applied to similar problems of different magnitude? Has the proposing vendor ever done work on the scale required?

Technical Requirements What technical resources will be required?

Budget

Budgets are standard, critical parts of proposals because all organizations want to spend as little money as possible to solve their problems and want to fully understand what they are getting for their money. Thus, RFPs ask for a detailed general budget and sometimes request budget information for specific costs.

Third-Party Costs Some vendors subcontract with other companies to complete part of the larger task. An RFP might ask for a breakdown of third-party costs.

Cost of Provided Services Some vendors provide services instead of products, and services can't always be easily quantified. Consequently, an RFP may ask for a budget that details the services provided and the costs of those services.

Provisions

To ensure that there is no confusion about what a vendor will provide if its proposal is accepted, RFPs often ask that proposals include a specific list of what the vendor will deliver, provide, or do.

Personnel

RFPs ask that proposals indicate who would be working to solve the problem. This information serves two purposes: it allows proposal readers to know the level of personnel working on the problem, and it begins to establish accountability by showing who is responsible for solving the problem.

References

RFPs ask that proposals include references for the bidding company and perhaps even for the designated individuals so that proposal reviewers can evaluate the credentials of those who would be solving the problem. Have they worked on similar kinds of projects? Have other respectable organizations trusted them? Have those clients been pleased with their work?

Portfolio

RFPs often ask for examples of a proposing organization's previous work. In some cases, though, previous work is irrelevant or impossible to show, particularly for RFPs calling for products or services that are consumable.

Awards

Sometimes organizations are recognized for their outstanding work by other members of their professional community. Knowing that a proposing organization received an award from members of its peer community might carry some weight because it shows credibility. Thus, RFPs generally ask for this information.

Certification and Licensing

Many fields require individuals or organizations in those fields to hold certification or licenses from boards or agencies that maintain standards within those fields. Therefore, RFPs that seek individuals or organizations to do work that requires licensing or certification ask that licensing or certification be verified in the proposal.

ANALYZE THIS

The preceding section lists most of the components that are commonly found in RFPs. However, few RFPs contain all of them; the authors of RFPs usually choose the specific components they need. Find two examples of RFPs in an area of interest to you, and compare the components that each contains. To find the examples, you might do a Google search using the terms "RFP" and "Engineering" or some other field. Then write a one- to two-page comparison and analysis of the two RFPs, considering the following questions:

- What features and components do the two RFPs have in common?
- How are the two RFPs different? Does one contain fewer sections?
- How do the subject and the audience of each RFP influence the choice of components?
- Do either of the RFPs contain sections or components that aren't mentioned in this textbook? Do those sections seem important or useful?
- Do either (or both) of the RFPs suggest to you a project you might undertake now or in the future?

Ethics and Professionalism in RFPs

Ethical and professional considerations affect RFP writers. Because RFPs request help with problems and thus initiate professional relationships, RFP writers must realize that if one party acts unethically, the relationship will not likely succeed. An RFP sets the framework for the relationship; for it to benefit both parties, RFP writers should consider these issues:

- *The posting of an RFP indicates that a proposal will be accepted* and someone will be invited into a professional relationship with the requesting organization. In some cases, however, no proposals are accepted. An RFP might not draw any proposals that offer proper solutions, funding for the project might be cut, or the problem might be solved in another, unanticipated way (e.g., closing the department with the original problem). In unavoidable cases like these, RFP writers need to pull their RFPs and explain to respondents why no proposal was accepted. However, it is better policy not to post an RFP in the first place unless the project is certain to be completed.
- *RFPs should never ask for the impossible*; they need to ask for legitimate responses.
- *RFPs should disclose as much as possible about the problem* so that additional problems and issues are not presented to vendors at a later date. Vendors who agree to a contract based on an RFP and a proposal are not obligated to address issues not previously discussed.
- *RFPs should provide sufficient time for organizations to compose proposals.* An organization may be in a hurry to solve a problem, but asking respondents to provide detailed information in an unreasonable amount of time leads to flawed proposals and strained relationships.
- *RFPs should not solicit proposals that have little chance of being reviewed*; doing so is unethical because it wastes people's time. In those instances when RFPs must be posted despite someone's having an inside line on the contract, the RFP writer should discourage respondents from writing proposals. For instance, you might be required to post an RFP for a government agency looking for a vendor to supply staplers. For twenty years that agency has contracted with a single company, but the contract has expired, and government regulations require that an RFP be posted publicly to determine the best company to serve

as the new vendor. Your department has been happy with the stapler support provided for the past twenty years and has every intention of awarding the contract to the same vendor. In such a situation, the RFP is really for a **sole-source contract** and might be crafted to read something like this:

Sole-source contract
a contract predetermined to go to a specific vendor despite the contract's appearance of being open to bidders

> Government agency seeks proposals from vendors to supply staplers, stapler support services, and stapler maintenance program. Qualified proposals will indicate a minimum of nineteen years of previous experience supplying staplers, support, and maintenance to this same organization.

ANALYZE THIS

Shortly after Hurricane Katrina struck the Gulf Coast of the United States in September 2005, FEMA began to distribute funds to various companies and organizations in an attempt to solve many of the hurricane-related problems. Some of these funds were awarded as sole-source contracts, with no proposals or bidding considered from other organizations. These sole-source contracts awarded by FEMA drew criticism, and FEMA's failure to solicit proposals became a media story in itself.

Do some research on FEMA's use of sole-sourcing to address Hurricane Katrina–related problems. What is your opinion of FEMA's use of sole-sourcing and of sole-sourcing in general? What are the benefits of this strategy? What are the disadvantages? What ethical or political ramifications are connected to sole-sourcing?

Technology and RFPs

Technology plays a particularly important role with RFPs. Because RFP writers typically want their RFPs to reach as many potential vendors as possible, they need a medium that reaches many people.

World Wide Web

Posting RFPs on the Web makes access easy and widespread. Thus, writers need to consider web design when composing RFPs. But they must also recognize that the Web makes RFPs available to many unqualified respondents as well.

Databases and Lists

Most professional organizations maintain databases, Listservs, bulletin boards, and other forms of information dissemination. Posting RFPs through these media is likely to put RFPs in the hands of qualified readers.

Communication Technologies

Many organizations subscribe to services that track RFP postings and alert them when an RFP is posted that might interest them. These subscriptions cost money but do help. Other companies pay these same services to post their RFPs to ensure that the right organizations see them. In addition, these services sometimes provide templates for organizations to use when composing RFPs.

PDFs

Because many RFPs are disseminated electronically, it is a good idea to make RFPs available in portable document format (PDF). Because nearly all computers have a PDF reader, posting an RFP as a PDF practically ensures that readers can see the RFP as intended. In addition, PDFs are printable, allowing readers to make hard copies.

Forms

Because RFPs frequently ask for information in particular formats, many RFP writers provide templates for respondents to fill out, thereby ensuring that respondents provide the right kinds of information in a standardized fashion. Templates are often created as PDFs, which can be printed without losing the intended format.

IN YOUR EXPERIENCE

RFPs aren't always formal; they can sometimes be a simple e-mail or some other informal request. Take a moment to consider whether anyone has ever solicited a suggestion from you to solve a problem or if you have ever asked anyone for help with your problems. Think about all kinds of requests—a roommate having problems with a boyfriend or a girlfriend, a friend with a money problem, a classmate who has to miss class. Consider what was being requested, and then write a more formal RFP for that situation.

PSA

WRITING A PROPOSAL

Whether internal or external, sales or research, solicited or unsolicited, proposals have one main goal: acceptance. To be successful, proposal writers must address a number of rhetorical problems, particularly in the early stages of writing—the Plan and Research phases.

- Because proposals are problem-solving documents, proposal writers must be able to offer feasible, realistic, well-defined solutions.
- Proposal writers must reach their audience. A proposal that fails to address audience needs and expectations is likely to be denied.
- Proposal writers must be able to persuade their audience; proposals have to be convincing.
- Proposal writers must display the credentials that allow readers to take the proposal seriously.

Thinking through the problem-solving approach can help proposal writers plan a well-conceived proposal. The PSA helps writers understand their purpose and identify the problem to be addressed in the proposal. The PSA also provides proposal writers with guidelines for finding the information needed to write the proposal.

PSA

RFP Guidance

If the proposal is being written in response to an RFP, that RFP establishes the rhetorical context for the proposal writer. It identifies the audience and the purpose of the proposal and provides specific information about how to write the document. Proposal writers must carefully read and analyze the RFP as they work through the Plan phase of the PSA. Some RFPs are detailed, explaining exactly what a proposal should contain, whereas others offer little detail about what the organization wants from the proposal. In either case, writers must gain as much rhetorical guidance from the RFP as possible, asking the following questions:

1. What problem does the RFP present that needs to be solved?
2. What clues does the RFP offer about who the proposal audience will be? Does the RFP provide any information, either directly or indirectly, about that audience? What is the audience apt to want from the proposal?

3. Based on the RFP, what should the purpose of the proposal be?
4. What solutions can the proposal offer in response to the problems identified in the RFP?
5. What information is readily available, and what information will need to be gathered in order to write the proposal?

Components of Proposals

See chapter 12 for more about memos.

See chapter 22 for more about formal reports.

Most proposals are organized in a way that is similar to that of reports and manuals: they include front matter, body, and end matter. But because proposals are driven by their contexts, purposes, and classifications, not all proposals contain the same subordinate parts or use the same format. The outline that follows can be adjusted to meet the needs of any specific proposal. For instance, shorter, less formal proposals might adopt a memo format, beginning with a standard memo heading, whereas longer, more formal proposals might look more like a formal report. Your task as a writer is to make rhetorical decisions about which parts and approaches are most effective for the kind of proposal you are writing. Fortunately, as mentioned earlier, proposal writers often have guidance from RFPs to help make some of those rhetorical decisions.

Front matter
- Title page/cover
- Letter of transmittal
- Executive summary
- Table of contents
- List of visuals

Body
- Introduction
  - Statement of purpose
  - Background/statement of problem
  - Scope
  - Organization
- Approach
- Requirements
- Plan of action
- Qualifications/experience
- Budget
- Proposed schedule
- Conclusion/recommendations

End matter
- Bibliography
- Glossary
- Appendixes

It is important to note that proposal writers should never simply rehash the RFP. Proposals that closely adopt the language of the RFP often say little more than was in the original RFP. Experienced proposal readers are looking for substance. Take the time to propose and to persuade by providing new, important, detailed, relevant information.

The sections that follow include numerous examples of proposal components. A full proposal is included at the end of this chapter (Figure 20.9), but like most proposals, it does not include all of these parts.

Front Matter

Title Page/Cover Like any other technical document, proposals need to convey specific information quickly so that readers can determine whether the document is relevant to their needs. A proposal title page conveys three critical pieces of information.

Title of the Proposal A title identifies a document by explaining its purpose. Proposal titles may also indicate the intended audience. Consider these examples:

- A Proposal to Install a G-31 Communication Network System for Philips Northtell International
- Proposed Upgrade Requirements for GlowLode Marketing Software
- Proposed Pollution Prevention
- Statistical Analysis of Internet Crimes against Children: A Proposal Written for the Oklahoma State Division of Child Welfare
- Proposed Research in Wetlands Maintenance

Byline Bylines identify the writer(s) of the proposal. They can refer to a single author, a group of authors, or a company or organization.

Date Like all documents, proposals become artifacts in company records. Providing the date when the proposal was submitted relates the proposal to the time when the problem was identified, when the RFP was posted, and when proposals were due to be filed.

Other Elements If a single organization has posted numerous RFPs or has several divisions posting RFPs, proposal writers should include an audience address on the cover or the title page (see Figure 20.3). This is particularly necessary when the requesting organization is not familiar with the proposing organization.

Unlike manuals, instructions, and reports, proposals traditionally do not include cover images. However, companies may choose to include their logo to help readers understand who submitted the proposal; recognizable company logos help establish credibility. As imaging software is becoming more prevalent and easy to use, many proposal writers *are* including images to enhance the appeal of their proposals, but they should be sure to avoid images that are distracting or that offer nothing more than fluff.

Letter of Transmittal Letters of transmittal—sometimes called cover letters—identify writers' intentions to officially submit a proposal. Not all proposals include letters of transmittal; they are often unnecessary in short, solicited, internal proposals but are common in external proposals, both solicited and unsolicited. Letters of transmittal preface proposals by introducing writers, their organizations, and their purposes. The letters adhere to the standards of formal business letters and are generally no more than one page in length. They should contain several kinds of information.

See chapter 13 for more about writing letters.

EXPLORE

Locate two proposal covers: one that integrates an image and one that relies solely on text. Which is more effective? In what ways are the designs similar? In what ways are they different?

A Proposal to Monitor North Pacific Fisheries Seasonal Restriction Policies

For:
The Alaska Fisheries Research Consortium
1048 Lower Point Road
Seward, AK 99664

Submitted by:
Cory Aist
Department of Fisheries Biology and Aquaculture
University of Atlantis, Offshore Extension

July 28, 2007

FIGURE 20.3 A cover page including an audience address

Context for Submission of the Proposal Letters of transmittal explain *why* writers are submitting the proposal, including which RFP is being addressed and what problem is being solved. Some organizations may have publicized dozens or even hundreds of RFPs; the cover letter allows the reader to immediately place the proposal in the correct context so that it is reviewed appropriately.

Summary Letters of transmittal offer brief summaries of proposal objectives and parts, particularly those that writers don't want readers to miss. Remember that readers are often busy and may not pay close attention to an entire proposal. The letter of transmittal should begin to persuade readers to take interest in the proposal. The letter probably won't persuade readers to accept the proposal, but it might persuade them to read it carefully.

Date of Action If the proposal is time sensitive, the transmittal letter can point out any time constraints or simply provide a brief overview of the time the project will take, identifying key dates such as anticipated beginning and completion dates.

Gratitude It is a good idea to thank the audience for time spent considering the proposal. And if the readers have assisted in any way during the proposal process—for instance, by providing information or by explaining details of the RFP—writers should acknowledge that assistance. This professional courtesy might remind readers of who the proposal writers are and might enhance their credibility.

Although letters of transmittal appear first in the proposal, they are generally the last part of the proposal written. Because they contain information that summarizes the contents of the proposal, writers must first write the proposal to know what information to include in the transmittal letter. Figure 20.4 shows a sample transmittal letter.

Atlas Global Network Systems
111 Josephine B. Baker Blvd.
St. Louis, MO 63113

October 2, 2005

Ms. Hannah Cone
Pacific International Design
201 Union St.
Seattle, WA 98102

Dear Ms. Cone:

Context → In response to your September 13, 2005, request for proposals asking for designs for a prototype Personal Area Network (PAN) system, we submit the enclosed proposal. We are grateful for the opportunity to present this information and are appreciative of the time and effort Pacific International Design is devoting to reading our proposal.

Summary {

As you know, individuals worldwide have begun to carry numerous personal information devices as a regular means of accessing, storing, and transmitting information: cellular phones, personal digital assistants (PDAs), digital watches, pagers, handheld video game devices, personal music players, digital cameras, and even portable DVD players. Although all of these devices allow for information mobility, few of these devices interact with each other (though some strides have been made to integrate cellular phones with music players, PDAs, and game-playing devices on a limited scale). We believe that by networking these personal information devices within a PAN, the devices themselves become more useful and more efficient users and producers of information.

You will find in the enclosed proposal detailed plans for establishing a PAN that links directly with networked servers in order to route information quickly to and from individuals, regardless of the devices being used. The proposal identifies the benefits of instant Internet and World Wide Web access not just through wireless laptops or PDAs but any personal information device. Such a network allows for rapid dissemination of information not just to, say, workers in the field, but from them, as well as the rapid transmittal of that information worldwide.

Date of action → As you will see from our proposal, we believe we can establish a reliable and cost-effective series of networked personal information devices within the next eighteen months and that our prototype network will not only serve Pacific International Design's needs, but will change the way personal information devices are employed.

Again, thank you for reviewing our proposal. We look forward to hearing from you and establishing a working relationship between Pacific International Design and Atlas Global Network Systems.

Sincerely,

W. M. Stagler, Director

FIGURE 20.4 A letter of transmittal

Executive Summary Sometimes called an abstract, the executive summary is one of the most important parts of a formal proposal because it offers readers an easy-to-understand synopsis of the entire proposal, allowing readers to determine whether they should continue reading. Summaries provide readers, who are often from upper-level management, with a quick overview in language they can understand. Proposals often address problems involving technical details that management-level readers do not know about; the technical details are for the workers who will do the actual problem solving. Executive summaries, on the other hand, are for readers with little time and usually less technical knowledge or familiarity with jargon. Thus, executive summaries are generally brief, ranging from a few sentences to a paragraph or two, but never more than that. They can take a narrative form and/or use bulleted lists to highlight key points.

Summaries should be thought of as persuasive elements. In composing them, writers should reflect on the following:

- Does the summary provide a brief overview of the problem being addressed?
- Does the proposed solution in the summary emphasize key points without going into detail?
- Does the summary identify the benefits the reader gains by adopting the proposal?

Like letters of transmittal, executive summaries are one of the first parts of a proposal to be read, but writers usually cannot compose them until the entire proposal is finished. Summaries must be clear and convincing.

IN YOUR **EXPERIENCE**

Assume for a moment that this textbook is a detailed proposal, one that provides possible solutions for solving technical communication problems and includes lots of technical details, like how to write a proposal or how to employ the problem-solving approach. Imagine that you are trying to persuade a reader to examine the entire book as a proposed method for solving communication problems. How would you summarize the key points of the proposal/book? Write the executive summary that should be included when the proposal/book is transmitted to readers.

Table of Contents Readers need guidance to access information quickly and efficiently, and tables of contents direct them to the specific information they seek. For instance, a finance officer might be particularly interested in budget items, whereas a technician might be most interested in the mechanical aspects of a proposal. Each reader needs a quick way to locate the desired information. Even a reader who reads the entire proposal might later need to refer to specific parts.

Tables of contents should reflect the level of detail in the proposal. Primary section headings should always be listed, and secondary and tertiary headings as well if they are crucial section dividers. Tables of contents that are too general do not effectively guide readers through the information in the proposal. Figure 20.5 shows an overly general table of contents.

Tables of contents should provide as much detail as possible about the proposal and about where specific information is located. This does not mean that a table of contents should account for every page in the proposal, but the location of every part

Table of Contents

Introduction 2
Proposed Solution 6
Conclusion 32
Appendix 47

FIGURE 20.5 A very general table of contents

or section of the proposal should be identified. Figure 20.6 presents a more detailed and helpful table of contents.

Table of Contents

FIGURE 20.6 A detailed table of contents

ANALYZE THIS

In reviewing the general table of contents in Figure 20.5, what do you learn about the actual contents of the proposal and the way you might locate specific information? Do you know what information is contained on pages 2 through 5? or on pages 6 through 31? Is the conclusion really fifteen pages long? List the kinds of things that are not conveyed by this general table of contents. What might you as a reader want to find in the proposal but have difficulty doing so because of the generality of the table of contents?

List of Visuals Many proposals contain visuals—graphs, charts, tables, and illustrations—to clarify information for readers. Budgets may be presented in charts; schedules may appear in tables; processes and procedures may be presented as flowcharts; photographs or illustrations may be used to clarify points or to provide examples. In the actual document, graphs, charts, and illustrations are referred to as figures and are listed in consecutive numerical order. Tables are simply called tables but are also numbered.

Sometimes visuals convey critical information that readers need to understand easily and quickly. Like tables of contents that guide readers through proposals, lists of visuals guide readers in locating all the visual elements in the proposal. However, a generic list of visuals, like the one found in Figure 20.7, is of very little help. Lists of visuals should include at least the titles of the figures and the tables, and these titles should correspond exactly to the titles used in the document. Figure 20.8 shows a more helpful list of visuals.

See chapter 8 for more about using visuals.

List of Figures and Tables

FIGURE 20.7 A generic list of figures and tables

List of Figures and Tables

FIGURE 20.8 A detailed list of figures and tables

Body

Introduction Like all good introductions, proposal introductions inform and persuade readers. They should include four primary elements:

- A statement of purpose
- A description of the background and/or a statement of the problem
- An account of the scope of the proposal
- An explanation of how the proposal is organized

Writers can divide the introduction into sections that address each of these elements or can build all of them into a single introductory statement. Writers who divide the introduction should use headings that clearly identify what is addressed in each section.

Statement of Purpose Simply put, the statement of purpose tells readers the goal of the proposal. Statements of purpose are generally a few sentences, identifying why the proposal has been written and what it tries to do. Statements of purpose function like a thesis statement for the proposal. Consider this example:

> The purpose of this proposal is to identify that Gaines Green Gadgets (G3) is best suited to provide personal digital assistant medical-record software to Alliance Medical Personnel. G3's medical-record software packages, when employed with

G3's wireless network, allow medical personnel to share critical medical records, regardless of their global location. The proposal provides details of the medical-records system and the wireless network that will best serve Alliance Medical and that can be provided in a timely, cost-effective manner.

Statements of purpose do not always begin with "The purpose of this proposal is . . . ," although this phrase draws the reader's attention. Proposal writers might find a statement of purpose redundant if it repeats information contained in the letter of transmittal and executive summary, but there are two important points to remember:

1. *Reiterating a key component of the proposal helps emphasize it.* A savvy writer plays down the notion of repetition but highlights reiterated points by varying sentence structure and presentation. Smart rhetorical choices allow writers to stress the same points without using the same words, structure, or organization.
2. *Many readers won't read an entire proposal.* Reiterating a key point in three or more parts of a proposal means that more readers are likely to encounter it and understand that point.

Background/Statement of the Problem The background segment of the introduction is more substantial than the statement of purpose. This portion of the introduction discusses the problem that the proposal addresses and is critical to the success of the proposal; it sets the context in which the proposed solution functions. If the problem being discussed isn't significant for the reader, then the reader will not be interested in the solution. Or if the proposed solution does not relate directly to the problem, then the proposal will seem inconsistent and will not be received well. The statement of the problem must identify all facets of the problem and must show that the writer fully understands the context in which the problem and the solution are taking place.

Proposals—particularly unsolicited proposals—frequently have to convince readers that a problem exists that requires a solution. Proposals offering third-party expertise can give readers a different view of their situation and perhaps persuade them that the writers have a clear view of the problem. But proposal writers must remember that sometimes the "problem" is actually an opportunity. For instance, a particular company might have overlooked a potential client group simply because that company has not had the technological services to identify or reach this group. In that case a proposal can identify a new opportunity for the company.

Because people and organizations generally don't want to hear that they have problems, writing the statement of the problem can be a rhetorical challenge for proposal writers. Especially in an unsolicited proposal, it's a good idea to present the information positively. In fact, a good proposal writer approaches this section not as a simple statement of a problem, but as a statement of an opportunity to solve a problem. This does not mean that writers should hide information or present inaccuracies in order to present a positive message; doing so would be a serious breech of professional ethics. But readers are more likely to be receptive to a positive opportunity to improve.

Scope Because problems often require multiple solutions, some proposals focus on solving only part of the problem. In such cases writers include a section in the introduction that specifies the parts of the problem the proposal addresses. This section identifies the parameters of the proposal: exactly what the proposal does and, in some cases, does not do. For instance, a proposal might recommend the use of a particular software, but it might not address upgrading the computer systems in order to accommodate that software. The scope section can be as brief as a sentence or as long as a page or more, depending on the length of the proposal as a whole. And again, not all proposals require a scope section.

Organization This section of the introduction tells readers about the order in which information will be presented. The section can be as short as a sentence, or it can be a longer, detailed overview of the entire proposal. Generally, however, proposal writers explain their organizational scheme concisely. Consider this example:

> The remainder of this proposal provides a detailed explanation of the methods we propose to use, a week-by-week budget of all anticipated expenses, a likely materials list, a roster of necessary personnel, and a time line of projected progress. The proposal concludes with the expected results at completion.

Approach The remaining sections in the body portion of the proposal tell about the solution(s) and provide details about what is being proposed. For some proposals it is necessary to give a detailed account of the approaches or methods used to solve the problem. With research and grant proposals, for example, the methodologies used are central to whether the proposal is seen as valid. In fact, providing detailed explanations of and rationales for proposed research may be one of the most important parts of such a proposal. However, not all proposals will include an approach section.

Requirements If a proposal has particular needs, writers must identify all of them so that readers can determine whether the requirements are feasible. These sections should not merely list requirements but should include a rationale explaining why they are necessary to solve the problem. Often a vendor may have specific needs in order to complete a task or service. For instance, if a vendor proposes to install network software for a company, the proposal might include required network upgrades so the system can handle the software. And the company might be required to provide the vendor with access to network administrators or to restricted network areas. In other cases a proposal might require a company to supply needed equipment or additional personnel. Sometimes, too, a proposal cannot solve the problem without economic resources in advance. Countless types of requirements may exist, but not all proposals have requirements. Therefore, a requirements section may not always be necessary.

Plan of Action Readers want to be assured that proposal writers or their organizations can and will complete the proposed task. To persuade readers, proposals usually include a plan-of-action section that addresses the methods by which the task will be completed. In other words, this section explains how things will get done. The plan of action serves two rhetorical purposes in persuading readers to accept the proposal:

1. *It shows that the proposed method of solving the given problem is valid.* It details that method to persuade readers that it is the most feasible and efficient.
2. *It establishes credibility* by showing that those responsible for the proposal know how to accomplish the task. The plan of action demonstrates the professional know-how to deliver results.

The level of detail required in a plan of action is determined by the complexity of the solution and the problem. Some plans of action are simple: "We will deliver the hash browns every Tuesday morning." Others require a more thorough explanation. There should always be a narrative description of the plan; and sometimes a visual element, such as a flowchart, can help walk readers through the plan. The narrative portion is best organized chronologically: first we plan to do this, then this,

See chapter 7 for more about chronological organizational strategies.

and finally this. Plans of action should address the entire solution, from inception to completion.

Qualifications/Experience In many instances proposal writers need to show that they are qualified to solve the problems properly. Qualifications can be demonstrated in a number of ways.

Company Background This section can include a brief history of the company. How long has it been in business? How competitive is it in its field of work?

Personnel This section can identify how many people work for the organization and how many of those will be directly involved with the work described in the proposal. In some cases the people who work for the company are critical, particularly if they are recognizable or have positive reputations. When personnel do matter, proposal writers should provide short, biographical statements or abbreviated résumés of key individuals.

Experience Because experience is a key factor in persuading readers to accept a proposal, this section should identify past experience solving problems like those addressed in the proposal. The logic is simple: someone who has previously completed a similar task should have the know-how to do it again. This section can provide a résumé of relevant experience.

Providing information about qualifications and experience presents important ethical decisions. Proposal writers should never falsify or exaggerate qualifications. Doing so can lead to a negative relationship with the hiring organization, which will also probably see to it that other organizations find out about your questionable work ethics. Providing inaccurate information is unethical and unprofessional and ultimately bad for business.

Portfolio To emphasize and illustrate problem-solving experience, writers sometimes include a portfolio of previous work. Portfolios allow readers to view past work and assess the quality of that work. However, there is no need to include *all* of the work that has been done. Examples should be limited to relevant, recent work.

Equipment Noting the ability to use and access specific equipment can help readers recognize problem-solving abilities. For instance, a company that makes customized training videos might want to identify that it owns six professional-grade cameras and editing equipment, which allow multiple-angle shooting and in-house editing. If equipment needs to be purchased, rented, or maintained, writers should note this in the budget section of the proposal as well.

Licenses In some fields writers must show that they or their organizations are licensed to do the proposed work. If the work falls under local or national licensing regulations, writers must provide confirmation of those licenses. In addition, identifying how long a license has been held can sometimes suggest valuable experience.

Awards In some fields outstanding work is recognized through local, national, and international awards. Sharing recognition for exemplary work can help build credibility. However, providing too many accolades can become tiresome for readers. Proposal writers should be professional, not boastful.

References Sometimes companies don't want to hear self-praise but instead want to know what clients have thought about the proposing company's work. Lists of references help readers locate others familiar with a company's work in case they want to discuss previous experience.

Budget For many readers the budget section of a proposal is the most important because, for many companies and organizations, it all boils down to money—is the solution cost-effective? Budgets are difficult to write because they are *proposed* budgets that are expected to show readers the *exact* costs of implementing the proposed solution. But anticipating and estimating all associated costs can be tricky. Nonetheless, because the budget offered in the proposal is likely to shape the budget that is agreed on in a contract, the proposed budget must be as realistic as possible. In other words, it's not smart to lowball a budget in order to secure a contract because you'll be expected to adhere to that budget later. In addition, lowballing a budget to gain a contract is unethical.

The budget portion of your proposal should itemize all expected expenses, including materials, managerial expenses, personnel, and equipment. Furthermore, the budget should detail both direct and indirect costs. Direct costs are those that specifically affect the proposed goals and solutions—for example, materials, equipment, and personnel. Indirect costs refer to those not tied directly to the proposed goals and solutions—for example, a water cooler at a job site.

Detailed budgets persuade in two primary ways. First, when budgets offer great detail, readers sense that the proposing organization is not asking for more money than is actually necessary to complete the task. Certainly, a low budget is appealing, but a budget that shows detail is more persuasive. In addition, a detailed budget demonstrates an understanding of the problem and the solution, showing that the proposal writers are aware of everything that is involved in solving the problem. Well-written budgets evoke confidence.

Budgets are often reported in tables, charts, and graphs. However, as discussed in chapters 8 and 9, visuals can clarify information as well as obscure it—either intentionally or unintentionally. Proposal writers should be sure that visuals are clear and accurate. Hiding a bad budget behind fancy graphics is bad professional ethics.

For more about using visuals, see chapters 8 and 9.

Proposed Schedule Most problems need to be solved within a given time period. Even without time constraints, however, most organizations want to know when a project will be completed. Thus, proposals need to explain when the proposed work will commence and when it will be completed. In larger, multiphased projects, the schedule should also identify the beginning and ending of all phases. In addition, schedules should account for dates when progress reports will be submitted, when goods or services will be delivered, when equipment maintenance will take place, and when other key activities will occur. Projected payment schedules should be included if applicable.

Proposed schedules can be written as a detailed narrative but are most frequently presented as lists, graphs, charts, or time lines, which can improve readability. Charts, tables, and network diagrams make it easier and more efficient to quickly find and review particular dates and times.

Regardless of format, schedules must be realistic. They should not promise to accomplish more than is possible in a given time, nor should they project taking longer than is reasonable. Schedules should anticipate delays but should not overcompensate.

Conclusion/Recommendations Like any good document, a proposal needs closure. The conclusion restates the problem or opportunity, summarizes the proposed solution, and reminds readers why the solution is beneficial for them. The conclusion is one last opportunity to persuade readers to accept the proposal.

Many proposal readers don't have the time, or in some cases the interest, to read all of a proposal and may turn to the conclusion first as a way to quickly determine what the proposal is about and what it suggests. Thus, writers should think of the audience for the conclusion as a more general one and should consider these suggestions:

- *Indicate what should be done first.* If the problem is time sensitive, be sure to identify when the first action must be taken.
- *Highlight important details in bulleted or numbered lists*, as this example does:

  We recommend the following:
  1. Purchasing hash browns from Potatoes-R-Us
  2. Making hash browns a breakfast choice at least five days per week
  3. Serving hash browns for lunch and supper in addition to breakfast

See chapter 9 for more about design elements.

- *Use good design elements*, emphasizing important information through placement on the page, use of white space, font, boldface, and so on.
- *Provide quantitative assessments of the proposal*—for example, what the problem costs, what the solution costs, what the solution ultimately saves. Provide numbers and details if applicable.

End Matter

See chapter 6 and appendix B for more on research, sources, and citations.

Some proposals require materials that aren't crucial to the success of the proposal but that do provide further or clarifying information. These materials are placed in bibliographies, glossaries, and appendixes.

Bibliography Proposal bibliographies document outside sources used in the proposal; they are common in research and grant proposals. Bibliographies employ a uniform bibliographic style, preferably one that is consistent with the area of work presented in the proposal (e.g., APA, Chicago, CSE, or MLA).

Glossary Proposals are often read and vetted by individuals with varying levels of expertise. This means that proposal writers must make careful rhetorical choices about language and terminology. In some cases they must use language that technicians understand, at the risk of talking over the heads of others. As a result, many proposals include glossaries of key terms to help readers understand unfamiliar terminology, jargon, abbreviations, and acronyms. Glossaries are particularly useful with proposals for transnational audiences.

Glossaries are necessary only if a proposal contains numerous terms that need to be defined. If a proposal has only a few difficult terms, it might be more efficient to include definitions internally or as footnotes on the pages where the terms are used. However, if a glossary *is* used, it is necessary to somehow identify the terms that are to be found there, usually in one of two ways:

See chapter 15 for more about writing definitions in workplace documents.

1. Terms listed in the glossary can be italicized or highlighted in the text of the proposal. With this approach the glossary should be explained in the introduction, along with the method of identifying glossary terms in the text.
2. Terms listed in the glossary can be marked with an asterisk (*) in the text. Writers can also identify this strategy in the introduction or can explain it in a footnote the first time the mark is used.

Appendixes Appendixes include any additional information that proposal readers might want to see, such as the details of data or information presented in the proposal. For instance, a proposal might indicate that seven of ten recently conducted surveys found an increase in customer satisfaction. The appendix, then, might include copies of those original ten surveys for readers to examine. Appendixes often include the following:

Surveys

Maps

Charts

Previous correspondence

Reports that have influenced the proposal

Documents that establish the background or context of the problem

Spreadsheets or other databases

Images of locations, equipment, damage, and so on

Statistics

All information in appendixes is ancillary; the proposal should never depend on it. Information that is critical to the argument made in the proposal should be included in the front matter or the body of the proposal. Appendixes are usually accessed only by experienced or technical readers.

Use of Technology

As discussed earlier, technologies like PDF files, Listservs, and the Internet affect how writers both find proposal opportunities and disseminate proposals. And World Wide Web resources help writers find information to use in preparing proposals. However, one of the most significant technological tools available to proposal writers is electronic templates. Microsoft, for instance, offers a number of free proposal templates that can be downloaded from the Microsoft Office website for use with Microsoft Word. Microsoft also offers other features for proposals, as well as advice for writing them. Nonetheless, even though templates can make things a bit easier for proposal writers, templates also limit the possibilities available to them. A savvy proposal writer relies more on the ability to make good rhetorical choices than on template formats.

Coffee Rings and Pretzel Crumbs

Whether you are trying to persuade readers to fund your research, buy your product, or hire your service, your writing is all that proposal readers will have to use in evaluating you and often the only chance you will have to persuade them. Thus, the writing in your proposal must be flawless. Even if your information is valuable, your solution useful and feasible, your credentials impeccable, a poorly written proposal can easily prevent your success. Companies read proposals to determine whether they are going to invest money; they need to be confident that their money will be well spent. A sloppy proposal usually suggests sloppy work, and most companies and organizations do not want to invest in sloppy work.

Imagine being on an airplane, opening the tray table in front of you, and finding coffee rings and pretzel crumbs. Little indications of sloppiness can undermine your confidence in the airline: if they can't take care of the little stuff, can you trust them to take care of the big stuff? Overlooking coffee rings and pretzel crumbs signifies a lack of professionalism and a lack of attention to details.

PSA

In a similar way, proposal readers are likely to see sloppy writing as a sign of larger inefficiencies. Consequently, you should always approach proposal writing with diligence and professionalism, understanding that your proposal is a representation of you and/or your company. It is imperative that you spend ample time concentrating on the Review phase of the problem-solving approach. Careful revision and editing can help ensure that the proposal you send out represents your highest caliber work. Figure 20.9, which follows, shows a proposal that takes all of these points into consideration.

Kaufman & Associates
873 SW 85th Street
Providence, RI 02905

401.383.2100
www.kaufmanandassociates.com

Title → **PROPOSAL**
ACCOUNTING AND PAYROLL SERVICES

Date → May 14, 2008

PREPARED FOR Jakob Israel, General Manager
Coastal Computer Repair

Byline → **PREPARED BY** Gary A. Olsen, Vice President
Accounting and Payroll Services
Kaufman & Associates

Executive summary[a] → **DESCRIPTION:**
Coastal Computer Repair is in need of an accounting and payroll service to manage daily accountancy and payroll. Currently those tasks fall to an overburdened accountant manager. With only limited resources and staff, accountant and payroll functions have suffered. Outsourcing these duties to Kaufman & Associates will free the accounts manager to focus on growth aspects of this thriving company.

Proposal number[b] → Proposal Number: 138-2008

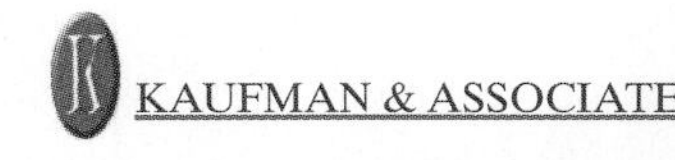

[a]Including the executive summary with the cover information is a time- and space-saving device used instead of placing the summary after the letter, which is more standard.

[b]A number and date system identifies which proposal this is.

FIGURE 20.9A

Letter of transmittal[a]

KAUFMAN
&
ASSOCIATES

May 14, 2008

Mr. Jakob Israel, General Manager
Coastal Computer Repair
6785 NW 134th Ave.
Providence, RI 02906

Dear Mr. Israel:

It was a pleasure to speak with you last week about the accounting challenges Coastal Computer Repair faces. Enclosed you will find a brief proposal outlining how Kaufman & Associates can offer a cost-effective solution to alleviate your in-house accounting burden.

As you know, Kaufman & Associates specializes in payroll and accounting services and has over thirty years of experience working in the Providence area. We take pride in providing clients with confidential, accurate, and professional service.

The enclosed budget is based on our earlier conversation and can easily be adjusted as we customize services to fit your needs.

I look forward to discussing this proposal with you soon. Please contact me should you have any questions about the proposal.

Sincerely,

Gary A. Olsen, Vice President

873 SW 85TH STREET • PROVIDENCE, RHODE ISLAND • 02905
401.383.2100 • WWW.KAUFMANANDASSOCIATES.COM

[a]This letter is an opportunity to clarify why the proposal is being submitted and what solution is being proposed.

FIGURE 20.9B

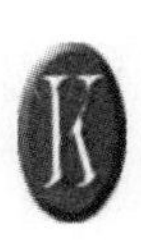

Table of contents[a]

TABLE OF CONTENTS

KAUFMAN & ASSOCIATES

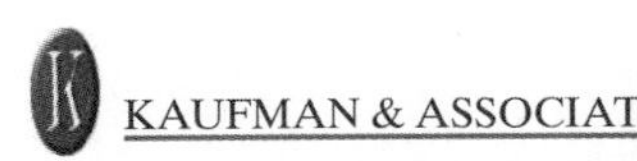

[a]Notice the identification of subsections, not just primary sections, in order to improve reader access.

FIGURE 20.9C

1

INTRODUCTION

PURPOSE

Purpose[a]

Coastal Computer Repair is in need of an accounting and payroll service to manage its daily financial accounting and its payroll system. By outsourcing these tasks to Kaufman & Associates, Coastal Computer Repair can redirect its current accounting resources toward much-needed attention to company growth and diversification. This proposal offers solutions to those problems.

CLIENT BACKGROUND

Background[b]

Jakob Israel established Coastal Computer Repair in 2002 as a small, part-time income-earning initiative. However, Coastal Computer's reputation for efficient, inexpensive home computer repair resulted in more business than Mr. Israel had anticipated. In 2003, Coastal Computer hired two service technicians and incorporated. By the end of 2003, Coastal Computer had branched out to repair and consultation for corporate network systems, elevating its status to one of the most widely contracted independent repair and consulting services in the Providence area. In 2004, Coastal Computer employed more than a dozen repair technicians, four computer engineers, six office staff, and one accountant. In the next four years, Coastal Computer doubled its number of employees, yet all company accounting and payroll was still managed by one accountant; Pam Gilbert.

In this same time period, Coastal Computer increased its client base by more than 300%, now contracting with over 900 individual, corporate, and government clients.

By 2008, Ms. Gilbert's responsibilities had exceeded her resources and time, resulting in a backlog of data entry, delayed billing, payroll not disbursed on

Introduction

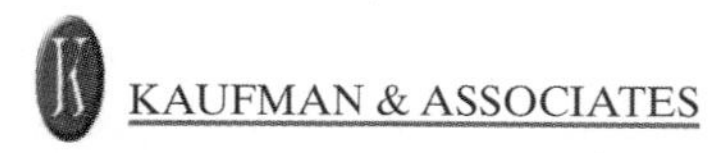

[a]Notice the direct statement of purpose and the concision of the purpose section.

[b]Even in providing background information, this section takes on a persuasive approach.

FIGURE 20.9D

2

time, and late tax filing. Facing the impending continued growth of Coastal Computer, Ms. Gilbert offered Mr. Israel two solutions: either hire at least two more accountants or outsource the accounting and payroll needs. Those needs require a service to do the following:

- Provide professional payroll and accounting services
- Provide services at an equivalent or lower cost than an additional in-house accountant would require
- Guarantee on-time payroll disbursement
- Guarantee accurate accounting records

Based on this background, Kaufman & Associates was asked to prepare a proposal addressing Coastal Computer Repair's needs.

Scope[a]

SCOPE

This proposal demonstrates that Kaufman & Associates' Accounting and Payroll Services can alleviate Coastal Computer Repair's immediate and long-term accounting and payroll problems dependably and cost-effectively. This proposal does not address issues of future investment, though Kaufman & Associates' Investment Services can address such issues in the future.

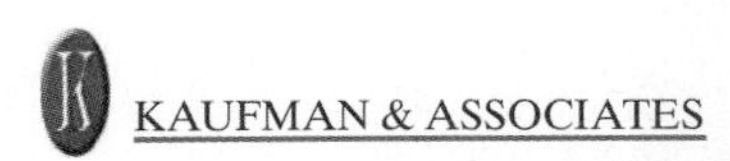

[a]The scope section identifies what the proposal does and does not do.

FIGURE 20.9E

Approach[a]

3

PROPOSED SOLUTION

Kaufman & Associates has specialized in accounting and payroll services in the Providence area for more than thirty years. K&A focuses on working with growing mid-sized businesses. All of K&A's accountants are CPAs, and K&A provides Bar Associated tax lawyer services should the need arise.

Given the immediate needs of Coastal Computer Repair, K&A CPAs recommend implementing the following action:

ACCOUNTING SERVICES

Kaufman & Associates will provide

- Ledger maintenance
- Account balancing
- Daily, quarterly, and yearly accounting reconciliations
- Year-end tax statements and summaries
- Quarterly tax payment reports

PAYROLL SERVICES

Kaufman & Associates will provide

- Standard pay-period check determination and origination
- Withholding allotments
- Benefit summaries

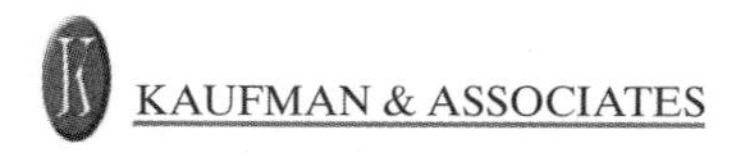

[a]This section provides details of exactly what is being offered as a solution.

FIGURE 20.9F

4

ACCOUNTS PAYABLE SERVICES

Kaufman & Associates will provide
• Payment of all bills owing
• Budget plans
• Term negotiations

FIGURE 20.9G

5

BUDGET/COST SUMMARY

The following budget/cost summary is an estimate for the proposed accounting, payroll, and accounts payable services. These numbers are estimates only.

INITIAL ACCOUNT SETUP	First-Year Costs
Conversion from Coastal Computer Repair to K&A (time period to cover January 2008 to present)	$1,500
Production of Ql and Q2 Reports	600
Production of Current Standing Report	1,000
Total setup costs	$3,100
ONGOING MONTHLY COSTS	
Daily account reconciliations ($1,200/mo.)	$14,400
Biweekly payroll ($600/mo.)	7,200
Total ongoing monthly costs	$21,600
ONGOING QUARTERLY/YEARLY COSTS	
Quarterly tax documents ($600/qtr.)	$2,400
Quarterly filings ($400/qtr.)	1,600
Quarterly reports ($600/qtr.)	2,400
Year-end summaries	400
Year-end tax filing summaries	400
Maintenance of employee benefit reports	300
Total ongoing quarterly/yearly costs	$7,500
Total setup costs	$3,100
Total ongoing monthly costs	$21,600
Total ongoing quarterly/yearly costs	$7,500
Total first-year costs	$32,200

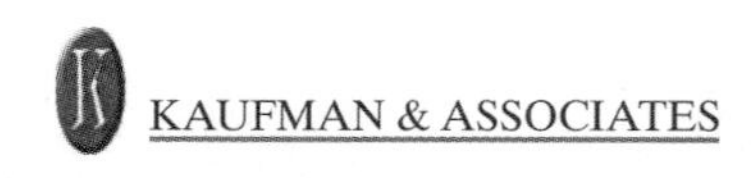

Budget[a]

[a]Notice the organization of the budget into three primary parts, the design used to present the budget, and the summary of the total costs at the end. Notice, too, the statement at the beginning of the budget section identifying these figures as estimates.

FIGURE 20.9H

6

Schedule[a]

SCHEDULE

July 1, 2008	Contracts signed
July 7, 2008	Initial consultation meeting with Jakob Israel and Pam Gilbert
July 14, 2008	Submit transfer plan to Pam Gilbert
July 21, 2008	Initiate accounting transfer
July 28, 2008	All accounting and payroll responsibilities transfer to K&A

Qualifications[b]

QUALIFICATIONS

Kaufman & Associates has served the Greater Providence area for more than thirty years. K&A offers accounting and legal services through its Board Certified CPAs and Bar Associated lawyers.

PERSONNEL

Kaufman & Associates has on staff over 30 Certified Public Accountants. Your account will be overseen by an account manager and attended to by a team of up to three additional accountants. Coastal Computer Repair payroll and accounting will be managed by Karen Westland, who has worked for K&A for nine years. She may be reached at

Karen Westland, Accounts Manager
Kaufman & Associates
873 SW 85th Street
Providence, RI 02905

401.383.2100 x274
karen_westland@kaufmanandassociates.com

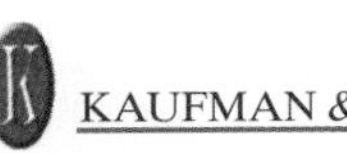

KAUFMAN & ASSOCIATES

[a]Notice the format in which the schedule is presented.

[b]Notice the organizational strategy of subdividing the section into multiple points of qualification, like Personnel, Licensing, and References.

FIGURE 20.9I

7

Kaufman & Associates also has on staff five Bar Associated tax attorneys. All legal matters pertaining to your account will be addressed by Matthew Salzberg, who has been a Bar-approved attorney since 1999 and has worked with K&A since 2002. You may contact Mr. Salzberg at

Matthew Salzberg
Kaufman & Associates
873 SW 85th Street
Providence, RI 02905

401.383.2100 x301
m_salzberg@kaufmanandassociates.com

Résumés for these and other K&A professionals are available upon request.

License[a]

LICENSING
All Kaufman & Associates accountants are licensed Certified Public Accountants.
All Kaufman & Associates tax attorneys hold JD degrees and are Bar approved.

Credentials and licensing information for any K&A employee is available upon request.

References[b]

REFERENCES
References for Kaufman & Associates accountants are available upon request.

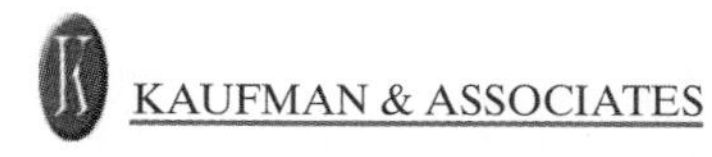

[a]The primary licensing qualifications are listed.

[b]This proposal doesn't provide direct reference information; instead, it offers to supply that information to clients who wish to see it.

FIGURE 20.9J

8

RECOMMENDATIONS

Recommendations[a]

Based upon the needs of Coastal Computer Repair, we recommend that Kaufman & Associates be contracted to manage all aspects of Coastal Computer's accounting and payroll. K&A can provide a dedicated account manager, ensuring accuracy and efficiency, as well as legal support should any need arise. By outsourcing all accounting and payroll services to K&A, Coastal Computer's in-house accountant will be better able to attend to issues of growth management that are evolving at Coastal Computer Repair.

Kaufman & Associates offers a number of benefits for Coastal Computer:

- Cost-effective accounting services
- Priority attention to deadlines
- Professional tax preparation and reporting
- Dedicated payroll and accounting professional committed to Coastal Computer
- Confidential service

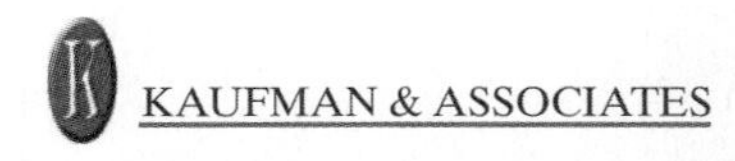

[a]Notice the distinction between the narrative of the recommendations and the bulleted list that clearly identifies particular points.

FIGURE 20.9K

SUMMARY

- Workplace writers use proposals to solve problems.
- Internal proposals try to persuade an audience that a problem exists, that it is understandable, that the problem should be fixed, and that it will escalate if not addressed.
- Internal proposals can be written informally or formally.
- External proposals work to persuade readers outside the writer's organization to opt for the solutions that the proposal offers.
- Proposals can be solicited or unsolicited.
- Requests for proposals (RFPs) are formal appeals put out by organizations requesting that interested vendors submit formal proposals in response to a particular set of needs.
- Reading RFPs carefully can help writers craft their proposals in response to the RFPs.
- Information provided in RFPs includes a title and/or brief description of the RFP, contact information, a proposal schedule, background information, the proposal selection process, a project summary, and the project details.
- Information requested in RFPs includes company/organization information, a proposed solution, details of the proposed solution, a budget, provisions, personnel, references, portfolio, awards, and certification and licensing.
- Writers must consider ethical and professional issues when writing RFPs.
- Technology plays a particularly important role with RFPs.
- Sales proposals attempt to persuade another organization that the writer's company can best supply a needed product or service.
- Research and grant proposals attempt to persuade a sponsoring organization to support research or a project.
- Proposal writers must be able to offer solutions, must reach and persuade their audiences, and must display the credentials that allow readers to take the proposals seriously.
- Proposal front matter includes a title page, a letter of transmittal, an executive summary, a table of contents, and a list of visuals.
- The proposal body includes an introduction, the approach, requirements, a plan of action, qualifications/experience, a budget, a proposed schedule, and a conclusion/recommendations.
- The proposal end matter includes a bibliography, a glossary, and appendixes.

CONCEPT REVIEW

1. What are the differences between internal and external proposals?
2. What forms might internal proposals take?
3. Why must proposals be persuasive?
4. In what ways are proposals problem-solving devices?
5. What is a solicited proposal?
6. What is an unsolicited proposal?
7. What is a sole-source contract?
8. Why might a company issue an RFP?
9. What information should an RFP provide for readers?
10. What information should an RFP request from proposal writers?
11. What are some ethical implications of posting RFPs?
12. What kinds of technologies affect RFP writing and releasing?
13. In what ways do RFPs relate to other documents?

14. What is the primary function of a sales proposal?
15. What is the primary function of a research or grant proposal?
16. What are the basic components of a proposal?
17. What role can minor errors play for readers of proposals?
18. In what ways do proposals relate to other documents?
19. What role do various technologies play in writing proposals?

CASE STUDY 1

The Search for South Central Louisiana State University's New Mascot

In 2005 the National Collegiate Athletic Association (NCAA) Executive Committee adopted a new policy to prohibit NCAA colleges and universities from displaying hostile or abusive mascots, nicknames, or imagery at NCAA championships.

> "The NCAA objects to institutions using racial/ethnic/national origin references in their intercollegiate athletics programs," said NCAA President Myles Brand. "Several institutions have made changes that adhere to the core values of the NCAA Constitution pertaining to cultural diversity, ethical sportsmanship and nondiscrimination. We applaud that, and we will continue to monitor these institutions and others. All institutions are encouraged to promote these core values and take proactive steps at every NCAA event through institutional event management to enhance the integrity of intercollegiate athletics related to these issues."[2]

This new policy will affect the athletic program at South Central Louisiana State University (SCLSU), whose mascot is the Fighting Indians. The university president does not want to change the mascot name but realizes that ultimately this must be done in order to preserve the athletic department's competitiveness in NCAA-related tournaments and events. Thus, in conjunction with a university committee, SCLSU has posted an RFP soliciting the help of an outside PR agency to develop a new mascot that represents the spirit of the university and the region while still complying with the new NCAA policy.

Imagine that you work for a new, small PR firm that wants to land this account. Alone or in teams, develop a sales proposal for SCLSU that reflects the university's needs and demonstrates your company's ability to complete this project satisfactorily. Refer to the sections of this chapter that discuss sales proposals and the different parts of a proposal. You will need to supply fictional information that relates to the needs and desires of the client.

Included on the next page is the RFP posted by the SCLSU athletic program. We have removed the deadline for submission so that your instructor can determine when your proposal should be due.

---

[2] NCAA News Release, http://www.ncaa.org/wps/wcm/connect/NCAA/Media+and+Events/Press+Room/News+Release+Archive/2005/.

Request for Proposals
South Central Louisiana State University Athletic Program seeks proposals from public relations firms and advertising agencies to assist in developing and marketing a new mascot that represents the spirit of SCLSU and the dynamic region of south central Louisiana. The contracted agency will develop a new mascot name and image as well as new colors, slogans, and possibilities for marketing the mascot to the communities inside and outside SCLSU. Successful proposals will address specifically how the new mascot will comply with the NCAA policy prohibiting NCAA colleges and universities from displaying hostile and abusive racial/ethnic/national origin mascots, nicknames, or imagery. Successful proposals will describe how the proposed mascot will reflect the ideals of the university and this region of Louisiana, how the mascot will appear visually on uniforms and university paraphernalia, and how long it will take to implement the change.

Background
In 2005 the presidents and chancellors who serve on the NCAA Executive Committee adopted a new policy to prohibit NCAA colleges and universities from displaying hostile and abusive racial/ethnic/national origin mascots, nicknames, or imagery at any of the eighty-eight NCAA championships. The SCLSU mascot, the Fighting Indians, was deemed to be offensive by the NCAA Executive Committee. Despite its thirty-five-year heritage as the SCLSU mascot, President Victoria Vallencourt, in conjunction with the SCLSU Athletic Program, decided to abide by the NCAA ban and seek a new mascot to represent the university's athletic teams.

Selection
A university committee consisting of a representative of the president's office (who will chair the committee), three representatives from the university athletic association, three students (including one student athlete), three faculty representatives, and two representatives from the local community will review submitted proposals. The committee will recommend which proposal should be accepted and will deliver that recommendation to President Vallencourt. The committee will have three weeks following the closing date for submission of proposals to make its recommendation.

Schedule

Final submission of proposals	(to be determined)
Initial proposal review	(TBD)
Interviews with vendors	(TBD)
Contracts to be tendered	(TBD)

Inclusions
1. Summary of the proposed solution.
2. Information about the proposing vendor company, including indications of previous experience with mascot development and marketing, a brief company history, and a list of relevant personnel.
3. A detailed proposed solution suggesting one or more possible mascot replacements, a marketing plan for making the mascot visible to the university and local communities, and a design plan for integrating the mascot into the SCLSU culture.
4. A time line of how long it will take to shift the university from one mascot to another, including a task list of what will be involved in such a shift.
5. A detailed budget of what the change will cost over the next year, two years, and five years.
6. A portfolio of previous mascot designs.

Completed proposals should be submitted to

Victoria Vallencourt,
President South Central Louisiana State University
Box 1, Route 8
St. Nachitibideaux, Louisiana 71333

CASE STUDY 2

Proposing Updates to the Greta Engineering Web Page

Cartwright, Nevada, is a midsize town located about fifty miles from Las Vegas. Like other communities close to Las Vegas, Cartwright has profited from Las Vegas's growth. Between 1990 and 2000, the Las Vegas population increased by just over 85 percent, most of the growth occurring away from Las Vegas's famous resort and gambling areas. In fact, Las Vegas has been rated as one of the best places in the United States to live, and such publicity has drawn a new population to the Las Vegas area, primarily comprised of young professionals. With the increase in population, Las Vegas's construction industry grew rapidly, too, as did other industries associated with community growth. Cartwright's proximity to Las Vegas has also provided many of Cartwright's local businesses with opportunities to expand their coverage areas into Las Vegas, particularly those businesses associated with building and construction.

Two years after you graduate from college, you accept a job with Greta Engineering, a midsize engineering firm in Cartwright. Greta Engineering provides engineering consulting to private and government clients in a number of engineering fields, including electrical engineering, security engineering, construction management, mechanical engineering, civil engineering, and water management. Greta is not a large firm; it employs just eleven engineers, a company lawyer, two bookkeeping and billing accountants, and four clerical staff, one of whom is the secretary for William Greta, the company's owner. You are the first person Greta Engineering has hired in ten years. Unlike other firms in Cartwright, Greta Engineering has not actively sought to expand its coverage area to include Las Vegas, because, as William Greta explains it, "There's plenty of work to be done in Cartwright, and growth isn't always a good thing." The rest of the employees at Greta Engineering seem to accept this business philosophy and don't seem eager to do anything more than they already do.

In your first few months at Greta Engineering, as you become acclimated both to your new work environment and to your new community, you begin to notice a number of ways that Greta Engineering could actively attract more business in Cartwright without looking to Las Vegas. Cartwright's growth, in fact, includes a number of exciting (and potentially lucrative) projects that would benefit the community, yet Greta Engineering never seems to be in the running for those projects, despite William Greta's regular submission of proposals and bids for such projects. Instead, Greta Engineering seems to contract over and over with the same government divisions and the same private companies, never expanding its client base and never changing the kinds of projects it engages.

You begin to wonder if perhaps the reason Greta Engineering doesn't win more bids on new projects is the company reputation. At home you decide to review Greta Engineering's web page and compare it to other engineering firms' web pages. You discover that the Greta page hasn't been updated since the late 1990s. Many of the internal links don't work, and even the contact information for the firm—like phone numbers—is obsolete. You decide that the web page can only be thought of as dilapidated and in need of substantial revision.

You decide that as the newest member of the Greta Engineering team, you should propose a major overhaul of the web page, but you face a difficult task in making such a proposal. William Greta and the rest of the company seem comfortable in their present situation and are rarely enthusiastic about any changes that might affect that comfort. As the new person, you might be seen as an agitator. Despite the shortcomings you've noticed about Greta Engineering, you enjoy working for the company and living in Cartwright.

Your task, you decide, is to write an unsolicited, internal proposal urging William Greta to hire a web designer to redesign and update the Greta Engineering web page. You conclude that the best way to make this argument is to identify the numbers of community-related projects that Greta Engineering is missing out on because newer companies in town just don't know anything about the firm. You also decide to show how other, more active firms in Cartwright use the web to make themselves visible to contracting companies.

For this assignment, write the unsolicited, internal proposal to William Greta. Use examples from engineering firm websites to develop your own ideas of how Greta Engineering might design a new website. And remember that you face a few challenges in convincing William Greta, so be sure to address those.

CASE STUDIES ON THE COMPANION WEBSITE

Proposing New Computers for Barksdale Accounting

In this case you assume the role of a young CPA working for a small accounting firm. You and your coworkers must work each day on out-dated computers and have difficulty convincing the company president to purchase new ones. You must produce a proposal that helps persuade the president of the necessity of new computers.

Banking on Productivity: Seeking Food Service Bids for a Corporate Cafeteria

You are the manager of operations at Bank of America Corporate Center and are given the task of finding a food service company that can staff a new corporate cafeteria, supply it, and provide the bank's employees with menus that will make them want to eat at work. This case asks you to research the food service industry and the ways that other companies have handled situations like this one. You also must create and distribute a request for proposals to these food service companies.

JMG Construction and Lifeline Safety: Writing a Proposal That Can Save Lives

In this case you are a licensed safety engineer who works at Lifeline Safety, a consulting firm that helps large construction companies meet safety standards set by the Occupational Safety and Health Administration. You are asked by your supervisor to prepare an unsolicited proposal for JMG Construction, a firm in your area that has recently suffered a number of work-related accidents on its job sites. In writing this proposal, you must carefully analyze your audience and its needs in order to compose the most effective document.

Testing the Waters: Establishing an Environmental Ethos

You are a state biologist that conducts water quality tests and other experiments on the water in remote areas of south Florida. You notice that the boat engines you use are environmentally friendly but could be better. Since it is your job to protect the environment, you feel compelled to convince your boss that he should replace or outfit the boat engines with greener technology.

VIDEO CASE STUDY

DeSoto Global

Collaboration and Writing a Request for Proposals

Synopsis

DeSoto Global's automobile navigation systems are popular with North American markets because they all feature DeSoto Global's patented Tracker1 system, which allows DeSoto Global to locate stolen units through a GPS tracker chip embedded in each unit. Dora Harbin reports that Japan has launched an aggressive campaign to combat increased car theft. Of greatest interest to DeSoto Global is the campaign's desire to use more efficient vehicle-tracking devices—like DeSoto's Tracker1 system. The snag for DeSoto is that in order for the Tracker1 system to work, it needs to be tied into a local security company with connections to local law enforcement agencies.

WRITING SCENARIOS

1. You are probably aware of problems other students face on your campus. These problems may be bureaucratic problems, such as difficulties with registration processes or with student financial aid forms. Or they may be logistic problems, such as difficulties in getting around a large campus in a timely fashion, arranging feasible class schedules, or navigating school web pages. Or they may be social problems, such as conflict resolution in the residence halls or the failure of the school food service to provide late-night meals. No matter where you go to school, you are apt to see problems that school administrators aren't aware of.

   Identify such a problem (or an opportunity) that should be addressed. Write an unsolicited, formal proposal (you may want to collaborate and work in a small group) to the school administrator(s) who have the authority to address the problem. Use the proposal as an opportunity to bring the problem to the attention of the administrator(s), explain the details and conditions of the problem, and provide a possible solution. Once you have written the proposal, deliver it to the administrator(s).

2. Chances are good that you have recently purchased or are considering purchasing a piece of electronic equipment: a handheld video game unit, MP3 player, cell phone, digital camera, and so on. Learn all you can about the device from available sales materials, and then write an unsolicited formal sales proposal directed at your classmates, identifying why it might be a problem for them *not* to have such a device, what opportunities are available to them if they purchase it, and how they might acquire it.

3. Pull up the U.S. government web pages at www.usa.gov and search for "rfp." Select one RFP that seems particularly interesting, and open it. What form does it take? What information does it provide? What information does it request? How long and how detailed is it? Consider whether the RFP might have been written differently or whether it was written as it is for a particular reason. Then write an analysis of the RFP.

4. The U.S. Department of Energy maintains a web page for science-related grants. Look through the web page at http://www.science.doe.gov/grants/. Look specifically at a minimum of five grant solicitation notices, and compare the similarities in those RFPs. In a memo to your instructor, describe the kinds of consistencies that you found among the grant proposal requests.

5. Proposals often become the focus of media attention. Visit the website of your favorite news organization, and search for an article describing some type of proposal. After you've carefully read the article, write a short memo to your instructor describing the proposal and comparing it to the proposals discussed in this chapter. Does the article seem to define *proposal* in the same way this chapter does? Does the article make reference to any of the specific topics or sections of proposals covered here?

6. Research where RFPs for your field of interest might be located. Then become familiar with those RFPs. Select one RFP that seems particularly interesting to you, and identify how it provides information and how it requests information. Once you have identified its various parts, identify what problem the RFP seeks to have solved. Create an outline of a solicited proposal that might be written in response to the RFP.

7. Learn more about the sole-source contract requirements of various organizations. What are the benefits and drawbacks of sole-sourcing for both proposal solicitors and proposal writers? Consider, in particular, the ethical implications

for companies that are required to post RFPs even when they intend to use a particular vendor. After reading more about sole-sourcing, write a short document that explains the benefits and drawbacks of sole-sourcing.

8. Each year more than 300,000 college students around the country enroll in technical communication classes in their pursuit of degrees in many different disciplines. What should those 300,000 technical writing students be taught about ethics and proposal writing? Should there be one set of ethical standards for proposal writing? What might such standards look like? Write general ethical guidelines or a code of ethics (see chapter 4) for technical writing students who are learning to write proposals.
9. Examine the memo that follows. Then assume the position of Shaia Noteworthy, the senior engineer at ACME Naval Engineering, and write a memo to Dingus Dorfman, explaining what action you want him to take next and why.

MEMORANDUM

TO: Shaia Noteworthy, Senior Engineer, ACME Naval Engineering
FROM: Dingus Dorfman, Engineer
SUBJECT: Oceanic Design RFP
DATE: March 19, 2007

At your request, I've reviewed the Oceanic RFP posted 28 February 2007 at rfpsrus.com, and although I have some concerns, I do think we should submit a proposal for the navigation aids program. I am a bit concerned with two primary parts of the RFP: first, we certainly don't have any experience in this area, although I think getting the experience through projects like Oceanic's will better position us down the road to pursue such contracts. Second, because we haven't done anything like this before, I've got no idea how we'd propose a budget for something like this. However, I think if we fluff up how we present our other naval engineering experience, we can probably mask our inexperience. The budget's a bit more difficult, but I think if we take the numbers from one of our old proposals and change the wording a little, we might be able to get by with that.

There's no question that this contract could lead to bigger projects, so I really think we should pursue it. Let me know if you want me to start drafting the proposal; Oceanic's deadline is three weeks from now.

10. Think about a piece of equipment you want—for fun, for a hobby, for your home, for your work space. Now imagine that you have been given a one-time-only

grant to cover the cost of this piece of equipment. But the grant is not unlimited; it will allow you to purchase a good piece of equipment, but it won't let you buy the top of the line. In order to make the best purchase, you have decided to ask vendors to submit proposals to persuade you about which piece of equipment is the most feasible for you to purchase. Write and design that RFP in the way that you want it released. Limit it to no more than two pages.

11. Formal proposals can contain a number of different parts and can vary in length, depending on the proposal's goals. Using the information found in this chapter, design a generic template or instructive outline for a formal proposal.
12. Assume that you work for BigRequest.com, an online RFP and proposal contractor. The company collects RFPs and distributes them to subscribers. It also collects proposals submitted in response to those RFPs, sorts them, evaluates them, and provides RFP-posting clients with evaluations and recommendations. BigRequest.com requires that all proposals be submitted in electronic form and has sometimes required a standardized online form, which clients have found useful when reviewing proposals. To make standard submission forms more available to clients, your supervisor has asked you to develop an online template for formal proposals. The prompts you use must ask respondents to submit particular information in each section. For this assignment you don't have to produce a functioning online form, but your design should show what questions you would ask, in what order, and how they would appear on the form.
13. Select one of the technologies important to RFP writers, and consider the ways workplace writers might come to use it. Use the Internet and other research tools to learn as much about that technology as you can. Then imagine yourself in a workplace that does not have access to it. Write an unsolicited informal proposal that identifies why not having that technology can be seen as a problem (or an opportunity), how your company might solve the problem, and what would be gained in solving that problem. Address the proposal to a fictitious supervisor of a fictitious company.
14. Microsoft Word provides downloadable proposal templates. Find other sources for downloadable proposal templates, and then download one template from one of those sources and one from the Microsoft selection. Be sure the two you select are comparable—two sales proposals, for instance. Compare the form and structure of the two, and then compare them to what this chapter has described as good proposal writing. Write a memo to your instructor that tells what you have found in your analysis.
15. Take any one of the writing assignments you have completed for this chapter, and present it in three different formats: a print version using word processing software, a PDF version for distribution via the Internet, and a web version with internal hyperlinks.

PROBLEM SOLVING IN YOUR WRITING

PSA

Research is a key component of the problem-solving approach. Using any one of the assignments you completed for this chapter that required you to conduct some level of research, write a memo to your instructor explaining how you (a) determined the types of information and sources necessary to complete the assignment, (b) conducted the research and gathered the information to complete the assignment, (c) organized and grouped the information to fit the needs of your project, and (d) assessed the information and decided what was appropriate to use and whether further research was necessary.

21

Informal Reports

CHAPTER LEARNING OUTCOMES

After completing this chapter, you will be able to do the following:

- Distinguish between **informal and formal reports**
- Recognize that informal reports **solve problems** and serve **different purposes**
- Understand that informal reports can be **written or spoken**
- Know that informal reports can be composed for **internal and external audiences**
- Recognize that informal reports can be written as **memos, letters, e-mails, or other genres**
- Know how and when to compose **progress reports**
- Know how and when to compose **laboratory reports**
- Know how and when to compose **directives**
- Know how and when to compose **incident reports**
- Employ the **problem-solving approach** when composing informal reports
- Understand the major **ethical issues** in informal report writing

DIGITAL RESOURCES

On the Companion Website www.prenhall.com/dobrin:

- Case 1: Reporting Changes in the Book Business at Pearman College
- Case 2: New Incident Reporting Procedures for Arctic Blast Heating and Air Conditioning
- Case 3: Construction Delays at the Oklahoma State Penitentiary
- Case 4: Game Over? Issuing Reports Regarding Faults in the New Z-Sphere Video Game System
- Video Case: Writing a Progress Report at DeSoto Global
- PowerPoint Chapter Review, Test-Prep Quiz, Exercises and Activities

REAL PEOPLE, REAL WRITING

MEET ETZER DAROUT • Research Associate with Scripps Research Institute

What types of writing do you do in your workplace?

As a synthetic chemist, I write short reports describing the experiments I perform, along with any information I am able to extract from these experiments. It is important to keep accurate records of my work for anyone who wishes to duplicate or learn about what I have done. I also write (usually with coauthors) longer scientific articles for publication and a variety of other less formal documents.

Who are the typical readers of your workplace documents?

I work with other chemists, biologists, and an administrative staff, who are my primary readers. When writing to fellow scientists, we share a common language and therefore use specific terminology to describe our work and ideas to one another. However, in writing documents to administrators, such as an e-mail to order chemicals or supplies, I must keep in mind that they are not scientists, and therefore, I use terminology they will understand.

What role does technology play in the writing you do?

Technology plays a major role in the writing I do. Presentations and documents, such as publications, are done using document software. Information is transferred primarily by e-mail. I use Internet sources, such as online databases, to research and stay up-to-date on topics of interest. I also rely greatly on technology to store and transport data, depending heavily on computers, CD-ROMs, and flash drives.

What role does ethics play in the writing you do? Do you have particular ethical issues or responsibilities you must consider?

Scientific research and the way it is validated rely on honesty and ethics. When other scientists and society in general read my work, they must be able to trust the data and results. It is important to give an unbiased and accurate description of what is observed in the experiments. The falsification of data damages one's reputation as a scientist and may put into question the integrity of science in general.

What advice about workplace writing do you have for readers of this textbook?

Workplace writers should be objective and accurate. In relaying information to others, whether it is through writing or other forms of communication, it is important to be coherent and precise. As part of your preparation, think about your target audience and exactly what you want to convey to them.

INTRODUCTION

The Olympic Games are an international gathering of many of the world's best athletes, who compete in summer and winter events every four years. To no one's surprise, these games capture worldwide attention and interest, generating millions of dollars in revenue for host cities and millions more in television contracts, licensing fees, and sponsorships. What may be surprising about the Olympic Games, however, is the extent of the preparation that takes place on the part of the cities that host them. In addition to spending millions of dollars simply to bid for the games, cities must spend even more if they are fortunate enough to be chosen to host them. Cities have to make sure they can accommodate the tens of thousands of fans, tourists, and athletes: they must provide hotels, restaurants, shopping, transportation, and other amenities; and they often must build the stadiums and other venues where the games will actually take place. It takes an enormous amount of time and money just to plan the security for the international visitors and athletes.

Because of the complexity of planning for the Olympic Games, host cities are typically given seven years to prepare. But they are not simply awarded the games and then left to their own devices to get things ready. Rather, a host city's planning activities are overseen by the International Olympic Committee (IOC) to ensure that every detail is planned properly. During the seven years of preparation, a host city is required to provide various progress reports to the IOC, showing exactly how the planning is going. For example, in planning the 2010 winter games in Vancouver, the Vancouver Organizing Committee (VANOC) submitted six different progress reports to the IOC between July 2005 and November 2007. Figure 21.1, provides information from a progress report submitted to the IOC in November 2005.

We are pleased to report that excellent progress is being made to help us deliver the 2010 Winter Games:

- We have started Olympic and Paralympic Winter Games venue construction on schedule with the expressed goal of early completion.
- Our Canadian sponsorship and marketing program is enjoying early success.
- Planning is now underway for the design of the 2010 Paralympic Winter Games emblem to be introduced in 2006.
- The entire senior management team is in place as well as a number of key vice presidents and managing directors. This team is supported by more than 150 VANOC employees in dozens of functional areas.
- We are pleased to report that Dena Coward has been appointed as VANOC's Director of the 2010 Paralympic Winter Games, effective January 2006.

FIGURE 21.1 Portions of a Vancouver Organizing Committee progress report

This chapter treats progress reports like the VANOC document as one type of *informal* report, as opposed to *formal* reports, which are discussed in chapter 22. Informal reports are "informal" because they tend to be shorter than formal reports; they often take the form of workplace genres such as memos, letters, and e-mails; and they sometimes lead to the composition of a larger, more complex document. That is, an informal report might function as a piece of a longer document. For instance, the VANOC progress report in Figure 21.1 is one brief progress report that contributes to longer and more complex recommendation reports, feasibility reports, and evaluation reports (these are covered in chapter 22). Furthermore, informal reports are often delivered orally through spoken presentations. Figure 21.2 shows the Vancouver Olympic Committee reporting information orally. In short, informal reports are not "informal" because they are incomplete, sloppy, or casual in tone and presentation.

See chapter 22 for more on formal reports.

FIGURE 21.2 Vancouver Olympic Committee reporting information through a spoken presentation

Like other workplace documents, informal reports provide information to specific audiences so they can understand, explore, analyze, or act in some particular manner. Informal reports can explain, inform, persuade, and/or recommend. They consider audience needs and backgrounds. They often involve research. They can use visuals and principles of document architecture. They may be written collaboratively. And they solve problems.

ANALYZE THIS

Look carefully at the VANOC progress report in Figure 21.1. It provides information about the status of many aspects of the 2010 Olympic Games. Locate particular passages from this report that refer to activities that would themselves probably need workplace documents. For example, the first bulleted point discusses plans for "venue construction" for the games. Documents that might be necessary for someone planning to build new venues like these would include contracts, blueprints, and site surveys. Examine how this progress report implies the writing of various other documents.

TYPES OF INFORMAL REPORTS

Informal reports are common workplace documents, although they can cover many different subjects and appear in many different forms. Informal reports are important in many different workplace and rhetorical situations, including the following:

- Announcing the status of a project to a manager
- Describing a problem to a client
- Introducing a proposal or a plan to investors
- Detailing an investigation to authorities
- Explaining a new policy to employees
- Providing results from field or lab work to colleagues

Informal reports can be written for many different purposes, such as these:

- To inform
- To explain
- To analyze
- To persuade
- To recommend

And they can be written for both internal and external audiences:

- Internal audiences, such as coworkers, supervisors, or executives
- External audiences, such as clients, customers, or outside agencies

Informal reports can take on the layout and design characteristics of formal reports (e.g., covers, tables of contents, sections and subsections, and visuals), but informal reports can also be composed in different genre types, such as these:

- Memos
- Letters
- E-mails
- Presentations

Various workplace situations create the need for writers to compose informal reports of various types.

Progress Reports

Progress reports update audiences on the status of an ongoing project, typically detailing where the project is in relation to its overall goals and objectives. Progress reports allow writers to communicate with supervisors or managers, who are interested in the progress of projects that affect them and that may need to be managed in relation to other projects going on at the same time. Progress reports are usually written during varying stages of completion, whether the progress is good or bad, and they typically provide their audiences with projections of future activities and expected dates of accomplishment. Figure 21.3 is an example of an informal progress report produced in memo format. It serves many purposes and is directed to a specific audience.

Although only a few pages long, this informal report reviews past information, explains the progress of the project (including project delays), provides its audience with expected dates of completion, and invites its audience to ask for additional information about the project itself. It also lets the audience know when to expect the final, more formal report. Thus, this informal report is only one document, which leads to the production of other documents.

MEMORANDUM

DATE: June 15, 2006
TO: Marsha Pace MP
FROM: Anthony Wittman
RE: Progress Report on Quartz Etch-Rate Project

This memo provides an update on the quartz etch-rate project. It includes a general overview of the project, a review of its scope and objectives, and details on the progress made on each objective.

Project Overview
Subject. The project develops a method for obtaining quartz etch rates from the Polyflow vertical quartzware cleaner. A report on the findings will be delivered to you and the engineering staff.
Purpose. The purpose of the project is to obtain quartz etch-rate data for future reference. Instructions will be provided that ensure accurate results when procedures are followed correctly.

Progress Overview
Obtaining the necessary supplies for the tests provided some initial difficulties, but once these difficulties were overcome, the project began to run smoothly. The project is now approximately 75% completed but is currently running 6 weeks behind schedule, according to the checkpoint dates I provided in my January 9, 2006, memo. However, it will take less time to analyze and summarize the data results than I originally believed because we are using new computer software that I was unaware of when the proposal was created. Time saved by the new software improves our schedule by 2 weeks. Therefore, you will have the final report by December 3, 2006. The three main phases of the project are these:

- Obtaining test materials, researching quartz etch, developing an etch procedure, writing recipes for the Polyflow, and training for new software
- Performing the etch-rate tests
- Analyzing etch-rate data using statistical process control, authoring preliminary report for approval, revising preliminary report and submitting it for departmental review, and completing the final report and updating tool specification to include new etch procedure

The first objective is completed. The second objective is 75% complete. All etch-rate tests will be completed by the end of this month. I have begun to collect data from the etch tests and input it into the computer for statistical process analysis. Most of the time has been spent on the first phase of the project.

Objective 1: Material and information acquisition
Work completed: Immediately upon your approval of the project, I ordered the quartz chips for the etch tests from Phoenix Corp. As you know, there was difficulty getting the purchase order approved because of recent budget cutbacks. It was necessary to get approval for the purchase of these chips from the fabrication director, Carlos Hinojosa.This delayed the process by 3 weeks. Your assistance in obtaining approval was invaluable. Then, Phoenix Corp. was approximately 3 weeks late shipping the test disks. During this time I did additional research on quartz cleaning and developed numerous etching recipes on the Polyflow, discussed surface testing with the development branch, and trained on the new test software to help the project run more smoothly.

Objective 2: Performing etch-rate tests
Work completed: The etch-rate tests have so far been flawless. A new method for measuring thickness and roughness of the disks was created. This allowed some disks to be in all phases of the testing process at once. We created detailed instructions that let anyone repeat the procedure. Several operators performed etch tests using only the written instruction; they repeated the tests without error.
Work remaining: 29 of the 38 necessary etches have been performed.The remaining 9 tests will be done by June 30.

Objective 3: Analysis of data and report
Work completed: Data from completed etch tests are currently being analyzed through our software.
Work remaining: I must write and submit for approval the final project report and then add the etch-rate procedure to the Polyflow operation specification. According to the revised schedule, the report will be submitted on December 3, 2006.
I foresee no future problems with the project that would delay its completion. If you would like to see data from the testing or information about our new software, please let me know. And if you have any suggestions or concerns, I would be happy to speak with you soon.

FIGURE 21.3 A progress report in memo format

This progress report can be understood as a problem-solving document in which the writer had to make many rhetorical choices. It seeks to provide its audience—a supervisor—with needed information, thereby directly addressing a stakeholder in the project. In addition, because the project is behind schedule, the writer provides his audience with accurate information about how this problem is being solved. The report also reassures the audience that the project is indeed running smoothly at the current moment and will be finished within a reasonable amount of time. And the report details exactly how much work has been completed on each of the three project objectives, as well as how much work still remains. The report is precise in the information it provides.

Progress reports like this one are common types of informal reports and can be written in a variety of genres, depending in large part on the preference of the audience. However, regardless of format, progress reports describe the details of a project over time; that is, they provide information relative to a projected time frame. For reports that detail the status of multiple goals, objectives, or tasks, like the one included here, writers may wish to organize their reports in a corresponding order.

Project overview

Progress overview

Objective 1 (repeated for each objective)

  Work completed

  Work remaining

Conclusion

For projects that do not have distinct objectives or tasks, writers may want to follow a strictly chronological pattern of past, present, and future work. The conclusions of informal progress reports do not need to be lengthy; their purpose is simply to provide readers with an honest assessment of the project's status.

ANALYZE THIS

In the informal report about the quartz etch-rate project, a writer informs his supervisor about the progress of his project. Even though you do not work at this company or know any of the individuals involved in this project, the document provides some sense of the rhetorical situation.

Look back at the first two phases of the rhetorical problem-solving approach, and explain in greater detail the kinds of problems this report addresses and solves, the audience needs, the report's purposes, and the persuasive strategies the writer uses, including why he may have chosen the memo format. If you were the audience of this document, are there any additional questions you would have for the writer?

Lab Reports

Generally speaking, the purpose of a lab report is to document the procedures, methods, and results of a lab experiment and explain their significance. Lab reports are common in the sciences and engineering but should do more than simply present data. They should demonstrate the writer's understanding and comprehension of the data, the underlying conceptions behind the data, and the principles and reasons behind the experiment itself. Although they vary widely, many lab reports include the following components, even though the reports might not use these exact headings or include all of these elements:

1. *Title page.* This should include the author's name and the names of any others who collaborated on the project, the name of the experiment, and the date(s) the experiment was performed.
2. *Abstract.* This describes four crucial aspects of the report: the purpose of the experiment, the method or theory underlying the experiment, key findings, and major conclusions. Although there is no set word limit on abstracts, they should be one paragraph only and should not exceed approximately 250 words. Abstracts allow readers to decide whether they want to read the whole report.
3. *Introduction.* This identifies the experiment undertaken, the objectives of the experiment, its importance, and any background needed for audiences to understand the experiment.
4. *Materials.* This section identifies, and in some cases describes, any equipment or other materials that were used in the experiment.
5. *Procedures.* This section narrates how the experiment occurred and how others could reproduce the results the writer found. These sections are usually written as first-person narratives, although not always. Most writers describe their procedures in chronological order, explaining all steps in the order in which they happened and including specific and detailed information.

See chapter 8 for more on creating and incorporating visuals into documents.

6. *Results and discussion.* This is one of the most important aspects of a lab report and is sometimes separated into two distinct sections. Results may be presented as calculations, tables, or figures, depending on the nature of the experiment, but writers must label and describe any visuals clearly. In discussing the results, writers should analyze them and consider their ramifications, as well as any errors that affected the experiment. Writers often focus on these concerns:
   - What do the results clearly indicate?
   - What is the significance of these results?
   - How do expected results compare with obtained results?
   - What experimental errors occurred?
   - How do results relate to experimental objectives?
   - How do results compare/contrast with those of similar experiments?
   - Were there any limitations to the experiment design?
7. *Conclusions.* This section presents any conclusions that the experimenter drew from the results and discussion. Conclusions must be supported with evidence, referring to specific results and observations from the experiment. The writer might wish to summarize reasons for any inconsistencies between actual and expected results and might also provide recommendations for correcting the underlying problems.
8. *References.* This section documents any reading that the experimenter has done that aided in the experiment. Individual disciplines have their own style guides for citing references.
9. *Appendix.* This section includes any raw data, calculations, graphs, and other quantitative materials that were integral to the experiment but were not reported in any of the other sections. In other sections of their lab reports, writers often refer to appendixes to highlight important information.

Lab reports usually cover a great deal of information and may deviate somewhat from the sections presented here. However, these components can be used for other informal reports in other workplaces; that is, these guidelines for lab reports are typical of the sciences and engineering but are not exclusive to them. For an example of a full lab report, see Figure 21.8 on pages 652–654.

See chapter 12 for a short lab report written in memo format.

EXPLORE

Locate a copy of a short lab report from an online source. This can be from any field, such as physics, chemistry, biology, engineering, or computer science. Which of the eight components does the lab report include? Which does it not include? Who do you think is the audience for the lab report? What is its purpose?

Directives

Directives are informal reports—sometimes in e-mail, memo, or letter format—that apprise audiences of policies or procedures that they should be aware of and follow. In many cases writers discuss where these new policies and procedures came from, why they were put into place, what the specifics are, how readers should comply, and where they can find additional information. In some situations writers may place this information under separate headings to allow audiences to follow it more easily.

Figure 21.4 shows a presidential directive released by the White House press secretary. The directive makes its readers aware of how the government will develop, maintain, and use information to handle those individuals suspected of terrorist activities or links. Although the directive is not written in strict memo format, it does have many characteristics of standard memos, such as subject line and date at the top. There is no one universal format for writing directives. In fact, many agencies, companies, and organizations have developed their own formats that must be followed. However, regardless of the format of a directive, writers must follow the problem-solving approach, reflecting on the rhetorical situation at hand and recognizing the specific information the audience requires. Policy or procedural changes that radically alter the way employees go about their daily jobs should be accompanied with detailed reasons and, if necessary, face-to-face meetings.

IN YOUR EXPERIENCE

In a brainstorming activity come up with two or three different directives you have encountered in the past—perhaps documents that explained new ways to pay for tuition, new policies for filling out time cards at work, or new procedures for securing season tickets. You may not remember the specifics of the documents, but describe what they looked like, what they said, how they were delivered to you, and what made you pay attention to them—or not.

Incident Reports

An incident report is another kind of informal report that is common in workplaces. Such reports document events in the workplace, such as accidents, emergencies, and problems. Their main purpose is to inform audiences about the events, although they can also explain causes and handling. Sometimes an incident report is not simply informative; it can also serve as a kind of warning, implying that the reader should take a certain course of action to avoid future complications.

Like other informal reports, incident reports can be written as memos, letters, or e-mails; however, most companies, agencies, and organizations have standardized forms to be used for incident reports—consider the standardized forms that police officers must use for traffic tickets or accident reports. Figure 21.5 is an example of an incident report form created by the Occupational Safety and Health Administration

President George W. Bush

Homeland Security Presidential Directive/Hspd-6

September 16, 2003

For Immediate Release
Office of the Press Secretary

Subject: Integration and Use of Screening Information

To protect against terrorism it is the policy of the United States to (1) develop, integrate, and maintain thorough, accurate, and current information about individuals known or appropriately suspected to be or have been engaged in conduct constituting, in preparation for, in aid of, or related to terrorism (Terrorist Information); and (2) use that information as appropriate and to the full extent permitted by law to support (a) Federal, State, local, territorial, tribal, foreign-government, and private-sector screening processes, and (b) diplomatic, military, intelligence, law enforcement, immigration, visa, and protective processes.

This directive shall be implemented in a manner consistent with the provisions of the Constitution and applicable laws, including those protecting the rights of all Americans.

To further strengthen the ability of the United States Government to protect the people, property, and territory of the United States against acts of terrorism, and to the full extent permitted by law and consistent with the policy set forth above:

(1) The Attorney General shall establish an organization to consolidate the Government's approach to terrorism screening and provide for the appropriate and lawful use of Terrorist Information in screening processes.

(2) The heads of executive departments and agencies shall, to the extent permitted by law, provide to the Terrorist Threat Integration Center (TTIC) on an ongoing basis all appropriate Terrorist Information in their possession, custody, or control. The Attorney General, in coordination with the Secretary of State, the Secretary of Homeland Security, and the Director of Central Intelligence shall implement appropriate procedures and safeguards with respect to all such information about United States persons. The TTIC will provide the organization referenced in paragraph (1) with access to all appropriate information or intelligence in the TTIC's custody, possession, or control that the organization requires to perform its functions.

(3) The heads of executive departments and agencies shall conduct screening using such information at all appropriate opportunities, and shall report to the Attorney General not later than 90 days from the date of this directive, as to the opportunities at which such screening shall and shall not be conducted.

(4) The Secretary of Homeland Security shall develop guidelines to govern the use of such information to support State, local, territorial, and tribal screening processes, and private sector screening processes that have a substantial bearing on homeland security.

(5) The Secretary of State shall develop a proposal for my approval for enhancing cooperation with certain foreign governments, beginning with those countries for which the United States has waived visa requirements, to establish appropriate access to terrorism screening information of the participating governments.

This directive does not alter existing authorities or responsibilities of department and agency heads to carry out operational activities or provide or receive information. This directive is intended only to improve the internal management of the executive branch and is not intended to, and does not, create any right or benefit enforceable at law or in equity by any party against the United States, its departments, agencies, entities, officers, employees or agents, or any other person.

The Attorney General, in consultation with the Secretary of State, the Secretary of Homeland Security, and the Director of Central Intelligence, shall report to me through the Assistant to the President for Homeland Security not later than October 31, 2003, on progress made to implement this directive and shall thereafter report to me on such progress or any recommended changes from time to time as appropriate.

GEORGE W. BUSH

FIGURE 21.4 A presidential directive about homeland security

Violence Incident Report Forms*

A reportable violent incident should be defined as any threatening remark or overt act of physical violence against a person(s) or property whether reported or observed.

1. Date:__________________
   Day of week:__________________
   Time:__________________
   Assailant: Female ____ Male ____

2. Specific Location:
   ____

3. Violence directed towards: ____ Patient ____ Staff ____ Visitor ____ Other
   Assailant: ____ Patient ____ Staff ____ Visitor ____ Other
   Assailant's Name:__________________
   Assailant: ____ Unarmed ____ Armed (weapon)

4. Predisposing factors: ____ Intoxication ____ Dissatisfied with care/waiting time
   ____ Grief reaction
   ____ Gang related ____ Prior history of violence
   ____ Other (Describe) __________________

5. Description of incident:
   ____ Physical abuse
   ____ Verbal abuse
   ____ Other

6. Injuries:
   ____ Yes
   ____ No

7. Extent of Injuries:
   ____
   ____

8. Detailed description of the incident:

9. Did any person leave the area because of incident?
   ____ Yes ____ No ____ Unable to determine

10. Present at time of incident:
    ____ Police __________________
    Name of department
    ____ Hospital security officer

11. Needed to call:
    ____ Police __________________
    Department
    ____ Hospital security

12. Termination of incident:
    Incident diffused ____ Yes ____ No
    Police notified ____ Yes ____ No
    Assailant arrested ____ Yes ____ No

13. Disposition of assailant:
    ____ Stayed on premises
    ____ Escorted off premises
    ____ Left on own
    ____ Other __________________

14. Restraints used:____ Yes ____ No
    Type:__________________

15. Report completed by:__________________ Title:__________________
    Witnesses:__________________
    Supervisor notified:__________________ Time:__________________

Please put additional comments, according to numbered section, on reverse side of form

***This form was taken from: Guidelines for Preventing Workplace Violence for Health Care And Social Service Workers, OSHA Publication 3148, 1996.**

FIGURE 21.5 A report form for violent incidents

(OSHA) for documenting violent incidents. Notice that in addition to the boxes and blanks that simply need to be checked off or filled in with factual information like dates and times, this form also includes a section where the writer must provide a detailed description of the incident.

Even though many incident report forms are prefabricated, those who compose them must often do a good deal of actual writing. In such instances, it is important that writers be specific, accurate, and ethical because it is possible that the incident report will later be used (perhaps even in court) to help decision makers take some kind of action.

IN YOUR EXPERIENCE

Have you ever had to document or report an incident of some sort, for either personal, legal, or employment reasons? What was the incident? Did you write the report, or did you provide information to someone else who wrote it? Was there a standard form for you to use? What kinds of questions did the form ask? How was the form constructed for ease of use and thoroughness of reporting? As well as you can, recall what the incident was in detail, and describe the means by which the incident was reported.

COMPOSING INFORMAL REPORTS

PSA

Because reports—whether informal or formal—help solve problems, the problem-solving approach provides workplace writers with strategies to compose these reports. You might also want to review the chapters that provide specific information about composing e-mails and memos, letters, formal reports, or oral presentations (see chapters 12, 13, 22, and 23 respectively). However, a number of important considerations stretch across these different genres. For example, informal reports must include introductory material, a body, and a conclusion.

- *Introduction.* Regardless of the kind of informal report, an introduction typically must accomplish three main tasks: announce the project, experiment, situation, or problem that the report will discuss; explain the role of the writer; and identify the purpose of the report, forecasting what it will cover.
- *Body.* The body of an informal report provides readers with the details of the project, experiment, situation, or problem. This section may be made up of many different subsections, such as those that describe how and why things have happened, what problems have arisen, and what will and should happen in the future. Each kind of informal report necessitates that writers divide the body section in particular ways.
- *Conclusion.* A conclusion typically reminds the audience of important dates and information, asks the audience to take some particular course of action, asks for a response, indicates a willingness to discuss the report in greater detail, and/or provides the audience with relevant contact information.

As writers consider the problem-solving approach in relation to informal reports, these concerns are crucial:

- Who is my audience? What are the needs of this audience?
- What is my purpose in writing to this audience?
- What kind of informal report will best suit audience needs and my purpose?

- Do I have the proper information or research to write the report?
- What tone and style should the informal report use for this audience?
- What genre will the audience expect for this informal report?
- How should I use layout/design principles and visuals in the report?
- What revisions should I make in the draft of the report?
- How should I transmit this document to my audience?

The answers to these questions change with each new rhetorical situation, causing the writer to make new decisions with every report.

For more on spoken presentations, see chapter 23.

Figure 21.6 offers another example of a progress report, although it is a spoken presentation, as is often the case with reports. The report's title and language make it sound and appear formal, but it remains a progress report, which is a kind of informal report. Because it is a written transcript of a spoken presentation, it does not include the headings and sections that would appear in a document composed for readers. However, like the earlier progress report in Figure 21.3, it was composed for a particular audience, it has a purpose, it describes problems and their causes, it updates audiences on the status of the project (i.e., the search for WMDs), and it relates a chronological progression of activities.

ANALYZE THIS

Because the search for WMDs was highly politicized, David Kay's rhetorical situation required him to address his audience in careful and sensitive ways. Explain how his speech reflects the characteristics of a progress report. How is it divided into distinct sections? How would you characterize the speaker's tone? What makes it appropriate or inappropriate for the situation at hand? How would you describe the speaker's style—his use of words and sentence structure? How does it fit with his purpose and his audience? What seems to be the purpose(s) of this spoken progress report?

ETHICAL ISSUES IN REPORT WRITING

The ethical issues related to informal reports do not differ much from those that arise with other kinds of workplace documents. However, there are two in particular that should be reviewed here.

Disclosure

Disclosure means that writers tell audiences everything they need to know about a project, experiment, situation, or problem; it means that nothing of importance or relevance is left out. There is an old saying that the truth is in the details, a reality that sometimes leads writers to cover up details in order to keep from telling readers things they will find disagreeable. Because progress reports and lab reports are often connected to projects that receive a great deal of monetary funding and support, it is not uncommon for those involved to hide problems and hindrances from supervisors, managers, and those who fund such projects. The workers who compose the progress reports often do not want to disappoint those in charge. However, it is ethically necessary that writers provide all relevant details about problems, delays, budgetary failures, and any other unpredicted changes to the project or situation.

[Statement by David Kay on the activities of the Iraq Survey Group (ISG) before the House Permanent Select Committee on Intelligence, the House Committee on Appropriations, the Subcommittee on Defense, and the Senate Select Committee on Intelligence]

October 2, 2003

Thank you, Mr. Chairman. I welcome this opportunity to discuss with the Committee the progress that the Iraq Survey Group has made in its initial three months of its investigation into Iraq's weapons of mass destruction (WMD) programs. I cannot emphasize too strongly that the Interim Progress Report, which has been made available to you, is a snapshot, in the context of an ongoing investigation, of where we are after our first three months of work. The report does not represent a final reckoning of Iraq's WMD programs, nor are we at the point where we are prepared to close the file on any of these programs. While solid progress—I would say even remarkable progress considering the conditions that the ISG has had to work under—has been made in this initial period of operations, much remains to be done. We are still very much in the collection and analysis mode, still seeking the information and evidence that will allow us to confidently draw comprehensive conclusions to the actual objectives, scope, and dimensions of Iraq's WMD activities at the time of Operation Iraqi Freedom. Iraq's WMD programs spanned more than two decades, involved thousands of people, billions of dollars, and were elaborately shielded by security and deception operations that continued even beyond the end of Operation Iraqi Freedom. The very scale of this program when coupled with conditions in Iraq that have prevailed since the end of Operation Iraqi Freedom dictates the speed at which we can move to a comprehensive understanding of Iraq's WMD activities.

We need to recall that in the 1991–2003 period the intelligence community and the UN/IAEA inspectors had to draw conclusions as to the status of Iraq's WMD program in the face of incomplete, and often false, data supplied by Iraq or data collected either by UN/IAEA inspectors operating within the severe constraints that Iraqi security and deception actions imposed or by national intelligence collection systems with their own inherent limitations.The result was that our understanding of the status of Iraq's WMD program was always bounded by large uncertainties and had to be heavily caveated. With the regime of Saddam Hussein at an end, ISG has the opportunity for the first time of drawing together all the evidence that can still be found in Iraq—much evidence is irretrievably lost—to reach definitive conclusions concerning the true state of Iraq's WMD program. It is far too early to reach any definitive conclusions and, in some areas, we may never reach that goal. The unique nature of this opportunity, however, requires that we take great care to ensure that the conclusions we draw reflect the truth to the maximum extent possible given the conditions in postconflict Iraq.

We have not yet found stocks of weapons, but we are not yet at the point where we can say definitively either that such weapon stocks do not exist or that they existed before the war and our only task is to find where they have gone. We are actively engaged in searching for such weapons based on information being supplied to us by Iraqis. Why are we having such difficulty in finding weapons or in reaching a confident conclusion that they do not exist or that they once existed but have been removed? Our search efforts are being hindered by six principal factors:

- From birth all of Iraq's WMD activities were highly compartmentalized within a regime that ruled and kept its secrets through fear and terror and with deception and denial built into each program.
- Deliberate dispersal and destruction of material and documentation related to weapons programs began preconflict and ran trans- to postconflict.
- Post-OIF looting destroyed or dispersed important and easily collectable material and forensic evidence concerning Iraq's WMD program. As the report covers in detail, significant elements of this looting were carried out in a systematic and deliberate manner, with the clear aim of concealing pre-OIF activities of Saddam's regime.
- Some WMD personnel crossed borders in the pre-transconflict period and may have taken evidence and even weapons-related materials with them.
- Any actual WMD weapons or material is likely to be small in relation to the total conventional armaments footprint and difficult to near impossible to identify with normal search procedures. It is important to keep in mind that even the bulkiest materials we are searching for, in the quantities we would expect to find, can be concealed in spaces not much larger than a two-car garage.
- The environment in Iraq remains far from permissive for our activities, with many Iraqis that we talk to reporting threats and overt acts of intimidation and our own personnel being the subject of threats and attacks. In September alone we have had three attacks on ISG facilities or teams. The ISG base in Irbil was bombed and four staff injured, two very seriously; a two-person team had their vehicle blocked by gunmen and only escaped by firing back through their own windshield; and on Wednesday, 24 September, the ISG Headquarters in Baghdad again was subject to mortar attack.

What have we found and what have we not found in the first 3 months of our work? We have discovered dozens of WMD-related program activities and significant amounts of equipment that Iraq concealed from the United Nations during the inspections that began in late 2002. The discovery of these deliberate concealment efforts has come about both through the admissions of Iraqi scientists and officials concerning information they deliberately withheld and through physical evidence of equipment and activities that ISG has discovered that should have been declared to the UN. Let me just give you a few examples of these concealment efforts, some of which I will elaborate on later:

FIGURE 21.6 Excerpt from a spoken progress report about weapons of mass destruction

- A clandestine network of laboratories and safehouses within the Iraqi Intelligence Service that contained equipment subject to UN monitoring and suitable for continuing CBW research.
- A prison laboratory complex, possibly used in human testing of BW agents, that Iraqi officials working to prepare for UN inspections were explicitly ordered not to declare to the UN.
- Reference strains of biological organisms concealed in a scientist's home, one of which can be used to produce biological weapons.
- New research on BW-applicable agents, Brucella and Congo Crimean hemorrhagic fever (CCHF), and continuing work on ricin and aflatoxin were not declared to the UN.
- Documents and equipment, hidden in scientists' homes, that would have been useful in resuming uranium enrichment by centrifuge and electromagnetic isotope separation (EMIS).
- A line of UAVs not fully declared at an undeclared production facility and an admission that they had tested one of their declared UAVs out to a range of 500 km, 350 km beyond the permissible limit.
- Continuing covert capability to manufacture fuel propellant useful only for prohibited SCUD variant missiles, a capability that was maintained at least until the end of 2001 and that cooperating Iraqi scientists have said they were told to conceal from the UN.
- Plans and advanced design work for new long-range missiles with ranges up to at least 1,000 km—well beyond the 150 km range limit imposed by the UN. Missiles of a 1,000 km range would have allowed Iraq to threaten targets throughout the Middle East, including Ankara, Cairo, and Abu Dhabi.
- Clandestine attempts between late-1999 and 2002 to obtain from North Korea technology related to 1,300 km range ballistic missiles—probably the No Dong—300 km range antiship cruise missiles, and other prohibited military equipment.
- In addition to the discovery of extensive concealment efforts, we have been faced with a systematic sanitization of documentary and computer evidence in a wide range of offices, laboratories, and companies suspected of WMD work. The pattern of these efforts to erase evidence—hard drives destroyed, specific files burned, equipment cleaned of all traces of use—are ones of deliberate, rather than random, acts. For example, on 10 July 2003 an ISG team exploited the Revolutionary Command Council (RCC) Headquarters in Baghdad. The basement of the main building contained an archive of documents situated on well-organized rows of metal shelving. The basement suffered no fire damage despite the total destruction of the upper floors from coalition air strikes.Upon arrival the exploitation team encountered small piles of ash where individual documents or binders of documents were intentionally destroyed. Computer hard drives had been deliberately destroyed. Computers would have had financial value to a random looter; their destruction, rather than removal for resale or reuse, indicates a targeted effort to prevent Coalition forces from gaining access to their contents.
- All IIS laboratories visited by IIS exploitation teams have been clearly sanitized, including removal of much equipment, shredding and burning of documents, and even the removal of nameplates from office doors.

Although much of the deliberate destruction and sanitization of documents and records probably occurred during the height of OIF combat operations, indications of significant continuing destruction efforts have been found after the end of major combat operations, including entry in May 2003 of the locked gated vaults of the Ba'ath party intelligence building in Baghdad and highly selective destruction of computer hard drives and data storage equipment along with the burning of a small number of specific binders that appear to have contained financial and intelligence records, and in July 2003 a site exploitation team at the Abu Ghurayb Prison found one pile of the smoldering ashes from documents that was still warm to the touch.

I would now like to review our efforts in each of the major lines of enquiry that ISG has pursued during this initial phase of its work....

FIGURE 21.6 Continued

Accuracy

Accuracy in reporting is a kind of disclosure, although it differs in referring to the degree to which writers are truthful in their writings. Accuracy includes recording and reporting truthful lab results, correctly explaining the extent of problems and hindrances to projects, and providing the precise details of a situation, as in an incident report. In the workplace it is not uncommon for writers to stretch the truth in order to report results that they believe their audiences wish to hear. However, ethics requires accuracy.

Almost every workplace has an established set of ethical codes or conduct codes by which its employees must abide. When you report for your first full-time job or when you join a professional organization, you will not be expected to know everything about ethics but will be shown the particular ethical standards of your workplace or organization. As an example, look at Figure 21.7, which includes portions of the code of ethics of the Ecological Society of America. Even though this is an organization dedicated specifically to the promotion of ecological sciences, notice how many of the items refer to issues of *reporting* information, including writing, research, problem solving, professionalism, disclosure, and accuracy.

Preamble:
This code provides guiding principles of conduct for all members of the Ecological Society of America and all ecologists certified by the Society. It is the desire and purpose of the Society to support and encourage ecological research and education, and to facilitate the application of ecological science in the management of ecological systems. Towards these ends, this Code is intended to further ecological understanding through the open and honest communication of research; to assure appropriate accessibility of accurate and reliable ecological information to employers, policy makers, and the public; and to encourage effective education and training in the disciplines of ecological science. Individuals aware of breaches of this Code are encouraged to refer to the Society's procedures for addressing violations of the Code, and to communicate with the Society's Executive Director, who will explain the code and process.

General:
All members of the Ecological Society of America and all ecologists certified by the Society should observe the following principles in the conduct of their professional affairs:

1. Ecologists will offer professional advice and guidance only on those subjects in which they are informed and qualified through professional training or experience. They will strive to accurately represent ecological understanding and knowledge and to avoid and discourage dissemination of erroneous, biased, or exaggerated statements about ecology.
2. Ecologists will not present themselves as spokespersons for the Society without express authorization by the President of ESA.
3. Ecologists will cooperate with other researchers whenever possible and appropriate to assure rapid interchange and dissemination of ecological knowledge.
4. Ecologists will not plagiarize in verbal or written communication, but will give full and proper credit to the works and ideas of others, and make every effort to avoid misrepresentation.
5. Ecologists will not fabricate, falsify, or suppress results, deliberately misrepresent research findings, or otherwise commit scientific fraud.
6. Ecologists will conduct their research so as to avoid or minimize adverse environmental effects of their presence and activities, and in compliance with legal requirements for protection of researchers, human subjects, or research organisms and systems.
7. Ecologists will not discriminate against others in the course of their work on the basis of gender, sexual orientation, marital status, creed, religion, race, color, national origin, age, economic status, disability, or organizational affiliation.
8. Ecologists will not practice or condone harassment in any form in any professional context.
9. In communications, ecologists should clearly differentiate facts, opinions, and hypotheses.
10. Ecologists will not seek employment, grants, or gain, nor attempt to injure the reputation or professional opportunities of another scientist by false, biased, or undocumented claims, by offers of gifts or favors, or by any other malicious action.

FIGURE 21.7 Ecological Society of America's code of ethics

ANALYZE THIS

Reread the code of ethics for the Ecological Society of America (Figure 21.7). Locate particular instances of ethical standards that deal specifically with communication and issues related to writing, speaking, research, professionalism, honesty, and accuracy. Compare these standards with the ethical code of behavior that students at your university are expected to follow in their work. What are the similarities and the differences?

SAMPLE LAB REPORT

Figure 21.8 gives an example of a full lab report.

Examination of Protozoan Cultures to Determine Cellular Structure and Motion Pattern

Abstract

Protozoans are unicellular eukaryotes with either plant-or animal-like characteristics. Through careful observation, we analyzed various protozoan cultures in order to identify characteristics associated with cell structure and movement of these one-celled organisms. We found that Protists exhibit certain characteristics that allow them to be categorized into different groups, mainly determined by their locomotion patterns. Despite differences in locomotion and the varying plant-like and animal-like organelles, all protists share key characteristics and functions that allow them to feed, grow, and reproduce--processes essential for survival and common to complex organisms

Introduction

Unicellular eukaryotes belong to the kingdom Protista, and are often referred to as "protists" or "protozoans." The name "protozoan" means "first animal," but eukaryotes may display either plant or animal-like characteristics, or a combination of both. Although unicellular, they have a nucleus and membrane-bound organelles, making them functionally complex despite their small size. Each small protist is a self-supporting unit, carrying out all the processes for survival in just one cell. They thrive on moisture and can be found on moist soil and in fresh and marine bodies of water. There are about 30,000 known species of protozoans, commonly classified according to their movement patterns as sarcodines—moving with false feet called pseudopodia or, flagellates—moving with whip-like structures known as flagella, ciliates—moving with short hairs known as cilia, and sporozoans—with no movement. They all have varying shapes, sizes, and survival strategies. For example, some may "hunt" small particles of food such as bacteria or algae; whereas others may be parasitic, inhabiting larger organisms. Despite their differences, all protists have several characteristics in common. In addition to a nucleus or nuclei to house their genetic material, most protists have mitochondria for metabolic functions, and vacuoles for digestion and excretion. With the help of these and other cellular structures, protists may feed, grow, and reproduce.

In this lab we observed select examples of protists in order to identify their cellular structures, and determine to which group of protista they belong based on their form of movement. We also made drawings of our observations using light and dissection microscopes to practice proper microscopy skills, including making wet-mount slides and cell sizing. By observing, drawing, and classifying protista, we learned about the cell structure and movement patterns of these one-celled organisms. We also learned about the differences and similarities of various protist cells.

Since we will observe how protists move, it will be interesting to figure out patterns of locomotion. For example, what happens when the protist encounters an obstacle? Does motion change when the organism is feeding? How does motion relate to where the organism lives? What characteristics do the protists exhibit: plant, animal, or both? Do the plant/animal characteristics influence motion patterns?

Methods

Three protists were chosen for observation. See the list of protists below to choose three samples. For each of the protists, a pipette was used to extract a few drops of culture from the culture jar. The drops of culture were placed on a clean microscope slide and covered with a slide cover slip. Using a light microscope, each protist was examined at different magnifications until the best field of view was found for identifying cellular structures. The color, shape, and motion cellular structures was noted. Each of the protists was drawn and the drawings were labeled. Field-of-view, magnification, and cell size was noted on the drawings, along with the organism's name and protist group.

Protists available for observation:
Euglena
Paramecium
Difflugia
Blepharisma
Didinium
Amoeba
Stentor
Spirostomum
Vorticella
Volvox
Bursaria

FIGURE 21.8A Example of a lab report

Results

All protists that were selected had features in common, but they all moved differently. The example protists were: *Euglena*, *Paramecium, and Amoeba*. *Euglena* moved with a flagellum and so is classified as a flagellate (see Fig. 1). *Paramecium* moved with cilia and so is classified as a ciliate (see Fig. 2). Finally, *Amoeba* moved with a pseudopod, and so is a sarcodine (see Fig. 3). All three protists had a nucleus, as expected, but the *Paramecium* had two nuclei, a micronucleus and a macronucleus. The *Paramecium* and *Amoeba* both had food and contractile vacuoles, but these were lacking in the *Euglena*. All protists had animal-like characteristics in terms of their movements and feeding patterns. Of the three, *Euglena* was the only one that had chloroplasts, an organelle common in plants.

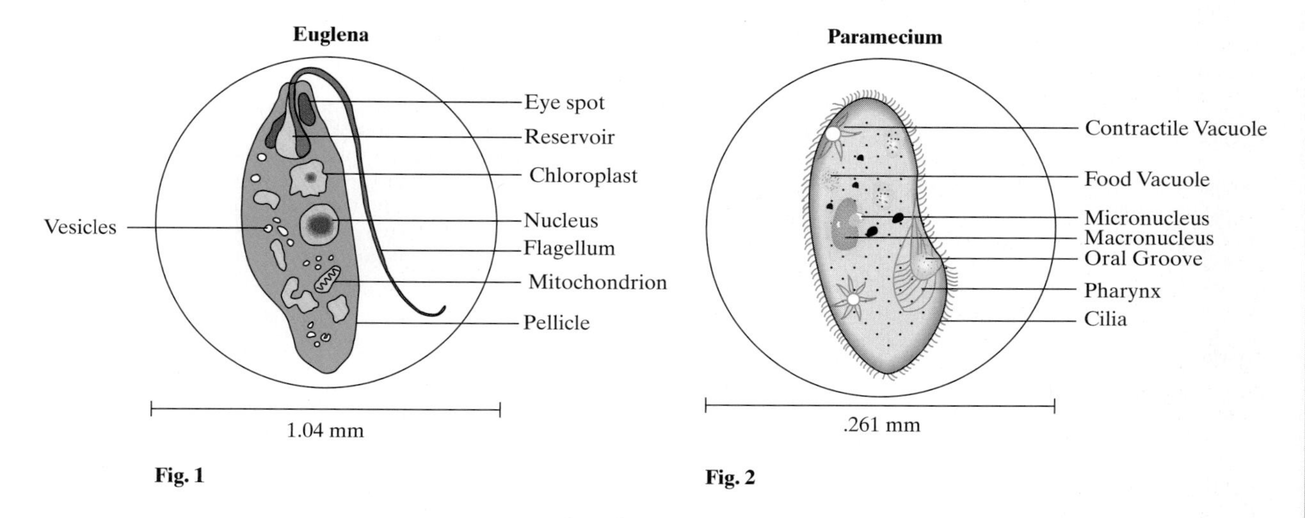

Fig. 1

Fig. 2

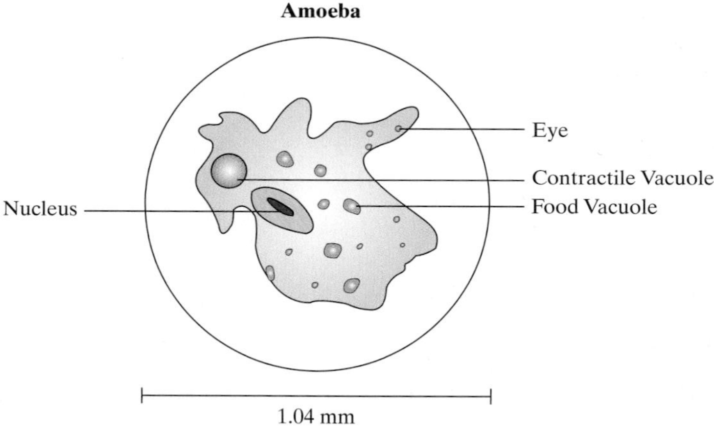

Fig. 3

Discussion

Protists seem to share certain characteristics even when they are classified into different groups. Their organelles are a mixture of animal and plant structures, but they all have nuclei, a feature which distinguishes Protists from other unicellular organisms. The protists' motion was consistent with their locomotion organ: cilia, flagella, or pseudopod. This motion was very clear under the light microscope, but interactions of protists with others in the culture jar were better observed using the dissection scope. The *Amoeba* moves by extending part of its cell. This extruding part is the pseudopod, and allows the *Amoeba* to drag itself from one place to another (see Fig. 3). Its movement is slow, and changing directions is just a matter of extending a pseudopod in a new direction. *Amoebas* do not seem to have a particular shape, with the exception of the pseudopodia that consistently protrude from the cell. This shapeless but ever shifting quality of the *Amoeba's* shape allows it to surround, engulf, and ingest its food by a process called phagocytosis.

FIGURE 21.8B

Paramecia are smaller than *Amoebas*. They move with the help of microscopic hair-like structures called cilia, which act like oars to push them through the water. They swim by rotating slowly and changing directions often. If the *Paramecium* comes upon an obstacle, it stops, swims backwards, and then angles itself forward on a slightly different course. Cilia help the *Paramecium* move as well as feed. When the *Paramecia* feed, it does so by drawing its food into a funnel-shaped opening called the oral groove that is lined with cilia (see Fig. 2). The oral groove is like a mouth, taking food in with the help of cilia, which direct and move the food inward.

The *Euglena* moves rapidly, using its flagellum to propel itself through the water rather quickly, shifting directions with whip-like movements. Unlike the *Amoeba* and the *Paramecium*, the *Euglena* has plant-like characteristics. It is sometimes referred to as a "plant-like" protist. The organelle that gives it this plant-like quality is the chloroplast (see Fig. 1), a green organelle responsible for carrying out photosynthesis in plants. The *Euglena* senses light with a light-sensitive organelle called the "eyespot," which directs the organism to a light source strong enough for photosynthesis to occur. Since it can undergo photosynthesis, *Euglena* is able to make its own food just like plants.

The three protists examined in this lab are examples of protists that use specialized structures for locomotion. Although the *Euglena* has some "plant-like" characteristics, all protists mentioned above, exhibit animal-like movements. These protists exemplify the animal-like and motile types of protozoans. As compared to other protists, the animal-like features of the protists we observed allow them to be motile. Their motility comes in handy for moving about their environment and finding food. They may be contrasted to another class of protist, the sporozoans. Sporozoans have no form of locomotion and are primarily parasitic, ingesting their food by absorption through their cell membranes. No matter what type of locomotion a protist uses, all protists must be able to carry out the metabolic functions of multicellular organisms. Based on the observations in this lab, protists are very small yet highly complex. They have all the organelles necessary for a variety of functions such as digestion, excretion, reproduction, respiration, and movement. Protists are self-supporting "one cell factories" churning out all the processes that are usually carried out by a highly-organized network of cells.

Conclusion

In this lab I learned about the structure and function of the smallest eukaryotic organisms, the unicellular protists. Although very tiny, these organisms are very complex, housing all the necessary life tools in one single cell. This shows that the complexity of an organism is not necessarily related to its size. I also learned to identify and classify different types of protists. I was able to observe locomotion patterns as well as other characteristic features. In doing so, I gained useful microscopy skills such as making wet mount slides, finding the proper magnification for viewing, and drawing microscope observations with all the proper labels.

FIGURE 21.8C

SUMMARY

- Workplace reports inform, explain, persuade, and recommend. They solve problems, consider audience backgrounds and needs, usually involve research, and often are written collaboratively.
- Reports differ from other texts in terms of their complexity, length, and the kinds of problems they solve.
- Informal reports are important to many different situations in the workplace.
- Informal reports can be written for many different purposes, for both internal and external audiences, and in different genre types, such as e-mails, memos, letters, and presentations.
- Typical informal reports include progress or status reports, lab reports, directives, and incident reports.
- Informal reports have identifiable sections: an introduction, a body section, and a conclusion.
- The problem-solving approach should guide the writing of an informal report.
- Ethical issues in report writing include disclosure and accuracy.

CONCEPT REVIEW

1. What are some of the purposes that informal reports serve?
2. What kinds of audiences are informal reports composed for?
3. What genres do writers usually use for informal reports?
4. What are progress reports, and what purposes do they serve?
5. When might someone compose a progress report? Give an example.
6. What are laboratory reports, and what purposes do they serve?
7. When might someone compose a laboratory report? Give an example.
8. What are directives, and what purposes do they serve?
9. When might someone compose a directive? Give an example.
10. What are incident reports, and what purposes do they serve?
11. When might someone compose an incident report? Give an example.
12. Which aspects of the problem-solving approach are relevant for informal reports?
13. What does *disclosure* mean as an ethical concern?
14. What does *accuracy* mean as an ethical concern?

CASE STUDY 1

E-mail at Agriflex: Writing Directives for Company Employees

As the chief compliance officer at Agriflex Corporation, imagine that you are responsible for ensuring that Agriflex's 2,500 employees comply with internal policies and regulations. However, your staff has conducted research on daily operations at Agriflex and has given you a short, informal presentation about lost productivity among workers who are misusing the company's e-mail system in a variety of ways. You are concerned but do not wish to do anything drastic at this point. Instead, you decide that the first step in solving this problem is simply to remind company employees about the proper use of Agriflex's e-mail system. You jot down the following points:

- E-mail should be professional and courteous.
- E-mail must not contain any illegal, libelous, or offensive statements.
- All statements meant to harass—sexually or otherwise—are prohibited.

- E-mail is for business purposes, not for personal use.
- All e-mail is company property.
- The company has the right to access e-mail sent to or from every company computer.
- The company has the right to retrieve e-mail stored on its servers that users have deleted from their e-mail programs.
- Employees who violate e-mail policies are subject to disciplinary measures or termination.
- The full e-mail policy is available in the employee handbook, Section 13.5.

You now must compose two documents: first, a one- to two-page directive for Agriflex's 2,500 employees (everyone from senior executives to clerical staff) to make them aware of the current problem, the e-mail policy, and the points you have listed. Second, you will need to compose a memo to your secretary, Christine Schneider, who will distribute the directive to the 2,500 employees. You must explain to her how you want the directive distributed—by letter, e-mail, memo, or presentation—and why you are choosing this means of distribution.

Remember that your purpose in writing the directive is to inform employees about this policy and that not everyone misuses the e-mail system. You must decide on the appropriate information to include, the appropriate tone, the appropriate genre, and the appropriate means of distribution.

CASE STUDY 2

Reporting Progress on the 2014 Winter Olympic Games in Sochi, Russia

You work as an assistant to the executive director of U.S. Figure Skating. In 2007 it was announced that the 2014 Winter Olympics will be hosted by Sochi, Russia. The executive director asks you to put together a progress report on the activities of the host city, which will be distributed to potential sponsors of the figure skating team. The executive director realizes there are too many facets of the Olympic Games for you to report on, so you are asked to focus on the progress of Olympic Village, venues for figure skating events, transportation around Sochi, lodging in Sochi, and security for the games. You will need to conduct research on these aspects of the 2014 winter games and create a progress report that is clear and appropriate for audience needs. Use layout/design principles and visuals that you believe are necessary for this audience.

CASE STUDIES ON THE COMPANION WEBSITE

Reporting Changes in the Book Business at Pearman College

As vice president of business affairs at a small college, it is your job to oversee the campus bookstore. For years you have done everything in your power to help improve the relationship between the bookstore and college professors, but you finally decide to make some large-scale changes. In this case you must decide on ways to report those changes to various stakeholders.

New Incident Reporting Procedures for Arctic Blast Heating and Air Conditioning

In this case you assume the role of an owner of a heating and air conditioning company. Because you oversee many people who encounter various kinds of problems on a daily basis, you decide it is necessary to develop a standardized incident report form. You must also consider ways to train employees to use this form.

Construction Delays at the Oklahoma State Penitentiary

In this case you serve as the general manager who oversees all facets of a large construction company's work on new prisons. Your crew faces many problems with one particular project, and you must report these problems in a progress report to your supervisors. Although you know that you cannot simply cover up these problems, that you must tell the truth, you also must decide how these problems are going to be solved so that you can show your supervisors that the situation is under control.

Game Over? Issuing Reports Regarding Faults in the New 2-Sphere Video Game System

As an employee of Mega-Rom Electronics, you must produce both internal and external informal reports to help your company maintain a positive image with various stakeholders in the wake of problems with the company's new Z-Sphere Video Game System.

VIDEO CASE STUDY

DeSoto Global

Writing a Progress Report at DeSoto Global

Synopsis

DeSoto Global has secured a contract with Shobu Automotive and is negotiating with two other automotive manufacturers. However, the company has not been successful yet in partnering with a security company to tie DeSoto's Tracker1 auto-theft tracking system in with local law enforcement. Nichols explains that she wants a progress report that explains any successes and setbacks in negotiations with security companies. She needs to report this information to DeSoto executives, who in turn will soon be meeting with investors to discuss the company's progress in Asia. Despite the fact that Nichols is asking only for an informal progress report, information in that report will be used by those at the top of the company.

WRITING SCENARIOS

1. Choose a project currently going on at your campus—a building project, a policy change, a proposed change in curriculum, or any other project currently being implemented. As a group, research this project: its goals and objectives, its history, its current status, and its future outcomes. Then compose a progress report for students that makes them aware of how the project is coming along. Assume that the report will be distributed through e-mail.
2. Reread the oral progress report presented in Figure 21.6. What do you believe were the main purposes of this report? Be sure to identify particular sections of the report to support your assertions. From what you know about this report and its audience, why do you think it was presented as an oral report rather than a written one? Compose your response in a document type assigned by your instructor.
3. Consider recent experiences that have happened in your life and that would be worthy of an incident report. Choose one of these, and write an incident report to the individual(s) who most need to know about it. After composing the report, discuss in a small group what makes the incident report effective as a written document instead of an oral one. Are there any reasons ever to give an oral incident report without documenting it in some written format?

4. This chapter discussed the Vancouver Organizing Committee's progress report for the 2010 winter Olympics. Research another significant Olympics report, current or past, and explain its rhetorical situation as thoroughly as possible: the writer, the audience, subject matter (problems), purpose, tone, and persuasive strategies. Compose your responses in one of the note-taking formats described in chapter 6 or another format suggested by your instructor.

5. In a local or national newspaper, find a news story about an emergency, disaster, or problem that interests you. Read the story, and collect as many details as you can. Then compose an incident report in memo format for an audience of your choosing that you believe needs to be informed about this. Use an appropriate tone and style for the audience and your purpose.

6. From what you already know about your future profession and what you can find out through research, describe in a memo to your classmates the kind of informal reports that are most common in this profession. In a document type assigned by your instructor, explain why these particular kinds of informal reports are so important in your particular profession.

7. Examine again the second case study in this chapter. Given the audience, purpose, and content of the progress report, what are the specific ethical concerns that the writer would have to consider? Is there anything about the rhetorical situation that the writer would have to be especially careful to address in order to be ethical? Explain your responses in a memo to your classmates.

8. Lab reports are often written by those in the natural sciences—biology, chemistry, and physics, for example—who deal with objective and unbiased information and data gathered through scientific methods. What particular ethical concerns must those doing lab work and writing lab reports be aware of? Record your responses in a set of notes, and be prepared to discuss your answers in class.

9. Write a progress report to the instructor of this course, discussing your progress so far this semester. Remember that progress reports do more than simply explain where things are now; they also provide overviews of progress, describe problems that have been encountered, and forecast future activity. You'll need to decide on an organizational pattern.

10. Find a paper or an essay (not a lab report) that you have written for another course—English, history, philosophy, or psychology, for example. Review what you wrote and why. Then take the information in the paper or essay, and reformat it in the style of a lab report. Finally, write a memo or a letter to your instructor, explaining the strengths and limitations of the lab report format for this subject matter. Does the format work for subjects other than science or engineering? Why or why not? Do you find this format more comfortable or more constraining? What other kinds of disciplines besides science and engineering could you see benefiting from the lab report format? Explain.

11. Locate a progress report, lab report, directive, or incident report of any kind. Take notes on how the document was designed for its audience and purpose; in particular, how is it organized, what genre is used, what is its tone, and does it use any particular layout and design elements? Are the design elements appropriate, given the document's audience and purpose? Explain briefly in a set of informal notes.

12. Of the four kinds of informal reports discussed in this chapter—progress reports, lab reports, directives, and incident reports—which do you believe generally best lends itself to the following technologies: e-mail, websites (Internet and intranet), and videoconferencing? Conversely, which generally lends itself least to these technologies? Write your responses in memo format to your classmates.
13. Look again at the first case study in this chapter. Explain how and why each of the following technologies would or would not work in that particular rhetorical situation: e-mailing the directive, sending it as a hard-copy memo, writing it in letter form, posting it on a website, or delivering it in an oral presentation. With each of these different means of distribution, what are the advantages and disadvantages? Reiterate why you chose the means of distribution that you did. Compose your responses in a document type assigned by your instructor.
14. Research two or three companies that provide software and hardware to companies and organizations—IBM, Microsoft, Macromedia, for example. What are some of the recent technological developments that would make writing and distributing informal reports more efficient and effective in workplace contexts? In a memo to your instructor, explain some specific information about these new software and hardware products that shows how they can change the production and distribution of documents such as progress reports, lab reports, directives, and/or incident reports.

PROBLEM SOLVING IN YOUR WRITING

For either case study in this chapter or any of the writing scenarios you were assigned by your instructor, outline the particular points in the problem-solving approach that influenced how you developed your answer. Explain in detail why these particular components were important in your thinking and writing.

22

Formal Reports

CHAPTER LEARNING OUTCOMES

After completing this chapter, you will be able to do the following:

- Recognize the major differences between **informal and formal reports**
- Understand how formal reports are **problem-solving documents**
- Establish **criteria** as standards of judgment in formal reports
- Know how to write **recommendation reports**
- Know how to write **feasibility reports**
- Know how to write **evaluation reports**
- Understand **audience needs** when composing formal reports
- Recognize the need for **research** in formal reports
- Understand the role of **visuals** in formal reports
- Recognize that many formal reports are written **collaboratively**
- Know the appropriate means to **distribute** formal reports to audiences
- Know the three main components of formal reports: **front matter, body, and end matter**
- Know the individual **sections** of these main components
- Understand the **ethical issues** involved in formal report writing

DIGITAL RESOURCES

On the Companion Website www.prenhall.com/dobrin:

- Case 1: In Violation of Federal Regulations: An Internal Review Reveals Problems at Berkshire Med-Tech Corporation
- Case 2: Swiss Sweets: Time for an Upgrade and a New Look?
- Case 3: Flawed Tests and Faulty Parts: An Engineer Develops Solutions for a New Device
- Case 4: Maintaining the Greens: Acquiring a New Maintenance Shed for Hampshire Greens Golf Course
- Video Case: Writing a Progress Report at DeSoto Global
- PowerPoint Chapter Review, Test-Prep Quiz, Exercises and Activities

REAL PEOPLE, REAL WRITING

MEET SOFIA PEREZ • Marketing Analyst for Sabroso Foods Limited

What types of writing do you do as a marketing analyst for Sabroso Foods?

My company packages snack foods from U.S.-based companies and distributes them abroad, mainly in Spanish-speaking countries. Much of my work as a marketing analyst involves researching the different foods and beverages available in our network and determining if and how they can be sold worldwide. A wide range of documents are part of this process, including many e-mails, memos, letters, proposals, and informal and formal reports.

Who are the typical readers of your workplace documents?

My readers vary with the purpose of the document. Many of the e-mails and letters I send are written to vendors in Spain, Mexico, and Central America, and my goal is often to determine if they could sell our products in their stores and vending machines. Most of the longer documents I create are written to our company managers and administrators, who make decisions about how we should market, advertise, package, and distribute snack foods abroad. To aid these executives, I create proposals and reports that inform them about different products, markets, and vendors.

What different types of reports do you write in your workplace?

I write a number of different reports that make recommendations about or evaluate our company's plans to distribute snack foods. For example, I recently completed a recommendation report in which we identified several Spanish vendors who might be interested in carrying a line of cornmeal crackers. The report recommended three vendors, based on the size of their market, our previous relationships with them, and a projected sales price for the final product. Like many of the reports I create, this report was written with a team of researchers, analysts, and sales managers. This is just one example of the kinds of reports I write on a regular basis.

What is your process for writing a report?

Many of my reports, including the recommendation report I just mentioned, can be long and complex and require lots of research and collaboration before any writing gets done. I'd say I devote about twice as much time researching and planning as I do actually writing. I am often able to enlist the help of others within my company to do the research, and sometimes we will each write different sections of a longer report. We usually work together to revise and edit the final document to make sure our style, tone, and message is consistent throughout the document.

What have you learned about writing reports as a marketing analyst?

I've learned that complex reports are often written to many different readers, each of whom has different needs. Because most of my reports are written primarily for managers and executives, it is important to write an introduction that gives them the important facts, details, and information right up front. They are often busy and may not read an entire report before making a decision. However, the other components of a report are also important, since an executive may forward the report to someone else in the company who will read it for a different purpose.

What advice about workplace writing do you have for the readers of this textbook?

Learn to work with others. So much of the writing I do involves collaborating with people inside and outside my organization, and being open and receptive to others' input is crucial to success. If you are writing with a team, try to recognize and understand each person's perspective, since each may contribute something new and important to the document you are creating.

INTRODUCTION

In December 2007 the general public was given access to a formal report that brought together the worlds of professional sports, medicine, law, and government. Succinctly dubbed "The Mitchell Report" by the media because its lead author was former Senator George Mitchell, the official title is "Report to the Commissioner of Baseball of an Independent Investigation into the Illegal Use of Steroids and Other Performance-Enhancing Substances by Players in Major League Baseball." The purpose of the report was to investigate how and why steroids and human growth hormones became a problem in major league baseball, to link particular players to the use of performance-enhancing drugs, to explore the health and legal effects of these drugs, and to make recommendations to the Major League Baseball organization about how to handle past, present, and future use of these drugs by players.

The Mitchell Report is a complex and lengthy document (409 pages), which was debated widely after its release. Because the report focuses on controversial and illegal behavior by professional athletes, the report generated a great deal of public attention. In fact, speculation about the report's contents was rampant even in the months before its release, which spawned interest and anticipation in many people (mainly journalists and sports fans) who would otherwise probably not have been particularly excited about a lengthy report. The report led to a range of news conferences, press releases, lawsuits, legal inquiries, and a lengthy congressional hearing featuring testimony by major league baseball players, including the famed Yankees pitcher Roger Clemens (see Figure 22.1).

Although most formal reports do not garner this sort of public attention, tens of thousands of such reports are produced each year. Even though most are not anxiously awaited by large numbers of people, formal reports are often the documents that seek to solve some of the most serious and complex problems that face companies, organizations, agencies, and institutions. Just because we rarely hear the media discuss formal reports, we should not think that they are rare occurrences. In fact, it is highly likely that you will someday be involved in the writing of a formal report in your workplace.

This chapter covers what formal reports are, what different types exist, who they are written for, when they are needed, and the kinds of problems they can explore and solve. Formal reports tend to be longer than informal reports, more thorough and more detailed, formatted in distinct ways (not written as memos, for example), and often composed for multiple audiences.

By using the term *formal reports*, as distinct from *informal reports*, we are not suggesting that formal reports are all the same, requiring all the same components

FIGURE 22.1 Baseball player Roger Clemens at a congressional hearing on performance-enhancing drugs

or including the same kinds of subject matter. Instead, the difference between formal and informal reports centers mainly on the kind of problems that need to be solved and different audience expectations. Formal reports typically work to solve large, complex problems and, thus, require more work than informal reports. Of the many varieties of formal reports, this chapter focuses on three in particular: recommendation reports, feasibility reports, and evaluation reports. As problem-solving and rhetorical documents, these three share some commonalities, yet each does something different—that is, solves a different problem in a unique way—and writers must be attuned to their distinct facets.

All formal reports require writers to be familiar with the particular sections and parts of such reports (e.g., title pages, background, summaries) and to recognize what audiences need and expect when reading a formal report. However, despite the specific parts and pieces that make up formal reports, the type of critical thinking and research that goes into writing a formal report is actually not all that different from much of the thinking and research that people do in their everyday lives. There may be a difference in degree but not in kind.

For instance, in the last few months, you may have engaged in one or more of the following activities:

- Purchased a new car, computer, or other product after carefully weighing the pros and cons of different brands or models
- Made a careful decision about whether to go on a trip, vacation, or weekend getaway during a busy time in your life
- Reflected on the outcomes—positive and/or negative—of a big decision that you had made previously, such as choosing a certain college, taking on a new job, getting involved in a particular activity

When we consider which is the best of many choices and offer our thoughts on why that choice is best, we are making recommendations. When we decide whether something is possible—or is a good idea—based on various factors and contingencies in our lives, we are deciding on the feasibility of a plan. And when we think about the goods and the bads that have derived from decisions and choices we made in the past, we are engaged in evaluation. Understanding how we make recommendations, consider feasibilities, and evaluate choices in our own lives is an important part of writing formal reports.

In most of these examples, we have served as both writer and audience; our conclusions were composed *by* us but also meant solely *for* us. When writing formal reports, workplace writers engage in these same kinds of critical thinking activities, but the audience will more typically be external clients, internal colleagues, managers, executives, a more general public, or some combination of these, as was the case with the Mitchell Report. Thus, formal reports—whether recommendation, feasibility, or evaluation reports—must be composed accurately and ethically with the intended audience in mind at all times.

IN YOUR **EXPERIENCE**

List and describe any recommendations, feasibility assessments, or evaluations you have made in your own life during the last few weeks. Did you simply think about these things, or did you discuss them with others? Were you required to do any kind of writing and/or research to arrive at a solution or conclusion? Explain your answers in a format prescribed by your instructor.

CRITERIA

criteria the standards or guidelines used to make a decision about something

Before we focus on how to write different kinds of formal reports, it is important to first discuss a crucial characteristic of formal reports, one that lies at the core of all formal reports and determines their success or failure. This crucial characteristic is *criteria*, which are the standards we use to make recommendations and evaluations. They are the guidelines or principles by which we judge something. Whether we know it or not, we use criteria every day when we assess things.

For example, if a friend asks you to recommend a good restaurant in your city, you couldn't answer that question appropriately unless you first understood the set of criteria that friend was using. The concept of "good restaurant" is open to interpretation, and a good restaurant in one situation may not be good in another. To recommend a good restaurant, you'd have to ask questions such as "How much money do you want to spend?" "How much time do you have to eat?" "Are you going to dine with family or friends?" "Is this a date?" A good restaurant for a few work colleagues who want to grab a quick bite for lunch is completely different from a good restaurant for someone who wants to impress a date. In short, you'd have to know more about what your friend wanted in a restaurant—that is, the criteria.

Situations are not always this simplistic, nor are criteria always subject to good taste or general impressions. Criteria often depend on the particular problem that needs to be solved as well as the audience needs and expectations. But the one thing that all formal reports have in common is that they require writers to develop criteria by which they will make evaluations and/or recommendations. Sometimes it takes detailed research to develop the necessary and appropriate criteria.

Let's look at a second example. Suppose that you recently took a job in the purchasing department of Agriflex, a large agricultural firm in South Dakota. The president of the company wants to provide company vehicles to thirty-five of the company's top executives, who often travel a great deal between the office and the many farms and ranches in South Dakota and surrounding states. You have been asked to write a report that recommends three different vehicles that executives can choose from to use as their company vehicle. The president of the company has limited the cost of each vehicle to no more than $45,000.

In situations like this, there are sometimes obvious, explicit criteria, such as the $45,000 limit on per-car cost. However, there also tend to be implicit criteria that must be developed with critical thinking and research. In this particular example, you could develop other criteria by looking carefully at the rhetorical situation. For instance, because the executives will put a lot of miles on these vehicles, you might consider good gas mileage as one criterion. And since they'll be driving the vehicles over South Dakota farmlands, you might consider vehicles that are durable and can function well in that terrain and climate. In addition, these are top-level executives, so the vehicles should provide the comforts and luxuries that executives expect.

However, you cannot develop these criteria in any detail without conducting research. Good gas mileage, durability, and luxury may be criteria, but they are too vague to be useful. What does "good" gas mileage mean? How do you define "durability" or "luxury"? To refine the criteria, you will have to research appropriate sources. For gas mileage, for example, you can look at government sources such as www.fueleconomy.gov, which provides the gas mileage ratings of vehicles in all classes: small cars, family sedans, coupes, convertibles, small SUVs, large SUVs, and so on. You might decide that one criterion will be that the vehicle gets the average or better fuel mileage of all vehicles in its class, according to www.fueleconomy.gov. To develop a list of luxury options suitable for company executives, you would probably have to survey or interview company executives to determine what best suits their needs and desires.

Every formal report requires writers to develop at least three or four criteria to be used as standards of judgment for evaluation or recommendation. These criteria are unique to each problem and situation and are not determined arbitrarily or haphazardly; rather, they are the outcome of critical thinking and research. We will discuss criteria further as this chapter continues.

RECOMMENDATION REPORTS

At their most basic, recommendation reports answer the question What option should be chosen? Or, out of many options, which is best? Recommendation reports derive from situations in which a company, group, or organization has a recognized problem or need that can be addressed or solved by making some kind of choice—for instance, purchasing a product, hiring new workers, reorganizing the company, selling certain assets, or opening new stores. When companies and organizations are faced with difficult choices, those who write recommendation reports are trying to help the company or organization make the best choice, which typically involves comparing and contrasting the different choices, drawing conclusions, and making the actual recommendations.

The formal report entitled "The Power of the Internet for Learning" was composed by the Web-Based Education Commission—made up of politicians, educators, and business executives—for an audience of diverse individuals, such as the president and congressional representatives. The report's purpose is to look at the various ways Internet technologies can be used to help students become better learners from prekindergarten through twelfth grade and to make recommendations about how best to implement Internet technologies in the nation's schools. A portion of the 185-page report is included in Figure 22.2.

See the Companion Website for a full copy of this report.

There is no doubt that this is a formal report; it assesses a great deal of information, includes a professionally enhanced cover, and tackles serious subject matter. In addition, the report should be considered formal because it is being presented to the president and the Congress of the United States. Such audiences expect a document that is detailed, thorough, clear, and professional, with a visually appealing cover, an easy-to-follow organizational structure, and coherent recommendations supported with good reasons. Furthermore, tackling a problem or issue like the role of the Internet in the future of education surely necessitates more than quick, uninformed recommendations in a memo or e-mail.

Not all recommendation reports are as long as this one, but they all tend to work toward the same problem-solving goals:

- They understand the different facets of the problem at hand.
- They recognize the audience of stakeholders who are influenced by the recommendation report.
- They recommend a choice(s) that best fit the criteria and suit the needs of the company, organization, or stakeholders themselves, depending on the nature of the problem.

These real-world examples illustrate when workplace writers might need to compose a recommendation report:

Based on the findings of our work, the Commission believes a national mobilization is necessary, one that evokes a response similar in scope to other great American opportunities—or crises: Sputnik and the race to the moon; bringing electricity and phone service to all corners of the nation; finding a cure for polio.

Therefore, the Commission is issuing a call to action to

Make powerful new Internet resources, especially broadband access, widely and equitably available and affordable for all learners. The promise of high quality web-based education is made possible by technological and communications trends that could lead to important educational applications over the next two to three years. These include greater bandwidth, expansion of broadband and wireless computing, opportunities provided by digital convergence, and lowering costs of connectivity. In addition, the emergence of agreement on technical standards for content development and sharing will also advance the development of web-based learning environments.

Provide continuous and relevant training and support for educators and administrators at all levels. We heard that professional development—for preK–12 teachers, higher education faculty, and school administrators—is the critical ingredient for effective use of technology in the classroom. However, not enough is being done to assure that today's educators have the skills and knowledge needed for effective web-based teaching. And if teacher education programs do not address this issue at once, we will soon have lost the opportunity to enhance the performance of a whole generation of new teachers, and the students they teach.

Build a new research framework of how people learn in the Internet age. A vastly expanded, revitalized, and reconfigured educational research, development, and innovation program is imperative. This program should be built on a deeper understanding of how people learn, how new tools support and assess learning gains, what kinds of organizational structures support these gains, and what is needed to keep the field of learning moving forward.

Develop high quality online educational content that meets the highest standards of educational excellence. Content available for learning on the Web is variable: some of it is excellent, much is mediocre. Both content developers and educators will have to address gaps in this market, find ways to build fragmented lesson plans into full courses, and assure the quality of learning in this new environment. Dazzling technology has no value unless it supports content that meets the needs of learners.

Revise outdated regulations that impede innovation and replace them with approaches that embrace anytime, anywhere, any pace learning. The regulations that govern much of education today were written for an earlier model in which the teacher is the center of all instruction and all learners are expected to advance at the same rate, despite varying needs or abilities. Granting of credits, degrees, availability of funding, staffing, and educational services are governed by time-fixed and place-based models of yesteryear. The Internet allows for a learner-centered environment, but our legal and regulatory framework has not adjusted to these changes.

Protect-online learners and ensure their privacy. The Internet carries with it danger as well as promise. Advertising can interfere with the learning process and take advantage of a captive audience of students. Privacy can be endangered when data is collected from users of online materials. Students, especially young children, need protections from harmful or inappropriate intrusions in their learning environments.

FIGURE 22.2 Excerpt from a recommendation report

- The United States is increasingly threatened by the possibility of bioterrorism and does not have sufficient plans to account for and respond to such threats. The U.S. General Accounting Office (GAO) investigates and releases a report to Congress about bioterrorism, recommending how different governmental agencies can best prepare for these threats.
- A large bank and brokerage firm finds itself dealing with more and more clients who rely on it to help them make financial decisions. The bank must find a way to cheaply, quickly, privately, and accurately disseminate recommendations to clients about their personal investments and finances. The bank ends up releasing recommendation reports via secure websites.
- A city in Ontario, Canada, finds itself facing extreme and unexpected population growth. A planning committee in this city assesses the situation and recommends strategic responses to an audience of city council members.

EXPLORE

Choose a large corporation like Ford, GM, Bank of America, Microsoft, Coca-Cola, or ConAgra. Briefly research the chief executive officer (CEO) of this corporation to find out what major recommendation(s) or decision(s) this person has made for the company in the last few years.

What recommendation(s) or decision(s) were made? What were the other choices, if known? Can you find any recommendation reports that may have influenced this person? If not, why do you think such reports are not available to you? What information were you able to find about the decision(s) this person made?

FEASIBILITY REPORTS

Feasibility reports study a situation or a problem and determine whether a plan to address that situation or problem is feasible. These reports help readers decide whether it is possible (e.g., technologically, economically, or practically) to implement the plan. Feasibility reports do not always give readers a clear-cut yes or no; they may provide a maybe or suggest that more research be done to decide whether the plan is feasible. Feasibility reports, then, are similar to recommendation reports in making a type of recommendation to readers. However, feasibility reports study whether a particular plan can be implemented successfully, whereas recommendation reports provide the best choice among others, often in a compare-contrast format.

Anyone who considers whether it is possible—financially, practically, or safely—to take the family on a vacation to a specific place is engaged in a kind of feasibility study. Discussing the topic with family members, reviewing family finances, making some phone calls, and visiting some websites may allow a person to decide rather quickly whether to implement a vacation plan. In the workplace, however, feasibility reports often deal with much weightier subjects, require a great deal of research and collaboration, and consider a diverse range of audiences. For example, a feasibility report written by the Center for Agribusiness and Economic Development (associated with the College of Agricultural and Environmental Sciences at the University of Georgia) studied the feasibility of an advanced ethanol plant in Georgia. The introduction of this report, shown in Figure 22.3, identifies the audience and the purpose of the forty-nine-page document.

See the Companion Website for a full copy of this report.

This particular feasibility report treats a specific scientific and agricultural topic and is written to specialists with particular knowledge of such issues, as is often the case with formal reports. The audience is members of the Georgia Agricultural Commodity Commission for Corn and the Georgia Cooperative Development Center, and the report describes the economic feasibility of an ethanol production facility in the state of Georgia. Even though this issue might not make major news headlines, such a production facility would likely involve great sums of money for those involved. Therefore, a careful and thorough study of the situation is necessary. Individuals who read reports that involve large amounts of technical information and/or large sums of money expect that such reports are well researched, systematic, comprehensive, and professionally rendered. Feasibility reports would also be necessary and expected in the kinds of problem-solving and rhetorical situations represented in these real-world examples:

IN YOUR **EXPERIENCE**

Consider a problem or an issue that affects your campus or local community. Briefly write about how this problem or issue might be addressed or solved, and then write further about how and why such a plan would need a feasibility study and report before being implemented.

Introduction

The Georgia Agricultural Commodity Commission for Corn in association with the Georgia Cooperative Development Center contracted the Center for Agribusiness and Economic Development (CAED) to do a feasibility study on the production of ethanol in central Georgia. This study was to follow up a study that had been done two years previously, and was to expand the knowledge base of the feasibility of ethanol production in Georgia. The concern of the corn growers in Georgia was that, given the price of feedstock corn is typically higher in Georgia than in the Midwest, a Georgia based ethanol plant will operate at a competitive disadvantage to Midwest-based ethanol plants. To offset this competitive disadvantage, the Georgia Agricultural Commodity Commission for Corn was interested in examining other ethanol production options besides the typical conventional dry-grind ethanol plant model that is currently the industry standard. The study was to evaluate the potential for using alternative feedstocks as well as implementing a dry fractionation process of the corn prior to conversion to ethanol. Dry fractionation reduces the corn kernel to three parts—the bran, the germ and the endosperm. Fractionation would allow bran from the feedstock to fire the boiler for the plant, thus saving on energy costs. In addition, the dry fractionation process was theorized to increase the protein level of the co-product dried distillers grains and solubles (DDGS) from around 23 to 28 percent to around 40 to 42 percent. It was also thought to be able to increase the efficiency of the plant as well, by running just the endosperm through the plant, therefore increasing the cost efficiency per unit. The corn germ would be sold as a by-product.

The CAED subcontracted out the tasks leading to the estimation of the production costs and capital expenditures. They chose the consulting firm Frazier, Barnes and Associates to perform these functions. Frazier, Barnes and Associates (FB & A) are well known for their work in the ethanol industry and had an interest in looking at this type of "advanced" ethanol plant. FB & A also performed the initial feasibility analysis that led to the construction of the Commonwealth Agri-Energy plant in Hopkinsville, KY, that is in its second year of operation and is currently producing between 23 and 30 million gallons annually.

The current industry standard plant is a dry-grind facility where the entire corn kernel is run through the conversion process. The estimated capital cost for a 30 million gallon per year conventional plant is about $52.5 million with annual operating costs of about $48.9 million. In comparison, the estimated capital cost of an advanced fractionation plant with co-firing capacity is about $70 million with annual operating expenses of about $46.8 million.

FIGURE 22.3 Excerpt from a feasibility report

- Researchers in the Pacific Northwest witness a substantial decline in the number of salmon and steelhead in regional rivers. The Army Corps of Engineers assesses the feasibility of a measure to help increase the survival of juvenile fish through the lower Snake River Project, which includes a series of locks and dams.
- Automobile traffic between the metropolitan areas of Austin and San Antonio, Texas, is increasing at harmful rates, placing an undue burden on the Interstate 35 transportation corridor. The Texas Department of Transportation conducts a feasibility study to see whether a commuter rail is practical and cost-effective for this particular area of the state.
- Overwhelming medical evidence shows tobacco use as the number one cause of preventable cancer and cardiovascular death. Research shows that smoking is responsible for about 30 percent of cancer-related deaths and about 90 percent of lung cancer deaths. Clinical researchers at the U.S. National Institutes of Health propose conducting tests and writing a feasibility report on an intensive behavioral and pharmacologic tobacco cessation program for radiation oncology patients and their families and/or companions.

EVALUATION REPORTS

Evaluation reports usually provide a judgment—or a researched opinion—about the worth, value, or effectiveness of something. Did something work the way it was supposed to? Did the company make the correct decision? Has someone been a worthwhile and valuable employee? Recommendation reports and feasibility reports typically focus on future courses of action that may be taken by a company, organization, or group, whereas evaluation reports tend to look back at the past to explore whether previous actions or decisions were effective and appropriate. Evaluation reports are sometimes referred to as assessment reports or performance reports because they may assess how a plan or solution worked or how a plan, solution, or person performed over a given period of time.

Evaluations and evaluation reports are probably the most common kinds of reports and, in some sense, reflect the kinds of evaluative thinking many of us engage in all the time. How often do we evaluate and assess people, objects, and ideas—that new neighbor or work colleague, the mileage we get with a new kind of gasoline, a new cell phone, a new video game, or a new route home from school or work? Evaluation reports extend this habitual assessment and evaluation into detailed, comprehensive, written documents for diverse audiences.

One particular evaluation report derives from Duke University's initiative to encourage innovative uses of technology in learning by giving all entering freshmen Apple iPods. The excerpt in Figure 22.4 contains portions of the first two sections of this fifteen-page evaluation report: a summary and an evaluative overview of findings.

Summary

As part of a university initiative to encourage creative uses of technology in education and campus life, Duke distributed 20 GB Apple iPod devices, each equipped with Belkin Voice Recorders, to over 1600 entering first-year students in August 2004.

The Center for Instructional Technology (CIT) coordinated an evaluation of the academic uses of iPods, drawing on course-level feedback; student and faculty focus groups; a broad survey of first year students and faculty; and discussions and feedback among staff, administrators, and important campus stakeholder groups. This evaluation focused on the feasibility and effectiveness of the iPod as a tool for faculty and student use. The primary purpose of this evaluation was to assist project stakeholders and Duke decision makers in determining what iPod uses were most fruitful and to help shape future Duke academic technology initiatives. This report summarizes the main findings of this collaborative assessment effort.

Evaluation Findings

At least fifteen fall courses with a total enrollment of 628 unique students and an estimated thirty three spring courses with a total enrollment of over 600 students incorporated iPod use. . . . As expected, foreign language and music courses integrated the device, but its use also extended to other social science and humanities courses. . . . Recording was the most widely used feature for academic purposes, with 60 percent of first-year students reporting using the iPod's recording ability for academic purposes.

Benefits of Academic iPod Use

- Convenience for both faculty and students of portable digital course content, and reduced dependence on physical materials
- Flexible location-independent access to digital multimedia course materials, including reduced dependence on lab or library locations and hours
- Effective and easy-to-use tool for digital recording of interviews, field notes, small group discussions, and self-recording of oral assignments
- Greater student engagement and interest in class discussions, labs, field research, and independent projects
- Enhanced support for individual learning preferences and needs

Barriers to and Problems Encountered with Academic iPod Use

- Significant challenges in integrating multiple systems for content storage, access, sharing, and distribution with one another and with existing technology infrastructure
- Absence of systems for bulk purchase or licensing of commercial MP3 audio content for academic use
- Difficulties in locating commercial sources and obtaining licenses for content from independent and international publishers in appropriate formats
- Inherent limitations of the device (e.g., no mechanism for input other than synchronization, lack of instructor tools for combining text and audio)

FIGURE 22.4 Excerpt from an evalutation report

The summary hints at the report's complexity; it shows that it is written for multiple audiences, particularly stakeholders who have the power and ability to make decisions based on the findings in the report. In addition, the summary lets us know that this evaluative report is based on research, mainly primary research with students and faculty. Finally, and perhaps most importantly, the report does not simply close down conversation on the subject. It looks back at the effectiveness of iPod use by students and faculty over the course of an academic year, but these evaluations are meant to help make plans for the future. Evaluation reports would also be necessary in the following real-world scenarios:

- An engineering firm has stockholders that need to know about the company's recent (i.e., annual) successes and failures. The firm composes an annual report that discloses and assesses the firm's performance.
- Hazardous contaminants are found in soils and groundwater in Texas and different bioremediation techniques are used to treat these problems. A study is then done to assess these techniques, and an evaluation report is written to explain the effectiveness of the treatments.
- The Securities and Exchange Commission works to ensure fairness and well-being in financial tradings and investments in the United States. Because the commission's cases involve so much money, the Office of the Inspector General of the United States regularly evaluates the commission's ethical standards and integrity—from high-level managers to low-level staff members—and produces evaluative or performance reports on the commission.

There are distinctions among the three kinds of reports, but sometimes the lines between them blur. For instance, recommendation reports are often written on the basis of previous evaluation reports, and feasibility reports are often written after workers have been given a particular choice or plan that was suggested in a recommendation report. Thus, these three kinds of documents are often produced in cooperation with one another. As we have said repeatedly, technical documents often affect other documents, and many formal reports are actually combinations of different types, such as the Mitchell Report. In fact, it is not uncommon for a single formal report to include an evaluation of something in the past, a recommendation of a future plan, and a feasibility study of that plan—all in one.

EXPLORE

What kinds of evaluation reports have been written at your university during the last year or two? Find at least one, read through it, and summarize the details. What in particular makes it an evaluation report? Convey your responses in a document type assigned by your instructor.

GUIDELINES FOR COMPOSING FORMAL REPORTS

Like other workplace documents, formal reports are specific to their particular problems and rhetorical situations. Therefore, there is no one single formula to follow when you are producing a formal report. However, you should use problem-solving and rhetorical strategies in all cases. The rest of this chapter provides and discusses a sample formal report to help you understand how formal reports solve problems and how writers make rhetorical choices that best suit the problem and the audience needs and expectations. This sample report, produced by the Army Corps of Engineers, is entitled "Improving Salmon Passage: Lower Snake River Juvenile Salmon Migration Feasibility Report/Environmental Impact Statement." The complete report is over

To see the entire salmon passage report, visit the Companion Website.

fifty pages long; the sample in this chapter contains some of the most relevant of those pages, which appear in Figures 22.5–22.17 on pages 679–92.

Address Audience Needs

PSA

Like other workplace writings, formal reports are driven by particular problems and their audiences. The Plan component of the problem-solving approach reminds you to review the rhetorical situation: define the problem, identify the audience and stakeholders, and understand your purpose by understanding what these individuals need and expect.

Workplace writers must recognize that some formal reports will be used by different audiences with different levels of expertise and different reasons for wanting or needing the document itself. In the report on improving salmon passage, the Army Corps of Engineers addresses not only state and federal legislators, wildlife experts, and engineers, but also concerned citizens of the region (see Figure 22.7 on page 681.) Consequently, the report provides a great deal of background information and uses a writing style accessible to people with different levels of expertise.

See chapter 2 for more on audiences and chapter 5 for more on transnational audiences.

ANALYZE THIS

Reread the letter presented in Figure 22.7. Given the content of this letter and the purpose of the report, why do you think the author addressed the letter to "Concerned Citizen"? What impact does this have? Is it effective or not? Explain your responses in a written format assigned by your instructor.

Conduct Necessary Research

Most formal reports are detailed and comprehensive documents that attempt to answer questions an audience might have about the problem or issue at hand. Therefore, formal reports are typically well researched and documented, providing accurate information and facts to support any recommendations or evaluations. Workplace writers need to know which kinds of research are necessary for their individual projects. For instance, the Duke University study of iPods in academic studies relied on student and faculty interviews and surveys, whereas the study of salmon migration in the Pacific Northwest relied on scientific studies and experiments as well as data from civil engineering (i.e., dam construction). Figure 22.12 on page 686 describes the research processes the writers used in the salmon passage report.

See chapter 6 for more information about research.

Include Helpful Visuals

One thing that makes formal reports "formal" is that they often provide readers with visual aids and graphics that make reports more readable and usable. There is no set number of visual aids, nor are there particular types of visuals that a report must include. Workplace writers include visuals that actually help readers make sense of the information being presented; visuals and graphics should not be used gratuitously or indiscriminately. Remember that visuals do not simply include pictures or photographs but charts, graphs, and tables as well. Several pages in the "Improving Salmon Passage" report illustrate the effective use of visuals.

See chapter 8 for more on using visuals in technical documents.

For instance, the visual on the cover of the report (see Figure 22.5 on page 679) is not entirely necessary for audiences to understand the content of the report, but it helps establish the credibility of the report because of its professional caliber. Carefully produced high-quality visuals, combined with high-quality writing, can help persuade readers and make a report stand out. However, no fancy visuals can make up for poor writing and planning.

Using visuals in a report is part of the Plan stage in the problem-solving approach, which focuses on choosing persuasive strategies. In this sense persuasion involves letting an audience *see* information as well as read it. For example, the salmon report details the movement of salmon through different dams on the lower Snake River. One section of the report describes each of these dams and provides readers with aerial photos of them all (see Figure 22.11 on page 685) to help readers visualize and better understand information about the dams and their relationship to the salmon.

Collaborate, Draft, and Revise

Formal reports are often long, complex documents; many are fifty to one hundred pages in length, although many others are shorter or longer. Because of their length and complexity, formal reports are often composed collaboratively by teams or committees, with each member having a particular role or responsibility in producing the report. The introduction to the "Improving Salmon Passage" report (see Figure 22.8 on page 682) demonstrates this collaboration. The third paragraph in the introduction credits the many "scientists, engineers, and economists" who collaborated; one can only imagine the many e-mails, memos, informal reports, and presentations that took place over the course of a six-year-long study. In addition, we can safely assume there were numerous writers, photographers, and layout and design staff involved. Workplace writers should expect that their work on formal reports will be collaborative; recommendations, feasibility studies, and evaluations usually are team efforts not often left to a single individual.

See chapter 7 for more on drafting and chapter 10 for more on revising and rewriting.

Like many other documents, formal reports are not necessarily composed in linear fashion. They can be, and likely should be, written "inside out," meaning that writers often begin in the middle with body sections or with research and data. In addition, formal reports are often written in multiple drafts, revised many times until finalized. Notice on the cover of the salmon report (Figure 22.5) that the word *Final* is prominent, implying that previous drafts have existed. In fact, the introduction (Figure 22.8) refers to an earlier draft that was extensively reviewed. We can conclude that the final version was the result of considerable revision and rewriting.

Distribute Professionally

Formal reports are important documents that often result from much care and attention. Therefore, most workplace writers who compose formal reports want their finished products to reflect their careful work and to appear professional and finished. As mentioned in the Distribute stage of the problem-solving approach, how formal reports are produced and disseminated to their audiences is important to consider. Formal reports are typically sent as hard copies, printed on quality paper, and illustrated with color, when applicable. This practice, however, does not mean that formal reports should never be sent as e-mails or uploaded onto web pages; in fact, many companies and organizations have standard methods for making reports available electronically, and some audiences may prefer and expect this form of distribution. As technologies progress to make electronic documents more professionally rendered and accessible, you can expect that this form of distribution will increase, particularly since printing and binding costs for lengthy, professional documents can be quite high.

See chapter 3 for more about using technology to distribute documents.

ANALYZE THIS

Choose three of the visuals or graphics included in the salmon report, and discuss with classmates whether these visuals are necessary for the report. How do they contribute to the audience's understanding? Are they gratuitous? Do they make the report more interesting and appealing? You will have to read portions of the report to understand the contexts of these visuals.

STANDARD PARTS OF FORMAL REPORTS

Most formal reports include the three standard parts—front matter, body, and end matter—each of which can be broken down into different sections. But every report responds to its own rhetorical situation, which calls for a specific and unique structure. Therefore, writers must be attuned to what their audiences need and, consequently, which sections their reports must include. In the descriptions that follow, you will notice some similarity to the sections of manuals, proposals, and informal reports discussed in chapters 19, 20, and 21, respectively.

Front Matter

Front matter includes all the information that identifies the report, the writer(s) of the report, and the components of the report.

Front Cover

Front covers can use images and pictures to help make them appealing to audiences. However, some companies, organizations, and institutions have stock front covers that must be used in all formal reports, usually including such things as company logos or seals. Beyond visuals, front covers generally include three primary pieces of information.

Titles The front cover should include the title of the report, something specific that tells readers exactly what they will be reading about. Note the full title of the salmon report:

> Improving Salmon Passage: Lower Snake River Juvenile Salmon Migration Feasibility Report/Environmental Impact Statement

Although this may not be the most provocative title ever written, it does give readers a clear understanding of the report's purpose.

Titles initiate a conversation with readers, identifying what a report seeks to accomplish. Thus, report titles should be clear in their statement of purpose and, when applicable, should identify the context for which a report is written. In other words, titles may also indicate the audience to which a report is directed.

Byline As a title begins a conversation with a reader, bylines let readers know with whom they are entering into that conversation. Report covers should include information about who wrote the report, whether a single author, a group of authors, or an organization or company. In some cases so many individuals may have collaborated in producing a report that it is more feasible to list a single organization.

Date Like other documents, reports become artifacts within company records. The date when a report was submitted places that report in a historical context for the reader, identifying when the report was produced in relation to the problem it attempts to solve.

Table of Contents

Tables of contents are necessary for most formal reports because these reports are typically lengthy and include many sections and subsections. A standard rule is that a report longer than ten pages should include a table of contents. Readers need to be able to navigate through formal reports easily, and a table of contents helps guide them to particular information. Writers should make sure that section headings, subheadings, and, if applicable, chapter titles appear in the table of contents as they appear in the report itself. Look at Figure 22.6 on page 680, and note that the table of contents of the salmon report lists even those sections that are only one page in length.

Lists of Tables, Figures, and Symbols

It is necessary to list all tables, figures, and symbols used in a report if the report includes more than five of these. If the lists of tables and figures are short (i.e., fewer than five each), they may be combined in one list, although a list of symbols should always be separate and composed in alphabetical order. If the report includes many tables and figures (i.e., five or more of each), writers should compose separate lists of tables and figures, again keeping the list of symbols separate.

Body

Again, all reports handle body sections a bit differently. Some may include all of the following, whereas others may include only a few. Writers must decide which sections are necessary to help solve the problem(s) at hand and meet audience needs and expectations.

Executive Summary

Some formal reports contain an executive summary, others do not, and ocassionally, executive summaries are seperate documents that precede or accompany a formal report. Although some people equate executive summaries and abstracts, they are not the same thing. Abstracts are highly generalized summaries of a report, typically a lab report, and are written in a few hundred words. Executive summaries, on the other hand, may be longer and more substantial, although no more than 10 percent of the length of the original. Thus, a one-hundred-page report could include a ten-page executive summary, although there is no rule that it *must* be this long. The overall purpose of the executive summary is to encapsulate the report into a shorter amount of space. Some reports may even be arranged according to an executive plan, in which case the conclusions and recommendations sections are moved to the front of the report, allowing a busy executive to see the most important information first.

As is always the case with workplace documents, audience is crucial. Workplace writers must decide when executive summaries are necessary. The salmon report, for example, does not include an executive summary, whereas the Mitchell Report does. In fact, the executive summary in the Mitchell Report is an exceptionally long forty pages (the entire report is 409 pages).

Introduction

The introduction indicates what kind of document the audience is reading: a recommendation report, feasibility report, or evaluation report. It also alerts readers to the report's overall purpose and, in some cases, may provide an overview of the contents of the report. The introduction should be clear, concise, and no longer than a page or two. Introductions in formal reports function much as abstracts do in lab reports (see chapter 21); they generally discuss what is to follow. Notice that the introduction to the salmon report (Figure 22.8) is relatively short, only one page out of a fifty-five-page document. It briefly discusses what the rationale for the report is and how the report came to be.

Description or Definition of the Problem

Because formal reports are problem-solving documents, many of them state the particular problem the report addresses in order to better explain and highlight the purpose of the report. Not all formal reports include this section, although many audiences appreciate it. Like the introduction, a description or definition of the problem does not need to be lengthy. If it takes writers many pages to describe the problem, they may need to write a different report altogether—one devoted solely to defining the problem. See Figure 22.9 on page 683 for a section that defines the problem in the salmon report.

Technical Background

Some reports require technical discussion to make the rest of the report meaningful to readers, particularly those readers who do not already have the necessary technical understanding. Technical background sections are not necessarily labeled as such. Rather, some reports simply provide appropriate sections of information that address the technical concerns that writers believe audiences will have. Depending on the audience, background information may be highly technical and sophisticated or simple and uncomplicated. The salmon report, as you may recall, was written for legislators and concerned citizens, many of whom would not have a strong background in fish populations and migrations, rivers, and dam operations. Therefore, one section of the report (Figure 22.10 on page 684) provides technical background information about how a dam works. The background information in this example is somewhat simplistic, geared for nonspecialists; it defines basic terms and presents readers with a visual aid. Even the section title, "How the Dams Operate," is simple and straightforward to help readers flow smoothly from one section to the next.

EXPLORE

The Mitchell Report, which was discussed earlier in this chapter, can be accessed on the Companion Website at www.prenhall.com/dobrin. Go to this web address, and skim through the table of contents to locate a section filled with technical information that some readers might not understand at first glance. How does or doesn't the report provide its audiences with technical background information so that they can fully understand the report's main points? Does the Mitchell Report do a better or worse job of explaining technical information than the salmon report? Explain in a short document assigned by your instructor, or discuss in a small group.

Criteria

As previously discussed, criteria are a necessary part of formal reports that present final recommendations or evaluations. All formal reports must make clear to audiences what criteria form the basis of the report's standards of judgment. Most reports include a criteria section, although not all do so explicitly. We do recommend that you include an explicit criteria section in all your formal reports in order to make clear what criteria you are using as well as how and why you chose those criteria.

Criteria can be formulated in different ways:

- *Numerical values:* Some criteria can be presented as maximum or minimum numerical values. Numerical values can help quantify how much something costs, how many hours something would take to accomplish, or even the number of square feet something might needed.
- *Yes/no values:* Some criteria can be presented simply as yes-or-no questions. In these instances, choices may be rejected if they do not garner sufficient yes answers on a list of requirements.
- *Ratings values:* Some criteria are set by others—perhaps outside ratings groups—who may assign comparative values, such as "good" or "best." For example, organizations such as *Consumer Reports* use their own criteria to assign comparative ratings to various products. A writer who believes such an organization to be credible may use its criteria and ratings as part of a formal report but should include an explanation and justification of the criteria nonetheless.

It is important that audiences accept and trust the criteria used. In most cases the more complex the problem is, the more complex and detailed the criteria should be—as well as the description and justification of those criteria. In all cases the criteria should be clearly and consistently presented throughout the report.

The salmon report provides a resource list of thirteen main categories that serve as a kind of criteria. All the options—called alternatives in this report—are evaluated in terms of how they will affect these thirteen categories. Thus, these categories are the standards by which the options are judged and the best option is ultimately recommended. Figure 22.15 on page 690 shows a table that indicates how the different options impact, these criteria categories.

Discussion of Options

In formal reports, primarily in recommendation reports, writers need to explain how they arrived at the different options or choices that the report is considering. This discussion may include brief descriptions of the options themselves to help readers better understand, later in the report, why one was chosen over the others. However, the discussion of options is not comparative at this point. In the salmon report the Army Corps of Engineers investigated four possible options, or alternatives, to the problem they faced (i.e., the decline of salmon populations in the Pacific Northwest). Even though the report clearly recommends one of the four options, the writers thoroughly discussed all four in order to better highlight for readers how and why they selected the one that they did. You can examine this discussion of options in Figure 22.14 on pages 688–89.

ANALYZE THIS

Look again at Figure 22.14. The authors compare the different options for solving the problem and have chosen to do this with a particular kind of layout and design. Given the purpose of this section, the audience, and the purpose of the report as a whole, what makes you think this layout and design is effective or ineffective? Explain your response with specifics from the figure. You may wish to refer back to chapter 9 before proceeding with this exercise.

Comparison

Another necessary component of most recommendation or feasibility reports is a comparison of the options or choices; evaluation reports typically are not comparative and therefore would not need such a section. A comparison section allows your audience to see your line of thinking, your rationale for the decision you are making. Comparison sections should usually use a category-by-category organization, rather than an option-by-option arrangement, and should identify which option is the best for each particular category. When that is not clear-cut, writers may have to provide more than one possible conclusion.

The salmon report looks at various solutions to the problem of diminishing salmon populations passing through the dams on the Snake River. The report does not look at each of the four options in full but instead looks at each in terms of its effects on various key environmental resources and economic factors. Look at the table of contents again (Figure 22.16) to see the list of categories affected. Because each of the four alternatives is examined in terms of each of these categories, the report has a highly comparative feel. Figure 22.13 on page 687 includes the section of the report that looks at the four alternatives in terms of their effect on one of the thirteen categories—Native American Indians.

Conclusions

The conclusions section of a formal report is a summary of the conclusions already reached in previous sections. Conclusions sections, however, must do more than simply restate what has already been said; they must also resolve any confusion or contradiction from previous sections. For example, the salmon report may highlight the benefits of various solutions, but the conclusion should make clear why one solution would be preferred in a given situation. Your conclusions section must resolve such contradictions in order to ultimately justify your recommendation or assessment, particularly when no one option is clearly the best.

In the case of evaluation reports, for example, a writer might review company sales and recognize that the company has increased its profits by 3.7 percent over the previous year but has also lost market share to its competitors. Thus, this writer must draw conclusions from a situation that is both good and bad to assess the company's past and, perhaps, to make recommendations for a future course of action.

Sometimes conclusions sections can be supplemented with visual aids, such as charts or tables, to clarify a decision. For instance, the salmon report provides a chart that grades each alternative as having a positive, minimal, or negative effect in each of its categories of study (see again Figure 22.15). The table is even color coded to allow readers to recognize patterns quickly and better understand why the report reaches the conclusions that it does. A chart like this is not adequate by itself but can be helpful when there are many different categories of comparison.

Final Recommendations or Opinions

This section of formal reports should clearly and concisely state the recommendation or evaluation offered. If a report must make several recommendations or evaluations, writers can provide a bulleted list. This section may not necessarily be titled "Final Recommendation" or "Final Opinions," but the title should clearly indicate that this section is the pinnacle of the report, the part that explains to readers what the report is ultimately suggesting about the problem. The salmon report titles this section "The Recommended Plan" (see Figure 21.16 on page 691).

End Matter

End matter refers to all information that follows the body of the report.

References

Chapter 6 discusses research and documentation in detail and identifies the leading style manuals of various disciplines.

Because most formal reports rely heavily on different forms of research, both primary and secondary, it is important that they include information about where these materials have come from. Reference sections, also known as works-cited pages or notes, are not specific to formal reports. Different disciplines, as well as different organizations and companies, require writers to format reference materials in specific ways.

Appendix

An appendix contains support material for the body section of a report or other document. A formal report may include more than one appendix, and each should be labeled alphabetically with a capital letter: Appendix A, Appendix B, and so on. However, an appendix may also have an additional name, such as Description of Equipment, Illustrations, Tables, or Figures.

Deciding whether to include something in an appendix or in the body of a report is not always easy or clear-cut. Although somewhat imprecise, one rule to follow is that information that breaks up the flow of the report belongs in an appendix. Again, because formal reports are often long and complex documents, they often include one or more appendixes.

Back Cover

Most of us do not think of our documents as having back covers, but formal reports often do include both front and back covers, both of which can influence a reader's trust in a report. In many instances budgetary and practical constraints prevent writers from focusing on front and back covers, although when documents will be read by important stakeholders, both front and back covers may be necessary. However, given the increasing likelihood of formal reports being distributed through e-mail and online environments, back covers are becoming less important.

Earlier we discussed the need for front covers to be visually appealing; a back cover is no different. It should also present different information about the report: its authors, the organization represented in the report, or additional information about the subject matter presented in the report. As always, audience needs will help dictate what kind of information is presented on the back cover. The salmon report includes a back cover that offers readers contact information for the U.S. Army Corps of Engineers (see Figure 22.17 on page 692).

ETHICAL ISSUES WITH FORMAL REPORTS

The same issues of disclosure and accuracy that were discussed in chapter 21 apply to formal reports as well as to informal reports. In addition, writers of formal reports should give special attention to honesty. As this chapter has stressed, formal reports are driven by audience needs and expectations. And in many cases these audiences are managers or clients, individuals with power over the careers of others. Thus, writers of formal reports are often pressured—directly and indirectly—to make recommendations, decisions, and/or evaluations that serve the immediate good of an individual manager or client, rather than the long-term good of the company, the audience, or stakeholders as a whole. Even though we typically don't like to contradict or even disappoint those for whom or with whom we work, honest answers are the ethical answers.

See chapter 4 for additional information about ethics.

ANALYZE THIS

Skim through either the Mitchell Report or the salmon passage report; read the table of contents and introductory information. Given your understanding of the problem at hand, the report's purpose, and its intended audience, what ethical concerns do you think the writers faced in composing the report? Explain your answers in a document type assigned by your instructor.

SAMPLE FORMAL REPORT

The figures that follow illustrate the key components of a formal report that were discussed in this chapter. All thirteen figures come from the "Improving Salmon Passage" report.

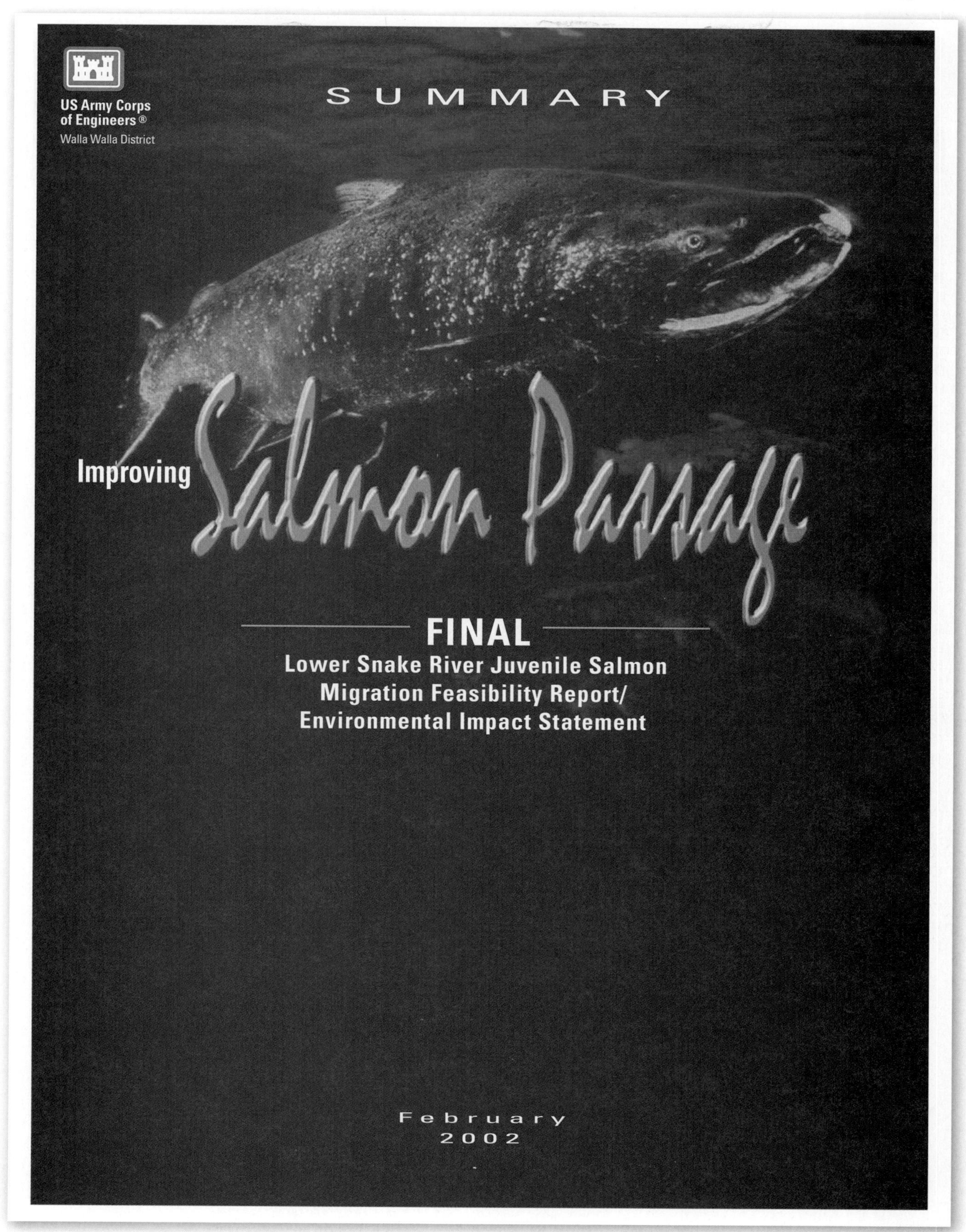

FIGURE 22.5 Front cover of the Salmon Report

FINAL

Lower Snake River Juvenile Salmon Migration Feasibility Report/ Environmental Impact Statement

Contents

FIGURE 22.6 Table of Contents in the Salmon Report

US Army Corps of Engineers®
Walla Walla District

Dear Concerned Citizen,

The U.S. Army Corps of Engineers (Corps), Walla Walla District's Final Lower Snake River Juvenile Salmon Migration Feasibility Report/Environmental Impact Statement (FR/EIS) represents more than 6 years of work by scientists, engineers, and technical staff. The Bonneville Power Administration, the U.S. Bureau of Reclamation, and the U.S. Environmental Protection Agency were cooperating agencies in the development of this report. Other Federal agencies, including the U.S. Fish and Wildlife Service and the National Marine Fisheries Service, provided essential input. Regional scientists, economists, and stakeholders also provided input.

The Corps operates four dams within a 140-mile stretch of lower Snake River: Ice Harbor, Lower Monumental, Little Goose, and Lower Granite. The Final FR/EIS explores four alternatives for improving salmon migration through those dams: continue the existing conditions at the dams, maximize transportation of juvenile salmon, make major system improvements (adaptive migration approach), and breach the dams. Based on a thorough evaluation of all alternatives, the Corps' recommended plan (preferred alternative) is a modified version of major system improvements (adaptive migration) that combines a series of structural and operational measures intended to improve fish passage through the lower Snake River.

This summary document presents an overview of the technical, environmental, and economic effects of the four alternatives. Salmon recovery has economic and environmental implications for the Pacific Northwest. Salmon are a national resource that must be protected and the dams are national investments. As stewards of both resources, we must ensure concerns are recognized and addressed. The decisions we make as a result of this study will have wide-ranging effects. Input from affected agencies, regional entities, tribes, and the public was vital to the development of this study. This active input from the region not only contributed to this study, but also contributed to regional processes that are taking other significant actions toward salmon recovery. These broad regional efforts are directed at reducing impacts associated with habitat, harvest, hatcheries, and hydropower. The Corps' recommended plan will complement these regional actions by assisting in increased salmon survival and aiding in overall salmon recovery.

We encourage you to take time to consider the data, analyses, and rationale found in our report that led to the selection of the recommended plan. Even with the uncertainties, this report and its associated documents contain the best information available to date. The information gained in this extraordinary study is sufficient to support the selection of Alternative 3—Major System Improvements (Adaptive Migration) as the recommended plan. The Corps considers this recommendation to be of critical importance.

For more information about available documents and other sources of information, please refer to the inside back cover of this summary.

In the spirit of the Corps, we say ESSAYONS, "Let Us Try."

Sincerely,

Richard P. Wagenaar

/signed/
Richard P. Wagenaar
Lieutenant Colonel, Corps of Engineers

FIGURE 22.7 Letter to audience of the Salmon Report

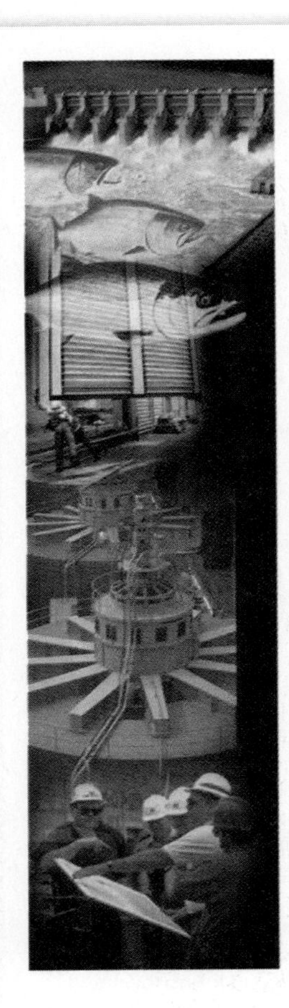

Introduction

This summary provides an overview of the analysis conducted during the Lower Snake River Juvenile Salmon Migration Feasibility Study (Feasibility Study). The results of this comprehensive analysis are documented in the Final Feasibility Report/Environmental Impact Statement (FR/EIS) and its 21 appendices. The Feasibility Study focused on the relationship between the four dams on the lower Snake River (collectively called the Lower Snake River Project) and their effects on juvenile fish traveling toward the ocean. However, as the past 6 years have shown, the technical considerations, potential implications, and interest in the Feasibility Study reach far beyond the immediate lower Snake River area. Local, regional, and national public interest in the study has been extremely high.

The genesis of this Feasibility study was the National Marine Fisheries Service (NMFS) *1995 Biological Opinion for the Reinitiation of Consultation on 1994-1998 Operation of the Federal Columbia River Power System and Juvenile Transportation Program in 1995 and Future Years* (1995 Biological Opinion). In 1998, NMFS issued a supplement to the 1995 Biological Opinion, and in 2000, it issued an updated Biological Opinion on Federal Columbia River Power System operations. The Corps' Feasibility study, and the resulting Final FR/EIS, respond to the reasonable and prudent alternative in these documents. Improvements in juvenile passage survival through the Lower Snake River Project, implemented as a result of this Feasibility Study, would be a step towards NMFS' regional survival and recovery goals for the salmon and steelhead species listed under the Endangered Species Act.

Many of the region's scientists, engineers, and economists have contributed to the Feasibility Study and other related regional processes. The Final FR/EIS includes the best availble information on the biological effectiveness, engineering components, costs, economic effects, and other environmental effects associated with four alternatives. It also reflects the extensive agency, peer, and public review process undertaken for the Draft FR/EIS. In the Final FR/EIS, the Corps identifies Major System Improvements (Adaptive Migration) as the recommended plan (preferred alternative) and explains the process for selecting that alternative.

FIGURE 22.8 Introduction in the Salmon Report

Defining the Problem

The decline of salmon and steelhead in Pacific Northwest rivers is a complex problem. It is not possible to point to one specific cause. The situation currently facing the salmon has been years in the making. The problem stems from a variety of interrelated sources that regional scientists are working hard to evaluate and understand. Historically, the runs have been affected by overfishing, poor ocean conditions, reduced spawning grounds, dams and reservoirs (Federal and non-Federal), and general habitat degradation. Several of these conditions continue today, along with predation, estuary destruction, and competition from hatchery fish and non-native fish.

Although many of these causes are known and the region has worked to correct some of them, the outstanding causes and their collective effect has resulted in the continued decline of some Columbia-Snake River Basin salmon and steelhead populations. Under the Endangered Species Act, NMFS listed the Snake River sockeye salmon as endangered in 1991. In 1992, Snake River spring/summer chinook and Snake River fall chinook salmon were listed as threatened. In 1997, lower Snake River steelhead were listed as threatened. By 1999, NMFS had placed another nine anadromous fish species throughout the Columbia River Basin on the Endangered Species List. Although this study focuses on the relationship between the Lower Snake River Project and the four listed lower Snake River stocks, defining the problem (and finding potential solutions) necessarily involves looking at the overall regional salmon decline and at causes above and beyond the four lower Snake River dams.

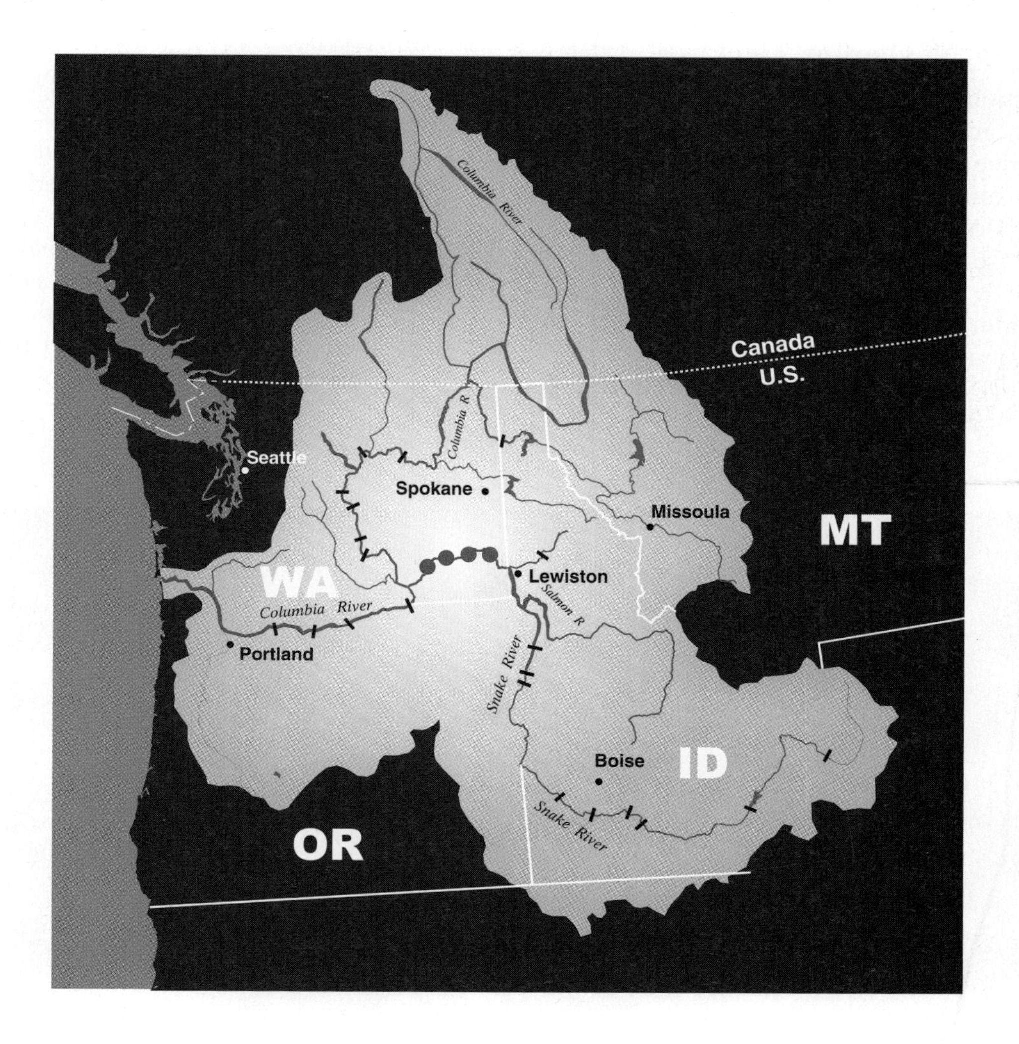

FIGURE 22.9 Defining the Problem in the Salmon Report

How the Dams Operate

Spillway

The spillway is a series of gates along the top of the dam that can open, allowing water to spill. Water is passed through the spillway to release excess flows: at times, to assist in juvenile fish migration, the corps voluntarily spills additional water through the spillways.

Navigation Lock

A navigation lock lifts and lowers boats and barges between the lower river level downstream of the dam and the higher reservoir level. Boats enter the lock, the gates close behind them, and the lock is slowly filled of drained until its water level is even with the destination water level. Then the gates are opened and the boats move from the lock to continue either upriver or down river.

Powerhouse

The powerhouse portion of the dam houses large generators for producing electricity. The water in the reservoir passes through turbine intakes in the powerhouse, rotating the turbines at 90 revolutions a minutes, and then passes into the river downstream of the dam.

Reservoir

Spanning the river, the dam forms a physical barrier that impedes the river's flow, forming an artificial lake or reservoir. Water polls behind each dam covering land that was previously exposed, allowing navigation and creating, opportunities for recreation, irrigation, and water supplies.

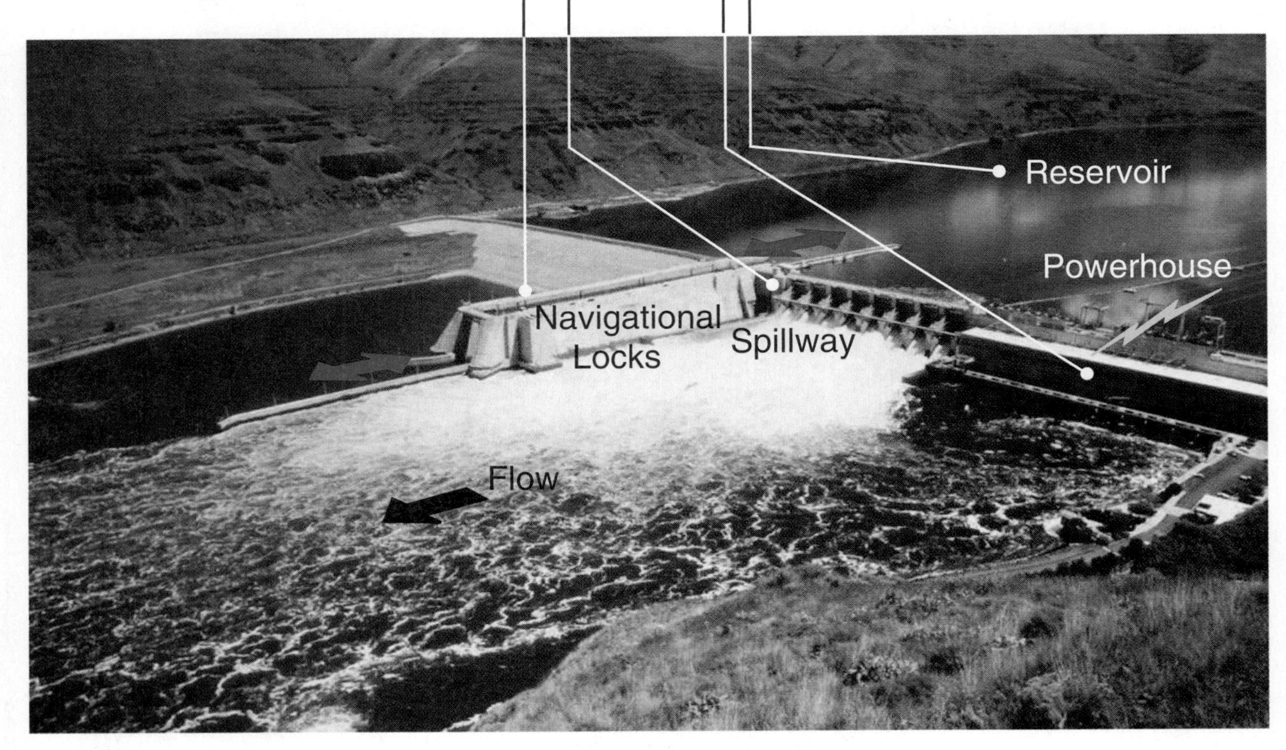

FIGURE 22.10 Description of how dams work in the Salmon Report

Ice Harbor Dam

Ice Harbor Dam, near river mile 10 (as measured from the Snake River's joining with the Columbia River), was placed in service in 1961. It is nearest to the point where the Snake River flows into the Columbia River. There are more than 4,000 acres of Corps-managed lands surrounding the dam and its reservoir, Lake Sacajawea. The reservoir extends 31.9 miles upstream. The dam has three 90-megawatt and three 110-megawatt generators, and a 90-foot-high, 86-foot-wide single-lift navigation lock. The spillway has 10 spillbays. Benefits are derived from the dam's hydroelectric power generation, seven developed recreation areas, navigation lock, wildlife habitat areas, irrigation water, fish passage facilities, and two port facilities.

Lower Monumental Dam

Lower Monumental Dam, near river mile 42, was placed in service in 1969. There are more than 9,100 acres of Corps-managed lands surrounding the dam and its reservoir, Lake Herbert G. West. The reservoir extends 28.7 miles upstream. The dam has six 135-megawatt generators and a 100-foot-high, 86-foot-wide single-lift navigation lock. The spillway has eight spillbays. Benefits are derived from the dam's hydroelectric power generation, six developed recreation areas, navigation lock, wildlife habitat areas, fish passage facilities, provision for irrigation water, and one port facility.

Little Goose Dam

Little Goose Dam, near river mile 70, was placed in service in 1970. There are more than 4,800 acres of Crops-managed lands surrounding the dam and its reservoir, Lake Bryan. The reservoir extends 37.2 miles upstream. The dam has six 135-megawatt generators and a 100-foot-high, 86-foot-wide single-lift navigation lock. The spillway has eight spillbays. Benefits are derived from the dam's hydroelectric power generation, seven developed recreation areas, navigation lock, wildlife habitat areas, fish passage facilities, three port facilities, and provision for irrigation water.

Lower Granite Dam

Lower Granite Dam, near river mile 107, was placed in service in 1975. Of the four dams, it is the farthest upstream. There are more than 9,200 acres of Corps-managed lands surrounding the dam and its reservoir, Lower Granite Lake. The reservoir extends 39.3 miles upstream. The dam has six 135-megawatt generators and a 100-foot-high, 86-foot-wide single-lift navigation lock. The spillway has eight spillbays. Benefits are derived from the dam's hydroelectric power generation, 13 developed recreation areas, navigation lock, wildlife habitat areas, fish passage facilities, water for six municipal and industrial pump stations, and three port facilities on Lower Granite Lake.

FIGURE 22.11 Dam descriptions with visuals in the Salmon Report

The Effects on Salmon

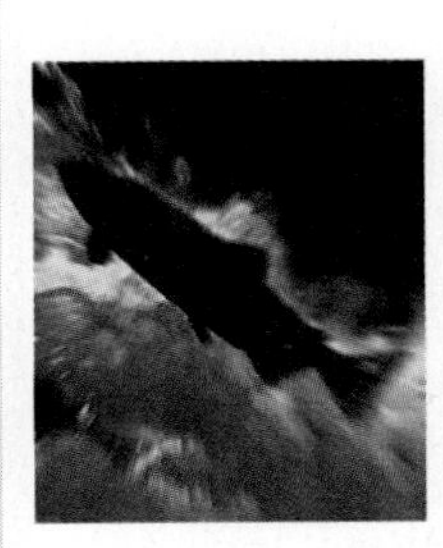

Background

Of the 12 anadromous fish stocks within the Columbia-Snake River System that are listed under the Endangered Species Act or that are candidates for listing, the Snake River stocks are: Snake River sockeye salmon, Snake River spring/summer chinook salmon; Snake River fall chinook salmon, and Snake River steelhead. Anadromous fish hatch in freshwater streams, rear in stream or lakes as juveniles, migrate downriver to the ocean, mature in the ocean, and then return upstream to spawn. This summary focuses on the effects of the alternatives on the juvenile lifestage of the listed salmon and steelhead stocks as they migrate downriver through the Lower Snake River Project. Conclusions about the effects of the alternatives on adult anadromous fish and species such as Pacific lamprey and American shad can be found in Chapter 5.5 of the Final FR/EIS; these effects are generally minimal.

Analyses Used

NMFS used two primary sets of analyses to help quantify the likely effects to the listed Snake River stocks—one developed by the Plan for Analyzing and Testing Hypotheses (PATH), and the other known as the Cumulative Risk Initiative (CRI).

The PATH analysis developed models that predict the likelihood of achieving survival and recovery of the listed Snake River stocks. The PATH model results were influenced by the effects of direct and indirect mortality. Direct mortality occurs while fish pass through the hydrosystem. Indirect mortality is assumed to occur after fish have left the hydrosystem, but is caused by having passed through the hydrosystem, including transportation. PATH defined indirect mortality in two general categories, differential delayed transport mortality and extra mortality. NMFS evaluation (Appendix A) of these two categories stated. "Debate about the importance of post Bonneville effects of dams has been highly contentious and data with which to estimate these parameters are generally poor."

The CRI analysis estimated the likelihood of extinction of listed fish stocks occurring within specified time periods. It compared how certain actions, including those outside of the hydrosystem, affect the chance of the selected stocks meeting the NMFS definition of acceptable risk of extinction criteria. The CRI analysis also evaluated the effects that a delay in implementing actions would have on the chances of specific stocks going extinct.

FIGURE 22.12 Research methods used in the Salmon Report

The Effects on Native American Indians

The FR/EIS discusses the following Native American Indian tribes and bands whose interests and/or rights may be affected by the proposed Federal actions described in the FR/EIS:

Confederated Tribes of the Umatilla Indian Reservation

Confederated Tribes and Bands of the Yakama Nation of the Yakama Reservation

Nez Perce Tribe of Idaho

Confederated Tribes of the Colville Indian Reservation

Wanapum Band

Confederated Tribes of the Warm Springs Reservation of Oregon

Shoshone-Bannock Tribes of the Fort Hall Reservation

Shoshone-Paiute Tribes of the Duck Valley Reservation

Burns Paiute Tribe of the Burns Paiute Indian Colony

The Spokane Tribe of the Spokane Reservation

Coeur d'Alene Tribe

Kalispel Indian Community of the Kalispel Reservation

Kootenai Tribe of Idaho

Northwestern Band of the Shoshoni Nation.

Five tribes—the Nez Perce Tribe, the Confederated Tribes of the Umatilla Indian Reservation, the Yakama Nation, the Confederated Tribes of Warm Springs Reservation of Oregon, and the Shoshone-Bannock Tribes of the Fort Hall Reservation—provided specific input because of their close cultural and economic links to the salmon and the lower Snake River: Impacts to tribal circumstances may be viewed in terms of tribal ceremonial, subsistence, and commercial harvest of salmon, and tribal access to lands significant to the tribes.

A Tribal Circumstances report was prepared by a private consultant in association with the Columbia River Inter-Tribal Fisheries Commission. The following alternative analysis was derived from that report.

Tribal salmon harvest numbers presented in that report were based on preliminary PATH data weight by its scientist extended by the drawdown Regional Economic Workgroup (DREW) Anadromous Fish Workgroup to represent all Snake River wild and hatchery stocks. Due to concerns associated with the weighting process, unweighted PATH results were used in all other analyses for this Feasibility Study.

Alternatives 1, 2, and 3—Existing Conditions, Maximum Transport of Juvenile Salmon, and Major System Improvements

According to the Tribal Circumstance report, Alternative 1 and 2 offer limited hope of salmon recovery within a timeframe considered reasonable by the five tribes represented. The report does not address Alternative 3, but the impacts of Alternative 3 are likely to compare closely with those for Alternative 2. There would be no change in tribal land use under any of these alternatives.

Alternative 4—Dam Breaching

According to the Tribal Circumstances report, this alternative would produce 2.4 times more tribal harvest of Snake River wild salmon and steelhead stock compared to Alternative 1 (2.6 times more harvest than alternatives 2). At the 50-years benchmark, estimated tribal wild and hatchery harvest would increase by about 1.7 million pounds. The Tribal Circumstance report concludes that only this alternative would redirect river actions toward significant improvements of the cultural and material circumstances of the tribes.

Approximately 14, 000 acres of previously inundated land would be exposed under this alternative. The Tribal Circumstances report states that the tribes would benefit from implementation of this alternative by gaining access to lands once used for cultural, material, and spiritual purposes.

FIGURE 22.13 Description of effects on Native Americans in the Salmon Report

Summary Comparison of the Four Final FR/EIS Alternatives

Alternative 1—Existing Conditions

Every FR/EIS has a starting points from which all other alternative are measured. Alternative 1 is the baseline or no action alternative under which the corps would continue operating the four lower Snake River dams according to their current configurations, including all fish passage programs now in operation. About 50 to 65 percent of the fish would be transported via truck and barge, While the remainder would migrate in river. This alternative does not mean that no further improvements would be made. The Corps, as part of its ongoing developments plans and in response to changes in agency requirements, plans to improve technology at the dams to promote fish passage. The Corps' current plan calls for turbine improvements, structural modifications to fish facilities at Lower Granite Dam, new fish barge, adult fish attraction modifications, trash boom at Little Goose Dam, modifications to fish separators, added cylindrical dewatering screens, and more or improved spillway flow deflectors.

Features

- No major changes to fish passage systems, Spill, juvenile transport
- Continued flow augmentation

- Slightly reduced extinction risks for listed stocks (Cumulative Risk Initiative [CRI])—Pre–1995 operations
- Continued juvenile fish passage for listed stocks
- Continued hydropower generation
- Continued navigational activity
- Continued irrigation and water supply
- No major economic impacts

Alternative 2—Maximum Transport of Juvenile Salmon

Most of the improvements planned for alternative I would also be included in Alternative 2. The emphasis in this alternative, however, is operating the existing facilities to maximize the passage of fish through the existing collectors into truck or barges for transport downriver. Voluntary spill to bypass fish would be minimized. The majority of the juveniles would be collected in the existing facilities and the transported past the dams. Under this alternative, there would be no need to modify spillway flow deflectors, because voluntary spill would be minimized. Some juvenile fish would still pass through the dam turbines.

Features

- Maximized juvenile fish transport with current systems
- Minimized voluntary spill
- Continued flow augmentation

Key Effects

- Slightly reduced extinction risks for listed stocks (CRI)—Pre–1995 operations
- Slightly reduced juvenile fish passage for listed stocks
- Continued hydropower generation
- Continued navigational activity
- Continued irrigation and water supply
- No major economic impacts
- Reduced total dissolved gases (voluntary spill)

FIGURE 22.14A Summary comparisons #1 and #2 in the Salmon Report

Summary Comparison of the Four Final FR/EIS Alternatives (*continued*)

Alternative 3—Major System Improvements (Adaptive Migration)

Alternative 3—Major System Improvements (Adaptive Migration) is the Corps' recommended plan (preferred alternative). This alternative would balance the passage of fish between in river and transport method to minimize risks and provide for the flexibility of adaptive migration. Alternative 3 would include all of the existing or planned structural configuration from Alternative 1 and most structural configurations found under Alternative 2—Maximum Transport of Juvenile Salmon. This alternative also includes major system improvements that would improve effectiveness and increase flexibility for optimized migration routes within seasons and years. Surface bypass collectors, behavioral guidance structures, and removable spillway wiers could be installed at one to four dams, if testing warrants, to maximize adaptive migration capabilities.

Features

- Testing of surface bypass system to optimize in river passage and transport
- Optimized voluntary spill
- Continued flow augmentation
- Operational modifications for flow augmentation and transportation

Key Effects

- Slightly reduced extinction risks for listed stocks (CRI)—Pre–1995 operations
- Slightly increased juvenile fish passage for listed stocks
- Continued hydropower generation
- Continued navigational activity
- Continued irrigation and water supply
- No major economic impacts
- Reduced total dissolved gases (voluntary spill)

Alternative 4—Dam Breaching

This alternative consists of breaching the four dams and creating a 140-mile stretch of river with near-natural flow. This would involve removing the earthen embankment section of each dam and eliminating reservoirs behind all four of the dams. Under this alternative, all facilities for transporting fish would cease to operate. A river with near-natural flow can be achieved removing only the embankment. The powerhouses, spillways, and navigation locks would not be removed , but would no longer be functional.

Features

- Removal of dam embankments
- Conversion of reservoirs into riverine environment
- Shutdown of navigation lock
- Shutdown of power generation
- End of juvenile fish transport program on the lower Snake River
- Reevaluation of fish and wildlife mitigation
- Expanded protection of cultural resources
- Modifications to some reservoir facilities
- Continued flow augmentation

Key Effects

- Moderately reduced extinction risks for all chinook and steelhead (CRI)—Pre–1995 operations
- Slightly reduced extinction risks for spring/summer chinook (CRI)—Pre–1995 operations.
- Moderately increased fish passage for listed stocks
- Loss of hydropower generation; raised electric rates
- Loss of navigational capacity; impact on other transportation system; increased transportation costs
- High sediment movement
- Impacts to irrigation and water supplies
- Short-term gain and long-term loss of jobs and income
- Change in recreation opportunities
- Reduced total dissolved gases (no voluntary of involuntary spills)
- Increased risk of major economic impacts

FIGURE 22.14B Summary comparisons #3 and #4 in the Salmon Report

Summary Comparison of the Effects of the Alternatives

Resource List	Alternative 2	Alternative 3	Alternative 4	
	Maximum Transport	Adaptive Migration	Dam Breaching (Short Term)	Dam Breaching (Long Term)
Aquatic Resources–Anadromous Fish	Minimal or notable change	Positive effect	Negative effect	Positive effect
Aquatic Resources–Resident Fish	Minimal or notable change	Minimal or notable change	Negative effect	Minimal or notable change
Water Resources				
Sediment	Minimal or notable change	Minimal or notable change	Negative effect	Negative effect
Temperature	Minimal or notable change	Minimal or notable change	Minimal or notable change	Minimal or notable change
Dissolved Gas	Positive effect	Positive effect	Positive effect	Positive effect
Contaminants	Minimal or notable change	Minimal or notable change	Negative effect	Minimal or notable change
Terrestrial Resources	Minimal or notable change	Minimal or notable change	Negative effect	Minimal or notable change
Air Quality				
Fugitive Dust Emissions	Minimal or notable change	Minimal or notable change	Negative effect	Minimal or notable change
Transporation Emissions	Minimal or notable change	Minimal or notable change	Minimal or notable change	Minimal or notable change
Replacement Power Emissions	Minimal or notable change	Minimal or notable change	Negative effect	Negative effect
Water Supply/Irrigation	Minimal or notable change	Minimal or notable change	Negative effect	Negative effect
Cultural Resources	Minimal or notable change	Minimal or notable change	Minimal or notable change	Minimal or notable change
Native Amrican Indians (Tribal values)	Minimal or notable change	Minimal or notable change	Positive effect	Positive effect
Transportation (Navigation)	Minimal or notable change	Minimal or notable change	Negative effect	Negative effect
Electric Power	Positive effect	Positive effect	Negative effect	Negative effect
Recreation and Tourism	Minimal or notable change	Minimal or notable change	Negative effect	Positive effect
Implementation/Avoided Costs (Economics)	Minimal or notable change	Negative effect	Negative effect	Negative effect
Social Effects	Minimal or notable change	Minimal or notable change	Negative effect	Negative effect

A positive effect (dark dot) Minimal or notable change in effect (gray dot) A negative effect (white dot)

Source: Condensed from Table 6-14 of Final FR/EIS, which also includes comparisons for lamprey, bull trout, traffic safety, geological resources, aesthetic resources, etc.

FIGURE 22.15 Summary comparisons in a table in the Salmon Report

The Recommended Plan
(Preferred alternative)

Based on a thorough examination of the best available biological, economic, social, environmental, and other related information, the Corps has selected a recommended plan (preferred alternative). The recommended plan is a modified version of alternative 3—Major System improvements (Adaptive Migration), with increased focus on adaptive migration capabilities. The alternative analysis and evaluation of impacts summarized in this document and described in detail in Chapter 5 of the Final FR/EIS include all components or actions contained in the recommended plan.

The recommended plan combines a series of the structural and operational measures described and evaluated in the FR/EIS for Alternative 3 that are intended to improve fish passage through the four lower Snake River dams. This alternative provides that maximum operational flexibility for juvenile fish passage; it optimizes in river passage when river conditions are best for fish and optimizes the juvenile transportation program when that operation is best for fish. It also allows for optimized combined passage when necessary for spread-the-risk operation or to conduct needed research. These improvements are not only intended to reduce direct mortality associated with dam passage, but also to reduce stress on juvenile fish, reduce total dissolved gas, and improve operational reliability.

FIGURE 22.16 Recommended plan in the Salmon Report

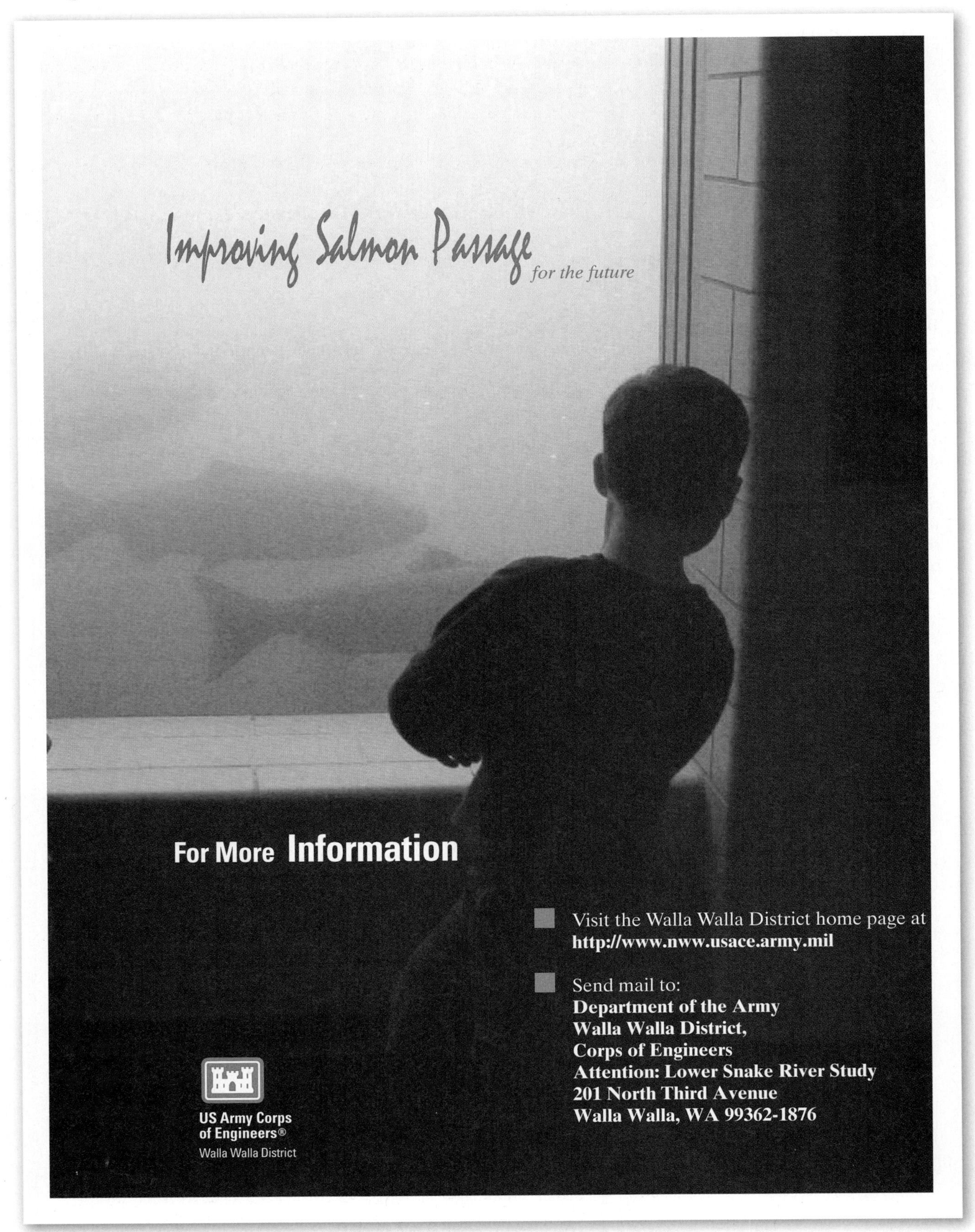

FIGURE 22.17 Back cover of the Salmon Report

SUMMARY

- Formal reports solve problems in ways that meet audience needs and expectations.
- Formal reports tend to be longer than informal reports, more thorough and detailed, formatted in distinct ways, and often composed for multiple audiences.
- The three main types of formal reports are recommendation reports, feasibility reports, and evaluation reports.
- Formal reports include criteria that serve as standards of jugdment by which writers reach their conclusions.
- Recommendation reports provide options for readers and recommend the best one to solve a problem(s).
- Feasibility reports help readers determine whether a particular plan is possible and will solve the problem(s).
- Evaluation reports provide a judgment—or a researched opinion—about the worth, value, or effectiveness of something or someone.
- Formal reports are driven in large part by their audiences.
- Most formal reports require research to answer all of the questions an audience might have about the problem or issue at hand.
- Formal reports use visuals to help solve problems and meet audience needs.
- Formal reports are often composed collaboratively.
- Formal reports are distributed to audiences as hard copies or electronically through e-mail or other online environment.
- Most formal reports include front matter, body, and end matter.
- Front matter typically includes a front cover, table of contents, and lists of tables, figures, and symbols.
- The body typically includes the introduction, description of the problem, background information, criteria, discussion of options, comparisons, conclusions, and final recommendation or opinions.
- The end matter typically includes references, appendixes, and a back cover.
- Ethics are central to all types of formal reports.

CONCEPT REVIEW

1. What are the main differences between formal and informal reports?
2. What is the purpose of a recommendation report?
3. What is the purpose of a feasibility report?
4. What is the purpose of an evaluation report?
5. What are criteria, and why are they necessary in formal reports?
6. In what ways are formal reports typically audience-centered documents?
7. Why do formal reports usually require research?
8. What role do visuals generally play in formal reports?
9. Why are formal reports often written collaboratively?
10. What are the three main components of formal reports?
11. What makes an appropriate title for a formal report?
12. Why is the front cover important in a formal report?
13. What purpose does the introduction of a formal report serve?
14. What is a description of the problem?
15. What is background information, and what purpose does it serve?
16. What are final recommendations or final opinions?
17. Why do formal reports often include references and appendixes?
18. What ethical issues might arise in formal report writing?

CASE STUDY 1

What's in a Name? A Recommendation Report to McAllen Independent School District

As the assistant superintendent of a school district in McAllen, Texas, you are asked by the superintendent to help come up with a new name for North McAllen High School, a name that honors an important person from the region. Ultimately, you will have to compose a recommendation report that considers at least four possible names and recommends one as the best choice. To get started, you will need to research McAllen, the region, and its people and cultures.

For this report, compose a criteria section, one that includes the actual criteria you would use as well as a justification for those particular criteria. In addition, develop a list of four possible names, a discussion of options in which you explain why the four options are good possibilities and how each relates to the criteria, and a conclusions section in which you recommend the name best suited for this high school. You do not need to complete the entire recommendation report, just the sections mentioned here. Your instructor may wish to change your focus to a high school in a different city or region of the country.

CASE STUDY 2

Where Should You Live? Evaluating America's Best Cities

According to *Money* magazine in 2007, these are the ten best places to live in the United States:

10. Suwanee, Georgia
9. Nether Providence, Pennsylvania
8. Chaska, Minnesota
7. Milton, Massachusetts
6. Papillion, Nebraska
5. Claremont, California
4. Lake Mary, Florida
3. Louisville, Colorado
2. Hanover, New Hampshire
1. Middleton, Wisconsin

You may be scratching your head: Georgia? Nebraska? Wisconsin? These cities and states may seem strange to some readers until they discover the criteria that *Money* used to determine these ten best places. For instance, the magazine considered the amount of criminal activity present, the quality of schools, economic opportunities, things to do, and a sense of community.

Criteria, as we have discussed, should not be arbitrary; they should depend on the particular problem that needs to be solved, as well as the audience needs and expectations. *Money* says this about its audience:

> *MONEY* Magazine is the nation's largest magazine of personal finance, with nearly 2 million subscribers and newsstand buyers and more than 7.5 million total readers. *MONEY*'s accessible, friendly and jargon-free articles are aimed at anyone who is responsible for making money decisions in his or her household—from seasoned investors to those who are just starting their financial lives. The magazine's

> mission is to help readers make better decisions, not just in their investment portfolio, but in all areas in which life and money intersect—their family, their home, their health, their spending and their future.*

In addition, the magazine gives this demographic information about its readership:

> Male/female: 65%/35%
>
> Median age: 48.2
>
> Graduated from college+: +50%
>
> Median household income: $84,778
>
> Professional/managerial: 39%
>
> Top management: 18%

Given the magazine's wide and diverse audience, as well as the age and education level of its audience, the criteria it used for the ten best places make a good deal of sense. The median age of 48 suggests that many of the magazine's readers probably have children in school, which probably explains why *Money* used quality of schools, things to do, and criminal activity as three of the criteria. The quality-of-schools criterion also makes sense because the audience tends to be college educated and probably values education. Furthermore, the median household income places readers of *Money* within the middle and upper-middle classes; therefore, it is sensible that economic opportunities would be a criterion. This audience would probably be interested in ways to increase economic worth. Thus, *Money's* ten best places to live make sense for the kind of person who reads *Money* magazine—the criteria fit this group of people.

Choose two different magazines that are not similar to each other. For example, don't choose *Vogue* and *Cosmopolitan*, or *Sports Illustrated* and *ESPN the Magazine*. Research as much as you can about the audiences of these two magazines. Then, based on your understanding of these audiences, develop the criteria you would use in creating a list of the ten best places to live for each of these two different audiences. In addition, conduct the research necessary to actually put together a list of the ten best places to live for each audience. Present your criteria, audience analyses, and findings in a formal report or in particular sections of a formal report assigned by your instructor.

---

* "Best Places to Live," http://money.cnn.com/magazines/moneymag/bplive/2007/.

CASE STUDIES ON THE COMPANION WEBSITE

In Violation of Federal Regulations: An Internal Review Reveals Problems at Berkshire Med-Tech Corporation

You work as a risk management advisor at Berkshire Med-Tech Corporation and are sent to one of the company's production facilities in Davenport, Iowa, to report on some recent problems there. In this case you are asked to both evaluate the problems there and make recommendations to help solve them.

Swiss Sweets: Time for an Upgrade and a New Look?

In this case you work in the marketing sector of a gourmet candy maker and are charged with researching the costs and benefits of updating the company's website and logo. This research should help lead to a feasibility report that examines whether the company pursues these updates.

Flawed Tests and Faulty Parts: An Engineer Develops Solutions for a New Device

As a biomechanical engineer at Watson Laboratories, Inc., you are a member of the research department that specializes in the development of technologies used to design and manufacture prosthetic limbs. You are asked by one of your supervisors to look into problems associated with one of the company's machines. In particular, you are asked to discover the reasons behind these problems and recommend solutions to solve them. In addition, you must consider the ethical issues that surround this particular situation.

Maintaining the Greens: Acquiring a New Maintenance Shed for Hampshire Greens Golf Course

In this case you are the operations manager at Hampshire Greens Golf Course and must produce a formal report for board members that solves a particular workplace problem: discovering the best means to handle certain maintenance issues that are crucial to the success of the golf course while catering to the needs and desires of different stakeholders.

VIDEO CASE STUDY

DeSoto Global

Writing a Progress Report at DeSoto Global

Synopsis

DeSoto Global has secured a contract with Shobu Automotive and is negotiating with two other automotive manufacturers. However, DeSoto has not been successful yet in partnering with a security company that can tie DeSoto's Tracker1 auto-theft tracking system in with local law enforcement. Jennifer Nichols wants a progress report that explains any successes and setbacks in the negotiations with security companies. She needs the progress report in order to develop her own formal report for the DeSoto executives, who will soon be meeting with investors to discuss the company's progress in Asia. In other words, Nichols uses the informal progress report as the basis of her formal report for those at the top of the company.

WRITING SCENARIOS

1. Locate a recent recommendation, feasibility, or evaluation report written by a company, organization, or institution that you recognize. You should be able to do this easily by using a web search engine. Analyze the report's use of research and visuals. What kind of research was conducted? What types of visuals are included? Why were these necessary, given the report's audience and its purpose? Are the research and the visuals effective? Why or why not? Compose your response in a document type assigned by your instructor.
2. Review the salmon report detailed in this chapter; the entire report is available on the Companion Website at www.prenhall.com/dobrin. Look carefully at the back cover. Given what you know about the report's audience and purpose, why do you think the writers of the report designed the back cover in this way? Discuss the use of images and words, as well as the rhetorical effects of the back cover—that is, how and why it works the way it does.
3. Imagine that you work as a financial advisor to a local investor who wants to open a chain restaurant—either Subway, McDonald's, or Papa John's Pizza. Ultimately you will need to write a recommendation report that tells your client which of these would be best in terms of its likely profit over a five-year period and its ease of daily operation. You know you will have to research and analyze how much money it takes to start up each of these chains, how many of each already exist in the area, and which chain has the best relationship with its franchisees. But what other criteria should you develop to use as the basis of your recommendation? Develop a list of other possible criteria, and then decide which are the most important for the problem at hand.
4. Locate a formal report written by employees in a company that you would like to work for when you graduate. In a one- to two-page summary, explain what kind of report it is, what problem it addresses, and how that problem is handled in the report. Also, discusse what stands out to you about this report and why.
5. Write a short evaluation report (5–10 pages) to the author of the report you located in Writing Scenario 1. Your report should evaluate the other report's use of visuals. What kinds are included? What makes them effective and necessary for the report's audience? Could they have been more effective? Are they integral or gratuitous? In making your evaluations, you'll have to develop some criteria that allow your report to be clear and sensible to the audience. Your report should include as many parts and sections of a formal report as you deem necessary for the task at hand.
6. You work for a large soft-drink company and are collaborating on a feasibility report exploring the possibility of the company's developing a new energy drink that contains over 300 milligrams of caffeine in a single serving. Your team leader asks that you write a three- to four-page technical background section that describes the effects of this amount of caffeine on the human body—kids, adults, the elderly. Your group leader also asks that you develop a table that lists the caffeine content in at least five competing energy drinks in order to set up some basis of comparison. If possible, this table should list the energy drinks in order of their market share—from highest to lowest. Compose these parts of the report, keeping in mind that your audience is comprised of nonspecialists in the areas of medicine and health. Be sure to consider layout and design issues, conduct necessary research, and document your sources.

7. In the previous writing scenario, what particular ethical concerns would all of the collaborators on this report face while composing this feasibility study? Consider the entire report, not just the sections described in the scenario. List at least three to five specific ethical concerns, and develop a memo to the collaborators that informs them of these concerns.
8. Research one real-life example of a company or organization that produced an unethical formal report. This can be a recent or an old report. Detail the specifics of the situation, particularly what could be construed as unethical and what the consequences of the unethical behavior were. Compose your findings in a one-page summary or document type assigned by your instructor.
9. Review the salmon report included in this chapter, or examine the entire document online on the Companion Website at www.prenhall.com/dobrin. Then list all of the ethical issues you believe the writers of the report faced. Compose your responses in note format, and be prepared to discuss them in class. You may wish to review chapter 4 on ethics before starting this assignment.
10. As was mentioned, the salmon report does not include an executive summary. Compose one for the president of an outdoor sports and recreation organization in the Pacific Northwest. You should begin by reading the full report carefully on the Companion Website located at www.prenhall.com/dobrin.
11. What technologies exist today that make writing and producing formal reports possible for the average workplace writer? Consider those technologies that allow workplace writers to produce not only words but also high-caliber visuals and sophisticated layout and design elements. You might refer to chapters 3 and 17. Record your responses in a memo format for your instructor.
12. Many formal reports are now available on websites, often as PDF files that can be accessed by just about anyone who wishes to see them. List three specific types of reports that you believe should be made available publicly via electronic technologies and three that you believe should not be made available publicly. Generally speaking, how does the availability of many formal reports affect your understanding of their role inside and outside workplaces? Summarize your responses, and be prepared to discuss them.
13. Despite the availability of various technologies to help workplace writers compose formal reports, what facets of formal report writing do you believe cannot be helped or improved by existing technologies? In a memo to your classmates, who are all likely to be workplace writers in some capacity, explain your responses to this question.

PROBLEM SOLVING IN YOUR WRITING

For one of the case studes in this chapter, or for any of the writing scenarios you were assigned by your instructor, outline the particular points in the problem-solving approach that influenced how you responded to it. Explain in detail why these particular components of the problem-solving approach were important in your thinking and writing.

23

Presentations

CHAPTER LEARNING OUTCOMES

After completing this chapter, you will be able to do the following:

- Deliver **formal and informal presentations**
- Recognize the different **audiences** for presentations
- Assess the **physical context** for various presentations
- Determine the **time and pacing** of a presentation
- Gather and evaluate the **information** you'll need for a presentation
- Choose the appropriate **visual aids** for a presentation
- Plan your **speech** for your presentation
- Organize the different **components** of your presentation
- Use appropriate **vocalization and body language** in a presentation
- Field **questions** from your audience

DIGITAL RESOURCES

On the Companion Website www.prenhall.com/dobrin:

- Case 1: Breaking the News: Presenting a Corporate Takeover at Pinnacle Accounting
- Case 2: Presenting a Very Positive State of the Hospital at Boise Presbyterian
- Case 3: Proposing a Bulkhead to Protect Carteret Technical Institute from Coastal Erosion
- Case 4: Mapping the City: Presenting the Changes to Fixed Food's Delivery Practices
- Video Case: Finding Employment at Cole Engineering—Water Quality Division
- PowerPoint Chapter Review, Test-Prep Quiz, Exercises and Activities

REAL PEOPLE, REAL WRITING

MEET MONTY HANSEN • PhD Candidate in Ocean and Resources Engineering

What kinds of presentations do you do as a PhD candidate in ocean and resources engineering?

I typically create and deliver presentations that summarize research that I have done or that I plan to do. My presentations usually consist of PowerPoint slides. I try to go light on text and heavy on graphics and animations to keep the presentation lively and interesting. I limit text to what is absolutely necessary as a prompt for myself.

Can you provide an example of a presentation you've recently done?

My most recent presentation was the defense for my master's thesis. I believe it was successful, based on the number and relevance of questions posed after the presentation. In my experience a successful presentation is one that is delivered to the audience in a manner that promotes enough understanding to generate relevant questions and lively discussion. In this case I ended up with a master's degree—another indication that the presentation was successful.

What are the most common physical spaces in which you deliver presentations, and how do those spaces influence your presentation?

I usually present in classrooms or conference rooms. I visit the space while I am preparing, as aspects of the room affect how I present the material. Light, sound, room size, location of projector, screen, podium, and electrical facilities become important factors in the success of the presentation.

What do you see as the most common problems in slide show presentations?

Most common is the use of excessive text on slides, usually accompanied by a speaker who drones monotonously from the slides. Another problem is the excessive use of PowerPoint enhancements such as fly-ins/-outs, appearing/disappearing arrows, and irrelevant audio, all of which can be distracting and possibly indicate thin content. I do not attempt to appear polished when I speak. I want to convey that I know the subject well, possibly understating the importance of the presentation itself.

What advice do you have concerning the actual delivery of oral presentations?

I know the theoretical dos and don'ts, but once I get in front of a crowd, the presentation assumes a life of its own. My pacing rule of thumb is two minutes per slide. I try to maintain my pace by checking a clock periodically. I will jiggle anything in my pocket, such as keys or change, so I make sure my pockets are empty. I have never heard a speaker who was too loud.

What advice about workplace communication do you have for the readers of this textbook?

Try to remember that audiences vary in experience, interest, and knowledge. All technical writers and presenters should spend lots of time thinking about who they're talking to before they write or speak. I try to deliver information that is inclusive, targeting those who have little background in my subject, while never dumbing things down for the audience.

INTRODUCTION

Delivering information orally is an essential part of most workplace environments. Sometimes presentations introduce, accompany, or follow a document or visual supplement; at other times the spoken word is the sole means of communication. You might speak briefly and informally to your colleagues and coworkers, or you might speak more formally to a group of managers, supervisors, clients, or the general public. Presenting information through presentations—regardless of their length, style, or formality—can be highly effective in solving workplace problems.

Consider the role of presentations in the Apple Computer, Inc. line of products. Each year at the MacWorld Conference and Expo, Apple CEO Steve Jobs begins with a keynote speech to thousands of attendees in which he introduces the newest Apple product—for example, the iPod, the iPhone, and in 2008 the MacBook Air notebook computer (see Figure 23.1). His presentations often involve multimedia displays in which he unveils new products in unique ways, such as his demonstration of the MacBook Air's thinness by sliding it into a manila envelope. Jobs's presentations have become so legendary that the buzz surrounding them often surpasses that of the products they introduce. In fact, the keynote speeches (known as "Stevenotes") are discussed for weeks on various tech blogs, and they become featured videos on YouTube, sometimes even becoming the subject of online games. Importantly, these presentations often cause substantial surges in Apple's stock prices.

FIGURE 23.1 Steve Jobs introduces the MacBook Air

Although Jobs's presentations seem relaxed and spontaneous, you can be sure they have been carefully designed and prepared. Most presentations involve more than simply standing up and talking; they involve a host of problem-solving activities. Effective workplace presenters plan their presentations before they begin to create them, and in many instances further research is necessary to adequately address the given problem(s). Creating or drafting the actual presentation may involve incorporating sophisticated multimedia applications. Thereafter, effective presenters test and solicit feedback on their presentations, revising them in the process to ensure effectiveness. Giving the presentation, the Distribute phase, is the most obvious—and sometimes stressful—aspect of the process and is quite different from releasing a print-based document.

PSA

One thing that makes presentations different from other forms of workplace communication is the degree of interaction between speakers and audiences. Most written communication is delivered to readers who are not in the same physical space with the writer and who may not respond to the information after reading it. Oral communication, on the other hand, is usually delivered face-to-face and often involves interaction between the speaker and the audience. Audience members sometimes make comments before, participate during, and ask questions after a presentation. Audiences sometimes interact nonverbally, too—they might nod, smile, or clap to show support or approval, or they might frown, grimace, or snore to show anger or boredom.

For many presenters this direct interaction causes stress: polls regularly show that fear of public speaking ranks as high as fear of snakes, heights, and death for many people. However, the guidelines in this chapter will help you become a comfortable and effective speaker in workplace environments. You may even come to view the interaction between you and your audience as an advantage because it allows you to continually assess and respond to audience needs, improving your chances of solving the problem at hand. Although your presentations may never reach the iconic status of those by Steve Jobs, you will become a more successful presenter with time, practice, and preparation.

ANALYZE THIS

Your technical writing class certainly involves presentations; in fact, it is likely that your instructor, your classmates, or perhaps even you recently presented something to your class. Analyze the degree of interaction between the audience and the presenter during a recent class presentation. How much interaction was there between the audience and the presenter? Did you serve as a participant or an observer in the presentation? Did the presentation seem at all stressful for you, your classmates, or your instructor? Why or why not? Write a short response to these questions, and then discuss your responses as a class or in a small group.

TYPES OF PRESENTATIONS

Oral presentations can be classified in two types—informal and formal. Although they often differ in scope, complexity, style, and format, both types require effective communication and the desire to respond to the audience's needs.

Informal Presentations

Most workplace presentations are informal. In fact, many people deliver oral presentations without realizing it; whenever they discuss or explain their ideas and thoughts without a great deal of preparation or rehearsal, they are giving an informal presentation. Sometimes informal presentations are *spontaneous*, that is, given without any advance notice. For instance, during lunch your supervisor might ask your opinion of a new product or marketing strategy that has just been announced. In spontaneous situations like this, you draw on your knowledge and expertise to say something productive and insightful, but you aren't normally expected to recite exact figures, dates, or details.

Informal presentations can also be *improvised*, when the speaker may refer to an outline or notes but creates the presentation along the way. In improvised situations you might be expected to provide a general overview or opinion of the subject and support it with rudimentary details or analyses. In general, most informal presentations are not accompanied by sophisticated written or visual aids, but they occasionally incorporate short documents or readily available images to help guide readers through the material.

Formal Presentations

Although they do not occur as frequently, formal presentations are important and can be the principal factor in determining whether ideas, products, or services are used and accepted. In fact, a presentation may actually *be* the service that a company provides—for example, a workshop or training seminar. Formal presentations are generally longer and more complex than informal presentations, although they can range from a few minutes to several days in length. Examples of formal presentations include briefings, sales meetings, product demonstrations, training seminars, workshops, panel and conference discussions, and speeches.

Presenters usually deliver formal presentations in set-aside spaces, which include offices or auditoriums reserved, or booked, in advance; a special room designated specifically for presentations; or even a space designed expressly for a particular type of presentation. Your college or university, for example, probably contains a number of specific spaces used for presentations, such as auditoriums, conference rooms, science labs, and perhaps even an outdoor amphitheater. These spaces have probably been carefully designed to accommodate formal presentations, even though they may serve other functions as well.

Most individuals giving formal presentations rely on various tools, ancillaries, and equipment to aid and reinforce their spoken words. Such aids range from low to high tech and can include a simple speaker's podium; a chalkboard or whiteboard; a model, mock-up, or example of a product or subject; a poster or chart; or a multimedia slide show presentation. Computer technology has become indispensable in formal presentations; many audiences now expect to see sophisticated visuals and graphics accompanying presentations, and studies show that audiences take a speaker more seriously and retain more information when it is delivered with multimedia equipment. According to MIT psychologist Stephen Pinker, this response is perfectly natural: "We are visual creatures. Visual things stay put, whereas sounds fade. If you zone out for 30 seconds—and who doesn't?—it is nice to be able to glance up at the screen and see what you've missed."[1]

See chapter 3 for more information on presentation technologies.

Fortunately, there are easy-to-use presentation software programs available, such as MS PowerPoint, Corel Presentations, Adobe Visual Communicator, and Apple iWork. These allow users to incorporate text, images, sound, and video into almost any presentation. And as chapter 3 notes, increases in bandwidth allow multimedia presentations to be delivered through the Internet; many organizations hold meetings and deliver their presentations across great distances, using videoconferencing services like GoToMeeting or GlobalConference. Because computer technology is such an important part of both formal and informal presentations, we discuss it throughout this chapter.

IN YOUR **EXPERIENCE**

Think about the last formal presentation you observed. Did it take place in a space that was specifically designed for that type of presentation, or was it a multiuse space? Did the presenter use effective equipment or ancillaries to aid in the delivery of the material? How might a different setting and equipment have changed your overall experience as a member of the audience? Discuss these formal presentations in a small group, noting similarities and differences in where they took place as well as what equipment was used.

Varied Presentations

Workplace writers often use combinations of informal and formal presentations to solve problems. They might even use several different presentations—with various levels of formality, delivered in different locations, and using an assortment of supplements and peripherals—to solve one problem. Imagine yourself as the sole salesperson for a small company that makes digital camera components. You notice that the company's new waterproof camera housing isn't selling well, and you realize that you aren't reaching your target audience for the housing—scuba divers. At your next meeting with your company's owner, you pitch the idea (in an informal presentation) that you should begin focusing your marketing on scuba shops rather than photography stores. Your presentation is very informal, uses no peripherals, and takes only a few minutes, but the owner agrees that you should target scuba shops.

---

[1] Stephen Pinker, *The Language Instinct: How the Mind Creates Language* (New York: Perennial Classics, 2000).

After compiling a list of the scuba stores in your area that might be interested in your camera housing, you begin visiting some of them. These visits are also fairly informal, although you distribute to shop owners a detailed flyer as well as a demo model of the camera housing for them to examine. Your presentations are not fully scripted, but you follow the outline of the flyer, and you gauge your level of technicality and detail by the shop owners' initial questions. One shop owner mentions that an underwater photography club meets in his scuba classroom each month, and he invites you to demo the camera housing at their next meeting.

Believing that club members are your best chance to increase sales, you prepare a more formal presentation. You develop a short PowerPoint presentation that describes the features of the housing and also includes various photographs and a short embedded video (see Figure 23.2). After presenting the information to the group, you pass around several demo models and field a variety of questions about the functions, features, price, and availability of the housing. Several days after the meeting the shop owner places an order for the camera housings, several of which have been requested by club members who attended your presentation.

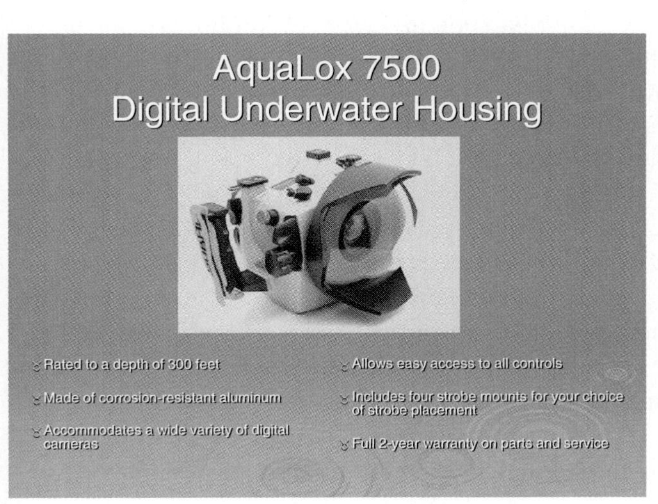

FIGURE 23.2 A PowerPoint presentation slide

At your next meeting with your company's owner, you describe your presentations and the results. Consequently, the owner encourages you to target and make presentations to other scuba stores and underwater photography clubs in surrounding areas. And as luck would have it, one of the club members you met sends you a picture of the camera housing being used to photograph a large sea lion (see Figure 23.3); you incorporate this image in your next presentation.

FIGURE 23.3 Photograph used in sales presentation

Varied Audiences

You might have noticed from this extended example that presentations can be delivered to a variety of audiences, which often determine the level of formality, the location or setting of the presentation, and also the materials and peripherals used to address them. With *all* audiences it is crucial to remember your ethical responsibility to speak truthfully and clearly.

Clients or Customers

Many workplace presentations are delivered to potential clients or customers, often addressing the primary benefits or features of a service or product. These presentations should be persuasive, focusing on what the audience gains from the product or service. They should include relevant facts and details but should focus on larger goals and purposes rather than the intricacies of the subject.

Colleagues

Some presentations are delivered to members of your own organization or individuals outside your organization who have similar jobs within your profession. Such presentations might discuss or examine a problem, or they might address a project or idea you've been working on. Presentations to colleagues often leave out background information and use technical language because colleagues already know something about the subject or topic. Sometimes, though, when you present to colleagues in different departments with different training and expertise, you need to carefully explain background and language to ensure understanding.

Supervisors and Employers

Presenting to supervisors and employers can be highly stressful because these individuals often determine whether your ideas will be implemented or your job will be secure. To diminish stress when speaking to supervisors or employers, you should spend extra time researching and preparing your presentations, anticipating questions and concerns the audience might have. In addition, because supervisors and employers are often busy, you should get to the main point quickly, front-loading the important facts and details they need to make a decision.

The Public

Occasionally, you may speak to public audiences outside your firm or company—perhaps civic or social organizations, governmental committees, customers or shareholders, or even the media. These audiences vary greatly, and you should learn as much about them as possible before speaking. With public audiences you should avoid industry jargon and technical details, focusing on how the information affects those in the audience. It is generally best to avoid negative remarks about any other individual or organization and to present yourself and your organization in positive ways.

ANALYZE THIS

Go to YouTube or some other video broadcasting website, and watch a portion of one of Steve Jobs's MacWorld keynote presentations. What can you learn about public presentations from the video? How might Jobs's presentation have been different had he delivered it to clients, customers, or colleagues? Write a short analysis of Jobs's methods of engaging a public audience.

PREPARING FOR A PRESENTATION

No matter how easy or spontaneous they look, most good presentations require careful planning, diligent preparation, and meticulous assessment of the situation. The first two aspects of the problem-solving approach—planning and researching—can be especially useful in preparing a presentation. Mark Twain once said, "It usually takes me more than three weeks to prepare a good impromptu speech." And although you may not have three weeks to prepare, a correlation exists between the amount of time spent preparing a presentation and the ease with which it is delivered.

Focusing on the Audience

The first thing to consider is precisely *why* you are creating a presentation. Although the answer may seem obvious in some situations (e.g., your boss asked you to speak at a meeting), you need to focus on the problem or issue that your presentation addresses, and that requires you to think carefully about your audience. Many unsuccessful presentations deliver global information about a topic, providing generalities that offer little substance, or they focus on the speaker's perspective and goals. In contrast, effective presentations focus on what the audience wants or needs to know. For example, in an employment interview (a type of presentation), when asked to speak about previous jobs, unsuccessful candidates might speak at length from their perspective—about

For more on job interviewing, see chapter 14.

what they've enjoyed about each job. Successful candidates, on the other hand, carefully select and highlight the skills and experiences from past employment that are most applicable to the position for which they're interviewing.

In other words, the goal or purpose of presentations is to provide solutions to audiences' problems. In his bestselling book on presentations, *I Can See You Naked*, Ron Hoff defines a presentation as "a commitment by the presenter to help the audience do something."[2] Thinking about a presentation in this light emphasizes the audience's, rather than the presenter's, interest. This is an important first step in solving problems.

The second step is to find out exactly who is part of the audience. Because different audiences require different presentation strategies, it is important to know a bit about the audience before you create a presentation. For example, if you were asked to deliver a sales presentation to a company that might purchase your product or service, you would need to determine whether the audience included other salespeople, technical specialists, managers, or the users of the product or service. Who is present at the presentation impacts how you make the presentation. Consequently, be sure that you inquire about the audience before developing the presentation, or perhaps even before you agree to do the presentation.

Here are some questions you will need to answer to help shape your presentation:

- What is the audience's purpose in attending your presentation? Do individuals expect to be entertained, informed, or persuaded to act?
- Will audience members use your information directly, or are they more likely to pass the information along to someone else?
- Will the audience include one person, a small group, or many people? Can you get an attendance list or an estimate of the number of attendees?
- Will the audience be uniform, consisting of one type of individual (e.g., salespeople or technical specialists), or will it be diverse, consisting of individuals with varying levels of expertise and interest?
- How much does the audience already know about the subject of your presentation? Will technical language and concepts be appropriate, or will you need to simplify and explain the technical aspects of your presentation?
- What beliefs about you, your company, or your product will audience members bring to the presentation? Are they likely to be excited by or hostile or neutral to your presentation?

Assessing the Physical Context

Presenters also need to determine where a presentation will take place; if you don't know, be sure to ask. For most formal presentations the location is determined beforehand. If possible, visit that location at least once before the presentation. If you have access to the space beforehand, consider rehearsing in it, either by yourself or with one or two colleagues or coworkers who can serve as a mock audience and can provide comments and suggestions.

Whenever possible, try to simulate the actual speaking conditions because familiarity with the physical context will make you more relaxed—and more effective—during the real presentation. When professional football teams prepare to play in a noisy or raucous stadium, they often practice with a public address system blaring crowd noises to simulate the real game situation. Although you may not need to pipe

---

[2] Ron Hoff, *I Can See You Naked* (Kansas City: Andrews and McNeel, 1992).

in the sounds of screaming football fans, you should try to practice in the actual setting or someplace similar. You might also videotape your rehearsal and review it before delivering the actual presentation.

Answering the following questions about the physical context can help you prepare for your presentation:

- Is the location difficult to get to or find? Will you need to provide transportation or directions for any member of the audience?
- Are parking spaces, restrooms, and refreshments easily accessible if they are necessary?
- Is the physical location appropriate for the audience? Is there sufficient seating or space for the anticipated number of attendees?
- Will you present from a lectern or podium, or will you be standing or sitting with nothing between you and your audience?
- Will all of the audience members, even those in the back, be able to see and hear your presentation, or should you try to arrange (if it is not already there) some method of broadcasting your voice and visuals?
- Does the room come equipped with any audiovisual equipment you might need, or will you be required to bring your own?
- If the room is equipped with the appropriate audiovisual equipment, will you need special permission or a password to use it? Will you need a key or code to access it?
- If you will be using someone else's computer, does it have the versions of software and hardware that you require? Older versions of some presentation software packages may not be compatible with a presentation created on a newer version.
- Will you have access to the location ahead of time to load, install, or set up any software, hardware, or other supplements you might use in your presentation?
- Will someone introduce you, or will the audience members be present before you arrive?

EXPLORE

Find a physical location that is designed specifically for presentations or one in which presentations are regularly given. Write a description of that physical space. Then write about the presentation types, strategies, and equipment that might best suit that physical space. For example, does this location suit longer, formal presentations, or is it better suited for brief, informal discussions? Would the space require you to broadcast your voice? Does it contain any presentation equipment, or would you need to provide your own?

Determining the Time Available

In many instances you will have a predetermined length of time to deliver your presentation, and you will know approximately how long your presentation takes if you rehearse it, as suggested earlier. If you are using slides, you should budget approximately one to two minutes for each one, although slides with detailed graphics, charts, or schematics take longer to explain or discuss. If you end up with additional time remaining, you can expand on your major points; don't simply extend the introduction or the conclusion just to fill the time.

You should always be certain to limit your presentation to the time allotted because other presenters may be impacted, audience members may have other activities scheduled, or the location may be reserved for another function. Nothing alienates an audience more quickly than a presentation that lasts longer than it should. In addition, although you want to deliver as much useful information as possible in the designated time, you should never plan to use *all* of the allotted time for the presentation; reserve some time for setup, interruptions, and questions after you speak. Even if no specific time has been allotted for a presentation, you should plan to deliver the important or significant information as quickly as possible. It's always better to finish a bit early than a bit late.

IN YOUR **EXPERIENCE**

In your experience as a student or an employee, you have probably delivered at least one presentation, whether informal or formal. Think about a past presentation, and consider how much preparing and planning you put into it. Did you consider the interests, composition, and prior knowledge of your audience? Did you carefully assess the physical location? Did you determine and then budget the time allotted for the presentation?

In a small group, discuss how you considered or failed to consider the audience, location, and available time for your presentation and how that attention affected the overall success or failure of your presentaton.

Gathering and Evaluating the Information

In preparing for a presentation, you'll also need to determine how much of the necessary information you have available and how much you might need to gather. Sometimes you'll be able to create your presentation entirely from your own personal knowledge and experience. At other times you'll be able to draw on information and material at your disposal—such as sales figures you've already compiled, reports and other documents you've written in the past, or physical objects in your possession—to augment your knowledge and experience. On other occasions you'll need to conduct research to gather more information to include in your presentation.

Aside from simply gathering facts and details, you might need to conduct other types of research that may be entirely new to you. Because most presentations contain visual aids, you may need to find or create images, charts, or graphs that are appropriate. You may even need to find or create audio or video to accompany your presentation. If your presentation is collaborative, part of the research process may involve discussing individual areas of expertise to discover who is best suited for what. For instance, you might find that one of your group members has created many computer presentations; as a group you might then decide to let that person determine the type of software to be used, assess the presentation location for support of that software, and design and create the slide show. Gathering and evaluating information early in the preparation process can save you a lot of time in the end.

Choosing and Creating the Appropriate Visual Aids

At some point in the preparation process, you will need to determine which tools and equipment you'll use in your presentation. In most cases the equipment takes the form of visual aids, which are influenced by the variables just discussed: audience, location, time available, and information to be included. For example, you probably wouldn't use sophisticated multimedia to deliver information to a few of your colleagues in an informal planning luncheon. But a more complex, formal conference presentation to a large group of peers would be more effective if it *did* include some sort of visual aid or display to help guide the audience through the material.

Generally speaking, the more formal, complex, or detailed the presentation, the more sophisticated the technology incorporated. Including technology can be especially important when the stakes are high; speakers are generally perceived as being more professional, more persuasive, easier to understand, and more interesting when they include visual aids. One study found that presenters who use visual aids are 43 percent more effective in persuading audience members to take a desired course of action than are presenters who don't use visuals. The study also revealed that typical presenters who used visual aids were as effective as better presenters who used no visuals and that audiences expect better presenters to incorporate higher-quality visual aids.[3]

The most effective visual aids are those that mesh seamlessly with the presentation. Visual aids should reinforce and strengthen information but should not overwhelm or take the place of quality information. Some presenters mistakenly include too many "bells and whistles" in their presentations when a simple, clear supplement would suffice. In presentations using computer software, this problem is sometimes called PowerPoint overload, which describes visual aids that are so busy that they impede audiences' understanding rather than support it.

To ensure that visual aids are appropriate and seamless, you should be certain that they serve one or more of the following four purposes:

- *To explain*: Visual aids work best when they help to explain or clarify difficult, complex, or abstract information. Use a visual aid whenever you think your audience will have trouble understanding what you have to say.
- *To emphasize*: Visual aids are also useful when you want to emphasize some aspect of your presentation. A textual or visual reinforcer of something you've said lets your audience understand that it is significant. For example, an instructor who writes something on the chalkboard is using a visual aid to emphasize something important.
- *To interest*: An appropriate visual attracts audience attention and keeps individuals interested in what you have to say. It allows them to look at something other than you, which helps them stay focused and can even alleviate some of your fears if you're nervous. Interesting visual aids need not be solely graphic; an appropriate quotation, fact, or statistic can capture audience interest through text.
- *To guide*: Audiences want to know where you are in your presentation and how a given point fits with the overall theme or subject of your talk. Some presenters distribute outlines or notes so that the audience can follow along; others use an introductory slide or graphic that provides an overview of the major points to be covered; still others rely on page or slide numbers to indicate how far into the presentation they are. Visual aids can also help keep *you* focused.

Types of Visual Aids

You should consider the advantages and disadvantages of each type of visual aid as you prepare your presentation. The most common types are discussed here.

---

[3] Jane Webster and Hayes Ho, "Audience Engagement in Multimedia Presentations," *Data Base* 28, no. 2 (1997): 63–77.

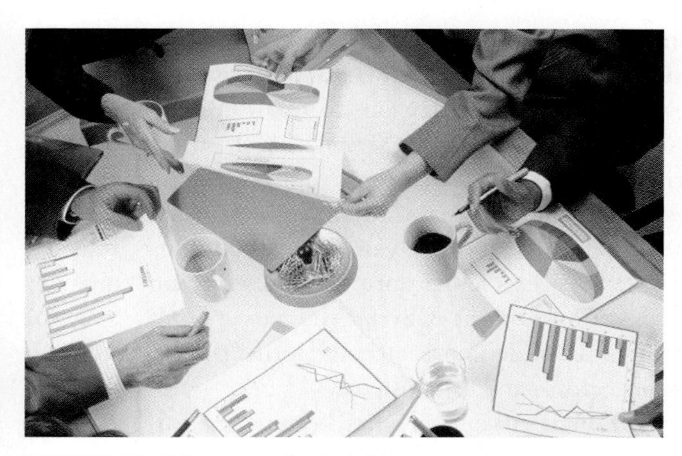

FIGURE 23.4 Presentation handouts

Handouts Many professional and technical presentations incorporate handouts, which can be duplicates of written material or slides, summaries of key points, supporting matter in graphs or charts or tables, or other supplementary information (see Figure 23.4). Handouts can reinforce the presentation. They are usually easy and inexpensive to prepare and can provide audiences with a space to take notes as well as something to take away from the presentation. However, some people can be distracted by handouts and can be tempted to read ahead, rather than focusing on what you have to say. If you do use handouts, they should be distributed to each member of the audience, and you should allow time for that distribution; it may take several minutes to circulate handouts to a large audience.

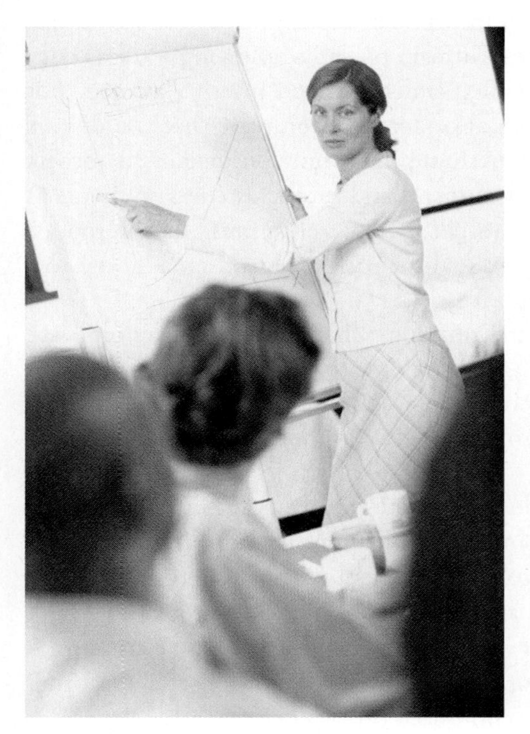

FIGURE 23.5 A flip chart

Posters and Flip Charts In presentations to smaller, more intimate groups, posters and flip charts can be effective methods of displaying visual information (see Figure 23.5). Posters can contain five to ten different boxes to show a procedure or a series of events in visual form. Flip charts usually consist of sheets of paper that are flipped forward or backward on an easel. Most can be written on by the speaker, which keeps audience members interested and involved in the presentation. Flip charts are especially useful in brainstorming meetings, when audience members can suggest ideas as the presenter writes them down. However, both posters and flip charts are difficult to see at a distance and are therefore inappropriate for large audiences and large locations.

Chalkboards, Whiteboards, and Smart Boards Presenters can use boards of various types to create or display information. These boards range from the basic chalkboards and erasable whiteboards available in most classrooms and offices to wireless smart boards, often called digital or interactive whiteboards (see Figure 23.6). Nearly all boards allow presenters to create graphics and charts on the fly, although sophisticated or highly detailed graphics are difficult to create using chalk or a pen. Most smart boards come equipped with touch pens, which combine the features of a computer mouse with those of graphic design software. Like posters and flip charts, boards are best suited to small groups because they are difficult to see at a distance.

Slides and Overhead Transparencies Both slides and overhead transparencies are easy to use, inexpensive, and readily available in most offices. Both can project text or images from a projector onto a blank wall or a screen. Although slides are quickly being replaced by digital photographs, most photo-processing stores can still create two-by-two-inch slides from film-based camera images. Overhead transparencies can be easily made by photocopying a printed page onto clear acetate paper. Slides can be useful in showing actual images of places or products; overheads can be useful in creating overlays of information (see Figure 23.7). However, when compared to computer projections, both slides and transparencies seem outdated and plain; they cannot include motion or sound and require a dark room to be viewed.

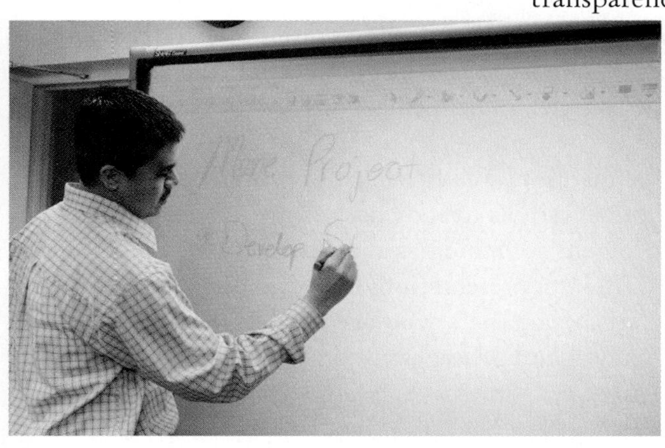

FIGURE 23.6 A digital smart board

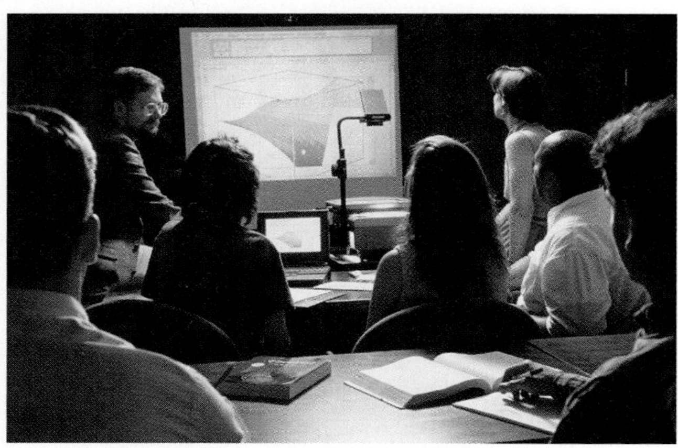

FIGURE 23.7 An overhead transparency

Computer Projections In today's technical world, computer projections are the primary visual aid used in presentations. In presentations to small groups, a monitor often suffices as a viewing screen; larger groups require a digital projector and screen or a high-resolution wide-screen television connected to a computer. Although web-based applications are widely used, most computer-projected presentations use some form of presentation software. Microsoft PowerPoint is by far the most popular program used to create slide shows, although other programs contain similar features and capabilities.

Presentation software has become increasingly user-friendly and powerful, enabling you to write the presentation while simultaneously creating slides; apply design templates and styles to the presentation; import content, graphs, and charts from other programs; and incorporate sound, video, and web links. In addition, many programs are platform portable, allowing for easy conversion to various word processing, print-ready, and Internet formats. For example, you can create a presentation for your company that you deliver to audience members in person through a digital projector, to others within the company as a PDF document, and to stockholders or clients as an XHTML document on the Internet. And wireless presentation rooms, like the one shown in Figure 23.8, allow presenters to link a laptop to a digital projector for convenient access.

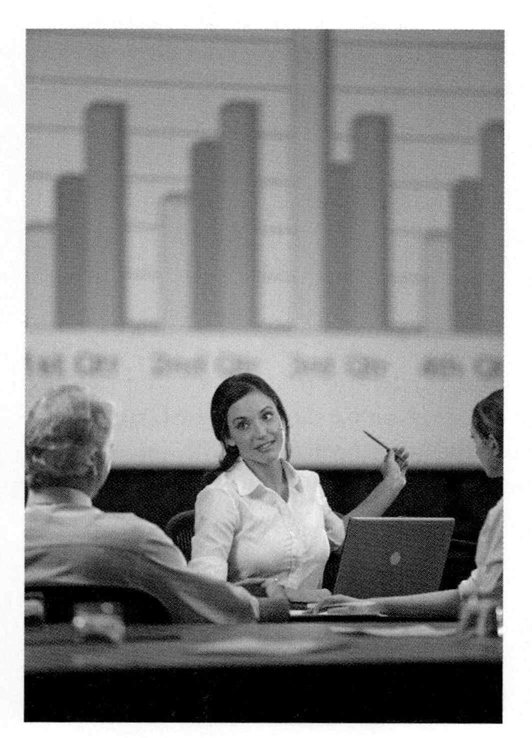

FIGURE 23.8 A computer presentation

Because these programs are constantly changing and developing, we do not offer here specific instructions on how to use them. However, most workplace presenters are familiar with and use such software, so you should learn how to use the program most common or applicable to your profession.

Computer projections do have some disadvantages. Of the visual aids we've mentioned here, computer projections have the highest initial cost—to purchase the hardware and software required to create and display them. And like slides and overheads, computer projections can't be easily delivered in outdoor or on-site locations that won't accommodate a computer and projector. Furthermore, unlike most of the other visual aids we've mentioned, computer projections can't be altered or written on during the presentation. And although the design templates and styles can be useful in the creation process, they do give you a limited number of choices. In addition, the physical space of computer slides sometimes requires presenters to break information into unnatural segments; complex and sophisticated topics cannot always be summarized in short, bulleted lists. Finally, computerized projections can be the most time-consuming of all the visual aids to create—especially if you include complex multimedia.

See chapter 5 for more about writing for transnational audiences.

Internet Presentations As we mentioned earlier in this chapter, some companies have begun broadcasting their presentations to audiences over the Internet. Internet presentations can incorporate different technologies and may consist of slides, video projections, or even animations. Internet presentations have two distinct advantages. First, they allow presenters to reach audiences who are not physically present; an expert in Iowa can deliver a presentation to colleagues or branch offices in New Jersey, Bangladesh, and New Zealand all at the same time. Second, these presentations can

FIGURE 23.9 A videoconferencing presentation

A clear heading signals the main idea

- Details or explanation should follow

FIGURE 23.10 An example of signaling

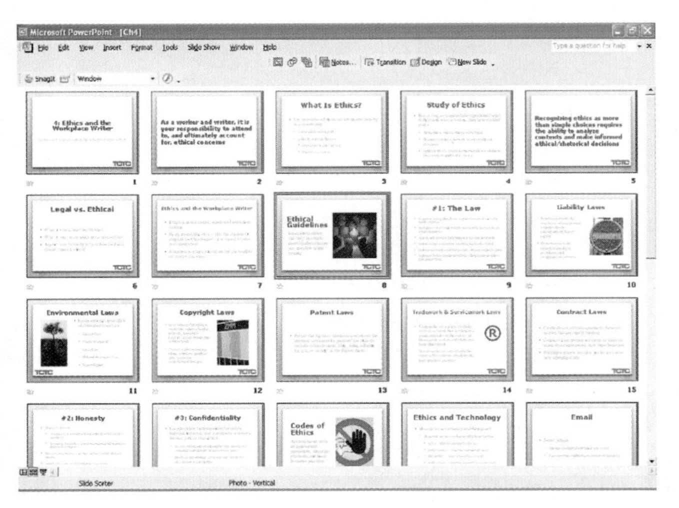

FIGURE 23.11 An example of segmenting

take place in real time, allowing users to interact with each other, or they can be stored and archived by users to view at their convenience. Some real-time Internet presentations take place using webcams or videoconferencing software, like the presentation depicted in Figure 23.9. Other presentations are placed online for later viewing—for example, Internet tutorials of products and services for customers and training seminars for employees. You should be aware of your viewers' technological limits if you do plan to deliver a presentation to them through the Internet. You should also follow the advice in chapter 5 if transnational audiences might view your Internet presentation.

Creating Effective Visuals

Although visuals can be effective supplements to a presentation, they must be chosen and created carefully to ensure that they enhance audience interest and understanding; this is especially true with computer projections and the slides that often accompany them. Even though it might be tempting to try to define what makes a good or bad slide presentation, it is more useful to think of what is effective or ineffective for each particular occasion. As with all forms of communication, effectiveness depends on the context; it is the product of circumstance. Consequently, there are few hard rules for creating slides. Nonetheless, you can be guided by five principles of effective visual design, which are drawn from the research of multimedia learning specialists Cliff Atkinson and Richard Mayer:[4]

Principle 1: Signal Visuals are effective when the material in them is organized and is signaled with a clear heading or description that explains the main idea being covered. Headings should be succinct and easy to find, and in most instances they should contain a subject and a verb in an active voice (see Figure 23.10).

Principle 2: Segment Visuals are effective when information is presented in bite-sized segments. One common problem with visuals is that presenters try to cram too much information into a single visual, which overwhelms and confuses the audience. You should try to keep an even pace from one visual to the next, without spending too long on any one visual. If you use presentation software, a viewing format allows you to see all of the slides at once to be sure you've used manageable segments in each slide (see Figure 23.11).

---

[4] Cliff Atkinson and Richard Mayer, "Five Ways to Reduce PowerPoint Overload," http://www.sociablemedia.com/PDF/atkinson_mayer_powerpoint_4_23_04.pdf.

The Problem-Solving Approach

- Plan
- Research
- Draft
- Review
- Distribute

Plan
-Define or describe the problem
-Establish your goal or purpose
-Identify stakeholders
-Consider document types

Research
-Determine sources
-Conduct research and gather information
-Organize and group the information you gathered

FIGURE 23.12 An example of narrating

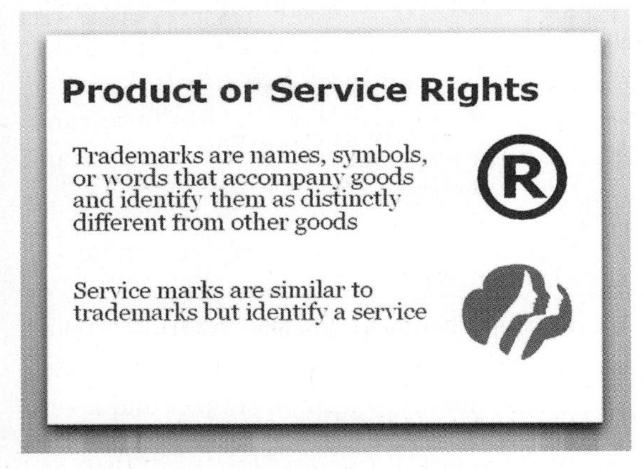

FIGURE 23.13 An example of showing and telling

Principle 3: Narrate Visuals are effective when the words are narrated by the speaker rather than being displayed as on-screen text. You can reduce the visual load on your audience by moving the narration off-screen. Although your visuals will probably include some text, you should not have large sections of text on a single visual and should usually avoid punctuated sentences. You can highlight keywords, phrases, facts, and figures instead. In a slide presentation such as PowerPoint, you can use the Notes feature to write out the narration and then summarize it on the slide itself (see Figure 23.12).

Principle 4: Show and Tell Visuals are effective when text is combined with images and pictures. Text generally tells an audience something, whereas images and pictures show them; combined, these two elements explain and reinforce your ideas (see Figure 23.13). Graphic elements also add interest and help keep audiences focused and alert. However, you should not overwhelm audiences with many different images in one visual or use an image in every visual. And you should be sure that visuals are appropriate and professional; avoid clip art that doesn't add anything to your presentation.

Principle 5: Streamline Visuals are effective when they are concise, simple, and to the point; remove any text or graphic that does not support your main idea. Because audiences can process only so much information, excluding extraneous information allows them to focus on the information that is most important in the presentation. This principle applies to text as well as to visuals, backgrounds, and logos; when in doubt, cut it out (see Figure 23.14).

Chapter Summary

- Technical and Professional Communication is communication about complex, highly detailed problems, issues, or subjects in the professional world
  - Helps audiences visualize and understand information so they can make informed and ethical decisions or take appropriate and safe actions
- Various forms of communication exist in workplaces, but writing is often the most common and most important.
  - This book asks students to understand technical and professional writing in three main ways: 1. Writing as Documents, 2. Writing as an Activity, and 3. Writing as Problem Solving.

Technical and Professional Communication

- Deals with complex professional problems or subjects
- Helps audiences visualize and understand information
- Allows readers to make informed and ethical decisions
- Considers writing as
  - Documents
  - An activity
  - Problem solving

FIGURE 23.14 An example of steamlining

Visuals and Transnational Audiences

See chapters 5 and 8 for more about using visuals when writing for transnational audiences.

Even though visuals can help clarify, emphasize, and organize your information, when preparing presentations for transnational audiences, you'll need to account for culturally different approaches to understanding visuals. Here are some points to remember:

- Different cultures interpret color differently.
- Some images might be regarded as insulting or inappropriate by some transnational audiences.
- Use of gestures can mean different things to different audiences.
- Familiar examples for one audience may not be common for another.
- Not all cultures read left to right; visuals should guide readers.
- Multiple examples can help reach multiple audiences.

Additional Suggestions

- Use text and visuals that can be read and seen at the back of the room. Most text in presentations should be 30-point font or larger.
- Instead of blocks of text, use bulleted or numbered lists that contain phrases and keywords. Most slides should contain no more than forty words and five lines or list items.
- Give each slide a heading or title, using a type size at least 10 points larger than the body text.
- Provide contrast between background and text; light backgrounds with darker text are usually easier to read.
- Be consistent with font style, sizes, spacing, and backgrounds.
- Use color sparingly to highlight main points and focus attention.
- Use graphs and charts rather than tables to quantify data; viewers often find the information in tables difficult to grasp.
- Avoid overcrowding and use plenty of white space.
- Try to use no more than fifteen slides in a presentation; more than that overwhelms an audience.
- Don't display a slide until you are ready to talk about it.
- Avoid the excessive use of enhancements in slide shows, such as fly-in text, noises and sounds with no real point, and meaningless or vague clip art.
- Avoid preset timings in your slides; a good presentation is flexible and may not fit an exact amount of time for each slide.

ANALYZE THIS

The presentation for Holland America Cruises that is shown in Figure 23.15 is an example of a professionally produced, visually dynamic series of slides. Analyze the presentation. What makes it effective or ineffective? Are the slides organized in a clear manner? Does the presentation seem to follow the five principles of visual design mentioned earlier? Does it follow the additional suggestions for slides? Which aspects or features of this presentation might you apply to your own presentations? Which aspects of the presentation would be difficult or unnecessary for you to apply?

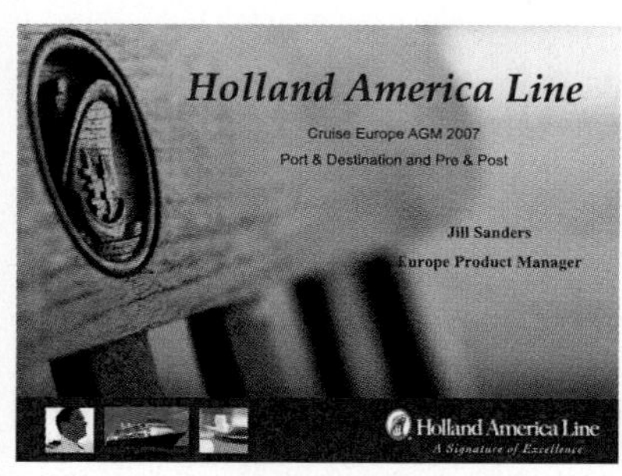

Cover slide depicting name and title

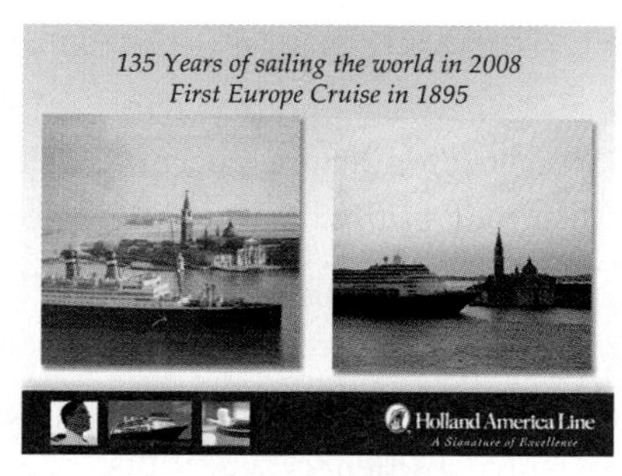

Introductory slide presenting the concept

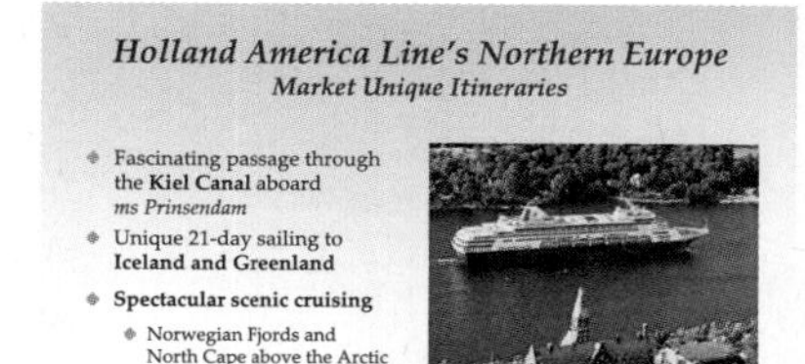

Body slide containing facts and image

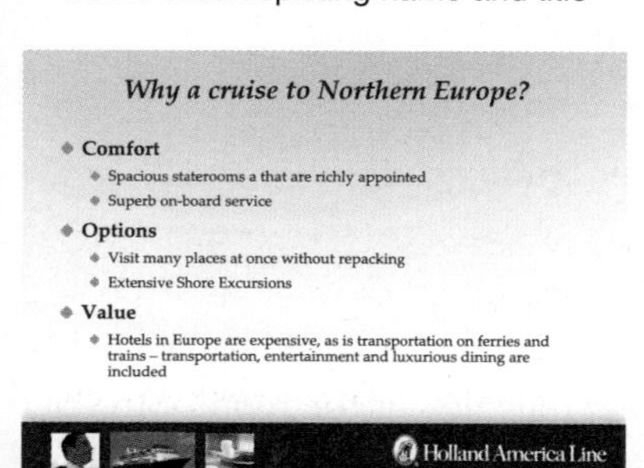

Body slide containing further detail

Body slide forecasting future slides

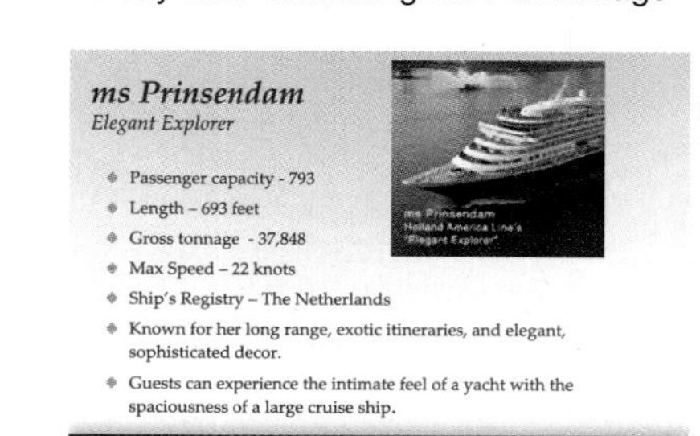

Body slide depicting first concept

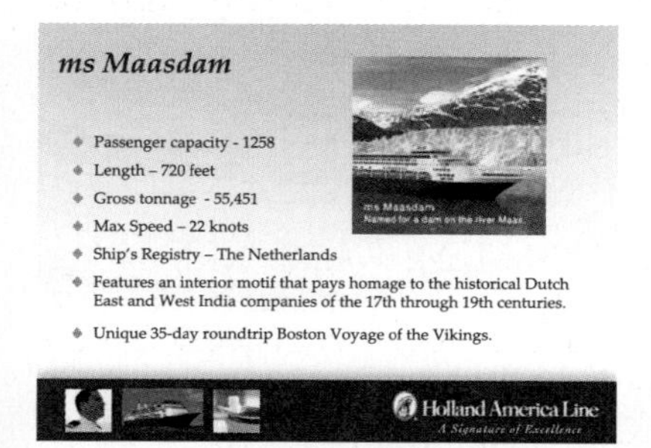

Body slide depicting second concept

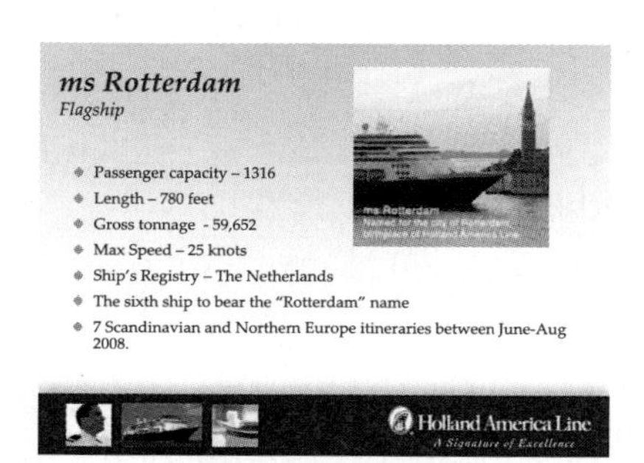

Body slide depicting third concept

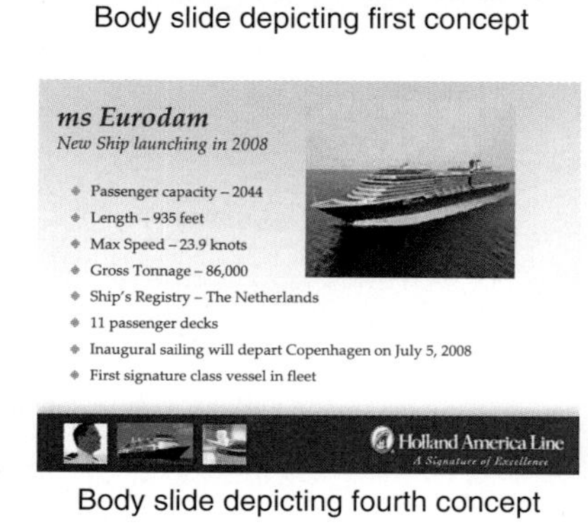

Body slide depicting fourth concept

Body slide presenting full image

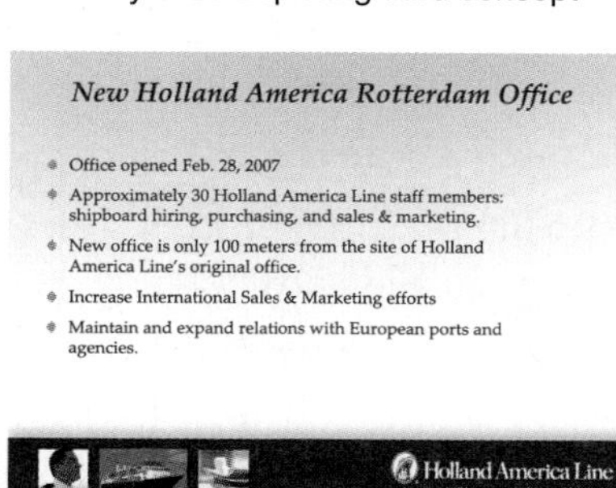

Body slide tying into previous slide

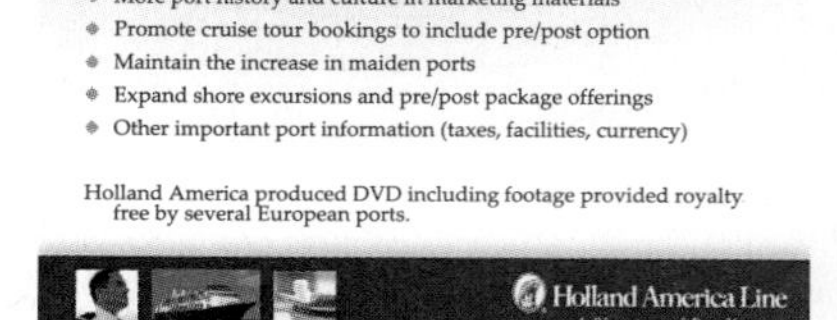

Concluding slide that summarizes

FIGURE 23.15 A professional slide presentation

PLANNING YOUR SPEECH

Planning a speech for a presentation is affected by the relative formality of the occasion. Generally speaking, formal presentations require more in-depth, specific planning than informal presentations do.

Choosing Your Written Format

There are three primary methods of preparing a speech: writing it out, creating an outline, or using notes to guide you. In some highly stressful, difficult situations, you might write out a formal speech that you would read word for word. Written speeches offer security to an anxious speaker, but reading a speech is usually the least effective approach because it offers little spontaneity and interaction with the audience. Many formal presenters prefer some form of outline, which can be a highly structured, numbered account of each topic, idea, and fact; or it can be a more loose collection of ideas in a rough order of importance or chronology. The third method is to prepare a series of note cards to guide you through the presentation; these can follow an outline organization or not.

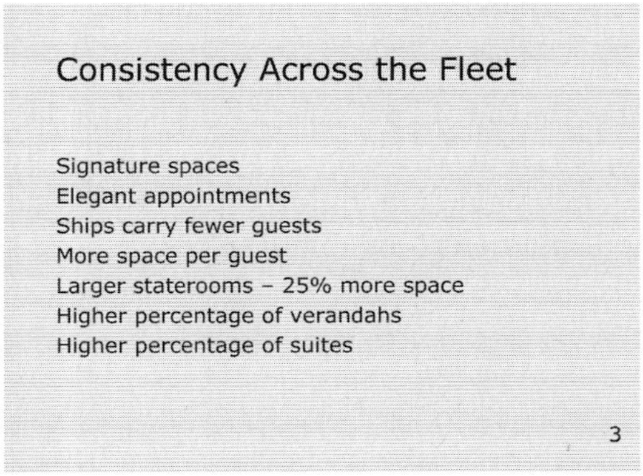

Using presentation software is advantageous because it allows you to convert existing slides into outline or note card form. Or you can create an outline and then convert it into a slide show presentation. If you are comfortable with your presentation topic and know it well, you may choose to convert the text in the presentation into an outline or note cards, as the example in Figure 23.16 demonstrates. However, you must be sure that you do not simply read all of the text in your notes word for word. If you are not comfortable with your topic or with public speaking in general, you can use a more detailed outline or note cards to guide you through the presentation. If you do not use a program that automatically inserts page numbers into your outline or notes, be sure to number them before your presentation; nothing is more frightening than losing your place when you are in front of an audience. Also, be certain to use large enough text and enough white space with your notes so that you can quickly glance at them to guide your speech.

Consistency Across the Fleet

Signature spaces
Elegant appointments
Ships carry fewer guests
More space per guest
Larger staterooms - 25% more space
Higher percentage of verandahs
Higher percentage of suites

3

FIGURE 23.16 Presentation text converted into a note card

Organizing Your Content

As with all other forms of workplace communication, you should think carefully about the organization of your presentation and should include an introduction, a body, and a conclusion. However, three primary organizational differences exist between a written document and an oral presentation.

Use of Detail

Most people can absorb more information when they see it than when they hear it; consequently, oral presenters must be careful not to overwhelm listeners. When delivering concepts, ideas, and facts, you should clarify them as much as possible. Because the audience can't go back to relisten to something you said, it must be clear the first time. And you should avoid any extraneous information, focusing just on the essentials needed to solve the problem. If you think the audience cannot absorb all the necessary information, prepare a handout or other document that audience members can take with them.

Use of Overlap

One way to make up for the lower retention rate of listeners is to repeat information to a greater degree than you would in a document. Although you should not be overly redundant, it is useful to provide more context and some restatement of important ideas in your presentation. Many presenters use the classic approach—"Tell 'em what you're going to say, tell 'em what you have to say, and then tell 'em what you just said"—in their introduction, body, and conclusion. This is often called a *shingle style* because it provides overlap in ideas, just as shingles are overlapped on a roof.

Use of Explicit Signals

Listeners often require more guidance than readers do in the flow of information. Specific types of language can signal the type of information that will follow; three common types are discussed here:

Advance Organizer An *advance organizer* is any information, such as an introduction, that tells listeners what they're about to hear. This statement—"In the next ten minutes I will explain how to use the automated tax return"—is an advance organizer because it provides structure for the information that follows.

Transition A *transition* is a verbal cue that a shift is being made from one idea to another. Transitions often refer to sequence or chronology. For example, statements such as these—"First I will discuss the log-in feature" and "The final thing you'll do is electronically sign your form"—show a transition from one point to the next.

Summary A *summary* provides an overview of what's been said, reviewing important details or ideas. Summaries often occur at the end of a presentation, although they can also be used to recap a section or a complex point. A statement such as this—"To recap, filing your automated tax return requires logging in, filling out the form, and electronically filing it"—signals that the presentation is coming to an end.

IN YOUR **EXPERIENCE**

For the presentations you've done in the past, you probably did something to prepare what you would say. Did you write out a formal speech? Did you use an outline? Did you create a set of note cards containing your main points? Did you print out the slides or notes from a software program? Was the method you used effective? Why or why not? Describe in writing how you prepared what you'd say in those presentations, and then discuss those approaches with your classmates. Which methods were most effective in which situations? Why?

DELIVERING A PRESENTATION

An effective presentation depends on two forms of communication that are just as important as the information and the visuals you use. To fully engage an audience, you must use effective vocalization and body language.

Vocalization

Vocalization refers to the way we say words. Most of us do a pretty good job of vocalizing in everyday conversation, but the anxiety created by a presentation can influence verbal delivery. By carefully practicing and paying attention to certain aspects of vocalization, you can be sure to keep listeners attuned and interested.

Use Adequate Volume

Finding the right volume for the space in which you deliver your presentation can be tricky, especially in a large room with poor acoustics. As Monty Hansen suggests in his interview at the beginning of this chapter, many people speak too softly during presentations. If you practice your presentation with an audience of friends or colleagues, ask one of them to sit in the back of the room to see whether you can still be heard. If you use a microphone, make sure that the volume is turned up enough that listeners can hear you but it's not so loud that your voice becomes distorted. If you haven't been able to practice your delivery in the actual room, ask members of the audience whether they can hear you. This is a simple, interactive icebreaker to begin a presentation.

Vary Pitch

Good pitch involves the sort of up-and-down inflections and emphasis that most of us use in an interesting conversation with someone else. Some speakers have a tendency to flatten their pitch into a monotone when delivering information, usually resulting in a bland, boring delivery. Others try too hard to sell their ideas by overemphasizing them with excessive pitch—you've probably seen used-car commercials that do this. Try to find a happy medium, raising and lowering your pitch in a natural way to add emphasis to important points.

Pace Yourself

The best speakers are those who use proper pacing and speed in their delivery; novice speakers often talk too quickly. You should slow down to make sure listeners can hear everything you say. And you can use the pace of your speaking to help clarify what you have to say—speak slowly when delivering complex, important, or difficult information, and pause at the end of each section or idea for emphasis. Above all else, don't forget to breathe; passing out during a presentation is not an effective delivery strategy.

Articulate Words Carefully

Careful and accurate pronunciation is also important during presentations. You should avoid rushing through tongue twisters and difficult-to-say words; enunciate them slowly and clearly so that listeners are certain to understand them. Accurate pronunciation also requires that you say words correctly, as they are meant to be said. For example, the phrase "nuclear weapon" should never be pronounced "new-cue-lar weapon." Check a dictionary if you're unsure of what is correct.

Avoid Filler Phrases

Unprepared or nervous speakers use fillers such as *uh*, *um*, *like*, and *you know* when searching for the right word or the next idea. These nonfluencies will detract from the effectiveness of your speech; it is better to simply remain silent as you think about what to say next. And you should not apologize if you mispronounce or can't remember something; it is better to simply say it again or move on, rather than drawing further attention to it. Planning and rehearsing helps minimize fillers and nonfluencies.

Body Language

Body language consists of all the nonverbal cues that you deliver through your physical presence. Audiences not only hear you but also see you. Good body language is important in capturing and keeping your audience's attention; successful speakers use body language to their advantage, mindful of effective characteristics. Body language is also culturally interpreted, so when presenting to a transnational audience, you must be sure to understand appropriate conventions.

Face the Audience

As simple as it sounds, simply facing the audience is an important aspect of body language because it lets audience members know that they are the focus of your presentation. You should keep your back erect and your head straight (but not stiff) and should lean slightly forward to show interest in your listeners. Avoid facing a screen, slumping, or looking down at your feet because these actions imply disinterest or uneasiness.

Make Eye Contact

Making eye contact with audience members allows you to continually assess whether your message is getting through; you can judge by their nonverbal responses. Keeping your eyes focused on your listeners also helps you avoid looking at your note cards or projection screen too much, which would distance you from your audience. Even when speaking to a large group, you should search for the eyes of those people who are engaged and speak directly to them. But be careful not to speak to only one individual, because that individual might become uncomfortable and others might feel excluded.

Use Movement and Gestures

There are several ways to use movement and gestures to establish a connection with your audience. Using your hands while you talk is a natural method of gesturing, and it can add emphasis to what you are saying. It can also give the audience an impression of action and animation. In addition, pointing at a graphic or an object in your presentation, whether with your hands or with a pointer, draws audience attention to that item. Even simply walking around can be an effective method of connecting with an audience. However, you should avoid excessive or meaningless movements and gestures; pacing back and forth, touching your hair or face, or tapping a pen on a table conveys nervousness and apprehension.

IN YOUR **EXPERIENCE**

Think about an effective presentation that you've witnessed in the recent past; this might even be one of your teacher's presentations in this class. How did the speaker use vocalization and body language to keep the audience engaged, interested, and attentive? Write down the specific aspects of vocalization and body language that made the presentation memorable. Then compare your list with that of one or two of your classmates, looking for similarities and differences among them.

Fielding Questions

A good presentation is more like a conversation than a lecture; interacting with the audience is something you should always deem important. Part of this interaction involves having dialogue with listeners and addressing the questions they have. Formal presentations usually address questions at the end of the talk, whereas some informal presentations allow continual discussion and interaction. You should let your audience know beforehand how you will address questions. Not only is this a good icebreaker, but it also avoids unwanted hand-raising and interruptions while you are speaking.

Thinking ahead about the sorts of questions the audience might ask will make the Q & A session much easier for you, especially if you are speaking about a controversial or highly important subject. Whenever possible, try to answer questions by referring to something specifically mentioned in the presentation—for example, you might refer to a graphic or a chart. And when you are asked a question, it is sometimes helpful to repeat it to ensure that you heard it correctly and to benefit others who may have missed it. This practice is especially useful when you are presenting to large groups and are using a microphone or are being recorded.

Of course, not all questions are easy to answer or understand. If the questioner rambles, asks a series of questions, or is unclear, you can restate the issue by saying, "What you seem to be asking is . . . ." If the question is unintelligible, simply ask the questioner to repeat it. You should try to answer any fair questions as specifically as possible, but some questions may be overtly hostile or biased. In such situations you can rephrase the question to more directly address your subject, or you can propose an alternative way of looking at it. If an audience member uses the opportunity to grandstand or to simply state a personal perspective, thank the individual for the input or offer to discuss the subject further after the session. Occasionally, you might be blindsided by a question you hadn't anticipated, and you may not have a quick response. Rather than making something up, it is better to say that you don't know the answer. If the question is a good one, you can offer to find the answer and share it later, perhaps through e-mail or at a later meeting. And if you do intend to answer the question later, be sure to write it down immediately to reassure the questioner that you intend to follow up.

When your time is up or you feel you've addressed all of the cogent questions, take a minute or two to summarize the main point(s) of your presentation. Thank the audience for joining you, and encourage them to e-mail any further questions to you if that seems appropriate.

SUMMARY

- The problem-solving approach can be useful in creating and delivering presentations.
- The degree of interaction between the speaker and the audience makes presentations different from most other forms of workplace communication.
- Presentations can be classified as informal or formal.
- Presentations can be delivered to many different audiences: clients or customers, colleagues, supervisors and employers, and the public.
- When preparing for a presentation, you should reflect on why you are presenting, identify your audience, assess the physical context, determine the time available, gather and evaluate the information, and choose and create the appropriate visual aids.
- Visual aids usually serve to explain, to emphasize, to interest, and to guide.
- Common visual aids include handouts; posters and flip charts; chalkboards, whiteboards, and smart boards; slides and overhead transparencies; computer projections; and Internet presentations.
- You should consider the principles of visual design as you create presentations: signal, segment, narrate, show and tell, and streamline.
- The three primary methods of preparing a speech for a presentation are writing it out completely, creating an outline of what you will say, and using note cards to guide you through the major points.
- The major differences between the audience for a document and the audience for a presentation are that listeners have less capacity for detail, require more overlap, and require more explicit signals.
- Effective vocalization involves using adequate volume, varying your pitch, pacing yourself, articulating your words, and avoiding filler phrases and nonfluencies.
- Effective body language includes facing the audience, making eye contact, and using appropriate movement and gestures.
- You should be prepared to field questions before, during, and especially after your presentation.

CONCEPT REVIEW

1. How can you use the problem-solving approach to create and deliver effective presentations?
2. How are presentations different from other forms of workplace communication?
3. What are the differences between informal and formal presentations?

4. What are the primary audiences for workplace presentations?
5. What questions might you ask to identify and connect with the audience for your presentation?
6. Which aspects of the physical context of a presentation should you consider?
7. How are time and pacing significant in a presentation?
8. What purposes might visual aids serve in a presentation?
9. What are the more common types of visual aids used in presentations?
10. What are the five principles of effective visual design in presentations?
11. What are the three primary methods of preparing what you will say in a presentation?
12. What are the three primary differences between a document and an oral presentation?
13. Which aspects of effective vocalization should you consider in a presentation?
14. How can you use effective body language in a presentation?

CASE STUDY 1

Senator Dereks and the UN Convention on Disabilities

You work as an assistant public relations advisor for state senator Owen Dereks, who recently took part in the United Nations Convention on the Rights of Persons with Disabilities. The convention's goal was to improve the worldwide medical treatment and social rights of persons with disabilities, and delegates from over 100 countries attended.

Senator Dereks has been invited to deliver a five- to 10-minute summary of the convention to a group of statewide industrial leaders who are interested in the UN discussion of the rights of persons with disabilities. The senator has asked you to help him prepare a short PowerPoint presentation for his talk to the industrial leaders. Fortunately, you were able to locate the PowerPoint presentation that was delivered to English-speaking delegates at the convention. You now need to view the presentation on the Companion Website and create a five- to 10-minute presentation geared toward industrial leaders who are interested in learning more about the convention's discussion of rights.

CASE STUDY 2

Trinity Furniture Creates a Web Tutorial

As a regional sales manager for Trinity Furniture, a retail furniture chain consisting of over fifty stores, you do presentations of different types nearly every day. Some of your presentations are informal discussions with individuals looking for furniture for their homes, some are more formal meetings with furniture manufacturers whose products you sell, and some are highly formal presentations delivered to Trinity executives and stockholders.

At a luncheon following one of your presentations to company executives, several of them approach you to say how much they admire your style, vocalization, and body language in delivering presentations. One executive adds that they've begun creating various tutorial videos to be placed on the company website for Trinity employees. He then asks you to create a five-minute informal presentation entitled "Delivering Presentations," to be placed on the company website for the benefit of other sales associates.

Drawing on this chapter's advice on delivering presentations, write out a five-minute informal speech to be delivered, videotaped, and placed on the Trinity Furniture website. Keep in mind that your audience consists of other sales associates who deliver their own presentations to various audiences for various purposes.

CASE STUDIES ON THE COMPANION WEBSITE

Breaking the News: Presenting a Corporate Takeover at Pinnacle Accounting

In this case you are the owner of a small accounting firm that decides to sell the company to a larger national accounting firm, which has given you and your employees fair offers. This case revolves around the kinds of presentations that you would need to give to your employees as they make decisions about whether to continue their work with the national firm or go somewhere else.

Presenting a Very Positive State of the Hospital at Boise Presbyterian

In this case you assume the role of the Director of Nursing at Presbyterian Hospital in Boise, Idaho. You are asked by the president of the hospital to report on the status of your division during a state-of-the-hospital address that he will present to the institution's board of trustees.

Proposing a Bulkhead to Protect Carteret Technical Institute from Coastal Erosion

As a chief engineer for Bogue Coastal Engineering, you are asked by your supervisor to prepare a presentation about a proposal your company has produced for a client. The difficulty with your presentation centers on the fact that you must address many different people, all of whom have different opinions and ideas about the project your company has proposed. Thus, this case asks you to think carefully about ways to successfully present the company's proposal, which includes some technical information, to multiple audiences that do not all understand technical jargon.

Mapping the City: Presenting the Changes to Fixed Food's Delivery Practices

In this case you are the logistics director for a mid-sized food courier service that has recently raised its rates to compensate for the implementation of a GPS system. You are to effectively present explanations and justification for the GPS system to diverse audiences.

VIDEO CASE STUDY

Cole Engineering–Water Quality Division

Finding Employment at Cole Engineering–Water Quality Division

Synopsis

Because of their increased success in a number of areas, Cole Engineering has recently begun hiring a number of new engineers. The Water Quality Division places a job ad and begins to interview possible additions to their team. This case shows the interview of Patrick Little with Asher Harris, Director of the Water Quality Division, and Cynthia Moore, Simulations Lab team leader. The interview is difficult, and the questions, at times, identify flaws in Patrick's application materials.

WRITING SCENARIOS

1. In a small group, create a short presentation that instructs your classmates in how to perform a basic but essential student activity, such as purchasing a textbook, registering for classes, or obtaining an e-mail account. Be sure to

conduct usability testing to ensure that your instructions work, and then present them to the class. Your instructor may assign specific activities to each group.

2. The World Wide Web is filled with websites containing advice about how to deliver effective presentations. Working in a small group, analyze these various websites. Then create a top ten list of things an effective presentation should do or contain. Compare your list with those of the other groups in your class.

3. Select one of the longer documents that you have already composed for this class, and convert it to a presentation, using whichever delivery format and visual aids you feel are appropriate. Deliver the presentation to an audience of your peers or classmates, giving them a copy of the document beforehand so that they can familiarize themselves with it. After your presentation, ask them to provide an analysis of it:
   - Did the presentation cover all of the important or appropriate information in the document?
   - Did the presenter choose an appropriate delivery format?
   - Did the presentation use graphics and images in an effective way?
   - Did the presenter use effective vocalization and body language?

4. It is likely that individuals in your field or intended profession deliver presentations on a regular basis. Conduct a short interview with someone working in your intended profession, asking about the individual's experiences, strategies, and advice concerning effective presentations. You may wish to take notes or record the interview (with permission). Then write a memo to your instructor, describing the interview and what you learned from it.

5. Find an example of a presentation that has been posted on the World Wide Web; many PowerPoint presentations can be found by doing a search for .ppt files. Analyze the chosen presentation to assess whether it would be effective according to the guidelines and criteria in this chapter.

6. Create and conduct a survey of ten to fifteen questions to determine the elements and features most viewers want in the presentations they view. Draw your survey questions from this chapter as much as possible. For example, you might ask respondents whether they prefer a particular type of visual aid. You might ask whether they prefer text, visuals, or some combination in presentations that use slide shows. You might also ask them whether they prefer formal written speeches or loosely organized discussions.

   If you plan to use this survey information to help you design a specific presentation (perhaps for one of the other assignments in this chapter), try to select respondents who you would consider your primary audience; if you want general information, try to select a diverse group of respondents. After you have completed your survey—we recommend surveying at least ten people—compile your results in a memo written to your instructor.

7. History is filled with examples of individuals and groups who used presentations and speeches to persuade others to think and act differently. Some presented in ways we consider ethical, such as Martin Luther King, Jr. or John F. Kennedy; others have used presentations and speeches to achieve unethical goals, such as Adolph Hitler and Osama Bin Laden. Visit your library and find a videotaped example of a presentation or speech by a historical or political figure. Analyze the presentation. What strategies did he or she use? How were vocalization and body language used? How did the speaker's goals and ethics influence the ways in which information was presented?

8. Find a presentation from a group whose beliefs, goals, or perspectives differ from your own. For example, you might examine a presentation posted online by an opposing political or social group or a competing company. Based on the presentation guidelines and advice given in this chapter and others in this book, analyze the presentation. Does it offer an accurate portrayal of the information? How does the presentation seek to persuade viewers? Does it present information about the organization or group ethically and fairly?
9. Presentations—particularly public presentations—are often delivered to diverse groups. As a result, presenters must consider and accommodate audience members who have special needs, such as those who are hearing or vision impaired. Create a code of ethics that presenters should consider when addressing diverse audiences that may include individuals with such impairments.
10. Look at the templates or slide designs of your favorite presentation software program. Choose five of the templates that you find interesting, and write a one- to two-paragraph analysis of each, identifying the overall style and feel of the design as well as its most appropriate subjects and audiences. Share your analyses with a classmate.
11. Within the next five to ten years, you will probably have to create and deliver several, perhaps many, different types of presentations. Write about the various types of presentations you expect to make in the future, focusing on the design choices and the visual aids that you think might be best for each type of presentation.
12. Color can be important in slide show presentations. Write about the ways in which you might use color in presentations. For example, if you use a light background in a slide show, what types and colors of images and text would work best? Do you prefer darker backgrounds or lighter backgrounds in your slide shows? Why? How do the physical spaces in which you deliver presentations affect your choices in colors?
13. Technology has certainly impacted contemporary workplace (and classroom) presentations. Can effective presentations still be done without the use of technology? Why or why not? Which technologies do you believe are essential to workplace communication? Which do you believe are outdated or unnecessary? In a memo to your instructor, address the role of technology in presentations.
14. Imagine that you have just graduated from college and have applied for a job in your field of study. The employer has asked you to create a five- to ten-minute presentation about yourself as a potential hire. Using presentation software, create a slide show that explains how your course work, activities, and work experience make you a good choice for the advertised job position. Then deliver the presentation to an audience of your peers or classmates. You may wish to consult chapter 14 to help you prepare for the presentation.

PROBLEM SOLVING IN YOUR WRITING

Select one of the documents or presentations you designed and created for this chapter. Consider how you employed the problem-solving approach in creating that document. Write a letter to your instructor identifying how you planned for the situation, conducted research to gather information, drafted the document or presentation, reviewed it, and finally distributed (or plan to distribute) the document or presentation.

APPENDIX A

HANDBOOK

STANDARD PROOFREADER'S MARKS

SYMBOL	MEANING	EXAMPLE
	Delete	Remove indicated characters or words
	Close up	Close up th e gap
	Delete and close up	Delete and close close the gap
	Insert	Insert word a or character t
#	Insert a space	Inserta # space
	Insert a comma	Insert a comma if you need one
	Insert an apostrophe	Youll need an apostrophe
	Insert a period	Insert a period
	Insert a semicolon	Insert a semicolon it might be useful
	Insert a colon	Insert a colon in these situations
or	Insert quotation marks	Insert double or single quotation marks
stet	Ignore the mark and keep the original	Keep the original stet
	Transpose	Transpose the order word or charatcer order
	Indent to the right/# of spaces	5 Indent 5 more spaces
	Indent to the left/# of spaces	5 Move 5 spaces to the right
	Align horizontally	Align horizontally
	Align vertically	Align vertically with surrounding text
	Begin new paragraph	Begin new paragraph
sp	Spell out	A 15 page report sp
sp	Incorrect spelling	Incorrect speling sp
cap	Set in CAPITALS	Set in capitals cap
lc	Set in lowercase	Set in Lowercase lc
ital	Italicize	Set in italics ital
bf	Make boldface	Set in boldface bf
rom	Make roman (plain text)	Set in *plain text* rom

ISSUES OF GRAMMATICAL CORRECTNESS

Sentence Fragments

A sentence fragment is an incomplete sentence that is punctuated to look like a sentence. A complete sentence must contain at least a subject and a verb (i.e., an independent clause). Workplace writers should avoid using sentence fragments because they can interrupt the flow of a document. They feel choppy, causing readers to stumble and alerting them to the fact that some element is missing.

Workplace writers often produce sentence fragments by simply forgetting to include a key word, like a verb or a noun. Fragments are most easily corrected by inserting the missing word(s) or repunctuating the fragment as part of the sentence that appears before or after the fragment.

Fragment: Conducted the field test.

Correction: The engineering firm conducted the field test.

Fragment: When the engineering firm completed the field test. The remaining phases of the project seemed feasible.

Correction: When the engineering firm completed the field test, the remaining phases of the project seemed feasible.

Run-on Sentences

Run-on sentences occur when writers punctuate two or more independent clauses as though they are one sentence. Run-on sentences can confuse readers by forcing too much information into a single sentence.

Run-on sentence: The World Wide Web Consortium has released updates to its XML protocol standards XML technologies may now be used in a variety of applications.

Correction: The World Wide Web Consortium has released updates to its XML protocol standards. XML technologies may now be used in a variety of applications.

Run-on sentence: The World Wide Web Consortium has released updates to its XML protocol standards, XML technologies may now be used in a variety of applications.

Correction: The World Wide Web Consortium has released updates to its XML protocol standards, and XML technologies may now be used in a variety of applications.

Run-on sentences can be corrected in four ways:

- Dividing the run-on into multiple, independent sentences

  Run-on: The biotech researcher is working with proteins he is trying to identify patterns of carbon atoms on the protein surface.

  Correction: The biotech researcher is working with proteins. He is trying to identify patterns of carbon atoms on the protein surface.

- Combining the two independent clauses with a comma and a coordinating conjunction

  Run-on: Informational goods have surpassed industrial goods as forces in world markets many executives have been forced to rethink their companies' positions in providing industrial goods.

  Correction: Informational goods have surpassed industrial goods as forces in world markets, and many executives have been forced to rethink their companies' positions in providing industrial goods.

- Combining the two independent clauses by making one subordinate to the other

  Run-on: The exclusive five-year contract includes impressive service agreements the agreement does not cover service on the entire airline fleet.

  Correction: The exclusive five-year contract includes impressive service agreements although the agreement does not cover service on the entire airline fleet.

- Combining the two independent clauses with a semicolon

  Run-on: MP3 is a compression format designed specifically for audio files it uses psychoacoustic models to eliminate components less audible to human hearing.

  Correction: MP3 is a compression format designed specifically for audio files; it uses psychoacoustic models to eliminate components less audible to human hearing.

Modifiers

Modifiers explain and clarify the meaning of parts of a sentence. Modifiers can be single words, phrases, or clauses in adjective or adverb form.

Single-word adjective modifier: <u>genomic</u> samples, <u>twelve</u> USB ports

Adjective phrase modifier: list <u>of parts</u>

Adjective clause modifier: The manual <u>that had been misinterpreted</u> was rewritten.

Single-word adverbial modifier: Chris submitted the proposal <u>late</u>. The test was <u>significantly</u> changed.

Adverbial phrase modifier: The parts will be delivered <u>in two weeks</u>.

Adverbial clause modifier: The report will be written <u>after the data have been gathered</u>.

Dangling and Misplaced Modifiers

Modifiers must have clear referents to modify. To correct a dangling modifier, writers should provide a logical referent. A misplaced modifier should be placed in close proximity to the relevant word.

Dangling modifier: Although unclear, the agency accepted the proposal.

Correction: Although the proposal was unclear, the agency accepted it.

Dangling modifier: Upon reading the report, a dangling modifier caught the CEO's attention.

Correction: Upon reading the report, the CEO identified a dangling modifier.

Misplaced modifier: The owner's manual came with the equipment that the technician had read.

Correction: The owner's manual, which the technician had read, came with the equipment.

Misplaced modifier: The engineer told the supervisor eventually the plans would be ready.

Correction: The engineer told the supervisor the plans would be ready eventually.

Pronouns

Pronouns are words that can replace or stand in for nouns. Different kinds of pronouns serve different purposes.

Interrogative pronoun: To whom was the progress report sent?

Personal pronoun: The company is committed to its existing contracts.

Relative pronoun: The design that the team chose was the most interactive.

Indefinite pronoun: The manager had to notify everyone about the policy change.

Agreement

In order for pronouns to be understood, they must clearly identify or reflect the nouns that they replace, that is, their antecedents. Pronouns must agree with their antecedents in both number and gender.

Incorrect agreement: The department purchased these computers because it is the best on the market.

Correct agreement: The department purchased these computers because they are the best on the market.

Writers should confirm gender when names are not gender specific or are culturally unfamiliar.

Verbs

Verbs are words that denote action and states of being (often called "to be" verbs).

Action: The IT technician replaced the hard drive.

State of being: The manual is incorrect.

Infinitives

Infinitives are the basic form of verbs. Most often, they are presented with "to" and then the root of the verb; they can function as nouns, adjectives, or adverbs.

Infinitive as noun: To write a proposal would be helpful.

Infinitive as adjective: Here are the documents to review.

Infinitive as adverb: The system shuts down to save power.

Gerunds

Gerunds are *–ing* forms of verbs that function as nouns.

Gerund: Editing requires careful attention to detail.

Gerund: The most interesting part of designing a web page is coding.

Participles

Participles are verb forms that can work as adjectives.

Participle: The cooling tower needed to be inspected.

Participle: The executives were impressed when they read the approved budget.

Agreement in Number and Person

Verbs must agree with their subjects in number and in person.

Incorrect number agreement: The keyboards delivers a quicker response.

Correct number agreement: The keyboards deliver a quicker response.

Incorrect person agreement: You is the winner.

Correct person agreement: You are the winner.

Voice

Verbs can appear in either the active voice, indicating that the subject acts through the verb, or in the passive voice, indicating that the subject receives the action of the verb.

Active voice: The design team approved the proposal.

Passive voice: The proposal was approved by the design team.

Active-voice sentences add clarity because they show who or what is acting. Passive-voice sentences can obscure meaning by de-emphasizing the role of the subject, or they can emphasize results. Active-voice verbs are generally preferable.

CORRECT USE OF PUNCTUATION

Periods

Periods end all complete sentences except for exclamations and direct questions.

Period to end a sentence: The medical examiner filed the report earlier this week.

See also the sections on sentence fragments and run-on sentences.

With Parentheses

When parenthetical information ends a sentence, the period belongs outside the closing parenthesis. However, when parentheses contain a complete sentence, the period precedes the final parenthesis.

Period outside parentheses: The data indicate that the torque limit was exceeded (see appendix C).

Period within parentheses: The RFP requests that proposals be submitted by July 28. (We may have difficulty meeting this deadline.)

With Quotations

When a quotation ends a sentence, the period belongs inside the quotation marks.

Quotation ending a sentence: The manual indicates that the pressure "should not exceed 400 psi."

With Abbreviations

Periods are generally used in the following abbreviations, but writers should consult the style manual for their particular field:

Pre–nominal titles: Mr., Mrs., Ms., Dr.

Post–nominal suffix: Jr., Sr.

Latin terms: etc., e.g., i.e.

Periods are not used in postal code abbreviations, such as AL, FL, ME, PA, WI.

Question Marks

Question marks indicate interrogatives.

Direct Questions

A question mark ends a sentence that is a direct question.

Direct question: When will the evaluation take place?

Indirect Questions

A question mark does not end a sentence containing an indirect question.

Indirect question: The customer asked who the supervisor is.

Shortened Questions

Question marks end questions that are not fully expressed.

Shortened questions: Will the announcement be made in the newspaper? the newsletter? the board meeting?

Questions within Sentences

Question marks are used with questions embedded in sentences.

Question within a sentence: The Baylor Group—or was it the Taylor Group?—conducted the initial survey.

Question Marks in Titles

If a title of a work contains a question mark, the question mark is always retained.

Title question mark within a sentence: The marketing report *How Can We Make Use of Online Sourcing?* has prompted our new sales strategy.

Question Marks with Quotations

Question marks are placed outside quotation marks unless the quotation itself is a question.

Question mark outside quotation marks: Will we need to review "Ten Suggestions for Upgrade"?

Question mark inside quotation marks: The survey asked, "Which product do you use?"

Exclamation Points

Exclamation points indicate excitement or urgency, but they do not strengthen the meaning of a sentence or boost the argument in a document.

Exclamation point at the end of a sentence: What a lesson we learned!

With Quotations

An exclamation point is placed outside quotation marks unless the quotation itself is an exclamation.

Exclamation point outside quotation marks: How useless to keep saying, "We're over budget"!

Exclamation point inside quotation marks: The speaker ended the presentation by shouting, "Show me the money!"

Commas

Commas are misused and overused, often because writers insert a comma to indicate a pause even though written pauses and spoken pauses are different.

Compound Sentences

Commas are used before a coordinating conjunction—*and*, *but*, *or*, *nor*, *for*, and *yet*—to link two independent clauses in a single compound sentence.

Comma used in a compound sentence: The data file needed to be saved, but it was larger than the flash drive memory.

When compound sentences are made up of two short independent clauses and the meaning of those clauses is clear, the comma is not necessary.

> *Compound sentence without a comma:* The supervisor retired and the manager resigned.

Series

Commas are used to separate three or more items in a list of words, phrases, or clauses.

> *Commas in a list of words:* The service technician ordered routers, connectors, and servers.
>
> *Commas in a list of phrases:* The applicant submitted a résumé, talked with the director, and toured the facility.

In informal writing, such as e-mails and casual letters, writers may opt not to include the comma that precedes the conjunction in a list.

Introductory Words, Phrases, and Clauses

A comma usually follows introductory phrases and always follows introductory dependent clauses.

> *Comma after introductory phrase:* Following the introduction of the new sales plan, the sales manager introduced the new sales force.
>
> *Comma after introductory clause:* After she learned the process, the intern was more confident in her ability.

If an introductory word is a transitional word like *however or therefore*, a comma follows.

> *Transitional word:* Therefore, we must reorganize the distribution plan.

Multiple Adjectives before a Noun

When a noun is preceded by two or more adjectives that could be joined by *and*, they are separated by commas.

> *Multiple adjectives with a comma:* The old, damaged phone in the conference room was replaced.

If a noun and a preceding adjective are seen as a unit, another preceding adjective is not followed by a comma.

> *Multiple adjectives without a comma:* The lecturer was a talented corporate speaker.

Direct Quotations

A comma precedes a direct quotation. If a comma follows the quotation, it precedes the quotation marks.

> *Comma with a direct quotation:* We reported, "These sixteen points must be considered by the full board."

Restrictive and Nonrestrictive Modifiers

Restrictive modifiers supply necessary, or essential, information within a sentence and are not set off by commas. Restrictive clauses often begin with *that*.

> *Restrictive clause:* The e-mail that announced the meeting was posted last week.
>
> *Restrictive phrase:* The manuals explaining the policies will be delivered on Tuesday.

Nonrestrictive modifiers supply extra, or nonessential, information within a sentence and should be set off with commas. Nonrestrictive clauses often begin with *which*.

> *Nonrestrictive clause:* The handouts, which were published in color, enhanced our presentation.
>
> *Nonrestrictive phrase:* Our supervisor, negotiating feverishly, secured the contract.

Determining whether a clause or phrase is restrictive or nonrestrictive is critical because the distinction can change the meaning of a sentence.

Parenthetical Words and Phrases

Parenthetical information is nonessential and interrupts the flow of the sentence. Parenthetical elements are set off by commas.

> *Parenthetical word:* The latest shipment, however, will not be delivered on time.
>
> *Parenthetical phrase:* Our competitors, of course, failed to anticipate our new sales drive.

Direct Address

Proper nouns used in direct address should be set off by commas.

> *Introductory direct address:* Christopher, your design is flawed.
>
> *Direct address within a sentence:* Your design, Christopher, is flawed.

Geographical Names

Commas are used to separate geographical elements and to set them off from the rest of the sentence.

> *Geographical names:* The firm opened in Muncie, Indiana, at the end of the year.

Dates

Dates written as month/day/year insert commas before and after the year.

> *Month/day/year format:* We signed the agreement on February 28, 2007, before beginning construction.

Dates written as day/month/year—a format commonly used in the U.S. military and other parts of the world—do not use commas.

> *Day/month/year format:* Please attend the opening reception on 15 June 2007.

Titles after Names

When a title follows a proper name, a comma separates the name from the title.

> *Name and title separated by a comma:* Ms. Leslie Darnell, accounts manager
>
> *Name and title separated by a comma:* Dante Hicks, PhD

Salutations in Personal Correspondence

Salutations in personal correspondence—letters or e-mails—are followed by a comma.

> *Salutation in personal correspondence:* Dear Teresa,

Semicolons

Semicolons are stronger than commas but weaker than periods.

With Independent Clauses

A semicolon can join two independent clauses to signal a close relationship between the information in the two clauses.

> *Two related independent clauses:* The software wasn't compatible with the operating system; the OS will have to be updated.

Connecting two independent clauses with just a comma would result in a run-on sentence.

In a Series

When a series of items contains internal punctuation, semicolons signal the division between items.

> *Items in a series with internal punctuation:* The home office has expressed interest in opening branch offices in Sydney, Australia; Rome, Italy; Rio de Janeiro, Brazil; and Allentown, Pennsylvania.

Colons

Colons are used primarily to introduce or to separate.

Lists, Quotations, Explanations, Appositives

Colons follow an independent clause to introduce lists, quotations, explanations, or appositives. What follows the colon explains or illustrates what precedes it.

> *Introducing a list:* All trainees will need to bring these items to the retreat: a new notebook, colored pens, a clothes hanger, and a gallon of water.
>
> *Introducing a quotation:* The director made his point clear: "We will raise the quarterly earnings, or we will find other work."

A colon does not follow an introductory word or phrase like *such as*, *including*, and *like*.

> *A list following an introductory word:* The RFP called for standard information such as background, experience, and budget.

A colon that follows a quotation always appears outside the closing quotation marks.

Salutations in Professional Correspondence

Salutations in professional correspondence are followed by a colon.

> *Salutation in professional correspondence:* Dear Ms. Marcum:

Quotation Marks

Quotation marks indicate exact words, either written or spoken.

Direct Quotations

Quotation marks identify exact spoken or written words. Misquoting is unprofessional; intentionally changing a direct quotation is unethical. Writers should identify the source of a quotation—both speaker and context—and should offer citation information when quoting from a written document (see Appendix B).

> *Direct quotation:* The safety inspector announced, "The plant will have to be closed."
>
> *Direct quotation:* The feasibility study identified that "75 percent of overstock can be moved to the north district warehouse" (p. 13).

Titles

Quotation marks identify the titles of works like reports, articles, essays, chapters of books, and special sections within a periodical. The titles of longer works, like books or periodicals, are italicized.

> *Title in quotation marks:* The acting director e-mailed everyone the article "Improving Productivity."

Special Terms

Quotation marks indicate that a word or phrase is used in an unusual or ironic way.

> *Quotation marks indicating special use:* The "irreplaceable" manager was let go.

Quotation marks should not be used to indicate emphasis.

Single Quotation Marks

Single quotation marks surround a quotation that appears within a quotation.

> *Single quotation marks:* Early in the meeting, the director announced, "I have been told that we will face a 'serious budget crisis' next year."

Punctuation with Quotation Marks

See individual marks of punctuation.

Italics and Underlining

Italics are used to emphasize and differentiate. Italics replaced underlining when word processing technology made italics widely available. Generally, underlining is now used only when italics are unavailable.

Titles

Italics indicate the titles of larger works, such as books, periodicals, newspapers, films, legal cases, comic strips, television shows, and CD-ROMs.

> *Book:* The speaker recommended the book *Making Important Business Decisions*.
>
> *Film:* After work we all went to watch *Pirates of the Caribbean*.
>
> *Legal case:* The company's wish to expand was delayed by the findings in *Wayne versus Delaware*.
>
> *Newspaper:* I have got to start reading the *Wall Street Journal*.
>
> *CD-ROM:* Perhaps you should order the Video Professor's CD-ROM *Wireless Networking*.
>
> *Abbreviated journal title:* The article about migraines appeared in *JAMA*.

The title of a document is not italicized on its cover or title page.

Specific Names of Ships, Trains, and Other Crafts

Italics indicate the proper names of vehicles such as ships, trains, aircraft, and spacecraft.

> *Ship name:* The *USS Kitty Hawk* has a displacement of approximately 80,800 tons (72,720 metric tons) at full load.
>
> *Aircraft name:* The *Concorde* made its maiden flight on 2 March 1969.

Words as Words

Words, figures, or letters used as words, figures, or letters appear in italics.

> *Word as word:* The word *failure* should not appear in the initial report.

Foreign Words

Foreign words used in English sentences are italicized unless they have been accepted as standard words in English. A dictionary can clarify a word's status.

> *Foreign word in English sentence:* The abbreviation i.e. stands for *id est*, which means "that is."
>
> *Foreign phrase accepted in English:* Hiring the manager's son was a fait accompli.

In E-mail

Some e-mail programs do not provide italicized fonts or underlining. In these cases writers should place an underscore character just before and after the emphasized word or phrase to indicate underlining or an asterisk before and after the emphasized word or phrase to indicate italics.

> *E-mail underscore:* I'm going to send you the book _Ten Lessons in Image Design_.
>
> *E-mail italics:* The report needs to define what is meant by *accessibility*.

On the World Wide Web

On the World Wide Web an underlined word or phrase indicates a hot link. Italics should be used to emphasize and differentiate in web documents so that readers do not misunderstand emphasis as a hot link.

Parentheses

Parentheses are used for explanations, digressions, afterthoughts, and supplementary information. Parenthetical information is generally not essential to the meaning of a text but may be useful.

> *Explanatory information:* The fourth-floor conference room (the small one with the new chairs) will be closed this week because of maintenance work.
>
> *Supplementary information:* Luis Jorge will be retiring this week (he's moving to Miami).

Abbreviations

Parentheses may follow a word or phrase to introduce an abbreviation that will be used subsequently.

> *Introduction of an abbreviation:* We have developed an automatic data handling (ADH) system that will handle your data collection more efficiently.

Numbers that are spelled out should not be followed by numerals in parentheses.

Run-in Lists

Parentheses enclose numbers or letters designating items in a list or series that is run into the text.

> *Numbers designating a series:* The report is organized in four primary parts: (1) introductory material, (2) data, (3) analysis and recommendations, and (4) appendixes.

With Other Punctuation

Parentheses do not affect the punctuation of the rest of the sentence. All other punctuation should follow the closing parenthesis.

Dashes

Dashes enclose, separate, and connect. Dashes can indicate sudden changes in thought and can offset parenthetical information. In typing, two hyphens indicate a dash, and most word processors automatically convert them into an "em dash" (a dash the width of a capital *M*). There should be no spaces before or after the dash.

Sudden shift: The figures came too late to include—this whole project has been late.

Parenthetical information: External storage devices—for example, flash drives, portable hard drives, and zip drives—may be necessary to transfer data.

Hyphens

Hyphens can link, or they can separate. Two hyphens together indicate a dash.

Compound Words

Hyphens link words together in compounds. A dictionary determines whether compounds are linked with a hyphen, written as a single word, or written as separate words. If a compound does not appear in the dictionary, it is written as separate words.

Linked with a hyphen: As we learned later, the initial rumor was only a half-truth.

Single word: The budget officer would have to earmark the funding for the project.

Separate words: The problem seemed to be with the differential gear.

Modifiers

Hyphens can link multiple-word adjectives that precede a noun; if they follow the noun they modify, the words are not hyphenated.

Hyphenated modifier: The investor had a reputation for collecting hand-carved artifacts.

Modifier without hyphens: The investor had a reputation for collecting artifacts that were hand carved.

Sometimes a hyphen is necessary to avoid ambiguity.

Ambiguous: We will have to use multiple degeneration techniques in order to collect accurate data.

Clear: We will have to use multiple-degeneration techniques in order to collect accurate data.

Modifiers using letters or numbers use hyphens.

Letter modifier: The pieces came together at an L-bracket connection.

Number modifier: The plans called for a 10-meter line of PVC pipe.

Prefixes and Suffixes

All compound words that begin with *cross-*, *ex-*, *self-*, and *great-* are hyphenated.

Word beginning with cross-: Cross-references are important to verify.

Word beginning with ex-: The ex-president of the company started a competing company.

Word beginning with self-: Recreational diving is a self-regulated industry.

Word beginning with great-: The new CEO is the great-grandson of the company's founder.

Compound words ending with *-elect* are hyphenated.

> *Word ending with -elect:* The president-elect will have a difficult time with the transition.

A dictionary or style manual can help with other prefixes and suffixes; most are not hyphenated.

With Sequences

Hyphens can be used in place of *to* (with a sense of *through*) with letters, numbers, and sometimes words. Printed materials often use an en dash, which is slightly larger than a hypen, for this purpose.

> *With numbers:* The information you will need is on pages 127-138.
>
> *With letters:* The supervisor divided the call lists alphabetically: A-K, L-S, and T-Z.

Apostrophes

Apostrophes indicate either possession or omitted letters.

Possessives

Possession is indicated by an apostrophe and an *s* added to singular nouns:

> engineer's office
>
> proposal's length
>
> Charles's advisor

Most singular nouns that end in *s* also add an apostrophe and an *s* to indicate possession. Exceptions vary according to style manuals. Plural nouns that end in *s* form the possessive by adding an apostrophe only. Irregular plural nouns add an apostrophe and an *s*.

> writers' organization
>
> women's movement

Joint possessives make the last noun in the list possessive.

> Juan, Diedra, Shelley, and Anne's presentation

Individual possessives make each noun possessive.

> Sean's and Clay's computers

Compound nouns form possessives by adding an apostrophe and an *s* to the last word of the compound.

> the officer of record's decision
>
> my brother-in-law's promotion

Contractions

Apostrophes identify missing letters or numbers within contractions.

> cannot = can't
>
> will not = won't
>
> it is = it's
>
> 2006 = '06

Contractions should be avoided in most formal documents but can be appropriate in less formal documents like e-mails.

Plurals

Apostrophes are sometimes used to create plurals of letters and numerals.

Plural letters: The misprint eliminated all of the *p*'s and *r*'s.

Plural numerals: How many 3's and 4's are in the URL?

Some style manuals form these plurals with just an *s*.

Brackets

Brackets generally indicate an insertion into a direct quotation to explain or fill in missing information.

> The report explained, "Ten percent of all profits [from last quarter] will be redirected to shareholders."

> The speaker announced, "[After the environmental impact study] has been completed, we will begin excavation."

Brackets are also used to enclose parenthetical information inside parentheses.

> The brochure touted the newest addition to the line of portable notebooks (six of which were released last year [including the X-tium 2.0]).

Ellipsis

Ellipsis points identify omitted words in a direct quotation. An ellipsis, or omission, is indicated by three periods separated by spaces (. . .).

> *Words omitted:* The report's final recommendation identifies that "the division will need to continue its examination of the relationship between sales and distribution . . . in order to streamline the customer notification process."

If the omission is one or more complete sentences, a period precedes the ellipsis points.

> *Full sentence omitted:* In the weekly staff meeting, the field supervisor told us to "consider all members of the research team as independent authorities. . . . They will be reporting directly to the research commission without confirming their data with the team or the manager."

Ellipsis points are not used at the beginning of a quotation.

CORRECT USE OF CAPITALIZATION

Capitals signal proper nouns, the first letter of the first word in a complete sentence, the first word in salutations and closings, and essential words in titles. Many dictionaries address capitalization conventions.

Proper Nouns

All proper nouns are capitalized, including names of people, places, and things.

Proper Nouns	Common Nouns
Atlantic Ocean	an ocean
University of Florida	a university
Britain/British history	history
Auntie Raul	my aunt
Chemistry 4351	chemistry
Supreme Court	the court

Days of the week, holidays, and months are considered proper nouns and are capitalized as such; the seasons are considered common nouns and are not capitalized.

Days of the week, holidays, and months: In the United States, Thanksgiving is celebrated on the fourth Thursday of November.

Names of seasons: The first day of winter occurs on the twenty-first day of December this year.

First Words

The first letter of the first word of a sentence is capitalized, as is the first letter of a quotation that is a complete sentence. Style manuals differ about capitalizing the first word of a complete sentence after a colon.

First word of a sentence: Always capitalize the first word of a sentence.

First word of a complete sentence in a quotation: The feasibility study argued, "If the company is to increase its profits, then spending cuts must be considered."

The first word in salutations and closings is also capitalized.

First word in a salutation: Dear Ms. Herrington:

First word in a closing: Sincerely,

Titles

Titles of Works

The first and all important words in a title are capitalized. Articles, conjunctions, and prepositions are not generally capitalized; however, style manuals differ.

Report title: "A Report to the Commissioner"

Book title: Of Mice and Men

Professional Titles

Titles that precede proper names are capitalized, as are titles that follow proper names in the addresses and signatures of letters.

Title before proper name: Director Montoya

Title after proper name: E. Montoya, Director of Financial Affairs

In other cases position titles are generally not capitalized.

Title in general use: E. Montoya, the director of financial affairs, will issue the report.

Title without proper noun: The financial officer will issue the report.

Letters

Letters used to indicate shape or used in a name are capitalized.

Letter indicating shape: The K-valve replaced the J-valve.

Letter in name: vitamin E

CORRECT USE OF ABBREVIATIONS

Abbreviations can increase efficiency and save space in documents. They are formed either by shortening a word or by using the first letter of each word in a string. Writers must make sure that readers understand what abbreviations represent.

Names and Titles

Abbreviations in personal names are followed by a period and a space.

J. P. Morgan

Some titles should always be abbreviated, such as *Mr.*, *Mrs.*, and *Ms.* Other titles used with last names alone are spelled out.

I asked Professor Kidd for a letter of recommendation.

Titles and degrees that follow a proper name are generally abbreviated.

Charlene Garcia, CEO

Hienrich Jansen, Jr.

Fadiya Birhanu, PhD

Style manuals differ on the use of periods in some abbreviations.

Dates and Time

Some time and date indicators are commonly abbreviated: a.m., p.m., AD, BC, BCE, and CE. (Only AD comes before the date—AD 1492.) Months and days are spelled out in formal documents, although they may be abbreviated in notes and charts. *May*, *June*, and *July* are not generally abbreviated.

The class will begin on Thursday, September 13, 2007, at 9:00 a.m.

Geographical Names

In a formal document, place names and geographical names are spelled out. A few conventional abbreviations—USA and UK—are exceptions. Abbreviations are appropriate in informal documents and when text formatting demands brevity.

Organization Names

Organization names that may not be familiar to readers should be spelled out the first time they appear in a document, followed by, in parentheses, the abbreviation that will be used. Some abbreviations of organization names are formed without any punctuation. An appropriate style manual should be consulted.

PBS	NAACP	UCLA	NCAA
NATO	FCC	NSF	GOP
IRS	AMA	CIA	FAA
AFL-CIO	NRC	FBI	IDF
IEEE	APA	DNC	NAFTA
STC	MLA	SEC	ATF

Measures

Terms of measure are spelled out in formal documents.

kilograms	miles
ounces	kilobytes

A few exceptions may always be abbreviated: Hz (Hertz) and other International System of Units (SI) measurements (daHz, hHz, etc.), mph, rpm, and dpi, for instance.

Specialized Terminology

Within the text of formal documents, specialized terminology is spelled out—for example, technical terms, Latin terms, or academic terms. Many of these terms are commonly abbreviated in parentheses, notes, documentation, and other instances that demand economic use of space, such as charts and tables.

In text: for example, all primary colors

In parentheses: (e.g., all primary colors)

COMMONLY MISSPELLED WORDS

A

acceptable
accidentally
accommodate
acquire
acquit
a lot
amateur
apparent
argument
atheist

B

believe
bellwether

C

calendar
category
cemetery
changeable
collectible
colonel (kernel)
column
committed
conscience
conscientious
conscious
consensus

D

daiquiri
definite(ly)
discipline
drunkenness
dumbbell

E

embarrass(ment)
equipment
exhilarate
exceed
existence
experience

F

fiery
foreign

G

gauge
grateful
guarantee

H

harass
height
hierarchy
humorous

I

ignorance
immediate
independent
indispensable
inoculate
intelligence
its/it's

J

jewelry
judgment

K

kernel (colonel)

L

leisure
liaison
library
license
lightning

M

maintenance
maneuver
medieval
momentum
millennium
miniature
miniscule
mischievous
misspell

N

neighbor
noticeable

O

occasionally
occurrence

P

pastime
perseverance
personnel
possession
proceed
principle/principal
privilege
pronunciation
publicly

Q

questionnaire

R

receive/receipt
recommend
referred
reference
relevant
restaurant
rhyme
rhythm

S

schedule
separate
sergeant
supersede

T

their/there/they're
threshold
twelfth
tyranny

U

until

V

vacuum

W

weather
weird

APPENDIX B

DOCUMENTATION

Most citation guides advise that sources should be documented twice: once in the text of the document and once in a works-cited, bibliography, or reference list at the end of the document. Chapter 6 provides a number of strategies for conducting research and documenting sources, as well as a list of style guides and the disciplines they most often serve. This appendix provides an overview of the approaches to source citation in three of the more widely used style guides: American Psychological Association (APA), Council for Science Editors (CSE, formerly the Council of Biology Editors—CBE), and Modern Language Association (MLA).

APA STYLE

The American Psychological Association (APA) style is most often used in the social sciences. The full APA manual provides information for both citing and writing documents. The information provided here is based on *Concise Rules of APA Style* (2005) and *Publication Manual of the American Psychological Association* (5th ed., 2001).

In-Text Citation

APA style addresses both in-text and parenthetical information within a document. APA generally identifies the name(s) of a cited work's author(s) and the date of publication. Page numbers of citations are included when the internal citation references specific information.

Paraphrased or Summarized Information

When summarizing or paraphrasing information from a work, you should put in parentheses, at the end of the passage, the author's last name, followed by a comma and the date of publication. If you identify the author of the work in the passage, then you should put the year of publication in parentheses soon after that identification.

> International business will become a necessary way of thinking about business (Stratton, 1967).
>
> According to Stratton (1967), international business will become a necessary way of thinking about business.

Specific Information

When citing a specific fact from a work, include a page number in the citation. APA style requires that you identify the page number with *p.* before the number.

> International business will become a necessary way of thinking about business (Stratton, 1967, p. 173).
>
> According to Stratton (1967), international business will become a necessary way of thinking about business (p. 173).

Direct Quotation

When citing a direct quotation from a work, include a page number in your citation. APA style requires that you identify the page number with *p.* before the number. For short quotations that appear within quotation marks, place the parenthetical citation after the quote but before the terminal punctuation.

The most interesting argument was that "the role of nepotism in American Business has generally been overlooked as a positive influence" (Bellow, 2004, p. 6).

Bellow (2004) argues that "the role of nepotism in American Business has generally been overlooked as a positive influence" (p. 6).

For long quotations, called block quotations, indent the entire quote, do not use quotation marks, and place the parenthetical citation after the terminal punctuation.

According to the preliminary OSHA report published in 2003,

> Workers experienced a high rate of fatigue following exposure to the gas leak. Tests conducted at Cedar Rapids Medical Facilities revealed no contaminants in the blood, urine, or breath of any of the employees tested. However, doctors involved in the study were quick to point out that long-term affects may not be immediately evident. (p. 134)

Works with Multiple Authors

When citing a work that has two authors, identify both names joined by an ampersand—&—in the parenthetical citation; join the names with the word *and* in the primary text.

The *Civil Engineering Handbook* clearly addresses the biological water treatment process (Chen & Liew, 2002).

According to Chen and Liew (2002), the art of estimation involves rounding numbers and understanding ratios and percentages.

For works written by three, four, or five authors, name all authors the first time the citation appears; however, in subsequent citations give only the first author's name, followed by *et al.*

First citation: There is traditionally a discrepancy between engineering standards that must be overcome (Stratton, Wormer, & Hoover, 1978).

Second citation: The engineer's responsibility must be guided by a strict code of ethics (Stratton et al., 1978).

First citation: According to Stratton, Wormer, and Hoover (1978), there is traditionally a discrepancy between engineering standards that must be overcome.

Second citation: According to Stratton et al. (1978), the engineer's responsibility must be guided by a strict code of ethics.

Multiple Authors with the Same Last Name

When citing different works whose primary authors have the same last name, use the primary authors' initials to distinguish between the similar names.

About.com, I. Smith (2001) claims, "is one of the greatest collections of information on diseases and conditions on the Net" (p. 1). However, P. Smith, Linkletter, and McClinton (2004) have explained that Net medical references can't always be trusted.

Works by an Organization or Corporation

If the author of a cited work is an organization, company, or corporation, use the organization's name as the author's name.

According to one set of guidelines, writers must consider audience members' disabilities when working to reduce bias in document language (American Psychological Association, 2001).

According to Microsoft Corporation (2003), writers must be consistent in their use of screen terminology.

Electronic Sources

Electronic documents should be cited as other documents are. If the author of a document is unknown, use an abbreviated version of the document title in the parenthetical citation. If the date of publication is unavailable, use *n.d.* for "no date."

> According to the policies manual, "Source packages should specify the most recent version number of the policy document with which your package complied when it was last updated" (Jackson & Schwartz, n.d.).

For documents in PDF format, include page numbers in the parenthetical citation. If the PDF does not list page numbers, identify paragraph numbers. If neither page numbers nor paragraph numbers are provided, identify section or heading information to direct readers to the citation.

> Peterman (1987) produced one of the finest sales catalogs available online (Catalog History section, para 9).

Personal Communication

Personal interviews, phone calls, letters, memos, and e-mails are cited by placing inside parentheses the words *personal communication*, followed by the date of the communication. Personal communications are not included in the reference list.

> Mary Benis (personal communication, August 5, 2006) identifies the implications of copyright issues in corporate documentation.

Reference List

APA style has a few basic rules for the reference list:

- Arrange entries alphabetically.
- For multiple works by the same author, list the works chronologically.
- List authors' names by last name, followed by initials. For works with six or more authors, list only the first six authors' names followed by *et al.*
- For titles of works other than journals, capitalize the first word of the title and the first word after a colon, if there is one. Other than that, capitalize only proper nouns.
- For journal entries capitalize all major words in the title.
- Italicize titles of major works like books or journals.
- Do not italicize, underline, or put quotes around shorter titles, such as journal articles or selections within a book.
- Use a hanging indent of five to seven spaces on all lines following the first line of an entry.

Books

Single Author Identify the author's last name and initials. After the author's name, place the year of publication in parentheses, followed by the title. Next, identify the place of publication and the publisher.

> Dobrin, S. I. (2008). *The Prentice Hall technical writer's handbook*. Columbus, OH: Prentice Hall.

Multiple Authors For works with multiple authors, separate two authors' names with an ampersand, and then follow the rules for a single-authored book. For works with three to six authors, separate the authors' names with commas, add an ampersand before the last author's name, and then follow the rules for a single-authored book.

Keller, C. J., & Weisser, C. R. (2006). *The locations of composition*. Albany: State University of New York Press.

Dobrin, S. I., Keller, C. J., & Weisser, C. R. (2007). *Technical communication in the twenty-first century*. Columbus, OH: Prentice Hall.

For works with more than six authors, list the first six names followed by *et al.*

Puccio, M., Connelly, L., Willis, A., Kent, R., Semenec, D., Kaufman, D., et al. (2001). *Youth violence and American culture*. Ft. Lauderdale, FL: Bully Press.

Same Author, Multiple Books If you cite two or more works by the same author, list them chronologically.

Pring, M. J. (1995). *Investment psychology explained: Classic strategies to beat the markets.* Indianapolis, IN: Wiley.

Pring, M. J. (2002). *Technical analysis explained: The successful investor's guide to spotting investment trends and turning points*. Boston: McGraw-Hill.

If you cite two or more works by the same author that were published in the same year, arrange them alphabetically by title, and include a letter in the parentheses with the date to identify an order. Be sure to include the corresponding letter in your in-text citation as well.

Krassman, M. (1979a). *Amphibian genetics*. New York: Henson.

Krassman, M. (1979b). *What frogs know and what they're not telling us*. New York: Statler & Waldorf.

Organization or Corporation as Author For works written by an organization or corporation, use the organization's name in place of the author's name. If the organization is credited as both author and publisher of the work, use the organization's name as the author, and use the word *Author* in place of the publisher.

American Medical Association. (2004). *American Medical Association family medical guide*. Indianapolis, IN: Wiley.

United States Center for Disease Control. (2005). *Reporting pathology protocols*. Atlanta, GA: Author.

Unknown Author If you cannot identify the author of a book, begin the entry with the title of the work.

Introduction to CSS. (2007). College Station, PA: College Printers.

Edited Book For books edited by a single editor, place the abbreviation *Ed.* in parentheses following the editor's name. For books edited by more than one person, use the abbreviation *Eds.*

Ziems, R. (Ed.). (2005). *Twenty great proposals*. New York: Technical.

Brown, R. A., Rech, J., & Rosenblum, S. (Eds.). (2004). *How to write policy that works*. San Francisco: Business Matters.

Chapter or Other Division in an Edited Book Begin with the name of the author of the chapter, followed by the date of publication in parentheses and the title of the chapter. Next, put *In* and the author or editor of the book, followed by the title of the book. In parentheses give the page numbers of the chapter and then the publication information.

Azar, T. (2002). Integrated cardiovascular imaging. In J. Spencer & J. Crowley (Eds.), *The future of cardiology* (pp. 273–294). Cleveland, OH: Carnegie Press.

Periodicals

Journal Article Identify the last name and initials of the author. Next, in parentheses put the date of publication, including a word like *Winter* or *Spring* if the journal includes it. Next, list the journal article title without italics or quotation marks. Add the volume number, set off by commas and italicized. If you include an issue number, put it in parentheses. Finally, identify the page numbers where the article can be found.

Vacanti, C. A. (2006). History of tissue engineering and a glimpse into its future. *Tissue Engineering*, *12*, 1137–1142.

Sanchez, R. (2004). Estimating pork futures in a regulated market. *The American Pork Industry Journal*, *42*(3), 47–68.

Magazine Article Begin with the last name and initials of the author, followed by the date of publication in parentheses. Next, include the article title, followed by the magazine title in italics. If there is a volume number, include it in italics, followed by the page numbers of the article.

Null, C. (2006, June). Stupid engineering mistakes. *Wired*, 45.

Newspaper Article List newspaper articles as you would journal or magazine articles. Be sure to identify the complete date, and use *p.* or *pp.* before the page number(s).

Wade, N. (2006, July 25). Scientists say they've found a code beyond genetics in DNA. *The New York Times*, p. A6.

Electronic Sources

Website When listing general websites, include the author's name when possible, followed by the date of publication in parentheses. If you cannot locate a publication date, put *n.d.* for "no date" in the parentheses. Put the title in italics, followed by a retrieval date—the date you accessed the web page. Precede the retrieval date with the word *Retrieved*. Follow the date with *from*, and list the URL as the site from which the information was taken.

Dawson, D. (n.d.). *Writing for corporate markets*. Retrieved May 20, 2006, from http://www.dawsonnet.com/writing.html

If you cannot locate the author of the page, begin the entry with the title.

Writing for corporate markets. (2005). Retrieved May 20, 2006, from http://www.dawsonnet.com/writing.html

For websites that are hosted by university programs, identify the institution and department or program that hosts the web page, followed by a colon and the URL for the page.

Design. (2004). Retrieved August 4, 2006, from University of Florida, Networked Writing Environment: http://www.nwe.ufl.edu/writing/help/web/authoring/design/

Online Periodical For articles that appear as electronic versions of print articles, list them as you would the print version, identifying the electronic version in brackets following the title.

Whitehead, S. P. (2005). Hypertextual linkage in the digital library [Electronic version]. *Digital Data*, *23*(2), 45–67.

For articles that appear only electronically or that are different from the print version of an article, list the article, followed by the retrieval date and the URL.

Harris, R. (2001). Citing web sources MLA style. *Virtual Salt*, *12*. Retrieved July 28, 2006, from http://www.virtualsalt.com/mla.htm

Database Articles When citing an article from an online database, provide the retrieval date, the name of the database, and an identification number for the article, if one exists.

Wickliff, G. A. (2006). Light writing: Technology transfer and photography to 1845. *Technical Communication Quarterly*, *15*(3), 293–313. Retrieved June 16, 2006, from Communication & Mass Media Complete, EBSCOhost.

E-mail APA style does not include e-mail citations in the reference list. Instead, e-mail should be identified in the text as a personal communication.

Online Forum or Discussion Board Posting When citing online forums, newsgroups, or discussion groups, provide the name of the poster, the date of the post in parentheses, the subject line or thread of the post, and the URL link for the post. If only a screen name is used for the post author, use the screen name. If there are any list identifiers, like message numbers, include that information in brackets following the subject or thread line.

Sanchez, Raul. (2008, June 1). Changes in transnational approaches to visuals [Msg 14]. Message posted to http://groups.earthlink.com/techcomm/messages/00014.html

Computer Software When citing computer software, include the software author's name, followed by the year the software was released in parentheses. Next, list the title of the software, followed by the words "Computer software" in brackets. Then, identify the location where the software was published, followed by a colon and the name of the software distributor.

Pearson, P. H., et. al. Technical Writing Connection [Computer software]. New York: Vallance.

Images

Images with Titles

From Library Databases When citing images like paintings that have been taken from a library database, list the artist's last name followed by a first initial, the date of the image in parentheses, the title of the image in italics with just the first letter capitalized, and the type or medium of the image listed in brackets. Then, identify the location where the original image is housed, followed by the city of that location. Next, identify the date when you accessed the image and the database from which you accessed it.

Matisse, H. (1902). *Notre-Dame, une fin d'après-midi (A Glimpse of Notre Dame in the Late Afternoon)* [Painting]. Albright-Knox Art Gallery, Buffalo, NY. Retrieved July 28, 2008, from Grove Art Online database.

From Free Web Pages When citing images like paintings that have been taken from a web page, list the artist's last name followed by a first initial, the date of the image in parentheses, the title of the image in italics with just the first letter capitalized, and the type or medium of the image listed in brackets. Then, identify the location where the original image is housed, followed by the city of that location. Next, identify when you accessed the image and the URL for the web page from which you accessed it.

Matisse, H. (1902). *Notre-Dame, une fin d'après-midi (A Glimpse of Notre Dame in the Late Afternoon)* [Painting]. Albright-Knox Art Gallery, Buffalo, NY. Retrieved July 28, 2008, from http://www.ibiblio.org/wm/paint/auth/matisse/matisse.notre-dame-am.jpg

From Printed Sources When citing images like paintings that have been taken from a print source, such as a book, list the artist's last name followed by a first initial, the date of the image in parentheses, the title of the image in italics with just the first letter capitalized, and the type or medium of the image listed in brackets. Then, identify the location where the original image is housed, followed by the city of that location. Next, identify the title of the print material from which you are referencing the image. Italicize titles like book titles, and then identify the author or editor of the print material. Next, list the place of publication, followed by the publisher of the print material and the date. Finally, list either the plate or page number for the image in parentheses.

> Matisse, H. (1902). *Notre-Dame, une fin d'après-midi (A Glimpse of Notre Dame in the Late Afternoon)* [Painting]. Albright-Knox Art Gallery, Buffalo, NY. *Matisse and Color*. By Jean Caude Dumond. New York: Parthenon Press, 2007 (121).

Images without Titles

The basic premise for citing untitled images is to use a self-created descriptor in place of the title.

From Library Databases When citing untitled images taken from a library database, list the artist's last name followed by a first initial. Then, in brackets, identify the image with a descriptive title, including the medium of the image as part of the descriptor. Following the bracketed title, include the date of the image in parentheses. Then, identify the location where the original image is housed, followed by the city of that location. Next, identify when you accessed the image and the database from which you accessed it.

> Landres, J. [Photograph of war refugees]. (1945). Canton War Museum, Canton, OH. Retrieved December 7, 2008, from Grove Art Online database.

From Free Web Pages When citing untitled images taken from a web page, list the artist's last name followed by a first initial. Then, in brackets, identify the image with a descriptive title, including the medium of the image as part of the descriptor. Following the bracketed title, include the date of the image in parentheses. Then, identify the location where the original image is housed, followed by the city of that location. Next, identify the date when you accessed the image and the URL for the web page from which you accessed it.

> Landres, J. [Photograph of war refugees]. (1945). Canton War Museum, Canton, OH. Retrieved December 7, 2008, from www.americanwarphotos.com/refugees1-2.jpg

From Printed Sources When citing untitled images taken from a print source, such as a book, list the artist's last name followed by a first initial. Then, in brackets, identify the image with a descriptive title, including the medium of the image as part of the descriptor. Following the bracketed title, include the date of the image in parentheses. Then, identify the location where the original image is housed, followed by the city of that location. Next, identify the title of the print material from which you are referencing the image. Italicize titles like book titles, and then identify the author or editor of the print material. Next, list the place of publication, followed by the publisher of the print material and the date. Finally, list either the plate or page number for the image in parentheses.

> Landres, J. [Photograph of war refugees]. (1945). Canton War Museum, Canton, OH. *Images of American War Stories*. By Paul Mitchell. New York: InSight Press, 1987 (96).

Other

Government Document First, identify the government agency that produced the document; in most cases abbreviate *United States* as *U.S.* If the document is identified by document identification numbers, include those in parentheses following the title.

U.S. Environmental Protection Agency. (2003). *EPA 2003 draft report on the environment* (Document #3245-876). Washington, DC: Author.

Pamphlet When listing a pamphlet or brochure, follow the rules for a book. However, following the title, you should include either the word *Brochure* or *Pamphlet* in brackets.

NJ Environmental Protection Agency. (2005). *A woman's guide to eating fish and seafood* [Pamphlet]. NJ: Author.

Report List reports according to the rules for books. However, if the report can be identified by an assigned identification number, include that number in parentheses following the title.

U.S. Department of State. (2005). *Trafficking in persons report* (Document #43899-00037). Washington, DC: Author (Government Document Service).

Interview If the title of the interview does not make clear that the work is an interview, include the word *Interview* and the interviewee's name in brackets following the title.

Boise, D. (2000). The Wired interview. *Wired, 8*(10), 46–49.

Olson, G. A. (2004). Confessions of an editor [Interview with Merry Perry]. *Editor's Monthly*, *47*, 105–107.

Personal interviews are not included in the reference list but are instead identified as personal communication in the text.

Letter Letters are not included in the reference list but are instead identified as personal communication in the text.

Memo Memos are not included in the reference list but are instead identified as personal communication in the text.

Presentation Begin with the last name and initials of the person delivering the speech, lecture, or presentation. In parentheses include the date the presentation was delivered. Then, list the title without italics or quotation marks. Finally, identify the kind of presentation (lecture, speech), followed by the location of the presentation.

Jobs, S. (2005, June 12). You've got to find what you love. Speech delivered at Stanford University, Stanford, CA.

CSE STYLE

The style of the Council for Science Editors (CSE), formerly the Council of Biology Editors (CBE), is most often used in biology, chemistry, geology, mathematics, medicine, and physics; it is the standard citation style for all science-related fields. The full CSE manual provides information for both citing and writing documents. The information provided here is based on *Scientific Style and Format: The CSE Manual for Authors, Editors, and Publishers* (7th ed., 2006).

In-Text Citation

CSE style addresses in-text citation and references at the end of a document. The in-text segment is divided into three approaches: the name-year system (also known as the "Harvard style"), the citation-sequence system (also known as the "Vancouver style"), and the citation-name style. The information found here addresses the first two of these approaches; you can consult the CSE style manual (particularly chapter 29) for more information about all three, as well as numerous examples. CSE style emphasizes reduced punctuation to make work at the keyboard more efficient.

Name-Year System

In this system writers provide full bibliographic information in an alphabetized reference list that appears at the end of the document. In the text, writers provide parenthetical information, including the author's name and the year of publication of the cited work.

> Lack of long-term strategy or vision for small businesses can limit a business's growth potential (Forrester, 2004), whereas overlooking immediate needs can squelch growth before a strategy can evolve (Wang, 2005).

This system provides readers with some context for the cited information, and it is relatively easy to add or remove sources from the document and the reference list. However, numerous parenthetical citations can be distracting for readers.

Citation-Sequence System

The citation-sequence approach directs readers to a reference list at the end of the work by providing sequential numbers with the in-text citations that correspond with entries in the reference list. The numbers are indicated as superscripts or within parentheses.

> High-end computer architectures, such as some experimental computer systems, Wallace notes[1], will likely improve efficiency, reliability, and usability [2–5].

> High-end computer architectures, such as some experimental computer systems, Wallace notes (1), will likely improve efficiency, reliability, and usability (2–5).

A source cited earlier in a document uses the same number originally attributed to that source.

> Wallace is also responsible for developing an experimental computer protocol [2–5].

The list of references follows the order of the works cited in the document.

The citation-sequence system only minimally disrupts the reader's attention and reduces the amount of space (ultimately, paper and ink as well) devoted to citation. However, readers must work harder to locate references and to match them with the in-text citations. Revising is also hampered because any citation that is added or deleted in the document changes the numeric order of citations both in the text and in the reference list. In addition, because the in-text citation does not identify authors or their works, they receive less recognition for their work and their direct or indirect contribution to the document.

Reference List

Including references in the CSE reference list is similar for both systems. The only difference is that the name-year system lists citations alphabetically, and the citation-sequence system lists them numerically in the order in which they were cited. Some general rules for both types of reference list include these:

- Do not italicize, underline, or place in quotation marks the titles of books.
- Capitalize only the first word and any proper nouns in book titles.

- Abbreviate periodical titles according to the rules listed in the manual.
- Do not italicize, underline, or place in quotation marks the titles of articles found in periodicals.
- Capitalize only the first word and proper nouns in the titles of articles found in periodicals.
- For book entries, end the citation by including the total number of pages in the book, followed by the abbreviation *p* with no period.
- For entries that identify a page range, use the abbreviation *p* with no period, followed by the first page number and then the last digit of the second number if the first digits are the same (i.e., 223–7, not 223–227).
- Identify dates by year/month/day, using three-letter abbreviations for the months (2007 Feb 28). No periods or commas separate the elements.

Books

Single Author First, identify the author's last name, followed without an intervening comma by the author's initials without periods for the initials. Next, list the title of the book, followed by the location and name of the publisher, the date of publication, and the number of pages in the book.

> Dobrin S I. The Prentice Hall technical writer's handbook. Columbus, OH: Prentice Halls; 2008. 677 p.

Multiple Authors List the authors' names as with the single author; do not use the word *and* to separate the names.

> 4. Weisser C R, Keller C J, Dobrin, S I. Technical communication in the twenty-first century. Columbus, OH: Prentice Hall; 2007. 740 p.

Organization or Corporation as Author Identify the organization in place of the author's name.

> 6. American Medical Association. American medical association family medical guide. Indianapolis, IN: Wiley; 2004. 323 p.

Edited Book Entries for edited books follow the same rules that other books follow but include the word *editor* or *editors* after the editors' names.

> Ziems R, editor. Twenty great proposals. New York: Technical Publishers; 2005. 282 p.

> 3. Brown R A, Rech J, Rosenblum S, editors. How to write policy that works. San Francisco: Business Matters, Inc.; 2004. 356 p.

Chapter or Other Division in an Edited Book Identify the author of the book chapter or division, followed by the title of the chapter. Write *In*, followed by a colon, the title of the book, and the name(s) of the editor(s) of the book. Next, provide publication information, followed by the page numbers on which the referenced work appears.

> Azar T, Integrated cardiovascular imaging. In: The future of cardiology. Spencer J, Crowley J, editors. Cleveland, OH: Carnegie Press; 2002. p 273–294.

Periodicals

Journal Article Identify the author's last name and initials, followed by the article title, which is not italicized or placed in quotation marks. Then list the journal name according to the CSE list of abbreviations. Identify the year, month if given, volume number, and page numbers of the article. If the journal paginates each issue within a volume separately, provide the issue number in parentheses following the volume number.

Vacanti C A. History of tissue engineering and a glimpse into its future. Tissue Eng 2006; 12: 1137–42.

12. Sanchez R. Estimating pork futures in a regulated market. Am Pork Ind J 2004; 42 (3): 47–68.

Magazine Article List the author's last name and initials, followed by the article title, which is not italicized or placed in quotation marks. Identify the magazine title according to CSE abbreviations, followed by the date of the issue in which the article appears and the page numbers of the article.

Null C. Stupid engineering mistakes. Wired 2006 Jun:45.

Newspaper Article Identify the author's last name and initials, followed by the article title, which is not italicized or placed in quotation marks. Provide the title of the newspaper in which the article appears, followed by the issue date, section, page(s), and column number. For articles appearing in newspapers that do not use section numbers, place a colon between the issue date and the page numbers.

5. Wade N. Scientists say they've found a code beyond genetics in DNA. New York Times 2006 Jul 25; sect A: 6 (col. 2).

Electronic Sources

Website Identify the name of the website, followed by the word *Internet* in brackets. Next, identify the location of the web page sponsor/publisher, followed by the name of the organization that produces the web page. Include a copyright date, if applicable. In brackets identify first the date the web page was updated and then the date you accessed the web page. Put *Available from:* and give the URL of the web page.

The Center for Disease Control [Internet]. Atlanta, GA: Center for Disease Control; c2006 [updated 2006; cited 2006 Jul 28]. Available from: http://www.cdc.gov/

Online Book Provide the author's last name and initials, followed by the title of the book, which is not italicized or placed in quotation marks. In brackets put the words *monograph on the Internet*. Next, provide the publication information, and then identify in brackets the date you accessed the book, followed by the page count for the book. If you have to estimate the page count, include the information in brackets. Finally, write *Available from:* and the URL where the book was located.

Mates B T. Adaptive technology for the Internet: Making electronic resources accessible to all [monograph on the Internet]. Chicago, IL: American Library Association; 2000 [cited 2006 Jun 15]. [about 150 p.] Available from: http://www.ala.org/ala/productsandpublications/books/editions/adaptivetechnology.htm

Online Periodical Provide the author's last name and initials, followed by the title of the article, which is not italicized or placed in quotation marks. Include the title of the periodical according to CSE abbreviations, followed by the words *serial on the Internet* in brackets. Next, include the date the article was published, followed in brackets by the date you accessed the article. Then, identify the periodical volume number, followed by the pages on which the article appeared, and finally *Available from:* and the URL of the article.

3. Wagner H J. Home office? It's in the yard. Wired [serial on the Internet]. 2006 Jul [cited 2006 Sep 4]. 7; 41–2. Available from: http://www.wired.com/news/technology/gizmos/0,71462-0.html?tw=wn_index_3

Database Articles Provide the author's last name and initials, followed by the title of the article, which is not italicized or placed in quotation marks. Provide the journal name, followed by the date of publication, the journal volume and issue, and the page numbers of the article. Then write *In*: and the name of the database, followed in brackets by *database on the Internet*. Identify the place of publication, followed by the host and the copyright date. In brackets put the date of access, then in brackets the length of the article. Finally, write *Available from*:, the URL to access the article, and the article identification number, when available.

Wickliff G A. Light writing: Technology transfer and photography to 1845. Tech Com Q 2006; 15 (3): 293–313. In: Communication & Mass Media Complete [database on the Internet]. Gainesville (FL): EBSCOhost [cited 2006 Jun 16]. [20 p.] Available from: http://weblinks2.epnet.com

E-mail Identify the last name and initials of the e-mail author, followed by the title/subject line of the e-mail. Put in brackets *electronic mail on the Internet*. Then write *Message to*:, followed by the recipient of the e-mail. Give the date the e-mail was sent, including the time, when available. In brackets identify when you accessed the e-mail. Finally, in brackets estimate the length of the e-mail.

5. Ohlinger M. Copyright concerns [electronic mail on the Internet]. Message to: Sid Dobrin. 2006 May 20, 1:35 am [cited 2006 May 22]. [About 2 screens].

Other

Government Document Identify the title of the document, followed by the place of publication and the publisher. Next, identify the publication date and the number of pages in the document, followed by *Available from*: and the source for the document.

EPA 2003 draft report on the environment. Washington: US Environmental Protection Agency; 2003. 250 p. Available from US EPA.

Map Follow the rules for listing other works, but indicate the type of map in brackets after the title.

Everglades, Florida [topographical map]. Washington: National Geographic Society; 1999. p 132.

Chart List charts as you would maps, indicating in brackets after the title what kind of chart is cited.

Annual budget expenditures [pie chart]. Columbus, OH: DataText; 2006. p 23.

MLA STYLE

The Modern Language Association (MLA) style is most often used for writing academic papers in the humanities. The full *MLA Style Guide* provides information for both citing and writing documents. The information provided here is based on the *MLA Handbook for Writers of Research Papers* (6th ed., 2003) and the *MLA Style Manual and Guide to Scholarly Publishing* (2nd ed., 1998).

In-Text Citation

MLA style provides author name and page citations for all works cited in the document. The approach varies depending on the kind of work cited, but for the most part MLA encloses citation information in parentheses following the cited material.

These citations are not designed to give readers full information but to direct them to specific information in the works-cited, or bibliographic, list. Chapter 6 of the handbook and chapter 7 of the style manual provide more information about this approach, as well as numerous examples.

Complete Work

Identify only the author's name when referring to an entire work by that author.

> The report was based entirely on Charles W. L. Hill's international business book.

Specific Pages

When identifying specific information from a source, put the author's last name and the page number of the source in parentheses with no internal punctuation and no abbreviation for page.

> The role of nepotism in American business has generally been overlooked as a positive influence (Bellow 6).

If the passage includes the author's name, include only the page number in parentheses.

> Bellow's research shows that the role of nepotism in American business has generally been overlooked as a positive influence (6).

Work with No Page Numbers

When a cited work does not have page numbers, use other identifying characteristics to provide accurate citation—a paragraph number, a screen number, or a section number. Use the abbreviation *par* or *pars* for *paragraph(s)*. If the parenthetical citation begins with the author's name, follow the name with a comma.

> The physics of wave formation might shed some light on ship construction (Elmore, par. 9).
>
> Elmore's explanation of wave physics might shed some light on ship construction (par. 9).

Multiple Works by the Same Author

When using more than one work by the same author, identify the specific work in the text, or include an abbreviated title in the parenthetical citation.

> Lawrence Gasman's book *Nanotechnology Applications and Markets* offers several interesting market approaches, including science-driven markets we have not considered (4–6).
>
> Lawrence Gasman offers several interesting market approaches, including science-driven markets we have not considered (*Nanotechnology* 4–6).

If you mention neither the title nor the author in the text, provide both in the parenthetical citation.

> There are several interesting market approaches, including science-driven markets we have not considered (Gasman, *Nanotechnology* 4–6).

Works with Multiple Authors

When citing works written by two or three authors, list all authors' names in the citation.

> The *Civil Engineering Handbook* clearly addresses the biological water treatment process (Chen and Liew 241).

When citing works written by four or more authors, include in the parenthetical citation the first author's name followed by *et al.*

> According to *Introduction to Electrical Engineering*, the art of estimation involves rounding numbers and understanding ratios and percentages (Ashby et al. 3).

Multiple Authors with the Same Last Name

If you use multiple sources written by different authors with the same last name, be sure to spell out their first names in the text. In the parenthetical citation, include the author's first initial before the last name.

> About.com, Dr. Ian Smith claims, "is one of the greatest collections of information on diseases and conditions on the Net" (1). However, Net medical references can't always be trusted (P. Smith 9).

Work by Organization or Corporation

When a work is produced by a company, organization, or corporation, you should use the name of that organization or company in place of the author's name in the parenthetical citation.

> According to one set of guidelines, writers must consider audience members' disabilities when working to reduce bias in document language (American Psychological Association 69–70).
>
> According to the *Microsoft Manual of Style for Technical Publications*, writers must be consistent in their use of screen terminology (Microsoft Corporation 4).

Electronic Sources

Electronic sources should be cited in text according to the rules that apply to print documents. If authors' names are not available, identify the title of the work, either in full in the text or in an abbreviated version in the parenthetical citation. Use page numbers when applicable; if page numbers are not available, use other document characteristics to identify the location of the citation, such as section numbers, screen numbers, or paragraph numbers. Do not include URLs in either the in-text or the parenthetical citations; do include URLs in works-cited and bibliography lists.

> According to the policies manual, "Source packages should specify the most recent version number of the policy document with which your package complied when it was last updated" (Jackson and Schwartz 4.1).

Works Cited

The MLA works-cited list is the second part of the MLA citation style. The works-cited section contains all of the citation details that cannot be identified in the in-text citations: publisher, date of publication, full titles, lists of authors, and so on.

Books

Single Author Include the author's last name, first name, middle name or initial when identified in the work, the complete title of the work (with all major words capitalized), the location of the publisher, the name of the publisher, and the year of publication.

> Dobrin, Sidney I. *The Prentice Hall Technical Writer's Handbook*. Columbus, OH: Prentice Hall, 2008.

Multiple Authors When listing books by two or three authors, include the first author's last name, first name, middle name or initial when identified in the work, and then the first, middle, and last names of the coauthors. Only the first author's name is inverted, and all authors' names are separated with a comma. Next, list the complete title of the work, the location of the publisher, the name of the publisher, and the year of publication.

Dobrin, Sidney I., Christopher J. Keller, Christian R. Weisser. *Technical Communication in the Twenty-First Century*. Columbus, OH: Prentice Hall, 2007.

Same Author, Multiple Books When the works-cited list includes more than one book by the same author, replace the author's name with three hyphens and a period in the second and subsequent listings. Arrange the entries alphabetically according to the title of the books.

Pring, Martin J. *Investment Psychology Explained: Classic Strategies to Beat the Markets.* Indianapolis, IN: Wiley, 1995.

________ *Technical Analysis Explained: The Successful Investor's Guide to Spotting Investment Trends and Turning Points*. Boston: McGraw-Hill, 2002.

Organization or Corporation as Author Follow the style for a single-authored book, but replace the author's name with the corporation's or organization's name.

American Medical Association. *American Medical Association Family Medical Guide*. Indianapolis, IN: Wiley, 2004.

Unknown Author If you do not know the author of a work, begin the entry with the title.

Introduction to CSS. College Station, PA: College Printers, 2007.

Edited Book List edited books as you would authored books. However, write *Ed.* after the editor's name or *Eds*. after multiple editors' names.

Ziems, Robert, Ed. *Twenty Great Proposals*. New York: Technical Publishers, 2005.

Brown, Royal Allen, Jota Rech, and Scott Rosenblum, Eds. *How to Write Policy That Works*. San Francisco: Business Matters, Inc., 2004.

Chapter or Other Division in an Edited Book Provide last name, first name, and middle name or initial for the author of the chapter or section in the edited book. Then include the title of the chapter or division in quotation marks. Next, provide the title of the book in which the chapter or section is found, followed by the editor's name in standard order and *Ed.* to indicate a single editor or *Eds*. to indicate multiple editors. Provide the publication information as you would for any other book. Finally, include the page numbers for the article or section.

Azar, Taraneh. "Integrated Cardiovascular Imaging." *The Future of Cardiology*. James Spencer and James Crowley, Eds. Cleveland, OH: Carnegie Press, 2002. 273–294.

Periodicals

Journal Article Identify the author of the article—last name, first name, and middle initial or name. Then identify the complete title of the journal article in quotation marks and the italicized title of the journal in which the article appears, followed by the volume number, year, and page numbers of the article. If the journal paginates each issue within a volume separately, place a period after the volume number, and add the issue number.

Vacanti, Charles A. "History of Tissue Engineering and a Glimpse into Its Future." *Tissue Engineering* 12 (2006): 1137–1142.

Sanchez, Raul. "Estimating Pork Futures in a Regulated Market." *The American Pork Industry Journal* 42.3 (2004): 47–68.

Magazine Article List the author's name in reverse order, followed by the article title in quotation marks and the name of the magazine in italics. Then list the issue date and the page numbers.

Null, Christopher. "Stupid Engineering Mistakes." *Wired*. June 2006: 45.

Newspaper Article List the author's last name, first name, and middle name or initial, followed by the article title in quotation marks and the title of the newspaper in italics. Then list the issue date and the page number of the article.

Wade, Nicholas. "Scientists Say They've Found a Code beyond Genetics in DNA." *The New York Times*. 25 July 2006. A6.

Electronic Sources

Website When citing an entire web page, first list the title of the web page in italics. If provided, identify the editor/author of the web page. Next, list the date the web page was published or updated and the organization or institution that sponsors the web page, if it is supported. Identify the date you accessed the web page to get the information you are citing. Finally, list the URL in angle brackets.

The Center for Disease Control. 2006. 28 July 2006. <http://www.cdc.gov/>

Article or Section of a Website List the author's last name, first name, and middle name or initial. Put the title of the work in quotation marks, and provide the web source information as you would for a full web page.

Harris, Robert. "Citing Web Sources MLA Style." *Virtual Salt*. 12 March 2001. 28 July 2006. <http://www.virtualsalt.com/mla.htm>

Online Book State the author's last name, first name, and middle name or initial, followed by the title of the book in italics. If print information is available, such as print publisher or publication date, include that as you would for a print book citation. If the book is published only electronically, identify the electronic publication date and publisher. Next, identify the date you accessed the book to get the information you are citing. Finally, list the URL in angle brackets.

Mates, Barbara T. *Adaptive Technology for the Internet: Making Electronic Resources Accessible to All*. Chicago, IL: American Library Association, 2000. 15 June 2006. <http://www.ala.org/ala/productsandpublications/books/editions/adaptivetechnology.htm>

Online Periodical List the author's last name, first name, and middle name or initial, followed by the title of the article in quotation marks. List the name of the publication in italics and the date of the article's publication. Identify all relevant publication information as you would for a print article, including volume and issue number as well as page number if available.

Wagner, Holly J. "Home Office? It's in the Yard." *Wired.* 28 July 2006. 4 September 2006. <http://www.wired.com/news/technology/gizmos/0,71462–0.html?tw=wn_index_3>

Database Articles Databases generally provide electronic access to print-based articles. To cite a database source, first provide information for the print article. Then list the database name, the subscription service, the library or other host location from which you accessed the information, the date you accessed the article, and the URL of the service.

Wickliff, Gregory A. "Light Writing: Technology Transfer and Photography to 1845." *Technical Communication Quarterly* 15.3 (2006): 293–313. Communication & Mass Media Complete. EBSCOhost. Univ. of Florida Lib., Gainesville, FL. 16 June 2006. <http://weblinks2.epnet.com>

CD-ROM Provide the author's last name, first name, and middle name or initial if available. List the title of the CD-ROM in italics, and write *CD-ROM*, followed by the place of publication, publisher, and date of publication.

Bawarshi, Anis. *Image Editing Software for the Technical Writer*. CD-ROM. Seattle: Tech-Write, 2006.

E-mail Indicate the author's last name, first name, and middle name or initial. If listed in the e-mail, identify the subject line in quotation marks. Then identify the recipient of the e-mail; if the e-mail recipient is the same as the document author, write *E-mail to the author.* Finally, identify the date on which the e-mail was sent.

Ohlinger, Monica. "Copyright Concerns." E-mail to the author. 20 May 2006.

Blanchard, Robert. "Weekly Staff Meeting." E-mail to Merritt Martin. 06 February 2006.

Other

Government Document Name the government agency that produced the document, and give the publication title and the edition or identification number, if available. Finally, identify the place and date of publication.

Unites States Environmental Protection Agency. *EPA 2003 Draft Report on the Environment*. Washington, DC, 2003.

Pamphlet Pamphlets are cited according to the rules used for books.

A Woman's Guide to Eating Fish and Seafood. New Jersey: NJ Environmental Protection Agency, 2005.

Report Reports are cited according to the rules used for books.

United States Department of State. *Trafficking in Persons Report*. Washington, DC, 2005.

Interview Identify the name of the person interviewed by last name, first name, and middle name or initial. If the interview has a title, include it in quotation marks. If no title exists, include the word *Interview*. Then give the information about where the interview was published.

Boise, David. "The Wired Interview." *Wired* 8.10, 2000: 46–49.

Olson, Gary A. Interview. *Editor's Weekly.* 28 July 2005: A3–A5.

If the interview is a personal interview rather than a published interview, first list the interviewee's last name, first name, and middle name or initial. Then write *Personal Interview,* followed by the date of the interview.

Bollea, Terrance Gene. Personal Interview. 20 May 2005.

Letter For letters written to the document author, identify the last name, first name, and middle name or initial of the letter author, followed by *Letter to the Author* and the date of the letter.

Maza, Jacob. Letter to the Author. 28 Feb. 2005.

If the letter was written to someone other than the author, provide that name instead.

Kaminsky, Melvin. Letter to Jerome Silberman. 11 Aug. 2004.

Memo Memos follow the same format used for letters.

Hirsch, Bud. Memo to Author. 11 Nov. 2002.

Feldman, Amy. Memo to William Bossing. 15 June 2006.

Presentation State the last name, first name, and middle name or initial of the speaker. Give the title of the speech or presentation in quotation marks, followed by the location and date of the presentation. If the presentation has no title, use a descriptor such as *Presentation*, *Speech*, or *Lecture* without quotation marks.

> Jobs, Steve. "You've Got to Find What You Love." Stanford University, Stanford, CA. 12 June 2005.

Map Maps follow the same citation rules used for books; however, following the title, include the word *Map*.

> National Geographic Society. Everglades, Florida. Map. Washington, DC: National Geographic Society, 1999: 132.

Chart Charts follow the same citation rules used for books; however, following the title, include the word *Chart*.

> Annual Budget Expenditures. Chart. Columbus, OH: DataText.

Images

Images with Titles

From Library Databases When citing titled images from library databases, begin with the artist's last name, then first name, and the title of the picture, followed by the date of the picture. Next, identify the museum where the image is housed, followed by the location of the museum. Then, list the name of the database used to retrieve the image, the library used to access the database, the date the image was accessed, and the URL for the database location of the image.

> Matisse, H. *Notre-Dame, une fin d'après-midi (A Glimpse of Notre Dame in the Late Afternoon)* [Painting]. 1902. Albright-Knox Art Gallery, Buffalo, NY. Grove Art Online Database. U of Florida Coll. 28 July 2008 <http://www.groveart.com>

From Free Web Pages When citing titled images from free web pages, identify the artist's last name and then first name, the title of the image, followed by the date of the image. Next, identify where the original image is housed, followed by the location of that place. Next, list the date you retrieved the image from the web page, followed by the URL of that web page.

> Matisse, Henri. *Notre-Dame, une fin d'après-midi (A Glimpse of Notre Dame in the Late Afternoon).* 1902. Albright-Knox Art Gallery, Buffalo, NY. 28 July 2008 <http://www.ibiblio.org/wm/paint/auth/matisse/matisse.notre-dame-am.jpg>

From Printed Sources When citing images found in printed sources, begin with the artist's last name, followed by the first name. Next, include the title of the image, followed by the date of the image. Then, include the location where the original image is housed, followed by the location of that place. Then, identify the title of the printed source, followed by the author or editor of the printed source, followed by publication information: place of publication, publisher, date, and page number of the image.

> Matisse, Henri. *Notre-Dame, une fin d'après-midi (A Glimpse of Notre Dame in the Late Afternoon).* 1902. Albright-Knox Art Gallery, Buffalo, NY. *Matisse and Color.* By Jean Caude Dumond. New York: Parthenon Press, 2007. 121.

Images without Titles

From Library Databases When citing untitled images from a database, begin with the artist's last name, then first name, followed by a descriptive name of the image. Do not italicize the description or place the description in quotes, but do include the medium of the image. Next, include the date of the image, followed by the location where the original image is housed and the location of that place. Then, name the database used to retrieve the image. Next, identify the library source used to access the database, followed by the date you accessed the image. Finally, list the URL for the database.

Landres, Jack. Photograph of war refugees. 1945. Canton War Museum, Canton, OH. Grove Art Online. U of Florida Coll. 7 December 2008. <http://www.groverart.com>

From Printed Sources When citing images taken from printed material like a book, begin with the artist's last name, then first name. Next, include a descriptive name of the image. Do not italicize the description or place the description in quotes, but do include the medium of the image. Next, include the date of the image, followed by the location where the original image is housed and the location of that place. Then identify the title of the print document, italicized or in quotes. Then, identify the print document's author, followed by place of publication, publisher, and date of publication. Finally, include the page number where the image was found in the print document.

Landres, Jack. Photograph of war refugees. 1945. Canton War Museum, Canton, OH. *Images of American War Stories*. By Paul Mitchell. New York: InSight Press, 1987. 96.

CREDITS

Chapter Opener Art Courtesy of www.istockphoto.com.

Chapter 3

Figure 3.1 Data from www.itu.int.

Figure 3.2 Data from http://www.zenker.se/Surprise/moore_intel.gif.

Figure 3.3 Data from http://www.microsoft.com/emea/presscentre/images/Word_NoGallery.jpg.

Figure 3.4 Data from http://office.microsoft.com/global/images/default.aspx?assetid=ZA101658571033.

Figure 3.5 Data from http://www.snapfiles.com/screenshots/adobeps.htm.

Figure 3.6 Data from http://en.wikipedia.org/wiki/IWeb.

Figure 3.7 ©2008 Adobe Systems Incorporated. All rights reserved. Adobe and InDesign are registered trademarks of Adobe Systems Incorporated in the United States and/or other countries.

Figure 3.8 Courtesy of WebSoft, Ltd.

Figures 3.10 Reproduced with permission of Yahoo! Inc. ©2008 by Yahoo! Inc. YAHOO! And the YAHOO! Logo are trademarks of Yahoo! Inc.

Figure 3.11 © Steve Chenn/Corbis.

Figure 3.12 Reprint Courtesy of International Business Machines Corporation, copyright 2008 © International Business Machines Corporation.

Chapter 4

Figure 4.2 © 2008 IEEE. Reprinted with permission of the IEEE.

Figure 4.3b Courtesy of Underwater Video Service. Photo by Charles Maxwell.

Figure 4.5 From *Newsweek,* August 3, 1981. ©1981 Newsweek, Inc. All rights reserved. Used by permission and protected by Copyright Laws of the United States. The printing, copying, redistribution, or retransmission of the Material without express written permission is prohibited.

Chapter 5

Figures 5.1b, 5.1c Courtesy of www.istockphoto.com.

Chapter 6

Figure 6.1 © John Zich/CORBIS NEWS. All rights reserved.

Figure 6.2 Reprinted by permission from Semi Karti, Center for Disease Control, http://www.cdc.gov.

Figure 6.3 Reprinted by permission from World Health Organization, www.who.int. All rights reserved.

Figure 6.4 Reprinted by permission from Weekly World News, www.weeklyworldnews.com/featurers/science/61485. © American media, Inc. All rights reserved.

Figure 6.7a CORBIS NEWS. All rights reserved.

Figure 6.7b © John Zich/CORBIS NEWS. All rights reserved.

Figure 6.8 Courtesy of TSA.

Chapter 7

Figure 7.8 Courtesy of Northwest Airlines, Inc.

Figure 7.12 Image courtesy of THE WEATHER CHANNEL, weather.com®.

Figure 7.15 Reprinted by permission from the Center for Responsible Nanotechnology, "A Technical Commentary on Greenpeace's Nanotechnology Report," September 2003, http://www.crnano.org/Greenpeace.htm.

Figure 7.16 Reprinted by permission from "United Nations Treaty collection," http://untreaty.un.org/English/guide.asp#protocols. © United Nations. All rights reserved.

Chapter 8

Figure 8.1 Courtesy of http://www.time.com/time/photogallery/0,29307,1662530_1446035,00.html.

Figure 8.3 Courtesy of http://ga.water.usgs.gov/edu/watercyclesummary.html.

Figure 8.4 © Frederic Larson/San Francisco Chronicle/Corbis.

Figure 8.5 Copyright © 1993, 2004 Steven C. McConnell. Used with permission.

Figure 8.6 Data from http://www.igd.com/cir.asp?menuid=50&cirid=1502.

Figure 8.14 U.S. Department of Labor; http://www.dol.gov/esa/minwage/chart.htm.

Figure 8.21 State of Delaware.

Figure 8.22 ©2008 Dell Inc. All Rights Reserved.

Figure 8.23 Reprinted by permission from Sword Marine Technology, Inc.

Figure 8.31 Technical Communication by William Horton. Copyright © 1993 by Society for Technical Communication (STC). Reproduced with permission of Society for Technical Communication (STC) in the format via Textbook via Copyright Clearance Center.

Figure 8.32 Reprinted from *Developing International User Information,* by Scott Jones et al., Digital Press, Copyright © 1992, with permission from Elsevier.

Figure 8.33b Courtesy of www.istockphoto.com.

Figures 8.37, 8.38, 8.39, 8.40 Used with permission from McDonald's Corporation.

Chapter 9

Figures 9.1, 9.2 Reprinted with permission of Rock the Vote. For more information about Rock the Vote, the nonprofit organization that engages young people in the political process, visit http://www.rockthevote.com.

Figure 9.6 Reprinted by permission of San Francisco Convention and Visitors Bureau.

Figure 9.9 Courtesy of www.science.gov.

Figure 9.10 Reproduced from "The Face of Recovery: The American Red Cross Response to Hurricanes Katrina, Rita, and Wilma," September 2007, with permission from the American Red Cross.

Figure 9.11 Reprint Courtesy of International Business Machines Corporation, copyright 2008 © International Business Machines Corporation.

Figures 9.16, 9.17 John Zoiner/Jupiter Images—Workbook Stock.

Case Study 2 Reprinted by permission. © 2004 Insurance Institute for Highway Safety. http://www.iihs.org/news/2004/iihs_news_070104.pdf.

Chapter 10

DILBERT: © Scott Adams/Dist. by United Feature Syndicate, Inc.

Figure 10.3 Information from http://www.westnet.com.au/products/broadband/adsl/technical.asp.

Figures 10.6, 10.7 Reprinted by permission from a class project in Preliminary Design, Professor Joseph King, Harvey Mudd College, 2002, http://emat.eng.hmc.edu/classes/e119/minilathe.htm.

Chapter 11

Figure 11.1 © Reuters/Corbis.

Figure 11.2, 11.3, 11.4, 11.5 Reprinted by permission from Sharon Walbridge, Karen R. Diller, and Janet Chisman, "Testing the Usability of a WebPac," Northwest Innovative Users Group Meeting, November 9, 1998, http://www.vancover.wsu.edu/fac/diller/usability/nwiug/usepresent.htm.

Figure 11.7 From http://www.irs.gov/pub/irs-pdf/fw4.pdf?portlet=3.

Chapter 12

Figure 12.1 Retrieved from http://i.a.cnn.net/cnn/2005/images/11/03/brown.emails.pdf.

Figure 12.2 Courtesy of Google, Inc.

Figure 12.7 Reproduced with permission of Yahoo! Inc. ©2008 by Yahoo! Inc. YAHOO! And the YAHOO! Logo are trademarks of Yahoo! Inc.

Figure 12.8 Courtesy of www.istockphoto.com.

Chapter 13

Figure 13.1 U.S. Food and Drug Administration. Sept 30, 2004.

Chapter 14

Figure 14.7 Nick Finck, Director of User Experience, Blue Favor.

Figure 14.8 From http./www.reslady.com/Randall-Frenett.html.

Chapter 15

Figure 15.1 © IAU/Martin Kommesser/Handout/epa/corbis CORBIS. All rights reserved.

Figure 15.2 Reprinted by permission from Remington's Pharmaceutical Sciences 18, ed. Alfonso R. Gennaro, et al., Philadelphia College of Pharmacy, 1990, p. 883. From Remington's Pharmaceutical Sciences 18, ed. Alfonso R. Gennaro, et al., Philadelphia College of Pharmacy, 1990, p. 883. Reprinted with permission.

Figure 15.3 From http://www2.preservationnation.org/dozen_distinctive_destinations/2008/. National Trust for Historic Preservation.

Figure 15.4 © Darrell Gulin/Corbis.

Figure 15.5 NASA, Glenn Research Center, http://www.lerc.nasa.gov/www/k-12/airplane/alr.html.

Figure 15.6 Courtesy of www.istockphoto.com.

Figure 15.7 US-VISIT procurement notice, DHS.

Figure 15.9 George J. Mitchell, "Report to the Commissioner of Baseball of an Independent Investigation into the Illegal Use of Steroids and other Performance Enhancing Substances by Players in Major League Baseball," December 13, 2007, http://files@mlb.com/mitchrpt.pdf.

Figure 15.10 New Technologies, Inc., "Computer Forensics Defined," http://www.forensics-intl.com/def4.html.

Chapter 16

Figure 16.1 J. Michael Griggs, Technical Director of the Loeb Drama Center, Harvard University. http://theatre.harvard.edu/space/loebex/index.html.

Figure 16.2 From http://www.boeing.com/commerical/757family/technical.html.

Figure 16.3 Copyright © I-O Data Device, Inc. All rights reserved.

Figure 16.4 Congressional Research Service, "Ricin: Technical Background and Potential Role in Terrorism," February 4, 2004, http://www.fas.org/irp/crs/Rs21383.pdf.

Figures 16.5, 16.6, 16.7 Thomas J. Martin, "Technical Report: Knee Brace Use in the Young Athlete, "*Pediatrics* 108, no. 2 (2001). Copyright © 2001 by The American Academy of Pediatrics. Used by permission.

Figures 16.8, 16.9 Mark A. Grace, "Field Guide to Requiem Sharks (Elasmobranchiomorphi: Carcharhinidae) of the Western North Atlantic" (NOAA Technical Report NMFS 153, U.S. Department of Commerce, 2001).

Figure 16.10 From http://www.yukonenergy.ca/community/education/power/.

Chapter 17

Figure 17.4 Courtesy of Pop Culture, LLC.

Figure 17.7 From http://www.si.edu. ©2008 Smithsonian Institution.

Figure 17.8 From www.usa.gov.

Figure 17.9 National Security Agency, http://www.nsa.gov/intro:htm.

Figure 17.10 Copyright Design Partnership of Cambridge. Reprinted with permission.

Figure 17.12 Reprinted by permission of the Modern Language Association of America. From www.mla.org/style_faq.

Chapter 18

FOX TROT: ©2001 Bill Amend. Reproduced with permission of UNIVERSAL PRESS SYNDICATE. All rights reserved.

Photo page 526 ©Douglas Keister/CORBIS. All rights reserved.

Figure 18.1 MapQuest and the MapQuest logo are registered trademarks of MapQuest, Inc. ©2008 by MapQuest, Inc. Used with permission.

Figures 18.8, 18.9 ©2007 Britax Child Safety, Inc.

Figure 18.14 Courtesy of David Clark Company Incorporated.

Figure 18.15 © 2007–2008 Dell Inc. All rights reserved. From http://support.dell.com/support/edocs/systems/ins1720/en/om_en/html/about.htm#wp1203619.

Figure 18.21 From www.loc.gov/jefftour/cutaway.html.

Figure 18.22 ©2003 Plain Dealer Publishing Company. Illustration by William Neff.

Figure 18.23 Courtesy of Techsonic Industries, Inc.

Chapter 19

NON SEQUITOR: ©1996 Wiley Miller. Dist. By UNIVERSAL PRESS SYNDICATE. Reprinted with permission. All rights reserved.

Figure 19.1 Copyright © Thomas Jefferson National Accelerator Facility. Reprinted with permission.

Figure 19.2 Reprinted by Permission of New York State Department of Motor Vehicles. Reprinted with permission.

Figure 19.6 Images courtesy of Altec Lansing, a Division of Plantronics, Inc.

Table 19.1 Adapted by permission of Cherryleaf from Justin Darley, "An Introduction to Single Sourcing: Part 1," January 2003, http://www.cherryleaf.com/news_single_sourcing.htm.

Chapter 21

Figure 21.1 From http://www.vancouver2010.com/resources/PDFs/IPCReport_05.pdf.

Figure 21.2 © Reuters/Corbis.

Figure 21.4 From http://www.whitehouse.gov/news/releases/2003/09/20030916-5.html.

Figure 21.5 From http://www.osha.gov/SLTC/etools/hospital/hazards/workplaceviolence/incidentreportform.html#*.

Figure 21.6 From http://www.cnn.com/2003/ALLPOLITICS/10/02/kay.report/.

Figure 21.7 Copyright Ecological Society of America.

Figure 21.8 From http://www.ncsu.edu/labwrite/res/labreport/res-sample-labrep.html.

Chapter 22

Photo page 661 Courtesy of www.istockphoto.com

Figure 22.1 © Brooks Kraft/CORBIS. All rights reserved.

Figure 22.2 U.S. Department of Education.

Figure 22.3 Reprinted with Permission from the Center for Agribusiness and Economic Development.

Figure 22.4 Yvonne Belanger, Head, Program Evaluation Academic Technology and Instructional Services, Perkins Library, Duke University.

Figures 22.5 to 22.17 U.S. Army Corps of Engineers, "Improving Salmon Passage," http://www.nww.usace.army.mil/lsr/final-fseis/study-kit/summary.pdf.

Chapter 23

Figure 23.1 Tony Avelar/AFP/Getty Images.

Figure 23.3 © Stuart Westmorland/CORBIS. All rights reserved.

Figure 23.4 © Tom & Dee AnnMcCarthy/CORBIS. All rights reserved.

Figure 23.5 © ImageShop/Corbis

Figure 23.6 © Najlah Feanny/CORBIS NEWS.

Figure 23.7 © George Disario/CORBIS.

Figure 23.8 Courtesy of www.istockphoto.com.

Figure 23.15 Courtesy of Holland America Line®.

INDEX